Haiduc · Zuckerman
Basic Organometallic Chemistry

Cover illustration

Decaphenylstannocene, $[\eta^5\text{-}(C_6H_5)_5]_2Sn^{II}$:
The first symmetrical main-group sandwich compound.
Reprinted with permission from the
Journal of the American Chemical Society.
Heeg, M.J.; Janiak, C.; Zuckerman, J.J.;
J. Am. Chem. Soc. **1984**, *106*, 4259–4261.

Ionel Haiduc
J. J. Zuckerman

Basic Organometallic Chemistry

Containing Comprehensive Bibliography

Walter de Gruyter
Berlin · New York 1985

Authors

Dr. Ionel Haiduc
Professor of Inorganic and Metallorganic Chemistry
Babes-Bolyaì University
Department of Chemistry
R-3400 Cluj-Napoca
Romania

Dr. Jerry J. Zuckerman
Professor of Inorganic Chemistry
The University of Oklahoma
Department of Chemistry
Norman, Oklahoma 73019
USA

Library of Congress Cataloging in Publication Data

Haiduc, Ionel.
Basic organometallic chemistry.

Bibliography: p.
Includes index.
1. Organometallic compounds. I. Zuckerman, Jerry J.
II. Title.
QD411.H297 1985 547'.05 84-27410
ISBN 0-89925-006-8 (U.S.)

CIP-Kurztitelaufnahme der Deutschen Bibliothek

Haiduc, Ionel:
Basic organometallic chemistry / Ionel Haiduc ;
J.J. Zuckerman. – Berlin ; New York : de Gruyter, 1985.
ISBN 3-11-007184-3

NE: Zuckerman, Jerry J.:

ISBN 3-11-007184-3 Walter de Gruyter · Berlin · New York
ISBN 0-89925-006-8 Walter de Gruyter, Inc., New York

Typesetting and Printing: Tutte Druckerei GmbH, Salzweg-Passau.
Binding: Lüderitz & Bauer, Berlin.
Cover design: Hansbernd Lindemann, Berlin.

Preface

This book had as its genesis Ionel Haiduc's text, *Chimia Compusilor Metalorganici*, published in the Romanian language by Editura Stiintifica in Bucharest in 1974. The English-language version has been completely updated and reedited.

The authors hope that this brief survey of the wonders of organometallic chemistry will win new recruits to our exciting area from among the students who use the text, and provide a useful pedagogical tool for the professors who teach from it.

Organometallic chemistry is almost as old as chemistry itself, with Cadet's fuming arsenical liquid, which contains methylarsines, having been synthesized in 1760, and Zeise's salt, which contains an olefin-transition metal π-bond, having been synthesized in 1827. The discovery of the silicones just before World War II, and the syntheses of ferrocene, $(\eta^5\text{-}C_5H_5)_2Fe$ in 1951, were important milestones in its modern development. The first chemotherapeutic agents were Ehrlich's organoarsenicals, which were used medicinally after the turn of the century, and tetraethyllead, which had been added to gasoline since the 1930's to improve engine performance until its deleterious side effects were recognized. Organotins currently find large-scale use in PVC-stabilization and as biocides. The application of organometallic compounds to organic synthesis, for example, of organozincs and mercurials, the Grignard and organolithium agents, and the more recently developed organocopper and -zirconium agents and certain metal carbonyls, began quite early. Organometallic compounds have also stimulated the application of physical methods and spectroscopy to chemistry, in general, by serving as ideal examples for study by vibrational, electronic, magnetic-resonance and mass spectroscopies, as well as by single-crystal X-ray diffraction techniques. Certain organometallic compounds have from time to time presented difficult problems for bonding theory, and the development of solutions to these has taken theoretical chemistry into new areas. This is fitting, since it was the organometallic derivatives that enabled Mendeleev to determine the oxidation states of several of the elements and hence guide their placement in his Periodic System. It was, in part, his prediction of the physical properties of tetraethylgermane that first convinced chemists of the value of the concept of periodicity. Organometallic derivatives are now known for almost every element, even the actinides (but not yet for the Group 0 gases).

This textbook emphasizes the synthesis, reactions and molecular structures of organometallic compounds. Less attention is paid to mechanism, spectroscopic properties and applications to catalysis or organic synthesis, which can be found in specialized monographs. *Basic Organometallic Chemistry* is intended mainly for the newcomer to the field, and, as the title indicates, includes only the basic facts and concepts of organometallic chemistry.

References to the primary literature have not been included. Instead, textbooks, monographs, and series containing organometallic information are listed in Chapter 4, which discusses the literature of the discipline. *The Professor's Edition, which contains a complete and up-to-date listing of the voluminous literature, divided up by chapter, at the end of the book, is being published simultaneously with the paperback student edition.* This version should prove useful to the advanced student and the research worker in the field, since it will constitute a unique guide to this information.

The time encompassed in the preparation of this text, a task conducted between the authors entirely by correspondence interspersed with only the briefest telephone conversations, witnessed an enormous accretion of new information concerning organometallic systems. Thanks to the publishers, we have been able to correct the text to delete statements now found no longer to be true and to add news of the latest discoveries in the discipline up to the middle of 1984. We are grateful for the chance to bring to our readers a freshly updated text, but we apologize in advance for the inevitable sins of omission and commission.

Lastly we must thank our wives (Rose E. Zuckerman typed the manuscript) for putting up with us during the course of preparing this book, and the members of our research groups and departmental colleagues who helped correct proofs.

Ionel Haiduc
Babes-Bolyai University
Cluj/Napoca
Romania

J. J. Zuckerman
University of Oklahoma
Norman, Oklahoma USA

Contents

Part I

1. Introduction 3

1.1. The Scope of Organometallic Chemistry 3
1.2. Some Historical Notes 4
1.3. The Classification of Organometallic Compounds 5

2. The Metal-Carbon Bond 7

2.1 Bond Types 7
Ionic Bonds 7
Sigma Covalent Bonds (Bicentric Bielectronic) 8
Electron-Deficient Bonds (Localized Polycentric Bonds) 8
Delocalized Bonds in Polynuclear Systems 8
Dative Bonds, with d-Orbital participation (Sigma Donor – Pi Acceptor) . 8
2.2. Metal-Carbon Ionic Bonds 9
2.3. Sigma (σ) Covalent Bonds (Bicentric Bielectronic) 12
2.4. Electron-Deficient Bonds (Polycentric Localized-Bonds) 14
2.5. Delocalized Bonds in Polynuclear Systems 16
2.6. Dative Bonds with Participation of *d*-Orbitals 18

3. Laboratory Techniques in Organometallic Chemistry 29

3.1. Synthesis and Isolation of Compounds 29
3.2. Analysis and Structural Characterization 33

4. The Literature of Organometallic Chemistry 35

4.1. Textbooks and General Monographs 36
4.2. Reviews 37
4.3. Primary Literature 38

Part II
Organometallic Compounds of Non-Transition Elements

5. Organometallic Compounds of Alkali Metals 45

5.1. Organolithium Compounds 45

5.1.1. Preparations 46
5.1.2. The Structure of Organolithium Compounds 48
5.2. Organometallic Derivatives of Sodium and the Heavier Alkali Metals 51

6. Organometallic Compounds of the Group II Elements 55

6.1. Organoberyllium Compounds 56
6.1.1. Types of Organoberyllium Compounds 56
Compounds of BeR_2 Type 56
Addition Compounds of the Type $BeR_2 \cdot nD$ (n = 1 or 2) 57
Anionic Compounds with Three- and Four-Coordinated Beryllium 57
Functional Derivatives of the Type RBeX 57
Hydrides of the Type RBeH 58
6.1.2. Diorganoberyllium Compounds, BeR_2, and their Derivatives 58
6.1.3. Organoberyllium Halides 61
6.1.4. Organoberyllium Hydrides 61
6.2. Organomagnesium Compounds 62
6.2.1. Disubstituted Derivatives, MgR_2 62
6.2.2. Grignard Reagents (Organomagnesium Halides), RMgX 63
6.3. Organometallic Compounds of Calcium, Strontium and Barium 66
6.4. Organozinc Compounds 67
6.4.1. Disubstituted Derivatives, ZnR_2 67
6.4.2. Organozinc Halides, RZnX 70
6.4.3. Organozinc Hydrides, RZnH 71
6.5. Organocadmium Derivatives 71
6.5.1. Disubstituted Derivatives, CdR_2 72
6.5.2. Monosubstituted Derivatives, RCdX 73
6.6. Organomercury Compounds 74
6.6.1. Disubstituted Derivatives, HgR_2 74
6.6.2. Monosubstituted Derivatives, RHgX 77

7. Organometallic Compounds of Group IIIA Elements 81

7.1. Organoboron Compounds 82
7.1.1. Types of Known Organoboron Compounds 82
Mononuclear Compounds with Three-Coordinated Boron 82
Heterocyclic Compounds with Boron as Heteroatom 83
Linear and Cyclic Polynuclear Compounds with Three-Coordinated Boron 83
Mononuclear Compounds with Four-Coordinated Boron 84
Cyclic Compounds with Four-Coordinated Boron 84
Carboranes 84
7.1.2. Trialkyl(aryl)boranes, BR_3, and Related Compounds 84
7.1.3. Halogeno-Organoboranes, R_nBX_{3-n} 87
7.1.4. Organoboron Acids, $R_nB(OH)_{3-n}$. 88
7.1.5. Esters of Organoboric Acids (Alkoxyboranes), $R_nB(OR')_{3-n}$. 89
7.1.6. Amino-Organoboranes, $R_nB(NR'R'')_{3-n}$, and Other Boron-Nitrogen Compounds 89

7.1.7. Sulfur-Containing Organoboron Compounds ... 91
7.1.8. Phosphorus-Containing Organoboron Compounds ... 92
7.1.9. Carboranes ... 92
7.2. Organoaluminum Compounds ... 97
7.2.1. Types of Organoaluminum Compounds ... 97
7.2.2. Trisubstituted Organoaluminum Compounds ... 98
7.2.3. Heterocyclic Compounds with Aluminium Heteroatoms ... 100
7.2.4. Mono- and Dinuclear Compounds with Four-Coordinated Aluminum ... 101
7.2.5. Cyclic Dimers, Trimers and Tetramers Containing Four-Coordinated Aluminum ... 102
Organoaluminum Halides ... 102
Alkylaluminum Hydrides ... 103
Organoaluminum Alkoxides, Thiolates and Amides ... 103
7.3. Organogallium Compounds ... 104
7.3.1. Trisubstituted Derivatives, GaR_3 ... 104
7.3.2. Diorganogallium Halides, R_2GaX ... 106
7.3.3. Cyclic Oligomers with Four-Coordinated Gallium ... 107
7.4. Organoindium Compounds ... 109
Trisubstituted Derivatives, InR_3 ... 109
Diorganoindium Halides, R_2InX ... 109
Organoindium Dihalides, $RInX_2$ ... 110
7.5. Organothallium Compounds ... 111
7.5.1. Trisubstituted Derivatives, TlR_3 ... 112
7.5.2. Disubstituted Derivatives, R_2TlX ... 113
7.5.3. Monosubstituted Derivatives, $RTlX_2$ ... 114
7.5.4. Monovalent Thallium Derivatives, TlR ... 114

8. Organometallic Compounds of Group IVA Elements ... 115

8.1. Organosilicon Compounds ... 116
8.1.1. Types of Organosilicon Compounds ... 116
Mononuclear Compounds with Four-Coordinated Silicon ... 116
Organic Heterocycles with Silicon as the Heteroatom ... 117
Linear and Cyclic Oligomers ... 117
Polymers ... 117
Organosilicon Compounds Containing Five- and Six-Coordinated Silicon 118
Silylenes, $:SiR_2$ and Free Radicals, $\cdot SiR_3$ ... 118
8.1.2. Tetraorganosilanes, SiR_4 ... 118
8.1.3. Organochlorosilanes, R_nSiCl_{4-n} ... 119
8.1.4. Organosilicon Hydrides (Organohydrosilanes), R_nSiH_{4-n} ... 121
8.1.5. Organoalkoxysilanes, $R_nSi(OR')_{4-n}$ ... 121
8.1.6. Organosilicon Carboxylates, $R_nSi(OCOR')_{4-n}$ ... 122
8.1.7. Organosilanols, $R_nSi(OH)_{4-n}$ ... 122
8.1.8. Organoaminosilanes, $R_nSi(NR'R'')_{4-n}$ ... 122
8.1.9. Other Organosilicon Functional Derivatives ... 123
8.1.10. Organofunctional Derivatives ... 123
8.1.11. Organic Heterocycles with Silicon as Heteroatom ... 124

8.1.12. Linear Oligomers, Cyclic Compounds and Polymers 127
Organopolysilanes 127
Organopolysiloxanes 129
Organoheterosiloxanes 134
Polyborosiloxanes 134
Polyphosphosiloxanes 134
Polymetallosiloxanes and Polysiloxymetalloxanes 135
Organopolysilazanes 136
Organopolysilthianes 139
Polycarbosilanes 140
8.1.13. Silylenes, $:SiR_2$ 142
8.1.14. Triorganosubstituted Free Radicals, $\cdot SiR_3$ 143

8.2. Organogermanium Compounds 143
8.2.1. Tetraorganogermanes, GeR_4 143
8.2.2. Heterocyclic Compounds with Germanium as Heteroatom 145
8.2.3. Organohalogermanes, R_nGeX_{4-n} 146
8.2.4. Organogermanium Hydrides, R_nGeH_{4-n} 147
8.2.5. Germanols, $R_nGe(OH)_{4-n}$ 147
8.2.6. Other Compounds with Ge-O Bonds 147
8.2.7. Organoaminogermanes, $R_nGe(NR'R'')_{4-n}$ 147
8.2.8. Linear Oligomers, Cyclic Compounds and Polymers 148
Organopolygermanes 148
Organopolygermoxanes 149
Organopolygermazanes 150
Organogermathianes (Organogermanium Sulfides) 151
8.2.9. Diorganogermylenes, $:GeR_2$ 152
8.2.10. Trisubstituted Organogermanium Free Radicals, $\cdot GeR_3$ 153

8.3. Organotin Compounds 153
8.3.1. Tetraorganostannanes, SnR_4 153
8.3.2. Organic Heterocycles with Tin Heteroatoms 155
8.3.3. Organotin Halides, R_nSnX_{4-n}, and Their Complexes 156
8.3.4. Organotin Hydrides, R_nSnH_{4-n} 158
8.3.5. Organotin Hydroxides, $R_nSn(OH)_{4-n}$ 159
8.3.6. Organotin Alkoxides, $R_nSn(OR')_{4-n}$ and Related Compounds 160
8.3.7. Organotin Amino-Derivatives, $R_nSn(NR'R'')_{4-n}$ 162
8.3.8. Organotin Thiolates, $R_nSn(SR')_{4-n}$ 164
8.3.9. Linear Oligomers, Cyclic Compounds and Polymers 164
Organopolystannanes 164
Organopolystannoxanes (Organotin Oxides) 166
Organostannazanes 167
Organotin Sulfides, Selenides and Tellurides 168
8.3.10. Stannylenes, $:SnR_2$, Free Radicals, $\cdot SnR_3$, and Related Compounds 169
Bis[bis(trimethylsilyl)methyl]tin(II), $:Sn[CH(SiMe_3)_2]_2$ 170
Dialkylstannylenes, SnR_2 171
Trisubstituted Organotin Free Radicals, $\cdot SnR_3$ 173

8.4. Organolead Compounds 173
8.4.1. Tetrasubstituted Derivatives, PbR_4 174
8.4.2. Heterocycles with Lead Heteroatoms 176
8.4.3. Organolead Halides, R_nPbX_{4-n} 177
8.4.4. Organolead Carboxylates $R_nPb(OCOR')_{4-n}$ 178
8.5.4. Organolead Hydrides, R_nPbH_{4-n} 178
8.4.6. Organolead Hydroxides, $R_nPb(OH)_{4-n}$ 178
8.4.7. Organolead Alkoxides, $R_nPb(OR')_{4-n}$ 179
8.4.8. Organolead Amides, $R_nPb(NR'R'')_{4-n}$ 179
8.4.9. Organolead Mercaptides, $R_nPb(SR')_{4-n}$ 180
8.4.10. Organolead Oligomers and Polymers 180
Organopolyplumbanes 180
Organoplumboxanes (Organolead Oxides) 181
Organolead Sulfides 181
8.4.11. Plumbylenes, $:PbR_2$ 181

9. Organometallic Compounds of Group VA Elements 183

9.1. Organoarsenic Compounds 183
9.1.1. Classification and Nomenclature 183
Compounds in which Arsenic is Bonded Only to Organic Groups (Homoleptic and Related Species) 183
Organoarsenic Hydrides, R_nAsH_{3-n} 184
Organoarsenic Halides 184
Oxygen-Containing Organoarsencicals 184
Sulfur-Containing Organoarsenic Compounds 185
Nitrogen-Containing Compounds 185
Compounds with Arsenic-Arsenic Bonds 186
9.1.2. Fully-Organosubstituted Compounds 186
Tertiary Arsines, AsR_3 186
Tetraorganoarsonium Salts, $[AsR_4]^+X^-$ 188
Organoarsenic Ylides (Alkylidenearsoranes), $R_3As=CHR'$ 190
Pentaorganosubstituted Derivatives (Pentaorganoarsoranes), AsR_5 190
Arsabenzene (Arsenine) 191
9.1.3. Organoarsenic Hydrides (Primary and Secondary Arsines) 191
9.1.4. Organoarsenic Halides 192
Mono- and Dihaloarsines, R_nAsX_{3-n} 192
Organotetrahaloarsoranes, $RAsX_4$ 194
Diorganotrihaloarsoranes, R_2AsX_3 194
Triorganodihaloarsoranes, R_3AsX_2 194
9.1.5. Oxygen-Containing Organoarsenic Compounds 195
Arsonous Acids, $RAs(OH)_2$, and Their Anhydrides, RAsO 195
Arsonous Acids Esters, $RAs(OR')_2$ 196
Arsinous Acids, R_2As-OH, and Their Anhydrides, $(R_2As)_2O$ (Diarsoxanes) 196
Arsinous Acid Esters, R_2As-OR' 196
Alkyl(aryl)tetraalkoxyarsoranes, $RAs(OR')_4$ 196

Dialkyl(aryl)trialkoxyarsoranes, $R_2As(OR')_3$ 197
Triorganodialkoxyarsoranes, $R_3As(OR')_2$ 197
Tetraalkylalkoxyarsoranes, R_4As-OR' 197
Arsonic Acids, $RAs(O)(OH)_2$ and Arsinic Acids, $R_2As(O)OH$ 197
Arsonic and Arsinic Acid Esters, $RAsO(OR')_2$ and $R_2As(O)(OR')$ 198
Tertiary Arsine Oxides, R_3AsO 198

9.1.6. Sulfur-Containing Organoarsenic Compounds 199
Organoarsenic Sulfides, RAsS 199
Thiodiarsines, $R_2As-S-AsR_2$ 200
Arsonous-Acid Thioesters, $RAs(SR')_2$ and Arsinous-Acid Thioesters, R_2As-SR' 200
Organoarsenic Sesquisulfides, $R_2As_2S_3$ 200
Arsine Sulfides, R_3AsS 201
Trithioarsonic Acids, $RAs(S)(SH)_2$, and Dithioarsinic Acids, $R_2As(S)SH$ 201

9.1.7. Nitrogen-Containing Organoarsenic Compounds 202

9.1.8. Organoarsenic Compounds Containing As–As Bonds 203

9.2. Organoantimony Compounds 205

9.2.1. Fully Organosubstituted Derivatives 205
Tertiary Stibines, SbR_3 206
Stibonium Salts, $[SbR_4]^+X^-$ 207
Pentaorganoantimony Derivatives (Stiboranes), SbR_5 208
Hexasubstituted Anions, $[SbR_6]^-$ 208
Stibabenzene (Antimonine) 208

9.2.2. Organoantimony Hydrides (Primary and Secondary Stibines), R_nSbH_{3-n} 208

9.2.3. Organoantimony Halides 209
Mono- and Diorganohalostibines, R_nSbX_{3-n} 209
Organoantimony(V) Tetrahalides, $RSbX_4$ 209
Diorganoantimony(V) Trihalides, R_2SbX_3 209
Triorganoantimony Dihalides, R_3SbX_2 210
Tetraorganoantimony(V) Halides, R_4SbX 211

9.2.4. Oxygen-Containing Organoantimony Compounds 211
Anhydrides of Stibonous and Stibinous Acids 211
Stibinous-Acid Esters, $RSb(OR')_2$ 212
Stibinous-Acid Esters, R_2Sb-OR' 212
Stibonic Acids, $RSbO(OH)_2$ 212
Stibinic Acids, $R_2Sb(O)OH$ 213
Organoalkoxystiboranes, $R_nSb(OR')_{5-n}$ 213
Triorganoantimony(V) Hydroxides, $R_3Sb(OH)_2$, and Stibine Oxides, R_3SbO 214

9.2.5. Sulfur-Containing Organoantimony Compounds 214
Dithiostibonous-Acid Esters, $RSb(SR')_2$ (Organoantimony(III) Dithiolates) 214
Thiostibinous-Acid Esters, R_2Sb-SR' 215
Triorganoantimony Dithiolates, $R_3Sb(SR')_2$ 215
Tetraalkylantimony Thiolates, R_4Sb-SR' 215

Stibine Sulfides, R_3SbS 215
9.2.6. Nitrogen-Containing Organoantimony Compounds 215
9.2.7. Organoantimony Compounds with Sb–Sb Bonds 216
9.3. Organobismuth Compounds 216
9.3.1. Fully Organosubstituted Bismuth Compounds 217
Tertiary Bismuthines, BiR_3 217
Tetraorganobismuthonium Salts, $[BiR_4]^+X^-$ 217
Pentaorgano-Substituted Derivatives, BiR_5 217
9.3.2. Organobismuth Halides 218
Triorganobismuth(V) Dihalides 218
9.3.3. Oxygen-Containing Organobismuth Compounds 219

Part III
Organometallic Compounds of Transition Metals

10. The Electronic Structure and Classification of Transition-Metal Organometallic Compounds 223
10.1 The Transition Metals 223
10.2. The Ligands 224

11. Compounds with Two-Electron Ligands 227

11.1. Metal Carbonyls 227
11.1.1. The Structure of Metal Carbonyls 229
11.1.2. Preparation of Metal Carbonyls 234
Titanium, Zirconium, Hafnium 235
Vanadium, Niobium, Tantalum 235
Chromium, Molybdenum, Tungsten 235
Manganese, Technetium, Rhenium 235
Iron, Ruthenium, Osmium 236
Cobalt, Rhodium, Iridium 237
Nickel, Palladium, Platinum 237
Copper, Silver, Gold 238
Heteronuclear Metal Carbonyls 238
11.1.3. Reactions of Metal Carbonyls 238
11.1.4. Metal-Carbonyl Anions 239
11.1.5. Metal-Carbonyl Hydrides and Related Hydrido Anions 243
11.1.6. Metal-Carbonyl Cations 246
11.1.7. Metal-Carbonyl Halides 247
11.2. Metal Thiocarbonyls 249
11.3. Metal Selenocarbonyls 250
11.4. Metal-Isocyanide Complexes 250
11.5. Metal-Carbene Complexes and Related Compounds 252
11.6. π-Olefinic Complexes 255

11.6.1. Monodentate Olefins 257
Chromium, Molybdenum, Tungsten 258
Manganese, Technetium, Rhenium 258
Iron, Ruthenium, Osmium 259
Cobalt, Rhodium, Iridium 260
Nickel, Palladium, Platinum 261
Copper, Silver, Gold 262
11.6.2. Bidentate Olefins 262
Chromium, Molybdenum, Tungsten 263
Iron, Ruthenium, Osmium 263
Cobalt, Rhodium, Iridium 264
Nickel, Palladium, Platinum 264
Copper, Silver, Gold 265
11.6.3. Tridentate Olefins 265
11.6.4. Bridge-Forming Olefins 265

12. Compounds with Three-Electron Ligands 269

12.1. π-Allylic Complexes and Related Compounds 269
Scandium, Lanthanides, Actinides 272
Titanium, Zirconium, Hafnium 272
Vanadium, Niobium, Tantalum 272
Chromium, Molybdenum, Tungsten 273
Manganese, Technetium, Rhenium 274
Iron, Ruthenium, Osmium 275
Cobalt, Rhodium, Iridium 277
Nickel, Palladium, Platinum 278
12.2. Cyclopropenyl Complexes 280
12.3. Metal-Carbyne Complexes 282

13. Compounds with Four-Electron Ligands 287

13.1. Butadiene Complexes and Related Compounds 288
13.2. Cyclobutadiene Complexes 291
13.3. η^4-Complexes of Cyclopentadiene, Cyclopentadienone, Fulvene and Heterocycles Derived from Cyclopentadiene 297
Cyclopentadiene 297
Cyclopentadienones 298
Fulvenes 299
Heterocycles 299
13.4. Complexes with Other Cyclic Dienes and Polyenes 300
Cyclohexadiene-1,3 300
Cycloheptadiene-1,3 302
Cycloheptatriene-1,3,5 303
Cyclooctatetrane 303
13.5. Trimethylenemethyl Complexes 304
13.6. Boron-Containing Four-Electron Ligands 305

14. Compounds with Five-Electron Ligands 307

14.1. Transition Metal, η^5-Cyclopentadienyl Complexes 307
Scandium, Yttrium, Lanthanides and Actinides. (Group III Elements) 313
Titanium, Zirconium, Hafnium 314
Vanadium, Niobium, Tantalum 315
Chromium, Molybdenum, Tungsten 317
Manganese, Technetium, Rhenium 319
Iron, Ruthenium, Osmium 320
Cobalt, Rhodium, Iridium 324
Nickel, Palladium, Platinum 326
Copper, Silver, Gold 327
14.2. Acyclic Pentadienyl η^5-Complexes 328
14.3. η^5-Complexes of Some Heterocyclic Ligands 328
14.4. η^5-Cyclohexadienyl Complexes 330
14.5. η^5-Cycloheptadienyl Complexes 331
14.6. Complexes of Carborane Ligands 332

15. Compounds with Six-Electron Ligands 335

15.1. η^6-Complexes of Benzene and its Derivatives 336
Titanium, Zirconium, Hafnium 336
Vanadium, Niobium, Tantalum 337
Chromium, Molybdenum, Tungsten 337
Manganese, Technetium, Rhenium 338
Iron, Ruthenium, Osmium 339
Cobalt, Rhodium, Iridium 340
15.2. η^6-Cycloheptatriene Complexes 340
15.3. Cyclooctatriene and Cyclooctatetraene as Six-Electron Ligands 341
15.4. Some η^6-Complexes with Heterocyclic Ligands 342

16. Compounds with Seven- and Eight-Electron Ligands 345

16.1. Seven-Electron Ligands 345
16.2. Eight-Electron Ligands 347

17. Organometallic Compounds Derived from Acetylenes 349

17.1. Complexes of Untransformed Acetylenes as Terminal (Dihapto) Ligands 350
Titanium, Zirconium, Hafnium 350
Vanadium, Niobium, Tantalum 351
Chromium, Molybdenum, Tungsten 351
Manganese, Technetium, Rhenium 352
Iron, Ruthenium, Osmium 352
Cobalt, Rhodium, Iridium 352
Nickel, Palladium, Platinum 352
17.2. Binuclear Complexes with Bridging Acetylene Ligands 353
17.3. Compounds Incorporating the Acetylene in a Polynuclear Cluster 354

17.4. Complexes Formed by Chemical Transformations of Acetylenes 356

18. Organometallic Compounds with σ-Metal-Carbon Bonds 359

Scandium, Yttrium, Lanthanides and Actinides (Group IIIB Elements) .. 364
Titanium, Zirconium, Hafnium 365
Vanadium, Niobium, Tantalum 366
Chromium, Molybdenum, Tungsten 366
Manganese, Technetium, Rhenium 368
Iron, Ruthenium, Osmium 369
Cobalt, Rhodium, Iridium 370
Nickel, Palladium, Platinum 371
Copper, Silver, Gold 373

Bibliography 377
Subject Index XVII

Part I

1. Introduction

1.1. The Scope of Organometallic Chemistry

Organometallic chemistry is at the same time an old and a new branch of chemistry. It is old because the first organometallic compound was prepared about 220 years ago: organic derivatives of many elements were known in the 19th century, but were investigated only in a handful of isolated laboratories; organometallic chemistry is new, since in the last 40 years organometallic compounds have become a subject of general interest, and the field is now recognized as an independent branch of chemistry.

Organometallic chemistry is the discipline dealing with compounds containing at least one direct metal-carbon bond. This bond can be simple covalent [as in lead tetraethyl, $Pb(C_2H_5)_4$] or π-dative [as in ferrocene, $Fe(\eta^5\text{-}C_5H_5)_2$] or even predominantly ionic [as in ethylsodium, $Na^+ C_2H_5^-$]. On this basis, compounds like metal alkoxides, [for example, aluminum triethoxide, $Al(OC_2H_5)_3$], metal amides, [for example $LiN(CH_3)_2$], chelate complexes, [for example, metal acetylacetonates], or the metal salts of carboxylic acids, are not recognized as organometallic compounds, although they contain both metal atoms and organic groups, since none of these systems contains a direct metal-carbon bond. These compounds are organic derivatives of metals, but not organometallic compounds.

The above definition can be further refined by adding that *organometallic chemistry deals with compounds in which an organic group is attached through carbon to an atom which is less electronegative than carbon.* Thus, organic compounds of the lighter halogens, and chalcogens are excluded from the field of organometallic chemistry, but not the organic derivatives of boron, silicon and arsenic since these are considered as metalloids. Phosphorus enjoys a somewhat privileged position: although it is slightly less electronegative than carbon, its compounds are not generally included in organometallic chemistry. Organophosphorus compounds are so diversified and numerous that it would be impractical to deal with them within the limits of organometallic chemistry and it is more efficient to have an independent branch of organophosphorus chemistry. Also, the non-metallic properties of phosphorus are so manifest that it would be hardly realistic to label its compounds "organo*metallic*". Perhaps in time an "organo*metalloidal* chemistry" will develop which will include the organic derivatives of elements having electronegativity slightly less or more than carbon (including phosphorus, sulfur, selenium, boron, silicon and arsenic).

Practically all elements, except at present the noble gases* form compounds with element-carbon bonds. Therefore, organometallic chemistry embraces the organic derivatives of the alkali and alkaline earth metals, the non-transition metals (main groups III–V), the transition metals (d-block elements, plus lanthanides and actinides) and some nonmetals (or metalloids) such as boron, silicon, antimony and tellurium.

1.2. Some Historical Notes

A brief glance at the history of organometallic chemistry can be instructive and interesting. There is no room here for a full treatment of this subject, but a few words about the development of organometallic chemistry are worthy of attention.

Several periods in the development of organometallic chemistry can be recognized. As mentioned above, the first organometallic compound prepared was a derivative of the dimethylarsenic group (called cacodyl); the product was in fact a mixture of tetramethyldiarsine, $Me_2As\text{-}AsMe_2$, and dimethylarsenous oxide, $(Me_2As)_2O$. An identification was made by Bunsen many years later (ca. 1848). The investigation of cacodyl marks the first period of organometallic chemistry. The next step was the synthesis of the alkylzinc compounds (Frankland 1849/50), and immediately following this the preparation of organic derivatives of mercury, antimony, tin, lead, cadmium, aluminum, magnesium and other main group elements, all before 1870. These were derivatives of non-transition metals, containing covalent, σ-bonds between the metal and carbon.

An independent development, which remained unknown to most chemists until much later, was the synthesis of metal derivatives in which an unsaturated organic compound (specifically an olefin) was bonded to a transition metal. Thus, the preparation of the platinum complex of ethylene, $K^+[C_2H_4PtCl_3]^-$, by W. C. Zeise in 1827 marked the birth of transition-metal organometallic chemistry. Although these compounds were investigated in some laboratories, and a few complexes of other olefins (for example, butadiene) with transition metals were also prepared, the importance of these compounds was not recognized until after 1950, when their structures were adequately explained.

The period 1870–1950 was dominated by the investigation of σ-bonded organometallic derivatives of the non-transition metals. These compounds were investigated intensively in the 20th century, but the emphasis was upon syntheses and applications as reagents in organic preparations. During this period the organometallic chemistry developed within the framework of classical organic chemistry, and aroused little interest from inorganic chemists. Structural investigations were scarce, but a great deal of preparative work was carried out, and some understanding of reactivity and mecha-

* The most likely would be a compound containing $Xe\text{-}CF_3$ groups, since the CF_3 group is comparable in electronegativity with a halogen. Evidence for the formation of $Xe(CF_3)_2$ has been reported recently.

nisms was also gained. This period is dominated by the names of Grignard, Schlenk, Kipping, Gilman, Ziegler, Rochow, Nesmeyanov and others who opened the way for the synthesis of organometallic derivatives of most metals.

A third dimension of organometallic chemistry before 1950 is the concentration on the chemistry of the carbonyls, treated for some time as coordination compounds. These were discovered by Mond who prepared first the nickel and iron carbonyls, and their chemistry was developed especially by Hieber after 1930. This work laid the foundation for further developments of transition metal organometallic chemistry to follow after 1950.

A turning point was the discovery of ferrocene in 1951 (Kealy & Pauson; Miller, Tebboth & Tremaine), developed further by E. O. Fischer and G. Wilkinson in an extensive chapter of organometallic chemistry. This discovery and the recognition of a new type of bonding between metals and organic unsaturated molecules prompted an enormous interest in these compounds, and resulted in the present state of organometallic chemistry. With the discovery of ferrocene, the classical period of organometallic chemistry drew to a close, and the modern era began. This was facilitated by the introduction of physical methods of investigation which afforded detailed information about the structure and bonding in organometallic compounds and made possible the understanding of their behavior. Of course, the development of the organometallic chemistry of the non-transition elements continued as well, but the modern period is dominated by the interest in transition metal derivatives.

New dimensions were added to organometallic chemistry by the recognition of the biological activity of some organometallic compounds which brought about a new field, *bio-organometallic chemistry*, by the understanding of the relationship between catalysis and formation of organometallic compounds, and more recently, – and in relation with their biological activity – by *environmental organometallic chemistry*.

The industrial interest in organometallic compounds must also be mentioned. Their application in organic synthesis was for a long time confined to laboratory uses, but now many industrial processes involve organometallic compounds. The use of Ziegler-Natta catalysts (organoaluminum-titanium compounds) for the polymerization of olefins, organosilicon compounds (silicone polymers and fluids), tetraethyl- and tetramethyllead as antiknock additives for gasoline, organotin biocides and polymer stabilizers, to name only a few important ones, should be cited. *Industrial organometallic chemistry* is expected to play a still more important role in the technology of the future.

1.3. The Classification of Organometallic Compounds

Organometallic compounds encompass a large diversity of types and structures, determined by the electronic configurations of the parent atom and their bonding type, which can be correlated with the place of the element in the Periodic Table. Accordingly, organometallic compounds can be classified as derivatives of main group elements (involving only *s*- and *p*-orbitals and electrons in bonding) and transition metal derivatives (involving *d*- and possibly *f*-orbitals as well).

The main group elements generally form metal-carbon σ-covalent bonds except for the alkali metals and alkaline earths, which form largely ionic compounds. Electron deficient structures are formed by elements like lithium, beryllium and aluminum. For transition elements the formation of π-complexes involving dative bonds is typical, and they can form normal covalent σ-metal-carbon bonds. The distribution of bonding type as a classification criterion for organometallic compounds is shown in Fig. 1.1. These bonding types are discussed in Chapter 2.

In this textbook organometallic derivatives of main group elements will be treated first and presented according to position in the Periodic Table. For transition metal derivatives it is more convenient to use a classification according to the ligand types, which can be better correlated with the electronic structure of the metals.

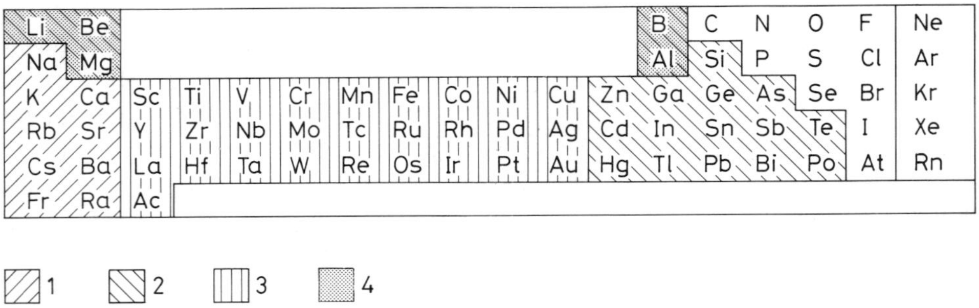

Li	Be											B	C	N	O	F	Ne
Na	Mg											Al	Si	P	S	Cl	Ar
K	Ca	Sc	Ti	V	Cr	Mn	Fe	Co	Ni	Cu	Zn	Ga	Ge	As	Se	Br	Kr
Rb	Sr	Y	Zr	Nb	Mo	Tc	Ru	Rh	Pd	Ag	Cd	In	Sn	Sb	Te	I	Xe
Cs	Ba	La	Hf	Ta	W	Re	Os	Ir	Pt	Au	Hg	Tl	Pb	Bi	Po	At	Rn
Fr	Ra	Ac															

1 2 3 4

Fig. 1.1. Classification of organometallic compounds.
(1) Ionic compounds, (2) Sigma-covalent compounds, (3) Sigma-covalent and pi-complex compounds, (4) Electron deficient compounds.

2. The Metal-Carbon Bond

2.1. Bond Types

The properties of organometallic compounds are decisively determined by the nature and stability of their metal-carbon bonds. The fact that some compounds are spontaneously inflammable in air, while others are perfectly stable in the presence of oxygen, is mainly due to the differences in the reactivities of their metal-carbon bonds. Factors including vacant orbitals or lone electron pairs at the metal, the possibility of electron delocalization, the bulk of the organic groups, exert a considerable influence upon the chemical behavior, and through this influence affect the reactivity of the metal-carbon bond, and can produce significant differences between the organometallic compounds derived from the same element.

The traditional types of bonds – ionic or covalent – can only in part explain the bonding between the metal atoms and the attached organic groups or molecules in organometallic chemistry. It is a peculiarity of some types of organometallic compounds that unsaturated organic molecules, capable of independent existence, can attach themselves to a metal atom which is in an apparently zero oxidation state. The occurence of bond types not found in other classes of chemical compounds, or discovered first in organometallic compounds, has contributed much to the general understanding of the nature of bonding.

Although any bond formed by overlap of atomic orbitals (or their hybrids) can be generally described as covalent, we will distinguish several possibilities. Thus, the following five metal-carbon bond types are found in organometallic compounds:

Ionic Bonds

These occur only in the compounds of alkali and alkaline earth metals. The metal is present in cationic form, M^{n+}, and the organic group is a carbanion; the negative charge of the carbanion can either be localized at a particular carbon atom (as in ethylsodium) or can be delocalized over several carbon atoms (as in sodium cyclopentadienide, Fig. 2.1).

$CH_3{-}\overset{\ominus}{C}H_2 \quad Na^{\oplus}$ — localized negative charge

$C_5H_5^{\ominus} \quad Na^{\oplus}$ — delocalized negative charge

Fig. 2.1. Carbanions with localized and delocalized charge.

The organometallic molecule in this case is formed by ion pairing. It can be assumed that some degree of covalency is found even in these cases, particularly if no donor solvent is present to produce ion-pair separation.

Sigma-Covalent Bonds (Bicentric Bielectronic)

These are typical for all non-transition elements but are also encountered in transition metal derivatives. These are the classical covalent bonds formed by pairing of two electrons of opposite spins. Thus, the bonds Sn-C, Pb-C, As-C and Al-C are σ-covalent bonds, with polar character of the type $M^{\delta+}$-$C^{\delta-}$, depending upon the electronegativity differences. It is worth mentioning that the presence of electron-attracting substitutents in the organic group (such as substitution of hydrogen by fluorine) increases the stability of the M-C bonds. Thus, the M—CF_3 and M—C_6F_5 compounds are considerably more stable thermally than their non-fluorinated analogues, especially if M is a transition metal.

Electron-Deficient Bonds (Localized Polycentric Bonds)

The possible formation of M···CH_3···M type bridges (for example, with M=Be, Al) can be understood in terms of an electron-pair bonding three atoms. Such bonds are weaker than the more common covalent bonds and form when the valence electrons available for bond formation are too few to fill all bonding orbitals with electron pairs.

Delocalized Bonds in Polynuclear Systems

The bonds between the lithium atoms in the Li_4 tetrahedra or Li_6 polyhedra in the polymeric lithium alkyls $(LiR)_n$ (n = 4 or 6) are delocalized metal-metal bonds: the organic groups (R) are attached simultaneously to several lithium atoms (usually three, on a polyhedral triangular face). In this case isolated Li—C bonds are not found. The formation of this type of bond can be explained in molecular orbital theory terms. A similar situation occurs in clusters present in metal carbonyls and in carboranes. In these molecules several atoms are held together by collectivizing the valence orbitals and electrons.

Dative Bonds, with d-Orbital Participation (Sigma Donor – Pi Acceptor)

A first type of dative bonding occurs when the carbon atom of a ligand acts as an electron-pair donor and the ligand molecule can accept back some of the electron density from the metal in its vacant molecular antibonding orbitals (back donation). In these interactions the metal may participate either with vacant *d*-orbitals (to accept the electron pair from the ligand) or occupied *d*-orbitals (for back donation). The typical ligands capable of such interactions include carbon monoxide (:C≡O), carbon monosulfide (:C≡S), isocyanides (:C=N—R), carbenes ($:CR_2$), cyanides ($:CN^-$), among others.

A second type of dative bonding involves an unsaturated molecule, which can release electrons from its π-molecular orbitals into vacant *d*-orbitals of the metal, while simultaneously accepting electrons from the occupied metal orbitals into the π^*-antibonding molecular orbitals. This type of bonding is of considerable importance in transition metal chemistry. Depending on their nature, such ligands can contribute a variable number of π-electrons: two (ethylene), three (the π-allyl group), four (cyclobutadiene), five (the cyclopentadienyl group), six (benzene), etc. (Fig. 2.2).

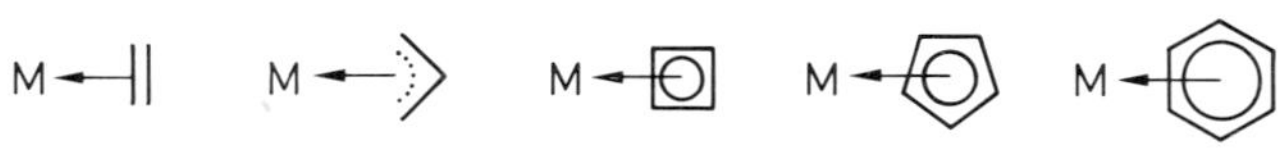

Fig. 2.2. Ligand types (donors of 2–6 electrons).

In order to participate in this type of bonding, each carbon atom must be sp^2-hybridized (thus possessing a π-electron in a p_z-orbital) and must be part of a planar carbon-atom network ("elplacarnet" = *el*emental *pla*nar *car*bon *net*work).

2.2. Metal-Carbon Ionic Bonds

Because of their polar character, metal-carbon bonds polarized as $M^{\delta+}—C^{\delta-}$ exhibit a certain degree of ionic character. However, when complete charge separation occurs, the organic group is present as a carbanion. In order to achieve this, it is necessary to have a sufficiently large electronegativity difference. The percent of ionic character (i) of a bond can be roughly calculated according to the relation given by L. Pauling:

$$i = 1 - e^{-0.25\,(X_A - X_B)^2}$$

where X_A and X_B are the electronegativities of atoms A and B. In Table 2.1 are given the electronegativities of all elements, the electronegativity difference $\Delta X = X_C - X_M$ for all element-carbon bonds, and the percent of ionic character (i) calculated according to Pauling's formula. The alkali and alkaline earth metal derivatives are characterized by the largest values of ΔX, and, therefore, exhibit the most pronounced ionic character of their metal-carbons bonds.

However, these values should not be taken too literally. In organolithium compounds the metal-carbon bond has a more covalent character than ΔX suggests, owing to the strong polarizability of the lithium atom. It is worth mentioning that in Group II A, beryllium and magnesium form more covalent metal-carbon bonds than calcium and the heavier congeners.

In the organometallic derivatives of the lower alkali metals (the best investigated is

Tab. 2.1. Element electronegativities (X_M), ΔX differences and the percent of ionic character of the M—C bonds.

Li 0.97 1.53 44	Be 1.47 1.03 23						H 2.1 0.4 4	← element ← electronegativity (x_M) ← difference $\Delta x = x_C - x_M$ ← % ionic character of the E—C bond				B 2.01 0.49 6	C 2.50 0 0	N 3.07 0.57 8	O 3.50 1.00 22	F 4.10 1.60 47	Ne — — —
Na 1.01 1.49 42	Mg 1.23 1.27 33											Al 1.47 0.93 19	Si 1.74 0.76 13	P 2.06 0.44 5	S 2.44 0.06 1	Cl 2.83 0.33 2	Ar — — —
K 0.91 1.59 46	Ca 1.04 1.46 41	Sc 1.20 1.30 34	Ti 1.32 1.18 29	V 1.45 1.05 24	Cr 1.56 0.94 20	Mn 1.60 0.90 18	Fe 1.64 0.86 16	Co 1.70 0.80 15	Ni 1.75 0.75 13	Cu 1.75 0.75 13	Zn 1.66 0.84 16	Ga 1.82 0.68 11	Ge 2.02 0.48 6	As 2.20 0.30 2	Se 2.48 0.02 0	Br 2.74 0.24 1	Kr — — —
Rb 0.89 1.61 48	Sr 0.99 1.51 43	Y 1.11 1.39 38	Zr 1.22 1.28 33	Nb 1.23 1.27 33	Mo 1.30 1.20 30	Tc 1.36 1.14 28	Ru 1.42 1.08 26	Rh 1.45 1.05 24	Pd 1.35 1.15 28	Ag 1.42 1.08 28	Cd 1.46 1.04 24	In 1.49 1.01 22	Sn 1.72 0.78 15	Sb 1.82 0.68 12	Te 2.01 0.49 6	I 2.21 0.29 2	Xe — — —
Cs 0.86 1.64 49	Ba 0.97 1.53 44	La* 1.08 1.42 39	Hf 1.23 1.27 33	Ta 1.33 1.17 29	W 1.40 1.10 26	Re 1.46 1.04 24	Os 1.52 0.98 22	Ir 1.55 0.95 20	Pt 1.44 1.06 24	Au 1.42 1.08 25	Hg 1.44 1.06 24	Tl 1.44 1.06 24	Pb 1.55 0.95 20	Bi 1.67 0.83 16	Po 1.76 0.74 13	At 1.90 0.60 8	Rn — — —
Fr 0.86 1.64 49	Ra 0.97 1.53 44	Ac 1.0 1.5 43	Th 1.11 1.39 38	Pa 1.14 1.36 36	U 1.22 1.28 33												

* These values can be used for all the lanthanides.

sodium) the organic groups bear almost a whole unit of negative charge, forming a carbanion. The carbanion center is the cause of the high reactivity and determines chemical behavior of these compounds. The negative charge can be localized in sp^3-, sp^2- or sp-carbon hybrid orbitals (Fig. 2.3).

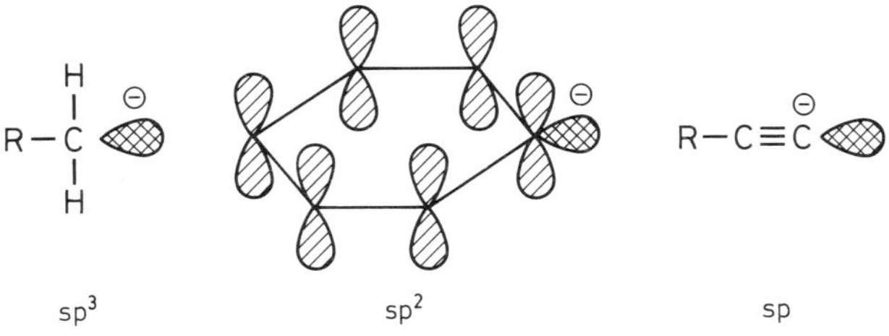

Fig. 2.3. Hybrid orbitals containing an electron pair in some typical carbanions.

The reactivity of the carbanions decreases in the order:

$$sp^3 > sp^2 > sp$$

owing to the increase in effective electronegativity of the carbon atoms in the same order. These differences are readily observed in the chemical behavior of organosodium compounds (for example, their ability to cleave ethers).

Stabilization of the carbanion, accompanied by a decrease in reactivity and often associated with formation of colored species, occurs when the negative charge can be delocalized over several carbon atoms. This takes place, for example, in the cyclopentadienyl, $C_5H_5^-$, anion mentioned above and in benzyl, $C_6H_5CH_2^-$, carbanion. In the $C_5H_5^-$ anion the formation of an aromatic electronic sextet has an additional favorable effect in the stabilization of the carbanion. The benzyl carbanion (Fig. 2.4) can delocalize its charge from the methylene group over the six-membered ring, resulting in an extended π-electron system.

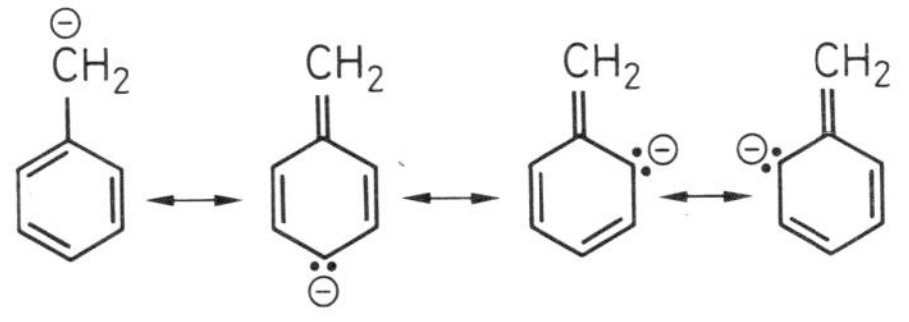

Fig. 2.4. Delocalization of the negative charge in the benzyl carbanion.

A less-common type of ionic organometallic compound is formed by addition of an electron released by an alkali metal to the electron system of an aromatic hydrocarbon (for example, naphthalene) without substitution of a hydrogen. The donated electron resides in a vacant π^*-antibonding molecular orbital of the aromatic hydrocarbon with the formation of a radical anion (for example, $C_{10}H_8^{\cdot-}$). This process is possible in strongly solvating media: the solvation energy of the alkali metal cation helps to stabilize the ionic product. Such addition compounds as $C_{10}H_8^{\cdot-}Na^+$ are colored and exhibit a high chemical reactivity.

Ionic metal-carbon bonds play a modest role in organometallic chemistry because of the relatively small number of metals (Groups I A and II A, excepting lithium and beryllium) able to participate in their formation.

Ionic organometallic compounds are extremely moisture-sensitive because of the high basicity of the carbanion, which abstracts a proton from water to form a hydrocarbon leaving an alkali metal hydroxide, and are reactive towards oxygen.

2.3. Sigma (σ) Covalent Bonds (Bicentric Bielectronic)

Covalent bonds (bicentric bielectronic) are found in organometallic derivatives of all elements, except those already mentioned as forming ionic bonds. The main group elements form compounds of the MR_n-type, called homoleptic compounds*, containing n metal-carbon σ-bonds, in which n is the typical valency of the element. The transition metals seldom form stable homoleptic compounds of the MR_n type; the instability of such compounds is of kinetic origin and generally arises from incomplete occupation of the *d*-orbitals. When electron pairs donated by π-acceptor ligands (able to form dative bonds of the type discussed in Section 2.6), for example, CO, PR_3, C_5H_5, etc., fill the *d*-orbitals, the kinetic stability of these transition metal compounds is considerably enhanced. Thus, $Ti(CH_3)_4$ is unstable at room temperature, but $(\pi\text{-}C_5H_5)_2TiR_2$ alkyls are stable.

Another cause of instability of σ-bonded transition metal organometallic compounds is the tendency of the organic group to be eliminated as an olefin (so-called β-elimination), with the formation of the metal hydride:

$$M—CH_2—CH_2—R \longrightarrow M—H + H_2C{=}CH—R$$

When the β-elimination is impossible because of the structure of the organic group (for example, in $—CH_2—SiMe_3$, $—CH_2—CMe_3$, $—CH_2—C_6H_5$) the σ-covalent compounds of the transition metals are stabilized.

The stability of σ-covalent organometallic compounds is determined by thermodynamic and kinetic factors. The thermodynamic stability is related to the heat of formation and can be judged with the aid of bond energies. Some bond energies (or dissociation energies) for several $M—CH_3$ bonds of nontransition metals are listed in Table 2.2.

* Homoleptic refers to all groups bound to a metal being identical; in heteroleptic compounds, two or more different groups are attached to a metal atom.

Tab. 2.2. Dissociation energies for the M—CH_3 bond (in kJ/mol).

Be	–	B	364	C	347	N	314
Mg	–	Al	276	Si	293	P	276
Zn	176	Ga	247	Ge	247	As	230
Cd	138	In	172	Sn	218	Sb	218
Hg	121	Tl	–	Pb	155	Bi	142

Some regularities can be observed. The M—C bond energies decrease descending a group, owing to the more diffuse character of s- and p-orbitals of the heavier elements, resulting in a less efficient overlap with the carbon sp^n-hybrid orbitals. The compounds with weak M—C bonds (for example, CdR_2, HgR_2, PbR_4, BiR_3) readily pyrolyze to deposit the metal. These compounds are thermodynamically unstable with respect to their decomposition to metals and hydrocarbons. They can, however, be isolated owing to kinetic stability, namely the lack of a decomposition mechanism with a sufficiently low activation energy.

The organometallic derivatives of the lighter elements (for example, BR_3, SiR_4, PR_3) have much higher thermal stabilities and show no tendency to decompose with release of the free element.

As mentioned above, the thermal stability of organometallic compounds is favored by an increase of the polar character of the metal-carbon bond. Thus, polyhalogenated, especially perfluorinated, groups form more stable compounds than analogous nonhalogenated groups. Also, aromatic groups tend to form more stable compounds than aliphatic.

All organometallic compounds are thermodynamically unstable with respect to oxidation. This is reflected in their high heats of combustion and arises from the formation of stable products (metal oxides, carbon dioxide and water). The oxidative stability observed for some organometallic compounds (for example, SiR_4, SnR_4, HgR_2, etc.) is again due to kinetic factors, namely the lack of oxidation mechanisms with low activation energy. Such kinetic stability is usually observed for compounds in which the central atom has no vacant orbitals of low energy. Conversely, such a mechanism is favored by the presence of vacant orbitals, and as a consequence the organometallic compounds of some elements (for example, AlR_3, ZnR_2, NaR, LiR) are spontaneously flammable in air. In other cases, the kinetic instability arises from the presence of lone pairs of electrons (for example, in SbR_3, BiR_3), which are also centers of reactivity. Thus, the tetravalent organometallic compounds of the elements from silicon to lead are insensitive to atmospheric oxygen and can be handled in the air, while the derivatives of the neighboring Groups III and V elements are much more sensitive to oxidation.

Some metal-carbon bonds readily hydrolyze according to the general reaction:

$$M{-}R + H_2O \longrightarrow M{-}OH + R{-}H$$

Hydrolytic stability depends upon the polarity of the M—C bond, and is favored by the presence of low-energy vacant orbitals on the metal. Vacant orbitals facilitate hydrolysis mechanisms of low activation energy through the possibility of coordinating the reagent (in this case water) with the metal in the transition state. Thus, compounds like ZnR_2, MgR_2, AlR_3, GaR_3 are water sensitive because of the polar character of their M—C bonds and the presence of vacant orbitals on the metal. The boron compounds of the type BR_3 are, however, hydrolytically stable, despite the presence of a vacant orbital at boron, because of the low polarity of the B—C bond.

Thus, kinetic factors play a significant role in determining the hydrolytic stability of organometallic compounds. Similar considerations are valid in judging the reactivity of organometallic compounds with other nucleophilic reagents (alcohols, amines, acids, etc.) possessing mobile hydrogen atoms.

2.4. Electron-Deficient Bonds (Polycentric Localized-Bonds)

Some organometallic derivatives of beryllium (BeR_2) and aluminum (AlR_3) are polymers despite the lack of lone electron pairs on the organic group able to bridge the metal with dative bonds. These compounds are associated by electron deficient or polycentric bonds.

Trimethylaluminum is, for example, a dimer, $[Al(CH_3)_3]_2$; the dimerization is achieved through CH_3 bridges between aluminum atoms:

$$\begin{array}{ccccc} H_3C & & \overset{H_3}{C} & & CH_3 \\ & \diagdown\, Al & \cdots\quad\cdots & Al\,\diagup & \\ H_3C & \diagup & \underset{H_3}{C} & \diagdown & CH_3 \end{array}$$

in which bridging bond distances are longer than in the terminal Al-CH_3 bonds. The formation of these bridging bonds can be explained as follows: the aluminum atom is hybridized sp^3, with four equivalent orbitals in tetrahedral orientation. Two orbitals are involved in the formation of normal, σ-covalent, electron-pair Al—CH_3 bonds, using two valence electrons of aluminum (Fig. 2.5).

On each $Al(CH_3)_2$ group thus formed there are then two sp^3-orbitals available for further bonding, but only one valence electron (see Fig. 2.5). On the other side, the carbon atom of a CH_3 group possesses a free sp^3-hybrid orbital containing a valence electron. By overlapping the sp^3-orbitals of the $Al(CH_3)_2$ groups with the sp^3-orbitals of the bridging CH_3 group in a way shown in Fig. 2.5, banana-shaped bonding orbitals are formed, each occupied by a pair of electrons which bonds three atoms in the Al···CH_3···Al bridge. These localized, three-center, bielectronic bonds are electron deficient by comparison with usual covalent bonds, and are weaker than two-center bonds because there is less electron density connecting the atoms. However, they are strong enough to produce the dimerization of several lower trialkyls of aluminum at

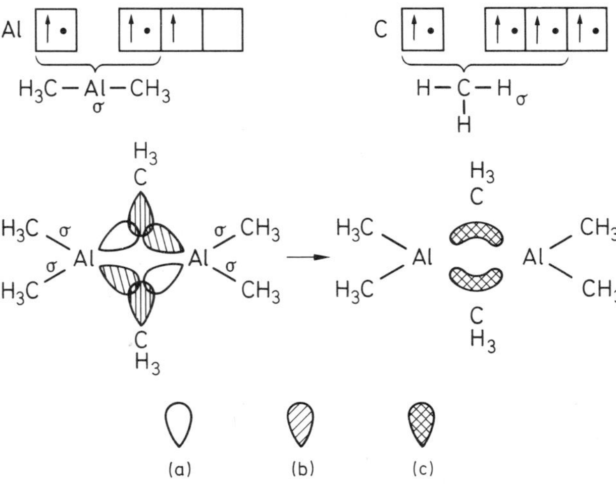

Fig. 2.5. The formation of three-center bonds in $[Al(CH_3)_3]_2$; (a) vacant orbitals; (b) orbitals occupied with one electron; (c) orbitals occupied with two electrons.

room temperature. The dissociation energies of the dimers are ca. 84 kJ/mol, which is comparable with the dissociation energy of some weak, electron-pair covalent bonds.

Triphenylaluminum also dimerizes. The participation of the phenyl group in bridge formation is probably achieved through a p_z-orbital of a phenyl carbon atom, possibly with some perturbation of the aromatic character of the ring (Fig. 2.6):

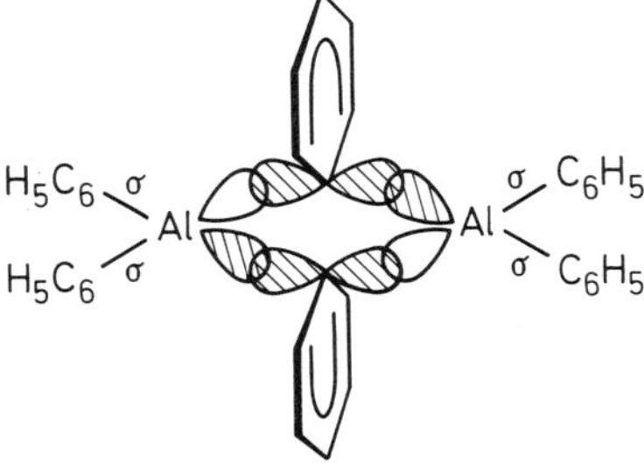

Fig. 2.6. The formation of bridging bonds in $[Al(C_6H_5)_3]_2$.

Additional examples of such three-center bonds are found in compounds such as $(\pi\text{-}C_5H_5)_2M(\mu\text{-}CH_3)_2Al(CH_3)_2$ (where M = Sc, Y, Dy, Ho, Er, Tm, Yb and μ- denotes the bridging methyls), $[Al(cyclo\text{-}C_3H_5)_3]_2$, and $[Al(C{\equiv}C{-}Ph)_3]_2$.
The diorganoaluminum hydrides, R_2AlH, are trimers, with electron deficient Al···H···Al bridging bonds:

R_2
Al
H H
R_2Al AlR_2
H

The bridging in dimethylberyllium can be explained in the same fashion. This compound forms chains in which the beryllium atoms are connected by two Be···CH_3···Be bridges, each metal atom being tetrahedrally surrounded by four carbon atoms. There are no normal σ-covalent bonds between beryllium and carbon because all four sp^3-orbitals of beryllium are involved in the formation of tricenter, bielectron bridges, as shown in Fig. 2.7:

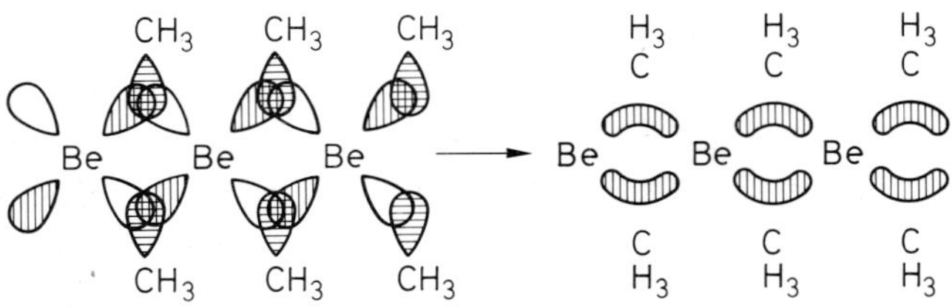

Fig. 2.7. The formation of three-center bonds in $[Be(CH_3)_2]_x$.

The formation of electron-deficient, polycenter bonds is confined to elements characterized by the following features: (a) a less than half-occupied valence shell (beryllium has 2 electrons and aluminum 3 electrons in a shell of four sp^3-orbitals which would require eight electrons to be completely filled); (b) an easily polarizable metal atom (due to a low charge: atomic radius ratio). The formation of this type of bond reflects the strong tendency of atoms to use as completely as possible their available valence orbitals and electrons in chemical-bond formation.

2.5. Delocalized Bonds in Polynuclear Systems

The organolithium compounds which form cluster tetramers $(LiR)_4$, hexamers $(LiR)_6$ and higher polymeric species, require an extension of the bonding theory. The formation of multiatom clusters through delocalized bonds means that we can no longer distinguish electron pairs binding pairs of atoms, and that we must consider several electrons as belonging to a group of atoms and holding the cluster together. The molecular orbitals of the polynuclear species thus formed from the atomic orbitals of the participating metal atoms will be occupied by spin-paired electrons. The number of electron pairs required to hold together a cluster of any size will be determined by the number of bonding molecular orbitals formed.

The structure of the $(LiR)_4$ tetramers (R = Me and Et) contains a tetrahedron of

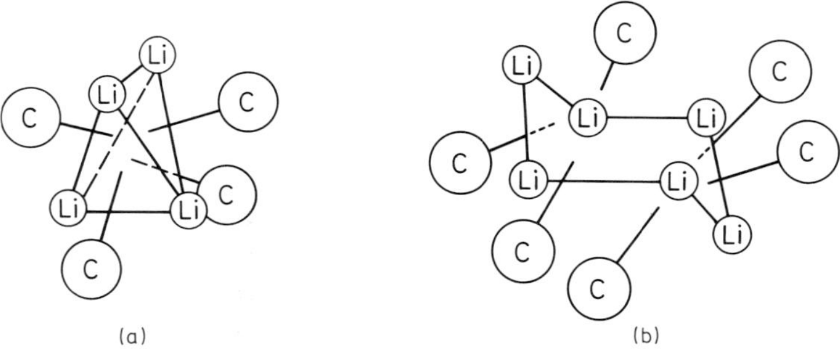

Fig. 2.8. The structure of $(LiR)_4$ (a) and $(LiR)_6$ (b).

lithium atoms (Li—Li distance = 256 pm≡2.56 Å) with the methyl groups located above the faces so that the carbon atoms are equidistant from all three lithium atoms (Fig. 2.8). The structure of $(LiC_2H_5)_4$ is somewhat more complicated, but also contains an Li_4 tetrahedron. For the $(LiR)_6$ hexamers a model has been proposed with the six lithium atoms in a chair-form hexagon, with the organic groups attached simultaneously to three lithium atoms.

The formation of Li_4 tetrahedra and Li_6 polyhedra can be interpreted by the sharing of lithium orbitals collectively. The carbon sp^3-orbital electrons are collectively shared, and delocalized in the cage formed by the lithium atoms. Alternately, the bond between the positively charged Li_n and the organic groups in carbanionic form, is essentially ionic.

In a similar manner, in carboranes (see Section 7.1.9), for example, in dicarba-closo-dodecaborane(12), $C_2B_{10}H_{12}$, as well as in the related anion $B_{12}H_{12}^{2-}$ (from which it formally derives by replacement of two B^- atoms with two isoelectronic carbon atoms), delocalized polycentric bonds serve to form an icosahedral cage (a regular figure with twelve vertices, composed of triangular faces). Each atom of this polyhedron forms an exo-cage bond with a hydrogen atom (B—H and C—H), leaving three sp^3-hybrid orbitals available for further bonding. Each boron atom also has two valence electrons, and each carbon atom has three. The collectivization of these hybrid orbitals and valence electrons to form molecular orbitals in the cage results in a super-aromatic molecule. Molecular-orbital calculations show that for each $B_nH_n^{2-}$ cluster, $2n + 2$ electrons are required, to fill the $n + 1$ bonding molecular orbitals. This can also be achieved by two carbon atoms each contributing three electrons, replacing two negatively charged boron atoms (B^-). This explains why only two carbon atoms can be introduced into the neutral carborane C_2B_{10} skeleton. Perhaps a three-carbon species based upon a C_3B_9 cluster would be stable in a cationic form, $C_3B_9H_{12}^+$, but so far such compounds have not been reported.

Delocalized bonds are also found in polynuclear metal carbonyl structures, and in other cluster-type organometallic compounds (see Chapter 11) in which metal atoms form polynuclear molecules via metal-metal bonding, sometimes with participation of carbon atoms. Transition metal bonding in polynuclear species is more complex because of the participation of d-orbitals in the formation of the delocalized electronic system.

2.6. Dative Bonds with Participation of *d*-Orbitals

Bonds between transition-metal atoms and unsaturated organic molecules are formed by electron donation in two opposite directions: from ligand to metal (direct donation) and from metal to ligand (back-bonding). To participate in such bonds the metal should have a partially occupied d-shell, and the unsaturated organic molecule must have vacant antibonding molecular orbitals.

The simplest example is the bond formed between carbon monoxide and metal atoms in metal carbonyls. To form such bonds, the metal should be in a lower oxidation state, preferably zero or ± 1, and must have vacant d-orbitals able to accept electrons donated by the coordinated CO molecule, and occupied d-orbitals able to donate d-electrons to the ligand. The carbon monoxide molecule has a lone pair of electrons in a molecular orbital localized at carbon (Fig. 2.9) which can be transferred

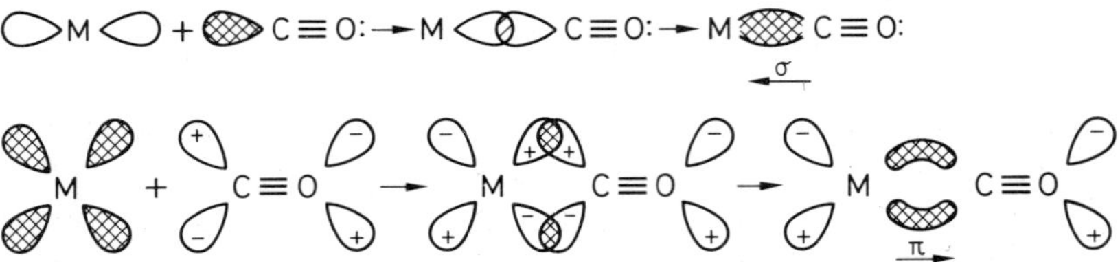

Fig. 2.9. The formation of dative bonds in metal carbonyls.

into a vacant orbital of the metal thus forming a σ-bond of the donor-acceptor type. The carbon monoxide molecule has vacant antibonding molecular orbitals which can accept electron pairs from the occupied *d*-orbitals of the metal to form a π-bond. The bond between a metal atom and a carbon monoxide molecule is then actually a double bond, formed by the superposition of a σ-bond (ligand-to-metal donation) and a π-bond (metal-to-ligand donation). Back-donation avoids the accumulation of an excess of electron density at the metal atoms and strengthens the bond between the metal and ligand.

The electron donation between metal and ligand also influences the strength of the bond between carbon and oxygen in the coordinated CO molecule. Since the metal electrons are accepted into the antibonding molecular-orbital of the CO molecule, the C—O bond order decreases, as can be shown by infrared spectroscopy. The CO bond stretching frequency gives information about the metal-ligand bond strength as shown by the ν_{CO} stretching frequencies in several neutral, anionic and cationic metal carbonyl species, compared with the magnitude of ν_{CO} in the CO molecule:

free CO: 2155 cm^{-1}

$[Mn(CO)_6]^+$ 2090 cm^{-1}		$Ni(CO)_4$	2128	2057 cm^{-1}
$Cr(CO)_6$ 2000 cm^{-1}		$[Co(CO)_4]^-$	1918	1883 cm^{-1}
$V(CO)_6$ 1973 cm^{-1}		$[Fe(CO)_4]^{2-}$	1790	1783 cm^{-1}
$V(CO)_6^-$ 1859 cm^{-1}				

Increasing the electron density at the metal decreases the CO stretching frequency and bond order, in agreement with donation to antibonding orbitals of the ligand, and the M—C bond strengthens.

The bonds between transition metal atoms and other ligands having a pair of electrons in a σ-type orbital are formed similarly: carbon monosulfide, :C=S, alkyl and aryl isocyanides, :C=N—R, and carbenes, $:CR_2$. In the latter case, the acceptor orbital of the ligand is a p_z-orbital:

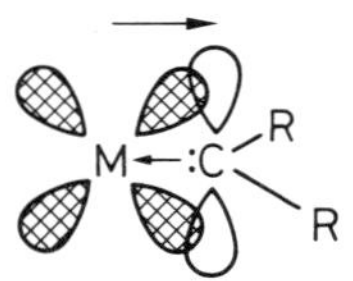

In most metal carbonyls (as well as in many other organometallic compounds) the transition metal tends to achieve a noble gas configuration, accepting as many electrons from the ligands as is required to complete the 18 electrons set in the valence shell, for example, an $(n-1)d^{10}ns^2np^6$ configuration. This empirical *18-electron rule* affords a simple guide to the composition of metal carbonyls and their substituted derivatives. The transition metals achieve an *effective atomic number* equal to that of the following noble gas.

Table 2.3 lists some typical metal carbonyls which obey the rule. Metals having even atomic numbers achieve the 18-electron configuration by bonding the number of CO molecules required to provide the necessary electron pairs. Metals having odd atomic numbers dimerize through metal-metal bond formation. Thus, manganese in the

Tab. 2.3. The formation of the 18-electron configuration in some metal carbonyls.

Compound	Metal	No. of valence electrons of the metal	CO ligand contribution	Electron balance in the metal valence shell
$Cr(CO)_6$	Cr	6	6×2	$6 + 12 = 18$
$Fe(CO)_5$	Fe	8	5×2	$8 + 10 = 18$
$Ni(CO)_4$	Ni	10	4×2	$10 + 8 = 18$
$(CO)_5Mn—Mn(CO)_5$	Mn	7	5×2	$7 + 10 + 1 = 18$
$(CO)_4Co—Co(CO)_4$	Co	9	4×2	$9 + 8 + 1 = 18$

group -$Mn(CO)_5$ and cobalt in the group -$Co(CO)_4$ have only 17 electrons in their valence shells, but each can achieve 18-electron shells after pairing the odd electron of each fragment by dimerization. Another possibility is to accept an additional electron (for example, from an alkali metal) with formation of $[Mn(CO)_5]^-$ and $[Co(CO)_4]^-$ anions.

There are exceptions to the 18-electron rule. Vanadium hexacarbonyl, $V(CO)_6$, has only 17 electrons (and is paramagnetic), but does not dimerize because the octahedral arrangement of the ligands is particularly stable. The neutral vanadium hexacarbonyl exhibits a marked tendency to accept an electron, however, to form the anionic species, $[V(CO)_6]^-$.

Some transition metals, particularly Rh, Ir, Pd can be satisfied with 16 electrons in their valence shells, and form carbonyls of a comparable stability to that of the 18-electron species.

The bond between a metal atom and an olefin (for example, ethylene in the simplest case) differs from the previous example in that the ligand can involve its π-orbitals both in donation and back-bonding. As shown in Fig. 2.10, the bonding π-orbital of ethylene is involved in electron donation and the antibonding π^*-molecular orbital in backbonding. Only the lobes of the d-orbitals participating in the bonding are shown. The cross-hatched orbitals are occupied and donate electrons; the other orbitals are vacant and accept electrons:

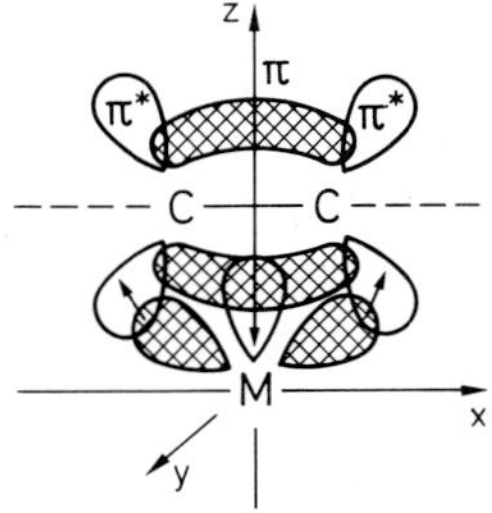

Fig. 2.10. The formation of metal-olefin bonds.

A better description is obtained if we combine the molecular orbitals of ethylene (both π-bonding and π^*-antibonding) with each of the various valence orbitals of the metal, as shown in Fig. 2.11: The s-, p_z-, $d_{z^2,\ x^2-y^2}$-orbitals of the metal can overlap with the occupied π-orbitals of the olefin, and participate in the ligand-to-metal donation. The p_x- and d_{xz}-orbitals of the metal overlap with the π^*antibonding molecular orbitals of the olefin, and participate in the back-bonding.

In this type of bonding the ethylene molecule uses its two electrons located in the π-molecular orbital, and, therefore, acts as a two-electron donor. Other unsaturated organic molecules or groups can contribute three electrons (π-allyl), four (butadiene and cyclobutadiene), five (cyclopentadienyl) or six electrons (benzene), simultaneously

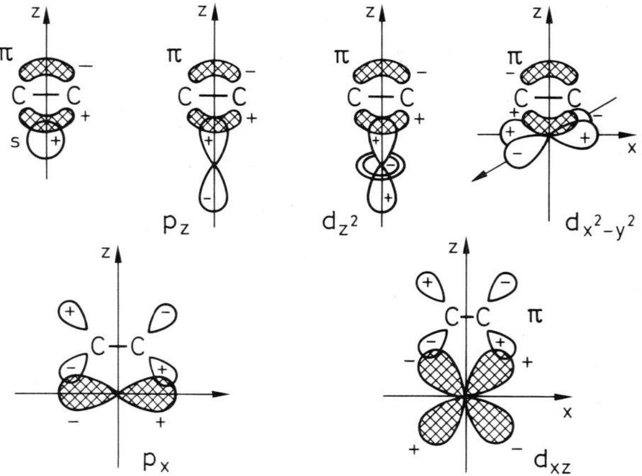

Fig. 2.11. Participation of various orbitals in the formation of π-olefin complexes.

accepting electron density from the occupied d-orbitals of the metal. The bonding in these systems can also be described with the aid of molecular-orbital theory. The molecular orbitals of the ligand having the appropriate symmetry will overlap with the atomic orbitals of the metal. Figure 2.12 shows the atomic orbitals of a transition-metal atom, and Figure 2.13 lists the molecular orbitals of the most important ligands encountered in organometallic chemistry:

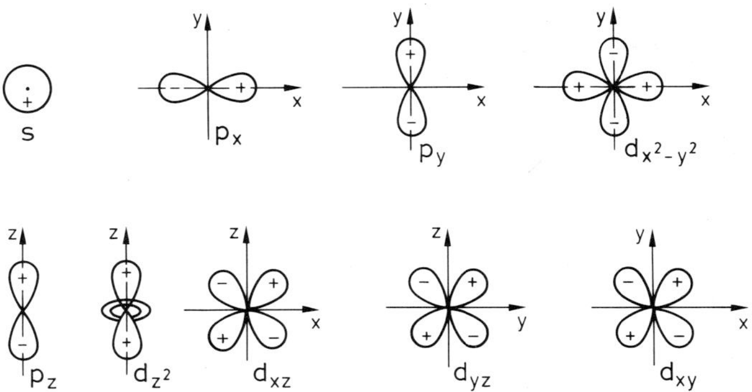

Fig. 2.12. Atomic orbitals of transition metals, participating in the bonding in π-complexes.

The π-orbitals of the ligand can combine with the s-, p_z- and d_{z^2}-orbitals of the metal to form metal-carbon bonds of σ-symmetry. When two lobes of the metal atomic orbitals (p_x-, p_y-, d_{xy}-, d_{yz}-)overlap two lobes of a ψ_2- and ψ_3- of the ligand, the bonds formed are of π-symmetry. The superposition of four lobes of the metal atomic orbitals ($d_{x^2-y^2}$- and d_{xy}-) with the orbitals of the ligand having two nodal planes, (such as ψ_3 for allyl, ψ_4 and ψ_5 for the other ligands) leads to formation of δ-bonds. The metal

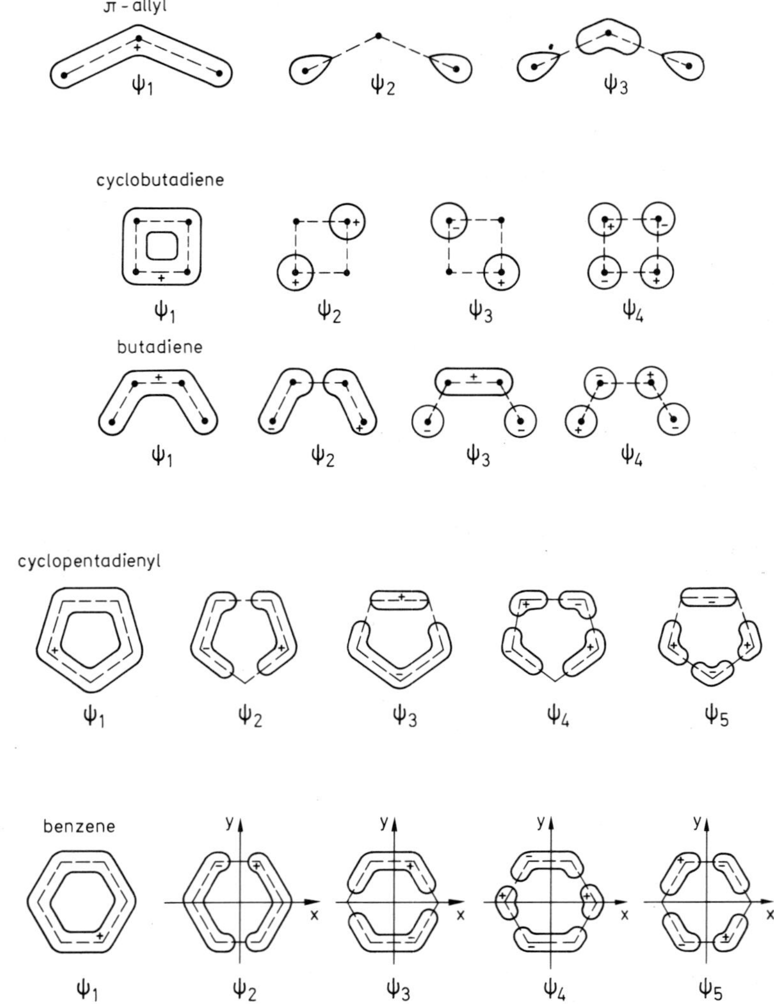

Fig. 2.13. The orbitals of organic ligands in π-complexes.

atom is placed at the center of the coordinate system and the unsaturated ligand is located on the z-axis, in a plane parallel to the xy-plane. Table 2.4 summarizes this discussion by listing the bonds (see Figures 2.12 and 2.13).

The 18-electron rule can also be used for unsaturated organic ligands: ligands combine to complete the effective atomic number of the next noble gas. For example, chromium needs twelve electrons to achieve a krypton-like electron configuration which it can obtain from two benzene molecules, each contributing six electrons, to form dibenzenechromium, $Cr(C_6H_6)_2$; or from a benzene molecule and three carbon

Tab. 2.4. Combination of the metal atomic orbitals with the molecular orbitals of the ligand.

π-Ligand*	Metal atomic orbitals σ-bonds	 π-bonds		 δ-bonds
	ns, np $(n-1)d_{z^2}$	p_x, p_y $(n-1)d_{xz}$ $(n-1)d_{yz}$		 $(n-1)d_{x^2-y^2}$ $(n-1)d_{x^2-y^2}$
allyl (3)	ψ_1	ψ_2		
cyclobutadiene (4)	ψ_1	ψ_2	ψ_3	
butadiene (4)	ψ_1	ψ_2	ψ_3	
cyclopentadienyl (5)	ψ_1	ψ_2	ψ_3	ψ_4
benzene (6)	ψ_1	ψ_2	ψ_3	ψ_4

* In parentheses is given the number of electrons contributed by the ligand to the 18-electron set in the valence shell.

monoxide ligands to form $C_6H_6Cr(CO)_3$. Iron needs 10 electrons, which can be obtained from two cyclopentadienyl groups, each contributing five electrons to form ferrocene, $Fe(C_5H_5)_2$. Other examples are given in Table 2.5.

Tab. 2.5. Formation of 18 electron configurations in some π-cyclopentadienylmetal carbonyls.

Organometallic compound	Metal	Number of valence electrons of the metal	Ligand contri-bution*	Electron balance
π-$C_5H_5Cr(CO)_3CH_3$	Cr	6	5 + 6 + 1	18
π-$C_5H_5Mn(CO)_3$	Mn	7	5 + 6	18
π-$C_5H_5Fe(CO)_2CH_3$	Fe	8	5 + 4 + 1	18
π-$C_5H_5Co(CO)_2$	Co	9	5 + 4	18
π-$C_5H_5Ni(CO)CH_3$	Ni	10	5 + 2 + 1	18

* The π-C_5H_5 group contributes five electrons, the CO groups two electrons each and a σ-bonded CH_3 group one electron.

Exceptions to the 18-electron rule include stable π-cyclopentadienylmetal compounds containing only 16 electrons in the valence shell, for example, titanium in $(\pi\text{-}C_5H_5)_2TiCl_2$ $[4 + (2 \times 5) + 2$ electrons]. Other exceptions are cobaltocene, $Co(\pi\text{-}C_5H_5)_2$ (19 electrons), and nickelocene, $Ni(\pi\text{-}C_5H_5)_2$ (20 electrons). These compounds are readily converted into those satisfying the 18-electron rule. More exceptions from the 18-electron rule can be found in the chemistry of d^8-systems, such as Ni(II), Pd(II), Pt(II), Au(II), Rh(I) and Ir(I).

The assignment of structures to π-ligand complexes on the basis of the 18-electron rule is a simple matter where the organic ligands use all available π-electrons for bonding with the transition metal atom, and if the number of π-electrons donated by the cyclic polyenes is equal to the number of carbon atoms in the ring. Indeed, this is the dominating tendency, for example, cyclobutadiene contributes four electrons, the cyclopentadienyl group five electrons, benzene six electrons, etc. However, many examples are known in which a ligand with several π-electrons does not use all of them in bonding with the metal. In other cases, some carbon atoms of the ligand can change their hybridization from sp^2- in which they can contribute an electron each to metal-ligand bonding, to sp^3- in which there is a saturated tetrahedral carbon, without π-electrons. The sp^3-hybridized carbon atoms are pushed out of the planar conjugated system and do not participate in the metal-ligand bond. This happens when the metal needs fewer electrons than the number offered by the ligand. As a result, many-electron ligands can exhibit various structures in transition metal complexes. Thus, cyclooctatetraene, COT, seldom uses its eight π-electrons, except with early transition metals which need many electrons to achieve the noble gas configuration.

Cyclopentadiene, C_5H_6, can act as a four- or two-electron donor, and the cyclopentadienyl group, $C_5H_5^-$, derived from it shows even more possibilities (Fig. 2.14). All these possibilities have been identified in transition metal complexes:

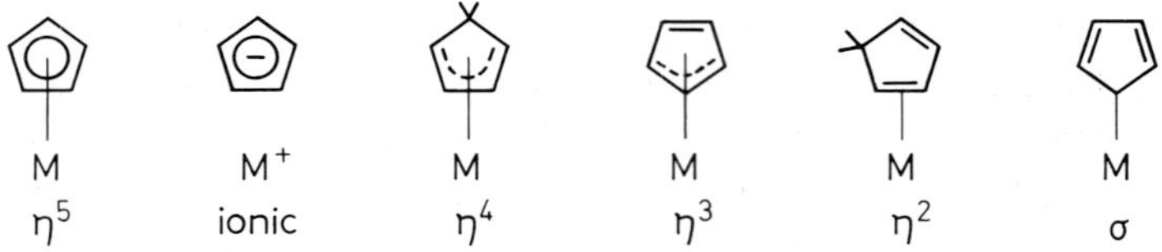

Fig. 2.14. Bonding possibilities of cyclopentadiene and cyclopentadienyl groups to metals.

In a similar manner, benzene and its substituted derivatives can act as six-, four-, or two-electron ligands, or the phenyl groups can be σ-bonded (Fig. 2.15):

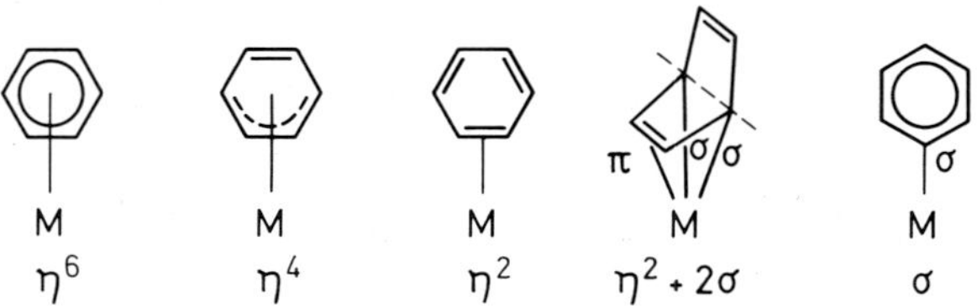

Fig. 2.15. Bonding possibilities of benzene to metals.

Similar possibilities are known for the seven- (Fig. 2.16) and eight-membered rings (Fig. 2.17):

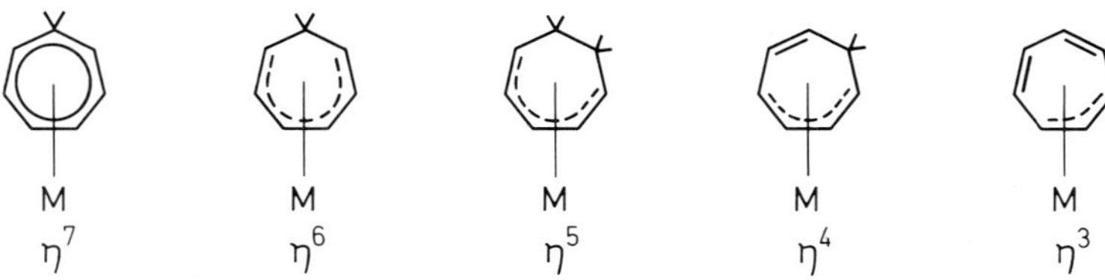

Fig. 2.16. Bonding possibilities of cycloheptatriene and cycloheptatrienyl group to metals.

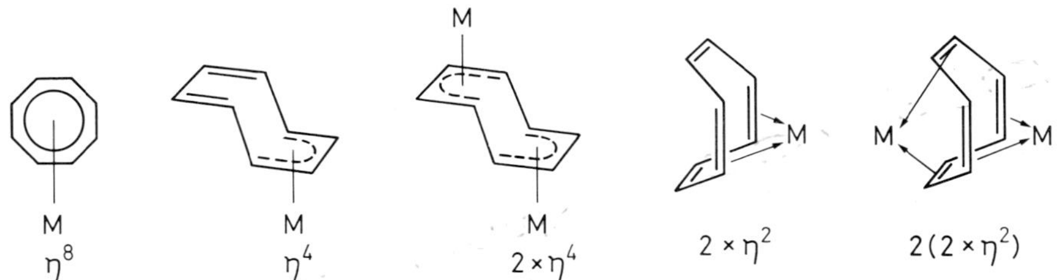

Fig. 2.17. Bonding possibilities of cyclooctatetraene to metals.

In order to indicate for each ligand the number of carbon atoms attached to a metal, a nomenclature system has been proposed, based upon the use of the Greek prefix *hapto-* (in Greek the word *haptein* means to fasten). Thus, monoolefins, in which the metal forms two equivalent bonds to the two carbon atoms are called di*hapto*-ligands (denoted as η^2 or h^2); the allyl group is tri*hapto-* (η^3), cyclobutadiene is tetra*hapto-* (η^4), etc.

Tab. 2.6. Possible combinations of various fragments in mixed ligand π-complexes.

Fragment	Number of electrons required for a noble gas configuration	π-Ligands which can be attached to the parent fragment
Metal-carbonyls		
$Cr(CO)_3$	6	η^6-C_6H_6, η^5-$C_5H_5 + \sigma$-R
$Mn(CO)_3$	5	η^5-C_5H_5
$Fe(CO)_3$	4	η^4-C_4H_4, η^4-C_4H_6
$Co(CO)_3$	3	η^3-C_3H_5
$Co(CO)_2$	5	η^5-C_5H_5
Metal cyclopentadienyls		
η^5-C_5H_5Cr	7	η^7-C_7H_7, 3CO + σ-R
η^5-C_5H_5Mn	6	3CO
η^5-C_5H_5Fe	5	η^5-C_5H_5
η^5-C_5H_5Co	4	η^4-C_4H_4
η^5-C_5H_5Ni	3	η^3-C_3H_5

There are many compounds known which contain several different ligands attached to the same metal atom. In the electron count the sum of the contributions of all ligands must be taken into consideration. Most common is the presence of cyclopentadienyl groups and carbonyl groups in the same molecule. Table 2.6 shows for several metals involved in various structural units in π-complexes, the number of electrons required by each fragment for completing the noble gas configuration. The last column shows the organic groups which can be used to form organometallic compounds with noble gas configurations.

The bonding capabilities of the $M(CO)_n$ (n = 3,4 and 5) fragments are determined by the combination of the available valence orbitals of the metal. A detailed analysis, based upon molecular-orbital theory can thus explain why of the two metal carbonyl fragments, $Mo(CO)_4$ and $Fe(CO)_3$, both requiring four electrons to satisfy the 18-electron rule, the former coordinates two isolated double bonds, while the latter coordinates a conjugated diene:

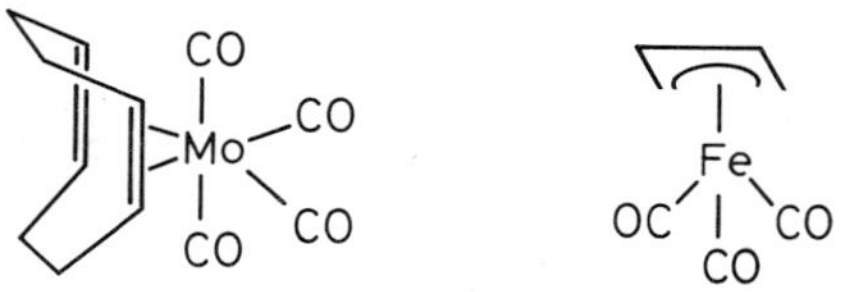

CO CO Mo CO CO

Fe OC CO CO

The orientation of the molecular orbitals of the fragments $M(CO)_4$ and $M(CO)_3$ is such that a better overlap is achieved in each case with a π-ligand as mentioned:

CO CO Mo CO CO

M OC CO CO

As a result, when an unconjugated diene (for example, 1,5-cyclooctadiene) is offered to an iron carbonyl fragment, the metal will force an isomerization, to produce a $Fe(CO)_3$ complex of the preferred conjugated diene, while molybdenum carbonyl will effect the reverse isomerization when offered a conjugated diene (for example, 1,3-cyclooctadiene):

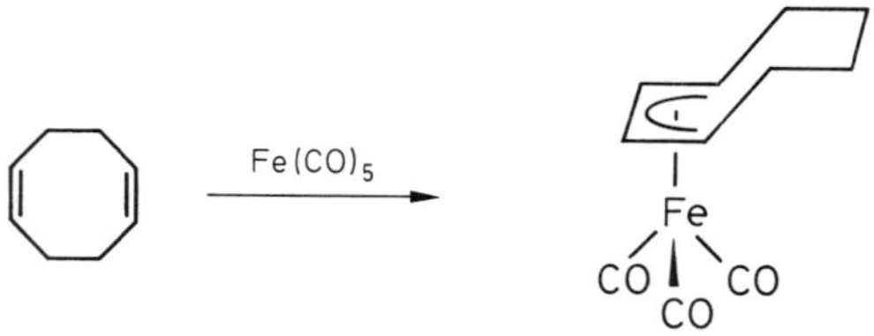

$Fe(CO)_5$

Fe CO CO CO

$Mo(CO)_6$

The structural diversity of organometallic compounds derived from cyclopolyenes is further favored by the various bonding possibilities of the ligand. An example is the *pentahapto*-cyclopentadienyl group, which can form both parallel (antiprismatic or staggered and prismatic or eclipsed) and bent structures when two η^5-C_5H_5 groups are attached to a transition metal atom, or can form triple-decker sandwiches, or can be attached to two metal atoms as η^1-, η^5-unit:

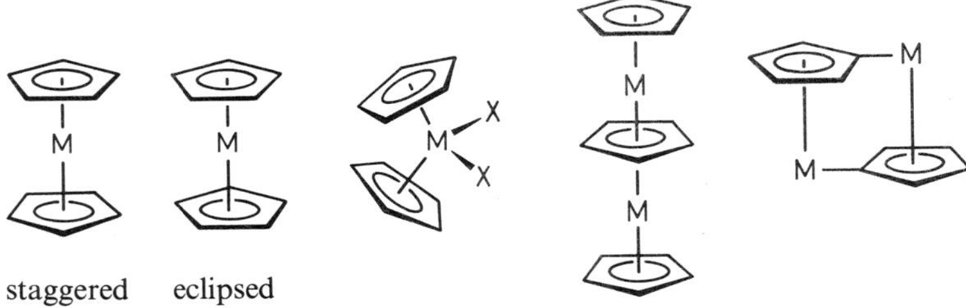

staggered eclipsed

Non-transition metals which participate only with s- and p-orbitals, also form monohapto-, dihapto- or pentahapto- derivatives, as well as mixed types, as shown in the following examples:

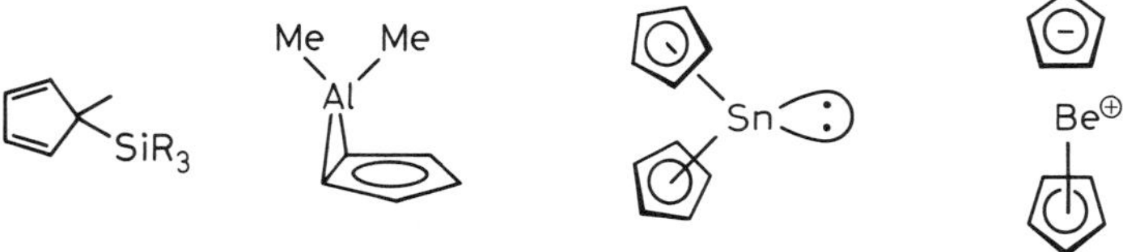

Thus some of the most-unexpected structures occur in the organometallic chemistry of the organic π-ligands.

3. Laboratory Techniques in Organometa Chemistry

3.1. Synthesis and Isolation of Compounds

Many organometallic compounds are sensitive to atmospheric oxygen and carbon dioxide, to moisture and even to light. Among these, the organolithium and organomagnesium compounds, often used as intermediates in the synthesis of organometallic derivatives, must be noted. The lower boron and aluminum alkyls are pyrophoric, and contact with atmospheric oxygen may lead to explosions. Finally, some organometallic compounds are stable towards air and water as solids (for example, many transition metal derivatives), but in solution are readily oxidized in contact with the atmosphere. There are classes of organometallic compounds which can be prepared and handled on the benchtop, but for most special precautions must be observed, such as the use of anhydrous and deoxygenated solvents, as well as anhydrous and inert atmospheres. The amount of materials used in research should be limited, especially when working with compounds whose properties have not been investigated. The amount of materials used in a reaction is often limited by a cost factor; usually organometallic reagents are expensive (for example, platinum metal carbonyls). Proper laboratory technique is of great importance and deserves a brief discussion here.

The laboratory techniques used in the classical period of organometallic chemistry are now considered primitive, but we can still admire the ingenuity of the founding fathers. The older techniques have largely been supplanted, but vacuum techniques devised by A. Stock are still used in modern laboratories for handling air- and moisture-sensitive compounds. The well-known series *Inorganic Syntheses* and *Organic Syntheses* can also be consulted for individual receipies.

The purification of solvents is extremely important in the work with organometallic compounds. They are normally degassed either by simple distillation before use, usually under inert atmosphere, or by bubbling a stream of dry nitrogen or argon through just before use. To remove the moisture – depending on their chemical nature – the solvents are stored and distilled over metallic sodium (ethers, hydrocarbons), lithium aluminum hydride (tetrahydrofuran, ethers), phosphorus pentoxide, calcium chloride (chlorinated solvents), molecular sieves, magnesium or calcium oxide and other dehydrating agents.

There are, in principle, two types of working procedures: *in vacuo* (in vacuum lines) or benchtop techniques under inert atmosphere. The vacuum-line technique is better, but it is somewhat cumbersome and slow, rather expensive, and for many purposes too sophisticated. Therefore, it is used only in special cases. Benchtop techniques are preferred, and an inert atmosphere is provided by passing a slow stream of inert gas over the solutions or by working in special air- and moisture-free spaces under controlled atmosphere (glove bags and boxes). The most frequently used inert gases are nitrogen and argon. The former is cheaper, but some organometallic compounds react with nitrogen. Argon is more expensive, but it is more efficient in preventing the diffusion of air toward the solution protected by the inert gas.

The inert gases are purified by passing through columns able to retain the oxygen (metallic copper heated to 600 °C, metal oxides in a lower valence state, or special products like the so-called "BTS catalyst" produced by BASF in West Germany). The moisture is retained in tubes or columns filled with solid dehydrating agents such as phosphorus pentoxide, magnesium perchlorate, molecular sieves, anhydrous calcium sulfate, anhydrous calcium chloride, or by passing the gas through concentrated sulfuric acid. Figure 3.1 shows such a nitrogen purification line schematically:

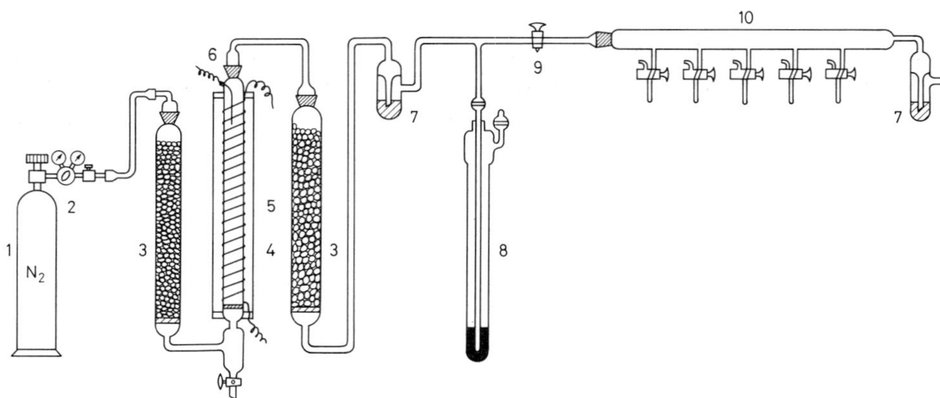

Fig. 3.1. The scheme of a nitrogen purification line
1 = Nitrogen cylinder; 2 = Pressure gauge; 3 = Drying column; 4 = Tube for deoxygenation; 5 = Active filling; 6 = Electric heating; 7 = Bubble counter; 8 = Safety valve with mercury; 9 = Stopcock; 10 = Distribution line.

The laboratory equipment used for the synthesis, isolation and purification of organometallic compounds is conventional with standard joints and gas inlets for introduction of the protecting inert gas. Figure 3.2 shows a three-necked flask equipped for introducing nitrogen (gas inlet); the calcium chloride tube prevents the contact of atmospheric moisture with solutions.

For inert atmosphere operations with amounts of materials from 1 to 100 grams, so-called Schlenk tubes (Fig. 3.3) are often used with appropriate devices for filtration, recrystallization, sublimation and other operations.

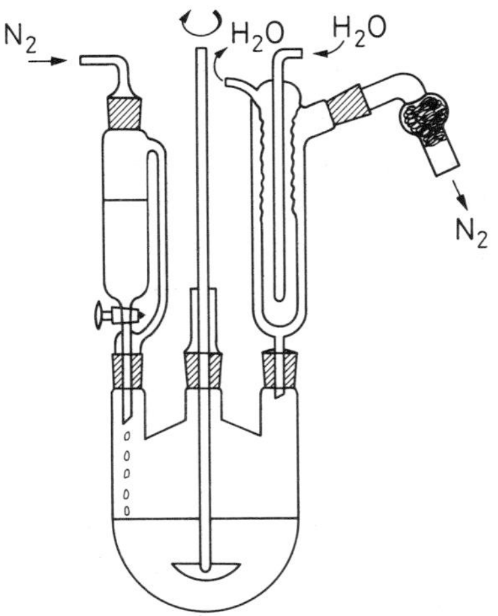

Fig. 3.2. Reaction flask for syntheses under inert atmosphere.

Fig. 3.3. Schlenk tubes.

In Fig. 3.4 a set-up for recrystallization and filtration in inert atmosphere is illustrated. Generally, it is not difficult to ensure anaerobic and anhydrous conditions during the reactions, but adequate care must be exercised during the isolation and purification operations.

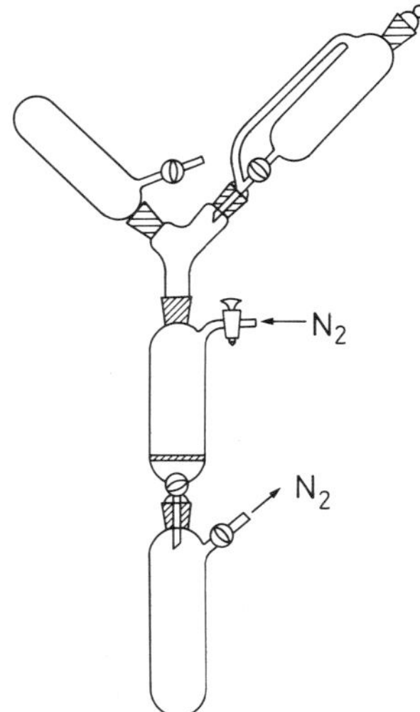

Fig. 3.4. Apparatus for filtration and recrystallization under inert atmosphere.

The synthesis of organometallic compounds is generally achieved in one of two ways: either by a direct reaction between the metal and an organic compound, or by first preparing a reactive organometallic intermediate, for example, an organolithium or organomagnesium reagent (by direct reaction) which is then reacted with the appropriate compound of a desired metal. For the direct reaction of metals with organic compounds, activated forms (usually freshly prepared by reduction) are recommended. A recent development is the use of a technique called *metal-atom synthesis* in which the metal to be reacted with an organic compound is evaporated first *in vacuo,* and the metal vapor (usually containing isolated atoms of the metal) is co-condensed on a cold wall at liquid nitrogen temperature with the organic reagent. This technique is very useful, since it has allowed preparations which otherwise could not have been possible. In some instances, electrochemical preparations of organometallic compounds can also be performed.

For the purification of organometallic compounds the usual methods can be employed, provided that the precautions already mentioned are observed during the operations. The purifications of liquids is done by distillation, in most cases under reduced pressure in order to decrease the boiling temperature, and thus minimize decomposition. This precaution is often required in organometallic chemistry, since high temperatures cause decomposition, molecular rearrangements or other reactions.

For the separation of solid substances, particularly transition metal compounds which are often colored, column chromatography is currently used by employing

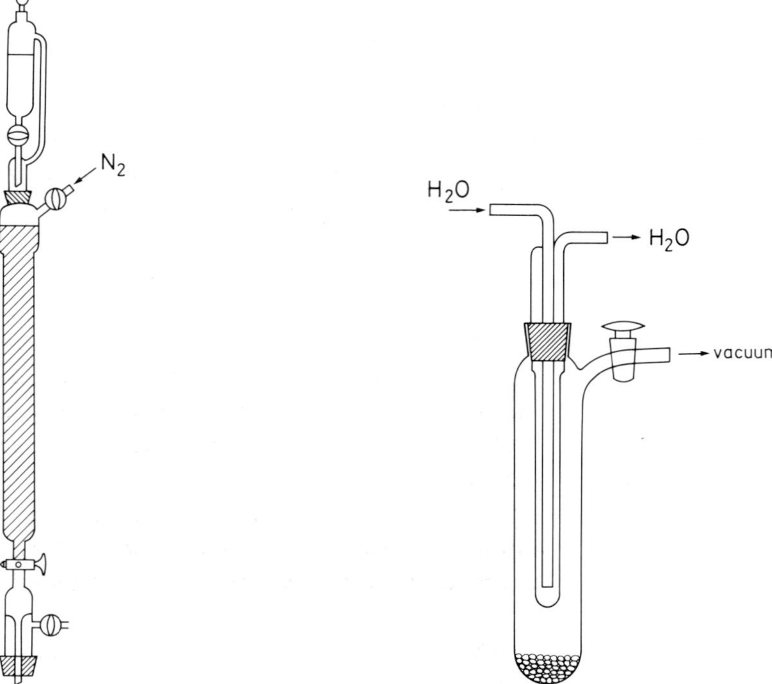

Fig. 3.5. Chromatographic column.

Fig. 3.6. Sublimation apparatus.

supports like silica gel, β-alumina or Fluorosil* (Fig. 3.5), and vacuum sublimation is employed if the stability and volatility of the organometallic compound permit (Fig. 3.6). Recrystallization is also an important procedure used for the purification of solid substances.

Generally, the laboratory operations used in organometallic chemistry resemble those used in organic chemistry, except for the additional precautions associated with the need to ensure an inert atmosphere.

3.2. Analysis and Structural Characterization

Organometallic chemistry is still at a stage in which synthesis plays an extremely important role. Every year thousands of new compounds are prepared and identified by analysis and characterized by spectroscopic and other physical methods. For this purpose all the classical and modern methods of organic and inorganic chemistry (elemental analysis, determination of physical properties such as melting point, boiling point, refractive index, density, spectroscopic characterization – or fingerprinting – and finally, structure determination by diffraction methods) are used.

The elemental analysis of organometallic compounds sometimes raises problems, mainly connected with the sampling and handling of air-sensitive compounds. Adequate methods are now available for most elements.

Among the spectroscopic methods, of particular importance is infrared spectroscopy, often supplemented by Raman spectroscopy, both giving vibrational information. This technique is important in the chemistry of transition-metal compounds containing metal-carbonyl bonds, since it affords a distinction between terminal and bridging carbonyl groups. Generally, the infrared spectrum plays a fingerprint role, since the spectra are rich. More sophisticated use of vibrational spectroscopy (correlated infrared and Raman) gives information about bond strengths, molecular symmetry and other features of interest.

Nuclear magnetic resonance is important in detecting the types of hydrogen atoms in organometallic molecules (proton magnetic resonance), the structural equivalence or non-equivalence of various atoms, and to estimate quantitatively the number of each kind of hydrogen atoms. It can be applied for the study of other nuclei (^{13}C, ^{11}B, ^{19}F, ^{29}Si, ^{31}P, ^{119}Sn, etc.). Nuclear magnetic resonance played a special role in the discovery and investigation of dynamic stereochemistry (fluxional behavior) in organometallic molecules.

Mass spectrometry is an important tool for investigation since it furnishes information concerning the molecular size (mass) and affords structural information which can be deduced from the molecular fragmentation under electron impact. The applications of this technique to organometallic chemistry are now numerous, but it is

* A mixture of magnesium oxide and silica.

limited to those compounds which are volatile and stable enough to survive vaporization into the mass spectrometer chamber.

No physical method is more important than X-ray diffraction which can be used when single crystals can be obtained for the investigation of molecular structure, and permits establishing the exact crystal and molecular structure, with bond distances and angles. Owing to the progress made in the computing techniques, this method (which requires extensive calculations) has become common.

Other techniques of structural investigation are used in certain cases, for example, electron diffraction (in the gas phase, applicable to simpler, volatile molecules), neutron diffraction (less used but affords the position of hydrogen atoms), photoelectron spectroscopy (gives information about the molecular-orbital energy levels in a molecule), Mössbauer spectroscopy (especially for organotin, organoiron and organoantimony compounds), electron spin resonance (for free-radical detection and investigation).

4. The Literature of Organometallic Chemistry

The explosive growth of the organometallic literature (original papers and communications, reviews, monographs and books) witnessed in recent years raises difficult problems both for the specialist and for the newcomer. The situation is made easier by the secondary literature (reviews and monographs) which selects (hopefully critically) and systematizes the most important facts and data, sending the reader to the primary publications only for detailed information. This chapter is intended as an initiation to the literature of organometallic chemistry, listing the general monographs and books and the specialized and general journals publishing such literature.

A detailed and systemic presentation of the literature on organometallic compounds of transition metals [1] and non-transition metals [2] is available and may serve as an excellent guide.

Searching the literature in organometallic chemistry will usually involve the following steps:

1. A book or monograph dealing with the subject. General information in a comprehensive treatise will suffice in most cases. For specific information a more detailed monograph (for example, the *Gmelin Handbook* series) will supply the information and the references to the original literature. Many recent inorganic chemistry textbooks provide a first entry into the field and may serve as a starting point.*
2. Primary journals are for specific details and for the most recent literature, not yet reviewed in the secondary literature. The current literature can be surveyed both directly, by regular consultation of selected journals, and/or with the help of *Chemical Abstracts,* in which Section 29 deals specially with organometallic compounds. The specialist follows the current primary journals to be continually informed about the most recent developments. This requires a great effort and is costly, but it may be a source of much satisfaction, pleasure and fresh ideas.

* Some inorganic chemistry textbooks which can be suggested for this purpose are:

F. A. Cotton and G. Wilkinson, *Advanced Inorganic Chemistry,* 4th ed., J. Wiley & Sons, New York, (1980).

J. E. Huheey, *Inorganic Chemistry. Principles of Structure and Reactivity,* 3rd ed., Harper & Row, New York (1983).

K. F. Purcell and J. C. Kotz, *Inorganic Chemistry,* W. B. Saunders Co., Philadelphia (1977).

J. C. Bailar, H. J. Emeléus, R. Nyholm, A. F. Trotman-Dickenson (eds.), *Comprehensive Inorganic Chemistry,* Pergamon Press, Oxford (1973), vols. 1–5.

T. Moeller, *Inorganic Chemistry: A Modern Introduction,* J. Wiley & Sons, New York (1982).

B. Douglas, D. McDaniel and J. Alexander, *Concepts and Models of Inorganic Chemistry,* 2nd ed., J. Wiley & Sons, New York (1983).

W. W. Porterfield, *Inorganic Chemistry, A Unified Approach,* Addison-Wesley, Reading, MA (1983).

4.1. Textbooks and General Monographs

The first comprehensive monograph on organometallic chemistry was published in 1937 by A. Krause and A. von Grosse [3]. In a voluminous tome these authors collected practically all the knowledge on organometallic chemistry available at the time and systematized all the literature published until 1935.

The rapid development of organometallic chemistry in the years following WWII made it impossible to cover the field exhaustively in a single volume. Therefore, the organometallic chemistry treatises published in the last 15 years, are either encyclopedic, multi-volume works directed to the specialist active in the field, or introductory handbooks, directed to a larger audience, particularly students making a first acquaintance with the field.

For rather detailed information, with references to the original literature, the volumes of Coates, Green, Wade and Aylett [4] is an indispensable work. Other textbooks and introductory level books on organometallic chemistry have been published [5–7]; some concentrate mainly on transition-metal chemistry [8–10]. Although not a textbook, the monograph edited by H. Zeiss [11] played an important role in the 1960's as a general presentation of the subject. Some elementary texts, written for the non-specialist or the undergraduate student, with the intent of stimulating interest in organometallic chemistry, are also mentioned here [12–14].

The industrial applications of organometallic compounds have been surveyed by J. L. Harwood [15] in a book based heavily on the patent literature.

Some recent attempts to systematize the organometallic literature in multi-volume monographs cover the field exhaustively or at least very comprehensively. One is the series edited by A. N. Nesmeyanov and K. A. Kocheshkov [16] published in Russian and partially translated into English. The publication of this series is still in progress.

The task of exhaustive coverage of organometallic chemistry has been taken more recently by the *Gmelin Handbook of Inorganic Chemistry*, published by the Gmelin Institute in Frankfurt, in German and English [17]. The coverage of organometallic chemistry in the *Gmelin Handbook* will require many volumes: the process of preparing and publishing these volumes is in progress and when complete, the Gmelin series will provide the definite and most comprehensive source of information on organometallic compounds. In the Gmelin Handbook the organometallic compounds are partially covered in the basic work (the 8th edition), partly in the New Supplement Series.

Excellent coverage of organometallic compounds is also given (in German) in the well-known (mainly to the organic chemists) Houben-Weyl treatise [18]. Volume 13 of this treatise deals with organometallic compounds and is planned in several parts, many of which are already published.

The six-part series *Comprehensive Organic Chemistry* edited by D. Barton and W. D. Ollis [19], includes a volume on *Sulfur, Selenium, Silicon, Boron Organometallic Compounds* edited by D. N. Jones and a new series entitled *Comprehensive Organometallic Chemistry* edited by G. Wilkinson, F. G. A. Stone and E. W. Abel has appeared in nine volumes [20].
This promises to become a major source of information in organometallic chemistry, since it covers over 30 000 literature references in more than 8 500 pages. The series indexes all organometallic structures studied by diffraction methods as well as review articles covering the organometallic literature.

A useful source of information, particularly for physical constants and quick literature reference is the multi-volume handbook of M. Dub [21]. This is a register of organometallic compounds, reported between 1937 and 1964, updated with supplements till 1968. For each compound are given the preparation methods, chemical reactions, physical properties and physical constants, as well as literature references. The work intended to continue the work of Krause and von Grosse [3].

Other registers of compounds are the books by H. C. Kaufmann [22], H. Hagihara, M. Kumada and R. Okawara [23] and Buckingham [24].

A new addition to the multivolume *Chemistry of Functional Groups* series edited by S. Patai is the series *Chemistry of the Metal-Carbon Bond* edited by F. R. Hartley and S. Patai which treats organometallic compounds from the carbon end of the bond [25].

Bio-organometallic chemistry is covered in a new monograph by J. S. Thayer [26].

4.2. Reviews

An important source of information is the review literature. Particular aspects of organometallic chemistry are reviewed when significant advances have been made, or when a given subject becomes of outstanding interest. Such reviews usually reflect the state of the art in a given area. Two publications are devoted exclusively to organometallic chemistry: *Advances in Organometallic Chemistry* (ed. by F. G. A. Stone and R. West) published in book form by Academic Press, New York. (20 volumes appeared between 1964 and 1983) and the journal *Organometallic Chemistry Reviews, Section A. Subject Reviews,* published by Elsevier, Amsterdam (8 volumes published between 1964 and 1971). In 1972 the latter was incorporated in the *Journal of Organometallic Chemistry,* but again in 1976 the series was reformated (in book form) under the title *Journal of Organometallic Chemistry Library.*

Reviews on organometallic chemistry are also published in the series *Advances in Inorganic Chemistry and Radiochemistry* and *Progress in Inorganic Chemistry*, or in the journals *Chemical Reviews, Uspekhi Khimii* (in English translation as *Russian Chemical Reviews*), *Angewandte Chemie* (translated as *Angewandte Chemie, International Edition in English*), *Chemical Society Reviews* (formerly *Quarterly Reviews of the Chemical Society*), *Accounts of Chemical Research, Inorganica Chimica Acta Reviews, Comments on Inorganic Chemistry, Reviews in Inorganic Chemistry, Coordination-Chemistry Reviews, Tetrahedron,* etc.).

The review literature surveys organometallic chemistry with a delay of 1–2 years, but usually in a systematic and critical manner. Reviews, are of great importance, since they offer the reader an overall view of a narrow field.

In addition to reviews on topical subjects, organometallic chemistry is reviewed periodically, element by element, in several independent publications. *Annual Surveys of Organometallic Chemistry* started in 1965 in book form (edited by R. B. King and D. Seyferth) and continued in 1968 in journal form: *Organometallic Chemistry Reviews. Series B. Annual Surveys* until 1972 when it was incorporated into the *Journal of Organometallic Chemistry.* More recently these annual surveys are being published in the *Journal of Organometallic Chemistry Library* series. Equally useful are the *Specialist Periodical Reports* published by the Royal Society of Chemistry, London, in several series. Two are of special interest for the organometallic chemist: *Spectroscopic Properties of Inorganic and Organometallic Compounds* (ed. by N. N. Greenwood and now by E. A. V. Ebsworth); and *Organometallic Chemistry* (ed. by E. W. Abel and F. G. A. Stone). Chapters on organometallic chemistry are also included in the *Annual Reports* of the Royal Society of Chemistry, London, but these are less comprehensive than the specialized reports just mentioned above. The Society also publishes *Index of Revievs in Organic Chemistry.* With the help of these publications the progress of the field can be followed.

Finally, to commemorate the publication of its 100th and 200th volumes, the *Journal of Organometallic Chemistry* issued special prospective books in 1975 and 1980 [27, 28].

4.3. Primary Literature

The primary literature (original papers) describing the new results of research is extremely voluminous now. This required the foundation of some specialized journals, publishing mainly organometallic literature. Thus, the *Journal of Organometallic Chemistry* (published by Elsevier, Amsterdam under the supervision of an international editorial board) started in 1964, is now a leading journal in the field. A second journal is *Synthesis and Reactivity in Inorganic and Metal-Organic Chemistry* (published by M. Dekker, New York). The journal *Organometallics in Chemical Synthesis* (also published by Elsevier) had a very short life and was absorbed by the *Journal of Organometallic Chemistry*.

In addition, many chemical journals published by learned societies or commercial publishers contain papers on organometallic chemistry. The American Chemical Society began the publication of *Organometallics* in 1982 which immediately became a leading journal.

The new journals, *Transition-Metal Chemistry, Journal of Chemical Research* and *Nouveau Journal de Chemie* also publish papers on organometallic chemistry, as do the *Journal of the American Chemical Society* and the *Journal of the Chemical Society, Dalton Transactions.*

[1] M. I. Bruce, *Adv. Organomet. Chem., 10* (1972) 274; *11* (1973) 448; *12* (1974) 388.

[2] D. R. M. Walton, J. D. Smith, *Adv. Organomet. Chem., 13* (1975) 453.

[3] E. Krause, A. von Grosse, *Die Chemie der Metall-Organischen Verbindungen,* Borntrager Verlag, Berlin (1937).

[4] G. E. Coates, M. L. H. Green, K. Wade, B. J. Aylett, *Organometallic Compounds,* 4th ed., Chapman & Hall, London (1979), ff. Vol. 1, Part 1: Groups I-III (by G. E. Coates, R. Snaith, K. Wade) 1983. Vol. 1, Part 2: Groups IV and V (by B. J. Aylett) 1979. Vol. 2. The Transition Elements (by M. L. H. Green) in preparation.

[5] E. G. Rochow, D. T. Hurd, R. N. Lewis, *The Chemistry of Organometallic Compounds,* J. Wiley & Sons, New York (1957).

[6] G. E. Coates, M. L. H. Green, P. Powell, K. Wade, *Principles of Organometallic Chemistry,* Methuen & Co., London (1968).

[7] J. J. Eisch, *The Chemistry of Organometallic Compounds, The Main Group Elements,* Macmillan Co., New York (1967).

[8] P. L. Pauson, *Organometallic Chemistry,* E. Arnold, London, (1967).

[9] R. B. King, *Transition Metal Organometallic Chemistry,* Academic Press, New York (1970).

[10] M. Tsutsui, M. N. Levi, A. Nakamura, M. Ichikawa, K. Mori, *Introduction to Metal-π-Complex Chemistry,* Plenum Press, New York (1970).

[10a] J.P. Collman, L.S. Hegedus, *Principles and Applications of Organotransition Metal Chemistry,* J.Wiley, New York (1980).

[10b] S.G. Davies, *Organotransition Metal Chemistry*, Pergamon Press, Oxford (1982).

[11] H. H. Zeiss (ed.) *Organometallic Chemistry,* Reinhold, New York (1960).

[12] E. G. Rochow, *Organometallic Chemistry,* Reinhold, New York (1964).

[13] O. Yu. Okhlobystin, *Tretyaia Khimiya: Elemento-organicheskie Soedineniya* [*The Third Chemistry: Organo-Element Compounds*] (in Russian), Nauka, Moscow (1965).

[14] F. R. Hartley, *Elements of Organometallic Chemistry* (Monographs for Teachers Series), The Chemical Society, London (1974).

[15] J. L. Harwood, *Industrial Applications of the Organometallic Compounds,* Chapman & Hall, London (1963).

[16] A. N. Nesmeyanov, K. A. Kocheshkov (ed.), *Methody Elemento-organicheskoi Khimii*

[*Methods of Organo-Element Chemistry*], Nauka, Moscow. The following volumes have appeared:

a) T. V. Talalaeva, K. A. Kocheshkov: Li, Na, K, Rb, Cs (1971).
b) S. T. Yoffe, A. N. Nesmeyanov: Be, Mg, Ca, Sr, Ba (1963).
c) N. I. Sheverdina, K. A. Kocheshkov: Zn, Cd (1964).
d) L. G. Makarova, A. N. Nesmeyanov: Hg (1965).
e) A. N. Nesmeyanov; R. A. Sokolik: B, Al, Ga, In, Tl (1964).
f) K. A. Andrianov: Si (1968).
g) K. A. Kocheshkov, N. N. Zemlyanskii, N. I. Sheverdina, E. M. Panov: Ge, Sn, Pb (1968).
h) A. N. Nesmeyanov, T. V. Nikitina, O. V. Nogina, E. M. Brainina, *et al.* Copper, scandium, titanium, vanadium, chromium, and manganese subgroups; actinides and lanthanides. Two volumes (1974) (the treatment is element-by-element).
i) E. V. Leonova, V. K. Syundyukova, F. S. Denisov, A. A. Koridze, *et al.* Cobalt, nickel, platinum metals (1978) (dealing with organo-Co, Ni, Ru, Os, Rh, Ir, Pd, and Pt compounds).
j) D. A. Bochvar, N. P. Gambaryan, R. A. Sokolik, L. P. Yur'eva, *et al.* Types of organometallic compounds of the transition metals. Two volumes (1975) (treating bonding, metal carbonyls, isocyanide, monoolefin, acetylene, quinone, cyclobutadiene, mono- and bis-cyclopentadienyl, and allyl metal complexes).
k) S. P. Gubin, N. A. Volkenau, L. G. Makarova, L. P. Yur'eva: π-Complexes of transition metals with arenes, σ-M—R compounds and acetylenides (two volumes, 1976).

[17] *Gmelin Handbuch der Anorganischen Chemie,* Gmelin Institute, Frankfurt.
The following volumes dealing with organometallic compounds have been published:
Achte Auflage (8th Edition, the Main Work):

a) Niobium. Part B 4 (1973) (includes organometallic compounds of the element).
b) Platinum. Part D (1957) (organoplatinum compounds are covered in an appendix).
c) Ruthenium. Supplement Volume (1970) (includes organometallic and carbonyl compounds).
d) Silver. Part B 5. Organosilver Compounds. Organosilver Salts (1975).
e) Tantalum. Part 2 (1971) (includes carbonyl and organotantalum compounds in the final chapter "Complex Compounds of Tantalum").

Ergänzungsbände (New Supplement Series):

Vol. 2/3. Organovanadium Compounds. Organochromium Compounds (1971, 1973).
Vol. 4. Transuranium Elements. Part C. The Compounds (1972) (includes organometallic derivatives).
Vol. 5. Organocobalt Compounds. Part I. Mononuclear Compounds (1973).
Vol. 6. Organocobalt Compounds. Part II. Polynuclear Compounds (1973).
Vol. 9. Perfluorohalogenoorgano Compounds of Main-Group Elements. Part 1. Compounds of Sulfur (1973).
Vol. 10/11. Organozirconium Compounds, Organohafnium Compounds (1973).
Vol. 12. Perfluorohalogenoorgano Compounds of Main-Group Elements. Part 2. Compounds of Sulfur, Selenium and Tellurium (1973).
Vol. 13. Boron Compounds. Part 1. Binary Boron-Nitrogen Compounds. B—N—C Heterocycles. Polymeric Boron-Nitrogen Compounds (1974).
Vol. 14. Organoiron Compounds. Part A 1. Ferrocene 1 (1974).
Vol. 15. Boron Compounds. Part 2. Carboranes 1 (1974).
Vol. 16. Organonickel Compounds. Part 1. Mononuclear Compounds (1975).
Vol. 17. Organonickel Compounds. Part 2. Mononuclear Compounds (concluded). Polynuclear Compounds (1974).
Vol. 18. Organonickel Compounds. Index for Parts 1 and 2 (1975).

Vol. 19. Boron Compounds. Part 3. Compounds of Boron with the Non-metals S, Se, Te, P, As, Sb, and Si and with Metals (1975) (includes organoderivatives).
Vol. 24. Perfluorohalogenoorgano Compounds of Main-Group Elements. Part 3. Compounds of Phosphorus, Arsenic, Antimony and Bismuth (1975).
Vol. 25. Perfluorohalogenoorgano Compounds of Main-Group Elements. Part 4. Compounds with Elements of Main-Group 1 to 4 (1975).
Vol. 26. Organotin Compounds. Part 1. Tin Tetraorganyls, SnR_4 (1975).
Vol. 27. Boron Compounds. Part 6: Carboranes 2 (1975).
Vol. 29. Organotin Compounds. Part 2. Tin Tetraorganyls, R_3SnR' (1975).
Vol. 30. Organotin Compounds. Part 3. Tin Tetraorganyls, Heterocycles and Spiranes (1976).
Vol. 34. Boron Compounds. Part 9. Boron-Halogen Compounds. 1 (1976).
Vol. 35. Organotin Compounds. Part 4. Organotin Hydrides (1976).
Vol. 36. Organoiron Compounds. Part Bl. Mononuclear Compounds (excluding ferrocenes) (1976).
Vol. 40. Organotitanium Compounds. Part 1. Mononuclear Compounds 1 (1977).
Vol. 41. Organoiron Compounds. Part A. Ferrocene 6 (1977).
Vol. 42. Boron Compounds. Part 11. Carboranes 3 (1977).
Vol. 43. Boron Compounds. Part 12. Carboranes 4 (1977).
Vol. 44. Boron Compounds. Part 13. Boron-Oxygen Compounds 1 (1977).
Vol. 45. Boron Compounds. Part 14. Boron-Hydrogen Compounds 1 (1977) (includes organodiboranes).
Vol. 46. Boron Compounds. Part 15. Amine-Boranes (1977) (including amine adducts of arylboranes).
Vol. 47. Bismuth-Organic Compounds (1977).
Vol. 48. Boron Compounds. Part 16. Boron-Oxygen Compounds 2 (1977).
Vol. 49. Organoiron Compounds. Part A. Ferrocene 2 (1977).
Vol. 50. Organoiron Compounds. Part A. Ferrocene 3 (1978).
Vol. 51. Boron Compounds. Part 17. Borazine and its Derivatives (1978).

After volume 52 the separate numbering of the supplement volumes ceased, and the new volumes are simply included in the main series as supplements. These include:

Organotin Compounds. Part 5. Organotin Fluorides, Triorganotin Chlorides (1978).
Organotin Compounds. Part 6. Diorganotin Dichlorides, Organotin Trichlorides (1978).
Organotin Compounds. Part 7. Organotin Bromides (1980).
Organotin Compounds. Part 8. Organotin Iodides, Pseudohalides (1981).
Organotin Compounds. Part 9. Triorganotin-Sulfur Compounds (1982).
Organotin Compounds. Part 10. Mono- and Diorganotin-Sulfur Compounds. Organotin-Selenium and Organotin-Tellurium Compounds (1983).
Boron Compounds. Part 19. Boron-Halogen Compounds 2 (1979).
Boron Compounds. 1st Supplement. Vol. 1 (1980) (Organoboron-Oxygen Compounds).
Boron Compounds. 1st Supplement. Vol. 2 (1980) (Organoboron-halides, Borazines, Other Heterocyclic Compounds).
Boron Compounds. 1st Supplement. Vol. 3 (1981) (Carboranes, Organothioboranes).
Boron Compounds. 2nd Supplement. Vol. 1 (1983) (Organoboron-Nitrogen, and -Oxygen Compounds).
Boron Compounds. 2nd Supplement. Vol. 2 (1982) (Organoboron-Halogen, Oxygen and Sulfur Compounds; Carboranes).
Organoiron Compounds. A 4. Ferrocene 4 (1980).
Organoiron Compounds. A 5. Ferrocene 5 (1981).
Organoiron Compounds. A 7. Ferrocene 7 (1980).
Organoiron Compounds. B 2. Mononuclear Compounds 2 (1978).

Organoiron Compounds. B 3. Mononuclear Compounds 3 (1979).
Organoiron Compounds. B 4. Mononuclear Compounds 4 (1978).
Organoiron Compounds. B 5. Mononuclear Compounds 5 (1978).
Organoiron Compounds. B 6. Mononuclear Compounds 6 (1981).
Organoiron Compounds. B 7. Mononuclear Compounds 7 (1981).
Organoiron Compounds. B 11. Mononuclear Compounds 11 (1981).
Organoiron Compounds. C 1. Binuclear Compounds 1 (1979).
Organoiron Compounds. C 2. Binuclear Compounds 2 (1979).
Organoiron Compounds. C 3. Binuclear Compounds 3 (1980).
Organoiron Compounds. C 4. Binuclear Compounds 4 (1981).
Organoiron Compounds. C 5. Binuclear Compounds 5 (1981).
Organocopper Compounds. Part 2 (1983)
Organogold Compounds (1980).
Organotitanium Compounds. Part 2 (1980).
Organoantimony Compounds. Part 1 (1981), Part 2 (1981), Part 3 (1982).
Uranium E 2. Coordination Compounds (including Organouranium Compounds) (1980).
Organoantimony Compounds. Part 1. Compounds of Trivalent Antimony with Three Sb—C Bonds (1981).
Perfluorhalogeno Compounds of the Main-Group Elements. Part 5. Compounds with Nitrogen (Heterocyclic Compounds) (1978).
Perfluorhalogeno Compounds of the Main-Group Elements. Part 6. Compounds with Nitrogen (Heterocyclic Compounds (continued)) (1978).
Perfluorhalogeno Compounds of the Main-Group Elements. Part 7. Aliphatic and Aromatic Nitrogen Compounds (1979).
Perfluorhalogeno Compounds of the Main-Group Elements. Part 8. Aliphatic and Aromatic Nitrogen Compounds (continued) (1980).
Perfluorhalogeno Compounds of the Main-Group Elements. Part 9. Aliphatic and Aromatic Nitrogen Compounds (continued) (1981).
Uranium C12 and C13 Uranium with Carbon (1983).

[18] *Houben-Weyl, Methoden der Organischen Chemie.*, E. Müller and O. Bayer, H. Meerwein, K. Ziegler, eds., G. Thieme Verlag, Stuttgart.

XIII/1. *CH-Acidität. Metallorganische Verbindungen der I. Gruppe des Periodensystems: Li, Na, K, Rb, Cs, Cu, Ag, Au* (by G. Bähr, P. Burba, H. F. Ebel, A. Luttringhaus, U. Schölkopf) 1970.

XIII/2a. *Metallorganische Verbindungen der II. Gruppe des Periodensystems (außer Hg): Be, Mg, Ca, Sr, Ba, Zn, Cd* (by G. Bähr, H. Gilman, H. O. Kalinowski, N. Nützel, G. F. Wright) 1973.

XIII/2b. *Quecksilber-Organische Verbindungen: Hg.* (by H. Leditschke, H. Straub, K. P. Zeller) 1974.

XIII/3. *Bor-Organische Verbindungen* (by A. Grassberger, R. Köster, P. J. Paetzold, G. Schimmel, G. Schmid, W. Siebert) 1982.

XIII/4. *Metallorganische Verbindungen der III. Gruppe des Periodensystems (außer Bor): Al, Ga, In, Tl, Sc, Y, La* (by G. Bähr, P. Burba, H. Lehmkuhl, K. Ziegler) 1970.

XIII/5. *Silicium-Organische Verbindungen* (by S. Pawlenko) – 1980.

XIII/6. *Germanium- und Zinn-Organische Verbindungen: Ge, Sn* (by G. Bähr, H. O. Kalinowski) 1978.

XIII/7. *Metallorganische Verbindungen von Blei sowie den Metallen der IV–VI. Nebengruppe des Periodensystems: Pb, Ti, Zr, Hf, V, Nb, Ta, Cr, Mo, W* (by G. Bähr, E. Langer, A. Segnitz) 1975.

XIII/8. *Arsen-, Antimon-, Wismuth-organische Verbindungen: As, Sb, Bi* (by S. Samaan) 1978.

XIII/9. *Metallorganische Verbindungen der VII. und VIII. Nebengruppe des Periodensystems: Mn, Tc, Re, Fe, Co, Ni* etc. (by H. F. Klein, E. Langer, H. Segnitz, K. von Werner) – in preparation.

XIII/10. *Metal-π-Komplexe* (by M. Herberhold, P. L. Pauson, M. Sauerbier, H. Werner) – in preparation.

[19] D. Barton and W. D. Ollis, eds., *Comprehensive Organic Chemistry,* Pergamon Press, Oxford (1979), 6 vols.

[20] G. Wilkinson, F. G. A. Stone, E. W. Abel, eds., *Comprehensive Organometallic Chemistry,* Pergamon Press, Oxford, 9 vols., (1982).

[21] M. Dub. *Organometallic Compounds,* 2nd ed., covering the literature from 1937 to 1964. Springer Verlag, New York:
Vol. 1. Compounds of Transition Metals (1966).
Vol. 2. Compounds of Germanium, Tin and Lead (1967).
Vol. 3. Compounds of Arsenic, Antimony and Bismuth (1968). Formula Index (1970). The literature from 1965 to 1968 is covered in Vol. 1, First Supplement (1975) and Vol. 2, First Supplement (1973).

[22] H. C. Kaufman, *Handbook of Organometallic Compounds,* Van Nostrand, Princeton (1961).

[23] H. Hagihara, M. Kumada, R. Okawara, eds., *Handbook of Organometallic Compounds,* W. A. Benjamin, New York (1968).

[24] J. Buckingham, *Dictionary of Organometallic Compounds,* 3 Vols. Chapman and Hall, London (1984).

[25] F. R. Hartley, S. Patai, *The Chemistry of the Metal-Carbon Bond,* Vol. 1, J. Wiley & Sons, Chichester (1982).

[26] J. S. Thayer, *Organometallic Compounds and Living Organisms,* Academic Press, New York (1984).

[27] C. Eaborn, ed., *Perspectives in Organometallic Chemistry, J. Organomet. Chem., 100* (1975).

[28] C. Eaborn, ed., *Further Perspectives in Organometallic Chemistry, J.. Organomet. Chem., 200* (1980).

Part II
Organometallic Compounds of Non-Transition Elements

5. Organometallic Compounds of Alkali Metals

The alkali metals loose their single valence electron to form organometallic derivatives with an ionic structure. Charge separation occurs with the formation of a carbanion from the organic part of the molecule, M^+R^- (see Section 2.2). This is valid to a much lesser extent for lithium, which forms organometallic compounds whose properties suggest a pronounced covalent character. In addition, these compounds are associated with the formation of clusters containing metal-metal bonds. The vacant orbitals of the metal, present both in the M^+ cations and in the covalent LiR compounds, favor strong solvation of the metal in donor solvents (ethers, amines or better, polyamines), and this is reflected in the chemical behavior of the alkali metal organometallic compounds in solution.

The first organometallic compounds of an alkali metal were isolated by Schlenk in 1914. Previously, Wanklyn in 1858 obtained addition products of zinc dialkyls with sodium and potassium alkyls (of the type $MR \cdot ZnR_2$). Organolithium compounds were first isolated in pure form in 1917 by Schlenk. The interest in alkali metal organometallic compounds has increased slowly but continuously from that time.

The organometallic derivatives of lithium will be discussed in greater detail than those of sodium and heavier alkali metals, consistent with their relative importance. The great reactivity of the alkali metal compounds and the small differences between the behavior of the sodium, potassium and the heavier alkali metal derivatives has limited interest to those of sodium, the most readily available of them.

5.1. Organolithium Compounds

Organolithium compounds are extremely important, owing to their use as reagents in preparative chemistry. They are readily prepared, exhibit high chemical reactivity and are soluble in hydrocarbons. They are today, along with the Grignard reagents, the most important alkylating reagents for metals and non-metals in synthetic organometallic chemistry and afford a large number of interesting and useful reactions in synthetic organic chemistry.

The organolithium compounds are of the type LiR. Despite the apparent simplicity of this formula, special problems of structure and bonding are inherent. As mentioned in Section 2.5, the LiR derivatives are seldom monomeric, but donor ligands, for example, diamines, may stabilize the monomeric form as adducts, LiR · D (D = donor molecule).

5.1.1. Preparations

The starting material in almost all syntheses of organolithium compounds is lithium metal. The use of inert atmospheres during the preparation, handling and use of organolithium reagents is necessary because of their reactivity toward oxygen, water and carbon dioxide.

When the isolation of the pure organolithium reagent is intended, it is convenient to use non-coordinating, inert solvents (benzene, petroleum ether or other hydrocarbons), and for this purpose the reaction of excess lithium metal with organomercury compounds is well suited:

$$2\,Li + HgR_2 \dashrightarrow 2\,LiR + Hg$$

The organolithium compound can be crystallized after removing the solvent and separating the mercury metal formed. In some cases the exchange between organomercury reagents and readily accessible organolithium compounds can be used for synthesis:

$$2\,LiEt + HgMe_2 \longrightarrow \underset{\text{insoluble}}{2\,LiMe} + HgEt_2$$

Usually, organolithium reagents are prepared for their application to other syntheses, and for this purpose other reactions are preferred. Of particular use is the direct synthesis from lithium metal and organic halides in hydrocarbon solvents:

$$2\,Li + RX \longrightarrow LiR + LiX$$

The use of ethers is avoided whenever possible, since organolithium reagents can cleave the C—O bond:

$$LiC_4H_9 + H_3C\text{—}CH_2\text{—}O\text{—}C_2H_5 \longrightarrow H_3C\text{—}\underset{\displaystyle Li}{\underset{|}{C}H}\text{—}O\text{—}C_2H_5 + C_4H_{10}$$

$$\downarrow$$

$$LiOC_2H_5 + C_2H_4$$

However, this undesired side-reaction can be sometimes used for preparative purposes, for example, in the synthesis of benzyllithium by cleavage of benzylmethylether. This reaction is useful since benzyllithium cannot be prepared directly (the reaction between benzyl chloride and lithium metal affords only the coupled product diphenylethane):

$$PhCH_2OMe + 2\,Li \xrightarrow{THF} PhCH_2Li + LiOMe$$

Of great utility, especially for the synthesis of organolithium reagents which are difficult to obtain by direct methods, is transmetalation (metal-metal exchange), used, for example, in the synthesis of vinyllithium:

$$Sn(CH{=}CH_2)_4 + 4\,LiPh \longrightarrow 4\,LiCH{=}CH_2 + SnPh_4$$
$$Pb(CH{=}CH_2)_4 + 4\,Li \longrightarrow 4\,LiCH{=}CH_2 + Pb$$

Another very useful reaction is lithium-halogen exchange in which common lithium alkyls are used to lithiate halogenated organic compounds. In this reaction the lithium atom becomes associated with the most electronegative organic group (aromatic, polyhalogenoaromatic):

$$LiR + R'X \longrightarrow LiR' + RX$$

The reaction has been used in the syntheses of organolithium derivatives of hexachlorobenzene (pentachlorophenyllithium), tetrachlorothiophene, pentachloropyridine and many other compounds. n-Butyllithium, which is commercially avaiable, is generally used:

$$C_6Cl_6 + n\text{-}BuLi \longrightarrow LiC_6Cl_5 + n\text{-}BuCl$$

An excellent solvent for this reaction is tetrahydrofuran, but in this case low temperature (below $-35\,°C$) must be used in order to avoid the C—O bond cleavage and LiCl elimination from the product.

Of comparable preparative versatility is the hydrogen-metal exchange (metalation) of aromatic compounds bearing electronegative substituents (halogen, OR, etc.) and heterocycles:

$$C_4H_4O + LiPh \longrightarrow C_4H_3O\text{-}Li + PhH$$

Lithium enters usually in the *ortho*-position to the electronegative substituent or heteroatom. Polyhalogenated organolithium reagents can be readily prepared by metalation (with n-butyllithium) of pentachlorobenzene, tetrachlorobenzenes, trichlorobenzene, pentafluorobenzene and tetrafluorobenzene, for example:

$$C_6Cl_4H_2 + LiBu \xrightarrow[-60\,°C]{THF} C_6Cl_4HLi + BuH$$

$$C_6F_5H + LiBu \xrightarrow[-60\,°C]{THF} C_6F_5Li + BuH$$

Acetylenes are readily metalated since the attached hydrogens even in the alkyl groups are more acidic, as shown by the following example (in which the organolithium reagent is converted to an organosilicon derivative, without isolation):

$$CH_3{-}C{\equiv}CH \xrightarrow{LiBu} CH_3C{\equiv}CLi \xrightarrow{LiBu} Li_3C{-}C{\equiv}CLi \xrightarrow{+\,Me_3SiCl} (Me_3Si)_3C{-}C{\equiv}C{-}SiMe_3 + (Me_3Si)_2C{=}C{=}C(SiMe_3)_2\,.$$

The formation of the polylithium compound in the reaction cited above deserves particular attention. Another interesting polylithium compound is tetralithium tetrahedrane C_4Li_4, formed by irradiation of dilithioacetylene.

$$2\ LiC{\equiv}CLi \xrightarrow{h\nu} C_4Li_4 \text{ (tetrahedrane, Li on each vertex)}$$

The synthesis of ylids through the reaction of quaternary amonium salts and organolithium compounds involves as an intermediate step the metallation of an $N{-}CH_3$ group:

$$N(CH_3)_4^+Cl^- \xrightarrow{LiPh} LiCH_2N(CH_3)_3^+Cl^- \xrightarrow[-LiCl]{} (CH_3)_3N^+{-}^-CH_2$$

It has been found in recent years that amine complexes of organolithium reagents of the type LiR·TMED (TMED = tetramethylethyleneduamine) are powerful metalating agents, and they have found considerable use. The metalation reactions are very numerous.

5.1.2. The Structure of Organolithium Compounds

Organolithium compounds are associated in solution, in the solid state and even in the vapor phase, not only to dimers and trimers, but tetramers and hexamers $(LiR)_6$ as well. The degree of polymerization depends upon the nature of the organic group and, sometimes, on the solvent, as illustrated by the data given in Table 5.2. It can be seen that hydrocarbons favor self-association, while solvating media decrease the association.

By mass spectrometry it is found that tetrameric and hexameric species are present in the vapor of ethyllithium, the predominant ions in the ionization chamber being $Li_4Et_3^+$ and $Li_6Et_5^+$.

The $(LiCH_3)_4$ tetramer consists of an Li_4 tetrahedron with a Li—Li distance of 228 pm (≡2.28 Å), with the methyl groups located above the center of the trigonal

Tab. 5.2. Association degree of organolithium compounds in solution.

R in $(LiR)_n$	Solvent	Association degree (n)
Me	THF, ether	4
Et	benzene, cyclohexane	6
iso-Pr	benzene, cyclohexane	4
n-Bu	benzene, cyclohexane	6
	ether, THF	4
sec-Bu	benzene, cyclohexane	2
tert-Bu	benzene, cyclohexane	4
$-CH=CH_2$	THF	3
Ph	THF, ether	2
$-CH_2Ph$	THF, ether	1

faces of the tetrahedron, simultaneously attached through polycentric bonds to three lithium atoms (Fig. 5.1):

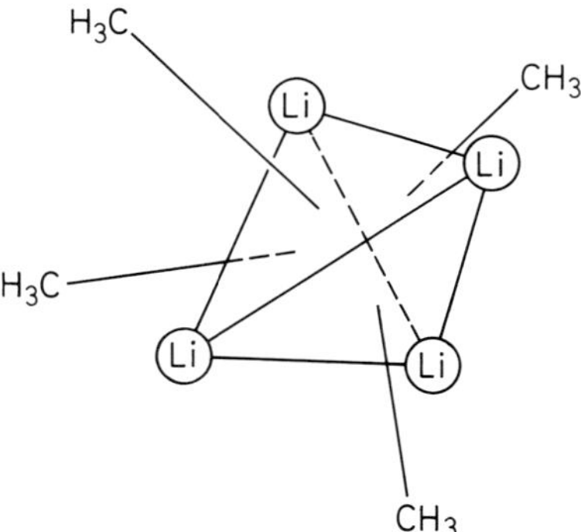

Fig. 5.1. The structure of $(LiCH_3)_4$.

Such molecules are electron deficient, and the bonding can be rationalized only by assuming molecular orbitals in which the four valence electrons of lithium are completely delocalized in the Li_4 tetrahedron.

The structures of di- or polyamine complexes of organolithium compounds show a completely different picture. The amine is coordinated through nitrogen to the lithium atom, thus supplying electrons for its vacant orbitals, and preventing association. Thus a monomeric structure has been established for $LiCPh_3 \cdot TMED$ in the solid state; the lithium atom is coordinated by two nitrogens and is attached to the triphenylmethyl group, as shown in Fig. 5.2a.

The TMED complex of phenyllithium is a dimer with phenyl bridges, as shown in Fig. 5.2b.

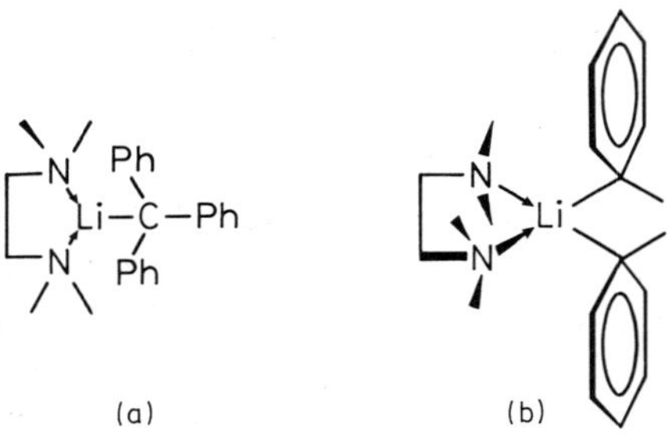

Fig. 5.2. The structures of (a) $LiCPh_3 \cdot TMED$ and (b) $(LiPh \cdot TMED)_2$.

The structures of the dilithium stilbene derivatives stabilized by chelation with tetramethylethylenediamine (TMED) or pentamethyldiethylenetriamine, are ionic in the solid state. In each compound the two chelated lithium atoms are located above and below the olefinic bond of the planar *trans*-stilbene molecule, as illustrated in Fig. 5.3:

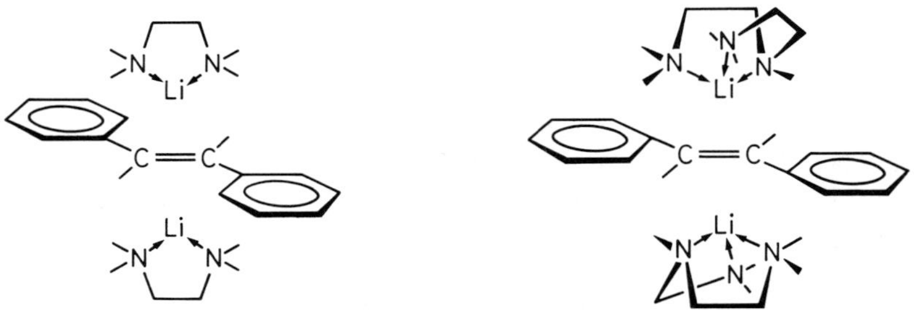

Fig. 5.3. The structures of two stilbene derivatives.

The triethylenediamine complex of benzyllithium has a polymeric structure (Fig. 5.4), consisting of chains made of alternating lithium atoms and diamine molecules.

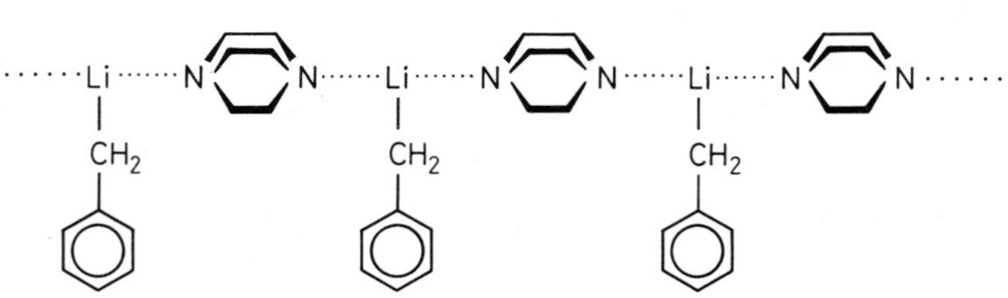

Fig. 5.4. The structure of $C_6H_5CH_2Li \cdot N(CH_2CH_2)_3N$.

5.2. Organometallic Derivatives of Sodium and the Heavier Alkali Metals

The organometallic derivatives of sodium, potassium are much less investigated, those of rubidium and cesium, differ from those of lithium in several respects: the former exhibit much higher chemical reactivity, and are non-volatile and insoluble in most organic solvents. This reflects their more pronounced ionic character. The increased chemical reactivity arises from the carbanion, R^-. These organoalkali-metal compounds react energetically with water, oxygen, carbon dioxide and most organic compounds except saturated hydrocarbons. As a consequence, they are more difficult to handle and seldom have been isolated as pure species. In most cases they are prepared only in solution and used in further reactions without isolation.

Organosodium compounds can be prepared by the reaction of organomercury compounds and metallic sodium in petroleum ether:

$$HgR_2 + 2\,Na \longrightarrow 2\,NaR + Hg$$

If this reaction is carried out in benzene, phenylsodium is formed because of the metalation of benzene (R=butyl, M=Na, K):

$$C_6H_6 + MR \longrightarrow MC_6H_5 + RH$$

In a similar manner, organozinc, -cadmium, and -lead compounds can be cleaved with an alkali metal.

The direct synthesis from organic halides and sodium metal cannot be used because of the coupling side reaction:

$$NaR + RX \longrightarrow R{-}R + NaX$$

but phenylsodium can, however, be obtained from chlorobenzene and finely divided sodium.

Hydrocarbons exhibiting a higher acidity, for example triphenylmethane, substituted acetylenes, cyclopentadiene, etc., can be directly metalated with sodium or potassium, either in liquid ammonia or tetrahydrofuran. Sodium hydride also metalates cyclopentadiene in tetrahydrofuran, to give sodium cyclopentadienide, $Na^+C_5H_5^-$, a compound of great utility in the synthesis of transition metal cyclopentadienyls.

A reaction with few analogies in the chemistry of other metals is the formation of organosodium and -potassium compounds by carbon-carbon bond cleavage in polyarylated ethanes, with sodium or potassium amalgam (R=aryl):

$$R_3C{-}CR_3 + 2\,M/Hg \longrightarrow 2\,M^+R_3C^- + 2\,Hg \qquad R{=}Ph,\ M{=}Na, K$$

Organometallic derivatives are also formed by addition of alkali metals to carbon-carbon double bonds. Since these derivatives can initiate the polymerization of olefins, the reactions play a role in polymer synthesis.

The sodium, potassium, rubidium and cesium alkyls are nonvolatile, insoluble and infusible solids which decompose at 100–200 °C. All are spontaneously flammable in air. Their structures are ionic, with pronounced charge separation. In those cases when the negative charge may become delocalized over several atoms (see Section 2.2), the organosodium compounds are colored, and the carbanion is more stable. Thus, benzylsodium is intensely yellow colored. Naphthalene and other polycyclic aromatic hydrocarbons readily form green solutions of anion radicals by accepting an electron furnished by the sodium atom. The reaction requires solvents with strong coordinating power toward the cation formed (for example, tetrahydrofuran or dimethoxyethane):

$$+ \text{Na} \longrightarrow \quad \text{Na}^{\oplus}$$

These compounds are unusual in that they are not substitution products (no hydrogen is replaced by sodium), but addition compounds. The electron provided by the alkali metal is accepted in a vacant antibonding molecular orbital of naphthalene.

The organosodium compounds are strong metalating reagents. Their reactivity toward hydrocarbons increases in the order of hydrocarbon acidity:

$$C_2H_6 < C_6H_6 < CH_3C_6H_5 < C_6H_5CH_2C_6H_5 < C_6H_5C{\equiv}CH$$

Solid ethylsodium contains double layers of alternating isolated Na^+ and $C_2H_5^-$ ions, and methylpotassium has an NiAs-type hexagonal structure, in which K^+ and CH_3^- ions also form alternating layers.

The sodium cyclopentadienide complex with tetramethylethylenediamine (TMED) consists of a chain structure of alternating $C_5H_5^-$ ions and solvated cationic groups $Na^+ \cdot$TMED, as shown in Fig. 5.5. Some degree of covalency is suggested by the relatively short Na—C distances in this compound:

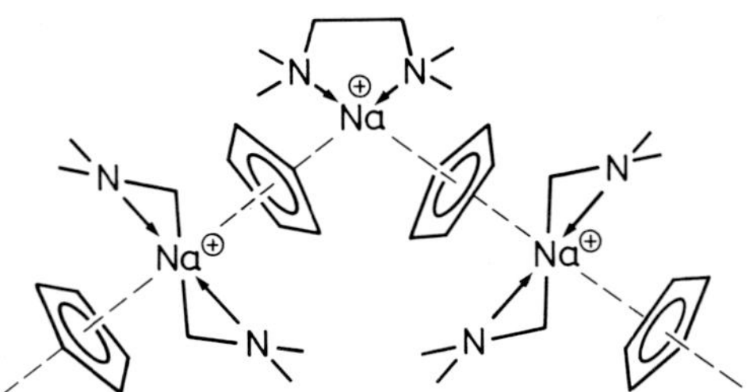

Fig. 5.5. The structure of $NaC_5H_5 \cdot$TMED.

The dipotassium tetramethylcyclooctatetraenide-diglyme solvate, a typical ionic compound, has the structure shown in Fig. 5.6. The potassium ions are solvated, and the eight-membered ring is planar, reflecting the delocalization of the double negative charge over the entire ring.

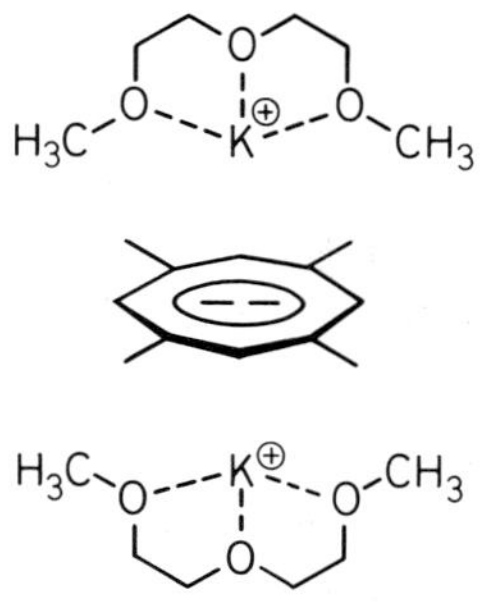

Fig. 5.6. The structure of $K_2(CH_3)_4C_8H_4 \cdot 2$ diglyme.

Solid biphenylrubidium bis(tetraglyme) contains a solvent-separated ion-pair in which each rubidium ion is spherically surrounded by ten oxygen atoms of the solvent molecule.

The examples cited demonstrate the important role of strongly coordinating solvents in the formation and stabilization of ionic organometallic compounds of the alkali metals. By favoring ion-separation, the solvation facilitates the formation of the organometallic compound.

6. Organometallic Compounds of the Group II Elements

The elements of Group II both in the main subgroup (Be → Ra) and in the B subgroup (Zn, Cd and Hg) use ns and np orbitals only. The (n − 1)d orbitals are completely filled and do not participate in chemical bonds; therefore, zinc, cadmium and mercury do not behave as transition metals and will be treated in this chapter.

The elements Be → Ra are strongly electropositive and form ionic compounds. This tendency increases for the heavier elements of the group. Only beryllium and magnesium form covalent, but rather polar, metal-carbon bonds. Their compounds are reactive and resemble the organolithium compounds. The heavier elements form metal-carbon bonds with a much more pronounced ionic character. Some organometallic compounds in which the organic group is anionic (for example, those containing the cyclopentadienyl anion, $C_5H_5^-$) have a clearly ionic character as in beryllocene, $Be(C_5H_5)_2$, and magnesocene, $Mg(C_5H_5)_2$.

The Group II elements can adopt sp-hybridization to form MR_2-type compounds. Two *p*-orbitals remain unoccupied, and they can accept electron pairs to form the adducts, $MR_2 \cdot D$ and $MR_2 \cdot 2D$ (D = donor molecule). These orbitals participate in sp^2- and sp^3-hybridizations, giving rise to trigonal and tetrahedral coordination geometries (Fig. 6.1).

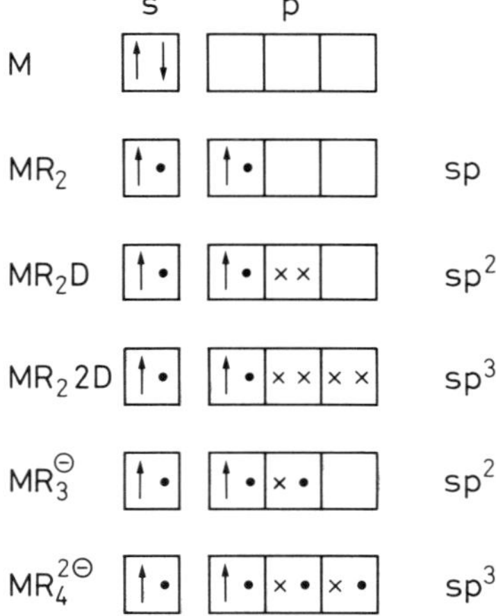

Fig. 6.1. The use of valence orbitals and electrons in the organometallic compounds of Group II.

The use of the p-orbitals results in the formation of $[MR_3]^-$ anions (for example, $[BePh_3]^-$), isoelectronic with the neutral organometallic compounds of the Group III elements, MR_3, or four-coordinated anions like $[MR_4]^{2-}$, isoelectronic with the neutral derivatives of the Group IV elements. In addition, by making full use of the p-orbitals, beryllium, and to some extent magnesium, participate in the formation of tricentric bonds (see the structure of dimethylberyllium in Section 2.4). The types of Group II metal compounds and their structures can be explained in terms of the tendency of the central metal atom to use completely its ns and np orbitals in bonding. For each case mentioned in Fig. 6.1 there are known examples.

The organometallic compounds of Group II elements were among the first organometallic compounds known, with the organozinc and organomercury compounds prepared in the middle of the 19th century by E. Frankland. Organomagnesium compounds were also obtained as early as 1859, by A. Cahours, but they became intensively used in laboratory only after the work of V. Grignard. The organometallic chemistries of these three elements flourished before WW I, because of their use in organic chemistry as intermediates (organozinc and magnesium compounds) or medicinal uses (organomercury derivatives).

6.1. Organoberyllium Compounds

The organometallic chemistry of beryllium had a slow development, although the first organoberyllium compound was prepared in 1873 by A. Cahours. For a long time this metal was scarce; in addition, its toxicity played a role. Recently, applications as polymerization catalysts and solid rocket propellants (!) were explored, but the high cost of beryllium rules out such uses on a large scale, especially since organoaluminum compounds are a powerful competitor in such applications.

6.1.1. Types of Organoberyllium Compounds

Beryllium is seldom di-coordinated, and – in the absence of steric hindrance – the association of organoberyllium compounds results in an increase of the coordination number of the metal. Therefore, a classification based upon structural criteria is not useful, and one based upon stoichiometry is preferred here.

Compounds of BeR_2 type

These can be polymeric (R = Me, Et), dimeric (R = iso-Pr) or monomeric (R = tert-Bu). The association is achieved through electron-deficient bonds.

Addition Compounds of the Type $BeR_2 \cdot nD$ (n = 1 or 2)

These are formed by coordination of donor molecules to beryllium dialkyls or diaryls. The donor can be an ether, amine, phosphine or a bidentate ligand such as bipyridyl, *ortho*-diamines, etc. In these compounds beryllium is three- or four-coordinated:

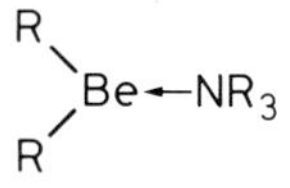

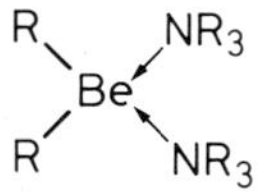

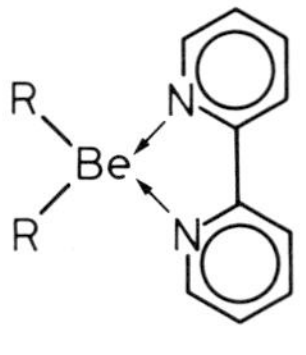

When the organic group itself contains a donor function in a suitable position, self-coordination may occur, with formation of spirocyclic compounds:

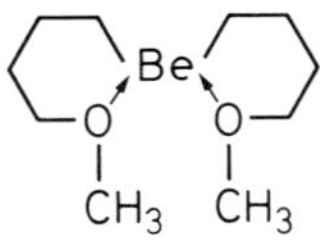

Anionic Compounds with Three- and Four-Coordinated Beryllium

These are compounds in which the beryllium central atom increases its coordination number by formation of more than two covalent bonds as in the anions $[BePh_3]^-$, $[R_2BeH]^-$, $[R_2Be—F—BeR_2]^-$ and $[BeR_4]^{2-}$.

Functional Derivatives of the Type RBeX

The organoberyllium halides are associated in the solid state and solvated or associated in solution. The RBeX formula thus does not reflect the actual structure. Even the bulky tert-BuBeCl is dimeric in benzene solution. These are some common features in the structures of RBeX compounds and the Grignard reagents (see further).

The functional derivatives RBeX in which X=OR, SR, SeR, NR_2, etc., are usually associated as cyclic dimers, trimers or cubane-type tetramers, through coordination bonds. The coordination number of beryllium in these oligomers is increased to three and four:

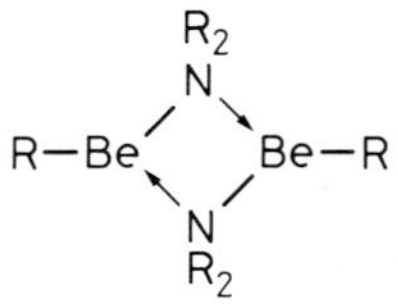

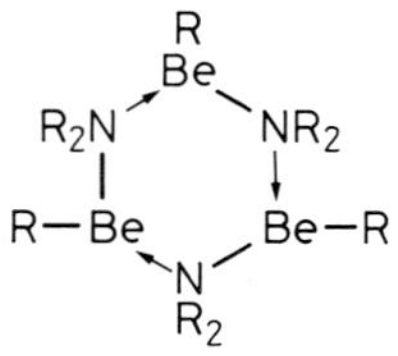

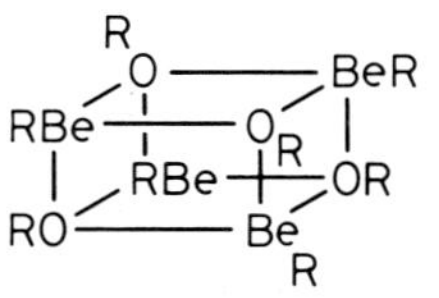

Hydrides of the Type RBeH

The simple (RBeH) or solvated ($RBeH \cdot OR_2$ and $RBeH \cdot NR_3$) hydrides are dimers formed via electron-deficient, three-center bonds:

$$R{-}Be(\mu\text{-}H)_2Be{-}R \qquad (R_2O)(R)Be(\mu\text{-}H)_2Be(R)(OR_2) \qquad (R_3N)(R)Be(\mu\text{-}H)_2Be(R)(NR_3)$$

6.1.2. Diorganoberyllium Compounds, BeR_2, and their Derivatives

Preparation. The reaction of anhydrous beryllium chloride and Grignard reagents in ether is used for the synthesis of BeR_2 compounds:

$$BeCl_2 + 2\,RMgX + Et_2O \longrightarrow BeR_2 \cdot OEt_2 + 2\,MgXCl$$

Ether adducts are usually formed, from which the donor molecule can be removed only with difficulty. Strong coordinating reagents like fluorides can displace the ether, but form fluoride-bridged dinuclear anions, in which beryllium is also three-coordinated:

$$2\,BeR_2 \cdot OEt_2 + KF \longrightarrow K^+[R_2Be{-}F{-}BeR_2]^- + 2\,Et_2O$$

When the organic group itself contains a donor function (—OR, —SR), intramolecular coordination leads to the formation of spirocyclic structures. In these syntheses dimethylsulfide can be used as solvent, since it does not coordinate to beryllium:

$$BeCl_2 + 2Cl(CH_2)_4OMe + 2Mg \rightarrow Be[(CH_2)_4OMe]_2 \text{ (spirocyclic, O}\rightarrow\text{Be)} + 2MgCl_2$$

$$BeCl_2 + 2Cl(CH_2)_3SEt + 2Mg \rightarrow Be[(CH_2)_3SMe]_2 \text{ (spirocyclic, S}\rightarrow\text{Be)} + 2MgCl_2$$

Organolithium reagents can be succesfully used, but an excess leads to trisubstituted derivatives:

$$BeCl_2 \xrightarrow[-2\,LiCl]{+2\,LiPh} BePh_2 \xrightarrow{+\,LiPh} Li^+[BePh_3]^-$$

Dimethylberyllium, treated with methyllithium in ether forms the salt $Li_2[BeMe_4]$, which contains a tetrahedral dianion.

Dicyclopentadienylberyllium (beryllocene), $Be(C_5H_5\text{-}\eta^5)_2$, is obtained from anhydrous beryllium chloride and sodium cyclopentadienide.

A preparative method which avoids the use of ethers in the synthesis of disubstituted derivatives involves the reaction of organomercury compounds with beryllium metal, on heating:

$$HgR_2 + Be \longrightarrow BeR_2 + Hg$$

Structure. The BeR_2 compounds can exhibit various structures, depending upon the nature of the organic substituent. Dimethylberyllium is polymeric in the solid state (Fig. 6.2), and its bonding was discussed in section 2.4:

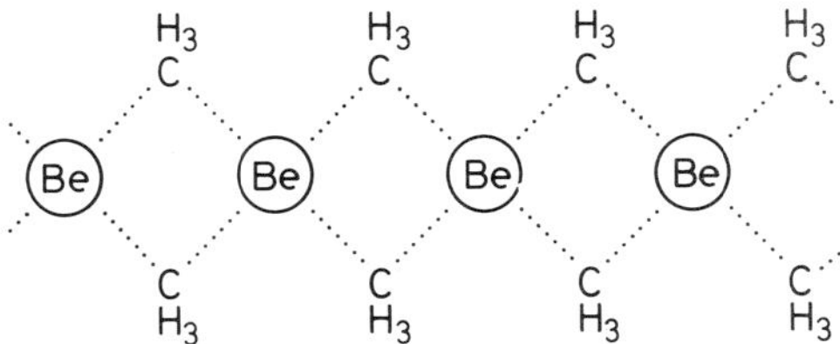

Fig. 6.2. The structure of $Be(CH_3)_2$ polymer.

In the vapor phase dimethylberyllium consists of a mixture of monomeric, dimeric and trimeric molecules in variable proportion, dependent upon temperature and pressure.

Methyl bridges are also present in the structure of methylberyllium borohydride, $[MeBeBH_4]_2$, a dimeric compound which also contains electron-deficient Be···H···B bridges:

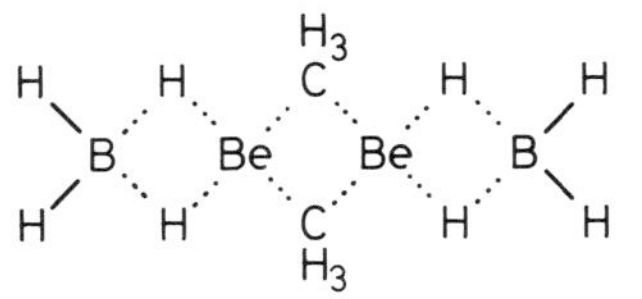

```
                 H3
 H      H        C       H       H
  \   .. ..    .. ..   .. ..    /
    B      Be        Be       B
  /   .. ..    .. ..   .. ..    \
 H      H        C       H       H
                 H3
```

The structure of this compound suggests that the vapor-phase dimers and trimers are built similarly.

The diethyl-, di-n-propyl, di-n-butyl and di-iso-propylberyllium alkyls are dimeric in benzene solutions; steric effects probably prevent formation of higher polymers. Di-tert.-butylberyllium is monomeric; its linear structure, in agreement with the expected sp-hybridization, contains a Be—C distance of 170 pm (≡1.7 Å), shorter than in dimethylberyllium (shown in Fig. 6.2), owing to the different nature of the two Be—C bonds:

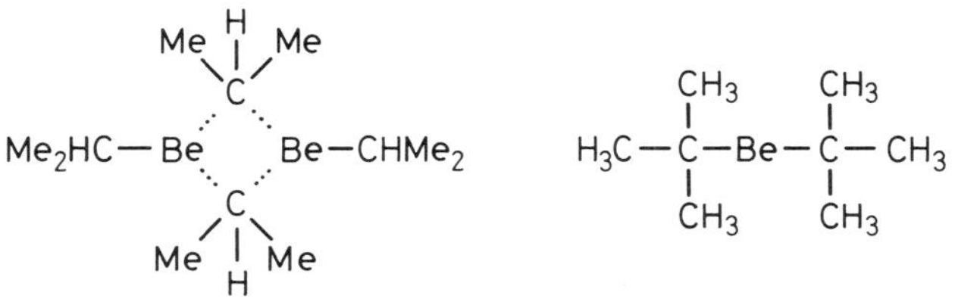

```
            H
        Me  |  Me                       CH3      CH3
          \ | /                          |        |
            C                    H3C — C — Be — C — CH3
          ..  ..                         |        |
Me2HC— Be        Be—CHMe2               CH3      CH3
          ..  ..
            C
          / | \
        Me  |  Me
            H
```

Dicyclopentadienylberyllium is an ionic compound and its molecular structure has been investigated in the solid and vapor phases. It consists of Be^{2+} ions sandwiched between two planar $C_5H_5^-$ anions (Fig. 6.3), at unequal distances:

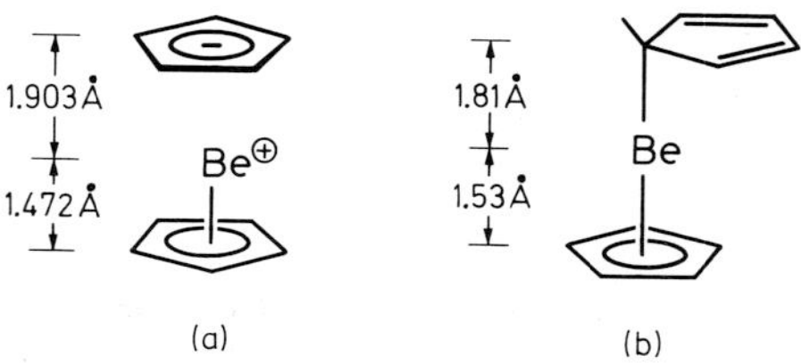

Fig. 6.3. The structure of $Be(C_5H_5)_2$. (a) in the vapor phase (b) in the solid state.

Structure (a) can be best described as containing an η^5-$C_5H_5Be^+$ cation (see Section 2.6) and a second, anionic cyclopentadienide ring.

In the solid state at $-120\,°C$ the compound has a structure with unequal distances, but this time containing one penta*hapto*- and one mono*hapto*-cyclopentadienyl ring.

Properties. The diorganoberyllium derivatives are chemically reactive and sensitive to oxygen and moisture. They are Lewis acids and form strong donor-acceptor bonds with nitrogen and somewhat weaker ones with phosphorus, oxygen and sulfur. They react with compounds containing active hydrogen attached to electronegative elements (N, O, P, S) with cleavage of one or both Be—C bonds:

$$BeMe_2 + NHMe_2 \xrightarrow{20\,°C} Me_2Be \cdot NHMe_2 \xrightarrow[-CH_4]{44\,°C} (MeBe{-}NMe_2)_3$$

$$BeMe_2 + NH_3 \xrightarrow{-80\,°C} Me_2Be \cdot NH_3 \xrightarrow[-CH_4]{10\,°C} (MeBe{-}NH_2)_x \xrightarrow[-CH_4]{50\,°C} (Be{-}NH)_x$$

$$2BeMe_2 + 2MeOH \longrightarrow (MeBe{-}OMe)_2 + 2CH_4$$

$$4BeMe_2 + 4PhCH_2OH \xrightarrow{Et_2O} (MeBe{-}OCH_2Ph)_4 + 4CH_4$$

The amino derivatives thus formed are cyclic dimers or trimers. The alkoxy derivatives, RBe—OR′, may be dimeric or trimeric, but in most cases they are tetramers containing a cubane-type Be_4O_4 cage, as in the $(MeBe{-}OSiMe_3)_4$ and thio, $(RBe{-}SR')_4$, derivatives.

Water reacts vigorously with beryllium dialkyls, by cleaving both Be-C bonds to yield $Be(OH)_2$ and the corresponding hydrocarbon.

The Lewis acid character of diorganoberyllium derivatives is manifest in the formation of numerous adducts with nitrogen donors like pyridine to give $BeR_2 \cdot 2py$, with dipyridyl (to form colored complexes), tetramethylethylenediamine, tetramethyl-*ortho*-phenylenediamine and with dimethoxyethane:

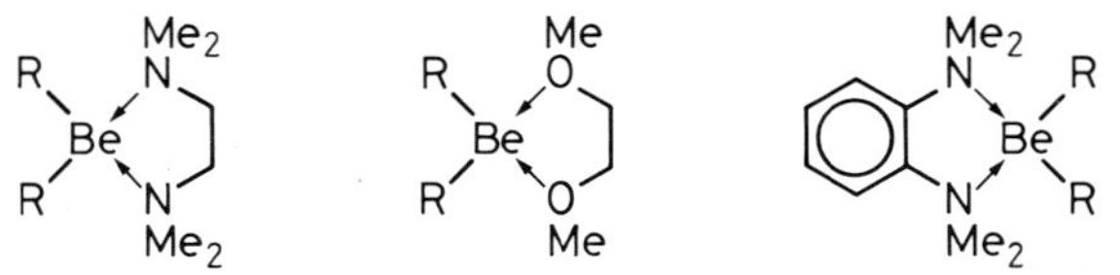

6.1.3. Organoberyllium Halides

The existence of RBeX halides had been questioned, but they can be demonstrated as intermediates in the formation of MeBeH when a mixture of $BeMe_2$ and $BeBr_2$ is treated with lithium hydride.

The organoberyllium halides can be prepared directly from alkyl halides and beryllium metal, in the presence of mercury(II) chloride as catalyst (X = Cl, Br):

$$Be + RX \longrightarrow RBeX$$

Less important reactions include cleavage of Be—N or Be—C bonds with suitable reagents:

$$MeBe{-}NMe_2 + HCl \longrightarrow MeBeCl + Me_2NH \cdot HCl$$
$$BeMe_2 + I_2 \longrightarrow MeBeI + MeI$$

Organoberyllium halides polymerize through halogen bridges, or through three-center $B \cdots CH_3 \cdots Be$ bonds. The halides are monomeric in ethereal solvents, owing to formation of solvates such as $RBeCl \cdot 2Et_2O$. The etherate of tert-butylberyllium chloride is dimeric in benzene solution, but monomeric in ether:

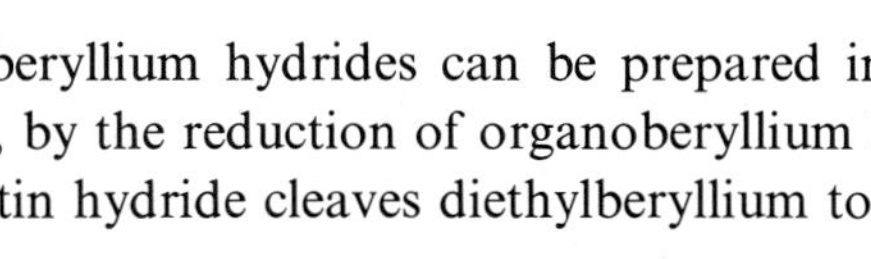

6.1.4. Organoberyllium Hydrides

Organoberyllium hydrides can be prepared in dimeric form as etherates, $(RBeH \cdot OEt_2)_2$, by the reduction of organoberyllium halides with lithium hydride in ether. Triethyltin hydride cleaves diethylberyllium to give the ether-free product:

$$2\,RBeBr + 2\,LiH \xrightarrow{Et_2O} (RBeH \cdot OEt_2)_2 + 2\,LiBr$$
$$2\,BeEt_2 + 2\,Et_3SnH \longrightarrow (EtBeH)_2 + 2\,SnEt_4$$

The etherates are also obtained by treating a mixture of $BeEt_2$ and $BeCl_2$ with $Na[HBEt_3]$ in ether:

$$BeEt_2 + BeCl_2 + 2\,Na[HBEt_3] \xrightarrow[-2\,NaCl]{OEt_2} (EtBeH \cdot OEt_2)_2 + BEt_3$$

Compounds of the type $M^+[R_2BeH]^-$, containing three-coordinated beryllium, are obtained in the reaction of beryllium dialkyls and metal hydrides. Organoberyllium hydrides are also formed in the pyrolysis of beryllium dialkyls.

6.2. Organomagnesium Compounds

The chemistry of organomagnesium compounds is dominated by importance of the Grignard reagents. Although diorgano derivatives, MgR_2, are known, they are much less investigated or used. In recent years the structures of the organofunctional $(RMgX)_n$ where X = OR, SR or NR_2, have been elucidated.

The first organomagnesium compounds were obtained by Cahours in 1859 and Wanklyn in 1866. The discovery of V. Grignard that RMgX reagents can be prepared from organic halides and magnesium metal in ether – for which he was awarded the Nobel prize – had enormous importance in organic and organometallic synthesis. These reagents are intensively used in laboratory and to some extent in industry in a great number of preparations.

The organometallic compounds of beryllium and magnesium are similar not only in types of compound, but also in their structure and chemical behavior.

6.2.1. Disubstituted Derivatives, MgR_2

Preparation. Disubstituted derivatives can be prepared by treating magnesium metal with organomercury compounds:

$$Mg + HgR_2 \longrightarrow MgR_2 + Hg$$

In ether the adducts $MgR_2 \cdot 2\,Et_2O$ are formed. A reaction of this type was used in the synthesis of an unusual magnesium heterocycle:

$$[\text{—}Hg(CH_2)_5\text{—}]_n + nMg \xrightarrow{THF} [\text{—}Mg(CH_2)_5\text{—}]_n \cdot 2n\,THF + nHg$$

The structure of the dimeric adduct, $(CH_2)_5Mg \cdot 2\,THF$, in the solid state is based upon a twelve-membered ring, containing two magnesium atoms.

The addition of dioxane to ethereal solutions of Grignard reagents precipitates adducts of magnesium halides, $MgR_2 \cdot$dioxane, leaving in solution the disubstituted compound, MgR_2, but this method does not afford pure compounds. A commercial procedure involves the reaction between magnesium metal, hydrogen and olefins at 100 °C to give MgR_2, as in organoaluminum chemistry (See Section 7.2).

Properties. The disubstituted derivatives, $MgMe_2$ and $MgEt_2$, have polymeric structures while the diethylether adduct, $Ph_2Mg \cdot OEt_2$, is monomeric, and contains four-coordinated magnesium.

The dicyclopentadienyl derivative, $Mg(C_5H_5\text{-}\eta^5)_2$ prepared from cyclopentadiene and magnesium metal at 500–600 °C, has a sandwich structure analogous to that of ferrocene, but is believed to consist of Mg^{2+} and $C_5H_5^-$ ions.

Diorganomagnesium compounds are very reactive, forming chelate complexes with bidentate amines (for example, tetramethylethylenediamine) and reacting with secondary amines to form substituted derivatives, $RMg{-}NR'_2$, which are cyclic dimers containing a four-membered, Mg_2N_2 ring:

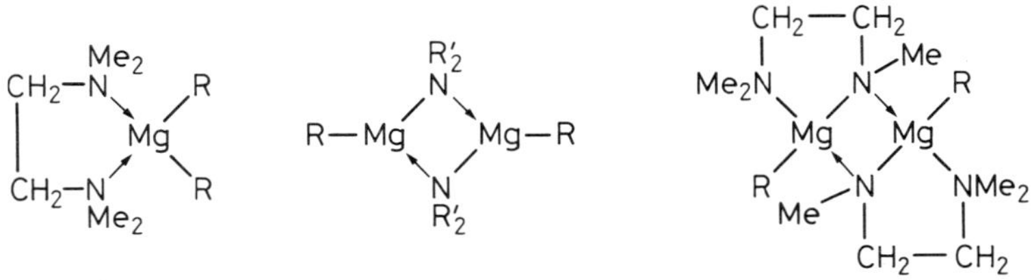

CH₂—CH₂
Me₂ R'₂ | |‿Me
CH₂—N R N Me₂N N R
Mg R—Mg Mg—R Mg Mg
CH₂—N R N R N NMe₂
Me₂ R'₂ Me | |
CH₂—CH₂

Alcohols can also cleave organic groups from magnesium to form tetramers, $(RMg{-}OR')_4$, with cubane-type structures; dimers and octamers have also been reported.

Derivatives with three-substituted, $Li^+[MgPh_3]^-$, and four-substituted, $Li_2^+[MgMe_4]^{2-}$, magnesium can be obtained with diorganomagnesium derivatives. The apparently five-coordinated $Li_3^+[MgMe_5]^{3-}$ may be an $MeLi \cdot [MgMe_4]^{2-}Li_2^+$ adduct.

A four-coordinated magnesium atom is at the center of the following spirocyclic structure:

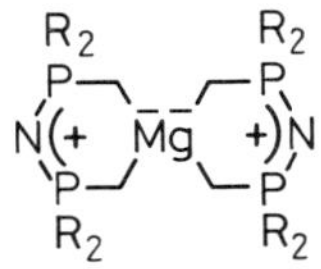

R₂ R₂
P P
N(+ Mg +)N
P P
R₂ R₂

6.2.2. Grignard Reagents (Organomagnesium Halides), RMgX

Preparation. Organomagnesium halides are relatively simply prepared by the reaction of organic halides and magnesium in anydrous ether or THF:

$$RX + Mg \xrightarrow{\text{ether}} RMgX$$

More recently, tertiary amines have been also used as solvents. Although the syntheses are apparently simple, the preparation of Grignard reagents in maximal yield is an art.

The reaction depends upon many conditions, including the quality of the magnesium metal used, the presence of activators or inhibitors, the purity and dryness of the solvent, etc. Because the surface of magnesium metal is always covered with a layer of magnesium oxide, the reaction can be difficult to initiate. Treatment with elemental iodine, dibromoethane,or simply by vigorously stirring the metal filings under a nitrogen atmosphere in the dry state, before introducing the reagents and the solvent can activate the reaction. Higher reaction temperatures are possible when tetrahydrofuran and dibutylether are used.

The reactivity of organic halides towards magnesium decreases in the order I > Br > Cl > F. The fluorides are virtually inert except when activated magnesium freshly prepared by reduction of an anhydrous halide with potassium metal is used. The formation of Grignard reagents can be accompanied by undesired coupling reactions with formation of hydrocarbons.

The Grignard reagents are sensitive towards moisture and oxygen. Usually, Grignard reagents are used immediately after preparation without isolation; however, the solutions can be stored for a long time and are commercially available.

Certain organomagnesium halides which are difficult to obtain directly are prepared by exchange. Thus, the reaction of hydrocarbons containing acidic hydrogen (acetylenes, cyclopentadiene) with ethylmagnesium bromide, results in the exchange of the organic groups attached to magnesium:

$$R{-}C{\equiv}CH \ + \ EtMgBr \longrightarrow R{-}C{\equiv}C{-}MgBr \ + \ EtH$$

$$C_5H_6 + EtMgBr \longrightarrow C_5H_5MgBr + EtH$$

This procedure can be used for the preparation of some organomagnesium fluorides.

Organolithium reagents can be converted to the corresponding Grignard reagents by reaction with anhydrous magnesium halides, but the reaction has little practical use, since the organolithium compounds themselves can be used for the same purposes as the Grignard reagents.

Structure. The structure of Grignard reagents is much more complex than indicated by the simple formula RMgX. In solution there is competition between the coordinating capacity of halogen and ether towards magnesium to give either simple monomeric etherates or halogen-bridged dimers:

$$RMgX(OR'_2)_2 \qquad [RMg(OR'_2)(\mu\text{-}X)]_2$$

The precipitation of dioxane adducts, $MgX_2 \cdot$dioxane, suggests that the following equilibrium occurs in solution:

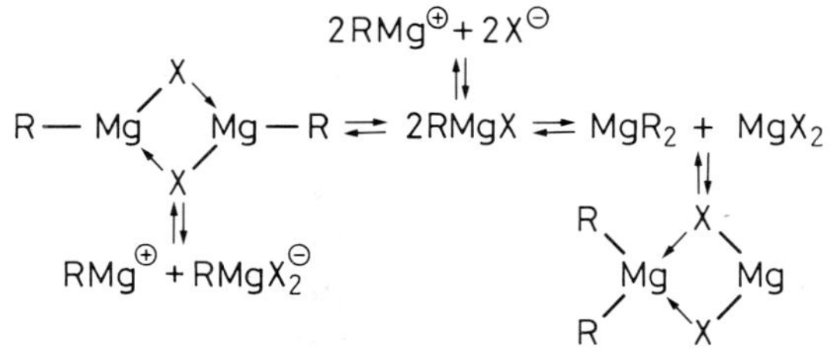

$$2RMg^{\oplus} + 2X^{\ominus}$$

$$R{-}Mg(\mu\text{-}X)_2Mg{-}R \rightleftharpoons 2RMgX \rightleftharpoons MgR_2 + MgX_2$$

$$RMg^{\oplus} + RMgX_2^{\ominus} \qquad R_2Mg(\mu\text{-}X)_2Mg$$

Monomeric species are predominant in strongly coordinating solvents or dilute solution. Dimerization and polymerization are favored by higher concentrations, weaker donor solvents or non-donating solvents like hydrocarbons. Structural investigation in the solid state can throw light upon the species potentially present in solution. Structures thus established are shown in Fig. 6.4:

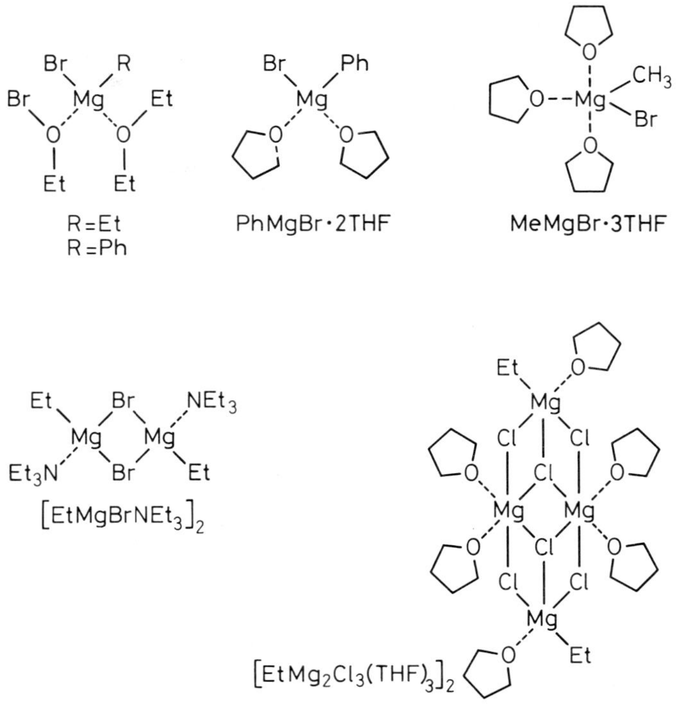

R = Et
R = Ph

$PhMgBr \cdot 2THF$

$MeMgBr \cdot 3THF$

$[EtMgBrNEt_3]_2$

$[EtMg_2Cl_3(THF)_3]_2$

Fig. 6.4. The structures of some organomagnesium halide solvates.

The crystallization of a compound with a certain structure is no guarantee that it exists in solution. During crystallization the equilibria can be shifted in favor of the species isolated which may merely be the one least soluble.

Properties. Grignard reagents enter into reactions with metal and non-metal halides, ketones, aldehydes, alcohols, etc. A few of these transformations are illustrated schematically in Fig. 6.5:

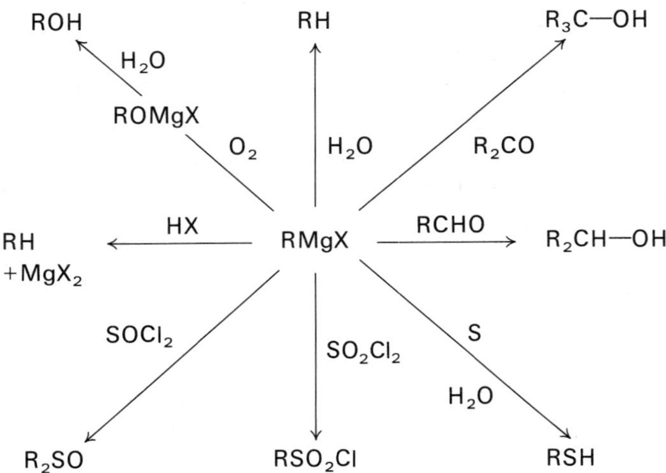

Fig. 6.5. Some reactions of Grignard reagents.

Simple transition metal salts (CuX_2, $FeCl_3$, $CoCl_2$, etc.) decompose common alkyl Grignard reagents, with formation of hydrocarbons (coupling), probably through intermediate free radicals.

6.3. Organometallic Compounds of Calcium, Strontium and Barium

These three elements are discussed together because of the relatively small number of known organometallic derivatives and of their modest importance. The metal-carbon bond in these compounds has a more pronounced ionic character than in their beryllium and magnesium analogues. These derivatives exhibit high chemical reactivity and are extremely sensitive to oxygen and moisture.

The organometallic compounds of the alkaline earths are obtained only with difficulty. The reaction of the metal with organomercury compounds fails, although diphenylbarium has been prepared in small yield by the reaction of diphenylcadmium and barium metal. The metals replace zinc in diethylzinc to form mixed compounds, formulated either as adducts $mMEt_2 \cdot nZnEt_2$ or as salts $M^{2+}[ZnEt_4]^{2-}$. The mixed compounds exhibit chemical behavior somewhat similar to that of alkali metal organic derivatives.

There is convincing evidence for the formation of organocalcium analogues of Grignard reagents. Thus, organic halides react (the reactivity order being RI > RBr > RCl) with calcium metal in ether or THF to form RCaX compounds of obscure structure.

The reaction of trityl chloride, Ph_3CCl, with calcium amalgam produces red colored $Ph_3CCaCl \cdot THF$, probably containing the Ph_3C^- anion. In these compounds the Ca—C bond is readily cleaved by reagents containing active hydrogen (HX where $X = NR_2$, PR_2, OR, SR, SeR). The disubstituted derivative, $Ca(CPh_3)_2 \cdot nTHF$, has been obtained by cleavage of hexaphenylethane with calcium amalgam.

Strontium metal likewise reacts with aryl iodides in a reaction catalyzed by mercury to give arylstrontium iodides.

The ionic cyclopentadienyls of alkaline earths can also be obtained by the reaction of the metal or metal hydride with cyclopentadiene.

6.4. Organozinc Compounds

Organozinc compounds played an important role as intermediates in synthetic organic chemistry before the discovery of the Grignard reagents. The first organozinc compounds were prepared by Frankland in 1849, and they are among the first organometallic compounds known. Interest in organozinc compounds has recently revived, and new applications as intermediates in organic synthesis have brought them again into use.

Monosubstituted functional derivatives, RZnX (X = halogen, OR, NR_2, SR, etc.), and disubstituted compounds, ZnR_2, are well-known, and three-coordinated $[ZnR_3]^-$ and four-coordinated $[ZnR_4]^{2-}$ derivatives are also described.

6.4.1. Disubstituted Derivatives, ZnR_2

Preparation. The original method of Frankland consisted in the disproportionation of alkylzinc halides, prepared without isolation, from zinc metal and alkyl iodides:

$$2\,Zn + 2\,RI \longrightarrow 2\,RZnI \longrightarrow ZnR_2 + ZnI_2$$

Zinc-copper alloy works better.

Better still, especially for aryl derivatives, is the reaction of organomercury compounds with zinc metal, in refluxing xylene:

$$HgR_2 + Zn \longrightarrow ZnR_2 + Hg$$

Zinc halides can be alkylated or arylated with the aid of Grignard reagents (a seldom used reaction), organolithium or aluminum compounds:

$$ZnR_2 \xleftarrow[\text{in } Bu_2O]{RMgX} ZnCl_2 \xrightarrow[\text{in } Et_2O]{LiR} ZnR_2$$

$$ZnCl_2 \xrightarrow{+ AlR_3} ZnR_2 + R_2AlCl$$

Alkali metal organometallics can react with disubstituted zinc derivatives to form higher coordinated anions. Thus, diphenylzinc reacts with phenyllithium to form $Li^+[ZnPh_3]^-$ and $Li_3^+[Zn_2Ph_7]^{3-}$ of unknown structure. Similarly, dimethylzinc forms with methyllithium in ether the compounds $Li_2^+[ZnMe_4 \cdot OEt_2]^{2-}$ and $Li_2^+[ZnMe_4]^{2-}$. In both $Li_2[ZnMe_4]$ and $Li_2[Zn(C{\equiv}CH)_4]$ a tetrahedral arrangement of organic groups about the metal is found.

The ability of zinc to form four Zn—C bonds is also illustrated by the synthesis of spirocyclic compounds:

$$2\ Me_3P{=}C{=}PMe_3 + ZnEt_2 \longrightarrow$$

Among diorganozinc compounds the structure of η^5-C_5H_5Zn—CH_3 is unique; it consists of isolated molecules in the vapor phase (a) and polymeric chains in solid state (b). In both cases the cyclopentadienyl group is bonded as penta*hapto*-ligand:

Properties. The disubstituted compounds, ZnR_2, are very sensitive to oxygen (the lower alkyls are spontaneously flammable and the higher alkyls fume in air). Unlike organomagnesium compounds, the organozinc compounds do not react with carbon dioxide, hence this gas served in earlier years of organometallic chemistry as a protective atmosphere. Organozinc compounds react with many functional derivatives (for example, with ketones) and participate in addition, reduction and enolization reactions.

Disubstituted organozinc compounds do not exhibit any tendency for polymerization: dimethylzinc and other dialkyls are monomeric with linear structures. Infrared spectroscopy suggests free rotation of the methyl groups about the Zn-C bonds in dimethylzinc. The zinc atom can make use of its available p-orbitals, however, by accepting lone electron pairs from donor molecules to form weak complexes. Simple ethers form relatively unstable complexes, but cyclic ethers (THF, dioxane, etc.) and chelating diethers such as dimethoxyethane increase the stability:

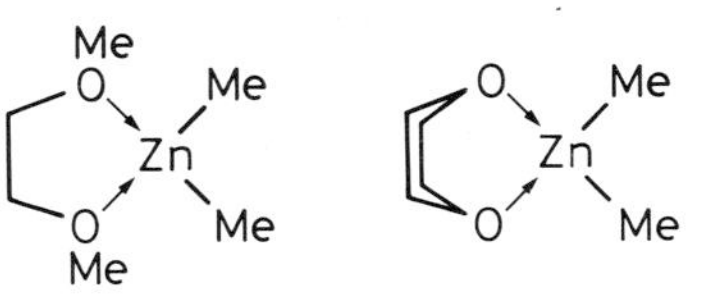

The amines form similar complexes, for example, $ZnMe_2 \cdot NMe_3$, $ZnPh_2 \cdot 2NMe_3$, as well as chelates with bidentate amines:

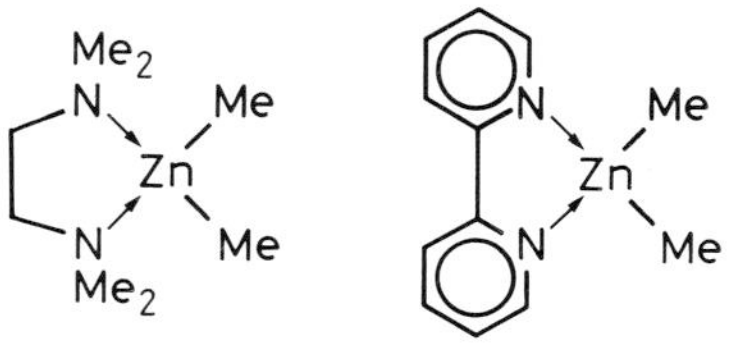

The complexes with phosphines, arsines and sulfides are unstable. Electron withdrawing groups at zinc increase the acceptor power of the metal and its tendency to form complexes. Thus, stable tertiary phosphine complexes such as $Zn(C_6F_5)_2 \cdot 2PPh_3$ can be obtained.

The disubstituted organozinc compounds are sensitive to the action of reagents containing active hydrogen. Water and alcohols cleave only one of the organic groups, forming, respectively, bridged dimers, $(RZn{-}OH)_2$, and tetramers, $(RZn{-}OR')_4$, with cubane structures. These tetramers are cleaved by pyridine to form cyclic dimers. When the alcohol contains a bulky organic group, cyclic dimers and trimers containing three-coordinated zinc are formed, such as $(EtZn{-}OCPh_3)_2$, $(EtZn{-}OCHPh_2)_3$, $(EtZn{-}OC_6F_5)_2$ and $(EtZn{-}OC_6Cl_5)_2$, etc:

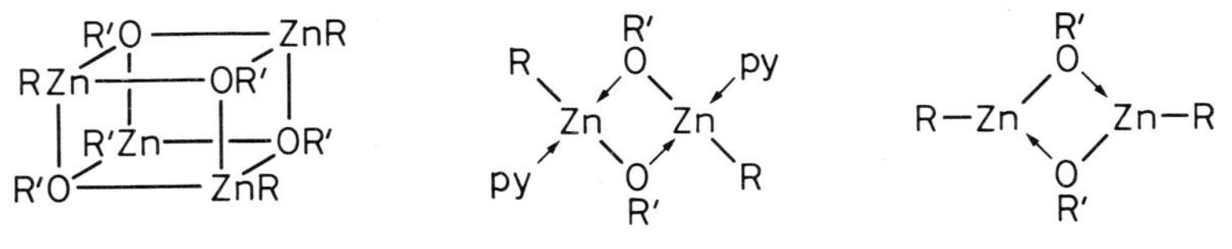

Disubstituted ZnR_2 compounds and thiols give insoluble polymers, $(RZn{-}SR')_n$ ($R = Me$), or oligomers ($n = 5$, 6, or 8) containing unusual Zn_nS_n polyhedral cages.

Primary and secondary amines can also cleave Zn—R bonds to form amino derivatives polymerized through coordination, as in the cyclic dimer, $(MeZn{-}NPh_2)_2$. The compound obtained from dimethylzinc and triphenylphosphinimine is a tetramer, $(MeZn{-}N{=}PPh_3)_4$, with a cubane-type structure:

Amino derivatives react with carbon dioxide, phenylisocyanate and phenylisothiocyanate to form addition compounds by insertion of the reagent molecule into the Zn—N bond:

$$EtZn{-}NR_2 \xrightarrow{+PhNCO} EtZn{-}N(Ph){-}CO{-}NR_2$$

$$EtZn{-}NR_2 \xrightarrow{+PhNCS} EtZn{-}N(Ph){-}CS{-}NR_2$$

$$EtZn{-}NR_2 \xrightarrow{+CO_2} EtZn{-}O{-}C(=O){-}NR_2$$

These products are associated. Thus EtZn—NPh—C(O)R (R = OMe, NPh_2) are cyclic trimers, and the compound with R = Me is a tetramer. With pyridine the trimers are degraded to dimers:

The examples cited reflect the coordinative unsaturation of the organozinc compounds.

6.4.2. Organozinc Halides, RZnX

Preparation. The formation of organozinc analogues of the Grignard reagents by a reaction of zinc metal and alkyl halides can be achieved only in strongly polar solvents, such as dimethylformamide, diglyme or dimethylsulfoxide:

$$Zn + RX \longrightarrow RZnX \qquad (X = Br)$$

For the synthesis of ketones from acyl halides, compounds with a lower reactivity than Grignand reagents are often useful. The latter are transformed into organozinc halides and are used without isolation:

$$ZnCl_2 + RMgX \xrightarrow{Et_2O} RZnCl + MgXCl$$

Organozinc halides can also be prepared by the reaction of disubstituted derivatives and zinc halides. The equilibrium:

$$ZnR_2 + ZnX_2 \rightleftharpoons 2\,RZnX$$

is strongly shifted towards the right.

Structure. The structure of organozinc halides in solution probably involves both coordinative solvation (in suitable solvents) and intermolecular association (in non-coordinating solvents). The RZnX derivatives are polymerized in the solid state. For example, ethylzinc iodide is a polymer consisting of macromolecular chains involving Zn_3I_3 rings with the metal four-coordinated and the iodine atoms three-coordinated.

6.4.3. Organozinc Hydrides, RZnH

Simple organozinc hydrides, RZnH, are difficult to obtain. Reducing diorganozincs with lithium alanate gives RZnH in THF, but conversion to RZn_2H_3 occurs:

$$ZnR_2 + LiAlH_4 \xrightarrow{THF} RZnH \xrightarrow{20\,°C} RZn_2H_3 \qquad R{=}Me, Ph$$

Diorganozinc derivatives, ZnR_2, react with zinc hydride in the presence of pyridine to form cyclic trimers:

$$ZnR_2 + ZnH_2 + py \xrightarrow{THF} (RZnH \cdot py)_3$$

$(RZnH \cdot py)_3$

6.5. Organocadmium Derivatives

Organocadmium compounds are less thoroughly investigated than those of zinc and magnesium, however, they can be used for the preparation of ketones from acyl halides. Unlike those of magnesium and zinc, organocadmium reagents do not react with the carbonyl group of the ketone formed.

Organo-disubstituted derivatives, CdR_2 and $CdRR'$, and monosubstituted, RCdX (X = halogen, OR, SR, etc.), are known. The coordinative unsaturation of cadmium makes possible the formation of adducts with donors, the association of the functional derivatives RCdX, and the formation of anionic species with three and four Cd—R bonds of the type $[CdR_3]^-$ and $[CdR_4]^{2-}$.

6.5.1. Disubstituted Derivatives, CdR_2

Preparation. The CdR_2 compounds are prepared from organomagnesium or lithium reagents and anhydrous cadmium halides:

$$CdX_2 + 2\,RMgX \longrightarrow CdR_2 + 2\,MgX_2$$
$$CdX_2 + 2\,RLi \longrightarrow CdR_2 + 2\,LiX$$

Adding hexamethylphosphortriamide (HMPT) which precipitates the magnesium halide as a complex, facilitates the isolation of the organocadmium derivative. Organolithium reagents are particularly suitable for the synthesis of aromatic compounds of cadmium, for example, phenyl and pentafluorophenyl derivatives.

The reaction of free cadmium (unlike zinc and magnesium) metal with organomercury derivatives cannot be used, since the organocadmium compounds are difficult to separate from the equilibrium mixture formed:

$$Cd + HgR_2 \rightleftharpoons CdR_2 + Hg$$

Divinylcadmium has been obtained by an exchange reaction between divinylmercury and dimethylcadmium.

The direct synthesis with alkyl iodides and cadmium metal in hexamethylphosphortriamide yields a complex:

$$2RI + 2Cd \xrightarrow{HMPA} CdR_2 \cdot 2\,HMPA + CdI_2$$

Bis(pentafluorophenyl)- and bis(pentachlorophenyl)cadmium can be prepared by thermal decarboxylation of organic salts:

$$Cd(OCOC_6F_5)_2 \longrightarrow Cd(C_6F_5)_2 + 2\,CO_2$$

and allyl derivatives by exchange between dimethylcadmium and the appropriate boron compounds:

$$3\,CdMe_2 + 2\,B(CH_2CR{=}CHR')_3 \longrightarrow 3\,Cd(CH_2CR{=}CHR')_2 + 2\,BMe_3$$

Thallium can also transfer organic groups to cadmium, as in the following preparation of bis(pentafluorophenyl)cadmium:

$$(C_6F_5)_2TlBr + Cd \xrightarrow{160\,°C} Cd(C_6F_5)_2 + TlBr$$

Properties. Dialkylcadmium derivatives are distillable, monomeric liquids, decomposing thermally at temperatures above 150 °C. They are less reactive than the analogous zinc compounds and are not spontaneously flammable in air, although they are oxidized to peroxides, $Cd(OOR)_2$. The Cd—C bond is easily cleaved by halogens, with formation of cadmium halides.

The reaction between CdR_2 derivatives and acyl halides is used in the synthesis of ketones:

$$2\,RCOCl + CdR'_2 \longrightarrow 2\,R{-}CO{-}R' + CdCl_2$$

The tendency to increase coordination number and reduce the coordinative unsaturation of CdR_2 compounds is reflected in the formation of anions containing three- and four-organic groups attached to cadmium. Thus, diphenylcadmium reacts with phenyllithium to form an unstable salt:

$$CdPh_2 + LiPh \longrightarrow Li^+[CdPh_3]^-$$

and the tetrahedral $[Cd(C{\equiv}CR)_4]^{2-}$ anion has been prepared:

$$Cd(SCN)_2 + 4\,KC{\equiv}CR + Ba(SCN)_2 \longrightarrow Ba[Cd(C{\equiv}CR)_4] + 4\,KSCN$$

The acidity of organocadmium compounds is much less than their organozinc analogues. Thus, the α,α-bipyridyl complex, $CdMe_2 \cdot$ bipy, is unstable and the dioxane complex, $CdMe_2 \cdot$ dioxane, is dissociated in solution.

Cadmium, like zinc, can form associated functional derivatives, RCdX, and react with compounds containing active hydrogen. Thus dimethylcadmium reacts with alcohols to form dimers, $(MeCd{-}OR)_2$ (R = tert-Bu), and tetramers, $(MeCd{-}OR)_4$, and with thiols to form insoluble polymers, $(MeCd{-}SR)_n$, or low-molecular weight oligomers ($n = 4$ when R = tert-Bu and $n = 6$ when R = iso-Pr). Triphenylphosphinimine forms a tetramer, $(MeCd{-}N{=}PMe_3)_4$, having a cubane-like structure.

6.5.2. Monosubstituted Derivatives, RCdX

Preparation. Organocadmium halides, RCdX, have been only recently isolated, although they have long been used in ethereal solutions for preparative purposes in organic chemistry. Adding anhydrous cadmium chloride to Grignard reagents in ether yields CdR_2 and RCdX, depending upon the reagent ratio. The RCdX derivatives do not play a role comparable to that of Grignard reagents, since it is more convenient to use solutions of CdR_2 compounds in organic preparations. They are, however, important for the preparation of the asymmetric derivatives, RCdR′, through reactions between RCdX and R′MgX compounds.

Organocadmium halides are obtained by redistribution reactions of cadmium dialkyls with dihalides:

$$CdR_2 + CdX_2 \longrightarrow 2\,RCdX$$

and by the reaction of Grignard reagents with cadmium halides:

$$CdX_2 + RMgX \longrightarrow RCdX + MgX_2$$

Properties. The halides, RCdX, are infusible crystalline solids, decomposing at ca. 100 °C, monomeric in dimethylsulfoxide solution. The alkoxy-, MeCd-OR, and thio-, MeCd—SR, derivatives are tetramers.

6.6. Organomercury Compounds

The first organomercury compound was obtained in 1853 by E. Frankland by the action of methyl iodide on mercury metal under sunlight irradiation. The number of organomercury compounds that were synthesised for pharmacological purposes is very large, but their role in chemotherapy has now been completely superceded. Use of organomercurial fungicides is also on the decline, owing to the toxicity to human beings and animals. Interest in the biological activity of organomercury compounds, particularly methylmercury species, has been focused in recent years, after the world-famous poisoning accident in Japan, known as Minimata disease, in which biomethylation of inorganic mercury salts was implicated. The great synthetic utility of organomercury compounds is based upon the ability of mercury to transfer organic groups to other metals and nonmetals.

The major types of organomercury compounds include HgR_2, RHgX and their addition compounds, but the acceptor ability of mercury in its organic derivatives is only moderate.

6.6.1. Disubstituted Derivatives, HgR_2

Preparation. One of the most versatile laboratory procedures for the disubstituted derivatives uses Grignard reagents:

$$HgX_2 + 2\,RMgX \longrightarrow HgR_2 + 2\,MgX_2$$

This method has recently been applied to the synthesis of bis(pentafluorophenyl)mercury, $Hg(C_6F_5)_2$. Organolithium reagents can be used equally succesfully, and the use of organoaluminum compounds facilitated by sodium chloride is also possible:

$$3\,HgCl_2 + 2\,AlR_3 + 2\,NaCl \longrightarrow 3\,HgR_2 + 2\,NaAlCl_4$$

Cyclopentadienyl derivatives of mercury are prepared by treating cyclopentadienyl sodium or thallium(I) with mercury(II) chloride.

Treating red mercury(II) oxide with triethylboron in aqueous alkali yields diethylmercury.

The original method of Frankland, based upon the reaction of alkyl and aryl halides or sulfates with sodium amalgam, is now seldom used:

$$2\,RX + Na + Hg \longrightarrow HgR_2 + 2\,NaX$$

Thermal decarboxylation of some mercury(II) carboxylates with elimination of carbon dioxide can be used for compounds containing rather electronegative organic groups. For example, bis(pentafluorophenyl)mercury:

$$Hg(OCOC_6F_5)_2 \longrightarrow Hg(C_6F_5)_2 + 2\,CO_2$$

Acetylenic derivatives can be mercurated directly with inorganic complexes:

$$2\,R{-}C{\equiv}CH + K_2HgI_4 + 2\,KOH \longrightarrow Hg(C{\equiv}CH)_2 + 4\,KI + 2\,H_2O$$

Disubstituted organomercury compounds are also prepared by reduction of organomercury halides, RHgX, with sodium, copper, alkali metal stannites and hydrazine hydrate, or by their disproportionation in reactions with tertiary phosphines, alkali metal iodides and other reagents:

$$2\,RHgX + 2\,Na \longrightarrow HgR_2 + Hg + 2\,NaX$$
$$2\,RHgX + 2\,PR'_3 \longrightarrow HgR_2 + HgX_2(PR'_3)_2$$
$$2\,RHgI + 2KI \longrightarrow HgR_2 + K_2[HgI_4]$$

The reduction with sodium iodide in ethanol or acetone is a standard procedure.

Unsymmetrically substituted diorganomercury derivatives, RHgR′, can be prepared by Grignard reactions (RHgX + R′MgX) or by redistribution of differently substituted symmetrical compounds:

$$HgR_2 + HgR'_2 \rightleftharpoons 2\,R{-}Hg{-}R'$$

The distribution is statistical when R and R′ are similar and non-statistical when R and R′ are different.

Structure. The monomeric diorganomercury compounds, HgR_2, are linear, reflecting sp-hybridization. Vibrational spectra show free rotation of the organic groups around the Hg—C bonds in dimethylmercury. In the linear structure of $Hg(C_6H_4CH_3\text{-}para)_2$ the two aromatic rings are in two planes forming a torsion angle of 60°.

Some organomercury compounds are formulated with non-linear C—Hg—C groupings. One such example is *ortho*-phenylene mercury prepared from *ortho*-dibromobenzene and sodium amalgam. First, a dimeric structure [Fig. 6.6(a)] was assigned, then an early X-ray investigation suggested a hexameric structure (b) in which all mercury atoms would be coplanar and the C—Hg—C bonds are colinear: Recently, a more accurate redetermination showed that this compound is trimeric and has structure (c). A trimeric structure was also found for the perfluorophenylene analogue, $[Hg(ortho\text{-}C_6F_4)_2]_3$.

The compound prepared from *ortho-,ortho′*-dilithiobiphenyl and mercury(II) chloride is not a heterocyclic monomer with bent C—Hg—C bonds as suggested earlier, but a tetramer (Fig. 6.7).

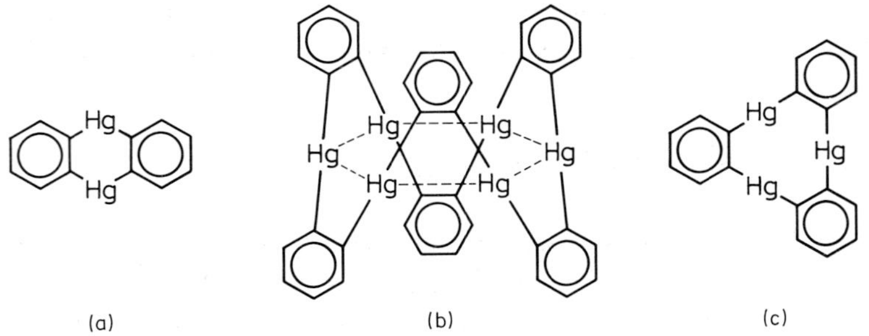

(a) (b) (c)

Fig. 6.6. The structures suggested for *ortho*-phenylenemercury.

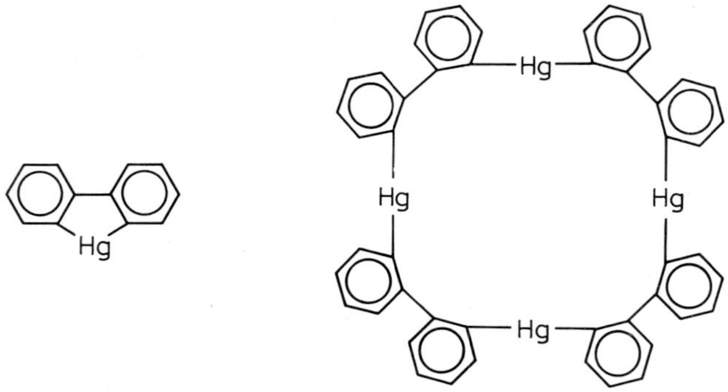

Fig. 6.7. The structure of *o,o'*-diphenylene mercury.

Properties. Dialkylmercury derivatives are very toxic, volatile liquids exhibiting moderate thermal stability. The analogous aromatic derivatives are stable solids; some are light-sensitive. The chemical reactivity of the diorganomercury compounds is much lower than with their zinc and cadmium analogues, for example, in not being sensitive to moisture and air.

The tendency of mercury to increase its coordination number is only moderate; sp-hybridization is more stable, for example, disubstituted derivatives do not form complexes with tertiary amines and phosphines. Perfluoro- and perchlorophenyl derivatives are exceptions with enhanced acceptor ability, and bis(pentafluorophenyl)mercury forms adducts with bipyridyl and tetraphenyldiphosphinoethane (diphos), while diphenylmercury does not:

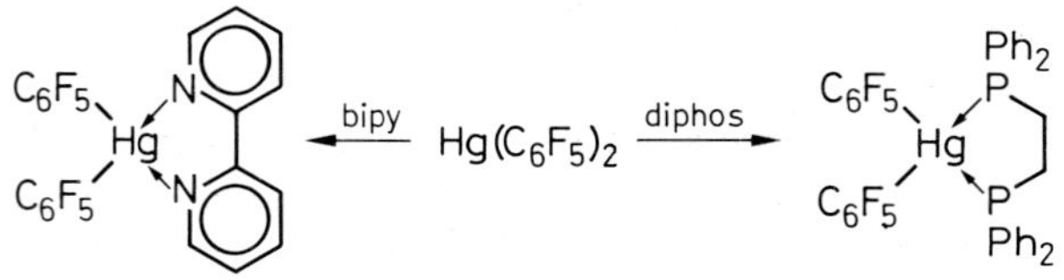

$$(C_6F_5)_2Hg(bipy) \xleftarrow{\text{bipy}} Hg(C_6F_5)_2 \xrightarrow{\text{diphos}} (C_6F_5)_2Hg(Ph_2PCH_2CH_2PPh_2)$$

The trifluoromethyl derivative, $Hg(CF_3)_2$, behaves simirlarly. Four-coordinated mercury has also been reported in the salts of the complex anion, $[Hg(CF)_2I_2]^{2-}$.

6.6.2. Monosubstituted Derivatives, RHgX

Preparation. Monosubstituted organomercury derivatives are readily obtained. Alkyl iodides react with mercury metal after photochemical initiation to form organomercury halides. The method can be applied to the synthesis of perfluoroalkyl derivatives:

$$RI + Hg \longrightarrow RHgI$$

By appropriate adjustment of the reagent ratio and reaction conditions, methods cited for disubstitution can also be used for the preparation of monosubstituted compounds, for example, the reaction of mercury(II) chloride with Grignard reagents:

$$HgCl_2 + RMgCl \longrightarrow RHgCl + MgCl_2$$

or the reaction of sodium amalgam with alkyl halides or sulfates:

$$3\,RX + 2\,HgNa \longrightarrow RHgX + HgR_2 + 2\,NaX$$

Organoaluminum halides also convert mercury(II) chloride into organomercury chlorides.

Organomercury halides are formed by redistribution of diorgano derivatives with mercury(II) halides:

$$HgR_2 + HgX_2 \longrightarrow 2\,RHgX$$

A specific synthesis of aromatic derivatives involves the reduction of diazonium and iodonium salts (R = aryl) with elimination of nitrogen or iodobenzene:

$$[R{-}N{\equiv}N]^+Cl^- + Hg \longrightarrow RHgCl + N_2$$
$$[R{-}N{\equiv}N]^+HgCl_3^- + Cu \longrightarrow RHgCl + N_2 + CuCl_2$$
$$[R{-}I{-}R]^+Cl + Hg \longrightarrow RHgCl + RI$$

The analysis of arylboronic acids is based upon the quantitative transfer of aromatic groups to mercury:

$$RB(OH)_2 + HgX_2 + H_2O \longrightarrow RHgX + B(OH)_3 + HX$$

The action of sulfinic acids on mercury(II) chloride was popular in the past:

$$RSO_2H + HgCl_2 \longrightarrow RHgCl + SO_2 + HCl$$

The compounds are also formed in reactions of mercury(II) halides with organic derivatives of tin, lead, antimony, bismuth, cadmium, thallium and other metals. The transfer of organic groups from these metals to mercury proceeds easily.

Mercuration is an important synthesis for monosubstituted, mainly aromatic derivatives. In this long-known reaction, aromatic compounds act on mercury(II) acetate or perchlorate. Aromatic amines, phenols, ethers, which are all more reactive than unsubstituted benzene, while nitro- and halogeno-substituted benzenes react very slowly:

$$RH + HgX_2 \longrightarrow RHgX + HX$$

The mercuration of benzene with mercury(II) acetate proceeds in acetic acid in an autoclave, and thiophene can be easily mercurated with an aqueous solution of mercury(II) acetate. The reaction is used to eliminate thiophene from benzene, since the dimercurated product is insoluble:

$$C_4H_4S + 2Hg(OCOCH_3)_2 \xrightarrow{H_2O} HO{-}Hg{-}C_4H_2S{-}HgOCOCH_3 + 3CH_3COOH$$

Furan forms a tetramercury derivative ($Ac = OCOCH_3$):

$$C_4H_4O + 4\,Hg(OCOCH_3)_2 \longrightarrow C_4O(HgOAc)_4 + 4CH_3COOH$$

Some aliphatic compounds can also be mercurated. Polymercury compounds, called "mercarbides" in the older literature, are formed, but their structure is unknown.

Mercuration products of ferrocene (for example, mono- and dimercurated derivatives), ruthenocene and η^5-cyclopentadienylmanganese tricarbonyl can be obtained, indicating aromatic properties of these π-complexes:

$$Fe(C_5H_4HgX)_2 \qquad Ru(C_5H_4HgX)_2 \qquad (C_5H_4HgX)Mn(CO)_3$$

Mercury salts undergo addition to olefins, in the presence of alcohols (oxymercuration):

$$H_2C{=}CH_2 + HgX_2 + ROH \longrightarrow ROCH_2CH_2HgX + HX$$

or amines (aminomercuration):

$$H_2C{=}CH_2 + HgX_2 + HNR_2 \longrightarrow R_2NCH_2CH_2HgX + HX$$

These two reactions have been widely applied to organic synthesis.

Properties. Monosubstituted organomercury compounds, RHgX, are crystalline solids, sometimes vacuum sublimable. They are water-soluble when $X = F$, NO_3, $1/2\,SO_4$, ClO_4, etc., with formation of RHg^+-cations.

The older literature contains references to organomercury hydroxides, RHgOH, (prepared by the reaction of the halides with silver oxide or potassium hydroxide in alcohol). These are actually mixtures of tris(organomercuryl)oxonium hydroxides, $[O(HgR)_3]^+OH^-$, and bis(organomercuryl)oxides, RHg-O-HgR, however.

The reaction of methylmercury bromide with sodium sulfide produces methylmercury sulfide, which can be converted to a tris(methylmercuryl)sulfonium salt (R = Me):

$$2RHgBr + Na_2S \longrightarrow RHg{-}S{-}HgR \xrightarrow{\text{decomp.}} HgR_2 + HgS$$

$$RHg{-}S{-}HgR \xrightarrow{H_2Cr_2O_7} [(RHg)_3S]_2^+\,[Cr_2O_7]^{2-}$$

The arylmercury dithiophosphates, $RHg\text{-}SP(S)(OR')_2$, represent another class of the little investigated mercury-sulfur derivatives.

The trimeric organomercury alkoxides, $(RHg{-}OR')_3$, have a cyclic structure:

$$\left.\begin{array}{l} RHgCl + NaOR' \\ RHgOH + R'OH \end{array}\right\} \longrightarrow (RHg{-}OR')_3$$

The trimethylsiloxy derivative, $(MeHg{-}OSiMe_3)_4$, is, however, monomeric in solution, unlike zinc and cadmium analogues, but in the solid state is tetrameric, with a cubane-type structure.

Tertiary phosphines disproportionate organomercury halides through intermediate salts:

$$RHgX + PR'_3 \longrightarrow [RHg\cdot PR'_3]^+X^- \xrightarrow{\text{dec}} HgR_2 + HgX_2(PR'_3)_2$$

($X = Cl$, BF_4^-, ClO_4^-, etc.).

7. Organometallic Compounds of Group III A Elements

The elements of Group IIIA (boron → thallium) form covalent organometallic compounds. The electronic structures of the Group IIIA elements, with two s-and one p-electron in their valence shell, lead to sp^2-hybridization and favor the formation of three normal covalent bonds in typical MR_3 compounds. Other modes of using the available valence orbitals and electrons are possible, however, as illustrated in Fig. 7.1:

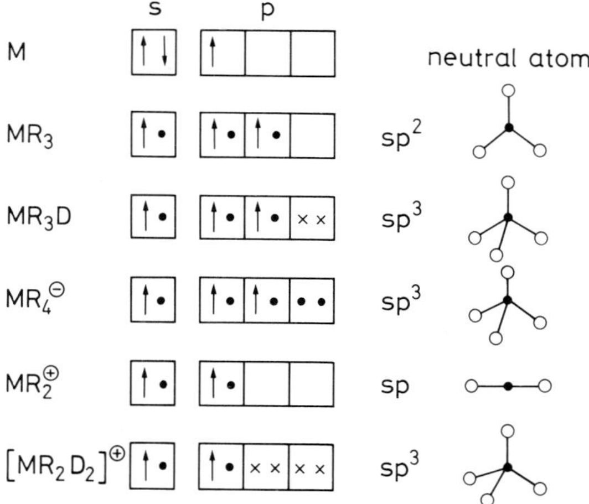

Fig. 7.1. The use of valence orbitals and electrons in the organometallic compounds of Group IIIA elements.

It can be seen that the vacant p-orbital remaining in the MR_3 compounds is available for accepting a donated electron pair. The formation of an additional bond is associated with sp^3-hybridization, and the $MR_3 \cdot D$ compounds are tetrahedral. Group IIIA elements can use all four sp^3-hybrid orbitals to form bonds to carbon. In this case the central atom bears a negative charge in the tetrahedral anions, $[MR_4]^-$, and becomes isoelectronic with a Group IVA element as a result.

Formation of $[MR_2]^+$ cations, isoelectronic with Group IIA, can be expected with Group IIIA elements. Such cations are particularly stable for Tl (R_2Tl^+ is isoelectronic with HgR_2). The boron-containing cations, R_2B^+, can be stabilized in the adduct $[R_2B \cdot 2D]^+$, in which the boron atom is again sp^3-hybridized.

This simplified explanation rationalizes the composition and structure of Group IIIA organometallic compounds. The possibilities illustrated in Fig. 7.1 will be exemplified with many compounds in the following pages. Not shown in Fig. 7.1 are the bonding peculiarities of organoaluminium compounds, which form three-center bonds (see Chapter 2); these will be discussed in the appropriate place.

Organometallic compounds of boron, aluminum and thallium, known since the last century, continue to attract interest. Organogallium and indium derivatives are much less investigated.

7.1. Organoboron Compounds

The first organoboron compound was prepared in 1860 by Edward Frankland. The present extremely voluminous literature of organoboron compounds is reflected in the volumes of the Gmelin Handbüch and other monographs and reviews.

While the hope of finding powerful propellants based upon organoboron compounds proved overly optimistic, and unexpected difficulties arose (high cost, toxicity, difficult handling, incomplete combustion, etc.), the scientific value of the research results stimulated was enormous. However, organoboron compounds are now used in the synthesis of thermostable polymers, polymerization catalysts and biologically active compounds. The high effective cross-section of the ^{10}B isotope in neutron capture and the preferential accumulation of some organoboron compounds in malignant tumors makes them attractive in the neutron-irradiation therapy for cancer. Sodium tetraphenylborate, $Na^+[BPh_4]^-$, is also used as an analytical reagent for heavier alkali metals.

7.1.1. Types of Known Organoboron Compounds

The great variety of organoboron compounds will be systematized here according to the nature of the coordination at boron atom and the number of boron atoms in the molecule.

Mononuclear Compounds with Three-Coordinated Boron

This class comprises trisubstituted derivatives of the type BR_3 (R = alkyl or aryl), and numerous functional derivatives of the types R_2BX and RBX_2, where the functional group X can be halogen, hydroxy, amino, mercapto or other groups:

R_2BCl	$RBCl_2$	chloroorganoboranes
$R_2B—OH$		borinic acids
	$R—B(OH)_2$	boronic acids
$R_2B—NR_2$	$R—B(NR_2)_2$	aminoorganoboranes
$R_2B—SR$	$R—B(SR)_2$	mercaptoorganoboranes

The organoboron hydrides, R_2BH and RBH_2, are dimers, formed through three-center hydrogen bridges, with structures derived from that of diborane:

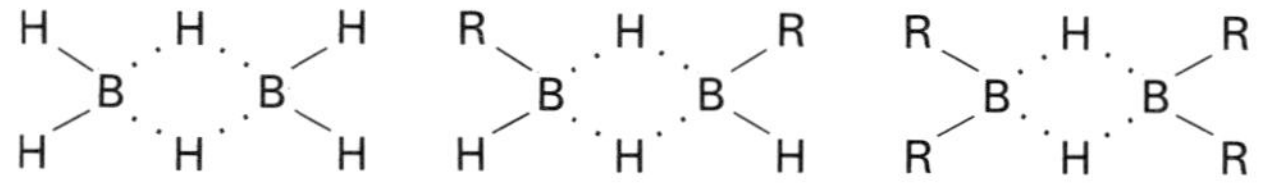

B_2H_6 $(RBH_2)_2$ $(R_2BH)_2$

Heterocyclic Compounds with Boron as Heteroatom

These are closely related to the compounds described under (a), but the organic substituents of boron are part of a cyclic system:

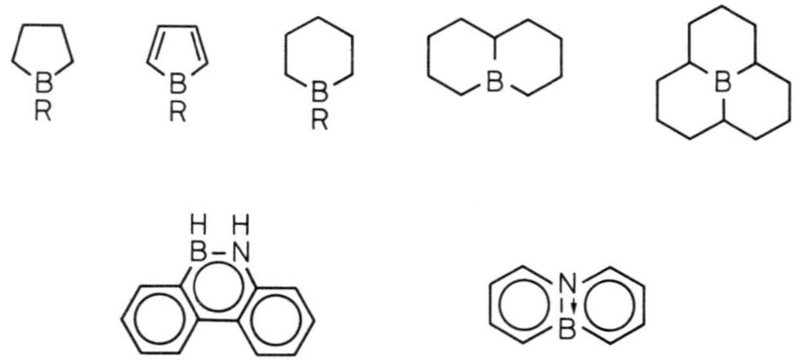

Linear and Cyclic Polynuclear Compounds with Three-Coordinated Boron

Dinuclear compounds, $R_2B—O—BR_2$ (diboroxanes), and $R_2B—NH—BR_2$ (diborazanes), are unstable toward disproportionation and have been little investigated, unlike the cyclic trimers containing an inorganic ring system with organic substituents of the types $(RBO)_3$, $(RBNR')_3$ or $(RBS)_3$:

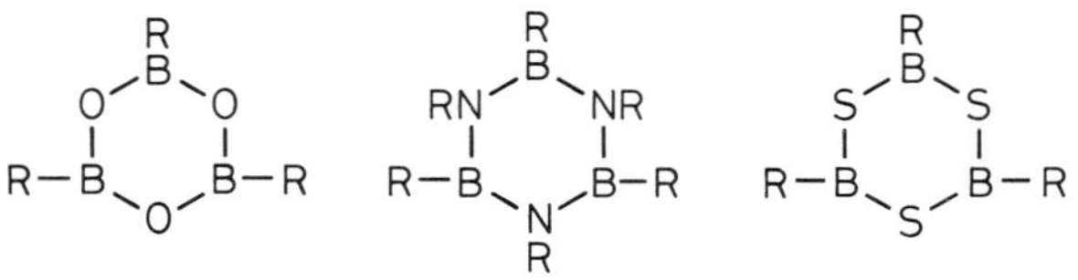

Mononuclear Compounds with Four-Coordinated Boron

The known species include anions, $[BR_4]^-$, neutral molecules, $R_3B \cdot NR'_3$, cations, $[R_2B \cdot bipy]^+$, chelated compounds $R_2B(acac)$, etc., including partially halogenated anions such as $[Ph_2BF_2]^-$ and $[PhBF_3]^-$:

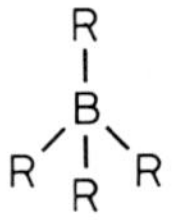

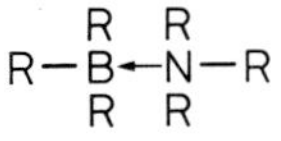

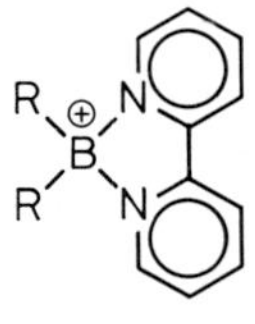

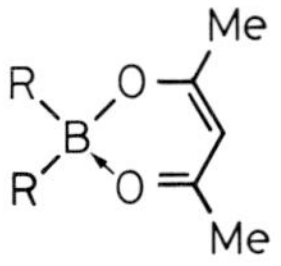

Cyclic Compounds with Four-Coordinated Boron

Oligomeric compounds are formed by association of some R_2BX derivatives in which functional groups, X, are able to donate an electron pair to the boron atom of a second molecule. The cyclization produces four-coordination, using all four valence orbitals of boron. The donor atoms can be nitrogen, phosphorus, or sulfur, as shown in the following examples:

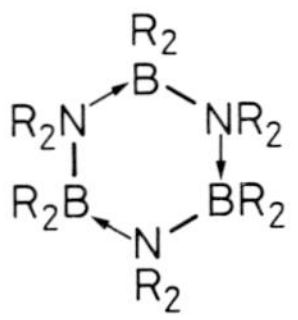

cyclotris-(aminoborine) 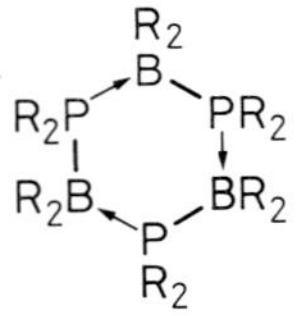 cyclotris-(phosphinoborine) 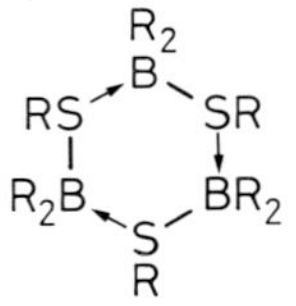 cyclotris-(mercaptoborine)

Carboranes

This is an unusual class of compounds formally derived from $B_nH_n^{2-}$ anions by replacing a negatively charged boron with an isoelectronic carbon atom, for example, in the series:

$$B_{12}H_{12}^{2-} \qquad B_{11}CH_{12}^{-} \qquad B_{10}C_2H_{12}$$

In $B_nH_n^{2-}$ anions the boron atoms form polyhedral clusters by the collectivization of available bonding orbitals and electrons. In *carboranes* the carbon atoms are introduced as heteroatoms in the molecular cluster skeleton.

7.1.2. Trialkyl(aryl)boranes, BR_3, and Related Compounds

Preparation. Trisubstituted boron derivatives are obtained by Grignard or organolithium reactions on boron halides, or better, from orthoboric esters (trialkylborates):

$$BF_3 \cdot OEt_2 + 3RMgX \longrightarrow BR_3 + 3MgFX + Et_2O$$
$$B(OMe)_3 + 3RMgX \longrightarrow BR_3 + 3Mg(OMe)X$$

Excess organomagnesium or organolithium reagent can give four-coordinated boron anions, $[BR_4]^-$, if R is an aromatic group. Arylation of $Na^+[BF_4]^-$ with excess Grignard reagent also yields $Na^+[BR_4]^-$.

Difunctional organometallic reagents can give heterocyclic derivatives, for example, borolanes:

$$Li(CH_2)_4Li + RBF_2 \longrightarrow C_4H_8BR + 2LiF$$

An interesting example is the synthesis of a heteroboracyclopentadiene (borole) ring system, containing four-coordinated boron (R = Ph):

$$R_4C_4Li_2 + Ph_2BCl \longrightarrow Li^{\oplus}[R_4C_4\bar{B}Ph_2] + LiCl$$

For industrial scale alkylations, organoaluminum reagents can be used in reactions with boron halides, alkoxides or the oxide:

$$BF_3 + AlR_3 \longrightarrow BR_3 + AlF_3$$
$$B(OR)_3 + AlR_3 \xrightarrow{100^\circ} BR_3 + Al(OR)_3$$
$$B_2O_3 + 2AlR_3 \longrightarrow 2BR_3 + Al_2O_3$$

Boron-carbon bonds can be formed from organotin compounds, for example (R = Ph):

$$R_4C_4SnR_2 + RBCl_2 \longrightarrow R_4C_4BR + R_2SnCl_2$$

$$C_5H_6SnR_2 + RBBr_2 \longrightarrow C_5H_6BR \xrightarrow{tert\text{-}BuLi} Li^{\oplus}[C_5H_5\bar{B}R]$$

The hydroboration of olefins, a process in which boron-carbon bonds are formed through addition of B—H groups to carbon-carbon double bonds, is an important reaction:

$$3R{-}CH{=}CH_2 + R_3N \cdot BH_3 \longrightarrow B(CH_2CH_2R)_3 + NR_3$$

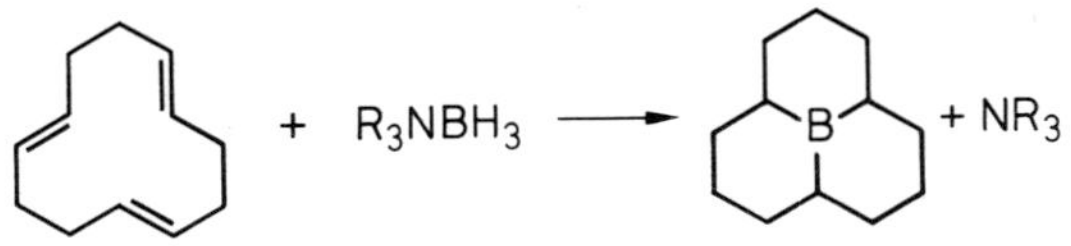

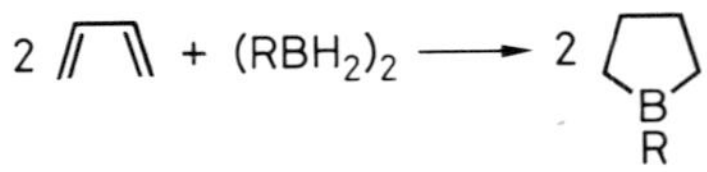

$$+ R_3NBH_3 \longrightarrow + NR_3$$

$$2 + (RBH_2)_2 \longrightarrow 2$$

The treatment of the products of hydroboration of butadiene with potassium hydride results in the formation of a heterocycle containing a transannular BHB bridge:

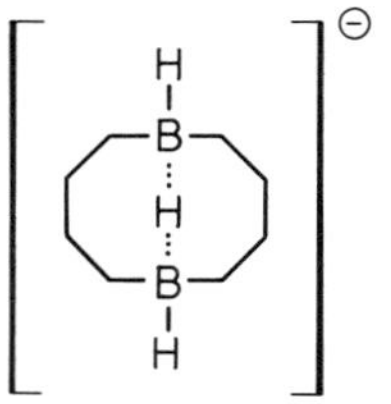

Hydroboration of 1,4-pentadiene, followed by heating and additional treatment with diborane, gives the hydrogen-bridged dimer:

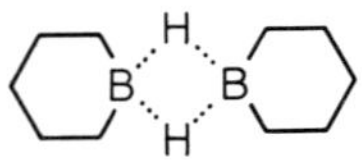

Hydroboration, H.C. Brown's Nobel prize winning discovery, has great potential for synthetic-organic chemistry.

Properties. Trisubstituted boranes are liquids (only BMe_3 is a gas with b.p. $-22\,°C$) and the aromatic derivatives are solids sensitive to the action of oxygen. Some are pyrophoric and also burn in chlorine or bromine vapor. The cleavage of boron-carbon bonds by oxidation with hydrogen peroxide (resulting in the formation of alcohols) or chromium(VI)oxide, CrO_3, (with formation of ketones) is of interest in synthetic organic chemistry.

Arylboranes react with sodium metal in liquid ammonia or ethers to form an anionic free radical R_3B^-, in which the unpaired electron occupies the vacant p-orbital not used in the formation of the three bonds to carbon.

Acetylacetone cleaves one bond from triorganoboranes to form chelates with four-coordinated boron:

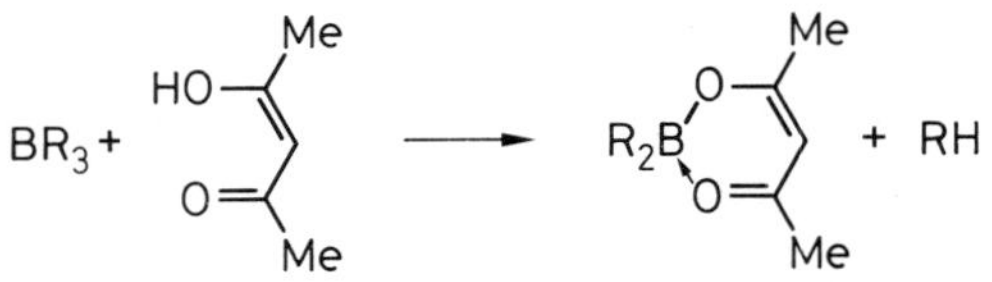

$$BR_3 + \longrightarrow R_2B + RH$$

Other chemical transformations of trisubstituted boranes are shown in Fig. 7.2.

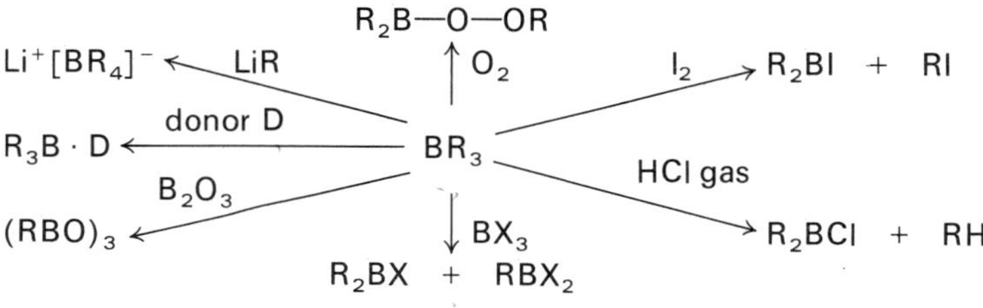

Fig. 7.2. Some reactions of triorganoboranes.

7.1.3. Halogeno-Organoboranes, R_nBX_{3-n}

Organoboron halides are formed on treatment of boron trihalides with organic derivatives of mercury, aluminum or tin; the latter are particularly versatile in the synthesis of aryl- and vinylboron halides:

$$2\,BCl_3 + HgR_2 \xrightarrow{180\,°C} 2\,R{-}BCl_2 + HgCl_2$$
$$3\,BCl_3 + AlR_3 \longrightarrow 3\,R{-}BCl_2 + AlCl_3$$
$$2\,BCl_3 + SnPh_4 \longrightarrow 2\,Ph{-}BCl_2 + Ph_2SnCl_2$$

Friedel-Crafts reactions (catalyzed by anhydrous aluminum halides) can also be used:

$$C_6H_6 + BCl_3 \xrightarrow[120\,°C]{AlCl_3} C_6H_5BCl_2 + HCl$$

as well as the redistribution between triorganoboranes and boron trihalides ($X = F$, Cl, Br):

$$BR_3 + BX_3 \longrightarrow R_2BX + RBX_2$$

or the reaction between organoboron acid esters and phosphorus pentachloride:

$$RB(OR)_2 + PCl_5 \longrightarrow RBCl_2 + (RO)_2PCl_3$$

Halogeno-organoboranes are important compounds in organoboron chemistry. They serve as starting materials in the synthesis of oxygen-, nitrogen- and sulfur-containing organoboron compounds or heterocyclic derivatives. Some reactions of halogeno-organoboranes are illustrated in Fig. 7.3:

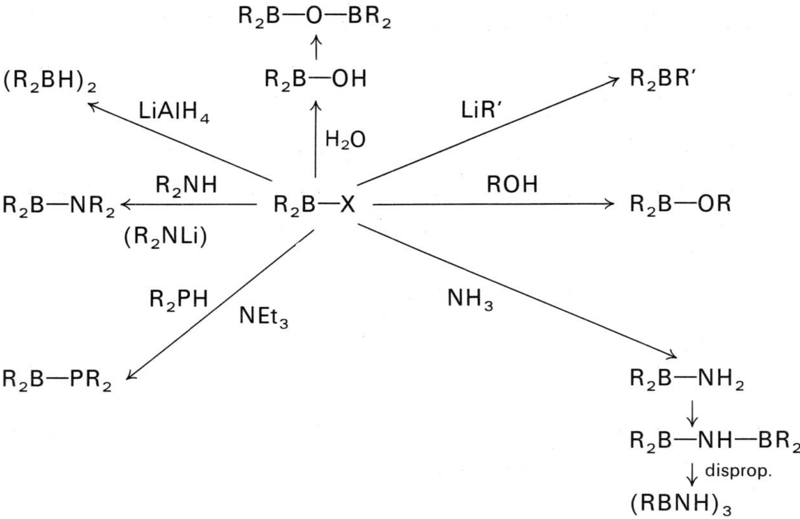

Fig. 7.3. Some reactions of halogenoorganoboranes (X = halogen).

7.1.4. Organoboron Acids, $R_nB(OH)_{3-n}$

Boronic, $R—B(OH)_2$, and borinic, $R_2B—OH$, acids are prepared by hydrolysis of the corresponding halides or esters. Borinic acids are best prepared by the hydrolysis of acetylacetone derivatives. All exhibit weak acidic properties and are readily converted to their anhydrides, sometimes spontaneously:

$$R_2BX \xrightarrow{H_2O} R_2B-OH \xrightarrow[-H_2O]{} R_2B-O-BR_2$$

$$RBX_2 \xrightarrow{H_2O} RB(OH)_2 \longrightarrow (RBO)_3 \text{ (cyclic B}_3\text{O}_3\text{ ring, each B bearing R)}$$

Cyclic boroxines, obtainable directly from boric oxide and triorganoboranes, serve as intermediates in the preparation of boronic acids by hydrolysis:

$$BR_3 + B_2O_3 \longrightarrow (RBO)_3 \xrightarrow{3\,H_2O} 3\,R—B(OH)_2$$

In solid phenylboronic acid, $Ph—B(OH)_2$, the molecules are associated as dimers.

7.1.5. Esters of Organoboric Acids (Alkoxyboranes), $R_nB(OR')_{3-n}$

Trialkoxyboranes react with Grignard or organolithium reagents to give these esters:

$$B(OMe)_3 + RMgX \longrightarrow RB(OMe)_2 + MgX(OMe)$$
$$B(OMe)_3 + 2\,RMgX \longrightarrow R_2B(OMe) + 2\,MgX(OMe)$$
$$B(OR)_3 + LiR \longrightarrow RB(OR)_2 + LiOR$$

An alternative preparation involves redistribution:

$$2\,BR_3 + B(OAr)_3 \longrightarrow 3\,R_2B{-}OAr$$

The esters are moisture-sensitive liquids with some of the aromatic derivatives crystalline solids, more stable hydrolytically and thermally. They serve as intermediates in the synthesis of other organoboron compounds.

7.1.6. Amino-Organoboranes, $R_nB(NR'R'')_{3-n}$, and Other Boron-Nitrogen Compounds

The compounds of the type $R_2B{-}NR'_2$ are monomeric when steric hindrance prevents oligomerization. In these cases the lone pair of electrons at nitrogen are donated *intramolecularly* to the boron atom, thus increasing the order of the B—N bond:

$$R_2B \overset{\leftarrow}{=} NR_2 \quad \text{or} \quad R_2\overset{\delta-}{B}{=}\overset{\delta+}{N}R_2$$

The structure of solid $Me_2B{-}NMe_2$ has been investigated at $-95\,°C$, and the molecule found to be planar with a B—N bond of 142 pm (≡1.42 Å) (calculated bond order of 1.57).

The partial double bond character of the B—N bond in these compounds is demonstrated by the isolation of *cis-trans* isomers of benzylmethylamino-benzylphenylboranes, where there is hindered rotation about the B—N bond:

$$(PhCH_2)(Me)N \rightleftharpoons B(Ph)(CH_2Ph) \qquad (PhCH_2)(Me)N \rightleftharpoons B(CH_2Ph)(Ph)$$

When the boron or nitrogen substituents are small, (for example, hydrogen) the nitrogen atom can donate its lone pair *intermolecularly*, resulting in the formation of cyclic dimers and trimers:

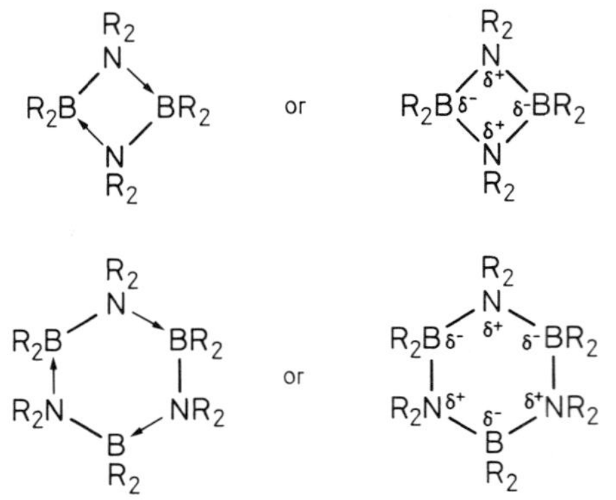

The dimer and trimer may also coexist in a temperature-dependent equilibrium.

Amino-organoboranes are obtained by reaction of diorganoboron halides and primary or secondary amines or their lithiated derivatives:

$$R_2B{-}Cl \;+\; R_2NH\,(\text{or } R_2NLi) \longrightarrow R_2B{-}NR_2 \;+\; HCl\,(LiCl)$$

Tris(diorganoboryl)amines, $N(BR_2)_3$, have also been prepared.

Mono-organoboron halides react with primary amines and ammonia to form cyclic borazines which have an extensive chemistry with many organosubstituted derivatives known:

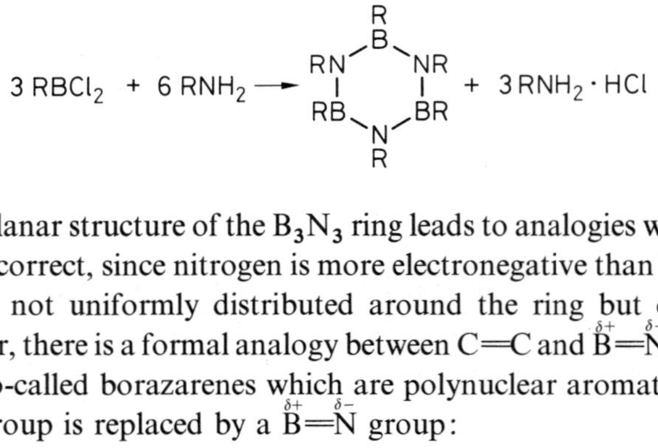

$$3\,RBCl_2 + 6\,RNH_2 \longrightarrow (RBNR)_3 + 3\,RNH_2 \cdot HCl$$

The planar structure of the B_3N_3 ring leads to analogies with benzene which are not entirely correct, since nitrogen is more electronegative than boron, and the π-electron cloud is not uniformly distributed around the ring but concentrates at nitrogen. However, there is a formal analogy between C=C and $\overset{\delta+}{B}{=}\overset{\delta-}{N}$ groups, which is invoked in the so-called borazarenes which are polynuclear aromatic compounds in which a C=C group is replaced by a $\overset{\delta+}{B}{=}\overset{\delta-}{N}$ group:

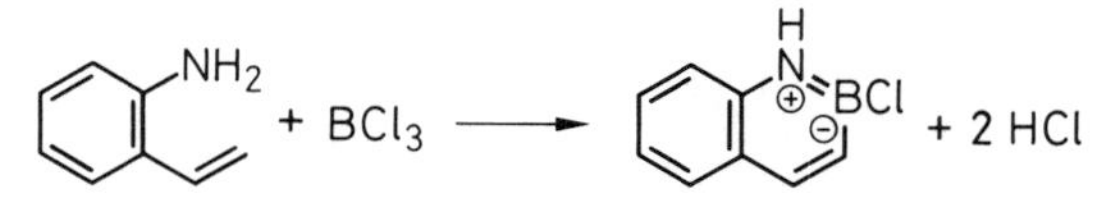

$$\text{o-aminostyrene} + BCl_3 \longrightarrow \text{(B-chloro-borazarene)} + 2\,HCl$$

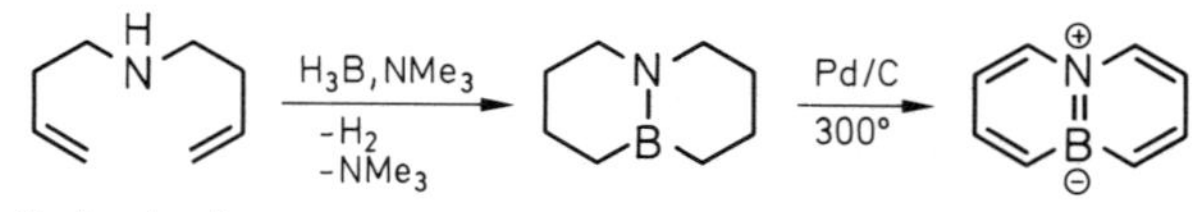

o-aminostyrene

$$\text{dibutenylamine} \xrightarrow[-H_2,\ -NMe_3]{H_3B,NMe_3} \text{(bicyclic aminoborane)} \xrightarrow{Pd/C,\ 300°} \text{(borazanaphthalene)}$$

dibutenylamine

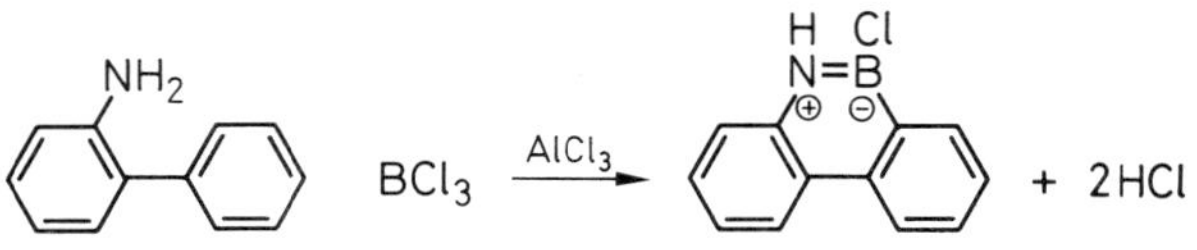

Among the other boron-nitrogen heterocycles known are the triazadiborolidines:

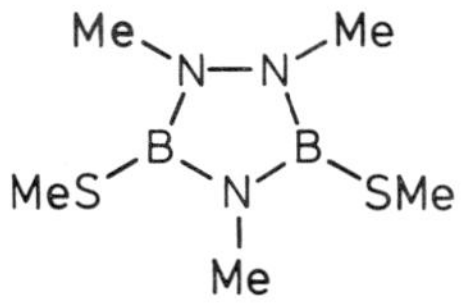

which can be prepared by treating the mercapto derivative, $B(SMe)_3$, with *sym*-dimethylhydrazine and methylamine:

$$2\,B(SMe)_3 + MeNHNHMe + MeNH_2 \longrightarrow (MeS)_2B_2N_3Me_3 + 4\,MeSH$$

The six-membered ring compound:

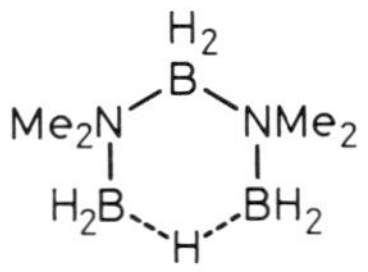

is obtained from $Me_2NHBH_2NMe_2BH_3$ (a unusual boron-nitrogen compound) and potassium in ether, followed by treatment with excess diborane.

7.1.7. Sulfur-Containing Organoboron Compounds

Organoboron compounds containing B—SH bonds eliminate hydrogen sulfide and undergo condensation with the formation of B—S—B bridges.

Mercapto-organoboranes are obtained from organoboron halides and mercaptides:

$$R_2B{-}Cl + NaSR \longrightarrow R_3B{-}SR + NaCl$$

$$RBCl_2 + 2\,NaSR \longrightarrow RB(SR)_2 + 2\,NaCl$$

or by elimination of volatile organosilicon halides in the reaction of Si—S compounds with organoboron halides:

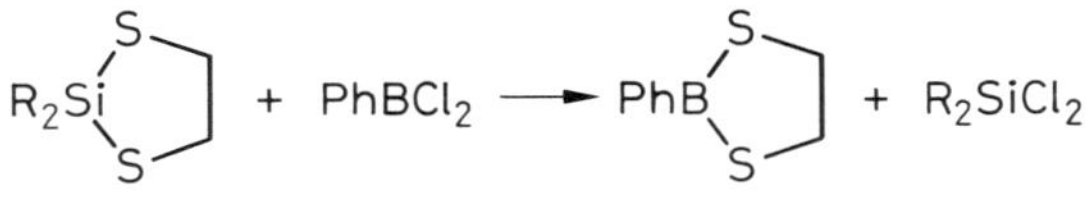

$$R_2Si(S_2C_2H_4) + PhBCl_2 \longrightarrow PhB(S_2C_2H_4) + R_2SiCl_2$$

Mercaptoboranes can also be obtained from lead mercaptides and organochloroboranes:

$$Ph_nBCl_{3-n} + Pb(SR)_2 \xrightarrow[-PbCl_2]{} Ph_nB(SR)_{3-n}$$
$$n = 1 \text{ or } 2$$

and five- and six-membered boron-sulfur rings, with organic substituents at boron, have also been prepared:

$$3RBX_2 + 3H_2S \longrightarrow (RBS)_3 \text{ (six-membered } B_3S_3 \text{ ring)} + 6HX$$

$$2RBX_2 + 2H_2S_2 \longrightarrow R{-}B\langle S{-}S, S\rangle B{-}R \text{ (five-membered ring)} + S + 4HX$$

The former may exhibit charge delocalization in the B_3S_3 ring.

7.1.8. Phosphorus-Containing Organoboron Compounds

The cyclic phosphinoborines show unexpectedly high thermal (to 500 °C) and chemical stability. Condensation of phosphines with organoboron halides yields the trimers, including when $R' = H$:

$$3R_2PH + 3XBR'_2 \xrightarrow{NEt_3} (R'_2B{-}PR_2)_3 + 3HX$$

With larger substituents the phosphinoborines are monomers in which electron donation from phosphorus to boron operates *intramolecularly*.

7.1.9. Carboranes

Carboranes are compounds formed by replacing one or two boron atoms in anionic boranes of the type $B_nH_n^{2-}$ with carbon heteroatoms.

Carboranes are generally obtained by reaction of boranes with acetylenes, for example, *ortho*-$B_{10}C_2H_{12}$, is prepared from decaborane and acetylene in the presence of diethylsulfide:

$$B_{10}H_{14} + HC{\equiv}CH \xrightarrow[90°C]{Et_2S} B_{10}C_2H_{12} + 2H_2.$$

The *meta*- and *para*-isomers are obtained by anaerobic pyrolysis of the *ortho*-isomer:

$$ortho\text{-}B_{10}C_2H_{12} \xrightarrow{465-500\,^\circ C} meta\text{-}B_{10}C_2H_{12} \xrightarrow{615\,^\circ C} para\text{-}B_{10}C_2H_{12}$$

Lower carboranes have trigonal bipyramidal ($B_3C_2H_5$), octahedral ($B_4C_2H_6$), pentagonal bipyramidal ($B_5C_2H_7$) and dodecahedral ($B_6C_2H_8$) structures, corresponding to the parent $B_nH_n^{2-}$ anion. Figure 7.4 shows some of these structures:

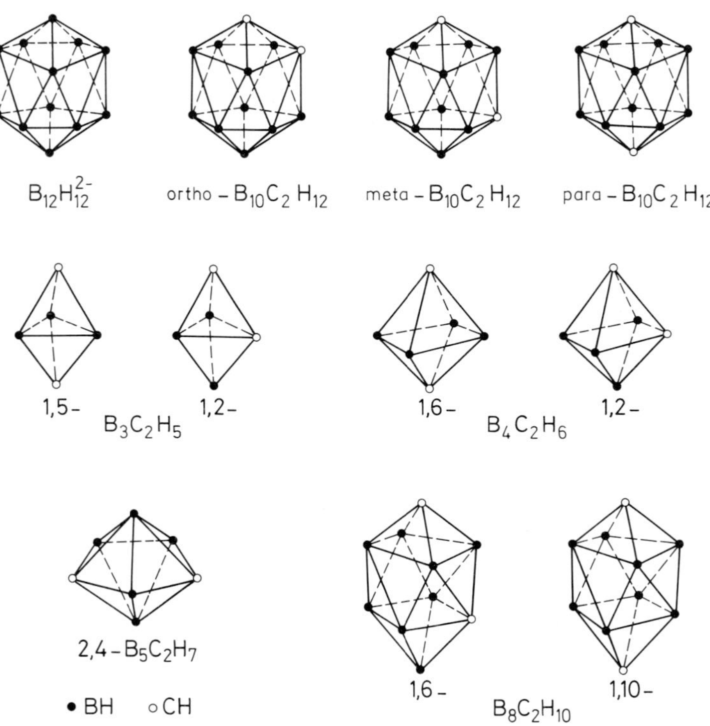

Fig. 7.4. Carborane structures: $B_{12}H_{12}^{2-}$ anion and the three carboranes derived from it; some lower carboranes.

The lower carboranes, 1,5-$B_3C_2H_5$, 1,6-$B_4C_2H_6$ and 2,4-$B_5C_2H_7$, are obtained as a mixture when gaseous acetylene and boranes (B_2H_6, B_4H_{10}, B_5H_9 and B_5H_{11}) are subjected to electrical discharges or high temperatures.

Multicarbon carboranes such as $C_4B_2H_6$ and the methyl derivative $Me_4C_4B_8H_8$ are also known.

In all carboranes the boron atoms form closed cages and the classical concepts of valence (trivalent boron, tetravalent carbon) are not applicable since both boron and carbon are bonded to a larger number of other atoms. A simplistic but acceptable explanation for the formation and the nature of bonding in these compounds based upon molecular orbital theory is a follows: the BH and CH groups gave each three identical sp^3-hybrid orbitals available for further bonding. Boron has two electrons and carbon three electrons available (one in each cross-hatched orbital):

The polyhedral arrangement of BH and CH groups with the hydrogens pointing outwards is formed by overlap and combination of the sp^3-hybrid orbitals with formation of molecular orbitals concentrated in the space inside the polyhedron. The electrons contributed by boron and carbon atoms are located in these molecular orbitals and are delocalized in the closed space of the B_mC_n cage. This process of shared collectivization of the available hybrid orbitals and valence electrons of the BH and CH groups can also involve elements other than carbon and boron, and the formation of heterocarboranes can be expected, including the non-metals, aluminum and gallium and the transition metals iron, cobalt and nickel. Known examples include $B_8C_2NH_{13}$ and $B_8C_2SH_{10}$, $B_9C_2H_{11}AsPh$ (related to $B_{10}C_2H_{12}$ by replacing a BH group by As—Ph), aluminum and gallium carboranes, the chromium complex of $CB_{11}H_{11}Ge$ as well as several heterocarborane derivatives, containing cobalt and iron in their skeletons (see Fig. 7.5):

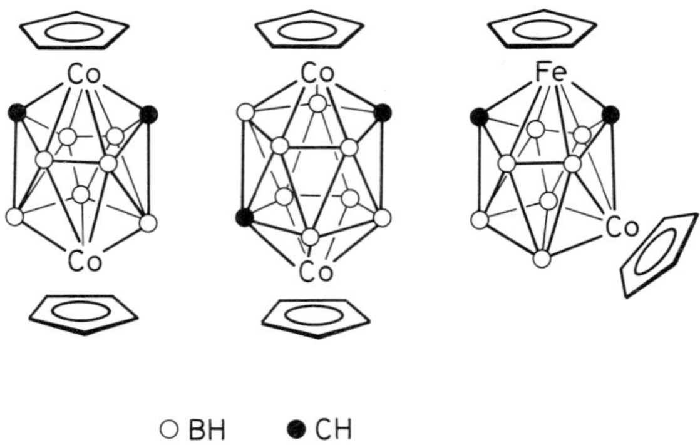

Fig. 7.5. Some heterocarborane structures containing transition metals.

Electron delocalization in the carborane cage results in aromatic character of the closed carboranes, but unlike benzene, in which the electron delocalization is planar, in closed carboranes the delocalization is in 3-dimensional space. For this reason the closed carboranes are sometimes referred to as superaromatic, which is also in agreement with their chemical behavior and stability. Indeed, in its reactions, the $B_{10}C_2H_{12}$ group behaves as an aromatic nucleus, and on this basis a whole chapter of organic and organometallic chemistry has been developed.

The best investigated carborane is *ortho*-$B_{10}C_2H_{12}$, whose external hydrogens can be readily substituted, for example by chlorine. More drastic conditions produce full chlorination, including at the carbon sites:

$$\underset{B_{10}H_{10}}{HC{-}CH} \xrightarrow{Cl_2} \underset{B_{10}Cl_{10}}{HC{-}CH} \xrightarrow{Cl_2} \underset{B_{10}Cl_{10}}{ClC{-}CH} \xrightarrow{Cl_2} \underset{B_{10}Cl_{10}}{ClC{-}CCl}$$

Metalation with organolithium reagents or lithium amides occurs only at carbon sites; the metalated derivatives can be converted into numerous other compounds, as shown in Fig. 7.6:

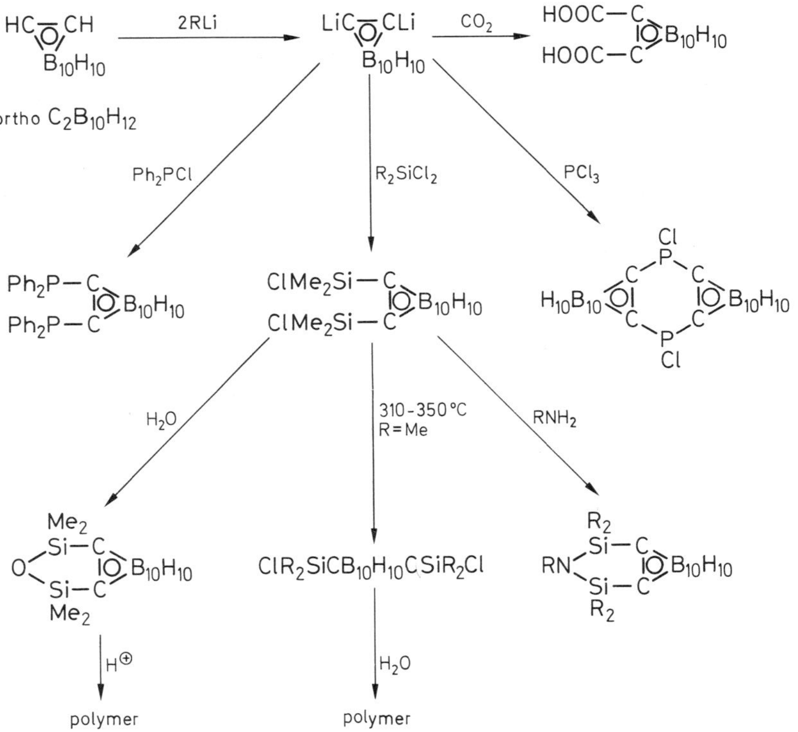

Fig. 7.6. Some reactions of icosahedral, closed carboranes.

The $B_{10}C_2H_{12}$ carboranes exhibit a remarkable thermal stability, illustrated by the high temperatures used in the isomerization of the *ortho*- and *meta*-derivatives. The high thermal stability is also conferred on the polymers with carborane nuclei backbones such as the thermally stable polymers (to 500 °C) containing *meta*-carborane units connected through siloxane bridges (known under the trade name *Dexil*) which are produced on a commercial scale for uses in space and aviation technology, as well as for stable supports in high temperature gas chromatography. *Dexil*-type polymers are obtained by condensation of methoxysilylcarboranes with chlorosilylcarboranes or chlorosilane, in the presence of iron(III) chloride ($R = CH_3$):

$$\left(\underset{R}{\overset{R}{\mathrm{Si}}}\!-\!CB_{10}H_{10}C\!-\!\underset{R}{\overset{R}{\mathrm{Si}}}\!-\!O\!- \right)_x \quad \textit{Dexil } 100$$

$$\uparrow \; ClR_2SiCB_{10}CSiR_2Cl/FeCl_3$$

$$x(RO)R_2SiCB_{10}H_{10}CSiR_2(OR) \xrightarrow{R_2SiCl_2} \left(-\underset{R}{\overset{R}{\mathrm{Si}}}\!-\!CB_{10}H_{10}C\!-\!\left(\underset{R}{\overset{R}{\mathrm{Si}}}\!-\!O \right)_2 \right)_x \quad \textit{Dexil } 200$$

$$\downarrow \; (ClR_2Si)_2O/FeCl_3$$

$$\left[-\underset{R}{\overset{R}{\mathrm{Si}}}\!-\!CB_{10}H_{10}C\!-\!\left(-\underset{R}{\overset{R}{\mathrm{Si}}}\!-\!O\!- \right)_3\!- \right]_x \quad \textit{Dexil } 300$$

No other synthetic or natural elastomer has the thermal stability of *Dexil*-type polymers.

The controlled degradation of higher carboranes affords the synthesis of novel, neutral or anionic carboranes with open structures by cleavage of fragments from the parent cage:

$$\textit{ortho}\text{-}B_{10}C_2H_{10} \xrightarrow[NMe_4^+ X^-]{KOH/MeOH} [NMe_4]^+[B_9C_2H_{12}]^- \begin{cases} \xrightarrow{NaH} Na_2B_9C_2H_{11} \\ \xrightarrow{HCl} B_9C_2H_{13} \end{cases}$$

The $B_{10}C_2H_{11}^{2-}$ anion is of particular interest since it has an open structure with a pentagonal face which results in a chemical behavior similar to that of the cyclopentadienyl anion, $C_5H_5^-$, that is, formation of transition metal π-complexes, analogous to the metallocenes (see Section 14.5).

The carboranes are probably the most fascinating chapter of organoboron chemistry, owing to their nonclassical structures and unusual chemical behavior. Their investigation expanded chemical bond theory and enriched it with new ideas about electron delocalization and bond formation.

7.2. Organoaluminum Compounds

Although the first organoaluminum compounds were synthesized in 1865 by G.B. Buckton and W. Odling, they became industrially important only in the last 25 years, after the discovery of their catalytic properties in stereospecific polymerization of olefins by the Ziegler-Natta method and the discovery of a simple and inexpensive industrial manufacturing procedure (the direct synthesis from aluminum metal, hydrogen and olefins – the Ziegler procedure). The readily available aluminum alkyls open the way for new syntheses of commercially important organometallic compounds, or in processes requiring alkyl-transfer reagents. In spite of their extreme sensitivity to air and moisture, aluminum alkyls are currently manufactured and handled on an industrial scale, not only as catalysts, but also as intermediates in organic synthesis, for example in the manufacture of higher alcohols and saturated acids.

7.2.1. Types of Organoaluminum Compounds

There are some formal similarities with organoboron compounds in composition, but there are remarkable structural differences. The tendency of aluminum to use completely its valence orbitals is manifest, and four-coordinated organoaluminum compounds are frequently encountered. Thus, even the compounds AlR_3 and R_2AlX dimerize, while the boron analogues are monomeric. The distinction between boron and aluminum seems to be mainly based upon the preference of the latter to form *intermolecular* donor-acceptor bonds; the boron atom uses the fourth available valence orbital mainly by *intramolecular* donation.

The main types of organoaluminum compounds can be classified as follows:

a) Trisubstituted organoaluminum compounds, AlR_3;
b) Aluminum heterocycles;
c) Mononuclear compounds with four-coordinated aluminum;
d) Cyclic dimers, trimers, tetramers and other oligomers with four-coordinated aluminum.

An uncommon situation occurs in the cyclopentadienyl derivative, $C_5H_5AlMe_2$, a compound in which aluminum (as an $AlMe_2$ group) is bonded in the gas phase to one side of the cyclopentadienyl ring:

H3C CH3
Al

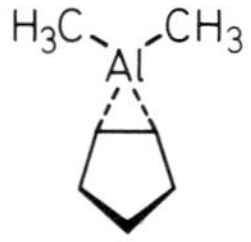

The unique bonding in this compound can be treated on the basis of *ab initio* molecular orbital calculations for the model compound $H_2AlC_5H_5$. The bonding between the aluminum atom and the cyclopentadienyl ring is di*hapto* and arises from an interaction of the $(a_1)\pi$-orbital of the ring with the 3s-orbital of aluminum, and of the $(e_{1x})\pi$-orbital of the ring with the $3p_x$-orbital of aluminum.

7.2.2. Trisubstituted Organoaluminum Compounds

The trisubstituted compounds, AlR_3, are dimers, formed through three-center, electron-deficient bonds. The dimerization of trimethylaluminum occurs through methyl bridges, and triphenylaluminum and dimethylphenylaluminum dimerize by sharing phenyl groups between aluminum atoms. These structures have been described in Section 2.4.

In the vapor phase, trimethylaluminum is monomeric and planar (D_{3h} symmetry), suggesting sp^2-hybridization at aluminum.

Compounds with bulky substituents producing steric hindrance to dimerization, for example, $Al(iso\text{-}Pr)_3$ and $Al(iso\text{-}Bu)_3$, are also monomeric.

Preparation. The preparation of organoaluminum compounds seldom uses Grignard reagents since ether adducts, $R_3Al \cdot OR_2$, result:

$$AlX_3 + 3\,RMgX \xrightarrow{Et_2O} R_3Al \cdot OEt_2 + 3\,MgX_2$$

With excess Grignard reagent, anionic species, $[AlR_4]^-$, are formed. Grignard reagents in hydrocarbon solutions, without ether, have also been used.

For aromatic compounds the reaction of organomercury compounds with excess aluminum metal is preferred:

$$2\,Al + 3\,HgR_2 \longrightarrow 2\,AlR_3 + 3\,Hg$$

The reaction which makes aluminum alkyls available on a large scale is the direct synthesis from aluminum-metal powder, olefins and hydrogen under heat and pressure:

$$Al + 3/2\,H_2 + 3\,RCH{=}CH_2 \xrightarrow[50-200\ bar]{110-60\ °C} Al(CH_2CH_2R)_3$$

The olefins can undergo addition to aluminum hydride, AlH_3, or lithium alanate, $LiAlH_4$, to form R_2AlH and AlR_3, or even $Li^+[AlR_4]^-$ (hydroalumination). The addition of organoaluminum hydrides to acetylenes gives the *cis*-product (R = iso-Bu):

$$R_2AlH + Me{-}C{\equiv}C{-}Ph \xrightarrow{50\ °C} \begin{matrix} Me & & Ph \\ & C{=}C & \\ H & & AlR_2 \end{matrix}$$

The addition of olefins to aluminum trialkyls may result in the increase of the substituent chain length, and finally in the polymerization of the olefin:

$$AlR_3 + H_2C{=}CH_2 \longrightarrow R_2AlCH_2CH_2R \xrightarrow{3n\,H_2C=CH_2} Al[(CH_2CH_2)_nR]_3$$

The reaction is of industrial importance, since after controlled oxidation and hydrolysis, higher alcohols and fatty acids are formed:

$$Al(CH_2CH_2R)_3 \xrightarrow{O_2} Al(OCH_2CH_2R)_3 \xrightarrow{H_2O} 3\,RCH_2CH_2OH + Al(OH)_3$$

Aluminum trialkyls, (for example, R = Et) result from the reaction between aluminum and magnesium metals with organic halides in ether to give an adduct:

$$2\,Al + 3\,Mg + 6\,RX + Et_2O \longrightarrow 2\,R_3Al \cdot OEt_2 + 3\,MgX_2$$

The acidic hydrogen atom in acetylenes cleaves the phenyl groups from triphenylaluminum and forms organoaluminum acetylenes:

$$AlPh_3 + HC{\equiv}C{-}Ph \xrightarrow{25-50\,°C} Ph_2Al{-}C{\equiv}C{-}Ph + PhH$$

The cyclopentadienyl derivatives of aluminum, gallium and indium, $R_2MC_5H_5$, are prepared straightforwardly from the dialkylmetal chlorides and sodium cyclopentadienide:

$$R_2AlCl + NaC_5H_5 \longrightarrow R_2AlC_5H_5 + NaCl$$

Properties. Triorganoaluminum derivatives are very sensitive to oxygen (most are pyrophoric in air) and to compounds with active hydrogen such as water, alcohols,

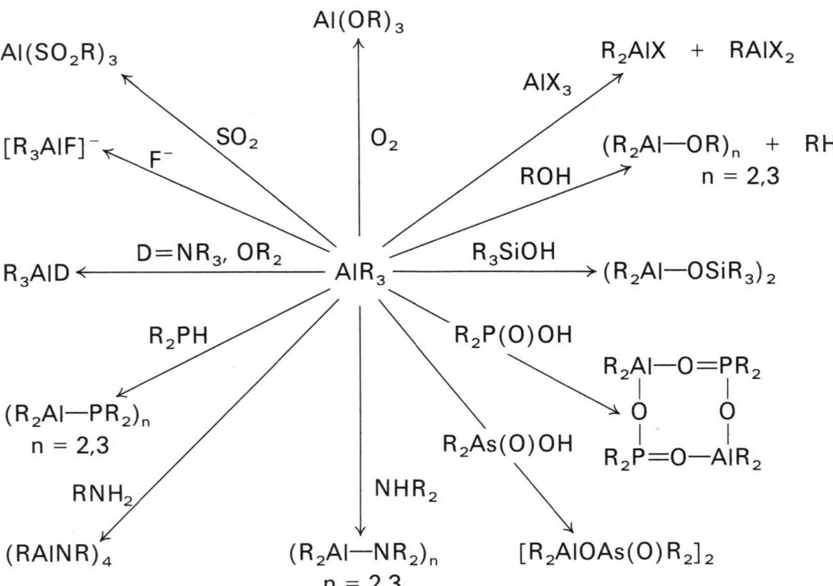

Fig. 7.7. Reactions of organoaluminum compounds.

amines, acids, etc. They hydrolyze to aluminum hydroxide and hydrocarbon, but controlled reactions with other active-hydrogen reagents can be of preparative value. Important reactions of triorganoaluminum derivatives are shown schematically in Fig. 7.7.

7.2.3. Heterocyclic Compounds with Aluminium Heteroatoms

Inclusion of aluminum into a ring may favor the preservation of the three-coordinated state and prevent dimerization. Two examples formed by alumination of aromatic nuclei on heating, with proton abstraction from the biphenyl nucleus by the leaving phenyl group are shown:

$AlPh_2$ $\xrightarrow{t}$ Al Ph + PhH

$MeN-AlPh_2$ $\xrightarrow{t}$ MeN–AlPh + PhH

A nearly planar six-membered ring, containing two aluminum atoms, has been prepared as a bis-THF adduct:

$$2\,Na^{\oplus}\left[\begin{matrix}Ph & & Ph\\ & C=C & \\ R_3Al & & AlR_3\end{matrix}\right]^{2-} \xrightarrow[-NaCl,\ -Me_3SiR]{+Me_3SiCl\ \text{in THF}} \begin{matrix}Ph & & Ph\\ & C=C & \\ R_2Al & & AlR_2\\ \cdot THF & & \cdot THF\end{matrix} \xrightarrow{-AlR_3\cdot THF}$$

R = Et

$$\begin{matrix} & R\ \ Al \leftarrow THF & \\ Ph & & Ph\\ \| & & \| \\ Ph & & Ph\\ & THF \rightarrow Al\ \ R & \end{matrix}$$

This reaction is interesting for it proceeds through three types of unusual organoaluminum compounds.

7.2.4. Mono- and Dinuclear Compounds with Four-Coordinated Aluminum

The presence of the vacant orbital in trisubstituted aluminums leads to the tendency to form a fourth bond, either by coordination of a neutral donor molecule of an amine, ether or phosphine, or by forming an additional Al—C bond, formally equivalent to the coordination of a carbanion, $:R^-$. In donor solvents triorganoaluminums are solvated, and the central atom becomes four-coordinated [$R_3Al \cdot D$ (D = ether, amine, etc.], with the donor ability toward aluminum decreasing in the order:

$$NMe_3 > PMe_3 > OMe_2 > SMe_2 > SeMe_2 > TeMe_2$$

Coordination of fluoride and hydride anions yields complex anions, such as $[R_3AlX]^-$, $[R_3AlH]^-$, $[R_2AlH_2]^{2-}$ and $[RAlH_3]^-$. A less-common example is the linear, symmetric, dinuclear anion containing a fluorine bridge, $[R_3Al—F—AlR_3]^-$, in the solid state. The anion in $K^+[MeAlCl_3]^-$ forms a distorted tetrahedral structure.

A nitrogen-containing analogue, $Na^+[MeN(AlR_3)_2]^-$, has been prepared by the reaction of $Na[R_3AlH]$ with methylamine.

Coordination can stabilize R_2Al^+ cations, otherwise incapable of free existence; *ortho*-phenanthroline and bis(phosphinimino)silanes are good ligands for this purpose:

Et_2Al(phen); $[Me_2Si(N{=}PMe_3)_2AlMe_2]^{\oplus}[AlMe_4]^{\ominus}$

In purely inorganic compounds, aluminum can achieve coordination numbers higher than four by using 3d-orbitals, for example, five in $AlH_3 \cdot$ bipy and six in $[AlF_6]^{3-}$ and $Al(acac)_3$, but in organometallic compounds this is observed only to a limited extent. Trimethylaluminum adducts with bipyridyl and tetramethyltetrazene contain five-coordinated aluminum:

R_3Al(bipy); $R_3Al(R_2N–N{=}N–NR_2)$

The four-coordinated $[AlR_4]^-$ anions form when aluminum halides react with an excess of Grignard or organolithium reagents. Acetylenic hydrogen reacts with lithium alanate, eliminating molecular hydrogen and forming tetrasubstituted compounds:

$$4\,RC{\equiv}CH + LiAlH_4 \longrightarrow Li[Al(C{\equiv}CR)_4] + 4\,H_2$$

Even benzene can be metallated by $Na^+[AlEt_4]^-$ in the presence of sodium ethylate to form tetraphenylalanate:

$$4\,C_6H_6 + Na[AlEt_4] \longrightarrow Na[Al(C_6H_5)_4] + 4\,C_2H_6$$

Cyclopentadiene, thiophene and furane eliminate hydrogen on heating with $M^+[AlH_4]^-$ to form tetrahedral organoderivatives. The salt $K^+[Me_3AlH]^-$ is formed by pyrolysis of $K^+[Me_3Al{-}SiH_3]^-$.

A four-coordinated, aluminum spiro-bicyclic system was prepared from a phosphorus diylid:

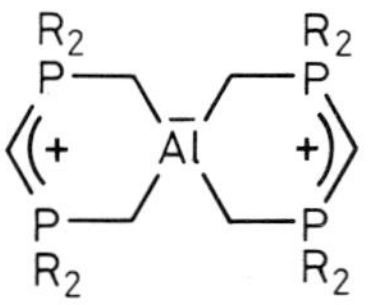

R2 R2
P P
(+ Al +)
P P
R2 R2

7.2.5. Cyclic Dimers, Trimers and Tetramers Containing Four-Coordinated Aluminum

The $R_2Al{-}X$ compounds in which X = halogen, OR, NHR, NR_2, PR_2, AsR_2, SR or SeR are associated, with cyclic structures in which aluminum is four-coordinated.

Organoaluminum Halides

These are dimers with halogen bridges as in the chlorides, $[R_2AlCl]_2$, and sesquichlorides, $R_3Al_2Cl_3$, but the fluorides, R_2AlF, are cyclic tetramers:

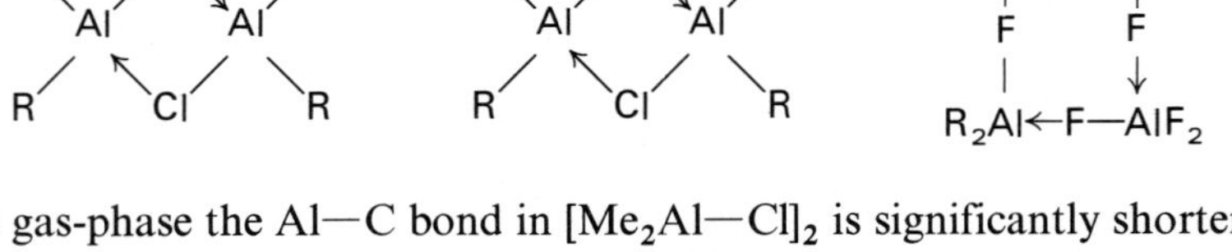

R Cl R
Al Al
R Cl R

R Cl Cl
Al Al
R Cl R

$R_2Al{-}F{\rightarrow}AlR_2$
↑ |
F F
| ↓
$R_2Al{\leftarrow}F{-}AlF_2$

In the gas-phase the Al—C bond in $[Me_2Al{-}Cl]_2$ is significantly shorter than the terminal Al—C bond in Al_2Me_6, and the Al—Cl bond is longer than the Al—Cl bridge in Al_2Cl_6.

The puckered, eight-membered ring, fluorine-bridged tetramer also survives in the vapor phase.

The direct reactions of alkyl halides with aluminum metal produce equimolecular mixtures of R_2AlX and $RAlX_2$ which are associated species:

$$3\,RX + 2\,Al \longrightarrow R_2AlX + RAlX_2 \longrightarrow R_3Al_2X_3$$

This reaction is of industrial importance in relation to the production of polymerization catalysts. A highly reactive form of aluminum, obtained by the reduction of halides with alkali metals, affords phenylaluminum halides by direct reaction.

Alkylaluminum halides are also formed in the reactions of aluminum trialkyls with aluminum and zinc halides, by substituent redistribution:

$$2\,AlR_3 + AlX_3 \longrightarrow 2\,R_2AlX$$
$$2\,AlR_3 + ZnX_2 \longrightarrow 2\,R_2AlX + ZnR_2$$

A precursor to the Ziegler synthesis is the reaction of aluminum metal, aluminum chloride, hydrogen and olefins:

$$Al + AlCl_3 + 3\,C_2H_4 + 1{,}5\,H_2 \longrightarrow (C_2H_5)_3Al_2Cl_3$$

Alkylaluminum Hydrides

These are also associated. Thus, in solution the R_2AlH compounds are trimers formed through Al···H···Al electron-deficient bridges, while dimers are found in the vapor phase (R = Me):

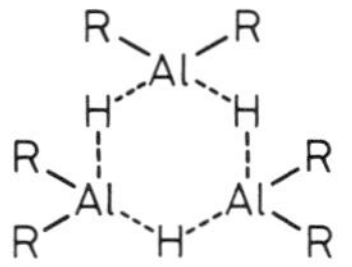

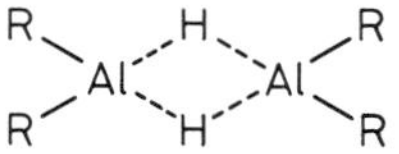

Organoaluminum Alkoxides, Thiolates and Amides

Organoaluminum derivates prepared from trialkyls with alcohols, thiols and amines are typical representatives of the cyclic oligomers with four-coordinated aluminum. Dimerization is common, but when the organic substituents of the functional groups (R in OR, SR or NR_2) are small, for example, CH_3, trimers can also form. With primary amines, cubane-type structures, as in $(RAl—NR')_4$ are formed, but hexamers, $(RAl—NR')_6$, heptamers, $(RAl—NR')_7$ and octamers, $(RAlNR')_8$ have also been confirmed.

In the solid dimer $(Me_2Al—NMe_2)_2$, the Al_2N_2 ring is planar. The dimer $(Me_2Al—NMePh)_2$ exists as a mixture of *cis*- and *trans*-isomers, with the two phenyl groups on the same or on different sides of the Al_2N_2 ring.

The trimer $(Me_2Al—NHMe)_3$ exists in two geometrical isomeric forms:

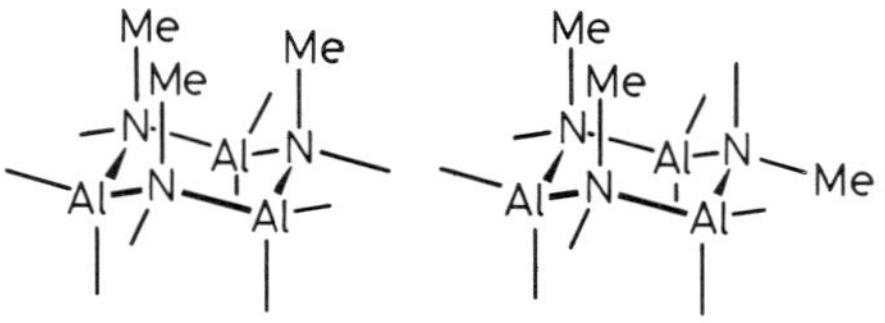

Other known structures include $(Me_2Al—NC_2H_4)_3$ and $(Me_2Al—OSiMe_3)_2$. The gas-phase structure of $(Me_2Al—OMe)_3$ contains a non-planar, six-membered ring. In the sulfur-containing cyclic dimer, $(Me_2Al—SMe)_2$, the S—Me groups are in *trans*-position in the gas phase:

7.3. Organogallium Compounds

The first representative, triethylgallium, was prepared in 1932. The compound types and chemical behavior are similar to those of the organoaluminums, with some differences in activity.

7.3.1. Trisubstituted Derivatives, GaR_3

Triorganogallium compounds are prepared from gallium metal and organomercury compounds, in a reaction catalyzed by mercury(II) chloride. Gallium trihalides can be alkylated and arylated by Grignard reagents, or organoaluminum or zinc compounds, as shown schematically in Fig. 7.8. In basic solvents they are obtained as solvates (etherates, etc.):

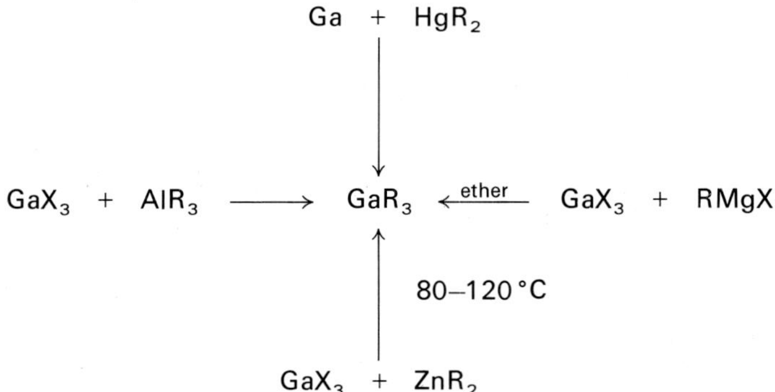

Fig. 7.8. Preparation of trisubstituted gallium derivatives.

Base-free trialkylgallium compounds, GaR_3, can be prepared by alkyllithium reagents. The 3 : 1 molar ratio between RLi and $GaCl_3$ must be strictly observed, because a deficit of RLi produces R_2GaCl or $RGaCl_2$ while an excess gives $Li[GaR_4]$.

Four-coordinated gallium is present in cyclic compounds prepared from trimethylgallium and phosphorus ylides:

The lower trialkylgallium derivatives are pyrophoric, while the higher members fume in air. Active hydrogen reagents, and even acetylenes, eliminate a hydrocarbon molecule and form the functional derivatives, R_2GaX, with associated cyclic or linear structures.

The trisubstituted derivatives, GaR_3, are monomeric in solution and the vapor phase, with the exception of trivinylgallium, which is dimeric. Solid triphenylgallium, unlike triphenylaluminum in the vapor phase, is monomeric. Trimethylgallium is monomeric in the vapor and the methyl groups are freely rotating at room temperature.

The cyclopentadienyl derivative, $C_5H_5GaMe_2$, forms chains of Me_2Ga units bridged by C_5H_5 rings in the solid state.

The typical reactions of GaR_3 derivatives are illustrated in Fig. 7.9:

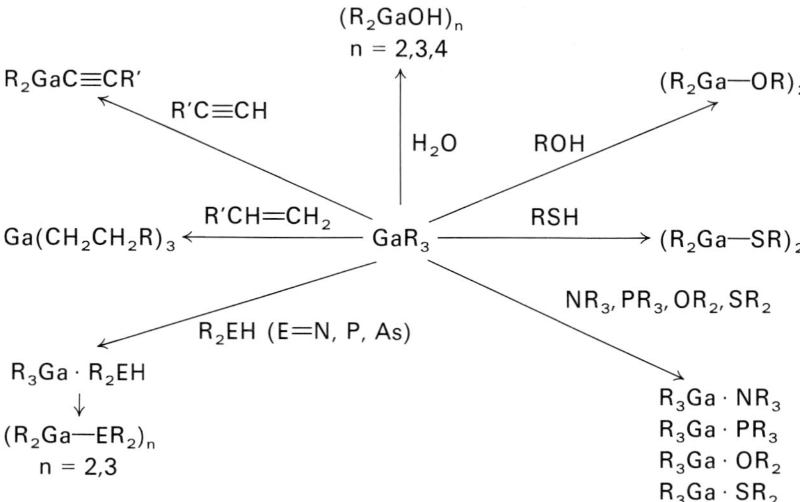

Fig. 7.9. Typical reactions of GaR_3 derivatives.

Ethers, amines, phosphines, thioethers form adducts in which the coordination number of gallium has been increased to four. The adducts of secondary amines, phosphines and arsines, whose stability decreases in the order N > P > As, eliminate a hydrocarbon molecule on heating, to form substitution products, $(R_2GaER_2)_n$, which are cyclic oligomers.

7.3.2. Diorganogallium Halides, R_2GaX

Diorganogallium monohalides are prepared by the alkylation of gallium trihalides with organolithium reagents, or better by cleavage of a Ga—R bond from GaR_3 compounds with hydrogen halides or halogens. They can also be obtained by redistribution between trisubstituted compounds and gallium trichloride (Fig. 7.10):

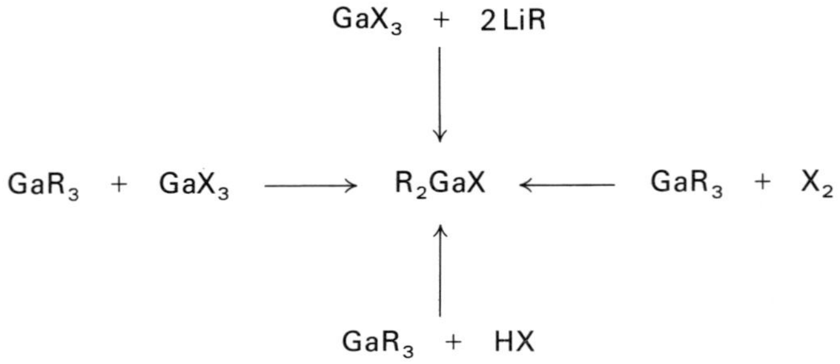

$$GaX_3 + 2\,LiR \longrightarrow R_2GaX$$

$$GaR_3 + GaX_3 \longrightarrow R_2GaX \longleftarrow GaR_3 + X_2$$

$$GaR_3 + HX \longrightarrow R_2GaX$$

Fig. 7.10. Preparations of R_2GaX compounds (X=halogen).

The monohalides are dimeric in the vapor phase and in solution, and contain halogen bridges connecting four-coordinated gallium. Only the dialkylgallium fluorides are trimeric and tetrameric:

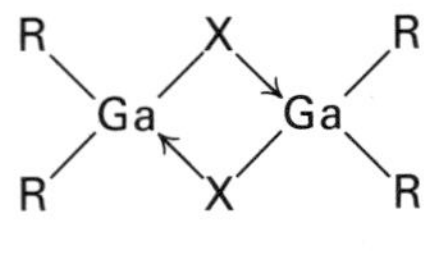

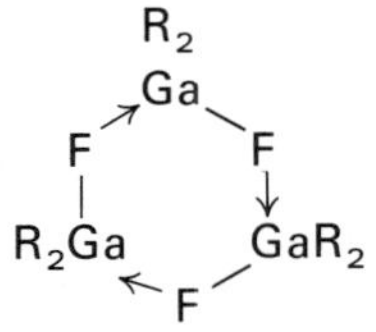

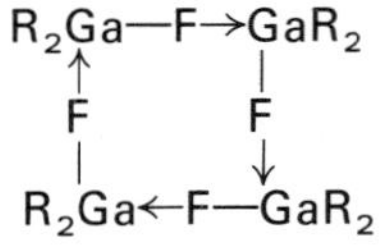

$$R_2Ga(\mu\text{-}X)_2GaR_2 \qquad (R_2GaF)_3 \qquad (R_2GaF)_4$$

In descending a main group of the periodic table a tendency to achieve higher coordination numbers and more ionic character of bonds is observed. This is illustrated by the five-coordinated adduct of Me_2GaCl with *ortho*-phenanthroline which adopts bipyramidal trigonal geometry in the solid state and in chloroform solution (a), but dissociation of chlorine occurs in water to form a cation (b):

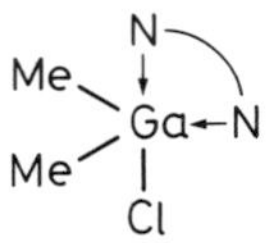

$$Me_2GaCl(N\text{—}N) \qquad [Me_2Ga(N\text{—}N)]^{\oplus}\,Cl^{\ominus}$$

(a) 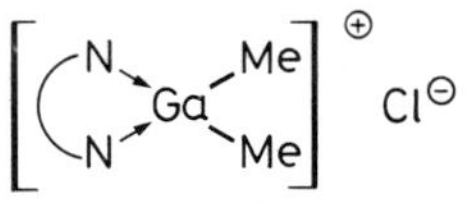 (b)

Monosubstituted organogallium dihalides, $RGaX_2$, are less stable. These can be prepared by alkylation of gallium trihalides with mild alkylating agents (for example, SiR_4, GeR_4, SnR_4 or ZnR_2), by cleavage of GaR_3 with anhydrous hydrochloric acid, by redistribution (for example, $PhGaX_2$ from $GaPh_3$ and GaX_3), or by addition of $HGaCl_2$ (obtained from $GaCl_3$ and $HSiMe_3$) to olefins (hydrogallation), as shown schematically in Fig. 7.11:

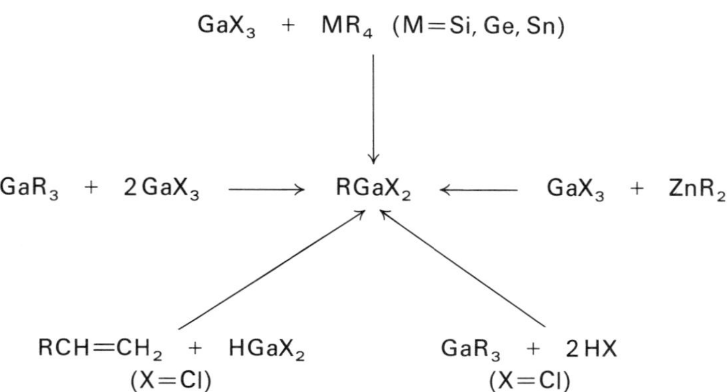

Fig. 7.11. Preparation of $RGaX_2$ derivatives.

Four-coordinated organogallium halide complexes, $[R_3GaX]^-$ ($R = Me$, Et; $X = F$ and $R = Me$, $X = Br$) and $[R_3Ga{-}X{-}GaR_3]^-$, are also known. The methylchlorogallate anions, $[Me_2GaCl_2]^-$ and $[MeGaCl_3]^-$, like their tetramethylarsonium salts are tetrahedral in the solid state.

7.3.3. Cyclic Oligomers with Four-Coordinated Gallium

Triorganogallanes react with water, alcohols, amines, and other active hydrogen reagents to give the functional derivatives, R_2GaX ($X = OH, OR, NR_2, SR, PR_2, AsR_2$) etc., which are associated in the solid state to glassy polymers and in solution to cyclic dimers, trimers and tetramers. Diorganogallium hydroxides, $(R_2Ga{-}OH)_n$, form as dimers, trimers and tetramers with the tetramer, $(Me_2Ga{-}OH)_4$, existing in the solid state as a puckered, eight-membered Ga_4O_4 ring:

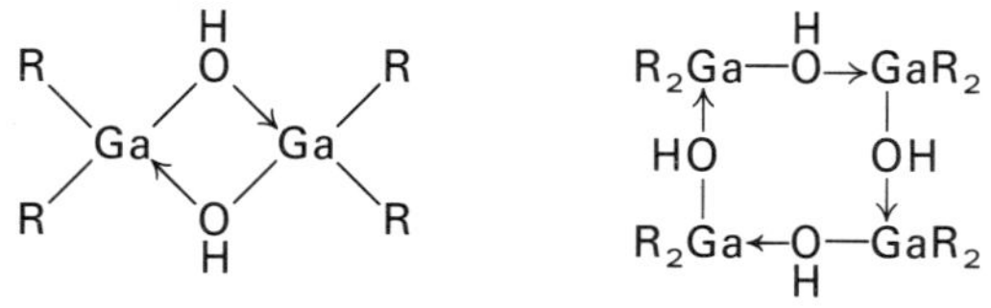

H
R O R
Ga Ga
R O R
H

H
$R_2Ga{-}O{\rightarrow}GaR_2$
HO OH
$R_2Ga{\leftarrow}O{-}GaR_2$
H

The alkoxides, $(R_2Ga{-}OR')_n$, are dimers; as are the mercapto, $(R_2Ga{-}SR')_n$, amino, $(R_2Ga{-}NR'_2)_2$, phosphino, $(R_2Ga{-}PR'_2)_n$, and arsino derivatives,

$(R_2Ga{-}AsR'_2)_n$, but trimeric phosphino- and arsino derivatives are also known, for example, $(Me_2Ga{-}PMe_2)_3$ and $(Me_2Ga{-}AsMe_2)_3$:

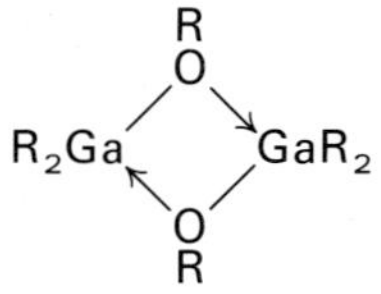

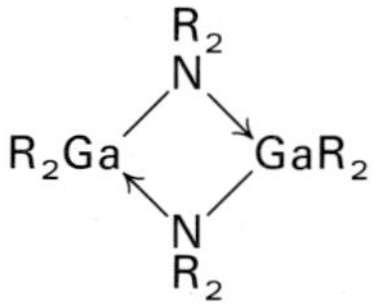

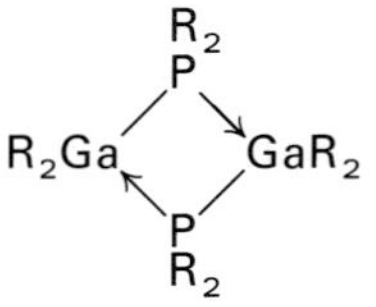

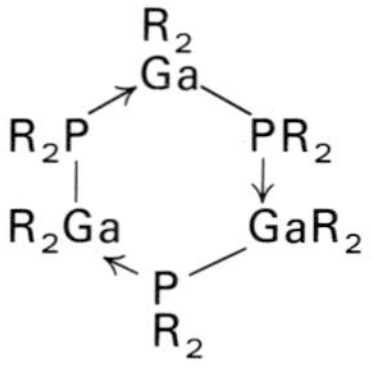

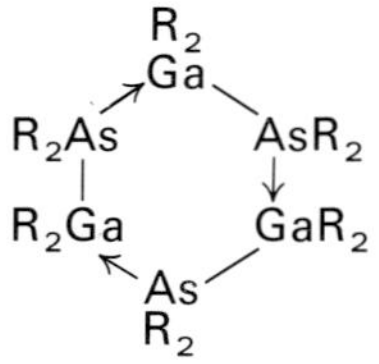

Eight-membered rings are formed by the reaction of gallium trialkyls with phosphinic, arsinic and sulfinic acids. Sulfinic acid dimers are also formed by sulfur dioxide insertion into the Ga—C bonds of triethylgallium:

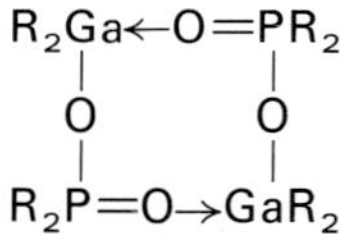

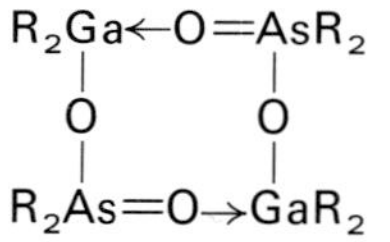

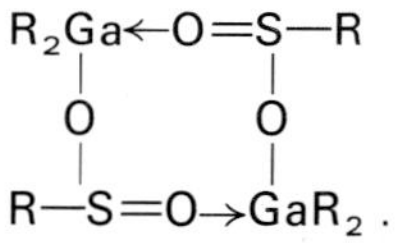

Organophosphorus thioacids behave differently; only monothiophosphinic acids form eight-membered ring dimers, while dithiophosphinic acids give rise to monomers, containing four-membered chelate rings:

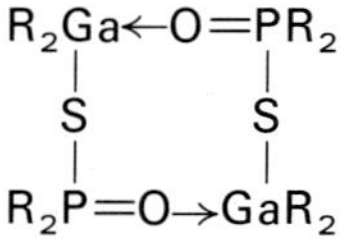

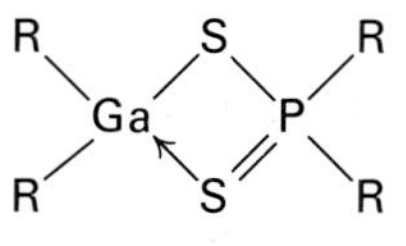

Chelates are also formed in the reaction of trimethylgallium with acetylacetone and salicylaldehyde:

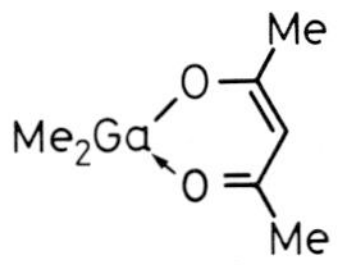

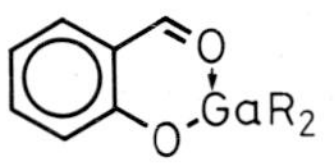

Diorganogallium cations are known only in four-coordinated species, as ammonia or diamine complexes. Thus, trimethylgallium etherate forms mono- and diammonia adducts with ammonium chloride. The latter is a salt containing the complex $[Me_2Ga(NH_3)_2]^+$ cation, in which ammonia can be replaced by ethylenediamine (en):

$$GaMe_3 \cdot OEt_2 \xrightarrow[-CH_4,\ -Et_2O]{NH_4Cl} \left. \begin{array}{c} Me_2GaCl \cdot NH_3 \\ + \\ [Me_2Ga(NH_3)_2]^+Cl^- \end{array} \right\} \xrightarrow{en} Me_2Ga \begin{array}{l} \leftarrow NH_2-CH_2 \\ \quad\quad\quad\ | \\ \leftarrow NH_2-CH_2 \end{array}$$

Four-coordinated gallium atoms forming four Ga—C bonds are present in cyclic compounds formed by the reaction of Me_2GaCl with the phosphorus ylide $Me_3P—\bar{C}H_2$:

$$Me_2GaCl + Me_3\overset{+}{P}—\overset{-}{C}H_2 \xrightarrow[-HCl]{} \begin{array}{ccc} Me_2\overset{-}{Ga} & —CH_2— & \overset{+}{P}Me_2 \\ | & & | \\ H_2C & & CH_2 \\ | & & | \\ Me_2\underset{+}{P} & —CH_2— & \underset{-}{Ga}Me_2 \end{array}$$

Indium and thallium analogues are formed similarly.

A phosphorus diylid yields a cyclic structure:

$$GaMe_3 \cdot OEt_2 + Me_3P{=}C{=}PMe_3 \longrightarrow Me_2\overset{\ominus}{Ga} \begin{array}{l} \diagup CH_2 - PR_2 \\ \quad\quad\quad\quad (+) \\ \diagdown CH_2 - PR_2 \end{array}$$

Organogallium compounds are mainly investigated for comparison with other elements of Group IIIA. Practical utility is limited by the high cost and rarity of the metal.

7.4. Organoindium Compounds

Organoindium chemistry is reminiscent of organoaluminum and organogallium chemistry, and this element will be only briefly discussed. The first organoindium compound, triethylindium, was prepared in 1934.

Trisubstituted Derivatives, InR_3

These are readily obtained by the reaction of indium metal with organomercury compounds, or by treatment of indium trihalides with organomagnesium, aluminum or lithium reagents.

Diorganoindium Halides, R_2InX

These are formed in reactions of indium trihalides with Grignard or organolithium reagents, or by cleavage of an In—R bond with halogens or haloforms.

Organoindium Dihalides, $RInX_2$

These are less-well known. Phenyl derivatives have been obtained by the action of the halogens upon triphenylindium.

Both mono- and dihalides, Ph_nInX_{3-n} (n = 1, 2), are prepared by oxidative arylation of indium(I) halides with diphenylmercury to give Ph_2InX and $PhInX_2$ which is associated through linear Ph_2In units linked through halogens, but $PhInI_2$ is unexpectedly a ionic compound, $[Ph_2In]^+[InI_4]^-$. The structures resemble the arylthallium more than the arylgallium halides.

Only a single monovalent organoindium compound is known, η^5-cyclopentadienylindium, η^5-C_5H_5In, prepared from indium trichloride and sodium cyclopentadienide, a reaction in which some tris(η^1-cyclopentadienyl)indium, $In(C_5H_5\text{-}\eta^1)_3$ is also formed. The solid state structure of $InC_5H_5\text{-}\eta^5$ suggests an ionic character with In^+ and $C_5H_5^-$ ions.

The properties of organoindium compounds are similar to those of the other Group III analogues. The lower trialkyls are pyrophoric, and all are readily oxidizable, react vigorously with water and active-hydrogen compounds.

The structure of trimethylindium in the vapor phase is monomeric and planar, but the solid state consists of tetrameric $In_4(CH_3)_{12}$ units formed through weak electron-deficient methyl bridges, in which each indium atom is surrounded by five CH_3 groups in a highly distorted trigonal-bipyramidal coordination. Three methyl groups lie in the equatorial plane at short distances (210 pm ≡ 2.1 Å), while the two axial bonds are longer (In—C 310 and 360 pm ≡ 3.1 and 3.6 Å). One methyl group is shared with another unit (Fig. 7.12):

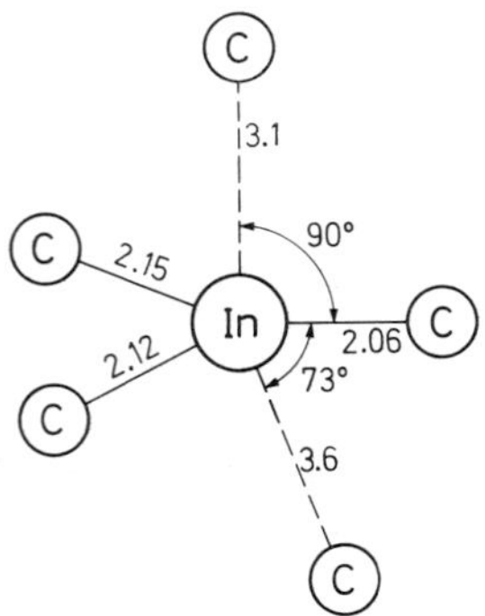

Fig. 7.12. The coordination geometry in solid $InMe_3$.

Triphenylindium on the other hand, is monomeric in the solid state, like triphenylgallium.

The halides Me_2InX and $(C_6F_5)_2InX$ form dimers in the vapor phase and in solution through halogen bridges with indium becoming four-coordinated.

The tendency for indium to increase its coordination number results in adduct formation, $R_3In \cdot D$ (for example, D = OR_2, SR_2, NR_3, etc.) and coordinative polymerization (oligomerization) of the functional derivatives, $(R_2In{-}X)_n$, where X = OR (n = 2 and 3), NR_2 (n = 2), PR_2 (n = 2 and 3) and AsR_2 (n = 2 and 3). Dimer formation is general, but when R is small cyclic trimers can also be formed:

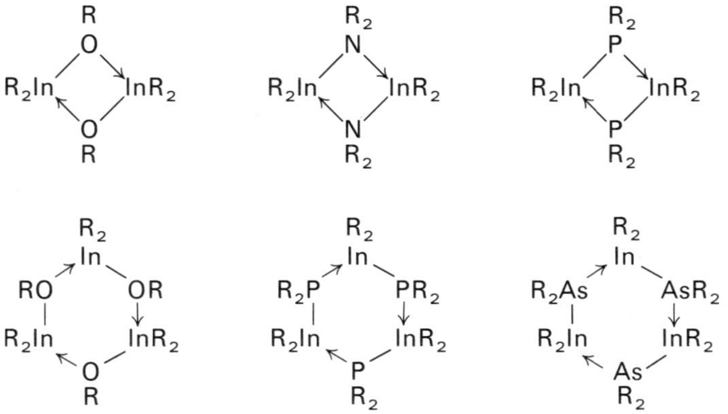

Four-coordinated chloroindates are formed in the reactions of Me_2InCl with trimethylantimony or -arsenic dichlorides:

$$Me_2InCl + Me_3SbCl_2 \longrightarrow [Me_4Sb]^+[MeInCl_3]^-$$

The dimer $[Me_2In{-}NMe_2]_2$ contains a four-membered, In_2N_2, ring.

Phosphinic acids form eight-membered coordination rings, but dithiophosphinic acids form monomeric chelates. A chelate with acetylacetone is also known. In each case indium is four-coordinated:

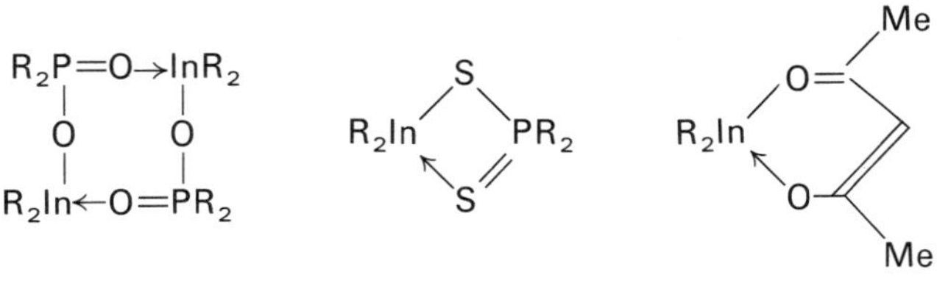

The formation of the five-coordinated compounds, $Me_2In(OAc)py$, and the six-coordinated $Me_2In(OAc)en$ reflects the tendency of the heavier metals to increase their coordination numbers.

The scarcity of indium and its high cost limit applications.

7.5. Organothallium Compounds

Organothallium compounds have been known since 1870, and are better investigated than those of gallium and indium. There are fundamental differences between organothallium chemistry and the organometallic chemistries of aluminum, gallium and indium, mainly in the unusual stability of the disubstituted derivatives, R_2TlX. The cations, R_2Tl^+, are isoelectronic with the organomercury compounds, R_2Hg, which are similarly stable and possess linear structures. The trisubstituted derivatives, TlR_3, particularly the trialkyls, are relatively unstable.

7.5.1. Trisubstituted Derivatives, TlR_3

Trisubstituted derivatives can be prepared from thallium(III) chloride and Grignard reagents in tetrahydrofuran; in ether the disubstituted derivatives form. Thallium(I) iodide reacts with organolithium reagents in the presence of alkyl iodides in ether through the intermediacy of organothallium(I) compounds which disproportionate to organothallium(III) compounds and thallium metal:

$$\begin{aligned} RLi + TlI &\longrightarrow TlR + LiI \\ 3\,TlR &\longrightarrow TlR_3 + 2\,Tl \\ 2\,Tl + 2\,RI &\longrightarrow R_2TlI + TlI \\ R_2TlI + RLi &\longrightarrow TlR_3 + LiI \end{aligned}$$

Thallium metal has been observed in the synthesis of triphenylthallium from thallium(I) chloride and phenyllithium, confirming the intermediate formation of monovalent thallium derivatives:

$$3\,TlCl + 3\,PhLi \longrightarrow 3\,TlPh \longrightarrow TlPh_3 + 2\,Tl$$

Since R_2TlX compounds are readily obtained and stable, their alkylation by organolithium reagents is convenient, affording unsymmetrically substituted compounds, R_2TlR':

$$R_2TlX + R'Li \longrightarrow TlR'R_2 + LiX$$

In liquid ammonia, sodium acetylides react with the ammonia adduct $TlCl_3 \cdot 4\,NH_3$ to form the anion $[Tl(C{\equiv}CR)_4]^-$, one of the few tetrasubstituted organothallium species containing four thallium-carbon bonds, others being the tetracoordinated anions $[Tl(C_6F_5)_4]^-$ and $[Tl(C_6F_5)_2(C_6Cl_5)_2]^-$ and the six-membered ring derivative:

$$(C_6X_5)_2TlBr + 2\,LiC_6X_5 \xrightarrow{NBu_4Br} [NBu_4]^+[Tl(C_6X_5)_2]^-$$
$X{=}F, Cl$

$$TlMe_3 + Me_3P{=}N{-}PMe_2{=}CH_2 \longrightarrow Me_2Tl^{\ominus}(CH_2{-}PMe_2{=}N{-}PMe_2{-}CH_2)^{\oplus} \text{ (six-membered ring)}$$

The trisubstituted derivatives with lower alkyl groups are pyrophoric; the others are also sensitive to oxygen, water and active hydrogen compounds. They pyrolyze more readily than their gallium and indium analogues, presumably via free radicals as suggested by the formation of biphenyl in the thermal decomposition of triphenylthallium.

The TlR_3 compounds show only weak acceptor properties; while $Me_3Tl \cdot NMe_3$ and $TlMe_3 \cdot PMe_3$ have been isolated, $TlMe_3$ does not coordinate arsines, and forms only a very weak adduct with ether. However, $Tl(C_6F_5)_3$ forms a stable etherate.

The trisubstituted compounds exchange substituents allowing unsymmetrically substituted compounds to be obtained by redistribution between TlR_3 and TlR'_3 species. Such exchanges do not take place with the more stable disubstituted compounds, R_2TlX.

7.5.2. Disubstituted Derivatives, R_2TlX

The action of Grignard reagents on thallium trichloride stops at the disubstituted product, but yields are diminished because of the reducing effect of the organomagnesium compound upon the trichloride. The bromides R_2TlBr are more readily obtained. The reaction of thallium trichloride with organolithium and organomercury reagents can also be used, but the reaction of thallium trihalides with arylboronic acids is less expected:

$$TlX_3 + 2\,RB(OH)_2 + 2\,H_2O \longrightarrow R_2TlX + 2\,B(OH)_3 + 2\,HCl$$

The disubstituted compounds, R_2TlX, are stable to 200–300 °C, and are little soluble in organic solvents, but dissolve readily in pyridine, owing to coordination. The halides are ionic in the solid state, for example, Me_2TlI and Me_2TlCl contain linear $[Me{-}Tl{-}Me]^+$ cations and halide anions. Other disubstituted, R_2TlX, compounds such as $(C_6F_5)_2TlX$ are dimerized in solution via halogen bridges.

The functional derivatives R_2TlX where X = OMe, OEt, SMe, SeMe, NMe_2, etc., which are obtained from the halides by alkali-metal derivatives of alcohols, thiols or amines, are also dimers with cyclic structures:

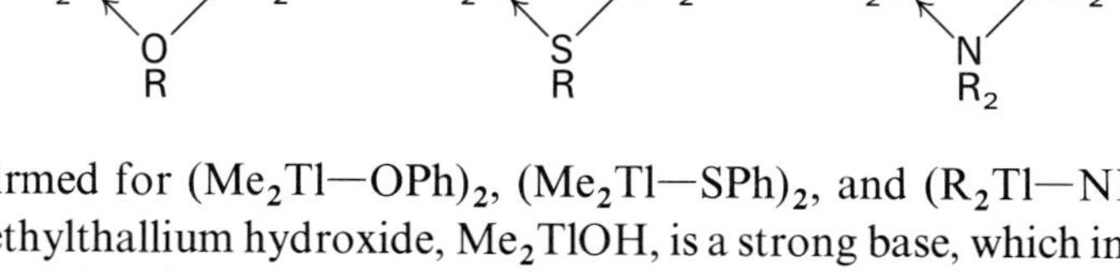

R R R₂
O S N
R_2Tl TlR_2 R_2Tl TlR_2 R_2Tl TlR_2
O S N
R R R_2

as confirmed for $(Me_2Tl{-}OPh)_2$, $(Me_2Tl{-}SPh)_2$, and $(R_2Tl{-}NR'_2)_2$.

Dimethylthallium hydroxide, Me_2TlOH, is a strong base, which in water dissociates into Me_2Tl^+ and OH^- ions.

The R_2Tl^+ ions form weak complexes, but chelates with β-diketones, salicylaldehyde, 8-hydroxyquinoline, etc., have been isolated. Unlike the aluminum and gallium analogues, the dimethyldithiophosphinic derivative, $[Me_2Tl]^+[Me_2PS_2]^-$, is an ionic compound.

The $Me_2Tl—NMe_2$ and $Me_2Tl—OMe$ derivatives undergo insertion reactions into the Tl—N and Tl—O bonds with double-bonded reagents like CO_2, CS_2, SO_2, SO_3, RNCS, RNCO, to form $Me_2Tl—C(X)Y$ derivatives, (X = O, NR, S and Y = NMe_2 or OEt).

7.5.3. Monosubstituted Derivatives, $RTlX_2$

Monoorganosubstituted halides cannot be isolated, but the carboxylates are stable. Aromatic compounds are formed in the reaction of thallium(III) carboxylates and organomercury compounds. The following redistribution reaction can also be used:

$$R_2Tl(COOOR') + Hg(OCOR')_2 \xrightarrow{20\,°C} RTl(OCOR')_2 + RHgOCOR'$$

Aromatic compounds can be subjected to a direct thallation reaction with thallium(III) carboxylates:

$$Tl(OCOR)_3 + RH \longrightarrow RTl(OCOR)_2 + RCOOH$$

7.5.4. Monovalent Thallium Derivatives, TlR

The only stable organic derivative of monovalent thallium is η^5-cyclopentadienylthallium, TlC_5H_5-η^5, which is obtained by treatment of a thallium(I) sulfate and bis(η^1-cyclopentadienyl)mercury in alcoholic alkalies. The product is insoluble in organic solvents and water. It sublimes *in vacuo*, however, and in the vapor the compound is monomeric with the thallium ion located above the C_5H_5 ring. In the solid state, on the other hand, the compound has an associated structure, consisting of alternating Tl^+ ions and $C_5H_5^-$ rings:

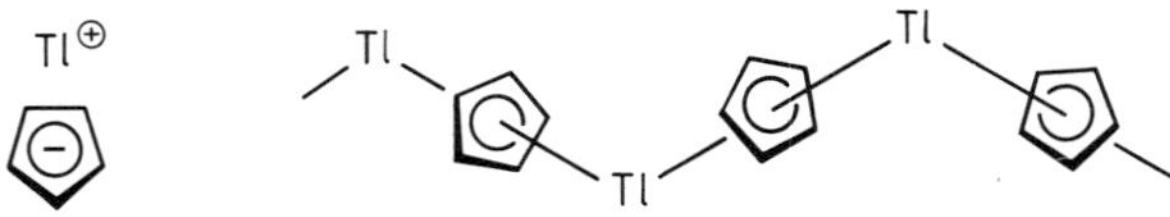

η^5-Cyclopentadienylthallium(I) is stable in air, and is important as a reagent for the synthesis of transition metal cyclopentadienyls (metallocenes), since it readily transfers C_5H_5 groups to other metals.

8. Organometallic Compounds of Group IV A Elements

The Group IVA elements form a great variety of organometallic compounds. The element-carbon bond has a pronounced covalent character. This bond is only moderately reactive, and the organometallic compounds of these elements are stable to heat, oxygen and moisture.

The Group IVA elements have in their valence shell two s-electrons and two p-electrons which undergo sp^3-hybridization to form four covalent bonds. This hybrid is particularly stable, and the central atom is four-coordinated in most cases. However, the participation of the d-orbitals expands the covalency of the central atom and five-, six- and seven-coordinated structures can be formed, beginning with silicon, in $[RSiF_4]^-$ (sp^3d-hybridization) or $[R_2SiF_4]^{2-}$ (sp^3d^2-hybridization). The tendency to achieve higher coordination numbers is favored by electronegative substituents like fluorine which contract the diffuse d-orbitals so that they can participate in bonding. This tendency increases descending Group IVA. Organotin and lead compounds, for example, provide many examples of increased coordination number, five to eight:

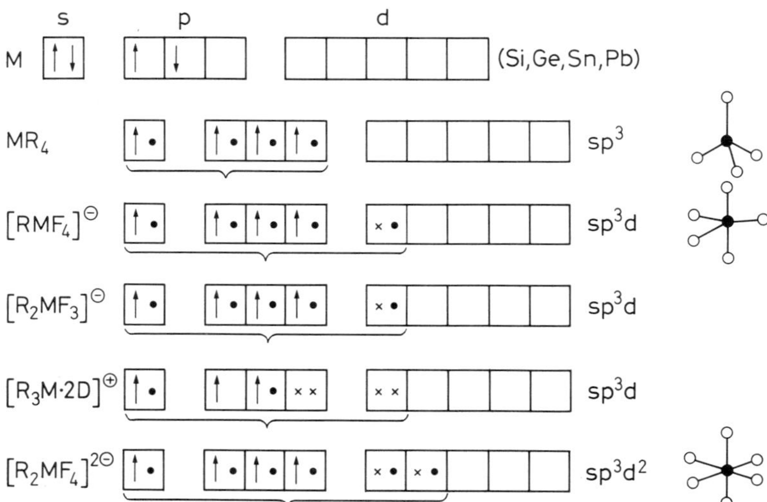

Fig. 8.1. The use of valence orbitals and electrons in the organometallic compounds of Group IV A elements.

The history of Group IVA organometallic chemistry began in the middle of the last century. Some development occured in the first half of our century, followed by an explosive increase in the literature of the field after 1950.

8.1. Organosilicon Compounds

Organosilicon chemistry is perhaps the most extensive chapter in organometallic chemistry, as illustrated by the more than 1000 papers published per year. The world literature on organosilicon compounds contains over 20,000 research papers and patents. Oligo- and polymeric organosilicon compounds have numerous industrial and commercial applications as oils, rubbers and resins. The world production of organosilicon compounds is of the order of tens of thousands of tons, and is increasing.

Organosilicon compounds were first prepared by C. Friedel and J. Crafts in 1863. The field was dominated in the first third of our century by the contributions of F.S. Kipping of Nottingham, England, whose work was initially based upon an analogy between silicon and carbon in the fourth group, but in 1935, after 30 years of investigation in the field, he came to the pessimistic conclusion that the reactions of organosilicon compounds are more limited than those of carbon, and that practical utility was unlikely. However, parallel work by K.A. Andrianov in the Soviet Union and J.F. Hyde and E.G. Rochow in the United States opened the way to useful organosilicon polymers, and as a result the silicone industry was born during WWII, stimulated by military needs and the unusual properties of the silicones.

8.1.1. Types of Organosilicon Compounds

Organosilicon compounds are generally four-coordinated monomers or linear or cyclic oligomers or macromolecular polymers.

a) Mononuclear Compounds with Four-Coordinated Silicon

Homo- and heteroleptic, tetrasubstituted SiR_4 (R = alkyl or aryl), and functional derivatives, R_nSiX_{4-n} (n = 1–3), are known:

R_nSiCl_{4-n}	organochlorosilanes
$R_nSi(OH)_{4-n}$	organosilanols (hydroxysilanes)
$R_nSi(OR')_{4-n}$	organoalkoxy(aroxy)silanes
$R_nSi(NR'R'')_{4-n}$	organoaminosilanes
$R_nSi(SR')_{4-n}$	organosilanethiols
R_nSiH_{4-n}	organosilicon hydrides

These compounds are monomeric and unlike similar Group IIIA derivatives are coordinatively saturated. They do not dimerize or form other oligomers by coordination polymerization.

b) Organic Heterocycles with Silicon as the Heteroatom

Silicon can participate as the heteroatom in many saturated and unsaturated organic heterocycles, as illustrated by the examples shown below:

SiR_2 $Si R_2$ $Si R_2$ $R_2 Si$ $Si R_2$

c) Linear and Cyclic Oligomers

The catenation of silicon atoms themselves or the alternation of silicon with oxygen, nitrogen, sulfur, or with other nonmetals yields linear and cyclic oligomers (in the following examples X is a functional group or an organic group):

-*linear* (unbranched):

$XR_2Si—(SiR_2)_{n-2}—SiR_2X$	polysilanes (n = 2 – 26)
$XR_2Si—(—O—SiR_2)_{n-2}—OSiR_2X$	polysiloxanes (n = 2, 3, 4 ...)
$XR_2Si—(—NH—SiR_2)_{n-2}—NHSiR_2X$	polysilazanes (n = 2, 3)
$XR_2Si—(—S—SiR_2)_{n-2}—S—SiR_2X$	polysilthianes (n = 2, 3)

-*cyclic*:

$(R_2Si)_n$	cyclosilanes (n = 3 – 36)
$(R_2SiO)_n$	cyclosiloxanes (n = 3 – 24)
$(R_2Si-NR')_n$	cyclosilazanes (n = 2 – 4)
$(R_2Si—S)_n$	cyclosilthianes (n = 2 – 4)

d) Polymers

Linear and branched polymers, for example, $R(R_2Si—O)_xSiR_3$ and the diols $HO(R_2Si—O)_xSiR_2—OH$, are known mostly in the siloxane series. The degree of polymerization can reach $10^3 - 10^5$. The tendency of the corresponding silicon-nitrogen compounds to form linear chains is very low, and long organosilicon-sulfur chains are practically unknown (but the structure of the inorganic SiS_2 is polymeric). Silicon-nitrogen and silicon-sulfur chains are unstable with respect to the corresponding rings and rearrange readily to form cyclic oligomers.

e) Organosilicon Compounds Containing Five- and Six-Coordinated Silicon

Examples include the tricyclic silatranes (triethanolamine derivatives), $R—Si(—O—CH_2CH_2—)_3N$, (with coordination number five at silicon), the diethanolamine derivatives of the type $R_2Si(—O—CH_2CH_2—)_2NH$ (also five-coordinated at silicon), the anions $[RSiF_4]^-$, $[R_2SiF_4]^{2-}$ and $[RSiF_5]^{2-}$, as well as adducts with bipyridyl and dihydroxybenzene, shown in Fig. 8.2:

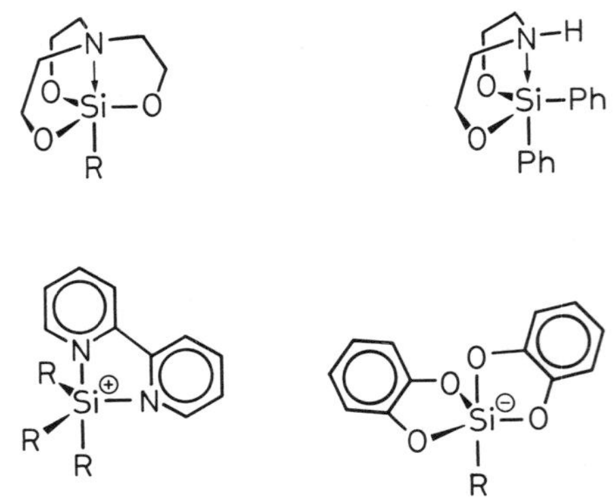

Fig. 8.2. Some compounds with five-coordinated silicon.

f) Silylenes, $:SiR_2$, and Free Radicals, $\cdot SiR_3$

Such species have been generated as unstable intermediates and exhibit rather interesting chemistry, which will be briefly summarized at the end of this Section.

8.1.2. Tetraorganosilanes, SiR_4

The first known organosilicon compound, tetraethylsilane, was prepared in 1863 by the reaction of silicon tetrachloride and diethylzinc, but alkyl or aryl halides and metallic sodium, organomagnesium or organolithium reagents are now used:

$$SiCl_4 + 4\,RCl + 8\,Na \longrightarrow SiR_4 + 8\,NaCl$$
$$SiCl_4 + 4\,RMgX \longrightarrow SiR_4 + 4\,MgClX$$
$$SiCl_4 + 4\,LiR \longrightarrow SiR_4 + 4\,LiCl$$

Mixed derivatives, containing different organic groups, can be obtained from organochlorosilanes with the same reagents.

The tetramethyl derivative, $SiMe_4$ (TMS), is the most famous for its use as a standard in nuclear magnetic resonance spectroscopy.

8.1.3. Organochlorosilanes, R_nSiCl_{4-n}

Chlorosilanes are intermediates in the synthesis of a large number of organosilicon derivatives. Methyl-, ethyl- and phenylchlorosilanes are produced in industrial quantities, mainly by direct synthesis. Halogeno derivatives of fluorine, bromine or iodine have only limited laboratory use.

Preparation. Organochlorosilanes can be prepared by Grignard reactions from silicon tetrachloride. This reaction has been used industrially for some time. However, difficultly separable mixtures of organochlorosilanes with various degrees of substitution (n = 1, 2, 3) are formed. The proportion of products depends upon the initial reagent ratio and reaction conditions.

The so-called "direct synthesis" (Rochow reaction) consists of a reaction between an alkyl or aryl chloride, in the gas phase with elemental silicon at 250–550 °C in the presence of a copper metal catalyst. This discovery by E. G. Rochow revolutionized organochlorosilane production:

$$nRX + Si \xrightarrow[Cu]{250-550\,°C} R_nSiX_{4-n}$$

The composition of the product depends upon the granularity and purity of the elemental silicon used, the presence, proportion and nature of the catalysts and promoters, the reaction temperature, the prior treatment of the contact mass, the contact time, the presence of inert gases, etc. The reaction with methyl chloride in the presence of copper metal as catalyst at 280–440 °C produces a mixture containing dimethyldichlorosilane, Me_2SiCl_2 (30–80%), methyltrichlorosilane $MeSiCl_3$ (10–40%), and in smaller proportions trimethylchlorosilane, Me_3SiCl, methyldichlorosilane, $MeSiHCl_2$, trichlorosilane, $HSiCl_3$, and silicon tetrachloride. The most sought after compound is dimethyldichlorosilane, the starting material for silicone rubber, oils and resins. The separation of this mixture is difficult, since the boiling points of the liquid products are very close:

	b.p. (°C)
$SiCl_4$	57.6
$MeSiCl_3$	66.0
Me_2SiCl_2	70.0
Me_3SiCl	57.7
$MeSiHCl_2$	40.7
$HSiCl_3$	31.8

The alkyl halide forms an unstable organometallic intermediate with the catalyst which decomposes to form free radicals. Simultaneously, the copper chloride produces surface chlorination of the elemental silicon; this surface reacts with the radicals, to form the alkylchlorosilanes:

$$CH_3Cl + Cu \longrightarrow CH_3CuCl \longrightarrow \cdot CH_3 + CuCl$$
$$CuCl + Si \longrightarrow Cu + [Si{-}Cl]$$
$$[Si{-}Cl] + \cdot CH_3 \longrightarrow \tfrac{1}{2}(CH_3)_2SiCl_2$$

The reactants are chemisorbed onto the solid silicon and a chain process occurs at active centers on the surface of the contact mass.

The hydrogen present in some by-products (for example, $MeSiHCl_2$) originates from the pyrolysis of the organic halide during reaction. The presence of hydrogen chloride with the methyl chloride, increases the $MeSiHCl_2$ in the product.

Organochlorosilanes can be prepared by the addition of chlorosilanes to unsaturated hydrocarbons at elevated temperature and pressure or in the presence of catalysts (hydrosilylation). Thus trichlorosilane, $HSiCl_3$ (readily prepared from gaseous hydrogen chloride and elemental silicon under conditions similar to those of the direct synthesis), can add to acetylene, ethylene, propylene and other olefins:

$$HC{\equiv}CH + HSiCl_3 \xrightarrow{Pt} H_2C{=}CH{-}SiCl_3$$
$$H_2C{=}CH_2 + HSiCl_3 \longrightarrow H_3C{-}CH_2{-}SiCl_3$$

Even silicon tetrachloride can add to acetylene at elevated temperatures to give chlorovinyltrichlorosilane:

$$HC{\equiv}CH + SiCl_4 \longrightarrow ClCH{=}CH{-}SiCl_3$$

Organochlorohydrosilanes, R_nSiHCl_{3-n}, can also take part in hydrosilylation.

Organohydrosilanes can undergo condensation reactions with organic halides at elevated temperatures and pressures to release hydrogen chloride and form organochlorosilanes. Thus trichloroethylene reacts with trichlorosilane at 550 °C to form dichlorovinyltrichlorosilane, and organohydrochlorosilanes react with vinyl chloride:

$$Cl_2C{=}CHCl + HSiCl_3 \xrightarrow{550\,°C} Cl_2C{=}CH{-}SiCl_3 + HCl$$
$$H_2C{=}CH{-}Cl + RSiHCl_2 \xrightarrow{600\,°C} H_2C{=}CH{-}SiRCl_2 + HCl$$

Thermal condensation is used in industry for the synthesis of phenyltrichlorosilane:

$$C_6H_5Cl + HSiCl_3 \xrightarrow{360\,°C} C_6H_5SiCl_3 + HCl$$
$$C_6H_6 + HSiCl_3 \xrightarrow{450\,°C} C_6H_5SiCl_3 + H_2$$

These reactions are not practical for laboratory preparations.

Properties. Organochlorosilanes are very reactive; the silicon-chlorine bond is readily attacked by nucleophilic reagents and such reactions are favored by the formation of five-coordinated intermediates which diminish the activation energy of substitution as in hydrolysis:

$$\equiv Si{-}Cl \xrightarrow{H_2O} \equiv Si(\leftarrow OH_2){-}Cl \longrightarrow \equiv Si{-}OH + HCl$$

Typical reactions of the monofunctional R_3SiCl chlorosilanes are shown in Fig. 8.3:

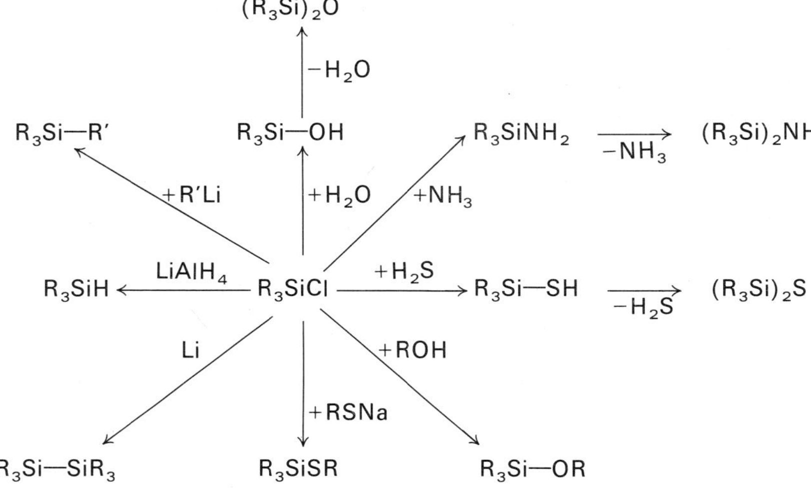

Fig. 8.3. Some typical reactions of organochlorosilanes.

The nature of the products obtained with chlorosilanes with more chlorine atoms is more complex, but the primary reactions are of the same type.

8.1.4. Organosilicon Hydrides (Organohydrosilanes), R_nSiH_{4-n}

These compounds are prepared by the reduction of organochlorosilanes with lithium alanate:

$$2\,R_2SiCl_2 + LiAlH_4 \longrightarrow 2\,R_2SiH_2 + AlCl_3 + LiCl$$

The hydrosilanes are useful in addition reactions to olefins.

8.1.5. Organoalkoxysilanes, $R_nSi(OR')_{4-n}$

Organoalkoxysilanes are obtained by reactions between organochlorosilanes and alcohols or phenols in the presence of tertiary amine acid scavengers or using metal alkoxides. Alternatively, tetraalkoxysilanes can be alkylated by Grignard or organolithium reagents:

$$R_nSiCl_{4-n} + (4-n)ROH \xrightarrow{NEt_3} R_nSi(OR)_{4-n} + nHCl$$
$$Si(OR)_4 + nRMgX \longrightarrow R_nSi(OR)_{4-n} + nMgXOR$$
$$Si(OR)_4 + nLiR \longrightarrow R_nSi(OR)_{4-n} + nLiOR$$

Organoalkoxysilanes can be used in the synthesis of siloxane polymers or in the preparation of silanols by hydrolysis. The alkoxysilanes are less reactive than chlorosilanes.

8.1.6. Organosilicon Carboxylates, $R_nSi(OCOR')_{4-n}$

These compounds can be readily prepared by the reaction between organochlorosilanes and salts or anhydrides of carboxylic acids. The best known are the acetoxysilanes which are volatile hydrolyzable liquids.

8.1.7. Organosilanols, $R_nSi(OH)_{4-n}$

Hydroxy-substituted organosilanes are obtained by hydrolysis of organohalogenosilanes, organoalkoxy-, organoamino- and organocarboxylatosilanes. Like orthosilicic acid, $Si(OH)_4$, to which they are related, the organosilanols are weak acids and undergo condensation to form siloxanes:

$$\gt Si{-}OH + HO{-}Si\lt \longrightarrow \gt Si{-}O{-}Si\lt + H_2O$$

Because of this, silanetriols, $RSi(OH)_3$, cannot be isolated. Silanediols, $R_2Si(OH)_2$, can be isolated only in media with neutral pH, since the acids favor condensation by acting as catalysts. Triorganosilanols, $R_3Si{-}OH$, are somewhat more stable, particularly when the organic groups are bulky, and some triorganosilanols can even be distilled without condensation. Alkali metal salts of silanols (silanolates) have also been isolated.

8.1.8. Organoaminosilanes, $R_nSi(NR'R'')_{4-n}$

Aminosilanes are obtained by the reaction of organochlorosilanes with primary or secondary amines. Those containing $Si{-}NH_2$ groups, $R_nSi(NH_2)_{4-n}$, undergo condensation with formation of silazanes. Compounds with more than two NH_2 groups at the same silicon atom cannot be isolated unless there are bulky organic groups at silicon:

$$\gt Si{-}NH_2 + H_2N{-}Si\lt \longrightarrow \gt Si{-}NH{-}Si\lt + NH_3$$

Derivatives of primary and secondary amines undergo condensation at elevated temperatures. Water, alcohols, phenols and acids cleave the silicon-nitrogen bond.

Transamination occurs on heating with an amine:

$$R_3Si{-}NHR + R'NH_2 \longrightarrow R_3Si{-}NHR' + RNH_2$$

The less volatile amine replaces the more volatile one in a redistribution equilibrium which is shifted by removal of one component from the system.

8.1.9. Other Organosilicon Functional Derivatives

Silicon-sulfur derivatives like organomercaptosilanes, $R_nSi(SR')_{4-n}$, are formed in the reactions of metal mercaptides (thiolates) with organochlorosilanes. These compounds are very sensitive to moisture and to compounds containing mobile hydrogen. Mixed derivatives which contain two different functional groups:

organoalkoxyaminosilanes	$R_nSi(OR)_m(NHR')_{4-m-n}$
organoaminohalosilanes	$R_nSi(NHR')_mX_{4-m-n}$
organoalkoxyhalosilanes	$R_nSi(OR')_mX_{4-m-n}$

are obtained by succesive substitutions using different nucleophilic reagents, though not all functional groups are compatible because of possible heterofunctional condensations:

$$-\overset{|}{\underset{|}{Si}}{-}Cl + HO{-}\overset{|}{\underset{|}{Si}}{-} \longrightarrow -\overset{|}{\underset{|}{Si}}{-}O{-}\overset{|}{\underset{|}{Si}}{-} + HCl$$

$$-\overset{|}{\underset{|}{Si}}{-}Cl + H_2N{-}\overset{|}{\underset{|}{Si}}{-} \longrightarrow -\overset{|}{\underset{|}{Si}}{-}NH{-}\overset{|}{\underset{|}{Si}}{-} + HCl$$

$$-\overset{|}{\underset{|}{Si}}{-}OH + R_2N{-}\overset{|}{\underset{|}{Si}}{-} \longrightarrow -\overset{|}{\underset{|}{Si}}{-}O{-}\overset{|}{\underset{|}{Si}}{-} + R_2NH$$

Organosilicon pseudohalides are obtained by the reaction of the halosilanes with the appropriate silver salt or thiourea for thiocyanates:

$$Et_3Si{-}Cl + AgNCO \longrightarrow Et_3Si{-}NCO + AgCl$$

$$Ph_3Si{-}Cl + SC(NH_2)_2 \longrightarrow Ph_3Si{-}NCS + NH_4Cl$$

8.1.10. Organofunctional Derivatives

In previous examples functional groups were attached directly to silicon, and reactions left the organic groups at silicon intact. Compounds in which a functional group is a part of the organic substituent are prepared by chemical transformations in the organic part of the molecule, without silicon participation. Thus, it is possible to introduce halogen-, nitrogen-, oxygen- or sulfur-containing functions in the organic group. Polyhalogenated organic groups, especially perfluoroalkyl, R_F, derivatives, exhibit enhanced thermal and chemical stability. Monomers of the type R_FSiRCl_2, which are used in the synthesis of special polymers, are synthesized from fluorinated organometallics, (for example, $C_6F_5SiMe_2Cl$ from C_6F_5Li and Me_2SiCl_2).

Chloroorganosilicon derivatives are obtained by photochemical chlorination of organosilanes. For example, chlorination of chlorotrimethylsilane yields $ClCH_2SiMe_2Cl$, $Cl_2CHSiMe_2Cl$ and $(ClCH_2)_2SiMeCl$. Chlorine in the chloromethyl substituent can be replaced with other groups, thus making possible the synthesis of a large number of organofunctional derivatives, as shown in Fig. 8.4:

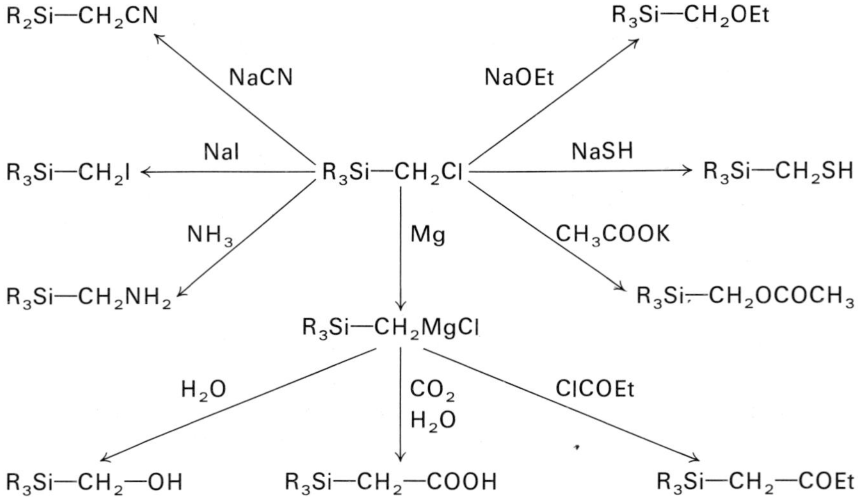

Fig. 8.4. Carbofunctional derivatives of silicon.

Perfluoro- and perchloroaromatic compounds such as $Si(C_6F_5)_4$, $C_6F_5SiMe_2Cl$, $C_6Cl_5SiMe_2Cl$, $Si(C_6Cl_5)_4$, etc., have been prepared.

8.1.11. Organic Heterocycles with Silicon as Heteroatom

The first silicon-containing heterocycle was prepared in 1915 by reacting the Grignard reagent obtained from $Cl(CH_2)_5Cl$ with silicon tetrachloride. Later the procedure was extended to the synthesis of the silacyclopentane and silacycloheptane heterocycles:

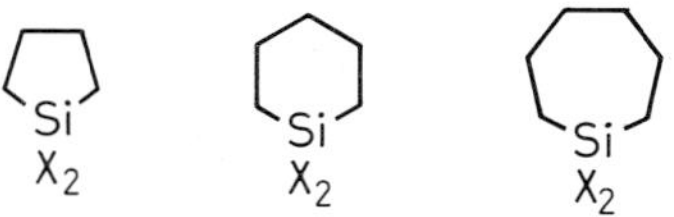

A more elaborate method is necessary for the preparation of a silacyclobutane:

$$Br(CH_2)_3SiMe_3 \xrightarrow[2.\ H_2O]{1.\ H_2SO_4} [Br(CH_2)_3SiMe_2]_2O \xrightarrow[NH_4Cl]{H_2SO_4}$$

$$\longrightarrow Br(CH_2)_3SiMe_2Cl \xrightarrow[ether]{Mg} \square\!\!-SiMe_2$$

Ring closure via intramolecular hydride addition in unsaturated silanes, such as $H_2C{=}CH{-}(CH_2)_nSiMe_2H$, can also be used. No cyclization occurs when $n = 0$ and 1, but ring formation takes place readily with $n = 2$.

Silacyclopropanes were first obtained as a part of polycyclic systems:

[Scheme: dibromo bis(cyclohexane-fused cyclopropyl) SiMe₂ compound —Mg/THF→ silacyclopropane-containing polycycle]

[Scheme: tetramethyl dibromo analogue (Me, Me, Br, Br, Me, Me, Si Me₂) —Mg/THF→ silacyclopropane (Me, Me, Me, Me, Si Me₂)]

The SiC_2 heterocycle exhibits extreme sensitivity to moisture and oxygen arising from strain in the three-membered ring.

Hexamethylsilacyclopropane (hexamethylsilirane) has been prepared by the bromination of dimethyldiisopropylsilane, followed by ring closure with magnesium:

$$Me_2Si(CHMe_2)_2 \xrightarrow{Br_2} Me_2Si(CMe_2Br)_2 \xrightarrow{Mg/THF} Me_2Si\langle CMe_2{-}CMe_2 \rangle$$

Other silacyclopropanes were prepared by the addition of dimethylsilylene ($:SiMe_2$) to olefins:

$$:SiMe_2 + RR'C{=}CRR' \longrightarrow RR'C{-}CRR' \text{ (bridged by } SiMe_2)$$

A spectacular achievement was the synthesis of silacyclopropenes (silirenes) – reactive, air- and moisture sensitive compounds – by either the addition of dimethylsilylene to dimethylacetylene:

$$:SiMe_2 + Me{-}C{\equiv}C{-}Me \longrightarrow \text{(Me)C=C(Me) with } SiMe_2 \text{ bridging (silirene)}$$

or by the reaction of hexamethylsilirane with bis(trimethylsilyl)acetylene:

$$\text{Me}_2\text{C—CMe}_2\text{ (bridged by SiMe}_2\text{)} + Me_3Si{-}C{\equiv}C{-}SiMe_3 \xrightarrow{-Me_2C=CMe_2} Me_2Si\text{(C(SiMe}_3\text{)=C(SiMe}_3\text{))}$$

Phenyl(trimethylsilyl)silirenes, which can be generated photochemically, dimerize on heating to disilacyclohexadienes:

$$Ph{-}C{\equiv}C{-}SiMe_2SiMe_3 \xrightarrow{h\nu} \text{Ph(C=C)SiMe}_3\text{, Si Me}_2 \longrightarrow$$

$$\longrightarrow \text{1,4-disilacyclohexadiene (Si Me}_2\text{; Ph, SiMe}_3\text{; Me}_3\text{Si, Ph)} + \text{1,4-disilacyclohexadiene (Si Me}_2\text{; Ph, Ph; Me}_3\text{Si, SiMe}_3\text{)}$$

Certain small rings exhibit uncommon reactions owing to ring strain. A spectacular example is disilacyclohexadiene-2,5:

$$\text{Me(C=C)Me with SiMe}_2{-}\text{SiMe}_2 \rightleftarrows \left[\text{Me–C(=SiMe}_2\text{)–C(=SiMe}_2\text{)–Me}\right] \xrightarrow{R-C\equiv C-R} \text{1,4-disilacyclohexadiene (Si Me}_2\text{; Me, R; Me, R)}$$

Another exotic heterocycle is silabenzene, which can be generated as an unstable intermediate; it can be trapped with hexafluorobutyne, to form bis(trifluoromethyl)silabarrelene:

$$\text{1-chloro-1-methylsilacyclohexadiene (Me, Si, Cl)} \xrightarrow[Et_2O]{(Me_3Si)_2NLi} \left[\text{silabenzene (Si–Me)}\right] \xrightarrow{CF_3C\equiv C-CF_3} \text{silabarrelene (CF}_3\text{, CF}_3\text{, Si–Me)}$$

The high reactivity of silabenzene is not surprising in view of silicon's reluctance to form $p_\pi - p_\pi$ double bonds or to participate in p_π-orbital conjugation.

Rings in which silicon is accompanied by a second heteroatom have also been prepared (R = Me):

$$ClCH_2SiR_2(CH_2)_3OH \xrightarrow{-HCl} \text{(cyclic } Si R_2 \text{, O)} \xleftarrow{-Br_2} BrCH_2SiR_2(CH_2)_3Br$$

$$BrCH_2SiR_2(CH_2)_3Br \xrightarrow{Mg + SiCl_4} \text{(ring: } SiCl_2 \text{, } Si R_2\text{)}; \quad \xrightarrow{Na_2S} \text{(ring: S, } Si R_2\text{)}; \quad \xrightarrow{RNH_2} \text{(ring: NR, } Si R_2\text{)}$$

Several polycyclic systems incorporating silicon heteroatoms, reminiscent of well-known organic heterocycles are known:

[Structures: dibenzo ring with O and SiR_2; dibenzo ring with N–H and SiR_2; fluorinated dibenzo ring with SiR_2 and SiR_2]

Biologically-active organosilicon compounds based upon heterocycles add a new dimension to this topic.

8.1.12. Linear Oligomers, Cyclic Compounds and Polymers

These compounds are formed by connecting Si—Si, Si—O, Si—N, Si—S or Si—C structural units in several possible ways to give classes of compounds known as polysilanes, polysiloxanes, polysilazanes, polysilthianes or polycarbosilanes, respectively.

Organopolysilanes

Organopolysilanes are based upon catenated silicon atoms forming linear and cyclic molecules, the simplest members of which are the disilanes, $R_3Si—SiR_3$, but linear molecules with longer chains (up to 24 silicon atoms) of the type $R_3Si—(SiR_2)_n—SiR_3$ or rings (up to 36 silicon atoms) of the type $(R_2Si)_n$ have been also synthesised.

Compounds with Si=Si double bonds have been isolated with bulky substituents, and evidence for the formation of transient tetramethyldisilene, $Me_2Si{=}SiMe_2$, and some of its reactions have been reported.

Organopolysilanes are obtained by the reactions of organochlorosilanes with sodium or lithium metal. The nature of the polysilanes formed is determined by the functionality of the starting chlorosilane. Thus, triorganochlorosilanes form disilanes, $R_3Si—SiR_3$, but diorganodichlorosilanes produce linear or cyclosilanes. The intermediate formation of silylenes, $:SiR_2$, (organic carbene analogues) which further polymerize, has been established:

$$2\,R_3SiCl + 2\,Na \longrightarrow R_3Si{-}SiR_3 + 2\,NaCl$$
$$nR_2SiCl_2 + 2\,nNa \longrightarrow (R_2Si)_n + 2\,nNaCl$$

Silicon homocycles containing four to six silicon atoms are isolated from this reaction:

$(R_2Si)_4$ — four-membered ring; $(R_2Si)_5$ — five-membered ring; $(R_2Si)_6$ — six-membered ring

with lithium in THF, cyclopolysilanes, $(SiR_2)_n$, containing medium-sized rings, with n = 6, 7, 8 and 9, can be obtained and even larger rings, with up to 36 silicon atoms, have been identified and separated by gas chromatography.

The structures of $(SiMe_2)_6$, $(SiPh_2)_4$ and $(SiMeBu^t)_4$, have been confirmed. The structure of bicyclic Si_9Me_{16} consists of two fused, six-membered polysilane rings:

[Structure: bicyclic Si_9 skeleton]

Cyclosilanes with silicon side chains, can also be prepared:

$$Me_3Si{-}SiMeCl_2 + Li(SiPh_2)_4Li \longrightarrow (Ph_2Si)_4Si(Me)(SiMe_3)$$

Cyclopolysilanes form anion-radicals by reaction with alkali metals, from which they accept an electron which is delocalized over the whole ring.

The cyclopolysilane rings are readily cleaved by halogens, with formation of linear oligomers, for example, $Cl(SiR_2)_nCl$ (n = 2–6) and by lithium metal to $Li(SiR_2)_nLi$. These oligomers can react with $LiSiR_3$ and $ClSiR_3$, respectively, to form fully-substituted organopolysilane chains, $R_3Si(SiR_2)_nSiR_3$, longer by two units than the parent polysilane fragment. Structures of the iso-pentane or bicyclic type are prepared using similar reactions and illustrate the possibility of branching the polysilane chain:

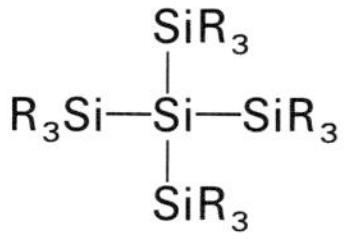

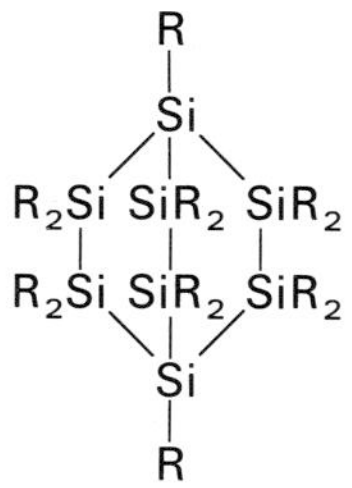

$$Si(SiR_3)_4 \qquad R-Si(SiR_2)_3Si-R \text{ (bicyclic)}$$

The polysilanes undergo aluminum chloride-catalyzed skeletal rearrangements, silyllithium-catalyzed redistribution reactions and photolytic degradation with loss of divalent silicon species, thermolysis and catalysis by noble metal complexes. Some examples are illustrated by the following reactions:

$$Me(SiMe_2)_5Me \xrightarrow{AlCl_3} Si(SiMe_3)_4$$

$$Ph_3Si(SiMe_2)_nSiPh_3 \xrightarrow{Ph_3SiLi} Ph_3Si{-}SiPh_3 + (SiMe_2)_5 + (SiMe_2)_6$$

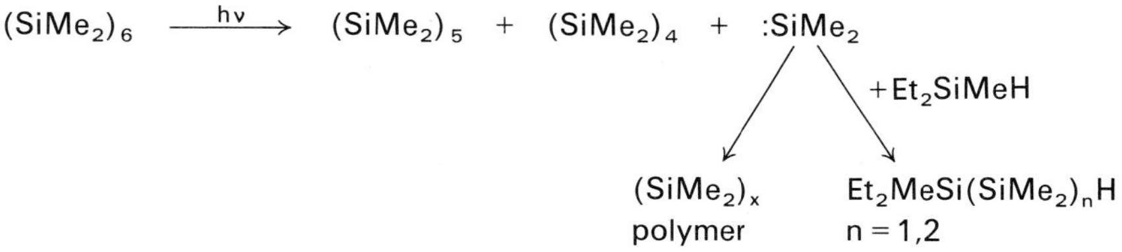

$$(SiMe_2)_6 \xrightarrow{h\nu} (SiMe_2)_5 + (SiMe_2)_4 + :SiMe_2$$

$$:SiMe_2 \rightarrow (SiMe_2)_x \text{ polymer}$$

$$:SiMe_2 \xrightarrow{+Et_2SiMeH} Et_2MeSi(SiMe_2)_nH \quad n = 1,2$$

Like other chlorosilanes, the chlorine-terminated polysilanes undergo nucleophilic attack and substitution with reagents such as water, ammonia, amines and hydrogen sulfide, as in dichlorotetramethyldisilane, $ClMe_2Si{-}SiMe_2Cl$, which reacts to form six-membered heterocycles:

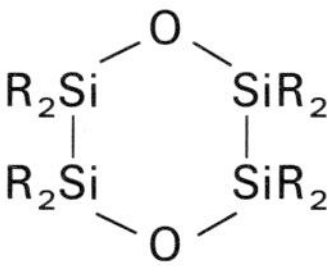

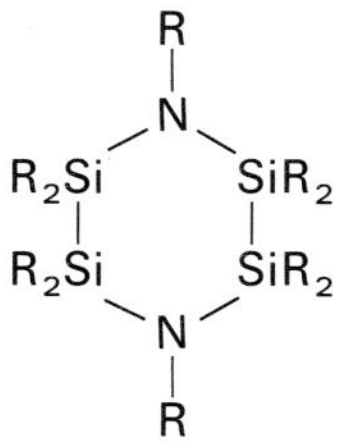

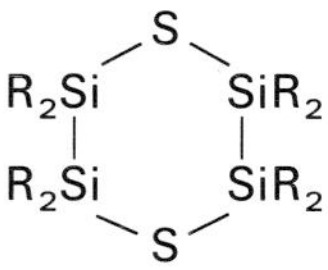

$$(R_2Si{-}SiR_2O)_2 \qquad (R_2Si{-}SiR_2NR)_2 \qquad (R_2Si{-}SiR_2S)_2$$

High-molecular weight, linear polysilanes $(SiMe_2)_x$, form from dimethyldichlorosilane and sodium metal along with the cyclic polysilanes.

Organopolysiloxanes

Compounds based on an —Si—O—Si—O— backbone are important industrial chemicals. The simplest are the disiloxanes, $R_3Si{-}O{-}SiR_3$, but linear organopolysiloxanes, $R_3Si—(—O—SiR_2—)_{n-2}—OSiR_3$, α,ω-difunctional derivatives $XR_2Si—(—O—SiR_2—)_{n-2}—OSiR_2X$ (X = Cl, OH, NH_2, NHR) and cyclosiloxanes, $(R_2SiO)_n$, are also well-known. There seems to be no limit for the value of n. When n is large, the high-molecular weight organopolysiloxanes have either a chain structure, $(R_2SiO)_x$, or a branched or polycyclic structure, $(RSiO_{1.5})_x$, or something intermediate. Structures are built up from mono- (M), di- (D) and tri- (T) functional units:

```
                    R              R
                    |              |
R3Si—O—          —Si—O—         —Si—O—
                    |              |
                    R              O

   M                D              T
```

The bond angle at oxygen in the cyclosiloxanes is significantly larger than tetrahedral, and this fact determines the conformation of the cyclosiloxane rings. In solid hexamethyldisiloxane, $Me_3Si—O—SiMe_3$, the Si—O—Si bond angle is 148° but, unexpectedly, in the hexaphenyl analogue, $Ph_3Si—O—SiPh_3$, the Si—O—Si bond angle is nearly linear (176.8°). In cyclosiloxanes the values are smaller: in cyclotrisiloxanes its value is ca. 120°, while in cyclotetrasiloxanes it is 140°. These data reflect great flexibility of the Si—O—Si bond angle, and suggest some ring strain in the cyclosiloxanes which facilitates the ring-opening polymerization of cyclosiloxanes.

The products of organopolysiloxane synthesis depend upon the functionality of the starting material and the reaction conditions. The reactions of Si—O bond formation can be classified as follows:

-heterofunctional polycondensation:

$$—Si—OH + X—Si— \longrightarrow —Si—O—Si— + HX$$
$$—Si—OH + RO—Si— \longrightarrow —Si—O—Si— + ROH$$
$$—Si—OH + R_2N—Si— \longrightarrow —Si—O—Si— + R_2NH$$
$$—Si—X + RO—Si— \longrightarrow —Si—O—Si— + RX$$
$$—Si—X + ROOC—Si— \longrightarrow —Si—O—Si— + RCOX$$
$$—Si—X + NaO—Si— \longrightarrow —Si—O—Si— + NaX$$
$$—Si—OR + R'COO—Si \longrightarrow —Si—O—Si— + RCOOR'$$

-homofunctional polycondensation:

$$—Si—OH + HO—Si— \longrightarrow —Si—O—Si— + H_2O$$
$$—Si—OOCR + RCOO—Si— \longrightarrow —Si—O—Si— + (RCOO)_2O$$
$$—Si—OR + RO—Si— \longrightarrow —Si—O—Si— + R_2O$$

These reactions require very different conditions; some occur only on heating and in the presence of catalysts, others are spontaneous even at room temperature.

The most frequently used is the hydrolysis of organochlorosilanes and, less often, of organoalkoxysilanes or other derivatives. Hydrolysis is a complex process, involving simultaneously substitution reactions:

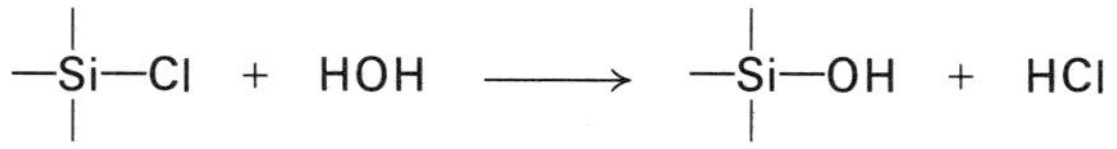

$$-\overset{|}{\underset{|}{Si}}-Cl + HOH \longrightarrow -\overset{|}{\underset{|}{Si}}-OH + HCl$$

and homo- and heterofunctional polycondensations:

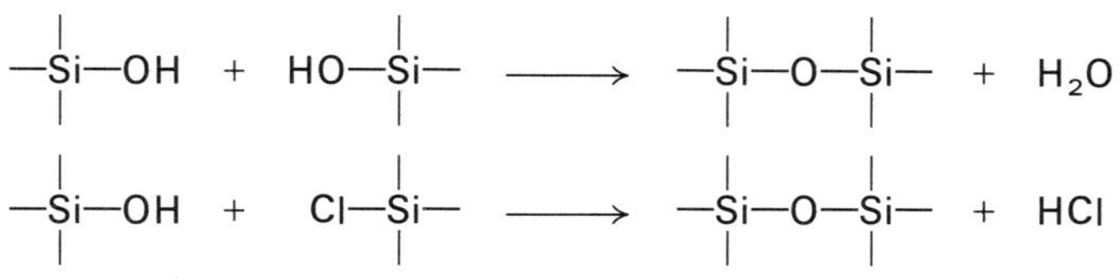

$$-\overset{|}{\underset{|}{Si}}-OH + HO-\overset{|}{\underset{|}{Si}}- \longrightarrow -\overset{|}{\underset{|}{Si}}-O-\overset{|}{\underset{|}{Si}}- + H_2O$$

$$-\overset{|}{\underset{|}{Si}}-OH + Cl-\overset{|}{\underset{|}{Si}}- \longrightarrow -\overset{|}{\underset{|}{Si}}-O-\overset{|}{\underset{|}{Si}}- + HCl$$

The length of the siloxane chain formed depends upon the functionality of the monomers, the reaction conditions, and the presence and proportion of monofunctional derivatives which can act as chain-terminating agents, (for example, $Me_3Si-O-SiMe_3$, able to furnish Me_3Si-terminal groups). The hydrolysis can sometimes be stopped at the silanol stage; this requires the removal of the acid formed in the reaction which catalyzes the polycondensation of Si—OH groups.

Monofunctional triorganochlorosilanes produce hexaorganodisiloxanes:

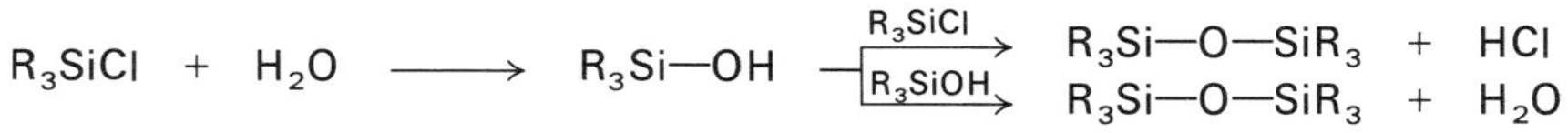

$$R_3SiCl + H_2O \longrightarrow R_3Si-OH \begin{cases} \xrightarrow{R_3SiCl} R_3Si-O-SiR_3 + HCl \\ \xrightarrow{R_3SiOH} R_3Si-O-SiR_3 + H_2O \end{cases}$$

Diorganodichlorosilanes can yield various products depending upon the reaction conditions. Thus dimethyldichlorosilane, with a deficiency of water, yields short-chain, linear α,ω-dichloropolysiloxanes $Cl(Me_2SiO)_nSiMe_2Cl$ ($n = 1$–5) in partial hydrolysis. If neutral pH is maintained during the reaction, by the presence of a neutralizing reagent, dimethylsilanediol, $Me_2Si(OH)_2$, and 1,1,3,3-tetramethyldisiloxanediol-1,3, $HO-SiMe_2-O-SiMe_2OH$, can be obtained. In excess water and without any precautions, the acidic medium favors polycondensation, and a mixture of cyclic polysiloxanes, $(Me_2SiO)_n$ ($n = 3,4,5,6 \ldots$), and linear polymers, $HO(SiMe_2-O)_xH$, results. The trimer content in this mixture is low, and the main component is the cyclic tetramer, $(Me_2SiO)_4$. The hydrolysis of organoalkoxysilanes proceeds similarly.

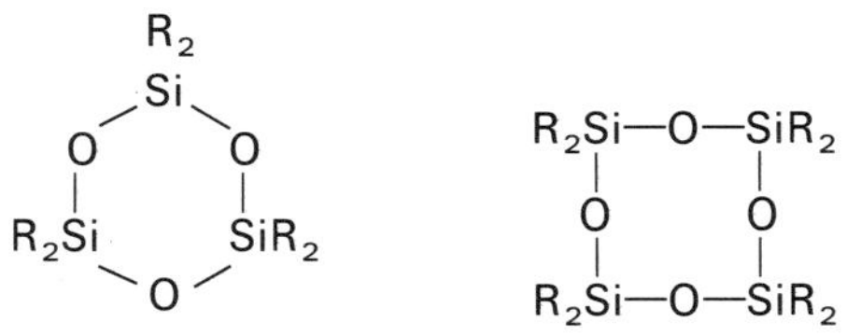

$(R_2SiO)_3$ [six-membered ring: R_2Si, O, SiR_2, O, SiR_2, O]

$(R_2SiO)_4$ [eight-membered ring: $R_2Si-O-SiR_2$, O, O, $R_2Si-O-SiR_2$]

In the hydrolysis of trifunctional organotrichlorosilanes, $RSiCl_3$, bulky groups favor the formation of polycyclic organopolysiloxanes of composition $(RSiO_{1.5})_4$, with adamantane-like structures. Cage polysiloxanes of composition $(RSiO_{1.5})_6$ or $(RSiO_{1.5})_8$ can also be isolated:

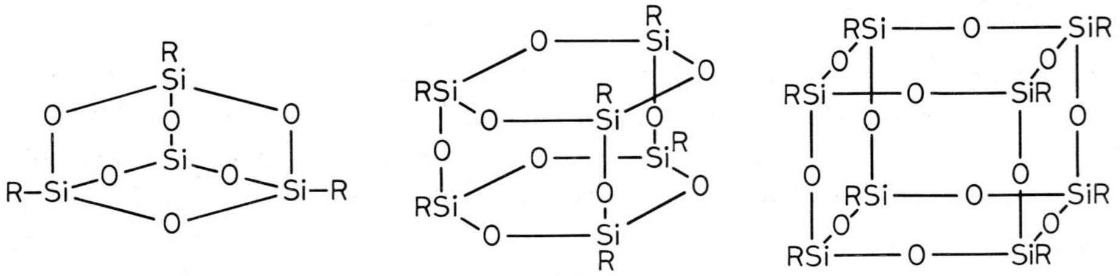

$(RSiO_{1.5})_4$ $(RSiO_{1.5})_6$ $(RSiO_{1.5})_8$

The hydrolysis of organotrichlorosilanes can also lead to monocyclic compounds with OH groups at silicon (for example, tetrahydroxycyclotetrasiloxanes, $[(HO)RSiO]_4$, and highly polymeric polycyclic structures, like those of the so-called "ladder polymers":

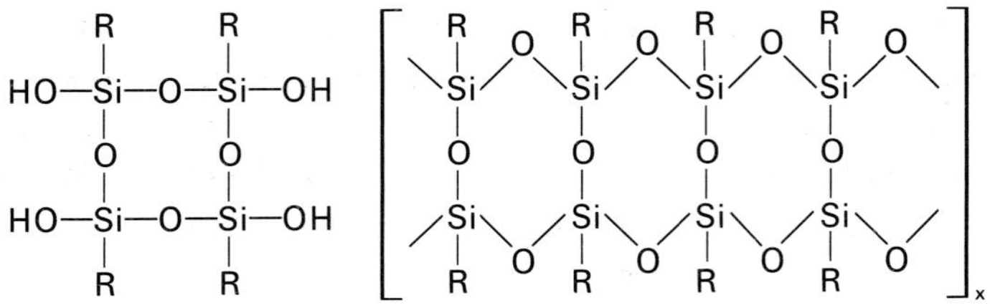

The hydrolysis of mixtures of chlorosilanes with different functionalities, for example, di- R_2SiCl_2, and trifunctional $RSiCl_3$ compounds, also leads to polymers with branched structures which are used in the manufacture of silicone resins.

Organopolysiloxanes are thermally stable compounds ($> 300\,°C$ and even higher if the contact time is short), but at elevated temperatures in the presence of alkaline catalysts they undergo depolymerization. The linear siloxanes form monocyclic compounds, $(R_2SiO)_n$ ($n = 3, 4, 5\ldots$), the trimer being predominant. The branched polysiloxanes containing di-, and tri- or tetrafunctional units depolymerize with formation of bicyclic siloxanes:

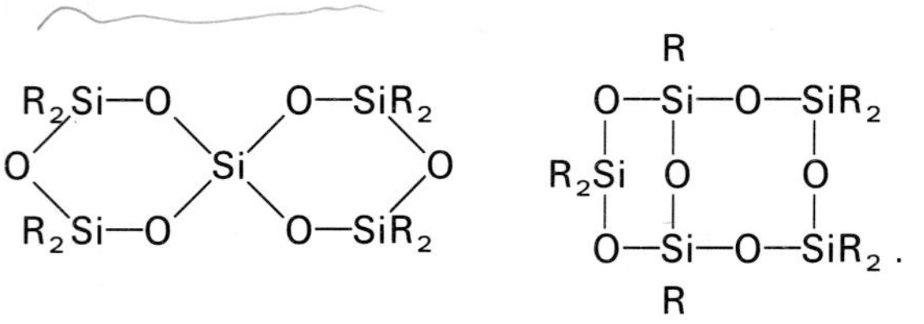

During the depolymerization a redistribution of siloxane building units occurs. An important redistribution is the polymerization of cyclosiloxanes into linear polymers known as silicone rubbers (R = Me):

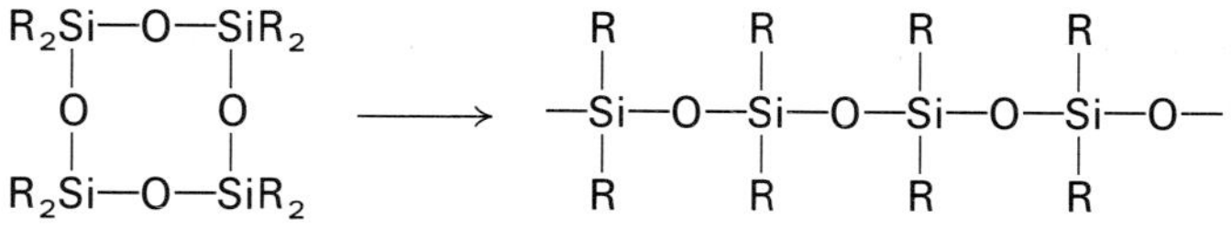

$$\begin{array}{ccc} R_2Si-O-SiR_2 \\ | \quad\quad | \\ O \quad\quad O \\ | \quad\quad | \\ R_2Si-O-SiR_2 \end{array} \longrightarrow -\overset{R}{\underset{R}{|}}\!\!\!\!\!Si-O-\overset{R}{\underset{R}{|}}\!\!\!\!\!Si-O-\overset{R}{\underset{R}{|}}\!\!\!\!\!Si-O-\overset{R}{\underset{R}{|}}\!\!\!\!\!Si-O-$$

Copolymerization of cyclosiloxanes with different organic substituents for example, $(R_2SiO)_n$ and $(RR'SiO)_n$ gives modified silicone rubbers (R = Me, R′ = $CH{=}CH_2$, $CH_2CH_2CF_3$, CH_2CH_2CN, etc.) in which the R′ groups are distributed statistically along the siloxane chain with their number regulated by the reaction stoichiometry.

Silicone oils, which are shorter siloxane chains terminated with SiR_3 groups, are synthesized by redistribution reactions between cyclosiloxanes and disiloxanes:

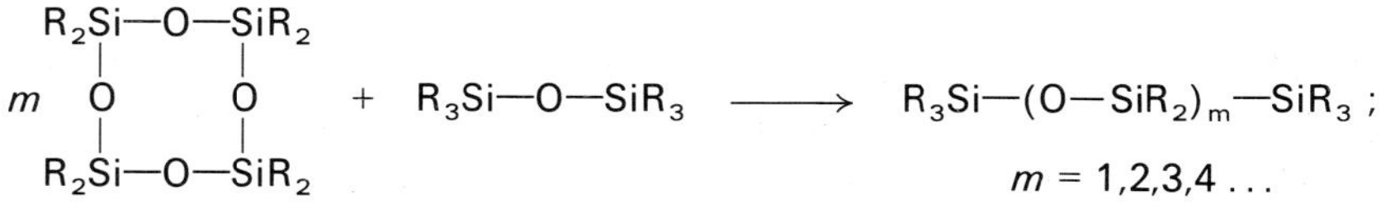

$$m \begin{array}{c} R_2Si-O-SiR_2 \\ | \quad\quad | \\ O \quad\quad O \\ | \quad\quad | \\ R_2Si-O-SiR_2 \end{array} + R_3Si-O-SiR_3 \longrightarrow R_3Si-(O-SiR_2)_m-SiR_3\,; \quad m = 1,2,3,4\ldots$$

The equilibrium is strongly shifted to the right.

Reagents able to open the ring (alkalies, acids, etc.) act as catalysts in these redistributions. The same reagents in stoichiometric amounts are able to cleave both the rings and the chains to form simple, low-molecular weight compounds, for example, silanolates, R_3SiONa and $NaOSiR_2-O-SiR_2ONa$ with alkalies. The redistribution between cyclosiloxanes and chlorosilanes catalyzed by Lewis acids, for example, $AlCl_3$ is used to synthesize halogen-terminated polysiloxanes:

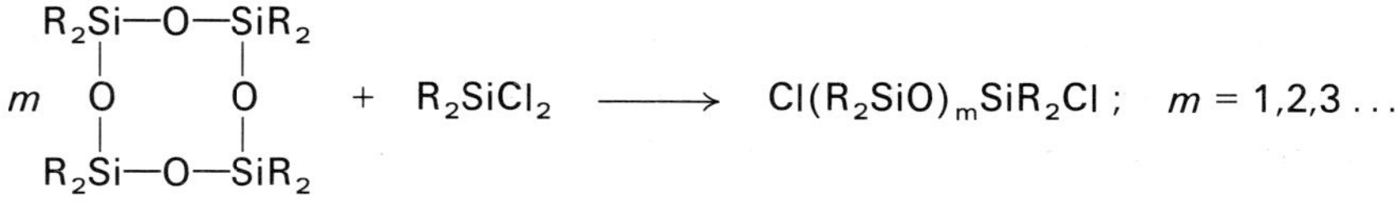

$$m \begin{array}{c} R_2Si-O-SiR_2 \\ | \quad\quad | \\ O \quad\quad O \\ | \quad\quad | \\ R_2Si-O-SiR_2 \end{array} + R_2SiCl_2 \longrightarrow Cl(R_2SiO)_mSiR_2Cl\,; \quad m = 1,2,3\ldots$$

The halogen-terminated, linear α,ω-polysiloxanes react like simple chlorosilanes with water, ammonia, amines, alcohols and other active-hydrogen reagents. Thus, hydrolysis affords α,ω-dihydroxypolysiloxanes, $HO(R_2SiO)_nSiR_2OH$, and cyclosiloxanes. With ammonia, the lower members (n = 1–5) undergo cyclization, and those with n > 6 produce linear α,ω-diaminopolysiloxanes, $H_2N(R_2SiO)_nSiR_2NH_2$:

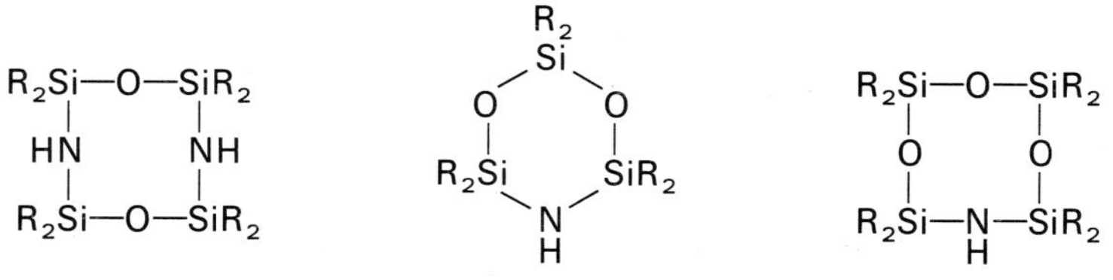

$$\begin{array}{c} R_2Si-O-SiR_2 \\ | \quad\quad | \\ HN \quad\quad NH \\ | \quad\quad | \\ R_2Si-O-SiR_2 \\ \\ n = 2 \end{array} \qquad \begin{array}{c} R_2 \\ Si \\ O \quad\quad O \\ | \quad\quad | \\ R_2Si \quad\quad SiR_2 \\ N \\ H \\ n = 3 \end{array} \qquad \begin{array}{c} R_2Si-O-SiR_2 \\ | \quad\quad | \\ O \quad\quad O \\ | \quad\quad | \\ R_2Si-N-SiR_2 \\ H \\ n = 4 \end{array}$$

Mixtures of dichloroorganosiloxanes and dichloroorganosilanes with ammonia give cyclosilazoxanes containing nitrogen atoms in the ring:

```
      R2
      Si                R2Si—O—SiR2          R2Si—O—SiR2
   HN    NH              |      |             |      |
   |     |               O      NH            HN     NH
  R2Si   SiR2            |      |             |      |
      O                 R2Si—N—SiR2          R2Si—N—SiR2 .
                             |                    |
                             H                    H
```

Organolithium or Grignard reagents can attach organic groups at the end of the siloxane chain; thus from C_6F_5Li and C_6Cl_5Li, oligomeric siloxanes terminated by perhalogenoaromatic groups, $C_6X_5(SiR_2O)_nSiR_2C_6X_5$, are formed.

Organoheterosiloxanes

These are compounds in which silicon has been partly replaced by other elements. Trialkylsiloxy derivatives of many elements, $M(OSiR_3)_n$ (where M = B, Al, Ga, Sn, Pb, P, As, Sb, Bi, V, Fe, etc., and n = the valence of the element M), are known. Heterocyclic siloxanes with boron, aluminum, phosphorus and arsenic, and related linear high polymers have also been synthesized.

Polyborosiloxanes

Trialkylsilylborates, $B(OSiR_3)_3$, form from the reaction of boric acid and trialkylchlorosilanes. Similarly, cycloborosiloxanes have been prepared by the reaction of phenylboric acid with dialkoxysilanes or dialkoxysiloxanes:

```
      R
      B                 R2Si—O—SiR2          R—B—O—SiR2
    O   O                |      |               |    |
    |   |                O      O               O    O
  R2Si   SiR2            |      |               |    |
      O                 R2Si—O—B—R            R2Si—O—B—R
```

Highly polymeric borosiloxanes are obtained by the condensation of oligomeric dihydroxypolysiloxanes with the esters of organoboric acids.

Polyphosphosiloxanes

Trialkylsilylphosphates, $(R_3SiO)_3PO$, are prepared by the condensation of trialkylalkoxysilanes with phosphoric acid. Cyclic phosphosiloxanes form from the reaction of dialkoxypolysiloxanes with methylphosphonic acid:

```
  O     R                                              O
   \\  /                                               ||
     P          R2Si—O—SiR2                            RP—O—SiR2
    / \           |      |                             |     |
   O   O          O      O                             O     O
   |   |          |      |                             |     |
R2Si   SiR2     R2Si—O—P=O                          R2Si—O—P=O
    \ /                 |                                  |
     O                  R                                  R
```

Hydrolytically unstable phosphosiloxane polymers have been obtained by the action of phosphoric anhydride on organoalkoxysilanes.

Polymetallosiloxanes and Polysiloxymetalloxanes

These two classes of highly polymeric compounds are related by the presence of metal atoms in the polymeric chain and by the presence of Si—O—M bonds. In polymetallosiloxanes, metal heteroatoms appear in polysiloxane chains; in polysiloxymetalloxanes the chain consists of alternating oxygen and metal atoms, with the organosilicon groups in the side chains:

```
  R     R           R     R
  |     |           |     |
—Si—O—Si—O—M—O—Si—O—Si—O—        polymetallosiloxanes
  |     |           |     |
  R     R           R     R
```

```
 OSiR3  OSiR3  OSiR3
  |      |      |
—M—O—M—O—M—O—                    polysiloxymetalloxanes
  |      |      |
 OSiR3  OSiR3  OSiR3
```

These polymers were synthesized in attempts to find materials able to stand high temperatures.

The aluminoorganosiloxane high polymers are soluble in organic solvents but are infusible solids (infusible polymers usually are also insoluble), and have a ladder structure:

```
[  R  O  R  O  R  O  R    ]
[ \Si/ \Si/ \Si/ \Si/     ]
[   |    |    |    |      ]
[   O    O    O    O      ]
[   |    |    |    |      ]
[ /Si\ /Si\ /Si\ /M\      ]
[  R  O  R  O  R  O       ]x
```

The cohydrolysis of organodichlorosiloxanes with titanium tetraalkoxides, or the condensation of diacetoxypolysiloxanes with $(R_3SiO)_2Ti(OR)_2$ produces high polymers containing titanium heteroatoms in a polysiloxane chain:

$$\left(\text{O}-\underset{\text{R}}{\overset{\text{R}}{\text{Si}}}-\right)_m\left(-\text{O}-\underset{\text{OR}}{\overset{\text{OR}}{\text{Ti}}}-\text{O}-\right)_n ; \left(-\text{O}-\underset{\text{R}}{\overset{\text{R}}{\text{Si}}}-\right)_m\left(-\text{O}-\underset{\text{OSiR}_3}{\overset{\text{OSiR}_3}{\text{Ti}}}-\text{O}-\right)_n$$

Organosiloxymetalloxanes containing aluminum and titanium are formed by partial hydrolysis of $Al(OSiR_3)_3$ and $Ti(OSiR_3)_4$, respectively:

$$-\underset{\text{OSiR}_3}{\overset{\text{OSiR}_3}{\text{Al}}}-\text{O}-\underset{\text{OSiR}_3}{\overset{\text{OSiR}_3}{\text{Al}}}-\text{O}- ; \quad -\underset{\text{OSiR}_3}{\overset{\text{OSiR}_3}{\text{Ti}}}-\text{O}-\underset{\text{OSiR}_3}{\overset{\text{OSiR}_3}{\text{Ti}}}-\text{O}-$$

Organopolysilazanes

The silazanes have an Si—N—Si—N backbone. In their reaction with ammonia, the monofunctional triorganochlorosilanes form disilazanes, $R_3Si{-}NH{-}SiR_3$, the difunctional organodichlorosilanes give cyclosilazanes, $(R_2SiNH)_n$ ($n = 3$ and 4), and the organotrichlorosilanes form polycyclic compounds, $[R{-}Si(NH)_{1.5}]_n$. The tendency for cyclization by silazanes is even stronger than for siloxanes, thus linear silazanes are thermodynamically unstable with respect to cyclosilazanes.

The ammonolysis of chlorosilanes is, like hydrolysis, a complex process consisting of several concurrent reactions:

-substitution:

$$-\overset{|}{\underset{|}{\text{Si}}}-\text{Cl} + 2\,\text{NH}_3 \longrightarrow -\overset{|}{\underset{|}{\text{Si}}}-\text{NH}_2 + \text{NH}_4\text{Cl}$$

-homo- and heterofunctional condensation:

$$-\overset{|}{\underset{|}{\text{Si}}}-\text{NH}_2 + \text{H}_2\text{N}-\overset{|}{\underset{|}{\text{Si}}}- \longrightarrow -\overset{|}{\underset{|}{\text{Si}}}-\text{NH}-\overset{|}{\underset{|}{\text{Si}}}- + \text{NH}_3$$

$$-\overset{|}{\underset{|}{\text{Si}}}-\text{NH}_2 + \text{Cl}-\overset{|}{\underset{|}{\text{Si}}}- \longrightarrow -\overset{|}{\underset{|}{\text{Si}}}-\text{NH}-\overset{|}{\underset{|}{\text{Si}}}- + \text{HCl}$$

Primary amines react likewise to form N-substituted silazanes, although the condensation of Si-NHR groups requires heating and the presence of a catalyst, more drastic conditions than for $Si{-}NH_2$:

$$-\overset{|}{\underset{|}{\text{Si}}}-\text{NHR} + \text{RNH}-\overset{|}{\underset{|}{\text{Si}}}- \longrightarrow -\overset{|}{\underset{|}{\text{Si}}}-\text{NR}-\overset{|}{\underset{|}{\text{Si}}}- + \text{RNH}_2$$

The disilazanes are hydrolyzed and decomposed by acids and alcohols. Metallation can be effected at nitrogen sites:

$$R_3Si{-}NH{-}SiR_3 + LiR \longrightarrow R_3Si{-}NLi{-}SiR_3 + RH$$
$$R_3Si{-}NH{-}SiR_3 + NaH \longrightarrow R_3Si{-}NNa{-}SiR_3 + H_2$$

N-Metallated silazanes are reactive compounds; with triorganochlorosilanes they give tris(organosilyl)amines:

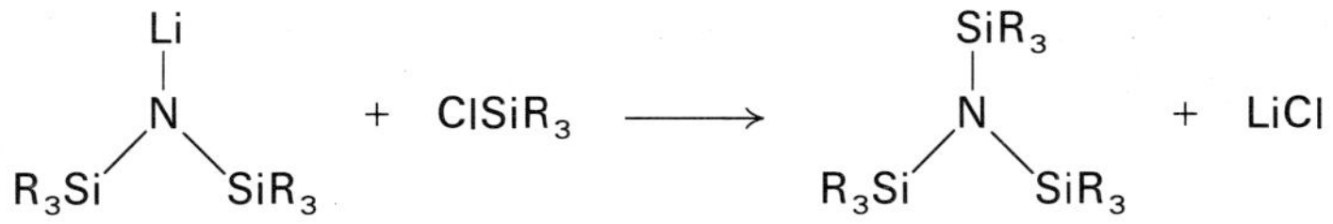

$$(R_3Si)_2N{-}Li + ClSiR_3 \longrightarrow (R_3Si)_2N{-}SiR_3 + LiCl$$

and with the halides of other elements they afford substituted silylamides of aluminum, titanium, iron, cobalt, vanadium, etc., for example:

$Al[N(SiR_3)_2]_3$ $(RO)_nVO[N(SiR_3)_2]_{3-n}$

$Fe[N(SiR_3)_2]_3$ $Cl_2OV{-}N(SiR_3)_2$

The bulky bis(trimethylsilyl)amino group $-N(SiMe_3)_2$ is a useful ligand in coordination chemistry, since it favors unusually low coordination numbers.

Cyclosilazanes, predominantly the trimer with some tetramer, are formed in the ammonolysis of diorganodichlorosilanes. Cyclosilazanes with more than four SiN units are unknown, but bicyclic compounds also form in some cases:

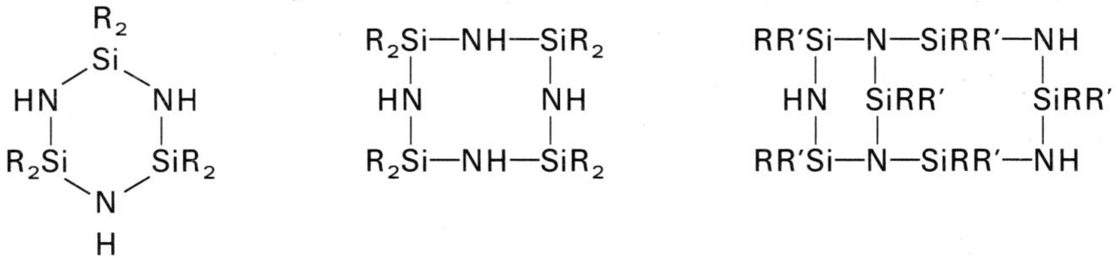

$(R_2SiNH)_3$ (six-membered ring: R_2Si, HN, NH, R_2Si, SiR_2, N–H)

$$\begin{array}{ccc} R_2Si{-}NH{-}SiR_2 \\ | \qquad\quad | \\ HN \qquad NH \\ | \qquad\quad | \\ R_2Si{-}NH{-}SiR_2 \end{array}$$

$$\begin{array}{l} RR'Si{-}N{-}SiRR'{-}NH \\ HN \quad SiRR' \quad SiRR' \\ RR'Si{-}N{-}SiRR'{-}NH \end{array}$$

Formally, organocyclosilazanes are polymers of the $p_\pi - p_\pi$ double-bonded compounds, $R_2Si{=}NR'$; these monomers have only a transient existence when generated by pyrolysis or photolysis of organosilyl azides; the monomer inserts into the $(Me_2SiO)_3$ and $(Me_2SiNMe)_2$ rings to yield expanded heterocycles:

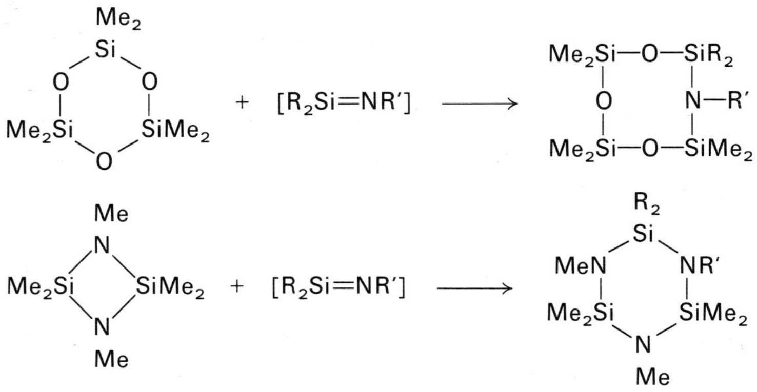

$$(Me_2SiO)_3 + [R_2Si{=}NR'] \longrightarrow \begin{array}{l} Me_2Si{-}O{-}SiR_2 \\ \;\; O \qquad\quad N{-}R' \\ Me_2Si{-}O{-}SiMe_2 \end{array}$$

$$(Me_2SiNMe)_2 + [R_2Si{=}NR'] \longrightarrow \text{cyclo-}(R_2Si{-}NR'{-}SiMe_2{-}NMe{-}SiMe_2{-}NMe)$$

The reactions of N-metallated amines or aminosilanes with chlorosilanes afford better control of the ring structure formed, making possible the preparation of cyclosilazanes and heterocyclosilazanes, including four-membered rings:

$$2\,RNLi_2 + 2\,R_2SiCl_2$$
$$R_2Si(NLiR)_2 + R_2SiCl_2 \longrightarrow$$

R
N
R_2Si SiR_2
N
R

$RN(SiR_2{-}NRLi)_2$ + R_2GeCl_2 → R_2Ge, RN, NR, R_2Si, SiR_2, N, R (six-membered ring)

$RN(SiR_2{-}NRLi)_2$ + $RBCl_2$ → R, B, RN, NR, R_2Si, SiR_2, N, R (six-membered ring)

$RN(SiR_2{-}NRLi)_2$ + $AsCl_3$ → Cl, As, RN, NR, R_2Si, SiR_2, N, R (six-membered ring)

$RN(SiR_2{-}NRLi)_2$ + R_2SnCl_2 → R_2Sn, RN, NR, R_2Si, SiR_2, N, R (six-membered ring)

Some nitrogen-rich heterocycles:

N=N, N, N, R_3Si, SiR_2, SiR_3

Me_2Si—$SiMe_2$, Me—N, N—Me, N, $SiMe_3$

illustrate exotic possibilities in silicon-nitrogen chemistry.

The structures of solid $(Me_2SiNH)_4$ and $(Me_2Si{-}NPh)_2$ confirm their cyclic nature, and of gas-phase $(Me_2SiNH)_3$ reveal that the six-membered Si_3N_3 ring is puckered, but that the deviation from planarity is small.

Attempts to polymerize organocyclosilazanes by heating cyclosilazanes with ammonium bromide (as an acidic catalyst) afford linear polysilazane with only a short chain length. Alkaline catalysts (which give good results in the polymerization of cyclosiloxanes) lead, unexpectedly, to the formation of a tricyclic compound:

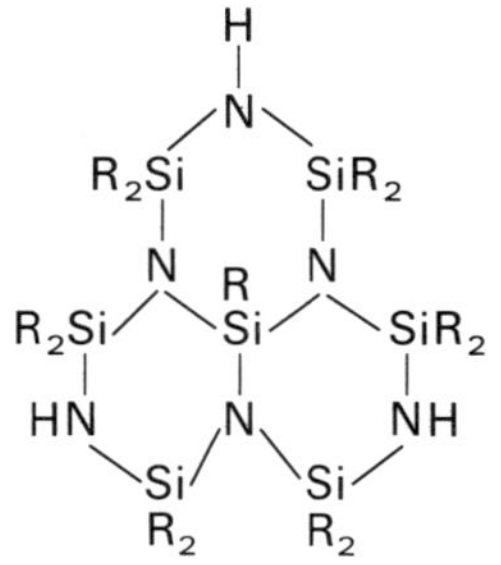

The ammonolysis of organotrichlorosilanes affords oligomers of the type $[RSi(NH)_{1.5}]_n$ (n = 4, 6, and 8) with polycyclic structures analogous to those of the related siloxanes.

Organopolysilthianes

The silthianes contain an alternating silicon and sulfur atom backbone. The Si—S bond is sensitive to moisture. Silthianes are obtained by the reaction of organohalosilanes and hydrogen sulfide (in the presence of an acid scavenger) or silver sulfide. Monofunctional chlorosilanes afford disilthianes, $R_3Si—S—SiR_3$, and organodichlorosilanes afford cyclosilthianes, $(R_2SiS)_n$ (n = 2 or 3). The four-membered, Si_2S_2 ring is formed preferentially, but the six-membered, Si_3S_3, ring can also be obtained. The tendency for cyclization is strong, and tricyclic tetramers, $(RSiS_{1.5})_4$, form in the reaction of organotrichlorosilanes and hydrogen sulfide:

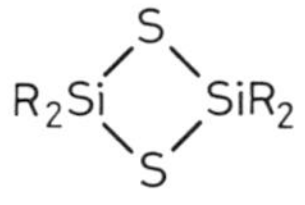

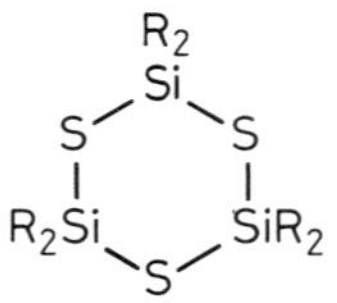

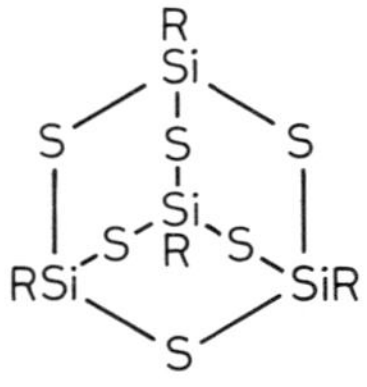

The reaction of dimethylsilane, Me_2SiH_2, with chalcogens affords six-membered rings, $(Me_2SiX)_3$, where X = S, Se or Te.

Solid *trans*-$(MePhSiS)_3$ adopts a twist-boat conformation and the tricyclic nature of solid $(MeSiS_{1.5})_4$ has been confirmed.

Linear silthiane polymers are thermodynamically unstable with respect to ring formation and the longest silicon-sulfur chain is the compound $Ph_3Si—S—SiMe_2—S—SiPh_3$, prepared from Me_2SiCl_2 and $[Ph_3Si—S^-\ [Et_2NH_2^+]$.

Cyclosilthianes undergo redistribution of Si—S and Si—N bonds in reaction with cyclosilazanes, leading to mixed Si—S—Si—N-heterocycles:

$$\text{cyclo-}(R_2SiS)_3 + \text{cyclo-}(R_2SiNR)_3 \rightleftharpoons \text{cyclo-}(R_2Si)_3(NR)_2S + \text{cyclo-}(R_2Si)_3(NR)S_2$$

Evidence for the transient existence of the parent monomer of cyclosilthianes, $R_2Si{=}S$, has been obtained in the reaction of thermally generated $Me_2Si{=}CH_2$ with thiobenzophenone:

$$(Me_2Si{-}CH_2)_2 \xrightarrow{610\,^\circ C} [Me_2Si{=}CH_2] \xrightarrow{Ph_2C=S} [Me_2Si{=}S] \longrightarrow (Me_2SiS)_n$$

The monomer is also formed in the pyrolysis of methylated cyclosilthianes, $(Me_2SiS)_n$ ($n = 2$ or 3), and can be trapped by hexamethylcyclotrisiloxane, to form an expanded eight-membered ring:

$$(Me_2SiS)_n \xrightarrow{\Delta} [Me_2Si{=}S] \xrightarrow{(Me_2SiO)_3} \begin{array}{ccc} Me_2Si{-}O{-}SiMe_2 \\ \vert \quad\quad\quad \vert \\ O \quad\quad\quad S \\ \vert \quad\quad\quad \vert \\ Me_2Si{-}O{-}SiMe_2 \end{array}$$

Polycarbosilanes

Many compounds with structures analogous to those of siloxanes in which the oxygen is replaced by $-CH_2-$, $-CH_2CH_2-$, or $-C_6H_4$-groups, have been prepared.. However, the syntheses are based upon completely different reaction types.

The simplest of these compounds have been obtained in a direct synthesis by passing methylene chloride over a silicon-copper alloy at 300 °C. The products are hexachlorodisilylmethylene, hexachlorocyclotrisilmethylene and a polymer:

$$Cl_3Si{-}CH_2{-}SiCl_3 \qquad \text{cyclo-}(Cl_2SiCH_2)_3 \qquad \left(-\overset{Cl}{\underset{Cl}{\vert}}\!\!Si{-}CH_2- \right)_x$$

Linear carbosilanes of specified chain length can be constructed:

$$R_3Si{-}CH_2Li \xrightarrow{ClCH_2SiR_2Cl} R_3Si{-}CH_2{-}SiR_2{-}CH_2Cl \xrightarrow[R_3SiCl]{Li} R_3Si{-}({-}CH_2{-}SiR_2)_2R$$

$$R_3Si{-}CH_2{-}SiR_2{-}CH_2Cl \xrightarrow{1.\ Li;\ 2.\ ClCH_2SiR_2Cl} R_3Si{-}(CH_2{-}SiR_2)_2{-}CH_2Cl \xrightarrow[R_3SiCl]{Li} R_3Si{-}({-}CH_2{-}SiR_2)_3R$$

$$R_3Si{-}(CH_2{-}SiR_2)_2{-}CH_2Cl \xrightarrow{1.\ Li;\ 2.\ ClCH_2SiR_2Cl} R_3Si{-}(CH_2{-}SiR_2)_3{-}CH_2Cl \xrightarrow[R_3SiCl]{Li} R_3Si{-}({-}CH_2{-}SiR_2)_4R$$

A compound with a highly branched neo-structure has been synthesised from the organolithium reagent, $R_3Si{-}CH_2Li$, and silicon tetrachloride:

$$4\,R_3Si{-}CH_2Li + SiCl_4 \longrightarrow R_3SiCH_2{-}Si(CH_2SiR_3)_3 + 4\,LiCl$$

A variety of polycarbosilanes have been identified in the pyrolysis products of tetramethylsilane and the methylchlorosilanes.

The properties of carbosilanes are similar to those of hydrocarbons. Their reactivity is generally low, the non-polar silicon-carbon bond being stable to moisture, oxygen and relatively high temperatures.

The $p_\pi - p_\pi$, double-bonded $R_2Si{=}CR'R''$ parent monomers of polycarbosilanes are produced by the gas-phase pyrolysis of dimethylsilacyclobutane and undergo addition of $HSiCl_3$, SiF_4 or $SiCl_4$ to the transient Si=C double bond or insertion into a cyclotrisiloxane ring with expansion:

$$\underset{H_2C{-}CH_2}{Me_2Si{-}CH_2} \xrightarrow{-H_2C=CH_2} [Me_2Si{=}CH_2] \xrightarrow{Cl_3SiH} Me_2HSi{-}CH_2SiCl_3$$

$$[Me_2Si{=}CH_2] \xrightarrow{(Me_2SiO)_3} \text{ring: } Me_2Si{-}O{-}SiMe_2{-}CH_2{-}SiMe_2{-}O{-}SiMe_2{-}O{-}$$

The unstable monomer $Me_2Si{=}CH_2$ can be frozen in an argon matrix at 10 K, and its spectral properties investigated.

Monomers stable enough to be isolated contain very bulky substituents, which protect the double bond and prevent dimerization:

$$(Me_3Si)_3Si{-}\underset{\underset{O}{\|}}{C}{-}R \xrightarrow{h\nu} \begin{array}{c} Me_3Si \quad\quad OSiMe_3 \\ Si{=}C \\ Me_3Si \quad\quad\quad R \end{array}$$

$R = C_{10}H_{15}$ (l-adamantanyl)

Silicon-carbon double-bond intermediates can also be generated by photolysis of 1-alkenyldisilanes or of trimethylsilyldiazomethane at 8 K in argon matrix, the latter to give 1,1,2-trimethylsilaethylene which is stable up to 45 K when it dimerizes to a mixture of *cis*- and *trans*-hexamethyldisilacyclobutane:

$$Me_3Si{-}CH{=}N_2 \xrightarrow[8\,K]{h\nu} [Me_2Si{=}CHMe] \xrightarrow{>45\,K} \begin{array}{c} Me \\ | \\ H{-}C{-}SiMe_2 \\ |\quad\quad | \\ Me_2Si{-}C{-}H \\ | \\ Me \end{array}$$

The unsubstituted silaethylene, $H_2Si{=}CH_2$, has been generated as a transient species, and some of its reactions investigated.

8.1.13. Silylenes, :SiR$_2$

Diorganosubstituted species, :SiR$_2$ (silylenes), in which silicon is formally divalent, are generated when hexamethylsilirane is heated, to 60–80 °C or by the photolysis of dodecamethylcyclohexasilane:

$$\left.\begin{array}{l} Me_2Si\langle\begin{array}{l} CH_2 \\ | \\ CH_2 \end{array} \xrightarrow{60-80^\circ} \\ (Me_2Si)_6 \xrightarrow{h\nu} \end{array}\right] \longrightarrow :SiMe_2$$

Dimethylsilylene inserts into hexamethylcyclotrisiloxane to expand the ring by an SiMe$_2$ unit and abstracts oxygen from dimethylsulfoxide to form monomeric dimethylsilanone, $Me_2Si{=}O$, which can be trapped by $Me_6Si_3O_3$:

$$:SiMe_2 \begin{cases} \xrightarrow{(Me_2SiO)_3} \begin{array}{c} Me_2Si{-}SiMe_2 \\ O \quad\quad O \\ | \quad\quad\quad | \\ Me_2Si \quad\quad SiMe_2 \\ O \end{array} \\ \xrightarrow{Me_2SO} [Me_2Si{=}O] \xrightarrow{(Me_2SiO)_3} \begin{array}{c} Me_2Si{-}O{-}SiMe_2 \\ | \quad\quad\quad | \\ O \quad\quad\quad O \\ | \quad\quad\quad | \\ Me_2Si{-}O{-}SiMe_2 \end{array} \end{cases}$$

Dimethylsilylene can be stabilized by coordination to transition metals. Thus, ultraviolet irradiation of $Me_2HSiNEt_2$ with iron pentacarbonyl yields a base-stabilized silylene complex:

$$Me_2HSiNEt_2 + Fe(CO)_5 \xrightarrow{h\nu} \underset{\substack{| \\ H}}{(CO)_4Fe}-\underset{\substack{| \\ NEt_2}}{SiMe_2} \xrightarrow{h\nu} (CO)_4Fe \leftarrow :\underset{\substack{\uparrow \\ HNEt_2}}{SiMe_2}$$

Reacting a disilane with diiron enneacarbonyl also yields an iron-silylene complex:

$$HSiMe_2SiMe_2R + Fe_2(CO)_9 \longrightarrow 2(CO)_3\overset{\substack{H \\ |}}{\underset{\substack{| \\ SiMe_2R}}{Fe}} \leftarrow :SiMe_2 + 3CO$$

$R = H, Me$

8.1.14. Triorganosubstituted Free Radicals, $\cdot SiR_3$

These reactive $\cdot SiR_3$ species are generated by the reaction of hexachlorodisilane with $LiCH(SiMe_3)_2$, followed by ultraviolet irradiation to give $\cdot Si[CH(SiMe_3)_2]_3$. The ESR spectrum of this radical shows coupling of the unpaired electron with the three equivalent protons at the neighboring carbon atoms.

8.2. Organogermanium Compounds

There are many similarities between silicon and germanium. They form analogous organometallic compounds, but there are important differences in the reactivity.

The first organic derivative, tetraethylgermane, was synthesized in 1887 by Winkler, only a year after he discovered the element. However, the organogermanium field lay dormant because of the rarity of the element. By 1950 *ca.* 200 compounds were known, and organogermanium compounds are no more a curiosity now.

8.2.1. Tetraorganogermanes, GeR_4

The tetrasubstituted derivatives, GeR_4, are prepared by the reaction of germanium tetrachloride with excess Grignard reagents. The tetraiodide is sometimes preferred while in the case of silicon it is rarely used. Hexaorganodigermanes, $R_3Ge\text{-}GeR_3$, are frequently formed as by-products, especially if residual magnesium metal is not removed.

Organolithium reagents show advantages over the Grignard reagents. The organometallic derivatives of zinc, mercury or tin are also able to transfer organic groups to germanium.

In the Wurtz-Fittig synthesis a mixture of an organic halide and germanium tetrahalide reacts with sodium metal:

$$GeCl_4 + 4\,RX + 8\,Na \longrightarrow GeR_4 + 4\,NaCl + NaX$$

Organogermanium hydrides are more reactive than their organosilicon analogues in such additions.

The germanium-carbon bond is more readily cleaved than the Si—C bond, and thus tetraorganogermanes are more reactive, especially in dealkylation and redistribution reactions which can be used to prepare organogermanium halides (with no analogy in organosilicon chemistry). The characteristic reactions of tetraorganogermanes are outlined in Fig. 8.6:

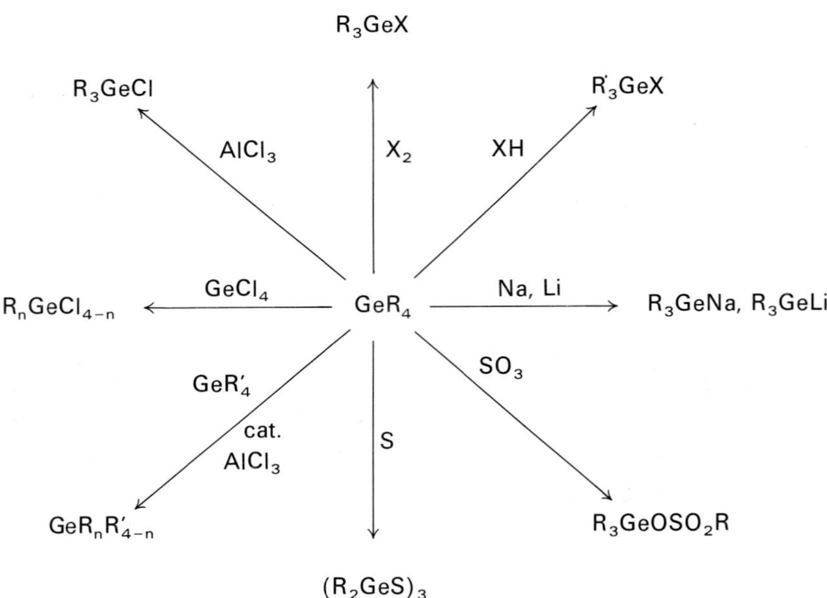

Fig. 8.6. Some typical reactions of tetraorganogermanes.

8.2.2. Heterocyclic Compounds with Germanium as Heteroatom

The small ring germanocyclobutanes react with sulfur, selenium, sulfur trioxide and carbenes:

$CH_2(CH_2Cl)_2 + 4Na + R_2GeCl_2 \longrightarrow R_2Ge\diamond \longleftarrow R_2Ge((CH_2)_3Cl)(Cl) + 2Na$

$R_2Ge\diamond \xrightarrow{S\,(or\,Se)}$ R_2Ge-S five-membered ring (or Se)

$R_2Ge\diamond \xrightarrow{SO_3}$ R_2Ge-O-SO_2 six-membered ring

$R_2Ge\diamond \xrightarrow{:CCl_2}$ R_2Ge-CCl_2 five-membered ring

Larger heterocycles can be prepared by Grignard reactions:

$$3XMg(CH_2)_5MgX + 2GeCl_4 \longrightarrow \text{(CH}_2)_5\text{GeCl}_2 + \text{(CH}_2)_5\text{Ge(CH}_2)_5 + 6MgXCl$$

or from organolithium compounds:

$$RC(Li){=}CR{-}CR{=}C(Li)R + R_2GeX_2 \longrightarrow \text{tetra-R-germole (Ge}R_2) + 2LiX$$

Addition of GeI_2 as a carbene-analogue to butadiene and acetylenes yields unsaturated heterocycles:

$$\text{(RC=CR)}_2(GeR_2)_2 \xleftarrow{RC\equiv CR} GeI_2 \xrightarrow{\text{butadiene}} \text{germacyclopentene (Ge}I_2)$$

Addition of trichlorogermane, $HGeCl_3$, to butadiene proceeds via $GeCl_2$, which is formed in an equilibrium:

$$HGeCl_3 \rightleftharpoons :GeCl_2 + HCl$$

which explains the formation of addition compounds in the above reaction:

$$5HGeCl_3 + 5\,\text{butadiene} \longrightarrow 3\,\text{germacyclopentene (Ge}Cl_2) + 2Cl_3Ge{-}CH_2{-}CH{=}CH{-}CH_3 + 3HCl$$

The direct synthesis with methylene chloride and elemental germanium in the presence of copper metal produces a germanium heterocycle:

$$CH_2Cl_2 + Ge \xrightarrow[Cu]{380°} \text{(}Cl_2Ge)_3\text{(CH}_2)_3 \text{ ring } + Cl_3GeCH_2GeCl_3 + MeGeCl_3$$

8.2.3. Organohalogermanes, R_nGeX_{4-n}

Organohalogermanes are prepared by direct synthesis from alkyl halides and elemental germanium in the presence of copper metal at temperatures between 300 and 350 °C. The reaction can be controlled to give $MeGeCl_3$ or Me_2GeCl_2, with higher temperatures favoring $MeGeCl_3$. In the presence of gallium and copper at 400 °C, methyl chloride and germanium give a mixture of 85% Me_3GeCl, 10% Me_2GeCl_2 and 5% $MeGeCl_3$. Bromopentafluorobenzene with a Ge/Cu alloy (60 : 40) gives a mixture of $C_6F_5GeBr_3$ and $(C_6F_5)_2GeBr_2$.

Redistribution reactions between tetraorganogermanes and germanium tetrahalides in the presence of aluminum chloride, are used to prepare trimethylchlorogermane:

$$GeR_4 + 3\,GeX_4 \longrightarrow 4\,RGeX_3$$
$$GeR_4 + GeX_4 \longrightarrow 2\,R_2GeX_2$$

Unlike in silicon chemistry, Ge—C bond cleavage by halogens or anhydrous hydrogen halides can also be used:

$$GeR_4 + X_2 \longrightarrow R_3GeX + RX \; (X = I, Br)$$
$$GeR_4 + HX \longrightarrow R_3GeX + RH\,.$$

Organogermanium iodides can be prepared by addition of germanium(II) iodide to organic iodides:

$$GeI_2 + RI \longrightarrow RGeI_3$$

Tetramethylgermane can be chlorinated with antimony pentachloride to form Me_3GeCl and Me_2GeCl_2. Germanium tetrachloride can be alkyated by tetraalkyltins to give $RGeX_3$ and R_3SnX. Methylgermanium fluorides, Me_nGeF_{4-n} (n = 1 – 3), are prepared by treating $(Me_3Ge)_2O$ or $(Me_2Ge_2O_3)_x$ with AsF_3.

In the gas phase Me_2GeF_2, $MeGeF_3$, Me_2GeBr_2 and $MeGeBr_3$, are monomeric and tetrahedral.

8.2.4. Organogermanium Hydrides, R_nGeH_{4-n}

The most stable are the triorganogermanium hydrides, which are prepared by the reduction of the corresponding halides with lithium alanate, or, unlike organosilicon analogues, with amalgamated zinc and hydrochloric acid.

The organogermanium hydrides are more stable towards alkalies than organosilicon hydrides. Thus R_3GeH compounds are not attacked by potassium hydroxide while the analogous silanes release hydrogen to form silanolates. The organogermanium hydrides also differ from silanes in their reactions with organolithium reagents where they react like triphenylmethane:

$$Ph_3SiH + RLi \longrightarrow Ph_3SiR + LiH$$

$$Ph_3GeH + RLi \longrightarrow Ph_3GeLi + RH$$

$$Ph_3CH + RLi \longrightarrow Ph_3CLi + RH$$

These differences are explained by the greater electronegativity of germanium making the Ge—H bond less polar.

8.2.5. Germanols, $R_nGe(OH)_{4-n}$

Few germanols, R_3GeOH, are known; usually $R = C_6H_5$ or C_6F_5. They are prepared by the hydrolysis of the corresponding halides which are more stable toward water than the analogous chlorosilanes.

Compounds with two OH groups at the same germanium atom are rare, but $Me_2Ge(OH)_2$ exists in aqueous solution.

8.2.6. Other Compounds with Ge-O Bonds

Organogermanium alkoxides, $R_nGe(OR')_{4-n}$, and carboxylates, $R_nGe(OCOR')_{4-n}$, are prepared by reactions similar to those used in organosilicon chemistry. The alkoxides are sensitive to hydrolysis and react with hydrogen sulfide and thiols:

$$R_2Ge(OR)_2 + H_2S \xrightarrow[-ROH]{} R_2Ge(SH)_2 \xrightarrow[-H_2S]{} (R_2GeS)_2$$

$$R_3GeOR + R'SH \longrightarrow R_3GeSR' + ROH$$

The Si—O bond cannot, on the other hand, be converted directly to the Si-S bond.

8.2.7. Organoaminogermanes, $R_nGe(NR'R'')_{4-n}$

Organoaminogermanes are obtained from chlorogermanes and amines. Compounds with Ge—NH_2 bonds are unstable and undergo condensation to form germazanes containing Ge—NH—Ge groups. The $Ge—NH_2$ group can be stabilized by bulky substituents, however, and thus $(iso\text{-}Pr)_3Ge—NH_2$ can be isolated from the reaction of $(iso\text{-}Pr)_3GeCl$ with KNH_2.

Germanium-nitrogen heterocycles have been prepared by transamination:

$$Me_2Ge(NEt_2)_2 + MeHN—CH_2—CH_2—NHMe \longrightarrow Me_2Ge\langle N(Me)—CH_2—CH_2—N(Me)\rangle + 2\,Et_2NH$$

Like in organosilicon chemistry, N-metallated amines are useful in the synthesis of Ge—N compounds:

$$2\,Et_3Si—NHLi + Me_2GeCl_2 \longrightarrow (Et_3Si—NH)_2GeMe_2 + 2\,LiCl$$

$$Me_3Si—NLi—CMe_3 + Me_3GeCl \longrightarrow Me_3Si—N(CMe_3)—GeMe_3$$

The action of reagents containing mobile hydrogen such as water, alcohols, acids and acetylenes cleaves the Ge—N bonds.

8.2.8. Linear Oligomers, Cyclic Compounds and Polymers

No long-chain, linear polygermoxanes (polysiloxane analogues) are known, and R_2GeO, R_2GeNH and R_2GeS adopt a cyclic form. Otherwise, the preparative methods and structural types are similar to those found in organosilicon chemistry.

Organopolygermanes

The simplest members of the series, the digermanes, $R_3Ge—GeR_3$, are prepared from triorganohalogermanes and alkali metals, or as by-products in the synthesis of tetraorganogermanes, with excess magnesium metal present:

$$3\,GeX_4 + 10\,RMgX + Mg \longrightarrow GeR_4 + R_3Ge—GeR_3 + 11\,MgX_2$$

The Ge—Ge bonds are thermally and hydrolytically stable, but are attacked by halogens and oxidizing agents; $Ph_3Ge—GePh_3$ resists attack by HX, and only phenyl groups are cleaved to give $X_2PhGe\text{-}GePhX_2$. Alkali metals cleave the Ge—Ge bond to form reactive organogermanides containing anionic organogermanium groups:

$$Ph_3Ge—GePh_3 + Na/K\ alloy \xrightarrow{THF} K^+GePh_3^- + Na^+GePh_3^-$$

Trigermanes are prepared from organodihalogermanes and alkali metal derivatives of triorganogermanes:

$$R_2GeCl_2 + 2\,R_3GeNa \longrightarrow R_3Ge{-}\underset{R}{\overset{R}{Ge}}{-}GeR_3 + 2\,NaCl$$

and a branched tetragermane has been synthesized from germanium(II) iodide:

$$GeI_2 + 3\,R_3GeLi \longrightarrow R_3Ge{-}\underset{Li}{\overset{GeR_3}{Ge}}{-}GeR_3 \xrightarrow{H_2O} R_3Ge{-}\underset{H}{\overset{GeR_3}{Ge}}{-}GeR_3$$

Organogermanium dihalides react with alkali metals to produce cyclotetra-, -penta- and hexagermanes:

$$R_2GeCl_2 \xrightarrow{\text{Na in xylene}} (R_2Ge)_4 \text{ (cyclotetragermane: } R_2Ge{-}GeR_2,\ R_2Ge{-}GeR_2\text{)}$$

$$R_2GeCl_2 \xrightarrow{\text{Li in tetrahydrofuran}} (R_2Ge)_5 + (R_2Ge)_6$$

From Me_2GeCl_2 and lithium metal, cyclogermanes, $(Me_2Ge)_n$, with n = 5, 6 and 7, have been obtained. Dodecamethylcyclohexagermane, $(GeMe_2)_6$ is isostructural with its silicon analogue.

Cyclic polygermanes are cleaved by halogens and alkali metals to afford linear polygermanes:

$$(Ph_2Ge)_4 + I_2 \longrightarrow I(Ph_2Ge)_4I$$

The longest $R(GeR_2)_nR$ chain contains seven germanium atoms.

Organopolygermoxanes

Triorganogermanium hydroxides eliminate water and form digermoxanes, $R_3Ge{-}O{-}GeR_3$. The synthesis of linear germoxanes with Ge—O—Ge chains is more difficult than in the case of siloxanes, since hydrolysis of organodihalogermanes yields cyclotri- and -tetragermoxanes:

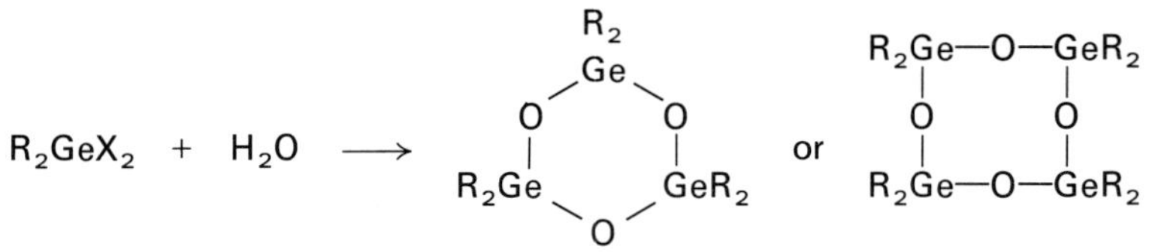

$$R_2GeX_2 + H_2O \longrightarrow (R_2GeO)_3 \text{ (cyclotrigermoxane) or } (R_2GeO)_4 \text{ (cyclotetragermoxane)}$$

The trimer and tetramer are interconvertible in a thermodynamically controlled equilibrium:

$$3\,(R_2GeO)_4 \rightleftharpoons 4\,(R_2GeO)_3$$

Unlike organosiloxanes, organocyclogermoxanes are not hydrophobic; some even dissolve in water to form the dihydroxide, $R_2Ge(OH)_2$. Germoxane rings are readily cleaved by hydrogen halides, alcohols and carboxylic acids or their anhydrides. An equilibrium redistribution with organogermanium dihalides affords halogen-terminated germoxane oligomers:

$$(R_2GeO)_n + R_2GeX_2 \rightleftharpoons X(R_2GeO)_nGeR_2X$$

This equilibrium favors linear oligomers.

The parent $R_2Ge{=}O$ monomer of cyclogermoxanes is formed as a transient species in the thermolysis of germanium-containing oxetanes. It polymerizes rapidly to cyclotrigermoxane, and inserts into C—O bonds:

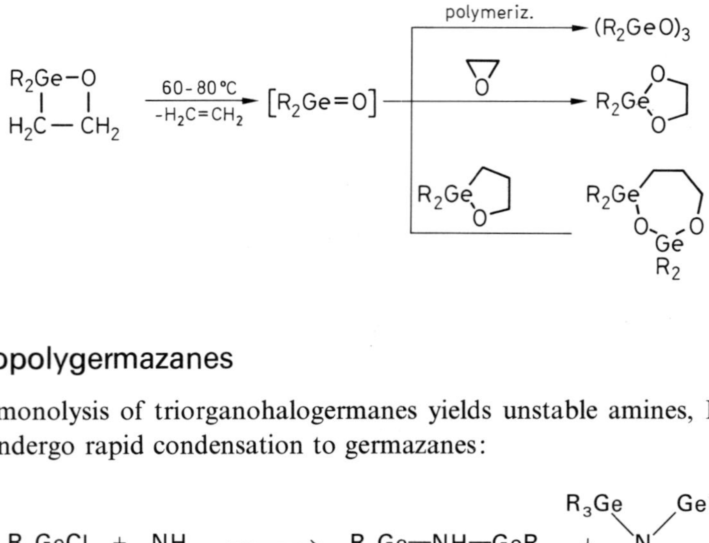

$$\text{R}_2\text{Ge—O—CH}_2\text{—CH}_2 \text{ (ring)} \xrightarrow[-H_2C=CH_2]{60-80\,°C} [R_2Ge{=}O]$$

$[R_2Ge{=}O]$ → (polymeriz.) $(R_2GeO)_3$; + ethylene oxide → 2,2-R_2-1,3-dioxa-2-germolane; + R_2Ge-oxetane → eight-membered ring R_2Ge–O–GeR_2–O (with $(CH_2)_3$)

Organopolygermazanes

The ammonolysis of triorganohalogermanes yields unstable amines, $R_3Ge{-}NH_2$, which undergo rapid condensation to germazanes:

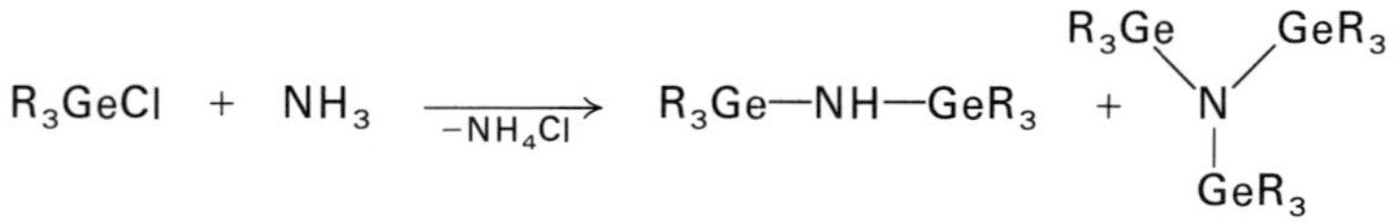

$$R_3GeCl + NH_3 \xrightarrow[-NH_4Cl]{} R_3Ge{-}NH{-}GeR_3 + N(GeR_3)_3$$

Organodihalogermanes react with ammonia and primary amines (R = Me) to form cyclotrigermazanes:

$$\text{(HN-GeR}_2)_3\text{ ring} \xleftarrow[-NH_4Cl]{+NH_3} R_2GeCl_2 \xrightarrow[-[RNH_3]Cl]{+RNH_2} \text{(RN-GeR}_2)_3\text{ ring}$$

Four-membered, germanium-nitrogen rings are prepared using a dimetallated amine:

$$2\,C_6F_5NLi_2 + 2\,Ph_2GeCl_2 \longrightarrow Ph_2Ge(N C_6F_5)_2GePh_2 + 4\,LiCl$$

Cyclization prevents Ge—N—Ge linear-polymer formation. Monomeric $Ph_2Ge{=}NMe$ has been detected as a transient species.

Organogermathianes (Organogermanium Sulfides)

Digermathianes, $R_3Ge{-}S{-}GeR_3$, form from triorganohalogermanes with hydrogen sulfide or silver and sodium sulfides. A heterocycle containing a Ge—S—Ge group forms by insertion of elemental sulfur into a Ge—Ge bond:

$$R_2Ge{-}GeR_2\ (\text{cyclic}) + S \longrightarrow R_2Ge{-}S{-}GeR_2\ (\text{cyclic})$$

Organodihalogermanes react with hydrogen sulfide even in aqueous solution to yield cyclic compounds. Elemental sulfur attacks tetraorganogermanes on heating to form cyclic compounds:

$$R_2GeX_2 + H_2S \xrightarrow[-HX]{} (R_2GeS)_3 \xleftarrow[-R_2S]{} GeR_4 + S_8$$

Four-membered, germanium-sulfur rings, $(R_2GeS)_2$, are formed from organodialkoxygermanes with hydrogen sulfide:

$$2\,R_2Ge(OR')_2 + 2\,H_2S \longrightarrow R_2Ge(S)_2GeR_2 + 4\,R'OH$$

Methyltribromogermane reacts with hydrogen sulfide in the presence of triethylamine to yield tetrameric $(RGeS_{1,5})_4$, with an adamantane-like structure:

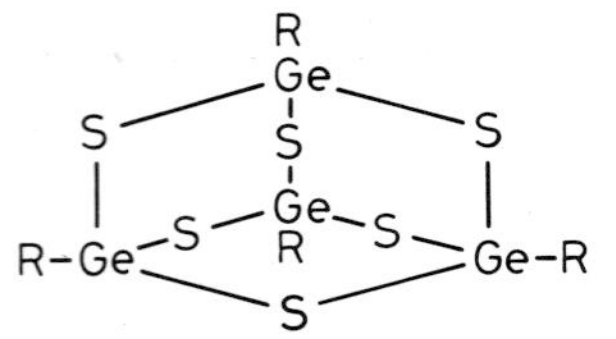

[Structure: adamantane-like $(RGeS_{1.5})_4$ cage with four R–Ge units bridged by six S atoms]

Cyclogermathianes are more stable toward moisture than their silicon analogues, but undergo redistribution of R_2GeS units with cyclosilthianes to form mixed heterocycles in a thermodynamically controlled equilibrium:

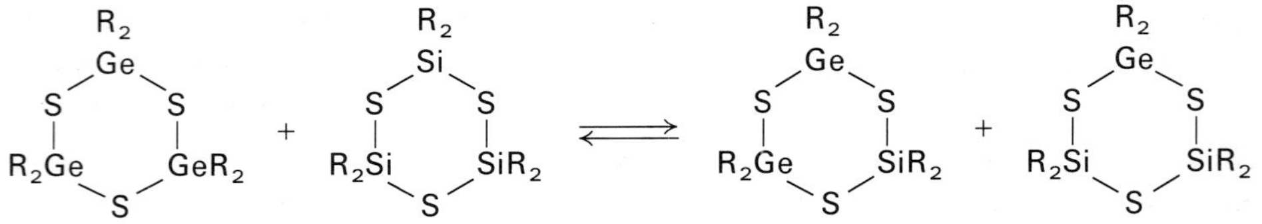

$$(R_2GeS)_3 + (R_2SiS)_3 \rightleftharpoons R_2Ge(SSiR_2)_2S + R_2Ge(SSiR_2S)\ldots$$

[Scheme: six-membered ring $(R_2GeS)_3$ + ring $(R_2SiS)_3$ ⇌ ring with one GeR_2 and two SiR_2 / R_2Ge + ring with one GeR_2, R_2Si, SiR_2]

8.2.9. Diorganogermylenes, :GeR$_2$

Divalent germanium is known in inorganic compounds, and can be stabilized with bulky organic substituents in :GeR$_2$ (germylenes). Thus, :Ge[CH(SiMe$_3$)$_2$]$_2$ has been prepared from Ge[N(SiMe$_3$]$_2$ and LiCH(SiMe$_3$)$_2$.

These species can be trapped as ligands in transition metal complexes. A chromium carbonyl complex of dimesitylgermylene has been obtained by ligand substitution. The coordinated germylene is a strong acceptor and can further coordinate pyridine:

$$(CO)_5Cr:GeCl_2\cdot THF \xrightarrow{RMgBr} (CO)_5Cr \leftarrow :GeR_2$$

$$\xrightarrow{.py} (CO)_5Cr \leftarrow :\underset{\underset{py}{\uparrow}}{GeR_2}$$

Germylene complexes can be prepared directly by reaction with metal carbonyl derivatives; the structure of solid (CO)$_5$Cr←:Ge[CH(SiMe$_3$)$_2$]$_2$ has been confirmed:

$$:Ge[CH(SiMe_3)_2]_2 \xrightarrow{+Cr(CO)_6} (CO)_5Cr \leftarrow :Ge[CH(SiMe_3)_2]_2$$

$$:Ge[CH(SiMe_3)_2]_2 \xrightarrow{+W(CO)_4NBD} trans\text{-}(CO)_4W\{:Ge[CH(SiMe_3)_2]_2\}_2$$

The reaction of metal carbonyl anions with diorganogermanium halides has the advantage of using more readily available starting materials:

$$Na_2Cr_2(CO)_{10} + R_2GeCl_2 \xrightarrow[-NaCl]{} (CO)_5Cr \leftarrow :GeR_2\cdot THF$$

8.2.10. Trisubstituted Organogermanium Free Radicals, $\cdot GeR_3$

Treatment of $GeCl_2$ with $LiCH(SiMe_3)_2$, followed by ultraviolet irradiation, gives $\cdot Ge[CH(SiMe_3)_2]_3$ as a stable species with a pyramidal structure and the unpaired electron located in an sp^3-hybrid orbital.

8.3. Organotin Compounds

The metallic character of tin is more pronounced than that of the lighter elements in Group IV A, and this is reflected in its organometallic chemistry. The bond to carbon is weaker, and Sn—C bond cleavage (for example, in redistributions) becomes more important. The tin atom achieves coordination numbers higher than four either by association or by chelation. Tetraorganotin compounds are four-coordinated monomers, but the functional derivatives R_3SnX and R_2SnX_2 dimerize or polymerize in the solid state and coordinate Lewis bases.

Organotin compounds are called either as substitution products of stannane, SnH_4, for example, $(CH_3)_2SnCl_2$-dichlorodimethylstannane, or $(C_6H_5)_3SnCl$-chlorotriphenylstannane by analogy with organic derivatives of silicon and germanium, or as salts of the metal, for example, dimethyltin dichloride, triphenyltin chloride.

Interest in organotins has been stimulated by the discovery of their use as stabilizers for polyvinyl chloride, and as pesticides, catalysts for polyurethane formation, as antioxidants, etc. An extensive review literature is available which summarizes the enormous development enjoyed by this chapter of organometallic chemistry in the last 25 years.

In addition to those mentioned for silicon and germanium, the tendency of tin to increase its coordination number beyond four gives rise to structural peculiarities. Thus, there are few complete analogies between silicon, germanium and tin compounds.

8.3.1. Tetraorganostannanes, SnR_4

Preparation. Tetrasubstituted derivatives are prepared from Grignard reagents and tin tetrachloride for symmetrical derivatives, or an organotin halide for unsymmetrical derivatives:

$$SnCl_4 + 4\,RMgX \longrightarrow SnR_4 + 4\,MgClX$$
$$R_2SnCl_2 + 2\,R'MgX \longrightarrow SnR_2R'_2 + 2\,MgClX$$

Organolithium reagents work even better, especially, for example, in the preparation of $Sn(C_6F_5)_4$.

Wurtz reactions, using tin tetrachloride, an organic chloride and sodium metal are less effective because sodium metal reduces the tetrachloride to metallic tin.

Organoaluminum compounds serve as alkylating agents in the presence of complexing agents (tertiary amines, ethers, even sodium chloride) required for the fixation of the aluminum chloride formed:

$$3\,SnCl_4 + 4\,AlR_3 + 4\,NaCl \longrightarrow 3\,SnR_4 + 4\,NaAlCl_4$$

Organometallic derivatives of sodium are employed mainly for the formation of Sn—C≡C—R groups, since sodium acetylides are readily available. Organothallium compounds can also transfer organic groups to metallic tin:

$$2\,(C_6F_5)_2TlBr + Sn \longrightarrow Sn(C_6F_5)_4 + 2\,TlBr$$

The addition of organotin hydrides to olefins is sometimes exothermic:

$$R_3SnH + H_2C{=}CHR' \longrightarrow R_3SnCH_2CH_2R'$$

The direct synthesis is not usually a method for tetrasubstituted derivatives, but it has been used in the preparation of $Sn(C_6F_5)_4$ by heating iodopentafluorobenzene with tin metal at 240 °C.

The acidic hydrocarbons, acetylenes and cyclopentadiene can replace amino groups to form additional Sn—C bonds:

$$Me_3Sn{-}NMe_2 + RH \longrightarrow Me_3Sn{-}R + HNMe_2$$

R = -C≡CH , (cyclopentadienyl, C–H)

Structure. The structures of tetraphenyltin $SnPh_4$ and tetrakis(pentafluorophenyl)tin are tetrahedral.

Properties. Tetraorganostannanes undergo redistribution of their substituents on heating in the presence of a catalyst:

$$SnR_4 + SnR'_4 \xrightarrow{AlCl_3} 2\,SnR_nR'_{4-n}$$

Some typical reactions of SnR_4 derivatives are illustrated in Fig. 8.7:

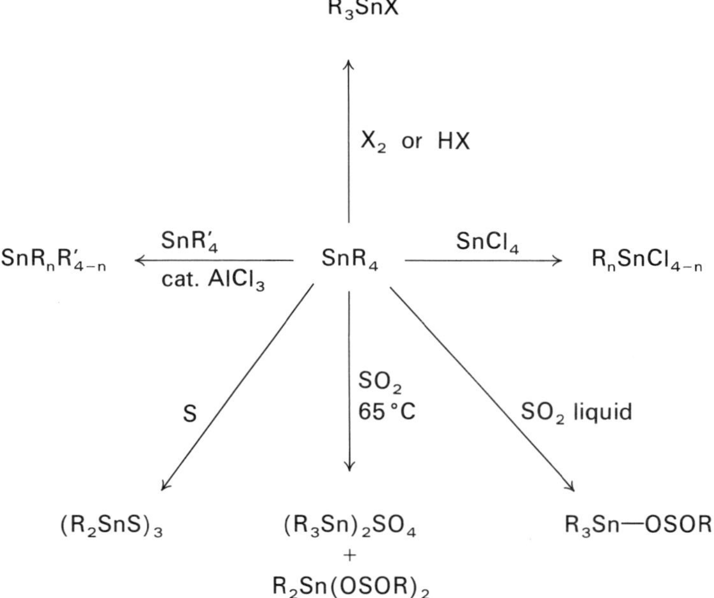

Fig. 8.7. Some general reactions of SnR_4 compounds.

8.3.2. Organic Heterocycles with Tin Heteroatoms

Tin-containing heterocycles are prepared by using Grignard reagents, organolithium compounds or organotin hydride addition reactions:

$R_2SnBr_2 + ClMg(CH_2)_5MgCl \longrightarrow$ [Sn, R_2 ring] $+ \; 2MgClBr$

[R, R, R, R, Li Li] $+ \; R'_2SnCl_2 \longrightarrow$ [R, R, R, R, Sn, R'_2] $+ \; 2LiCl$

[SnR_2] $\longleftarrow R_2SnH_2 \longrightarrow$ [SnR_2]

Tristannacyclohexane has been synthesized from organozinc reagents and magnesium coupling:

$$\mathrm{EtZnI + CH_2I_2 \xrightarrow[-EtI]{} IZnCH_2I \xrightarrow[-ZnICl]{Me_2SnCl_2} Me_2Sn(CH_2I)_2 \xrightarrow{Mg} Me_2Sn\langle CH_2SnMe_2CH_2\rangle SnMe_2}$$

Cyclic organotin compounds are intermediates in the synthesis of the corresponding heterocycles by treatment with boron or organoboron halides.

8.3.3. Organotin Halides, R_nSnX_{4-n}, and Their Complexes

Preparation. Redistribution occurs on heating a tetraorganostannane with tin tetrahalide

$$3\,SnR_4 + SnX_4 \longrightarrow 4\,R_3SnX$$
$$SnR_4 + SnX_4 \longrightarrow 2\,R_2SnX_2$$
$$SnR_4 + 3\,SnX_4 \longrightarrow 4\,RSnX_3$$

Aluminum alkyls replace only three chlorine atoms:

$$SnCl_4 + AlR_3 \longrightarrow R_3SnCl + AlCl_3$$

Organothallium reagents transfer organic groups to tin(II) chloride to form diorganotin dichlorides:

$$(C_6F_5)_2TlBr + SnCl_2 \longrightarrow (C_6F_5)_2SnCl_2 + TlBr$$

In the direct synthesis, tin metal and an organic halide, RX (X = I, Br, Cl in the order of decreasing reactivity), react at 60–180 °C:

$$Sn + 2\,RX \xrightarrow{\text{catalyst}} R_2SnX_2$$

Benzyl chloride acts on tin metal in boiling water to give tribenzyltin chloride, or in toluene to give dibenzyltin dichloride.

Organotin trichloride results from the addition of organic chlorides to tin(II) chloride in the presence of trialkylantimony catalysts:

$$SnCl_2 + RCl \xrightarrow{SbR_3\ \text{catalyst}} RSnCl_3$$

Organotin halides are synthesized by cleavage of organic groups by bromine or iodine:

$$SnR_4 \xrightarrow[-RX]{X_2} R_3SnX \xrightarrow[-RX]{X_2} R_2SnX_2$$

Aromatic groups are more readily cleaved, and saturated groups are less readily cleaved as the hydrocarbon chain lengthens. This makes possible the selective cleavage of organic groups from unsymmetrical compounds.

Inorganic tin is converted to organotin halides by the action of $HSnCl_3$ generated in ether on unsaturated esters:

$$SnCl_2 + HCl \xrightarrow{\text{ether}} HSnCl_3 \cdot Et_2O \xrightarrow{RR'C{=}CHY} Cl_2Sn(CRR'CH_2Y)_2 \text{ or } Cl_3Sn{-}CRR'CH_2Y \quad Y{=}COOR''$$

Structure. The structures of the organotin halides are interesting. In the vapor phase the triorganotin halides, R_3SnX (X = Cl, Br, I), are tetrahedral monomers. In the solid state, however, trimethyltin fluoride, Me_3SnF, is associated into a linear polymer in which tin is five-coordinated with trigonal-bipyramidal geometry:

$$-SnR_3-F\left[\rightarrow SnR_3-F\right]_n\rightarrow SnR_3-F\rightarrow$$

and the chloride has a similar structure. Likewise, trimethyltin thiocyanate has an associated structure with five-coordinated tin and SCN bridges. Triphenyltin chloride is, however, monomeric and tetrahedral in the solid state.

Diorganotin dihalides, R_2SnX_2, are also tetrahedral monomers in the vapor phase, and this structure is preserved in the solid state as suggested by their low melting points, but Me_2SnCl_2 and Et_2SnX_2 (X = Cl, Br and I) show intermolecular Sn ··· X interactions with the coordination geometry at tin intermediate between tetrahedral and octahedral.

The fluorides are highly associated and consequently are high melting and insoluble, as in Me_2SnF_2 which has a double fluorine-bridged structure containing six-coordinated tin:

$$\left[R_2Sn(\mu\text{-}F)_2R_2Sn(\mu\text{-}F)_2R_2Sn(\mu\text{-}F)_2R_2Sn(\mu\text{-}F)_2\right]_n.$$

The methyl groups are *trans*-.

Properties. Organotin halides form adducts with donor molecules or anionic complexes by coordination of additional halide ions. These compounds contain five- or six-coordinated tin and exhibit bipyramidal, trigonal or distorted-octahedral geometries, respectively, as in the pyridine, dipyridyl and dimethylsulfoxide adducts:

$$Ph_2SnX_2(py)_2 \qquad Me_2SnCl_2(\text{dipy}) \qquad Me_2SnCl_2(OSMe_2)_2$$

The complexes of Me_2SnCl_2 with phenanthrolines and 2,2′-dipyridyls have a *trans*-dimethyl structure.

Octahedral structures with *trans*-organic groups have also been identified in the anions $[Me_2SnF_4]^{2-}$ and $[Me_2SnCl_4]^{2-}$; $[Me_3SnCl_3]^{2-}$ is also octahedral. Five-coordinated, trigonal-bipyramidal geometries are adopted by the anions $[MeSnCl_4]^-$, $[Me_2SnCl_3]^-$, $[Ph_3SnCl_2]^-$ and $[Bu_3SnCl_2]^-$.

Methyltin chlorides undergo solvolysis in acidic solutions to form the solvated cations, $[Me_3Sn]^+$ and $[Me_2Sn]^{2+}$.

Internally-coordinated organotin halides with five-coordinated tin, for example,

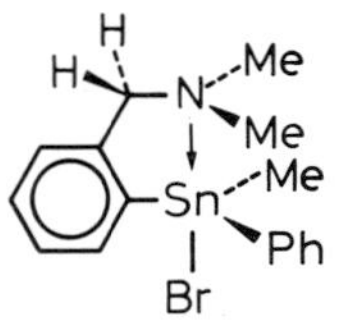

H H N Me Me Sn Me Ph Br

possess optical activity and are also fluxional.

8.3.4. Organotin Hydrides, R_nSnH_{4-n}

Organotin hydrides exhibit high reactivity and are used as intermediates in addition reactions and as reducing reagents in organic chemistry.

Their preparation is by the reduction of organotin halides with lithium alanate:

$$4\,R_3SnX + LiAlH_4 \longrightarrow 4\,R_3SnH + LiAlCl_4$$

but sodium borohydride can also be used:

$$2\,Me_3SnCl + 2\,NaBH_4 \longrightarrow 2\,Me_3SnH + 2\,NaCl + B_2H_6$$

Alkyltin hydrides are more stable than aryltin hydrides, and increased alkylation favors stability. Slow decomposition to SnR_4, tin metal and hydrogen is caused by impurities. Amines promote decomposition of the hydrides with formation of tin-tin bonds.

Organotin hydrides can add (hydrostannation) to compounds containing C≡C, C=O, C=N, and N=N bonds (Fig. 8.8).

Organotin hydrides reduce organic halides and other functional groups.

Organotin hydrides are tetrahedral monomers in the gas phase.

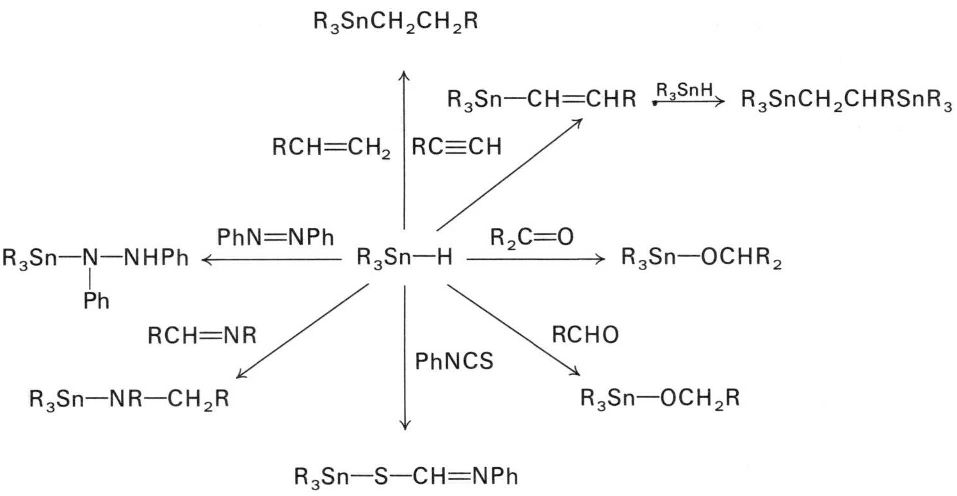

$$R_3SnCH_2CH_2R \xleftarrow{RCH=CH_2} R_3Sn-H \xrightarrow{RC\equiv CH} R_3Sn-CH=CHR \xrightarrow{R_3SnH} R_3SnCH_2CHRSnR_3$$

$$R_3Sn-N(Ph)-NHPh \xleftarrow{PhN=NPh} R_3Sn-H \xrightarrow{R_2C=O} R_3Sn-OCHR_2$$

$$R_3Sn-H \xrightarrow{RCH=NR} R_3Sn-NR-CH_2R$$

$$R_3Sn-H \xrightarrow{PhNCS} R_3Sn-S-CH=NPh$$

$$R_3Sn-H \xrightarrow{RCHO} R_3Sn-OCH_2R$$

Fig. 8.8. Addition of organotin hydrides to various compounds containing multiple bonds.

8.3.5. Organotin Hydroxides, $R_nSn(OH)_{4-n}$

Organotin hydroxides are obtained by alkaline hydrolysis of halides, but they are readily dehydrated:

$$R_3SnX + NaOH \xrightarrow[-NaX]{} R_3Sn-OH \xrightarrow[-H_2O]{} R_3Sn-O-SnR_3$$

Organotin hydroxides, like their halide precursors, undergo coordinative polymerization with increased coordination number at tin, Thus, trimethyl- and triphenyltin hydroxides are associated in the solid state, but Me_3SnOH is a cyclic dimer in solution:

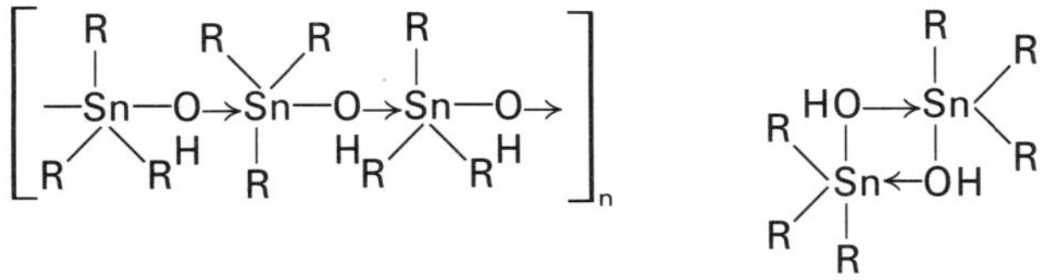

$$\left[-SnR_3(H)-O\rightarrow SnR_3-O(H)\rightarrow SnR_3-O(H)\rightarrow\right]_n \qquad R_3Sn(\mu\text{-}OH)_2SnR_3$$

Dimethyltin hydroxide nitrate is a dimer in the solid state with five-coordinated tin and bipyramidal trigonal geometry; $EtSn(OH)Cl_2 \cdot H_2O$ is also a dimer, but with six-coordination at tin:

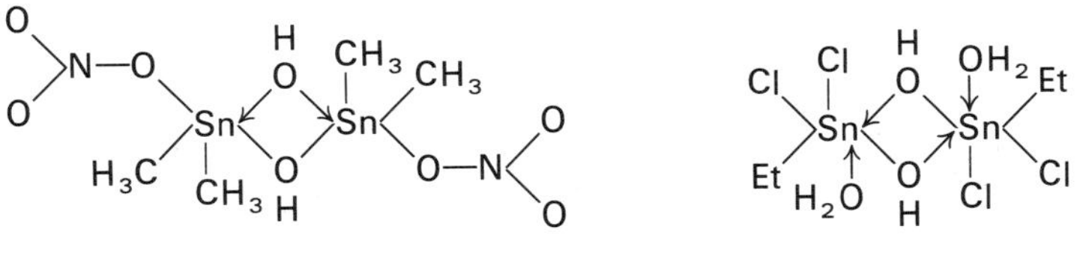

$$[(CH_3)_2Sn(ONO_2)(\mu\text{-}OH)]_2 \qquad [EtSnCl_2(H_2O)(\mu\text{-}OH)]_2$$

Organotin compounds with two hydroxy groups at tin undergo condensation, but $RSn(OH)_2Cl$ forms as a hydrolysis product of organotin trihalides. In alkaline solutions, octahedral $[Me_2Sn(OH)_4]^{2-}$ anions are formed with *trans*-methyl groups.

Organotin hydroxides are basic, reacting with organic acids to form organotin carboxylates and with dithiophosphoric acids to form dithiophosphates.

8.3.6. Organotin Alkoxides, $R_nSn(OR')_{4-n}$ and Related Compounds

Compounds containing Sn—OR bonds are obtained by treatment of organotin halides with alcohols in the presence of bases, or better with alkali metal alkoxides, or from distannoxanes and alcohols:

$$R_3SnX + R'OH + NR''_3 \longrightarrow R_3Sn{-}OR' + HX \cdot NR''_3$$

$$R_3SnX + NaOR' \longrightarrow R_3Sn{-}OR' + NaX$$

$$R_3SnOSnR_3 + 2R'OH \longrightarrow 2R_3Sn{-}OR' + H_2O$$

The elimination of a volatile alcohol by heating alkoxides with diols is used to synthesize organotin heterocycles:

$$R_2Sn(OEt)_2 + HO(CH_2)_nOH \longrightarrow R_2Sn\begin{matrix}O{-}(CH_2)_{n-1}\\ | \\ O{-}CH_2\end{matrix} + 2EtOH$$

The alkoxides undergo insertion of small molecules such as CO_2, SO_2, RNCS, RNCO, etc., into the Sn—OR bonds (Fig. 8.9):

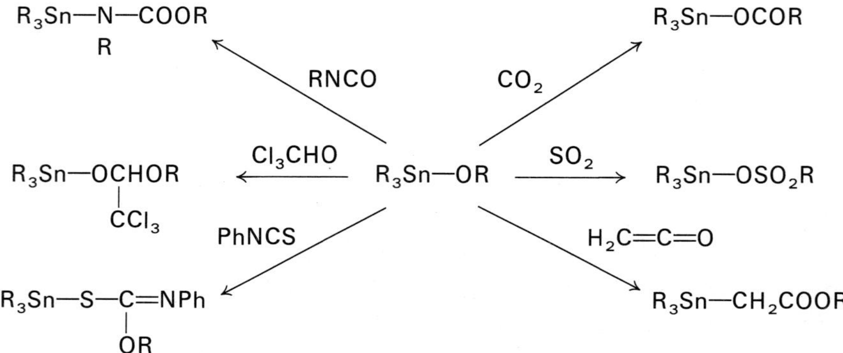

Fig. 8.9. Some insertion reaction of organotin alkoxides.

Alkoxy groups bonded to tin are replaced by acidic reagents, for example acetylacetone:

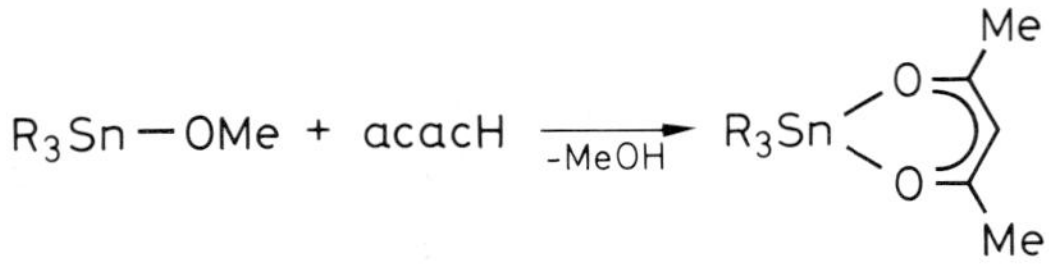

$$R_3Sn-OMe + acacH \xrightarrow{-MeOH} R_3Sn(acac)$$

An increase in the coordination number of tin occurs when another donor site is present in the organotin alkoxides, as in the oxinates (8-hydroxyquinoline derivatives) and β-diketonates:

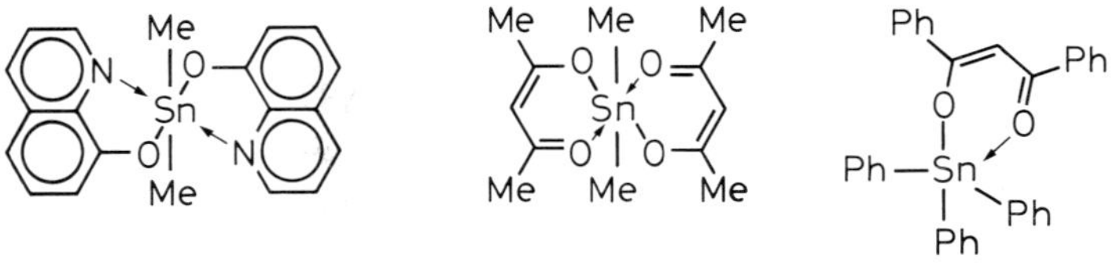

The presence of both alkoxy and acetylacetonato groups leads to dimer formation:

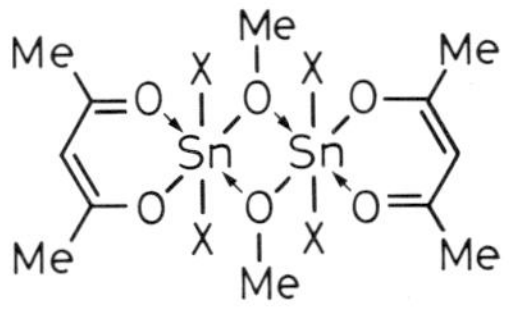

The structure of $Me_3Sn-OMe$ in the solid state contains planar Me_3Sn groups in trigonal-bipyramidal units bridged by O—Sn—O chains:

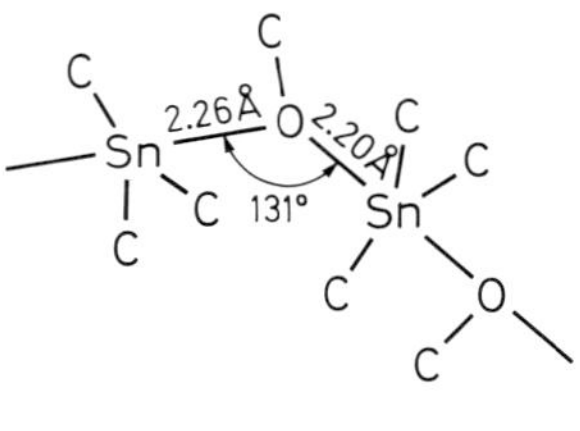

In solution Me_3Sn-OR is a tetrahedral monomer, but the $Me_2Sn(OR)_2$ and $MeSn(OR)_3$ derivatives are associated.

The stannatranes are five-coordinated organotin compounds derived from triethanolamine by reaction with $RSn(OEt)_3$ or $[RSnO(OH)]_x$ and $[(R_2SnO)]_x$. Diethanolamine forms similar compounds:

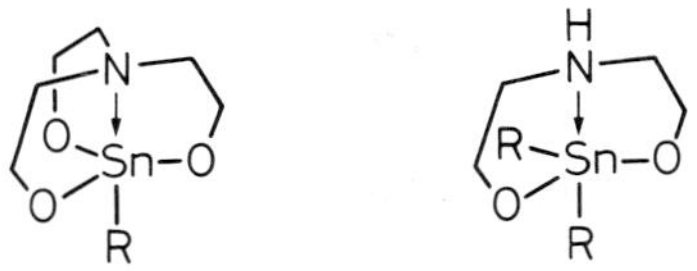

8.3.7. Organotin Amino-Derivatives, $R_nSn(NR'R'')_{4-n}$

Unlike their organosilicon and organogermanium analogues, the organotin halides do not react with ammonia or primary or secondary amines to form substitution products: only addition compounds (adducts) are formed instead. N-Metallated amines are necessary to form compounds containing Sn—N bonds:

$$R_3SnCl + LiNR'_2 \longrightarrow R_3Sn\text{—}NR'_2 + LiCl$$
$$R_3SnCl + R'_2N\text{—}MgX \longrightarrow R_3Sn\text{—}NR'_2 + MgXCl$$

The primary stannylamine, $Bu^t_3Sn\text{—}NH_2$, is prepared by Sn—Ph bond cleavage from $Bu^t_3Sn\text{—}Ph$ and KNH_2 in liquid ammonia.

Heterocycles containing Sn—N bonds can be prepared by transamination reactions in which a more-volatile amine is replaced by a less-volatile one:

$$R_2Sn(NEt_2)_2 + (R'NH)_2\text{(chain)} \longrightarrow R_2Sn(NR')_2\text{(ring)} + 2\,Et_2NH$$

N-Metallated silazanes yield six-membered, Si_2N_3Sn heterocycles by lithioamination:

$$R'SnCl_2 \quad (LiRN-SiR_2)_2NR \xrightarrow[-LiCl]{} R'_2Sn(NR-SiR_2)_2NR$$

Trimethyltin aziridine, $Me_3Sn\text{—}N(CH_2)_2$, is polymeric in the solid state and dimeric in solution:

$$[-SnMe_3-N(CH_2)_2\rightarrow]_n \qquad [Me_3Sn\text{—}N(CH_2)_2]_2$$

The Sn—N bond is sensitive to active-hydrogen reagents such as water, alcohols, acids, etc. (Fig. 8.10).

Small unsaturated molecules insert into Sn—N bonds (Fig. 8.11).

The organotin group can be removed from the product by cleavage of the Sn—C bond, leaving a purely organic derivative.

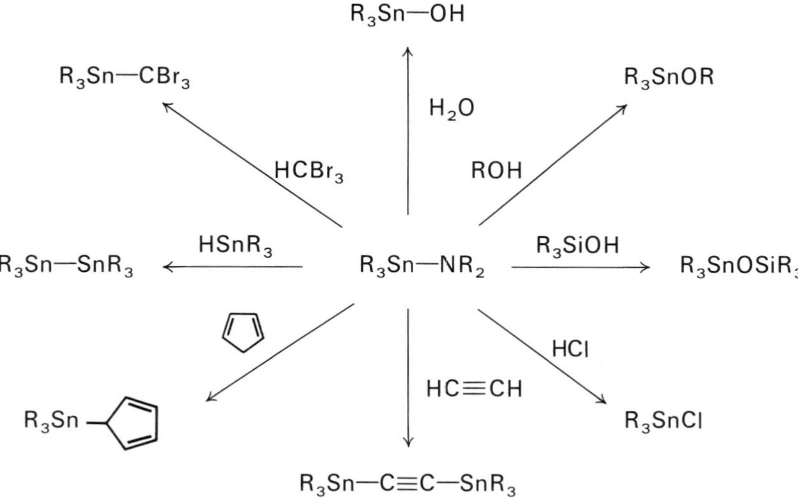

Fig. 8.10. Some reactions of tin-nitrogen compounds.

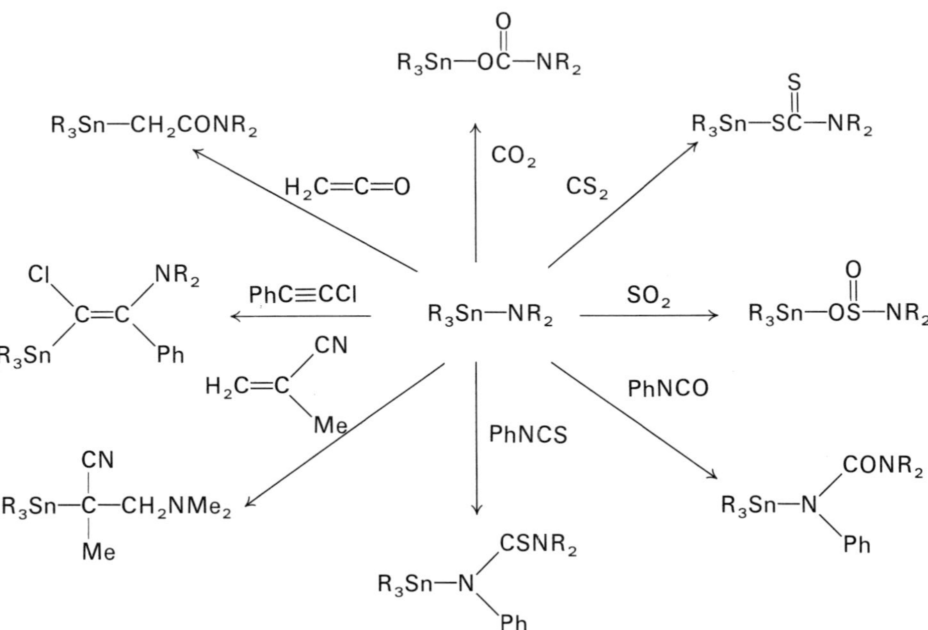

Fig. 8.11. Insertion reactions into Sn-N bonds.

8.3.8. Organotin Thiolates, $R_nSn(SR')_{4-n}$

Mercapto derivatives are prepared by treatment of organotin halides, oxides, alkoxides, hydrides or amines with thiols or alkali metal mercaptides:

$$R_3SnCl + NaSR' \longrightarrow R_3Sn{-}SR' + NaCl$$
$$R_3SnOSnR_3 + 2R'SH \longrightarrow 2R_3Sn{-}SR' + H_2O$$
$$R_3Sn{-}OR + R'SH \longrightarrow R_3Sn{-}SR' + ROH$$
$$R_3SnH + R'SH \longrightarrow R_3Sn{-}SR' + H_2$$

or by transfer of thiolato groups from silicon to tin:

$$R_3Si{-}SR' + R_3SnCl \longrightarrow R_3Sn{-}SR' + R_3SiCl$$

Organotin groups exhibit a great affinity for sulfur. Unlike their silicon and germanium analogues, organotin mercaptides hydrolyze only slowly. Tin can transfer SR groups to other elements by reaction with their halides:

$$ClHgSR \xleftarrow{HgCl_2} R_3SnSR \xrightarrow{BCl_3} B(SR)_3$$
$$R_3SnSR \xrightarrow{PCl_3} P(SR)_3$$

8.3.9. Linear Oligomers, Cyclic Compounds and Polymers

Organopolystannanes

Catenation is more limited for tin since Sn—Sn bonds are weak and reactive, but it is possible to prepare well-defined polystannanes with up to nine tin atoms in a chain.

Distannanes can be readily obtained by several ways:

$$2R_3SnCl + 2Na \longrightarrow R_3Sn{-}SnR_3 + 2NaCl$$
$$R_3SnLi + R_3SnCl \longrightarrow R_3Sn{-}SnR_3 + LiCl$$
$$R_3Sn{-}NR_2 + R_3SnH \longrightarrow R_3Sn{-}SnR_3 + R_2NH$$
$$R_3SnOSnR_3 + 2R_3SnH \longrightarrow 2R_3Sn{-}SnR_3 + H_2O$$
$$R_3SnOR + R_3SnH \longrightarrow R_3Sn{-}SnR_3 + ROH$$

Excess alkali metal gives organotin anions by cleavage of the Sn—Sn bond:

$$Me_3SnBr + 2Na \xrightarrow{liq.\ NH_3} Me_3Sn{-}SnMe_3 \xrightarrow{2Na} 2Na^+Me_3Sn^-$$

Linear polystannanes are obtained by reduction of organotin dihalides with sodium in liquid ammonia in the presence of monohalides:

$$R_2SnBr_2 \xrightarrow[-NaBr]{Na/NH_3\ liq.} NaR_2Sn{-}SnR_2Na \xrightarrow[-NaBr]{2\,R_3SnBr} R_3Sn(SnR_2)_2SnR_3$$

A branched-chain pentastannane has been synthesized:

$$4\,R_3SnLi + SnCl_4 \longrightarrow (R_3Sn)_4Sn \longleftarrow 4\,R_3SnCl + 8\,Li + SnCl_4$$

$$(R_3Sn)_3SnLi + R_3SnCl \longrightarrow (R_3Sn)_4Sn$$

Branched tetrastannanes have been prepared by condensing organotin trihydrides with a trimethylstannylamine:

$$RSnH_3 + 3\,Me_3Sn{-}NEt_2 \xrightarrow{20\,°C} R{-}Sn(SnMe_3)_3 + 3\,HNEt_2$$

A branched polytin chain includes a five-coordinated tin atom in solid $(Ph_3Sn)_3SnNO_3$:

$$Ph_2Sn(NO_3)_2 \cdot 3DMSO + AsPh_3 \longrightarrow Ph_3Sn{-}Sn(SnPh_3)_2(O_2NO) + \text{other prods.}$$

The decomposition of organotin hydrides in the presence of amine or alcohol catalysts produces cyclopolystannanes and hydrogen:

$$R_2SnH_2 \xrightarrow[R = PhCH_2]{DMF} (R_2Sn)_4 \qquad R_2SnH_2 \xrightarrow[R = p\text{-tolyl}]{DMF,\ 70°} (R_2Sn)_6$$

$$R_2SnH_2 \xrightarrow[R=Ph]{DMF} (R_2Sn)_5 \qquad R_2SnH_2 \xrightarrow[R=Ph]{MeOH} H(R_2Sn)_3H \xrightarrow{py} (R_2Sn)_6$$

Cyclopolystannanes result from the reaction between diorganotin amides and hydrides:

$$R_2SnH_2 + R_2Sn(NEt_2)_2 \longrightarrow 2/n(R_2Sn)_n + 2\,Et_2NH$$

The cyclic hexamer, $(Ph_2Sn)_6$, adopts a chair conformation.

Compounds of the composition SnR_2 were considered derivatives of divalent tin, but these compounds are in fact cyclic polymers, $(SnR_2)_n$, or ill-defined species containing SnR_2, SnR_3 or even SnR units, in polymeric networks whose overall composition is SnR_2. Elemental iodine cleaves the Sn—Sn bonds, and R_3SnI, R_2SnI_2 and $RSnI_3$ are identified. True organic derivatives of divalent tin have, however, now been prepared (see Section 8.3.10).

Organopolystannoxanes (Organotin Oxides)

Organotin oxides are different from organopolysiloxanes, both structurally and chemically.

Distannoxanes, $R_3Sn—O—SnR_3$, are formed in the hydrolysis of triorganotin halides by the condensation of the hydroxide intermediates. The structure of the distannoxanes is associated in the solid state:

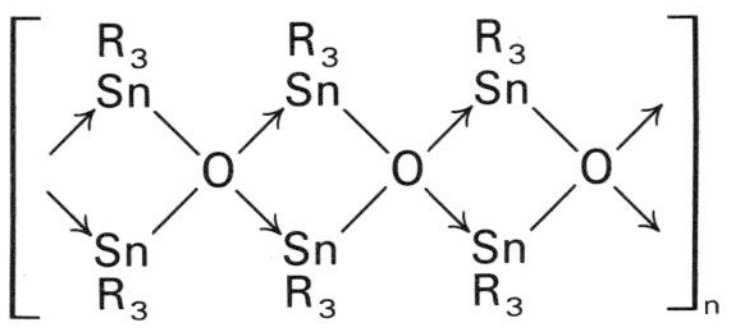

$$\left[\begin{array}{ccc} R_3Sn & R_3Sn & R_3Sn \\ O & O & O \\ SnR_3 & SnR_3 & SnR_3 \end{array} \right]_n$$

The hexaphenyl derivative, $Ph_3Sn—O—SnPh_3$, is monomeric because of the bulky phenyl groups, with a Sn—O—Sn angle of 137.8°.

A series of compounds of intermediate composition between R_2SnX_2 and R_2SnO are isolated from the hydrolysis of diorganotin dihalides and pseudohalides:

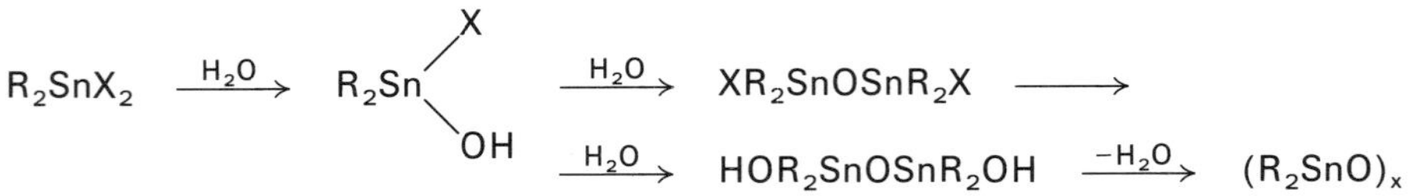

$$R_2SnX_2 \xrightarrow{H_2O} R_2Sn(X)(OH) \begin{array}{l} \xrightarrow{H_2O} XR_2SnOSnR_2X \longrightarrow \\ \xrightarrow{H_2O} HOR_2SnOSnR_2OH \xrightarrow{-H_2O} (R_2SnO)_x \end{array}$$

The distannoxane dihalides, $XR_2SnOSnR_2X$, are also obtained by heating a mixture of the oxide and halide:

$$1/x(R_2SnO)_x + R'_{4-n}SnX_n \longrightarrow X_{n-1}R'_{4-n}SnOSnR_2X\,.$$

These compounds are associated. The dihalides, $XR_2SnOSnR_2X$, are dimeric in solution, and $(Me_3SiO—SnMe_2)_2O$ and $(NCS—SnMe_2)_2O$ contain four-membered, Sn_2O_2 rings in the solid state with five-coordinated tin atoms:

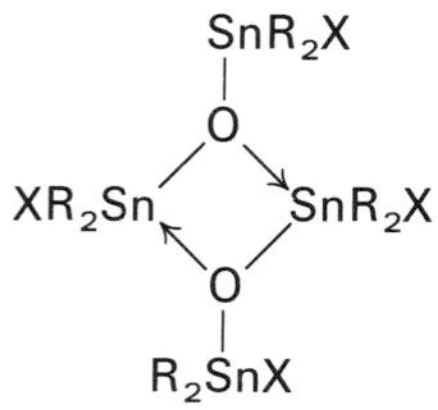

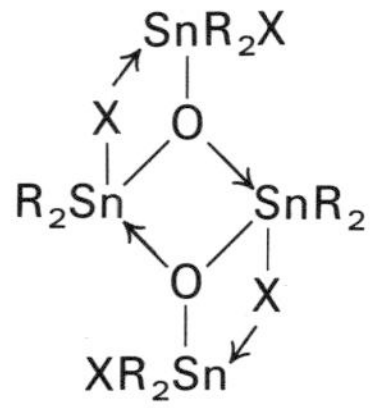

$$\begin{array}{c} SnR_2X \\ | \\ O \\ XR_2Sn \quad SnR_2X \\ O \\ | \\ R_2SnX \end{array} \qquad \begin{array}{c} SnR_2X \\ X \quad O \\ R_2Sn \quad SnR_2 \\ O \quad X \\ XR_2Sn \end{array}$$

In the oxides, $(R_2SnO)_x$, all the tin atoms are five-coordinated in polycyclic, macromolecular structures:

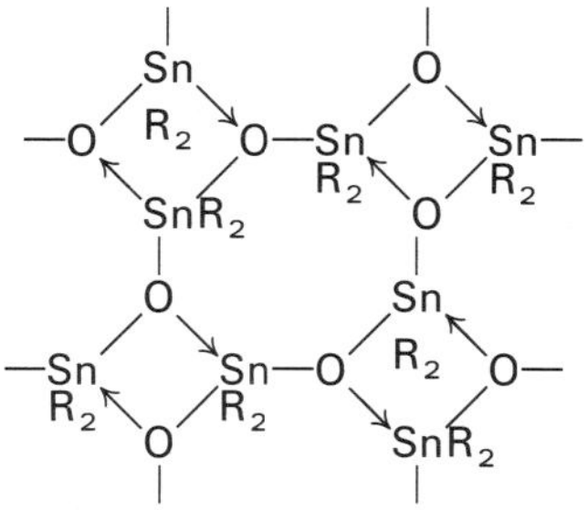

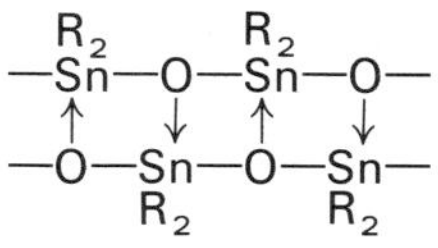

The *tert*-butyl derivative, $(Bu^t_2SnO)_3$, contains a planar Sn_3O_3 ring:

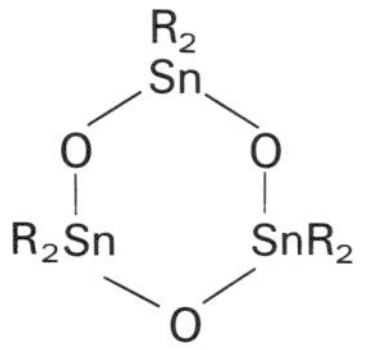

The hydrolysis products of the organotin trihalides have the composition $(R_2Sn_2O_3)_n$ and $(RSnOOH)_n$, of polymeric structure.

Organostannazanes

Distannazanes with Sn—N—Sn groups are prepared by the reaction of trimethyltin chloride and N-lithiated methylamine:

$$2\,Me_3SnCl \;+\; 2\,LiNHMe \xrightarrow[-LiCl]{} 2\,Me_3Sn{-}NHMe \xrightarrow[-MeNH_2]{} Me_3Sn{-}NMe{-}SnMe_3$$

or by transamination of aminostannanes with primary amines:

$$2\,R_3Sn{-}NMe_2 \;+\; EtNH_2 \longrightarrow Me_3Sn{-}NEt{-}SnMe_3 \;+\; 2\,Me_2NH$$

Tristannylamines are prepared by the reaction of trialkyltin chlorides and lithium nitride:

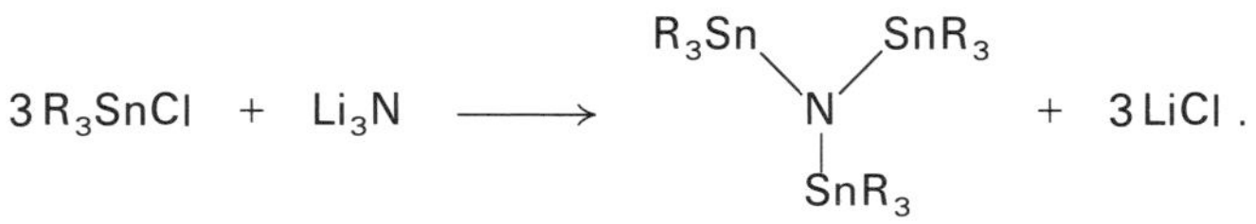

$$3\,R_3SnCl \;+\; Li_3N \longrightarrow N(SnR_3)_3 \;+\; 3\,LiCl\,.$$

Diorganotin amines form six-membered cyclostannazanes in transamination reactions:

$$3\,R_2Sn(NEt_2)_2 + 3\,EtNH_2 \longrightarrow (R_2SnNEt)_3 + 6\,Et_2NH\,.$$

R=Me

(six-membered ring: R_2Sn, EtN, SnR_2, N(Et), R_2Sn, NEt)

or by the reaction of di-*tert*-butyltin dichloride with potassium amide in liquid ammonia. Four-membered rings are formed in a sequence involving lithiation of the diamminostannanes:

$$R_2SnCl_2 \xrightarrow{RNHLi} R_2Sn(NHR)_2 \xrightarrow{BuLi} R_2Sn(NLiR)_2$$

$$R_2SnCl_2 \xrightarrow{KNH_2/liq.\ NH_3} (R_2SnNH)_3$$

$$R_2Sn(NLiR)_2 \xrightarrow{Me_2SnCl_2} R_2Sn(NR)_2SnR_2$$

Bulky di-*tert*-butyltin amides yield cyclodistannazanes in transamination reactions:

$$Bu^t_2Sn(NMe_2)_2 + RNH_2 \longrightarrow Bu^t_2Sn(NHR)_2 \xrightarrow{100-130\,°C} Bu^t_2Sn(NR)_2SnBu^t_2$$

The stannazanes are more sensitive to moisture than the corresponding silazanes.

Organotin Sulfides, Selenides and Tellurides

The Sn—S bond is more stable to hydrolysis than the Si—S and Ge—S bonds, and the organotin sulfides can be synthesized even in water. The triorganotin sulfides are prepared from the corresponding chlorides and sodium or silver sulfide, or by treatment of the oxides or hydroxides with hydrogen sulfide:

$$2\,R_3SnCl + Na_2S \longrightarrow R_3Sn{-}S{-}SnR_3 + 2\,NaCl$$

$$2\,R_3SnOSnR_3 + H_2S \longrightarrow R_3Sn{-}S{-}SnR_3 + H_2O$$

$$2\,R_3SnOH + H_2S \longrightarrow R_3Sn{-}S{-}SnR_3 + 2\,H_2O\,.$$

In solid $Ph_3Sn{-}S{-}SnPh_3$ the bond angle is nearly tetrahedral (107.4°).

Diorganotin halides react with sodium sulfide to form cyclic organotin sulfides, and diorganotin oxides can be converted to cyclostannathianes:

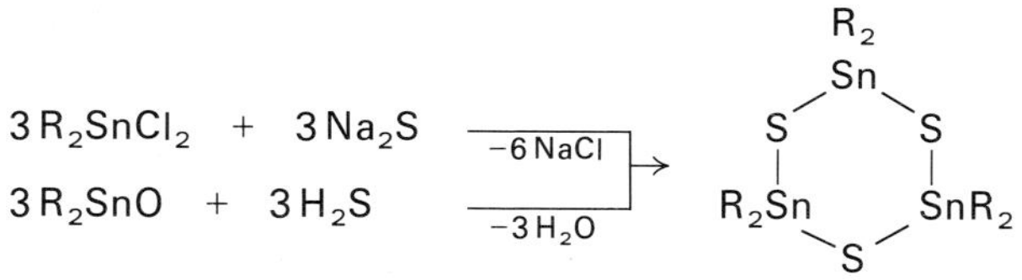

$$3\,R_2SnCl_2 + 3\,Na_2S \xrightarrow{-6\,NaCl} (R_2SnS)_3$$

$$3\,R_2SnO + 3\,H_2S \xrightarrow{-3\,H_2O} (R_2SnS)_3$$

With a deficit of sodium sulfide a linear oligomer is formed:

$$2\,R_2SnCl_2 + Na_2S \longrightarrow ClR_2SnSSnR_2Cl + 2\,NaCl\,.$$

With bulky *tert*-Bu groups, the reaction of diorganotin dichloride with sodium sulfide, selenide or telluride yields a four-membered, cyclic dimer:

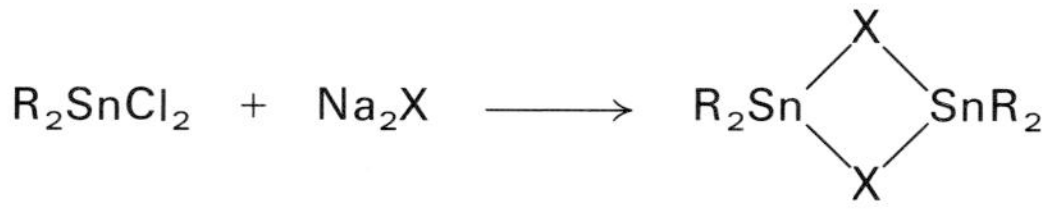

$$R_2SnCl_2 + Na_2X \longrightarrow R_2Sn(\mu\text{-}X)_2SnR_2$$

$R = Bu^t$; $X = S, Se, Te$

Dimethyltin dihydride reacts with elemental sulfur, selenium or tellurium to give the cyclic trimers, $(Me_2SnX)_3$ ($X = S, Se, Te$).

Tin-rich sulfur- and selenium-containing heterocycles have also been prepared:

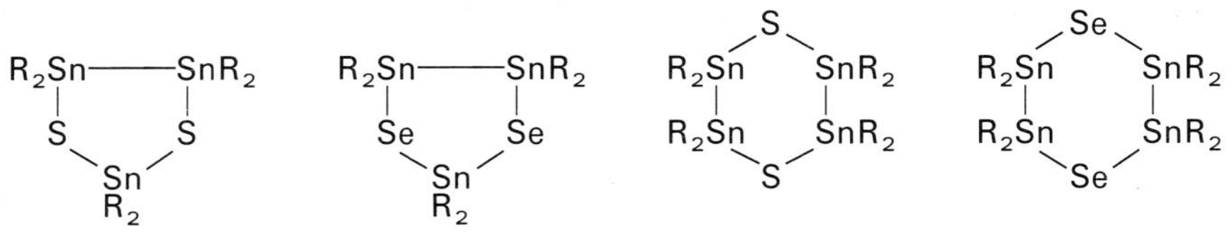

R₂Sn—SnR₂–S–SnR₂–S– (five-membered ring); R₂Sn—SnR₂–Se–SnR₂–Se– (five-membered ring); R₂Sn–SnR₂–S–SnR₂–SnR₂–S– (six-membered ring); R₂Sn–SnR₂–Se–SnR₂–SnR₂–Se– (six-membered ring)

The six-membered Sn_3S_3 ring in the trimer, $(Me_2SnS)_3$, adopts a nonplanar, twisted-boat conformation.

The tendency for cyclization in tin-sulfur chemistry is also manifest in the reaction of organotin trihalides with sodium sulfide, which gives tricyclic tetramers, $(RSnS_{1.5})_4$, with adamantane-like structures.

8.3.10. Stannylenes, $:SnR_2$, Free Radicals, $\cdot SnR_3$, and Related Compounds

There are few genuine organotin(II) compounds, of which the dicyclopentadienyl, $:Sn(C_5H_5\text{-}\eta^5)_2$, and bis[bis(trimethylsilyl)methyl], $:Sn[CH(SiMe_3)_2]_2$, derivatives are stable. Dialkylstannylenes are also formed as transient intermediates and have been trapped with suitable reagents. Transition metal complexes of stannylenes can also be generated, either directly from stable stannylenes, or by indirect routes, but tin can be better described as tetravalent in these complexes.

Dicyclopentadienyltin(II), $:Sn(C_5H_5\text{-}\eta^5)_2$, stannocene, is prepared from tin(II)chloride and sodium cyclopentadienide:

$$SnCl_2 + 2\,Na^+C_5H_5^- \longrightarrow :Sn(C_5H_5\text{-}\eta^5)_2 + 2\,NaCl$$

In the vapor or solid $:Sn(C_5H_5\text{-}\eta^5)_2$ is a pentahapto, π-complex with an angular sandwich structure; the mono-η^5-cyclopentadienyltin chloride is associated through weak chlorine bridges in the solid state. Pentamethyl-η^5-cyclopentadienyltin(II) cation, $(\eta^5\text{-}Me_5C_5)Sn:^+$, also has been obtained from $:Sn(C_5Me_5\text{-}\eta^5)_2$ and strong acids:

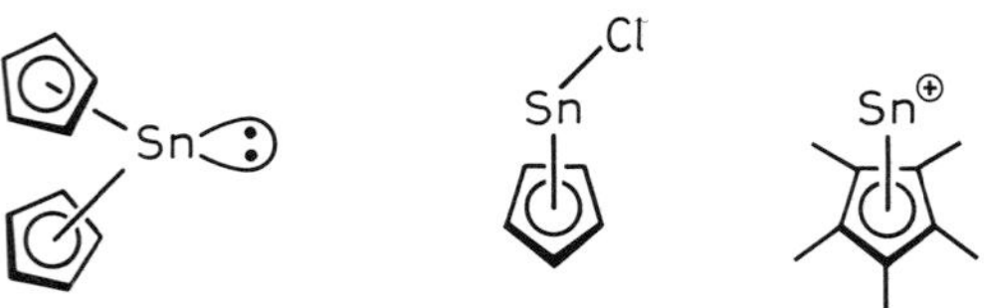

Di-η^5-cyclopentadienyltin(II) readily undergoes oxidative addition with halogens and some alkyl halides and forms transition metal complexes with metal carbonyls; it is oxidized by $SnCl_4$ to $(\eta^1\text{-}C_5H_5)_2SnCl_2$:

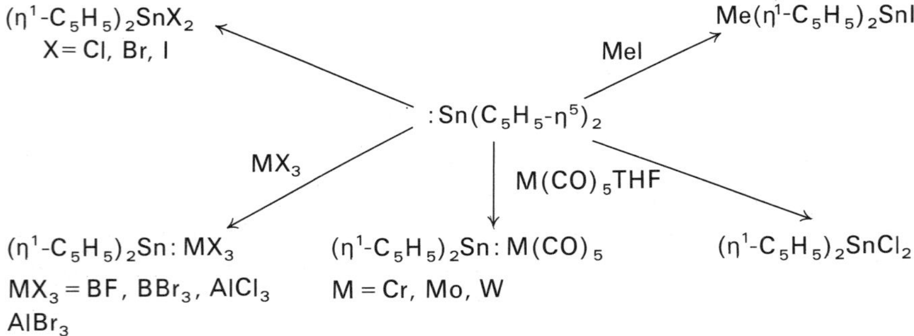

Bis[bis(trimethylsilyl)methyl]tin(II), $:Sn[CH(SiMe_3)_2]_2$

This orange material is obtained from tin(II) chloride and bis(trimethylsilyl)-methyllithium, or from tin(II) bis(trimethylsilyl)amide:

$$SnCl_2 + LiCH(SiMe_3)_2 \xrightarrow{0^\circ C} :Sn[CH(SiMe_3)_2]_2$$

$$Sn[N(SiMe_3)_2]_2 + LiCH(SiMe_3)_2 \longrightarrow :Sn[CH(SiMe_3)_2]_2$$

The compound is monomeric in solution but dimeric in the solid state, with an uncommon Sn—Sn bent double bond and an Sn—Sn distance of 276 pm (≡2.76 Å):

monomer dimer

The bulky $CH(SiMe_3)_2$ groups prevent the polymerization characteristic of stannylenes.

The reactions of $:Sn[CH(SiMe_3)_2]_2$ are typical of a diorganotin(II) compound: oxidative addition and complex formation with metal carbonyls to give, for example, the chromium complex $[(Me_3Si)_2CH]_2Sn:Cr(CO)_5$:

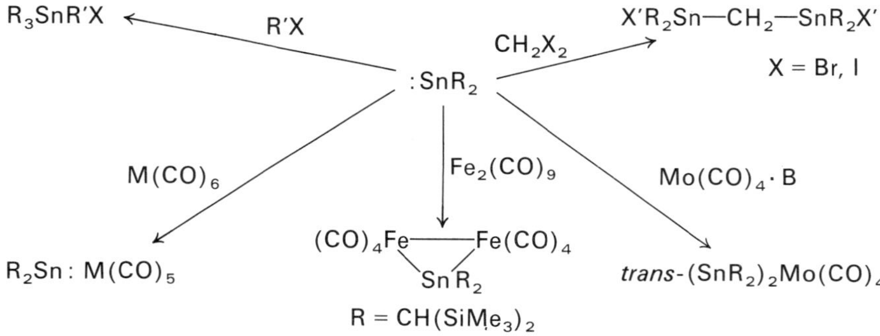

Dialkylstannylenes, $:SnR_2$

These are formed in several reactions as transient intermediates, and can be trapped by alkyl halides (oxidative addition). In the absence of a trapping agent, stannylenes polymerize to cyclopolystannanes. Stannylenes are formed by photolysis or pyrolysis of polystannanes, by thermolysis of tetraalkyldistannanes, dihalodistannanes, or pentaalkyldistannanes, or by decomposition of stannylene-transition metal complexes:

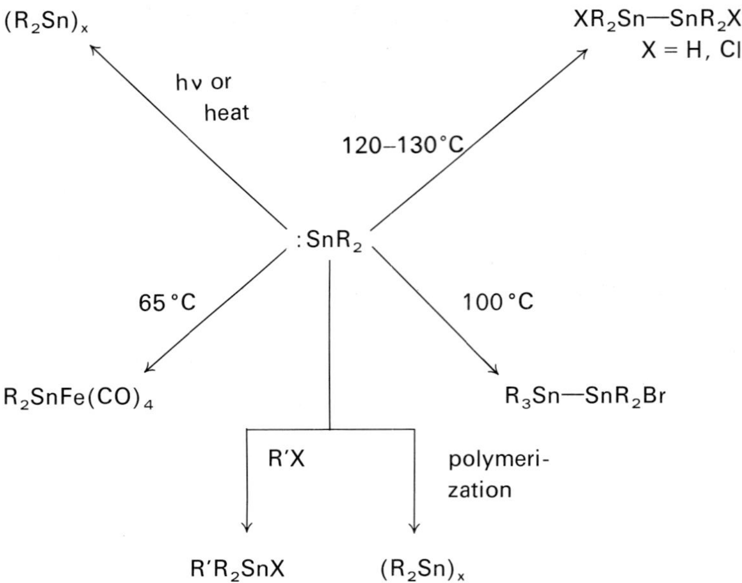

Dimethylstannylene forms in the reaction of 1,1-dimethyl-2,3,4,5-tetraphenyl-1-stannacyclopentadiene with acetylenes:

$$R_4C_4SnMe_2 + R'{-}C{\equiv}C{-}R'' \longrightarrow [:SnMe_2] + C_6R_4R'R''$$

$$[:SnMe_2] \xrightarrow{BuBr} BuMe_2SnBr$$

R = Ph

R' = R'' = COOMe; R' = Ph, R'' = H

Base-stabilized stannylene complexes can be obtained from diorganotin dihalides reacting with metal carbonyl anions in basic solvents:

$$R_2SnCl_2 + Na_2Cr_2(CO)_{10} \longrightarrow R_2Sn{:}Cr(CO)_5 \cdot THF + NaCl + Na[Cr(CO)_5Cl]$$

$$R_2SnCl_2 + Na_2Fe(CO)_4 \longrightarrow [R_2SnFe(CO)_4]_2 \xrightleftharpoons{2B} B{\rightarrow}SnR_2{:}Fe(CO)_4$$

B = THF, Py, MeCN, Et_2O, DMF

The structure of py · $Bu^t_2Sn:Cr(CO)_5$ involves coordination of the base to give four-coordinated tin:

$$C_5H_5N \rightarrow SnBu^t_2{:}\rightarrow Cr(CO)_5$$

The bonding is represented by I—III, but the ylid structure III is preferred:

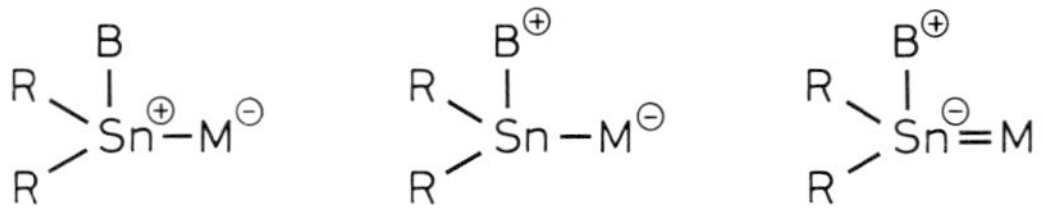

$R_2(B)Sn^{\oplus}{-}M^{\ominus}$ (I) $\quad R_2(B^{\oplus})Sn{-}M^{\ominus}$ (II) $\quad R_2(B^{\oplus})Sn^{\ominus}{=}M$ (III)

Tetrasubstituted organotin(II) derivatives are rare, but the tetrakis(η^5-cyclopentadienyl)tin(II) anion can be prepared from stannocene:

$$:Sn(C_5H_5\text{-}\eta^5)_2 + Sb(CH_3)_5 \longrightarrow [Sb(CH_3)_4]_2^+[Sn^{II}(C_5H_5\text{-}\eta^1)_4]^{2-}$$

A related anion, $[Me_2SnCl_2]^{2-}$, formed as a product of Me_2SnCl_2 with $B_{10}H_{10}^{2-}$, is isoelectronic with Me_2TeCl_2 and has the analogous structure:

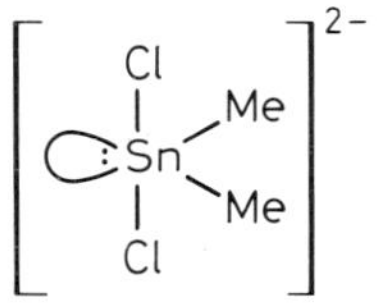

$$[(:)Sn(Cl)_2(Me)_2]^{2-}$$

Trisubstituted Organotin Free Radicals, $\cdot SnR_3$

The pyramidal $\cdot SnR_3$ can be prepared by irradiation of $:Sn[CH(SiMe_3)_2]_2$. These free radicals are unusually stable and can be isolated:

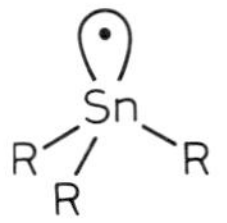

$$(\cdot)SnR_3$$

Hexaalkyldistannanes, $R_3Sn{—}SnR_3$, with bulky substituents dissociate reversibly at 180 °C (for R = 2,4,6-trimethylphenyl) or at 100 °C (for R = 2,4,6-triethylphenyl):

$$R_3Sn{—}SnR_3 \xrightleftharpoons{\Delta} 2\,R_3Sn\cdot$$

8.4. Organolead Compounds

Because of the lower metal-carbon bond strength, organolead derivatives decompose at moderate temperatures (100 to 200 °C), are slowly oxidized in air and are somewhat light sensitive. The Pb—C bond is stable to moisture, although those containing Pb—C_6F_5 groups are more readily hydrolyzed.

A peculiarity of organolead chemistry is the relative rarity of monoorganosubstituted derivatives. Apart from the carboxylates, $RPb(OCOR')_3$, iodides, $RPbI_3$, and the ill-defined and probably polymeric organoplumbonic acids, $RPbO_2H$, no other monoalkyl- or -aryllead derivatives are known.

The tetraalkylleads dominate organolead chemistry. The antiknock compounds $Pb(CH_3)_4$ and $Pb(C_2H_5)_4$ are manufactured in large, but decreasing quantities.

Lead forms the same types of compounds as other Group IVA elements: the tetra-substituted derivatives PbR_4 and R_nPbX_{4-n} and R_nPbX_{4-n} where X = halogen, H, OH, OR, NRR′, SR, etc., addition compounds with bases in which the lead atom has a coordination number greater than four and oligomers and polymers with Pb—Pb, Pb—O—Pb, Pb—S—Pb or Pb—N—Pb backbones.

The nomenclature of organolead compounds is similar to that of their tin analogues except that the salt-like nature of the compounds is reflected more frequently in names like diphenyllead dichloride, Ph_2PbCl_2, bis-(triphenyllead)sulfide, $Ph_3Pb{-}S{-}PbPh_3$, etc.

8.4.1. Tetrasubstituted Derivatives, PbR_4

Preparation. Because of the industrial interest in tetraethyllead, the synthesis of the tetrasubstituted derivatives has been investigated intensively.

Grignard and organolithium reagents are of most utility for laboratory purposes. Since lead(IV) chloride is unstable, lead(II) chloride is used as starting material. Organolead(II) derivatives, PbR_2, are intermediates which disproportionate to either PbR_4 or $R_3Pb{-}PbR_3$ as final products. The overall reactions are:

$$2\,PbCl_2 + 4\,RMgX \longrightarrow PbR_4 + Pb + 4\,MgClX$$
$$3\,PbCl_2 + 6\,RMgX \longrightarrow Pb_2R_6 + Pb + 3\,MgX_2 + 3\,MgCl_2$$
$$2\,PbCl_2 + 4\,LiR \longrightarrow PbR_4 + Pb + 4\,LiCl$$

The deposition of lead metal is avoided if an alkyl iodide is added. The organic group R must be the same as that in the Grignard or organolithium reagent (M = MgX or Li):

$$2\,PbCl_2 + 6\,MR + 2\,RI \longrightarrow 2\,PbR_4 + 2\,MI + 4\,MCl$$

The mechanism involves $R_3PbMgCl$ as an intermediate:

$$PbCl_2 + 3\,RMgCl \xrightarrow{THF} R_3PbMgCl + 2\,MgCl_2$$
$$R_3PbMgCl \xrightarrow{R'X} R_3PbR' \qquad R_3PbMgCl \xrightarrow{CH_2Cl_2} R_3Pb{-}CH_2{-}PbR_3$$

The use of the lead(IV) salts, $Pb(OCOCH_3)_4$ or $K_2[PbCl_6]$, offers no advantage over $PbCl_2$, however, these are used in the synthesis of acetylene derivatives:

$$K_2[PbCl_6] + 4\,LiC{\equiv}CR \longrightarrow Pb(C{\equiv}CR)_4 + 2\,KCl + 4\,LiCl\,,$$

Lead(IV) acetate is employed in the synthesis of tetraethyl- and tetramethyllead:

$$Pb(OAc)_4 + 4\,RMgCl \longrightarrow PbR_4 + 4\,MgClOAc$$

The large-scale availability of triethylaluminum stimulated its use in the synthesis of tetraethyllead. The reaction proceeds with deposition of lead metal, but in high yield when lead(II) acetate is used as starting material:

$$6\,PbX_2 + 4\,AlEt_3 \longrightarrow 3\,PbEt_4 + 3\,Pb + 4\,AlX_3\,.$$

With tributylaluminum, hexabutyldilead is obtained, and no tetrabutyllead is formed:

$$Pb(OAc)_2 + 6\,AlBu_3 \longrightarrow Bu_3Pb{-}PbBu_3 + Pb + 6\,Bu_2AlOAc$$

Addition of an alkyl iodide prevents the deposition of lead metal, and ensures a rich yield of tetraethyllead:

$$PbCl_2 + 4\,AlEt_3 + EtI \longrightarrow PbEt_4 + AlCl_2I$$

Trialkylboranes alkylate lead(II) oxide or hydroxide to give tetraalkyllead derivatives.

The direct synthesis with lead metal and an alkyl halide, which gives the organometal halides with silicon, germanium and tin, produces tetraalkyl-substituted derivatives of lead:

$$3\,Pb + 4\,RX \longrightarrow PbR_4 + 2\,PbX_2\,.$$

The addition of sodium tetraethylaluminate converts all the lead to organic products:

$$Pb + 2\,EtCl + 2\,Na[AlEt_4] \longrightarrow PbEt_4 + 2\,AlEt_3 + 2\,NaCl\,.$$

Lead-sodium alloy, first employed in 1853, reacts more efficiently with alkyl halides:

$$4\,PbNa + 4\,RCl \longrightarrow PbR_4 + 3\,Pb + 4\,NaCl$$

Ethyl iodide in alcoholic alkalies is electrolyzed with lead cathodes to prepare tetraethyllead. The complex $Na[Et_3Al{-}F{-}AlEt_3]$, melting at 35 °C, is used as an electrolyte with lead anodes:

$$4\,Na[Et_3Al{-}F{-}AlEt_3] + 3\,Pb \xrightarrow{e^-} 3\,PbEt_4 + 4\,NaEt_3AlF$$

The NALCO process (NALCO Chemical Corporation) produces tetraethyl- and tetramethyllead by the electrolysis of a Grignard reagent and an alkyl halide with lead anodes:

$$2\,RMgX + 2\,RX + Pb \longrightarrow PbR_4 + 2\,MgX_2\,.$$

Tetraphenyllead is obtained in low yield from phenyldiazonium tetrafluoroborate, $[Ph{-}N{\equiv}N]^+[BF_4]^-$, and metallic lead. The addition of organolead hydrides, R_3PbH, to olefins and Ziegler-type syntheses with lead, hydrogen and ethylene gives only small yields.

Properties. The properties of tetraorganolead derivatives reflect the weak Pb—C-bond, and Pb—C bond-cleavage reactions are used in the synthesis of organolead halides. Tetraorganolead compounds can also transfer organic groups to metals and nonmetals by distribution and act as alkylating agents. Small molecules like sulfur dioxide and trioxide insert into the Pb—C bond to form organolead sulfinates and sulfonates:

$$R_2Pb\text{—}OS(O)R \xleftarrow[\text{ether}]{SO_2,\ 0\,°C} PbR_4 \xrightarrow[\text{R.T.}]{SO_2\ \text{exces}} R_2Pb(OSOR)_2$$

$$PbR_4 \xrightarrow{SO_3} R_3Pb\text{—}OSO_2R \xrightarrow{SO_3} R_2Pb(OSO_2R)_2 \,.$$

The lead tetraalkyls decompose at > 100 °C, and only tetraethyllead can be distilled under atmospheric pressure.

Typical reactions of tetraorganolead compounds are shown in Fig. 8.12:

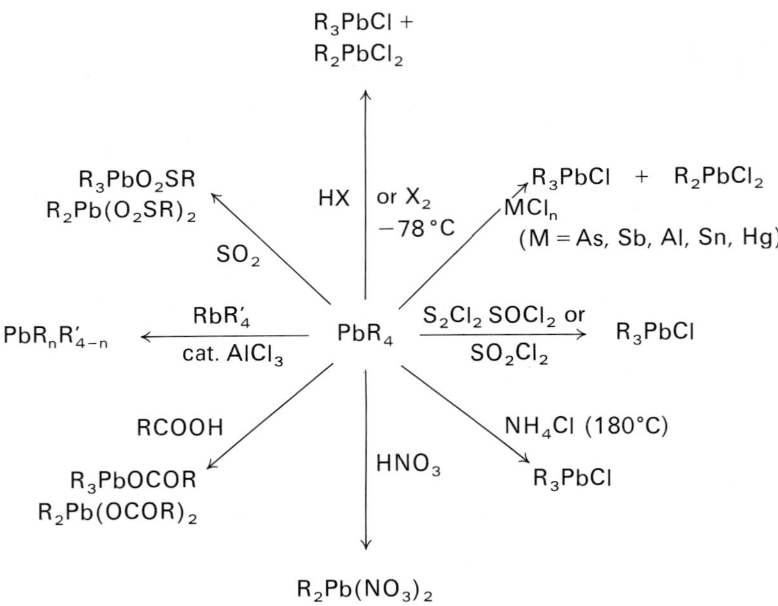

Fig. 8.12. Some typical reactions of PbR_4 derivatives.

8.4.2. Heterocycles with Lead Heteroatoms

Saturated heterocycles are prepared by Grignard reagents:

Pb R_2 Pb R_2

and organolithium reagents are used for the synthesis of tricyclic aromatic compounds:

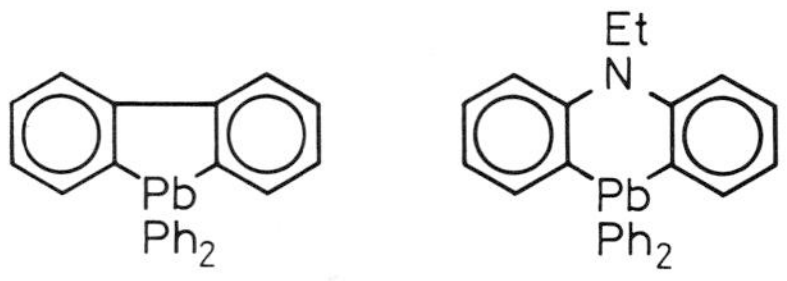

Spirocyclic compounds have been prepared with the aid of Grignard reagents:

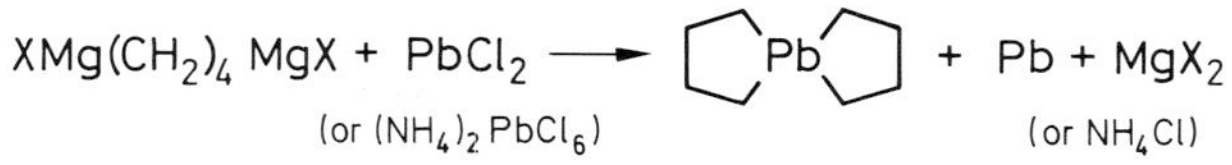

$$XMg(CH_2)_4MgX + PbCl_2 \text{ (or } (NH_4)_2PbCl_6) \longrightarrow \text{spiro-}Pb + Pb + MgX_2 \text{ (or } NH_4Cl)$$

8.4.3. Organolead Halides, R_nPbX_{4-n}

Derivatives with n = 3 and 2 are prepared from tetrasubstituted compounds by Pb—C bond cleavage. Cleavage with halogens or with the halides of other elements can be stopped at the R_3PbX or R_2PbX_2 stages. Dilead derivatives, $R_3Pb—PbR_3$, can be cleaved to give triorganolead halides, R_3PbX.

Monoalkyllead triiodides $RPbI_3$ are reaction products of lead(II) iodide with RI in the presence of an $SbMe_3$ catalyst.

Solid organolead halides contain lead in higher than four coordination, as in diphenyllead dichloride which is a chlorine double-bridged polymer:

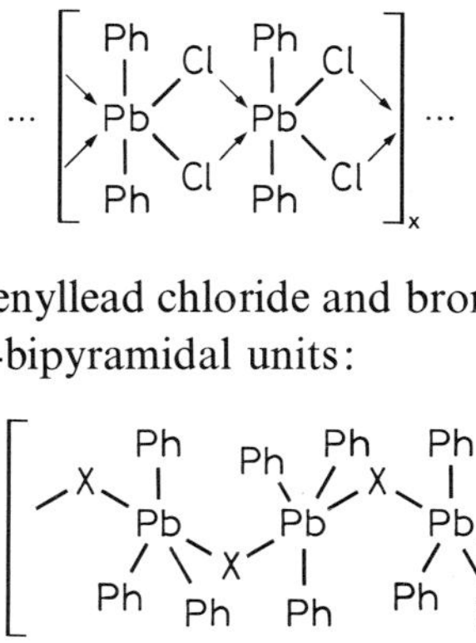

Triphenyllead chloride and bromide consist of halogen-bridged chains made up of trigonal-bipyramidal units:

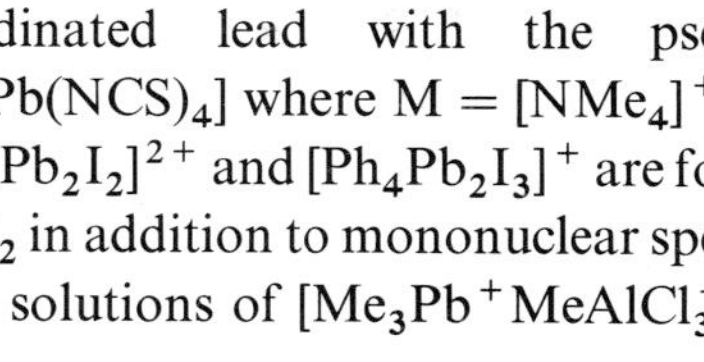

Anionic species containing five-coordinated lead atoms of the type $[R_3PbX_2]^-$ and $[R_2PbX_3]^-$, such as $[NMe_4]^+[Ph_3PbCl_2]^-$ and $Cs^+[Ph_2Pb(OAc)_3]^-$, and five- and six-coordinated lead with the pseudohalides like $M[Ph_3Pb(NCS)_2]$ and $M_2[Ph_2Pb(NCS)_4]$ where M = $[NMe_4]^+$ and $[AsPh_4]^+$ are known. The cationic species $[Ph_4Pb_2I_2]^{2+}$ and $[Ph_4Pb_2I_3]^+$ are formed in methanolic solutions of Ph_3PbX and Ph_2PbX_2 in addition to mononuclear species. The Me_3Pb^+ cations form in methylene chloride solutions of $[Me_3Pb^+MeAlCl_3^-]$ obtained from Me_3PbCl and $MeAlCl_2$.

8.4.4. Organolead Carboxylates, $R_nPb(OCOR')_{4-n}$

The only well-defined monoorganolead compounds belong to this class. Along with $RPb(OCOR')_3$, di- and triorgano-substituted lead derivatives, $R_2Pb(OCOR')_2$ and $R_3Pb(OCOR')$, are obtained by the cleavage of tetrasubstituted compounds with carboxylic acids for which $pK_a < 7$:

$$PbR_4 + R'COOH \longrightarrow R_3PbOCOR' + RH$$

These reactions are less vigorous than with halogens and, therefore, easier controlled. Organolead carboxylates are often preferred over the halides as starting materials owing to their ready availability through this synthesis.

Organolead dicarboxylates and mercury(II) carboxylates react to produce monoorganolead derivatives:

$$R_2Pb(OCOR')_2 + Hg(OCOR')_2 \longrightarrow RPb(OCOR')_3 + RHgOCOR'$$

The tetracarboxylates can also be used with diorganomercury derivatives:

$$Pb(OCOR')_4 + HgR_2 \longrightarrow RPb(OCOR')_3 + RHgOCOR'$$

Solid $Me_3PbOCOCH_3$ consists of associated chains of planar, trigonal-bipyramidal Me_3Pb groups bridged by acetato fragments.

8.5.4. Organolead Hydrides, R_nPbH_{4-n}

The hydrides are the least stable of those of the Group IVA elements. Triorganolead hydrides are obtained at low temperatures by reduction of the halides with $LiAlH_4$, but they decompose at 0°C:

$$4\,R_3PbH \longrightarrow 3\,PbR_4 + Pb + 2\,H_2$$

The hydrides reduce organic halides and add to olefins, acetylenes and isocyanates (hydroplumbation).

8.4.6. Organolead Hydroxides, $R_nPb(OH)_{4-n}$

The hydroxides are obtained by the hydrolysis of the corresponding halide in alcoholic alkali solutions or by wet silver oxide. These are ionic compounds which form weakly basic aqueous solutions and react with organic and inorganic acids to form the corresponding salts.

Hydroxylated dimethyllead species form in aqueous sodium perchlorate solutions:

$$\mathrm{Me_2Pb^{2+}} \underset{\mathrm{H+}}{\overset{\mathrm{OH-}}{\rightleftarrows}} \left[\mathrm{Me_2Pb}\overset{\mathrm{OH}}{\underset{\mathrm{HO}}{\rightleftarrows}}\mathrm{PbMe_2}\right]^{2+} \underset{\mathrm{H+}}{\overset{\mathrm{OH-}}{\rightleftarrows}} \mathrm{Me_2Pb(OH)_2}$$

pH <5   pH 5–8   pH 8–10

$$\underset{\mathrm{H+}}{\overset{\mathrm{OH-}}{\rightleftarrows}} \mathrm{Me_2Pb(OH)_3^-}$$

pH > 10

Solid triphenyllead hydroxide consists of associated zig-zag chains similar to those in the tin analogue.

8.4.7. Organolead Alkoxides, $R_nPb(OR')_{4-n}$

Triorganolead oxides react with alcohols to form alkoxides:

$$\mathrm{R_3Pb{-}O{-}PbR_3 + R'OH \longrightarrow R_3Pb{-}OR' + R_3Pb{-}OH}\ .$$

but a better procedure starts with the halides:

$$\mathrm{Me_3PbCl + NaOMe \longrightarrow Me_3PbOMe + NaCl}$$
$$\mathrm{Me_3PbBr + NaOSiMe_3 \longrightarrow Me_3Pb{-}O{-}SiMe_3 + NaBr}$$

The alkoxides undergo substitution and insertion reactions:

$$\mathrm{R_3Pb{-}S{-}\underset{\underset{S}{\|}}{C}{-}OR \xleftarrow{CS_2} R_3Pb{-}OR \xrightarrow{CO_2} R_3Pb{-}O{-}\underset{\underset{O}{\|}}{C}{-}OR}$$

$$\mathrm{R_3Pb{-}OR \xrightarrow{H_2C{=}C{=}O} R_3Pb{-}CH_2COOR}$$

$$\mathrm{R_3Pb{-}OR \xrightarrow{HC{\equiv}CPh} R_3Pb{-}C{\equiv}CPh}$$

8.4.8. Organolead Amides, $R_nPb(NR'R'')_{4-n}$

A rare organolead primary amide, $Bu^t_3Pb{-}NH_2$, is prepared from Bu^t_3PbI and KNH_2 in liquid ammonia. The amide hydrolyzes to give $Bu^t_3Pb{-}OH$ and the oxide $Bu^t_3Pb{-}O{-}PbBu^t_3$.

Derivatives of the type $R_3Pb{-}NR'_2$ result from lithioamination of triorganolead chlorides with N-lithiodiethylamine. The amides decompose at room temperature and are sensitive to moisture.

8.4.9. Organolead Mercaptides, $R_nPb(SR')_{4-n}$

The lead-sulfur bond is stable towards water. Mercapto derivatives are prepared from organolead chlorides, hydroxides or thiolates (R_3PbSNa is prepared from R_3PbCl and sodium sulfide):

$$R_3PbSNa + R'I \longrightarrow R_3PbSR' + NaI$$
$$2\,R_3PbCl + Pb(SR')_2 \longrightarrow 2\,R_3PbSR' + PbCl_2$$
$$R_3PbCl + R'SH \xrightarrow{py} R_3PbSR' + HCl \cdot py$$
$$R_3PbOH + R'SH \longrightarrow R_3PbSR' + H_2O\,.$$

Diorganolead derivatives, $R_2Pb(SR')_2$, are prepared similarly.

8.4.10. Organolead Oligomers and Polymers

Few polymers containing Pb—Pb, Pb—O, Pb—N or Pb—S units are known, but related compounds with lower molecular weights have been described.

Organopolyplumbanes

Diplumbanes are obtained by the reaction of lead(II) chloride and Grignard reagents:

$$3\,PbCl_2 + 6\,RMgX \longrightarrow 6\,[PbR_2] \longrightarrow R_3Pb\text{—}PbR_3 + Pb\,.$$

or by reduction of triorganolead halides with sodium in liquid ammonia.

The mechanism of the formation of diplumbanes via Grignard reactions is given by the following sequence:

$$PbCl_2 + 3\,Bu^tMgCl \xrightarrow{-60\,°C} Bu^t_3PbMgCl \xrightarrow[+Bu^t_3PbBr]{-30\,°C} Bu^t_3Pb\text{—}PbBu^t_3$$

Solid $Ph_3Pb\text{—}PbPh_3$ has a staggered conformation.
Simultaneous oxidation and hydrolysis with hydrogen peroxide and ice of Ph_3PbLi yields the branched pentaplumbane:

```
         PbPh3
           |
Ph3Pb—Pb—PbPh3
           |
         PbPh3
```

The PbR_2 species may be cyclic polyplumbanes.

Organoplumboxanes (Organolead Oxides)

Diplumboxanes or triorganolead oxides, $R_3Pb{-}O{-}PbR_3$, are obtained by hydrolysis of triorganolead halides, followed by condensation of the intermediate hydroxides:

$$2\,R_3PbCl \xrightarrow[-NaCl]{NaOH} 2\,R_3Pb{-}OH \xrightarrow[-H_2O]{} R_3Pb{-}O{-}PbR_3$$

The diorganolead oxides, R_2PbO, are amorphous, insoluble and infusible, suggesting a polymeric structure. The plumbonic acids, $RPbO_2H$, are also polymeric.

Organolead Sulfides

Triorganolead sulfides are obtained by the reaction of the halides with sodium sulfide:

$$2\,R_3PbX + Na_2S \longrightarrow R_3Pb{-}S{-}PbR_3 + 2\,NaX$$

The diorganolead sulfides are cyclic trimers. The phenyl derivative is prepared from diphenyllead diacetate and hydrogen sulfide:

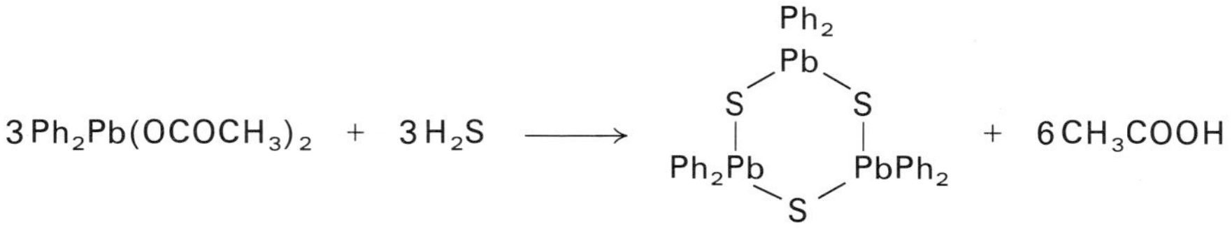

$$3\,Ph_2Pb(OCOCH_3)_2 + 3\,H_2S \longrightarrow (Ph_2PbS)_3 + 6\,CH_3COOH$$

(cyclic trimer: six-membered ring of alternating Ph_2Pb and S)

Heating $Ph_3Pb{-}PbPh_3$ with sulfur in benzene produces $Ph_3Pb{-}S{-}PbPh_3$ and $(Ph_2PbS)_3$. Carbon disulfide serves as a sulfur source in the conversion of triphenyllead hydroxide to the corresponding sulfide.

8.4.11. Plumbylenes, $:PbR_2$

Only the bis-η^5-cyclopentadienyllead(II) derivatives, $:Pb(C_5R_5\text{-}\eta^5)_2$, and the purple $:Pb[CH(SiMe_3)_2]_2$ are established monomeric derivatives of divalent lead.

Bis(η^5-cyclopentadienyl)lead(II) has an angular sandwich structure in the vapor phase, but a polymeric structure in the solid state:

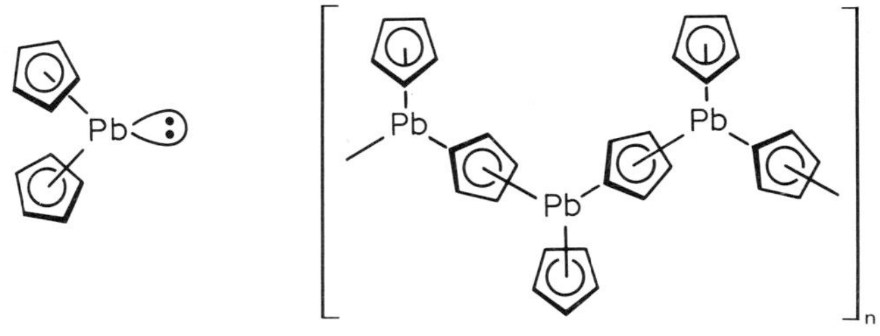

Pb Pb Pb Pb n

Cleavage of $:Pb(C_5H_5\text{-}\eta^5)_2$ with acids leads to monocyclopentadienyl derivatives, C_5H_5PbX.

The bis(trimethylsilyl)methyl derivative, $:Pb[CH(SiMe_3)_2]_2$, is prepared from $PbCl_2$ and $LiCH(SiMe_3)_2$ like the tin analogue and also forms metal carbonyl complexes.

9. Organometallic Compounds of Group V A Elements

The Group VA elements phosphorus, arsenic, antimony and bismuth form covalent organoelement compounds. The polarity of the M—C bond increases descending the group. The organobismuth compounds are the least stable and the most reactive.

Organophosphorus compounds are considered outside the scope of organometallic chemistry. Their great diversity, large number, and specific properties only partly imitated by arsenic would require many pages. Several books and monographs are available describing organophosphorus chemistry.

9.1. Organoarsenic Compounds

The first organoarsenical (and the first organometallic compound) was accidentally obtained by L.C. Cadet in 1760. The discovery of therapeutic activity of some organoarsenic compounds (Salvarsan, etc.) stimulated research, and many new compounds were synthesized.

Stable organoarsenic compounds form in two different oxidation states (III and V) and with one to six arsenic-carbon and other bonds to halogens, oxygen, sulfur and nitrogen.

9.1.1. Classification and Nomenclature

Organoarsenic(III) and (V) compounds will be dealt with within the same class, succesively, instead of dividing the chemistry into two parts according to the oxidation states. The main classes to be discussed in this chapter are:

Compounds in which Arsenic is Bonded Only to Organic Groups (Homoleptic and Related Species)

These are represented by four types of compounds:

-trisubstituted arsenic: AsR_3, tertiary arsines

-tetrasubstituted arsenic: $[AsR_4]^+$, tetraalkyl(aryl)-arsonium salts
$R_3As{=}CHR'$, organoarsenic ylides
-pentasubstituted arsenic: AsR_5, pentaorganoarsoranes

Disubstituted organoarsenic species, $[AsR_2]^+$ as in the diphenylarsinium perchlorate, $[AsPh_2]^+[ClO_4]^-$, prepared from diphenylchloroarsine and silver perchlorate in a polyether solvent, or in the six-membered heterocycle, arsabenzene or arsenine are also possible:

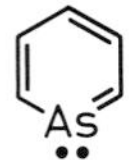

Organoarsenic Hydrides, R_nAsH_{3-n}

These are named as primary or secondary arsines:

$R{-}AsH_2$	primary arsines
$R_2As{-}H$	secondary arsines

No organoarsenic(V) hydrides are known.

Organoarsenic Halides

Both As(III) or As(V) derivatives containing from one to four arsenic-carbon bonds are known:

$R{-}AsX_2$	organodihaloarsines
$R_2As{-}X$	diorganohaloarsines
$R{-}AsX_4$	organotetrahaloarsoranes
R_2AsX_3	diorganotrihaloarsoranes
R_3AsX_2	triorganodihaloarsoranes
R_4AsX	tetraorganohaloarsoranes (usually X = F)

The tetrasubstituted derivatives, R_4AsX, are salts containing the arsonium cation, $[AsR_4]^+$; but when X = F, covalent $R_4As{-}X$ molecules form with five-coordinated arsenic.

Oxygen-Containing Organoarsenicals

There are a wide variety:

-arsenic(III) compounds:

$R{-}As(OH)_2$	arsonous acids
$RAs(OR')_2$	arsonous-acid esters
$(RAsO)_n$	arsonous-acid anhydrides or arsoxanes (also called arsenosoalkyls)

$R_2As{-}OH$	arsinous acids
$R_2As{-}OR'$	arsinous-acid esters
$R_2As{-}O{-}AsR_2$	arsinous-acid anhydrides (diarsoxanes)

-arsenic(V) compounds:

$R_nAs(OR')_{5-n}$	alkyl(aryl)alkoxyarsoranes
$R{-}As(O)(OH)_2$	arsonic acids
$R{-}As(O)(OR')_2$	arsonic-acid esters
$(RAsO_2)_n$	arsonic-acid anhydrides
$R_2As(O)OH$	arsinic acids
$R_2As(O){-}OR'$	arsinic-acid esters
$R_3As{=}O$	arsine oxides

Only a few halide, amide or hydride derivatives of arsonic and arsinic are known.

Sulfur-Containing Organoarsenic Compounds

Not every oxygen-containing type has a sulfur analogue, but there are similarities. The free thioacids cannot be isolated because of the instability of As—SH bonds, but their salts are known. The main classes are:

$(RAsS)_n$	alkyl(aryl)arsenic sulfides or thioarsonous acid anhydrides
$R_2As{-}S{-}AsR_2$	thiodiarsines (thioanhydrides of thioarsinous acids)
$R{-}As(SR')_2$	arsonous-acid thioesters
$R_2As{-}SR'$	arsinous-acid thioesters

-arsenic(V) compounds:

$RAs(S)(SH)_2$	trithioarsonic acids (known only as salts)
$(RAs)_2S_3$	organoarsenic sesquisulfides
$R_2As(S)SH$	dithioarsinic acids (known only as salts)
$R_3As{=}S$	arsine sulfides

Few functional derivatives of these thioacids (halides, amides, etc.) have been reported.

Nitrogen-Containing Compounds

Representatives of the following classes have been prepared:

-arsenic(III) compounds:

$R{-}As(NR'_2)_2$	diaminoorganoarsines
$R_2As{-}NR'_2$	aminodiorganoarsines
$(RAs{-}NR')_n$	arsazanes (cyclic or polymeric)
$R_2As{-}NR'{-}AsR_2$	diarsazanes

-arsenic(V) compounds:

$[R_2As(NH_2)Cl]^+$ aminoarsonium cations
$[R_2As{-}N{-}AsR_2]^+$ diarsazenium cations
$(R_2AsN)_n$ arsazenes (cyclic or polymeric), for example, with n = 3:

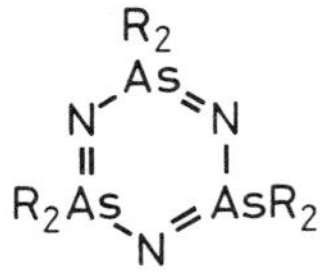

$R_3As{=}NR'$ arsinimines

Compounds with Arsenic-Arsenic Bonds

Only arsenic(III) derivatives are known:

$R_2As{-}AsR_2$ diarsines
$(RAs)_n$ cyclopolyarsines

9.1.2. Fully-Organosubstituted Compounds

Tertiary Arsines, AsR_3

Preparation. Grignard reagents act on arsenic trichloride to give triorganoarsines:

$$AsX_3 + 3\,RMgX \longrightarrow AsR_3 + 3\,MgX_2$$

Organochloroarsines are the starting material for the synthesis of unsymmetrical arsines, R_2AsR' or heterocyclic compounds (arsolanes, arsenanes) with di-Grignard reagents:

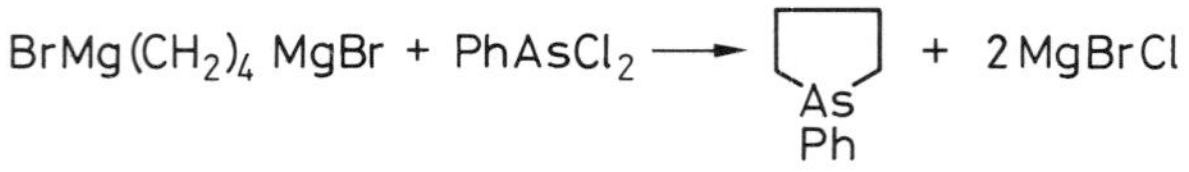

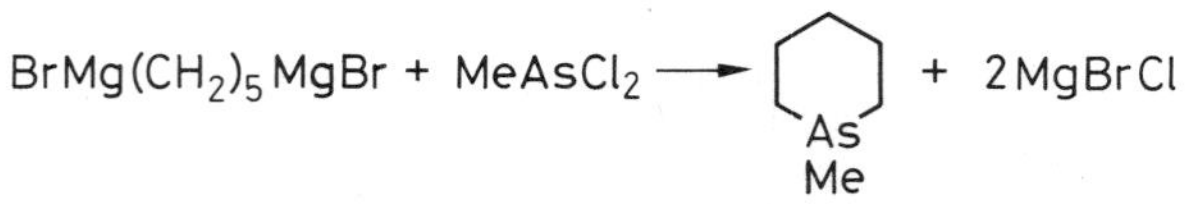

$$BrMg(CH_2)_4MgBr + PhAsCl_2 \longrightarrow \text{(arsolane, As–Ph)} + 2\,MgBrCl$$

$$BrMg(CH_2)_5MgBr + MeAsCl_2 \longrightarrow \text{(arsenane, As–Me)} + 2\,MgBrCl$$

Organomagnesium reagents also act on arsenic(III) oxide to give triarylarsines, but other organoarsinic by-products are formed.

Organolithium reagents are used for the preparation of trivinylarsines and for the synthesis of arsole derivatives:

$$\text{R}_4\text{C}_4\text{Li}_2 + \text{RAsCl}_2 \longrightarrow \text{R}_4\text{C}_4\text{AsR} + 2\,\text{LiCl}$$

Organomagnesium and lithium compounds are employed to synthesize the ditertiary phenylenediarsine "Diars" ligand used in transition metal coordination chemistry:

$$o\text{-C}_6\text{H}_4(\text{AsCl}_2)\text{Br} \xrightarrow[-\text{MgXCl}]{\text{RMgX}} o\text{-C}_6\text{H}_4(\text{AsR}_2)\text{Br} \xrightarrow[-\text{BuBr}]{\text{nBuLi}} o\text{-C}_6\text{H}_4(\text{AsR}_2)\text{Li} \xrightarrow[-\text{LiI}]{\text{R}_2\text{AsI}} o\text{-C}_6\text{H}_4(\text{AsR}_2)_2$$

The organosodium intermediates, $NaC{\equiv}CR$ and NaC_5H_5, are used to prepare the arsenic-containing acetylenic and η^1-cyclopentadienyl derivatives, $R_2As{-}C{\equiv}CR'$ and $As(C_5H_5\eta^1)_3$.

Organozinc and organomercury compounds are important when Grignard or organolithium reagents are difficult to obtain:

$$3\,\text{ZnR}_2 + 2\,\text{AsCl}_3 \longrightarrow 2\,\text{AsR}_3 + 3\,\text{ZnCl}_2 \ (e.g., \text{R} = \text{Me})$$
$$3\,\text{HgR}_3 + 2\,\text{AsCl}_3 \longrightarrow 2\,\text{AsR}_3 + 3\,\text{HgCl}_2 \ (e.g., \text{R} = 2\text{-furyl, 2-thienyl})$$

Organoaluminum compounds alkylate arsenic trichloride or trioxide to synthesize trialkylarsines:

$$\text{AsCl}_3 + \text{AlR}_3 \longrightarrow \text{AsR}_3 + \text{AlCl}_3$$
$$\text{As}_2\text{O}_3 + 2\,\text{AlR}_3 \longrightarrow 2\,\text{AsR}_3 + \text{Al}_2\text{O}_3$$

Organotin derivatives transfer organic substituents bearing functional groups (for example, ketones and esters):

$$\text{Ph}_2\text{AsCl} + \text{Et}_3\text{Sn{-}CH}_2\text{COMe} \longrightarrow \text{Ph}_2\text{As{-}CH}_2\text{COMe} + \text{Et}_3\text{SnCl}$$
$$\text{PhAsCl}_2 + \text{Ph}_2\text{Sn(CH}_2\text{COOMe)}_2 \longrightarrow \text{Ph{-}As(CH}_2\text{COOMe)}_2 + \text{Ph}_2\text{SnCl}_2$$

and organothallium compounds are used for the preparation of tris(pentafluorophenyl)arsine:

$$3\,(\text{C}_6\text{F}_5)_2\text{TlBr} + 2\,\text{As} \xrightarrow{190\,°\text{C}} 2\,\text{As(C}_6\text{F}_5)_3 + 3\,\text{TlBr}$$

Aminoarsines and cyclopentadiene give fluxional η^1-cyclopentadienylarsines:

$$\text{Me}_2\text{As{-}NMe}_2 + \text{C}_5\text{H}_6 \longrightarrow \eta^1\text{-C}_5\text{H}_5\text{AsMe}_2 + \text{HNMe}_2$$

haloarsines and η^1-cyclopentadienylsilanes also serve:

$$\text{Me}_2\text{AsCl} + \text{C}_5\text{H}_5(\text{H})\text{SiMe}_3 \longrightarrow \eta^1\text{-C}_5\text{H}_5\text{AsMe}_2 + \text{Me}_3\text{SiCl}$$

Tertiary arsines are prepared by Würtz reactions which do not require intermediate organometallic reagents to be prepared:

$$AsX_3 + 3\,RX + 6\,Na \longrightarrow AsR_3 + 6\,NaX$$

Metallated arsines, R_2AsM, are used to synthesize ditertiary or tritertiary arsines and prepare heterocycles:

$$2\,Ph_2AsK + Cl(CH_2)_nCl \longrightarrow Ph_2As(CH_2)_nAsPh_2 + 2\,KCl$$

$$3\,Me_2AsNa + (ClCH_2)_3C{-}CH_3 \longrightarrow (Me_2AsCH_2)_3C{-}CH_3 + 3\,NaCl$$

BrCH₂ CH₂Br (naphthalene) + $PhAs(MgBr)_2$ ⟶ (PhAs ring fused to naphthalene) + $2\,MgBr_2$

Structure. The solid tertiary arsines, trimethyl- and tri-p-tolylarsine are pyramidal because of the lone pair of electrons present in this oxidation state.

Properties. Tertiary arsines are weaker bases than amines, but are protonated in strong acids, for example, in anhydrous, liquid hydrogen chloride:

$$AsR_3 + HCl \rightleftharpoons [R_3AsH]^+Cl^-$$

The tertiary arsines are oxidized to arsine oxides by selenium dioxide, potassium permanganate or hydrogen peroxide. Trialkylarsines are oxidized by atmospheric oxygen with cleavage of an organic group and formation of dialkylarsinic acids, $R_2As(O)OH$, and react with elemental sulfur and selenium to form the oxidative-addition products, R_3AsX (X = S, Se). Halogens produce triorganoarsenic dihalides, R_3AsX_2.

Tertiary arsines are excellent ligands for transition metals in low-oxidation states.

Tetraorganoarsonium Salts, $[AsR_4]^+X^-$

Tertiary arsines react with alkyl halides to form arsonium salts:

$$AsR_3 + R'X \longrightarrow [R_3AsR']^+X^-$$

Electron-releasing groups at arsenic increase the reaction rate, and alkyl iodides are more reactive than the bromides and chlorides.

Spirocyclic arsonium cations are synthesized in this way:

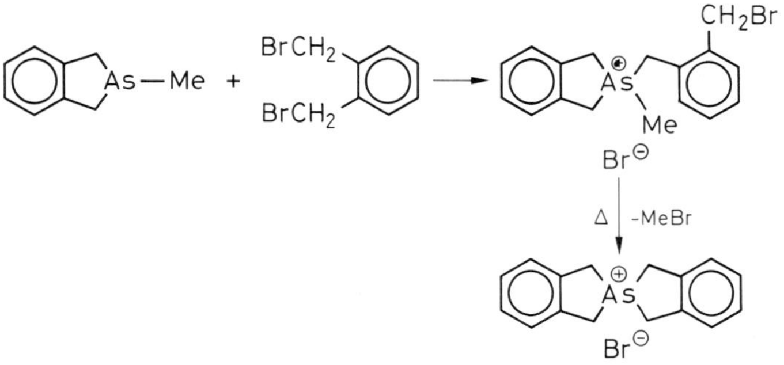

The reaction of aromatic arsine oxides with Grignard reagents yields tetraarylarsonium halides:

$$R_3AsO + RMgX \longrightarrow [AsR_4]^+X^- + MgO$$

Six-coordinated arsenic anions with six arsenic-carbon bonds are formed from arsonium salts with an organolithium reagent:

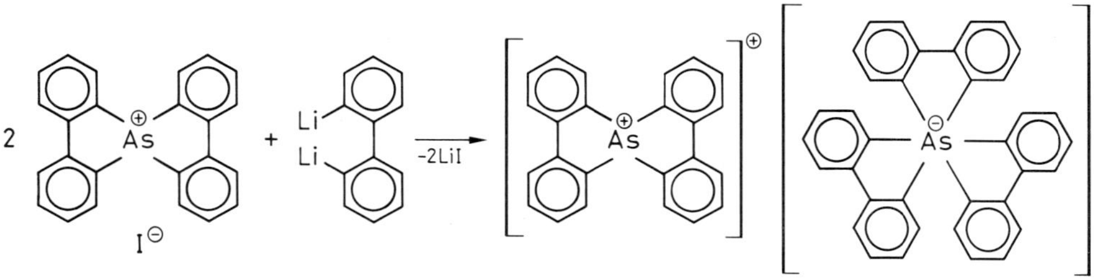

Heating of benzene, arsenic trichloride and anhydrous aluminum trichloride, followed by treatment with potassium iodide, precipitates tetraphenylarsonium iodide. A mixture of triphenylarsine, aluminum chloride and bromobenzene reacts on heating as well to give tetraphenylarsonium salts.

The zwitterion, $Me_3As^+{-}CH_2COO^-$ is one of the few naturally occuring organometallic compounds, having been isolated from the tail muscle of the western rock lobster.

Solid tetraphenylarsonium iodide and tetramethylarsonium bromide have a tetrahedral structure.

The arsonium salts are soluble in water and precipitate with numerous anions, a property useful in analytical applications. Thermal decomposition or electrolysis of arsonium salts leads to tertiary arsines:

$$[AsR_4]^+X^- \xrightarrow{\Delta} AsR_3 + RX$$

Organoarsenic Ylides (Alkylidenearsoranes), $R_3As{=}CHR'$

These active species are formally derivatives of pentavalent arsenic, and can be written in two mesomeric forms:

$$R_3As{=}CH{-}R' \longleftrightarrow R_3\overset{+}{As}{-}\overset{-}{C}HR'$$

They are prepared from quaternary arsonium salts and bases by deprotonation, for example, with LiR, NaH, NaOEt, $NaNH_2$, etc.:

$$[R_3\overset{+}{As}{-}CH_2R']X^- + \text{base} \xrightarrow[-HX]{} R_3As{=}CHR'$$

These reactive compounds are intermediates in organic syntheses.

Pentaorganosubstituted Derivatives (Pentaorganoarsoranes), AsR_5

Pentaphenylarsorane is obtained from tetraphenylarsonium bromide and phenyllithium:

$$[AsPh_4]^+Br^- + PhLi \longrightarrow AsPh_5 + LiBr$$

or from triphenylarsine oxide with phenyllithium:

$$Ph_3AsO + 2\,PhLi \longrightarrow AsPh_5 + Li_2O$$

Arsinimines react with 1,1′-dilithiobiphenyl to form a heterocyclic derivative with five-coordinated arsenic:

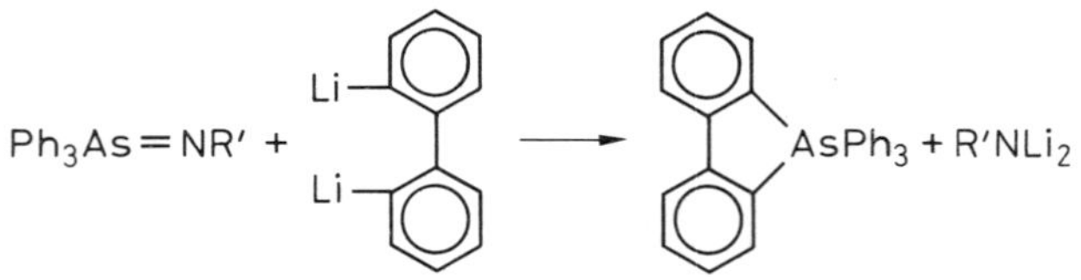

$$Ph_3As{=}NR' + \text{(2,2'-dilithiobiphenyl)} \longrightarrow \text{(biphenylene)}AsPh_3 + R'NLi_2$$

The spirocyclic arsonium salts react in a similar manner with phenyllithium:

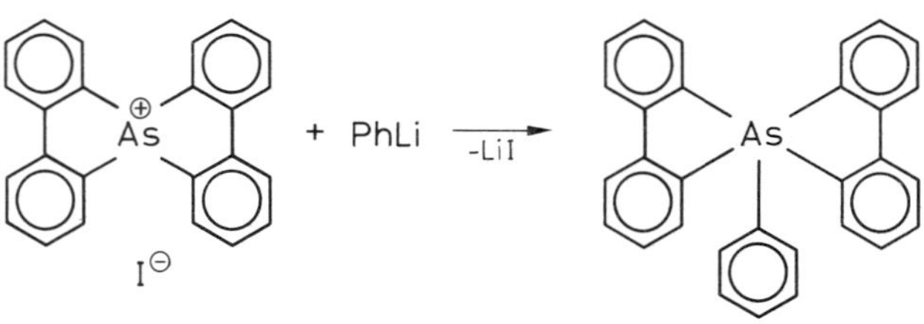

$$\text{(spiro-bis(biphenylene)}As^{\oplus})\,I^{\ominus} + PhLi \xrightarrow[-LiI]{} \text{spiro-bis(biphenylene)}AsPh$$

Solid pentaphenylarsorane has a trigonal-bipyramidal structure.

Trigonal-bipyramidal pentamethylarsorane, $AsMe_5$, is prepared by the reaction of Me_3AsCl_2 with MeLi.

Pentaphenylarsorane, $AsPh_5$, decomposes at 150 °C to triphenylarsine, biphenyl and benzene. The pentasubstituted derivatives are air- and moisture-sensitive, and are hydrolyzed to tetramethylarsonium hydroxide, while hydrogen fluoride forms tetramethylfluoroarsorane:

$$AsMe_5 + H_2O \longrightarrow [AsMe_4]^+OH^- + CH_4$$
$$AsMe_5 + HF \longrightarrow Me_4As{-}F + CH_4$$

Arsabenzene (Arsenine)

The six-membered heterocycle, arsabenzene, is prepared from a tin heterocycle and arsenic trichloride:

R R' / Sn Bu$_2$ $\xrightarrow[-Bu_2SnCl_2]{+AsCl_3/THF}$ R R' / As Cl $\xrightarrow{-R'Cl}$ R / As

The ring is planar in the vapor phase.

9.1.3. Organoarsenic Hydrides (Primary and Secondary Arsines)

Preparation. Primary arsines, $RAsH_2$, and secondary arsines, R_2AsH, are obtained by reducing mono- and diorgano-derivatives (halides, acids, salts) with zinc powder and hydrochloric acid, or with zinc amalgam. Lithium alanate, $LiAlH_4$, gives poorer results, but lithium borohydride, $LiBH_4$, reduces phenylarsenic tetrachloride, $PhAsCl_4$, and phenylarsenic dichloride, $PhAsCl_2$.

Alkali metal arsenides, prepared in liquid ammonia from arsine and an alkali metal, couple with alkyl halides:

$$AsH_3 + Na \xrightarrow{NH_3\,liq.} NaAsH_2 \xrightarrow{RX} RAsH_2$$
$$PhAsH_2 + Na \longrightarrow PhAsHNa \xrightarrow{RX} PhRAsH$$

Properties. The primary and secondary arsines are reactive, with the methyl and ethyl derivatives pyrophoric, and all are readily oxidized:

$$RAsH_2 \xrightarrow{[O]} (RAs)_n \xrightarrow{[O]} R_2As{-}O{-}AsR_2$$
$$RAsH_2 \xrightarrow{[O]} (RAs)_n \xrightarrow{[O]} (RAsO)_n \xrightarrow[+H_2O]{[O]} RAs(O)(OH)_2$$
$$R_2AsH \xrightarrow{[O]} R_2As{-}AsR_2 \xrightarrow{[O]} R_2As{-}O{-}AsR_2 \xrightarrow[+H_2O]{[O]} R_2As(O)OH$$

Halogenation leads first to substitution and finally to addition products:

$$RAsH_2 \xrightarrow[-2HX]{X_2} RAsX_2 \xrightarrow{X_2} RAsX_4$$
$$R_2AsH \xrightarrow[-HX]{X_2} R_2AsX \xrightarrow{X_2} R_2AsX_3$$

Hydrogen is replaced by alkali metals, to give arsenides:

$$RAsH_2 \xrightarrow[\text{or LiR}]{\text{Na metal}} RAsHNa \quad \text{or} \quad RAsLi_2$$

Phenyl groups are cleaved to give diorganoarsenides:

$$AsR_3 \xrightarrow{M} R_2AsM \qquad (M = Li, Na)$$

The addition of arsines to diacetylenes yields five-membered arsoles:

$$PhAsH_2 + RC{\equiv}C{-}C{\equiv}CR \longrightarrow \text{2,5-R}_2\text{-1-phenylarsole}$$

or seven-membered heterocycles:

$$PhAsH_2 + \text{(diyne)} \xrightarrow{cat} \text{1-phenylarsepine (As–Ph)}$$

Mixed ditertiary arsine-phosphines can be synthesized as ligands in coordination chemistry:

$$R_2AsH + H_2C{=}CH{-}PR'_2 \longrightarrow R_2As{-}CH_2CH_2{-}PR'_2$$
$$2\,R_2AsH + (H_2C{=}CH)_2PR' \longrightarrow R'{-}P(CH_2CH_2{-}AsR_2)_2$$

9.1.4. Organoarsenic Halides

Mono- and Dihaloarsines, R_nAsX_{3-n}

The chlorides are formed in redistribution reactions on heating mixtures of tertiary arsines with arsenic trichloride (unsatisfactory for R = Me):

$$AsR_3 + 2\,AsCl_3 \rightleftharpoons 3\,RAsCl_2$$
$$AsR_3 + RAsCl_2 \rightleftharpoons 2\,R_2AsCl$$
$$2\,RAsCl_2 \rightleftharpoons R_2AsCl + AsCl_3\,.$$

Organohaloarsines are prepared from organomercury compounds and arsenic(III) halides:

$$AsX_3 + HgR_2 \longrightarrow RAsX_2 + RHgX$$
$$AsX_3 + RHgX \longrightarrow RAsX_2 + HgX_2$$

Lead tetraalkyls alkylate arsenic halides:

$$PbR_4 + AsCl_3 \longrightarrow R_2AsCl + RAsCl_2 + R_3PbCl$$
$$PbEt_4 + RAsCl_2 \longrightarrow REtAsCl + Et_3PbCl$$

Organoaluminum and Grignard reagents (for example, for R = cyclohexyl and C_6F_5) are used, although the latter lead to tertiary arsines.

An $AsCl_2$ group can be substituted to an aromatic nucleus by heating benzene with arsenic trichloride:

$$C_6H_6 + AsCl_3 \longrightarrow C_6H_5AsCl_2 + HCl$$

Heterocyclic compounds are prepared by a similar dehydrohalogenation:

$$\text{(biphenyl-X)}AsCl_2 \xrightarrow{-HCl} \text{(tricyclic X, As)}AsCl$$

X = O, S, N

Organohaloarsines are also obtained by addition of arsenic trichloride to acetylene in the presence of aluminum trichloride which gives chlorovinylchloroarsines and tris(chlorovinyl)arsine:

$$3AsCl_3 + 6HC{\equiv}CH \xrightarrow{AlCl_3} ClCH{=}CH{-}AsCl_2 + (ClCH{=}CH)_2AsCl + As(CH{=}CHCl)_3 .$$

The direct synthesis of organohaloarsines occurs on heating elemental arsenic with organic halides:

$$2As + 3RX \xrightarrow{Cu/250-350\,°C} RAsX_2 + R_2AsX$$

$$4As + 6CF_3I \xrightarrow[\text{autoclave}]{220\,°C} CF_3AsI_2 + (CF_3)_2AsI + As(CF_3)_3 + AsI_3$$

$$2As + 3H_2C{=}CHX \xrightarrow[450\,°C]{Cu/Zn} H_2C{=}CHAsX_2 + (CH_2{=}CH)_2AsX \quad X = Cl, Br$$

Arsonic and arsinic acids are converted to organoarsenic(III) halides by reduction with sulfur dioxide in the presence of hydrochloric acid:

$$RAs(O)(OH)_2 \xrightarrow{SO_2/HCl} RAsCl_2$$

$$R_2As(O)OH \xrightarrow{SO_2/HCl} R_2AsCl$$

Hypophosphoric acid and PX_3 can also be used as reducing agents.

The conversion of primary and secondary arsines to halides has been mentioned above.

The halides are used for the synthesis of derivatives by halogen substitution. They hydrolyze readily and undergo halogen addition. The halides can be oxidized under hydrolytic conditions to the corresponding arsenic(V) acids, $RAsO(OH)_2$ and $R_2As(O)OH$.

Organotetrahaloarsoranes, $RAsX_4$

The chlorides are obtained by addition of chlorine to organodichloroarsines and the fluorides by the reaction of arsonic acids with sulfur tetrafluoride:

$$RAsCl_2 + Cl_2 \longrightarrow RAsCl_4$$
$$RAsO(OH)_2 + 3SF_4 \xrightarrow{70°C} RAsF_4 + 3SOF_2 + 2HF$$

These rare compounds are hydrolyzable and decompose on heating. Methyltetrachloroarsorane, $MeAsCl_4$, decomposes at 0°C.

Diorganotrihaloarsoranes, R_2AsX_3

The compounds are prepared by treatment of diorganohaloarsines with halogens, or arsinic acids with thionyl chloride:

$$R_2AsX + X_2 \longrightarrow R_2AsX_3$$
$$R_2As(O)OH + 2SOCl_2 \longrightarrow R_2AsCl_3 + 2SO_2 + HCl$$

They are more stable than organotetrahaloarsoranes, but decompose below 100°C, and are moisture-sensitive. Their structure may contain dihaloarsonium cations, $[R_2AsX_2]^+X^-$, but the trifluorides, R_2AsF_3, possess a trigonal-bipyramidal structure for R = Ph in solution, while the R = Me derivatives are associated.

Triorganodihaloarsoranes, R_3AsX_2

These derivatives are obtained by addition of halogens to tertiary arsines, but As—C bond cleavage interferes. Alternative routes involve the treatment of arsine oxides with hydrogen halides, or the reaction of diarylchloroarsines with aryldiazonium tetrachloroferrate:

$$R_2AsCl + [R{-}N{\equiv}N]^+[FeCl_4]^- \longrightarrow R_3AsCl_2 + FeCl_3 + N_2$$

Iodine monochloride is used as a chlorinating agent for ($R = C_6F_5$):

$$AsR_3 + 2ICl \longrightarrow R_3AsCl_2 + I_2$$

Difluorides are prepared by treatment of arsines with sulfur tetrafluoride, or by metathesis with silver fluoride:

$$AsR_3 + SF_4 \longrightarrow R_3AsF_2 + SF_2$$
$$R_3AsCl_2 + 2AgF \longrightarrow R_3AsF_2 + 2AgCl$$

Triphenylarsine oxide reacts with hydrofluoric-acid solution to give Ph_3AsF_2 and with concentrated HCl or HBr, to give Ph_3AsX_2.

The solid difluoride, Ph_3AsF_2, has a non-ionic, trigonal-bipyramidal structure with apical fluorines.

The dichlorides are weak electrolytes and the iodides are ionic in the solid state and in polar solvents, and contain the cation $[R_3AsX]^+$.

The dihaloarsoranes decompose on heating into the diorganoarsenic halides and organic halides. They can be reduced to tertiary arsines.

9.1.5. Oxygen-Containing Organoarsenic Compounds

The most important are the acids and their derivatives containing arsenic in the 3+ or 5+ oxidation states.

Arsonous Acids, $RAs(OH)_2$, and Their Anhydrides, RAsO

The hydrolysis of difunctional derivatives, $RAsX_2$ (X = halogen, OR, NR_2, CN, etc.), leads either to arsonous acids, $RAs(OH)_2$, or to their anhydrides, RAsO. The acids are formed when R is an unsubstituted aromatic group or has electron-releasing substituents, and the anhydrides are formed when R is aliphatic or an aromatic group bearing electron-withdrawing substituents. The two types are often formed together:

$$RAsO + H_2O \rightleftharpoons RAs(OH)_2$$

The reduction of arsonic acids, $RAsO(OH)_2$, with sulfur dioxide in hydrochloric acid, and the oxidation of primary arsines, $RAsH_2$, and polyarsines, $(RAs)_n$, also leads to arsonous acids or their anhydrides.

The anhydrides, RAsO, are trimers and tetramers with cyclic structures even in the vapor phase:

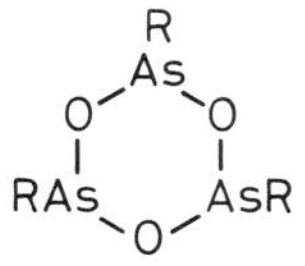

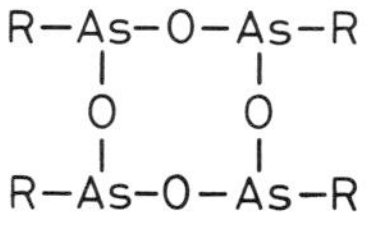

```
       R              R-As-O-As-R
      As                |     |
   O     O              O     O
   |     |              |     |
 RAs     AsR          R-As-O-As-R
     O
```

A solid, adamantane-like cage compound:

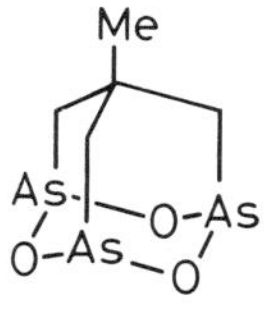

```
      Me
      |
   ___C___
  |   |   |
 As-  |  O-As
 /    |   /
O-As ---O
```

and an eight-membered ring, containing both methylene and oxygen bridges between the arsenic atoms are known:

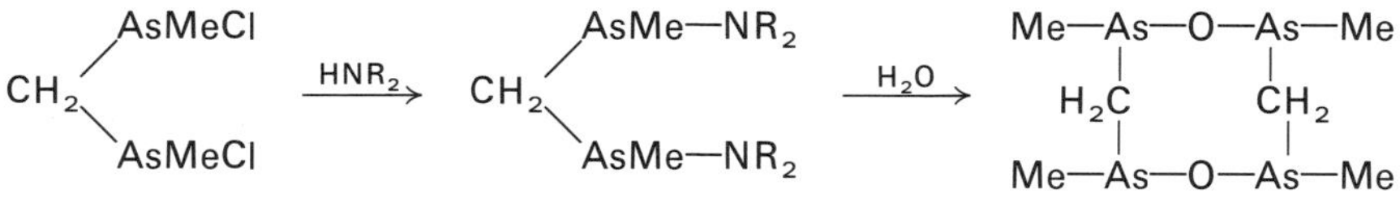

```
     AsMeCl               AsMe—NR2            Me—As—O—As—Me
    /          HNR2       /             H2O       |       |
CH2         -------> CH2           ------->     H2C     CH2
    \                     \                       |       |
     AsMeCl               AsMe—NR2            Me—As—O—As—Me
```

Arsonous Acids Esters, $RAs(OR')_2$

Organodichloroarsines react with sodium alkoxides, and alkylarsine oxides, RAsO, react with alcohols in the presence of anhydrous $CaCl_2$:

$$RAsCl_2 + 2\,NaOR' \longrightarrow RAs(OR')_2 + 2\,NaCl$$

$$RAsO + 2\,R'OH \xrightarrow[-H_2O]{CaCl_2} RAs(OR')_2$$

The moisture-sensitive arsonous esters are distillable *in vacuo*.

Arsinous Acids, $R_2As\text{-}OH$, and Their Anhydrides, $(R_2As)_2O$ (Diarsoxanes)

Hydrolysis of organoarsenic halides produces arsinous acids; when R is an aromatic group with electron-releasing substituents the anhydrides (diarsoxanes) are formed. The oxidation of diphenylarsine yields both $(Ph_2As)_2O$ and $Ph_2As(O)OH$.

The first organometallic compounds were contained in "Cadet's fuming liquid", a mixture of $Me_2As—O—AsMe_2$ ("cacodyl oxide") and $Me_2As—AsMe_2$:

$$As_2O_3 + 4\,KOOC—Me_3 \xrightarrow[-2\,K_2CO_3,\ -2\,CO_2]{} Me_2As—O—AsMe_2 + Me_2As—AsMe_2$$

Arsenic(III) oxide can be reacted with Grignard or organoaluminum compounds to give diarsoxanes:

$$As_2O_3 + 4\,RMgX \longrightarrow R_2As—O—AsR_2 + \text{other products}$$

$$As_2O_3 + 2\,AlR_3 \longrightarrow R_2As—O—AsR_2 + 2\,RAlO$$

Arsinous Acid Esters, $R_2As\text{-}OR'$

The title compounds form from diorganohaloarsines and sodium alkoxides. Their chemistry is like that of the arsonous esters.

Alkyl(aryl)tetraalkoxyarsoranes, $RAs(OR')_4$

These compounds are obtained from the corresponding tetrahalides and sodium methylate:

$$RAsX_4 + 4\,NaOMe \longrightarrow RAs(OMe)_4 + 4\,NaX$$

The more stable spirocyclic compounds form with diols:

$$RAsO(OH)_2 + \text{HO–CH}_2\text{CH}_2\text{–OH} \longrightarrow \text{spiro-R–As(–O–CH}_2\text{CH}_2\text{–O–)}_2 \longleftarrow RAs(OMe)_4 + \text{HO–CH}_2\text{CH}_2\text{–OH}$$

Dialkyl(aryl)trialkoxyarsoranes, $R_2As(OR')_3$

These compounds are obtained form diorganoarsenic halides, bromine and sodium methoxide:

$$R_2AsCl + 3\,NaOMe + Br_2 \xrightarrow{0°C} R_2As(OMe)_3 + 2\,NaBr + NaCl$$

Triorganodialkoxyarsoranes, $R_3As(OR')_2$

These compounds are prepared from tertiary arsines, bromine and sodium methoxide:

$$AsR_3 + Br_2 + 2\,NaOMe \longrightarrow R_3As(OMe)_2 + 2\,NaBr$$

Tetraalkylalkoxyarsoranes, $R_4As\text{-}OR'$

These compounds form on the addition of methanol to trimethylarsonium ylide:

$$Me_3As{=}CHSiMe_3 \xrightarrow[-Me_3SiOMe]{MeOH} Me_3As{=}CH_2 \xrightarrow{MeOH} Me_4As{-}OMe$$

The trigonal-bipyramidal product fumes in air and is pyrophoric.

Arsonic Acids, $RAs(O)(OH)_2$, and Arsinic Acids, $R_2As(O)OH$

The pentavalent organoarsenic acids are obtained by similar methods, leading to mono- or disubstitution.

The arylation of sodium arsenite with aryldiazonium salts occurs in buffered alkali ($pH = 8.8–9.2$):

$$Na_3AsO_3 + [R{-}N{\equiv}N]^+X^- \longrightarrow R{-}AsO(ONa)_2 + NaX + N_2$$

Yields are improved if the diazonium tetrafluoroborate is added to an alkaline solution of arsenite, or when an aromatic amine is diazotized in concentrated sodium nitrite in the presence of arsenic trichloride. Sodium arsinates, $R_2As(O)ONa$, form as by-products. Sodium arsonates react in alkaline medium with aryldiazonium salts to form the arsinates:

$$RAsO(ONa)_2 + R{-}N{\equiv}N^+X^- \longrightarrow R_2As(O)ONa + NaX + N_2$$

Phenols, aromatic amines and ethers can be arsenated directly with arsenic acid:

$$C_6H_5X + H_3AsO_4 \longrightarrow X{-}C_6H_4AsO(OH)_2 + H_2O\,.$$

Para-substitution is preferred. Thus, phenol gives *para*-hydroxyphenylarsonic acid, *para*-$HO{-}C_6H_4{-}As(O)(OH)_2$, and aniline yields the bis(*para*-aminophenyl)arsinic acid, $(p\text{-}H_2NC_6H_4)_2AsO(OH)$.

Organic halides react ($I > Br > Cl$) with sodium arsenite in alcohol to give both the arsonic or arsinic acid salts:

$$Na_3AsO_3 + RX \longrightarrow RAs(O)(ONa)_2 + NaX$$
$$RAsO(ONa)_2 + RX \longrightarrow R_2As(O)(ONa) + NaX\,.$$

Aromatic derivatives require sealed tubes and higher temperatures, as in the preparation of potassium phenylarsonate:

$$K_3AsO_3 + PhBr \xrightarrow{180-200\,°C} PhAs(O)(OK)_2 + KBr$$

Arsinic acids are prepared from arsenic trichloride and Grignard reagents in several steps:

$$AsCl_3 \xrightarrow{HNEt_2} Cl_2As{-}NEt_2 \xrightarrow{2\,RMgX} R_2As{-}NEt_2 \xrightarrow{H_2O}$$
$$\longrightarrow R_2AsOH \xrightarrow{H_2O_2} R_2As(O)(OH)$$

In situ hydrolysis of tetra- and trichloroarsoranes, obtained in a reaction of diazonium salts with arsenic chlorides, leads to arsonic and arsinic acids:

$$[R{-}N_2]^+Cl^- + AsCl_3 \longrightarrow [R{-}N_2]^+[AsCl_4]^- \longrightarrow$$
$$\xrightarrow[-N_2]{} RAsCl_4 \xrightarrow{H_2O} RAs(O)(OH)_2$$

$$[R{-}N_2]^+Cl^- + R'AsCl_2 \xrightarrow{Cu/EtOH} RR'AsCl_3 \xrightarrow{H_2O} RR'AsO(OH)$$

Arsonic and arsinic acids are stronger than carboxylic acids, but weaker than phosphonic and sulfonic acids. They are reduced to RAsO derivatives, and are converted to organochloroarsines by thionyl chloride. The sulfur-dioxide reduction of arsonic acids in hydrochloric acid gives organodichloroarsines. Drying leads to polymeric anhydrides, $(RAsO_2)_n$, of unknown structure.

Arsonic and Arsinic-Acid Esters, $RAsO(OR')_2$ and $R_2As(O)(OR')$

These compounds are prepared from the silver salts with alkyl chlorides, by oxidation of arsonous- and arsinous-acid esters with selenium dioxide, or by esterification of arsonic and arsinic acids with alcohols (the water is removed by azeotropic distillation).

Tertiary-Arsine Oxides, R_3AsO

The oxidation of tertiary arsines with aqueous potassium permanganate, hydrogen peroxide, selenium dioxide or elemental iodine gives arsine oxides. The aromatic-As bond resists oxidation, but in trialkylarsines one of the As—R bonds may be cleaved to form dialkylarsinic acids. Trialkylarsine oxides are prepared from the arsine with mercury(II) oxide in refluxing acetone, or by treatment of excess arsine with hydrogen peroxide. Arsine oxides, or the ill-defined hydroxides, $R_3As(OH)_2$, are obtained by hydrolysis of triorganodihaloarsoranes:

$$R_3AsX_2 + H_2O \longrightarrow R_3AsO + 2HX$$

Arsine oxides form as by-products in the arylation of arsenite by diazonium salts.

Tertiary-arylarsine oxides hydrate to form an equilibrium among three isomers in solution:

$$R_3AsO \cdot H_2O \rightleftharpoons [R_3\overset{+}{A}s\text{—}OH]OH^- \rightleftharpoons R_3As(OH)_2$$

Arsine oxides react with hydrohalic acids to form 1 : 1 adducts, believed to be hydroxotriorganoarsonium salts, in equilibrium with the two other isomers:

$$R_3AsO + HX \longrightarrow [R_3\overset{+}{A}s\text{—}OH]X^-$$

$$[R_3\overset{+}{A}s\text{—}OH]X^- \rightleftharpoons R_3AsO \cdots HX \rightleftharpoons R_3As(X)(OH) \rightleftharpoons [R_3\overset{+}{A}s\text{—}OH]X^-$$

With alkyl halides, tertiary arsine oxides undergo rearrangement:

$$EtPh_2AsO \xrightarrow[-EtX]{RX} Ph_2As\text{—}OR$$

9.1.6. Sulfur-Containing Organoarsenic Compounds

Organoarsenic Sulfides, RAsS

The oxides or organodichloroarsines react with hydrogen or sodium sulfide, primary arsines react with sulfur or thionyl chloride and cyclosilthianes with organodichloroarsines to give the title compounds:

$$RAsO + H_2S \longrightarrow RAsS$$

$$RAsX_2 + H_2S\ (NaSH) \longrightarrow RAsS$$

$$RAsH_2 + S \longrightarrow RAsS$$

$$RAsH_2 + SOCl_2 \longrightarrow RAsS$$

$$(R_2SiS)_3 + PhAsCl_2 \longrightarrow RAsS$$

The As—SH group in arsenic(III) compounds and the thioarsonous acids, $RAs(SH)_2$, undergoes rapid H_2S elimination to form the thioanhydrides, RAsS.

Like the oxygen analogues, RAsS compounds are associated in cyclic dimers to hexamers. Thus, two forms of the phenyl derivative, $(PhAsS)_n$, exist with n = 3 or 4:

$$\text{cyclo-}(RAsS)_3 \qquad\qquad \text{cyclo-}(RAsS)_4$$

The product from pentafluorophenyldichloroarsine and silver sulfide is the tetramer $(C_6F_5AsS)_4$.

Thiodiarsines, $R_2As\text{-}S\text{-}AsR_2$

Thiodiarsines are prepared from diorganochloroarsines and hydrogen sulfide, arsenic(III) sulfide and Grignard reagents, or from secondary arsines with elemental sulfur:

$$R_2AsCl + H_2S \xrightarrow{-HCl} R_2As{-}S{-}AsR_2 \xleftarrow{-H_2S} R_2AsH + S$$

$$As_2S_3 + RMgX \longrightarrow R_2As{-}S{-}AsR_2$$

The thioarsinous acids, $R_2As{-}SH$, are little studied.

Tetramethyldiarsine reacts with elemental sulfur to give solid $Me_2As({=}S)SAsMe_2$:

$$Me_2As{-}AsMe_2 \xrightarrow{+2S} Me_2As{-}S{-}As({=}S)Me_2$$

Arsonous-Acid Thioesters, $RAs(SR')_2$, and Arsinous-Acid Thioesters, $R_2As\text{-}SR'$

The title compounds are prepared from thiols or their salts, with arsonic, arsinic, arsinous acids or with organohaloarsines. The number of As—R bonds determines whether the product is $RAs(SR)_2$ or $R_2As{-}SR$:

$$RAsX_2 + 2\,HSR' \longrightarrow RAs(SR')_2 + 2\,HX$$
$$R_2AsX + HSR' \longrightarrow R_2As{-}SR' + HX$$

Organoarsenic Sesquisulfides, $R_2As_2S_3$

These five-membered heterocycles form in the reaction of arsonic acids with hydrogen sulfide, or carbon disulfide:

$$\text{R-As}\langle\text{S-S}\rangle\text{As-R (five-membered ring: R-As, S-S, As-R, bridging S)}$$

The selenium analogues result from reactions of organocyclopolyarsines with elemental selenium:

$$(AsPh)_6 + 9Se \xrightarrow{220\,°C} 3\ \text{Ph-As(Se-Se)(Se)As-Ph}$$

$$3(AsMe)_5 + 15Se \xrightarrow{180\,°C} 5\ \text{(MeAs)}_3\text{Se}_3 \text{ (six-membered ring: Me-As, Se, As-Me, Se, As-Me, Se)}$$

Arsine Sulfides, R_3AsS

Arsine sulfides are prepared form the oxides or triorganodihaloarsoranes and hydrogen sulfide, by addition of sulfur to tertiary arsines, or by the reaction of arsenic sulfide with Grignard reagents:

$$R_3AsX_2 + H_2S \xrightarrow{-HX} R_3AsS$$

$$R_3AsO + H_2S \xrightarrow[-H_2O]{} R_3AsS \xleftarrow[-H_2O]{} R_3As(OH)_2 + H_2S$$

$$R_3As + S \rightarrow R_3AsS \leftarrow RMgX + As_2S_3$$

Triphenylarsine can extract sulfur from $SPCl_3$, thiols, diorganodisulfides, or S_2Cl_2, to form Ph_3AsS.

Tertiary arsine sulfides are alkylated by dimethylsulfate or triethyloxonium tetrafluoroborate to give thioarsonium salts:

$$R_3AsS + (MeO)_2SO_2 \longrightarrow [R_3\overset{+}{A}s\text{—}SMe]MeSO_4^- \quad (R = Ph,\ C_6H_{11})$$

$$R_3AsS + Et_3O^+BF_4^- \longrightarrow [R_3\overset{+}{A}s\text{—}SEt]BF_4^- \quad (R = C_6H_{11})$$

Trithioarsonic Acids, $RAs(S)(SH)_2$, and Dithioarsinic Acids, $R_2As(S)SH$

These sulfur-containing acids cannot be isolated and are known only as salts; a sodium trithioarsonate results from the sesquisulfide with sulfur and sodium sulfhydride:

$$(PhAs)_2S_3 + S + 2\,NaSH \longrightarrow 2\,PhAsS(SNa)_2$$

Dithioarsinate salts are prepared from arsinates and hydrogen sulfide or sodium sulfide:

$$R_2As(O)ONa + 2H_2S \longrightarrow R_2As(S)SNa + 2H_2O$$

and can form transition metal complexes.

9.1.7. Nitrogen-Containing Organoarsenic Compounds

Organodichloroarsines and diorganochloroarsines react with ammonia to form the cyclic oligomers, $(RAsNH)_n$ and $R_2As{-}NH_2$, respectively. With secondary amines, $R_2As{-}NR'_2$, $RClAs{-}NR'_2$ and $RAs(NR'_2)$ are formed. Dimethylchloroarsine forms a diarsazane, $Me_2As{-}NMe{-}AsMe_2$, with methylamine. Arsoxanes and silicon-nitrogen compounds serve as intermediates in the synthesis of tris(dimethylarsin)amine:

$$3(Me_2As)_2O + 2NaNH_2 \longrightarrow 2N(AsMe_2)_3 + 2NaOH + H_2O$$
$$3Me_2AsCl + Me_3SiNHSiMe_3 \longrightarrow N(AsMe_2)_3 + 2Me_3SiCl + HCl$$

Ammonolysis of Me_2AsCl yields $[Me_2As{-}NH_3]^+Cl^-$, but Et_2AsCl produces $N(AsEt_2)_3$.

The four-membered hererocycles, $(RAs{-}NR')_2$, are obtained by transaminations of organoarsenic-dimethylamino derivatives or by thermal condensation:

$$MeAs(NR'_2)_2 + R''NH_2 \longrightarrow \begin{array}{c} Me{-}As{-}NR'' \\ | \quad\quad | \\ R''N{-}AsMe \end{array} \xleftarrow{170\,°C} MeAs(NHR'')_2$$

Aminochloroarsines and organomagnesium or -lithium reagents form substituted dialkylarsines:

$$Cl_2As{-}NR_2 + 2R'MgCl \longrightarrow R'_2As{-}NR_2 + 2MgCl_2$$
$$ClAs(NR_2)_2 + LiR' \longrightarrow R'As(NR_2)_2 + LiCl$$

which are obtained from organoarsenic halides and amines:

$$RAsCl_2 + 4HNR'_2 \longrightarrow RAs(NR'_2)_2 + 2[R'_2NH_2]^+Cl^-$$

The air- and moisture-sensitive organoarsenic(III) compounds are cleaved by water, alcohols or acids with liberation of the amine.

Nitrogen-bonded organoarsenic(V) analogues of similar phosphorus-nitrogen compounds form from diphenylchloroarsine and chloramine, or with a chloramine-ammonia mixture:

$$Ph_2AsCl \xrightarrow{ClNH_2} [Ph_2As(NH_2)Cl]Cl \xrightarrow[DMF]{} [Ph_2As(Cl){-}N{-}As(Cl)Ph_2]Cl$$

$$Ph_2AsCl \xrightarrow{ClNH_2/NH_3,\ 25\,°C} [Ph_2As(NH_2){-}N{-}As(NH_2)Ph_2]Cl \xrightarrow[DMF]{\Delta} (Ph_2AsN)_3 \xleftarrow{NH_3} Ph_2AsCl_3$$

Pyrolysis of diorganoarsenic azides, $R_2As{-}N_3$, produces cyclic oligomers $(R_2AsN)_n$:

$$(R_2AsN)_3 \text{ (six-membered ring)} \qquad (R_2AsN)_4: \; R_2As{=}N{-}AsR_2,\ N{=}R_2As{-}N{=}AsR_2 \text{ (eight-membered ring)}$$

Cleavage of pentaphenylarsorane with potassium amide in liquid ammonia yields $K^+[Ph_3As{=}N]^- \cdot NH_3$ and $K_2[PhAs(NH)_3]$.

The arsinimines, R_3AsNR', are prepared from tertiary arsines with Chloramine-T, $ClNHR'$ ($R' = para\text{-}CH_3C_6H_4SO_2$), chloramine, $ClNH_2$, or triarylarsines with organic azides:

$$AsR_3 + R'{-}N_3 \longrightarrow R_3AsNR' + N_2$$

Treatment of $[Ph_3As{-}NH_2]^+Cl^-$ with sodium amide in liquid ammonia also gives Ph_3AsNH.

Quasi-arsonium salts are obtained from tertiary arsines with amine-free chloramines:

$$AsR_3 + Cl{-}NR'_2 \longrightarrow [R_3\overset{+}{As}{-}NR'_2]Cl^-$$

9.1.8. Organoarsenic Compounds Containing As-As Bonds

Preparation. Compounds containing As—As bonds include the tetraorganodiarsines, $R_2As{-}AsR_2$, and cyclic $(RAs)_n$.

Diarsines are prepared by reduction of arsinic acids with hypophosphite, reduction of diarsoxanes, $R_2As{-}O{-}AsR_2$, with phosphorus acid or electrochemically. Hypophosphite reduction yields heterocycles containing As—As bonds:

$$R{-}As(O)(OH){-}(CH_2)_n{-}As(O)(OH){-}R \xrightarrow{H_3PO_2}$$

n = 1: $RAs{-}AsR$ / $RAs{-}AsR$ six-membered ring (two CH_2 bridges)

n = 2: eight-membered ring $RAs{-}AsR$ / $RAs{-}AsR$ with two $(CH_2)_2$ bridges

n = 3: five-membered ring $(CH_2)_3$ with $AsR{-}AsR$

n = 4: six-membered ring $(CH_2)_4$ with $AsR{-}AsR$

Diorganoarsenic iodides are coupled by mercury metal:

$$2\,R_2AsI + Hg \longrightarrow R_2As{-}AsR_2 + HgI_2$$

Condensation of diorganochloroarsines with secondary arsines also leads to diarsines:

$$R_2AsCl + HAsR_2 \longrightarrow R_2As{-}AsR_2 + HCl$$

Diarsines can be obtained from elemental arsenic by treatment with sodium in liquid ammonia and then with alkyl halides:

$$2\,As + 4\,Na \longrightarrow Na_2As{-}AsNa_2 \xrightarrow[-4\,NaX]{4\,RX} R_2As{-}AsR_2$$

Linear polyarsines with more than two arsenic atoms are rare:

$$RAs(SnR_3)_2 + 2\,R_2AsCl \longrightarrow R_2As{-}As(R){-}AsR_2 + 2\,R_3SnCl$$

$$As(SnR_3)_3 + 3\,R_2AsCl \longrightarrow R_2As{-}As(AsR_2){-}AsR_2 + 3\,R_3SnCl$$

Polyarsines, $(RAs)_n$, were formulated as dimers with As=As bonds, but they are cyclic oligomers. However, the double-bonded form, Ph—As=As—Ph, is trapped by coordination to transition metals as found in the complex $PhAsAsPhFe(CO)_4$.

Cyclic polyarsines are prepared by the reduction of arsonic and arsonous acids, organodihaloarsines or monoorganoarsenic oxides with hypophosphorous acid, dithionite, tin(II) chloride or sodium amalgam, by oxidation of primary arsines or by condensation of organodihaloarsines with primary arsines. Cyclopentaarsines form in the reaction of primary arsines, $RAsH_2$, with dibenzylmercury. A three-membered As_3-ring is formed as a part of a polycyclic cage:

$$MeC(CH_2AsI_2)_3 + Na/THF \longrightarrow$$

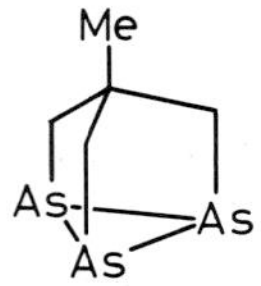

Me
As As
As

The solid $(RAs)_n$ compounds are polymerized. The cyclic structures of the phenyl (hexameric) and methyl derivative (pentameric) are puckered:

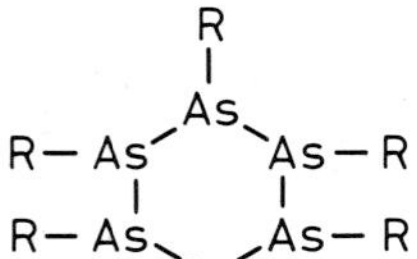

R=Ph

R=Me

Both the linear and cyclic polyarsines are cleaved by oxygen, and halogens; elemental sulfur inserts into As—As bonds:

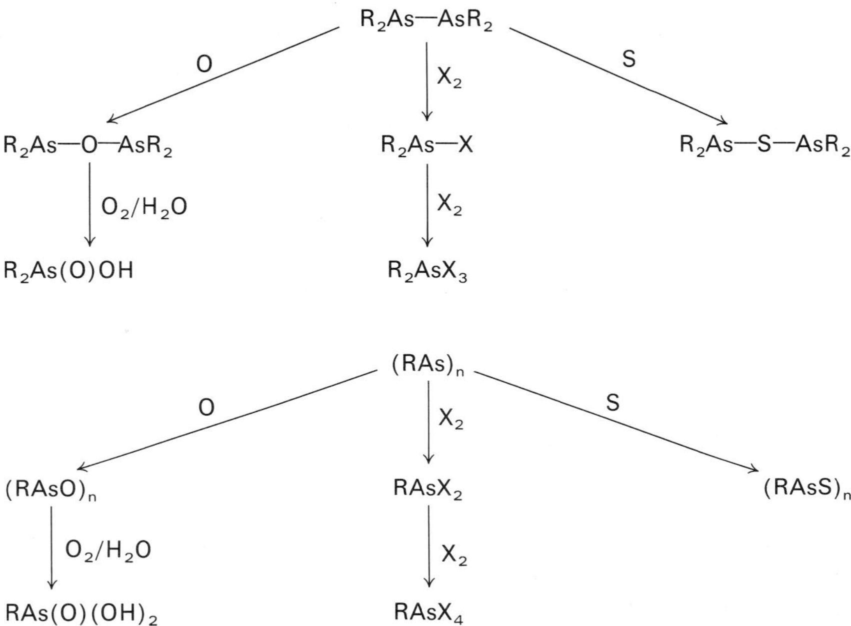

Polyarsines form transition metal complexes with metal carbonyls.

9.2. Organoantimony Compounds

Organoantimony compounds are analogous to organoarsenic compounds, and a similar systematization and nomenclature can be used. However, there are significant differences in the behavior of the two elements. Interest in organoantimony compounds arose after the chemotherapeutical properties of organoarsenic compounds were discovered.

The first organoantimony compound, a tertiary stibine, was prepared in 1850.

9.2.1. Fully Organosubstituted Derivatives

Fully substituted, homoleptic species include triorganostibines (tertiary stibines), SbR_3, tetrasubstituted cations (stibonium ions), $[SbR_4]^+$, and pentasubstituted compounds (stiboranes), SbR_5. Hexasubstituted anions, $[SbR_6]^-$, are also known.

Tertiary Stibines, SbR_3

Grignard alkylation is frequently employed for synthesis of tertiary stibines, but the lower alkyl derivatives can be difficult to separate from the solvent:

$$SbCl_3 + 3\,RMgCl \longrightarrow SbR_3 + 3\,MgCl_2$$

Dimagnesium compounds give heterocyclic stibines:

$$ClMg(CH_2)_5MgCl + MeSbCl_2 \xrightarrow[-MgCl_2]{} \text{(cyclic } C_5H_{10}SbMe\text{)}$$

Organolithium reagents yield arylstibines and the heterocyclic stiboles (R = Ph):

$$R_4C_4Li_2 + RSbCl_2 \xrightarrow[-2LiCl]{} R_4C_4SbR$$

Lithium cyclopentadienyl with Me_2SbBr yields the fluxional dimethylantimony η^1-cyclopentadienyls.

Organosodium reagents are employed for stibines containing acetylenic, $R_2Sb{-}C{\equiv}CR$, and cyclopentadienyl, $R_2Sb{-}C_5H_5\text{-}\eta^1$ and $Sb(C_5H_5\text{-}\eta^1)_3$, groups. Organosodium intermediates are also formed in the Wurtz-Fittig synthesis of triphenylstibine ($SbCl_3 + PhCl + Na$).

Organoaluminum compounds yield triorganostibines from antimony(III) oxide:

$$Sb_2O_3 + 2\,AlR_3 \longrightarrow 2\,SbR_3 + Al_2O_3$$

Reduction of triorganodihalogenostibines, R_3SbX_2, with zinc, lithium borohydride, or hydrazine hydrate also gives triorganostibines:

$$R_3SbX_2 + Zn \longrightarrow SbR_3 + ZnX_2$$

This reaction is used to purify stibines prepared with Grignard reagents by conversion to R_3SbX_2 which is isolated from the ether and then reduced.

Ditertiary stibines are prepared form secondary stibines, after metallation:

$$R_2SbH \xrightarrow{PhLi} R_2SbLi \xrightarrow[n=3-6]{Br(CH_2)_nBr} R_2Sb(CH_2)_nSbR_2$$

The direct synthesis from elemental antimony and organic halides, first used in 1850, is used for the preparation of tris(perfluoromethyl)stibine, $Sb(CF_3)_3$, from CF_3I at 165–170 °C, and tris(pentafluorophenyl)antimony, $Sb(C_6F_5)_3$, from $(C_6F_5)_2$ TlBr.

Tris(trifluoromethyl)stibine $Sb(CF_3)_3$ is trigonal pyramidal in the gas phase.

Trialkylstibines are air-sensitive; the lower alkyls are pyrophoric. The aromatic derivatives are more stable. The stibines reduce phosphorus pentachloride to elemental phosphorus. With halogens they undergo oxidative-addition reactions to form the dihalides, R_3SbX_2, and with alkyl halides to produce stibonium halides, $[R_3SbR']^+X^-$.

The tertiary stibines are cleaved by halogens, oxygen and lithium or sodium:

$$SbR_3 + 2M \xrightarrow[\text{liq. } NH_3]{\text{THF or}} R_2SbM + RM$$

Mercury(II) chloride cleaves triarylstibines to form R_2SbCl and RHgCl, but $CuCl_2$, $TlCl_3$, and $FeCl_3$ and the phosphorus, arsenic and antimony tri- or pentahalides oxidize them to R_3SbCl_2.

Tertiary stibines form complexes with transition metals in their lower oxidation states, but the stibines are weaker donors than the corresponding phosphines and arsines.

Stibonium Salts, $[SbR_4]^+X^-$

Tetraorganostibonium salts are prepared by the quaternization of tertiary stibines with alkyl halides. Aromatic stibines require trimethyloxonium tetrafluoroborate in liquid sulfur dioxide:

$$SbPh_3 + Me_3O^+BF_4^- \longrightarrow Ph_3SbMe^+ + BF_4^- + Me_2O$$

If the excess of halogen is avoided, the pentaorganoderivatives can be cleaved to stibonium halides with halogens, but this reaction is not suited as a preparative procedure:

$$SbR_5 + X_2 \longrightarrow SbR_4^+X^- + RX$$

Grignard arylation of triphenylantimony dichloride also yields a stibonium salt:

$$Ph_3SbCl_2 \xrightarrow[\text{2. HBr}]{\text{1. PhMgCl}} [SbPh_4]^+ Br^-$$

Tetraphenylstibonium bromide is obtained from a Friedel-Crafts reaction:

$$SbPh_3 + PhBr \xrightarrow{+AlCl_3} SbPh_4^+ Br^-$$

The solid salt, $[SbMe_4]^+[Al(OSiMe_3)_4]^-$, contains tetrahedral stibonium cations.

Five-coordinated, oxygen-containing insoluble $R_4Sb—OH$ derivatives form from tetraalkylstibonium halides and moist silver oxide. The alkoxides, $R_4Sb—OR$, are prepared from the halides and sodium alkoxides, and the reduction of stibonium halides with $LiAlH_4$ produces triorganostibines:

$$4[SbR_4]^+X^- + LiAlH_4 \longrightarrow 4SbR_3 + 4RH + LiX + AlX_3$$

Pentaorganoantimony Derivatives (Stiboranes), SbR_5

The aryl-pentasubstituted derivatives are more stable. They can be obtained by reactions of arylantimony(V) halides with organolithium reagents, organozinc and Grignard reagents:

$$Me_3SbBr_2 + 2MeLi \longrightarrow SbMe_5 + 2LiBr$$

$$[SbEt_4]^+Cl^- + EtLi\ (\text{or } ZnEt_2) \longrightarrow SbEt_5 + LiCl\ (\text{or } EtZnCl)$$

$$(H_2C{=}CH)_3SbBr_2 + 2H_2C{=}CHMgBr \longrightarrow Sb(CH{=}CH_2)_5 + 2MgBr_2$$

Mixed derivatives can be obtained with reagents containing different organic groups:

$$Et_3Sb(C{\equiv}CMe)_2 \xleftarrow{+LiC\equiv CMe} Et_3SbCl_2 \xrightarrow{+LiMe} Et_3SbMe_2$$

Pentaphenylantimony is obtained by the reaction of $[SbPh_4]^+Br^-$ and phenyllithium, by the reaction of antimony pentachloride and phenyllithium.

Solid pentaphenylantimony has an unusual square-pyramidal trigonal geometry unlike the bipyramidal geometry of $AsPh_5$ and other pentacoordinated compounds. Solid pentamethylantimony and $Sb(para\text{-}C_6H_4CH_3)_5$, however, have trigonal bipyramidal structures.

Hexasubstituted Anions, $[SbR_6]^-$

The highest degree of organosubstitution is achieved in anions produced from pentaphenylantimony and phenyllithium:

$$SbPh_5 + PhLi \longrightarrow Li^+[SbPh_6]^-$$

Stibabenzene (Antimonine)

This reactive dicoordinated organoantimony is obtained by dehydrohalogenation of a heterocyclic chloride:

$$C_5H_5SnBu_2 \xrightarrow[-Bu_2SnCl_2]{SbCl_3} C_5H_5SbCl \xrightarrow{-HCl} C_5H_5Sb$$

9.2.2. Organoantimony Hydrides (Primary and Secondary Stibines), R_nSbH_{3-n}

The unstable compounds R_2SbH and $RSbH_2$ are prepared by the reduction of the corresponding organoantimony halides, R_2SbX and $RSbX_2$, with alkali metal borohydrides or lithium alanate at low temperatures. The lower alkyl hydrides decompose slowly even on storage at $-70\,°C$.

9.2.3. Organoantimony Halides

Mono- and Diorganohalostibines, R_nSbX_{3-n}

Organoantimony(III) halides are seldom made by the Grignard synthesis but with tert-butylmagnesium chloride, the diorganoderivative, R_2SbCl, is obtained from antimony(III) chloride. To achieve partial substitution in SbX_3, a reagent of lower reactivity is required; organolead and organotin compounds are suitable for the chlorides, and organosilicon compounds are satisfactory for the fluorides:

$$SbCl_3 + PbR_4 \longrightarrow R_2SbCl + R_2PbCl_2$$
$$SbF_3 + 2\,[PhSiF_5]^{2-} \longrightarrow Ph_2SbF + 2\,[SiF_6]^{2-}$$

Pyrolysis of triorganodihalides, R_3SbX_2, the redistribution between inorganic trihalides and tertiary stibines and the reduction of stibonic and stibinic acids with sulfur dioxide and hydriodic acid in hydrochloric medium or with tin(II) chloride are used in the synthesis of organoantimony(III) halides:

$$R_3SbX_2 \xrightarrow{\Delta} R_2SbX + RX$$
$$SbX_3 + SbR_3 \longrightarrow R_2SbX + RSbX_2$$
$$RSbO(OH)_2 + SO_2 + HI \xrightarrow{HCl} RSbCl_2$$
$$R_2Sb(O)OH + SO_2 + HI \xrightarrow{HCl} R_2SbCl$$

The direct synthesis from alkyl halides and metallic antimony is used for the methyl derivatives, Me_2SbCl and $MeSbCl_2$, which can be separated only after conversion into the butyl derivatives, Me_nSbBu_{3-n}. Heating a chloromethylsilane with metallic antimony at 150–200 °C in the presence of tetraalkylammonium salts gives a mixture of $(RMe_2SiCH_2)_nSbCl_x$ (n = 2, x = 1, and n = 3, x = 2).

Organoantimony(V) Tetrahalides, $RSbX_4$

Only the aromatic derivatives, $ArSbCl_4$, can be prepared from diazonium salts and antimony(III) chloride, or by chlorination of aryldichlorostibines. The organoantimony tetrahalides are usually converted to stibonic acids.

The tetrahalides react with alkylammonium salts to form arylpentachloroantimonates, $[RNH_3]^+[RSbCl_5]^-$.

Diorganoantimony(V) Trihalides, R_2SbX_3

Dimethylantimony trichloride is prepared by chlorination of Me_2SbCl and bis(chlorovinyl)antimony trichloride by addition of acetylene to antimony pentachloride. Other alkyl derivatives are obtained by the chlorination of distibines,

$R_2Sb{-}SbR_2$. An ionic dimer of Me_2SbCl_3, $[SbMe_4]^+[SbCl_6]^-$, is prepared from $SbCl_5$ and Me_2InCl.

Aromatic derivatives are obtained from aryldiazonium salts with antimony chlorides ($SbCl_3$ or $SbCl_5$), from stibonic acids and hydrochloric acid, by the reaction of phenylhydrazine hydrochloride with antimony pentachloride in the presence of copper(II) chloride and oxygen, and by chlorination of diarylchlorostibines. The fluoride Ph_2SbF_3 is obtained by fluorination of Ph_2SbF with xenon difluoride and by treatment of $Ph_2Sb(O)OH$ with SF_4.

Solid Ph_2SbCl_3 is a dimer with chloride bridges of unequal length:

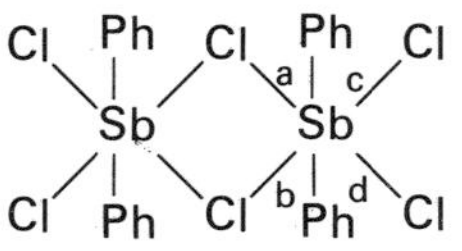

a = 283.9 pm = 2.84 Å
b = 262.0 pm = 2.62 Å
c = 238.8 pm = 2.39 Å
d = 234.6 pm = 2.35 Å

while Ph_2SbBr_3, $Ph_2SbClBr_2$ and Ph_2SbCl_2Br are monomeric, with trigonal-bipyramidal structures.

The tendency of antimony to increase its coordination number from five to six results in addition of halide and pseudohalide ions:

$$Ph_2SbCl_3 + [NMe_4]^+X^- \longrightarrow [NMe_4]^+[Ph_2SbCl_3X]^-$$
$$X = Cl, Br, N_3, NCS$$

The structure of the $[R_2SbX_4]^-$ anions is *trans*-R_2.

Triorganoantimony Dihalides, R_3SbX_2

The alkyl derivatives of antimony are generally unstable, but the trialkylantimony(V) dihalides are prepared by halogenation of trialkylstibine (with hydrogen evolution) or by reduction of phosphorus trichloride with a trialkylstibine with formation of elemental phosphorus. Aromatic derivatives, R_3SbX_2, precipitate on treatment of aromatic stibines with halogens. The reaction of antimony pentachloride with diphenylmercury also yields Ph_3SbCl_2 and PhHgCl. Triphenylantimony difluoride is formed in the reaction of Ph_3SbO with SF_4 or in the fluorination of $SbPh_3$ with XeF_2.

The dihalides decompose above their melting points with formation of R_2SbX. Hydrolysis yields $R_3Sb(OH)X$ and $R_3Sb(OH)_2$. With tertiary phosphines they form tetracoordinated anionic species by transfer of organic groups to phosphorus:

$$Me_3SbBr_2 + PR_3 \longrightarrow [R_3PMe]^+[Me_2SbBr_2]^-$$

Solid trimethylantimony dichloride, Me_3SbCl_2, tris(2-chlorovinyl)antimony-dichloride, $(ClCH{=}CH)_3SbCl_2$, and Me_3SbF_2 are trigonal bipyramidal with the organic groups equatorial and the halogens axial:

$$\begin{array}{c} X \\ | \\ R_2\text{Sb}-R \\ | \\ X \end{array}$$

Conductivity measurements in acetonitrile confirm their nonelectrolytic, covalent character.

Tetraorganoantimony(V) Halides, R_4SbX

Only the fluorides form molecular tetraorganoantimony derivatives, other halides being ionic (tetraorganostibonium) salts. Tetramethylantimony fluoride, Me_4SbF, is prepared from pentamethylantimony and KHF_2 or HF:

$$SbMe_5 \; + \; KHF_2 \text{ (or HF)} \longrightarrow Me_4Sb{-}F$$

Related pseudohalides are prepared by similar reactions (R = Me, Ph; X = N_3, CN, SCN):

$$SbMe_5 \; + \; HX \longrightarrow Me_4Sb{-}X$$

Solid Me_4SbF is polymeric with six-coordinated antimony.

9.2.4. Oxygen-Containing Organoantimony Compounds

Among the oxygen-containing organoantimony compounds are the following:

$R{-}Sb(OH)_2$	stibonous acids
$R{-}SbO$	organoantimony(III) oxides (stibonous acid anhydrides)
$R_2Sb{-}OH$	stibinous acids
$R_2Sb{-}O{-}SbR'_2$	distiboxanes (stibinous acid anhydrides)
$R{-}Sb(OR')_2$	stibonous acid esters
$R_2Sb{-}OR'$	stibinous acid esters
$R{-}SbO(OH)_2$	stibonic acids
$R_2Sb(O)OH$	stibinic acids
$R_nSb(OR')_{5-n}$	alkyl(aryl)alkoxystiboranes
$R_3Sb(OH)_2$	triorganoantimony(V) dihydroxides
R_3SbO	stibine oxides

Anhydrides of Stibonous and Stibinous Acids

The free acids, $RSb(OH)_2$ and $R_2Sb{-}OH$, are known only as their anhydrides, RSbO and $R_2Sb{-}O{-}SbR'_2$, respectively, which are prepared by alkaline hydrolysis of organoantimony(III) halides, or by *in situ* reduction of arylstibonic acids with sulfur dioxide followed by alkaline hydrolysis. Distiboxanes are also formed by thermal disproportionation of organoantimony(III) oxides:

$$4\,RSbO \longrightarrow (R_2Sb)_2O + Sb_2O_3$$

or by cleavage of triarylstibines with acids, followed by treatment with alkalis.

The insoluble monoorganoantimony(III) oxides are polymeric; the soluble, low-melting distiboxanes, $R_2Sb{-}O{-}SbR'_2$, are monomeric in the solid state.

Treatment of distiboxanes with carboxylic acids leads to diorganoantimony(III) carboxylates, $R_2Sb{-}OCOR'$.

Stibinous-Acid Esters, $RSb(OR')_2$

These derivatives are prepared from organoantimony dihalides and sodium alkoxides:

$$RSbCl_2 + 2\,NaOR' \longrightarrow RSb(OR')_2 + 2\,NaCl$$

or directly from dihalides and alcohols (diols) in the presence of bases:

$$RSbX_2 + \begin{matrix} HO\diagdown\!\diagup R' \\ | \\ HO\diagup\!\diagdown R' \end{matrix} \xrightarrow{NEt_3} R{-}Sb\begin{matrix} \diagup O\diagdown\!\diagup R' \\ | \\ \diagdown O\diagup\!\diagdown R' \end{matrix}$$

Stibinous-Acid Esters, R_2Sb-OR'

These derivatives are obtained similarly, from dialkylantimony halides and sodium alkoxides:

$$R_2Sb{-}Cl + NaOR' \longrightarrow R_2Sb{-}OR' + NaCl$$

Stibonic Acids, $RSbO(OH)_2$

Aliphatic derivatives are unknown, but stable aromatic derivatives are prepared from aryldiazonium salts with antimony halides, or by the precipitation of the $[Ar{-}N_2]^+[SbCl_4]^-$ salt from a hydrochloric solution of the diazonium salt on treatment with antimony(III) chloride followed by alkaline treatment producing nitrogen evolution and re-acidification to give the stibonic acid. The hydrolysis of arylantimony tetrachlorides, $RSbCl_4$, is also used.

The stibonic acids may be polymeric, but a hydrated six-coordinated structure, $[R{-}Sb(OH)_5]^-H^+$, is possible, analogous to the inorganic anion, $[Sb(OH)_6]^-$.

The aromatic stibonic acids form organoantimony(V) tetrachlorides, $RSbCl_4$, with concentrated hydrochloric acid used for the purification of the acids. With sulfur dioxide and hydrogen iodide in hydrochloric acid solution, the acids are reduced to aryldichlorostibines.

Stibinic Acids, $R_2Sb(O)OH$

Dimethylstibinic acid is prepared by the hydrolysis of dimethylantimony trichloride, Me_2SbCl_3, or by the wet oxidation of tetramethyldistibine, $Me_2Sb{-}SbMe_2$. Aromatic derivatives are formed as a mixture with stibonic acids in the reaction of arylhydrazines with antimony trichloride in the presence of copper(I) chloride, or from aryldiazonium salts with monoorgano-substituted antimony(III) compounds:

$$[R{-}N_3]^+Cl^- + R'SbCl_2 \longrightarrow RR'SbCl_3 \xrightarrow{H_2O} RR'Sb(O)OH$$

or by oxidation of triarylstibines with hydrogen peroxide in alkaline medium. Pure compounds are obtained by hydrolysis of diarylantimony(V)trichlorides.

Heterocyclic stibinic acids are formed by cyclodehydration of stibonic acids:

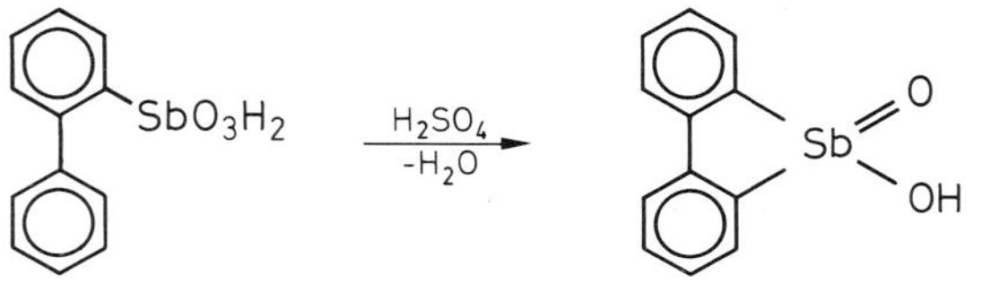

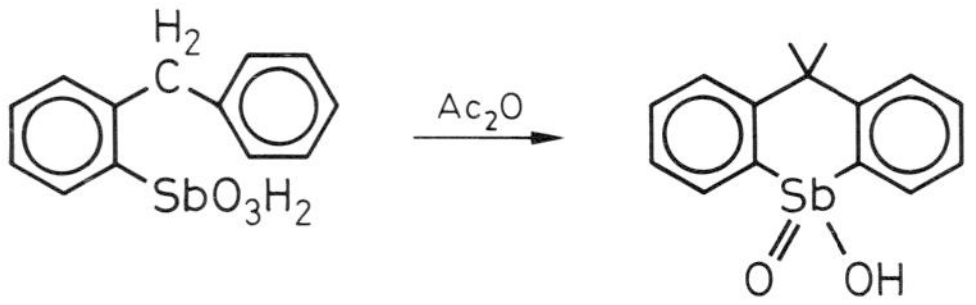

SbO_3H_2 $\xrightarrow[-H_2O]{H_2SO_4}$ Sb(=O)OH

H_2C ... SbO_3H_2 $\xrightarrow{Ac_2O}$ Sb(=O)OH

Stibinic acids are reduced with sulfur dioxide-hydrogen iodide in hydrochloric medium to diarylchlorostibines, and are converted with concentrated hydrochloric acid to diarylantimony trichlorides, R_2SbCl_3.

The polymeric stibonic and stibinic acid anhydrides, $RSbO_2$ and $R_2Sb(O){-}O{-}Sb(O)R'_2$, respectively, can be obtained by dehydration of the corresponding acids.

Organoalkoxystiboranes, $R_nSb(OR')_{5-n}$

Dialkyltrialkoxystiboranes (dialkoxyantimony trialkoxides), $R_2Sb(OR')_3$, are prepared from the corresponding trichlorides and sodium alkoxides at low temperature:

$$R_2SbBr_3 + 3\,NaOR' \xrightarrow{-40\,°C} R_2Sb(OR')_3 + 3\,NaBr$$

The $Me_2Sb(OMe)_3$ derivative is a dimer:

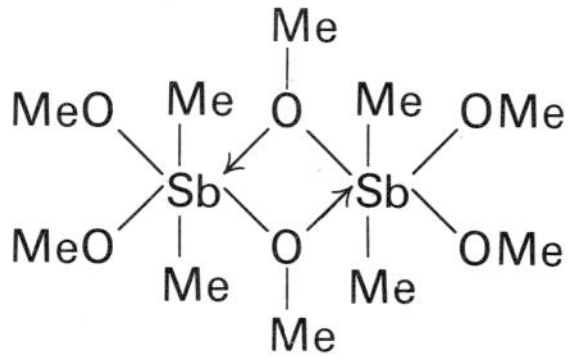

Me
MeO Me O Me OMe
Sb Sb
MeO O OMe
Me Me
Me

Triorganodialkoxystiboranes (triorganoantimony dialkoxides), $R_3Sb(OR')_2$, are prepared similarly from dihalides and alkali metal alkoxides.

Tetraorganoalkoxystiboranes (tetraorganoantimony alkoxides), R_4SbOR', are prepared by the cleavage of pentasubstituted derivatives with alcohols or phenols:

$$SbR_5 + R'OH \xrightarrow{90-100\,°C} R_4Sb{-}OR' + RH$$

Triorganoantimony(V) Hydroxides, $R_3Sb(OH)_2$, and Stibine Oxides, R_3SbO

These compounds are interconvertible. Trialkylantimony hydroxides are obtained by the hydrolysis of the trichlorides, but the trifluoromethyl derivative, $(CF_3)_3SbCl_2$, on hydrolysis yields the salt, $[(CF_3)_3SbCl_2(OH)]^- H_3O^+$, which can be converted with wet silver oxide into $Ag^+[(CF_3)_3Sb(OH)_3]^-$. Trimethylantimony(V) hydroxide, $Me_3Sb(OH)_2$, is dehydrated *in vacuo* to form trimethylstibine oxide, Me_3SbO.

Triarylantimony(V) hydroxides, are obtained by hydrolysis of dihalides in alkaline medium, or by oxidation of triarylstibines with hydrogen peroxide in acetone or with HgO in ether. They are dehydrated to form stibine oxides:

$$R_3SbCl_2 \xrightarrow{H_2O} R_3Sb(OH)_2 \xrightarrow[-H_2O]{} R_3SbO$$

The oxides, R_3SbO, are polymeric.

Triarylantimony(V) hydroxides, which exist as $[R_3Sb(OH)_3]^-$ anions in aqueous solution, are strong bases and precipitate metal hydroxides in reactions with their salts.

9.2.5. Sulfur-Containing Organoantimony Compounds

Organoantimony(III) sulfides RSbS and $R_2Sb{-}S{-}SbR'_2$, are prepared by the reaction of chloroorganostibines or the corresponding oxides with hydrogen sulfide. Arylantimony sulfides are obtained from the oxides with carbon disulfide in the presence of ammonia, or with dithiocarbamates. The RSbS compounds are polymeric.

Dithiostibonous-Acid Esters, $RSb(SR')_2$ (Organoantimony(III) Dithiolates)

These diesters are obtained from monoorganoatimony(III) oxides or dihalides with thiols. Heterocyclic esters are prepared from dithiols:

$$PhSbCl_2 + \text{HS–CH}_2\text{CH}_2\text{–SH} \xrightarrow{NEt_3} \text{Ph–Sb}\langle\text{S–CH}_2\text{CH}_2\text{–S}\rangle$$

Thiostibinous-Acid Esters, $R_2Sb\text{-}SR'$

These monoesters are formed in a similar manner:

$$\left.\begin{array}{l} 2\,R_2Sb{-}Cl \;+\; 2\,NaSR' \\ R_2Sb{-}O{-}SbR_2 \;+\; 2\,R'SH \end{array}\right] \longrightarrow 2\,R_2Sb{-}SR'$$

or by cleavage of triphenylstibine:

$$SbPh_3 \;+\; PhSH \xrightarrow{50\,°C} Ph_2Sb{-}SPh \;+\; PhH$$

Triorganoantimony Dithiolates, $R_3Sb(SR')_2$

These dithiolates are formed from triorganoantimony dihalides with thiols:

$$R_3SbCl_2 \;+\; 2\,R'SH \xrightarrow{NEt_3,\ -30\,°C} R_3Sb(SR')_2 \;+\; 2\,HCl$$

Tetraalkylantimony Thiolates, $R_4Sb\text{-}SR'$

These thiolates are obtained by cleavage of pentaalkyls with thiols:

$$SbMe_5 \;+\; RSH \longrightarrow Me_4Sb{-}SR \;+\; CH_4$$

Stibine Sulfides, R_3SbS

Stibine sulfides, R_3SbS, are prepared by oxidative addition of sulfur to tertiary stibines, by treatment of trialkylantimony hydroxides with hydrogen sulfide, or by reaction of triorganoantimony dichlorides with hydrogen sulfide in ammonia-alcohol solutions:

$$\begin{array}{ccccc} R_3Sb + S & \longrightarrow & R_3SbS & \longleftarrow & R_3Sb(OH)_2 + H_2S \\ & & \uparrow & & \\ & & R_3SbX_2 + H_2S & & \end{array}$$

9.2.6. Nitrogen-Containing Organoantimony Compounds

Organodiaminostibines $RSb(NR'_2)_2$, and diorganoaminostibines $R_2Sb{-}NR'_2$, are prepared from the corresponding halides and lithiated amines:

$$RSbX_2 \;+\; 2\,LiNR'_2 \longrightarrow RSb(NR'_2)_2 \;+\; 2\,LiX$$
$$R_2SbX \;+\; LiNR'_2 \longrightarrow R_2Sb{-}NR'_2 \;+\; LiX$$

The reaction of triorganoantimony dihalides with ammonia gives amino-stibonium salts, $[R_3Sb{-}NH_2]^+X^-$.

Triarylstibine imines, $R_3Sb{=}NR'$, are obtained from tertiary stibines and the sodium salt of N-bromacetamide or N-chlorosulfonamides and by treatment of triarylantimony dichlorides with sodium amide. Triorganostibines react with chloramine to form $(ClR_3Sb)_2NH$ derivatives.

9.2.7. Organoantimony Compounds with Sb-Sb Bonds

The older literature contains many references to compounds formulated as analogues of azobenzene, i.e., R—Sb=Sb—R. The insoluble $(RSb)_n$ compounds are, like the analogous arsenic compounds, cyclic oligomers or linear polymers. The solid tetramer, $(Bu^tSb)_4$, obtained from Bu^t_2SbLi and iodine or from Bu^t_2SbCl and magnesium in THF:

$$2Bu^t_2SbCl + Mg \xrightarrow{THF} \tfrac{1}{4}(Bu^tSb)_4 + SbBu^t_3 + MgCl_2$$

and the hexamer, $(PhSb)_6 \cdot C_6H_6$, prepared from phenylstibine are cyclic stibanes.

Other preparations of $(RSb)_n$ derivatives include the reduction of stibonic acids with sodium dithionite or hypophosphite, the decomposition of arylantimony hydrides, or the condensation of hydrides with aryldichlorostibines. The presence of halogen in the product is interpreted in terms of a polymeric structure with halogen terminal groups, $Cl(SbR)_nCl$.

Tetraorganodistibines, $R_2Sb—SbR_2$ are obtained by the action of organic free radicals upon metallic antimony mirrors, by reduction of dialkylantimony bromide with sodium in liquid ammonia or with magnesium in THF, or by the reaction of alkyl halides with antimony in the presence of sodium or lithium also in liquid ammonia. Aromatic derivatives are formed in the reduction of diaryliodostibines with sodium hypophosphite.

Tetraorganodistibines are cleaved by halogens and hydrogen halides. On heating above 200 °C they disproportionate to form tertiary stibines and metallic antimony.

9.3. Organobismuth Compounds

Bismuth is more metallic than its higher cogeners in Group V A. The bismuth-carbon bond is less thermally stable, and is readily cleaved. Several organoarsenic and organoantimony compound classes have no analogues in organobismuth chemistry. Thus organobismuth hydrides decompose at − 50 °C, as do the compounds containing Bi—Bi bonds, and the hydrolysis products of organobismuth halides are better described as hydroxides rather than acids. The organoarsenic and antimony acids have no analogues in organobismuth chemistry. For antimony a tendency to form six-coordinated hydroxides by addition of extra water to the acids is noted, and this trend becomes more pronounced for bismuth. Nitrogen- and sulfur-containing organobismuth compounds are little known. Only homoleptic and halide derivatives will be discussed.

Organobismuth compounds are extremely toxic.

9.3.1. Fully Organosubstituted Bismuth Compounds

Tertiary Bismuthines, BiR_3

Triethylbismuth, the first organobismuth compound was prepared in 1850 in a reaction between ethyl iodide and a bismuth-alkali metal alloy. Bismuth metal or bismuth tribromide reacts with diorganomercury compounds. Grignard or organolithium compounds are effective, and organosodium compounds lead to acetylenic derivatives. Organoaluminum reagents react with bismuth(III) oxide to form triorganosubstituted compounds. Other methods include the electrochemical synthesis of BiR_3 derivatives by electrolysis of the salt $Na^+[R_3AlOEt]^-$ with a bismuth anode, and the decomposition of the aryldiazonium salts, $[ArN_3]^+[BiCl_4]^-$, in the presence of metallic copper.

Solid triphenylbismuth is trigonal pyramidal with different angles of rotation of the phenyl groups about the Bi—C bonds.

The trialkylsubstituted bismuth derivatives are pyrophoric in air, decompose on distillation, and are readily cleaved by halogens. The aromatic derivatives are more stable in air. Strong acids cleave all organobismuth derivatives to inorganic bismuth(III) compounds, but weak acids lead to partial cleavage of aromatic groups to R_2BiX and $RBiX_2$ derivatives. The triorganobismuth compounds (bismuthines) are weaker donors than tertiary arsines and stibines, but some transition-metal complexes of triphenylbismuth are known.

Tetraorganobismuthonium Salts, $[BiR_4]^+X^-$

Only aromatic bismuthonium derivatives can be prepared from pentaphenylbismuth by bromine cleavage at $-78\,°C$:

$$BiPh_5 + Br_2 \xrightarrow{-78\,°C} [BiPh_4]^+Br^- + PhBr$$

Tetraphenylbismuthonium chloride is prepared by cleavage of pentaphenylbismuth with hydrogen chloride and can be converted to a more stable tetraphenylborate:

$$BiPh_5 \xrightarrow[-78\,°C]{HCl} [BiPh_4]^+Cl^- \xrightarrow{NaBPh_4} [BiPh_4]^+[BPh_4]^-$$
$$-PhH$$

The solid salt, $[BiPh_4]^+ClO_4^-$, contains tetrahedral bismuth.

Pentaorgano-Substituted Derivatives, BiR_5

Pentaphenylbismuth is prepared from phenyllithium with triphenylbismuth dichloride:

$$R_3BiCl_2 + 2\,RLi \xrightarrow{-75\,°C} BiR_5 + 2\,LiCl$$
$$R = Ph$$

The product decomposes at ca. 100 °C and is converted to tetraphenylbismuthonium salts with bromine in CCl_4 or HCl in ether. With excess of phenyllithium it forms a hexaphenylsubstituted anion, $[BiPh_6]^-$.

9.3.2. Organobismuth Halides

Organobismuth(III) mono- und dihalides R_nBiX_{3-n} have been obtained by redistribution between trisubstituted derivatives and bismuth(III) halides:

$$2\,BiR_3 + BiX_3 \longrightarrow 3\,R_2BiX \quad (X = Cl, Br, I)$$
$$BiR_3 + 2\,BiX_3 \longrightarrow 3\,RBiX_2$$

or by arylation of bismuth(III) bromide with tetraphenyllead. Organolead reagents are also used for the synthesis of vinylbismuth dichloride:

$$Pb(CH{=}CH_2)_4 + 2\,BiCl_3 \xrightarrow{CCl_4} 2\,H_2C{=}CH{-}BiCl_2 + (H_2C{=}CH)_2PbCl_2$$

and an organotin heterocycle is employed to prepare a halide which is the precursor of the unstable bismabenzene:

$$C_5H_6SnBu_2 \xrightarrow[-Bu_2SnCl_2]{BiCl_3} C_5H_6BiCl \xrightarrow{-HCl} C_5H_5Bi \longrightarrow \text{dimer}$$

Organobismuth halides are also obtained by cleavage of trisubstituted derivatives with hydrogen halides, iodine chloride, phosphorus or arsenic trichlorides, acyl chlorides, mercury(II) and thallium(III)chlorides.

The reactive organobismuth(III) halides are sensitive to moisture and alcohols and pyrophoric in air. Diarylbismuth halides are biologically active, showing strong sternutatory action, and are toxic.

The hydrolysis of dihalides leads to oxides, RBiO, but dimethylbismuth bromide forms a hydroxide, $R_2Bi{-}OH$. The halides react with sodium ethoxide and thiols to give ethoxy derivatives, $R_2Bi{-}OEt$, and thiolates, $R_2Bi{-}SR'$, respectively.

The anions $[Ph_2BiX_2]^-$ are formed by addition of halide or pseudohalide anions (X = Cl, Br, CN, SCN, N_3) to a diphenylbismuth halide.

Triorganobismuth(V) Dihalides

The R_3BiX_2 compounds are prepared by addition of halogens – in stoichiometric amounts and under controlled conditions to avoid Bi—C bond cleavage – to trisubstituted derivatives:

$$BiR_3 + X_2 \xrightarrow[CHCl_3]{0\,°C} R_3BiX_2$$

Sulfuryl chloride, sulfur monochloride, thionyl chloride or iodine trichloride are also used.

Solid triphenylbismuth dichloride is trigonal bipyramidal with organic groups equatorial and the halogens axial. Conductivity measurements in acetonitrile show no dissociation as $[R_3BiX]^+X^-$.

The thermal stability of the triorganobismuth dichlorides decreases in the order $F > Cl > Br > I$, the iodides decomposing as low as $-60\,°C$. The dihalides are reduced with hydrazine to triorganosubstituted compounds.

9.3.3. Oxygen-Containing Organobismuth Compounds

The triarylbismuth hydroxyhalides, $R_3Bi(OH)X$, are formed by treatment of the dihalides with aqueous ammonia, and the dihydroxides, $R_3Bi(OH)_2$, by treatment with wet silver oxide.

The solid, oxygen-containing salt, $[Ph_3Bi{-}O{-}BiPh_3]^{2+}(ClO_4^-)_2$, contains four-coordinated bismuth.

Part III
Organometallic Compounds of Transition Metals

10. The Electronic Structure and Classification of Transition Metal Organometallic Compounds

10.1. The Transition Metals

Transition metals are elements with partly filled d- or f-orbitals, either as atoms, as metals or in the zero or other positive or negative oxidation states. In the third period the 3d-level is being populated starting with scandium ($3d^1$) and ending with copper ($3d^{10}$). The elements from scandium to nickel ($3d^9$) are thus transition metals; copper is also considered a transition metal because it is d^9 in some derivatives. The general chemical behavior of copper justifies its inclusion among transition metals.

The electronic structure of transition metals is shown in Table 10.1. In the series from Sc to Cu ($Z = 21 - 29$) and from Y to Ag ($Z = 39$—47), the 3d-and 4d-levels are being occupied stepwise. In the lanthanide family ($Z = 57 - 71$), the 4f-level is being occupied by 14 electrons, followed by the filling of the 5d-level in the series from Hf to Au ($Z = 72 - 79$). These metals are grouped in three transition series, corresponding to the 3d-, 4d - and 5d-levels; the lanthanide (4f-level) and the actinide (5f-level) are inner-transition metals.

The formation of the organometallic derivatives of the transition elemens is dominated by the tendency of the metals to achieve the configuration of the next higher

Tab. 10.1. The Electronic Structure of Transition Metals

^{21}Sc $3d^1 4s^2$	^{22}Ti $3d^2 4s^2$	^{23}V $3d^3 4s^2$	^{24}Cr $3d^5 4s^1$	^{25}Mn $3d^5 4s^2$	^{26}Fe $3d^6 4s^2$	^{27}Co $3d^7 4s^2$	^{28}Ni $3d^8 4s^2$	^{29}Cu $3d^{10} 4s^1$
^{39}Y $4d^1 5s^2$	^{40}Zr $4d^2 5s^2$	^{41}Nb $4d^4 5s^1$	^{42}Mo $4d^5 5s^1$	^{43}Tc $4d^5 5s^2$	^{44}Ru $4d^7 5s^1$	^{45}Rh $4d^8 5s^1$	^{46}Pd $4d^{10} 5s^0$	^{47}Ag $4d^{10} 5s^1$
*	^{72}Hf $5d^2 6s^2$	^{73}Ta $5d^3 6s^2$	^{74}W $5d^4 6s^2$	^{75}Re $5d^5 6s^2$	^{76}Os $5d^6 6s^2$	^{77}Ir $5d^7 6s^2$	^{78}Pt $5d^9 6s^1$	^{79}Au $5d^{10} 6s^1$
n = 15	14	13	12	11	10	9	8	7

* Lanthanides $^{57-71}Ln\ 4f^{0-14} 5d^{0-1} 6s^2$

n = the number of electrons required to achieve a noble gas configuration noble gas by full occupation of (n − 1)d-, ns-and np-electron shells. This is achieved by accepting additional electrons from the ligands. As a result the formation of MR_n or MR_mX_n-type compounds (where R is a σ-bonded organic group) is not typical, because the formation of n covalent bonds (n = the valence of the metal) does not fill the (n − 1)d levels. The formation of compounds with organic ligands able to donate enough π-electrons to complete a noble gas configuration will be preferred. In Table 10.1 the number of electrons required by each transition metal atom is shown. The elements located at the beginning of a transition series require a larger number of electrons, and will usually not achieve the noble gas configuration. As a result the Group III metals form less-stable organometallic compounds, or prefer ligands able to donate a large number of electrons (for example, cyclooctatetraene, a potential 8-electron donor). The metals in the middle of the periods, between Groups V and VIII, are successful in achieving a noble-gas configuration with the aid of electrons from the ligands.

In Chapter 2 the bonds between transition metals and electron-donating organic ligands were discussed. The valence electrons are shared by the metal and the ligand, and back donation from occupied d-orbitals of the metal into vacant orbitals (usually antibonding) of the ligand plays an important role.

10.2. The Ligands

In principle, any unsaturated or aromatic organic molecule or radical can act as a π-ligand. A potentially planar network of sp^2-hybridized carbon atoms possessing unhybridized p_z-orbitals with transition π-electrons with a diameter when planar of ca. 210–230 pm (depending upon the size of the metal atom; 260 pm ≡ 2.6 Å for lanthanides and actinides) can bond to a single transition metal atom. This condition is satisfied only by a dozen examples, shown in Fig. 10.1.

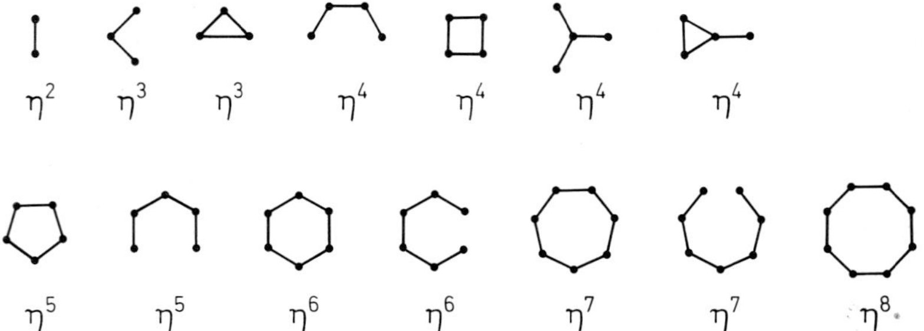

Fig. 10.1. Compact molecules able to bond to a single metal atom in a per*hapto*-fashion (η^n).

Units with larger dimensions or with branched structures can bond either partially to a single metal atom, or entirely to a set of two or more atoms, usually connected by metal-metal bonds. These units are illustrated in Fig. 10.2.

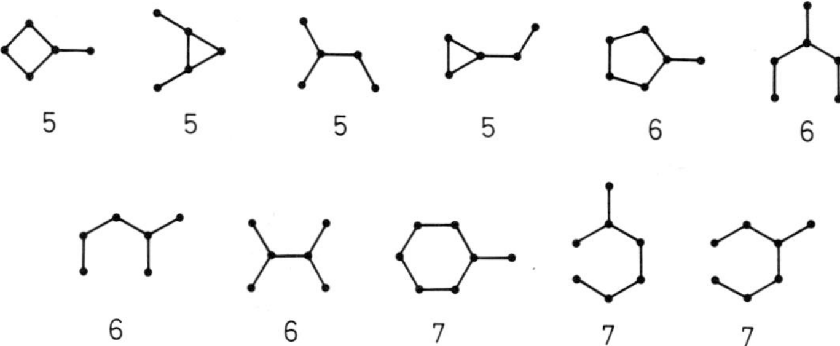

Fig. 10.2. Some branched molecules able to bond completely only to sets of two or more transition metal atoms.

Most of the examples shown in Fig. 10.1 are known to form transition metal complexes, but only few of those shown in Fig. 10.2 do so. Some form carbocation complexes, with the positive carbon uncoordinated, or bimetallic derivatives:

$\oplus$ CH_2 — Fe$(CO)_3$ $\oplus$ CH_2 — Fe — $(CO)_3$Fe — Fe$(CO)_3$

The simple monoolefins form only a single bond, while the polyolefins can use all or just a part of their p_z-orbitals and π-electrons to form bonds. The number of electrons accepted from the ligand depends upon the number required to achieve the next higher noble gas configuration (see the "effective atomic number" or the "18-electron rule" in Section 2.6).

The structure and chemical behavior of transition-metal organometallic compounds is determined largely by the ligand, but it is useful to compare the properties of different metal derivatives of the same ligand. It is thus more convenient to classify the transition-metal derivatives according to the nature of the ligand, or, more exactly, according to the number of electrons contributed by the ligand to attaining the noble gas configuration by the central atom. In the π-complexes known this number varies from two to eight. The ligands are classified as follows:

a) *Two-electron ligands*
carbon monoxide: (:CO), carbon monosulfide (:CS), carbon monoselenide (:CSe), organic isocyanides (:C=N—R), carbenes (:CR_2), cyanide (:CN^-), monoolefins or isolated double bonds.

b) *Three-electron ligands*
η^3-allyl (η^3-C_3H_5), carbyne ($\dot{:}$C—R), cyclopropenyl (η^3-C_3R_3) groups.

c) *Four-electron ligands*
cyclobutadiene, butadiene, cyclopentadiene (as η^4-C_5H_6), hexadiene-1,3 or other molecules containing a butadiene fragment, and the trimethylenemethyl radical.
d) *Five-electron ligands*
cyclopentadienyl (η^5-C_5H_5), cyclohexadienyl, carbollyl, etc.
e) *Six-electron ligands*
benzene and other aromatic molecules, borazine, $B_3N_3H_6$.
f) *Seven-electron ligands*
tropyllium cation (η^7-C_7H_7).
g) *Eight-electron ligands*
cyclooctatetraene.

The formation of a σ-metal-carbon bond, metal-metal bond or any other single covalent bond (for example, M—X, where X = halogen, OR, OH, SR, NRR′, etc.) contributes a single electron to the metal. Therefore, σ-alkyl and σ-aryl groups, as well as other groups attached to the metal through a single covalent bond, are considered in computing the electron balance as one-electron ligands.

Thus, the organometallic derivatives of transition metals will be classified here according to the nature of the ligands.

11. Compounds with Two-Electron Ligands

11.1. Metal Carbonyls

Metal carbonyls are an intensively investigated chapter in transition metal organometallic chemistry. Compounds in which carbon monoxide is attached through a metal-carbon bond to a transition metal atom in a low oxidation state (usually zero or ± 1) include binary compounds of the general formula $M_x(CO)_y$, heterobimetallic carbonyls (containing two different metals) of the type $M_xM'_y(CO)_z$, and many substituted derivatives. Substituted derivatives in which carbon monoxide is replaced by other organic groups, for example, allyl, cyclobutadiene, cyclopentadienyl, arenes, etc., will be discussed in the following chapters. Only mono- and polynuclear binary metal carbonyls will be presented here, with some metal carbonyl anions and cations, hydrides and halides. Polynuclear metal carbonyls containing heteroatoms in a cluster will also be briefly mentioned.

The first metal-carbonyl compound, iron pentacarbonyl, was synthesized in 1890 independently by L. Mond in England and M. Berthelot in France. The carbonyls of cobalt (1910), molybdenum (1910), chromium (1926), tungsten (1928), ruthenium (1936) and iridium (1940) were subsequently prepared. This early period was dominated by the work of W. Hieber and his coworkers. Since 1950 research on metal carbonyls has expanded considerably.

From the beginning, metal carbonyls raised difficult problems for bonding theory with their uncommon properties and stoichiometries (for example, their high volatility) for metal compounds. Their compositions were apparently not related to the formal valency, as it was then understood. Now the structures of many metal carbonyls are known, the nature of the metal-carbon bond is satisfactorily explained, and their composition can be understood in terms of the "effective atomic number rule", as shown in Chapter 2.

Binary metal carbonyls are readily formed by the transition metals of Groups VI – VIII (except palladium and platinum). In Table 11.1 the binary metal carbonyls known at this time are listed. A series of heteronuclear metal carbonyls, containing two more different metals are listed in Table 11.2. This list is not exhaustive, and many similar compounds are possible. The two tables illustrate the great diversity of compounds which can be obtained from transition metals and carbon monoxide only, in addition to the compounds formed using other ligands or by introducing main-group elements as heteroatoms in a polynuclear cluster.

Tab. 11.1. Binary Metal Carbonyls.*

Group IV	Group V	Group VI	Group VII	Group VIII			Group I B
$Ti(CO)_6$* $Ti_2(CO)_y$*	$V(CO)_6$ $V(CO)_{1-5}$* $V_2(CO)_{10}$* $V_2(CO)_{12}$*	$Cr(CO)_6$ $Cr(CO)_{3-5}$*	$Mn_2(CO)_{10}$ $Mn_4(CO)_{16}$ $Mn(CO)_5$*	$Fe(CO)_5$ $Fe_2(CO)_9$ $Fe_3(CO)_{12}$	$Co_2(CO)_8$ $Co_4(CO)_{12}$ $Co_6(CO)_{16}$ $Co(CO)_{1-4}$* $Co_2(CO)_7$*	$Ni(CO)_4$ $Ni(CO)_{1-3}$*	$Cu(CO)_3$* $Cu_2(CO)_6$*
		$Mo(CO)_6$ $Mo(CO)_{3-5}$*	$Tc_2(CO)_{10}$	$Ru(CO)_5$ $Ru_2(CO)_9$ $Ru_3(CO)_{12}$	$Rh_2(CO)_8$ $Rh_4(CO)_{12}$ $Rh_6(CO)_{16}$ $Rh(CO)_4$*	$Pd(CO)_{1-4}$*	$Ag(CO)_{1-3}$* $Ag_2(CO)_6$*
	$Ta(CO)_{1-6}$*	$W(CO)_6$ $W(CO)_{3-5}$*	$Re_2(CO)_{10}$	$Os(CO)_5$ $Os_2(CO)_9$ $Os_3(CO)_{12}$ $Os_4(CO)_{13}$ $Os_5(CO)_{16}$ $Os_6(CO)_{18}$ $Os_7(CO)_{21}$ $Os_8(CO)_{23}$	$Ir_2(CO)_8$ $Ir_4(CO)_{12}$	$[Pt(CO)_2]_n$ $Pt(CO)_{1-4}$*	$Au(CO)_{1,2}$*

* Species marked with an asterisk were identified only by low-temperature matrix isolation.

Tab. 11.2. Heteronuclear Metal Carbonyls.

Binuclear:	$MnRe(CO)_{10}$	$MnCo(CO)_9$	$ReCo(CO)_9$	$CoRh(CO)_7$
Trinuclear:	$Mn_2Fe(CO)_{14}$ $MnReFe(CO)_{14}$ $Re_2Fe(CO)_{14}$	$Mn_2Ru(CO)_{14}$ $Mn_2Os(CO)_{14}$ $Re_2Os(CO)_{14}$	$FeRu_2(CO)_{12}$ $Fe_2Ru(CO)_{12}$ $Ru_2Os(CO)_{12}$ $RuOs_2(CO)_{12}$	
Tetranuclear:	$Co_2Rh_2(CO)_{12}$ $Co_3Rh(CO)_{12}$ $Rh_3Ir(CO)_{12}$	$Co_2Rh_2(CO)_{13}$		
Hexanuclear:	$Re_2Fe_4(CO)_{24}$	$Co_2Rh_4(CO)_{16}$		

11.1.1. The Structure of Metal Carbonyls

Numerous spectroscopic and diffraction investigations have established the structures of metal carbonyls. Each carbon monoxide molecule contributes two electrons to the metal which achieves the electron configuration of the next higher noble gas (see Section 2.6). Metals of odd atomic number cannot form neutral, mononuclear metal carbonyls; these metals form dinuclear or polynuclear carbonyls containing metal-metal bonds, or metal-carbonyl anions. The only exception is vanadium, which forms paramagnetic $V(CO)_6$, a compound with only 17 electrons in the valence shell of vanadium.

Carbon monoxide can act as a ligand in several different ways:

a) terminal group, M—CO, with each carbon monoxide attached to a single atom;
b) bimetallic bridge, with CO connecting two metal atoms;
c) trimetallic bridge centered above a triangular face of a polyhedral cluster, connecting three metal atoms.

The three possibilities are illustrated in Fig. 11.1.

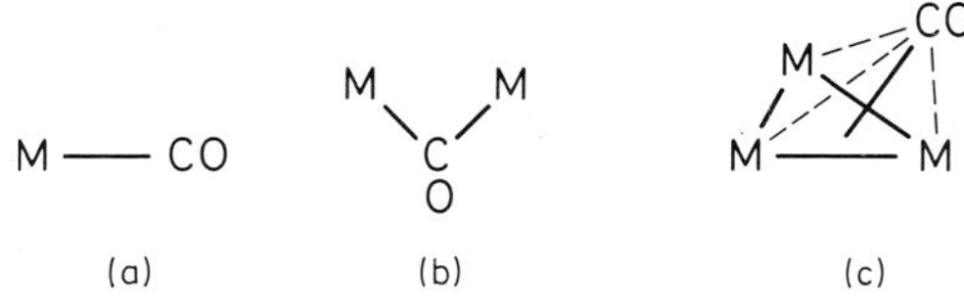

Fig. 11.1. Different ways of bonding of carbon monoxide in metal carbonyls.

Other types of bonding involve the participation of the π-system of the CO unit in bonding with one or two metal centers, as shown in Fig. 11.2.

The unsymmetrical bonding shown in Fig. 11.2 seldom occurs in binary metal carbonyls, but is identified in metal carbonyl anions or in substituted derivatives, for example, cyclopentadienylmetal carbonyls.

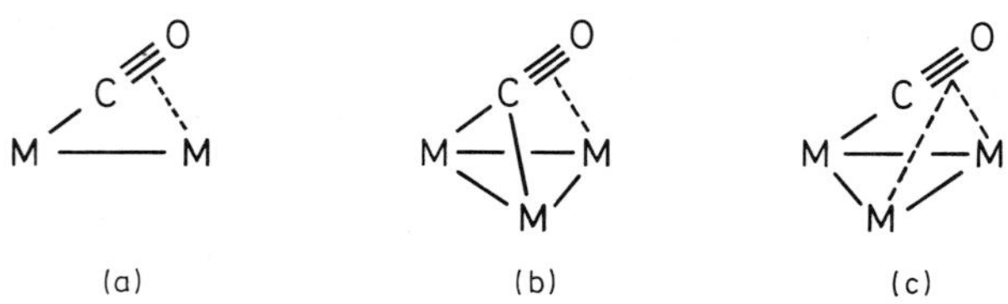

Fig. 11.2. New types of CO bonding in metal carbonyls.

Unsymmetrical M—CO ··· M bridges can occur in sets of compensating bridges, as in $Fe_3(CO)_{12}$ or $Fe_4(CO)_{13}^{2-}$ (Fig. 11.3) or as "semibridging" CO groups which may occur singly or in sets, as in $Fe_2(CO)_7 \cdot$ bipy or in $[FeCo(CO)_8]$:

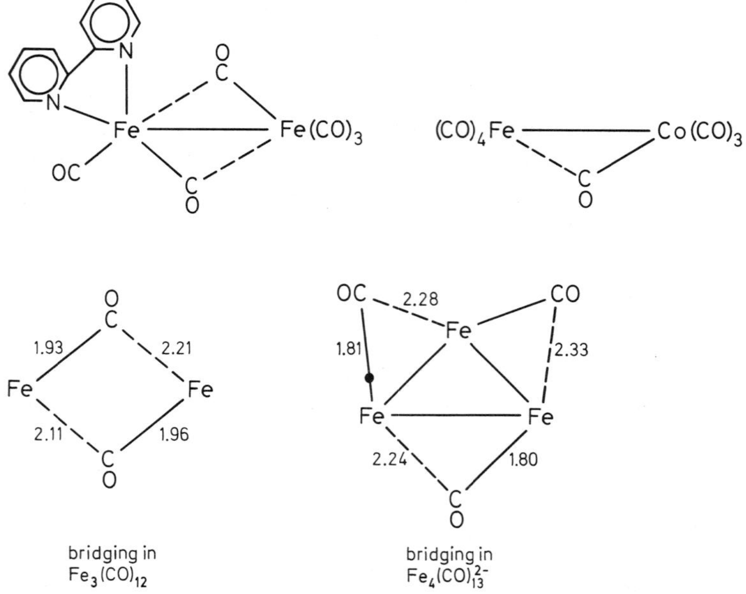

Fig. 11.3. Unsymmetrical carbonyl bridging (Bond lengths in Å).

Semibridging bonding is illustrated by the interaction shown in Fig. 11.4.

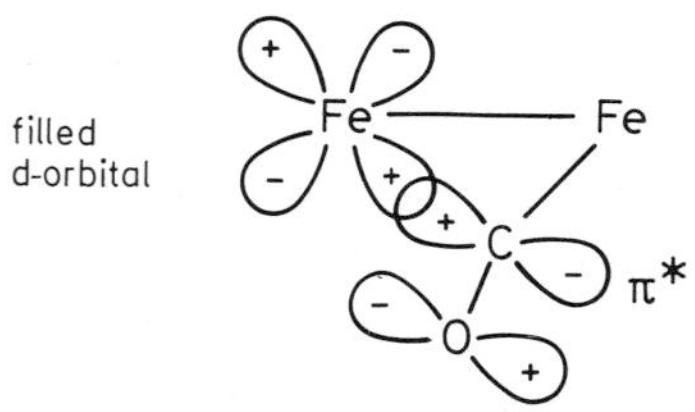

Fig. 11.4. Orbital interaction in unsymmetrical bridges.

Head-to-tail bridging of carbon monoxide, in which the coordinated CO molecule can further act as a donor through its oxygen:

$$M{-}C{\equiv}O \longrightarrow M'$$

has not been observed in binary metal carbonyls, but may occur when a strong acceptor of oxygen (for example, aluminum in organic compounds) is available, as illustrated in the following examples:

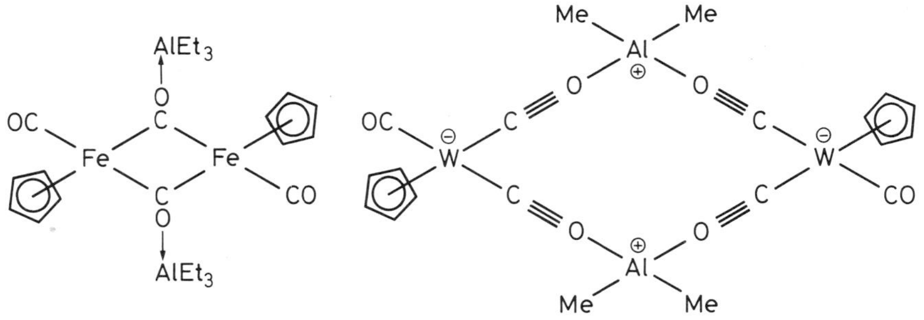

The mononuclear metal carbonyls adopt molecular geometries expected for a metal atom surrounded by n carbonyl groups: $Ni(CO)_4$ tetrahedral, $Fe(CO)_5$ trigonal bipyramidal, $Cr(CO)_6$ octahedral, and $V(CO)_6$ octahedral (Fig. 11.5).

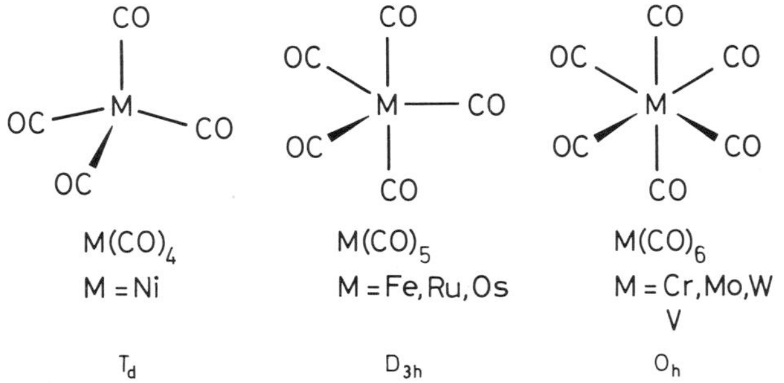

$M(CO)_4$ M = Ni T_d

$M(CO)_5$ M = Fe, Ru, Os D_{3h}

$M(CO)_6$ M = Cr, Mo, W, V O_h

Fig. 11.5. The structures of mononuclear metal carbonyls.

The structures of binuclear metal carbonyls such as $Mn_2(CO)_{10}$, $Fe_2(CO)_9$ and $Co_2(CO)_8$ could not be easily predicted, for example, $Mn_2(CO)_{10}$ is made up of two $Mn(CO)_5$ groups joined by a metal-metal bond (Fig. 11.6a) oriented in the solid with staggered CO groups. In the binuclear carbonyls of iron and cobalt, in addition to metal-metal bonds there are carbonyl bridges (Fig. 11.6b and 11.6c).

Dicobalt octacarbonyl can exist in another isomeric form, without bridging carbonyl groups (D_{3d} symmetry); a third isomeric form (D_{2d} symmetry), also without bridges, has been detected in a low-temperature matrix.

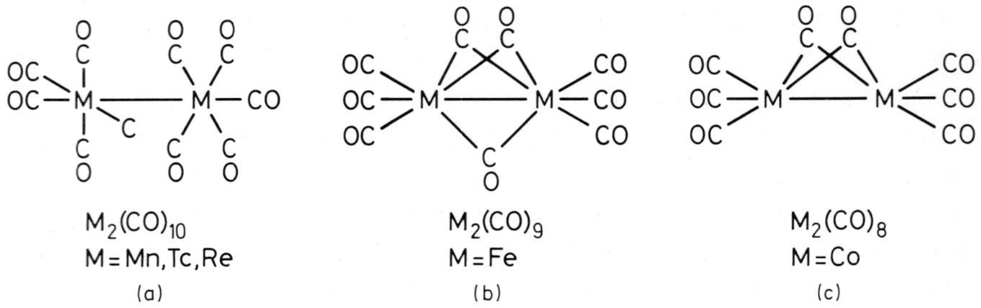

$M_2(CO)_{10}$
M = Mn, Tc, Re
(a)

$M_2(CO)_9$
M = Fe
(b)

$M_2(CO)_8$
M = Co
(c)

Fig. 11.6. The structures of dinuclear metal carbonyls.

In these three dinuclear carbonyls the metal achieves the next higher noble gas configuration: in the $Mn(CO)_5$ groups the metal receives 10 electrons from the five CO-coordinated molecules, thus making $25 + 10 = 35$ electrons. By sharing an electron in forming an Mn—Mn bond, each manganese atom achieves a noble gas configuration (36 electrons). In the $Co_2(CO)_8$ molecule each cobalt atom obtains the nine electrons needed for completing the noble gas configuration as follows: 6 from the three terminal CO groups; 2 from the two CO bridges; and 1 by metal-metal bond formation.

The trinuclear carbonyls, $Fe_3(CO)_{12}$, $Os_3(CO)_{12}$ and $Ru_3(CO)_{12}$, contain metal-metal bonded triangles (Fig. 11.7a and 11.7b). The structures are, however, different with and without bridges, despite analogous compositions. The structure of

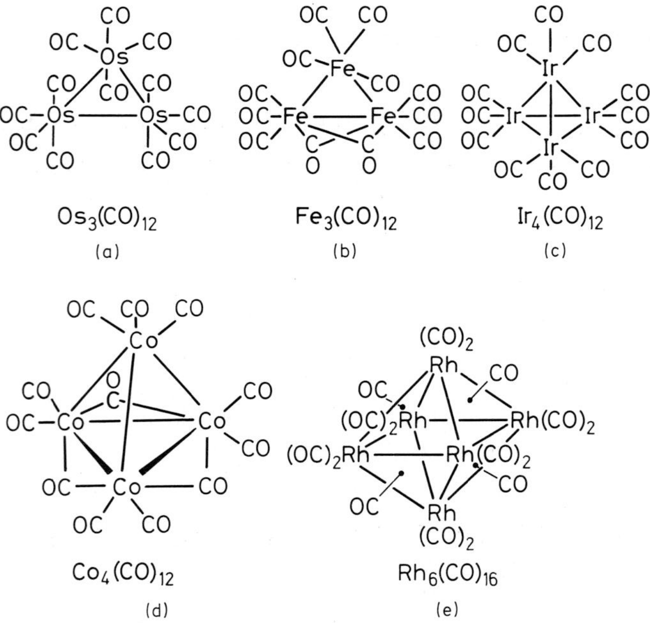

Fig. 11.7. The structure of polynuclear metal carbonyls.

$Fe_3(CO)_{12}$ can be deduced from that of $Fe_2(CO)_9$ by replacing a CO bridge in the latter with a $Fe(CO)_4$ bridge. The CO bridges in $Fe_3(CO)_{12}$ are unsymmetrical; another peculiarity of this molecule is its highly fluxional character with a rapid interchange of terminal- and bridging-CO groups. The ruthenium and osmium compounds are isostructural.

The structures of the tetranuclear carbonyls, $Co_4(CO)_{12}$ and $Rh_4(CO)_{12}$, are based upon a tetrahedral cluster of metal atoms (Fig. 11.7d). The triangle at the base of the pyramid has three CO bridges on the edges. The iridium compound, $Ir_4(CO)_{12}$, also has a tetrahedral structure, but without CO bridges (Fig. 11.7c).

The structure of the hexanuclear carbonyl, $Rh_6(CO)_{16}$, contains an octahedral cluster of rhodium atoms and four trimetallic bridges, in addition to two terminal CO groups at each rhodium atom (Fig. 11.7e).

The polynuclear carbonyls, $Os_5(CO)_{16}$, $Os_6(CO)_{18}$ and $Os_7(CO)_{21}$, contain only terminal-carbonyl ligands and are based upon the clusters shown schematically in Fig. 11.8.

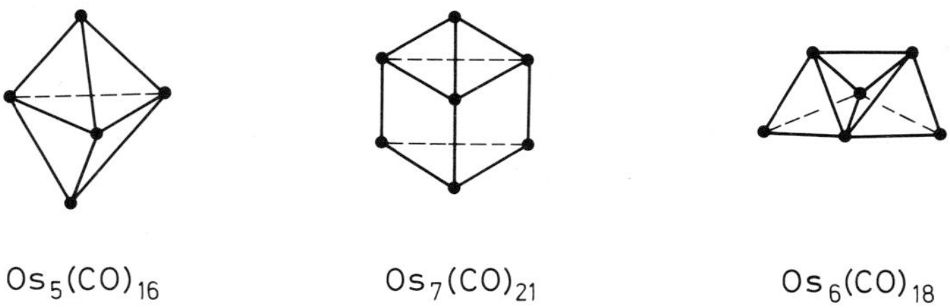

Fig. 11.8. Structures of three polynuclear osmium carbonyls.

Heterometallic polynuclear carbonyls containing non-transition metals have unusual cyclic structures determined by the presence of the carbonyl ligands only at the transition metal, for example, in $CdFe(CO)_4$ (tetramer) and $(bipy)CdFe(CO)_4$ (trimer):

```
(CO)4Fe—Cd—Fe(CO)4                 (CO)4
     |       |                       Fe
     Cd      Cd            bipy · Cd      Cd · bipy
     |       |                    |        |
(CO)4Fe—Cd—Fe(CO)4           (CO)4Fe      Fe(CO)4
                                      Cd
                                      ·
                                     bipy
```

The analogues $MFe(CO)_4$ (M = Zn, Hg, Pb) and $AgCo(CO)_4$ are also known. Some are polymers with chain structures.

11.1.2. Preparation of Metal Carbonyls

The preparation of metal carbonyls requires in most cases high pressures of carbon monoxide. Except for this condition which involves special equipment, the preparations are rather simple. In recent years normal pressure syntheses have been developed, and several compounds are now available commercially.

Iron and nickel metal react directly with carbon monoxide to form carbonyls. Since the oxidation state of the metal is zero in the metal carbonyls, and, as a rule, the syntheses start with a compound in a higher oxidation state, a reducing agent is necessary. The reduction of the metal compound is usually carried out under carbon monoxide pressure with such reducing agents as active metals (sodium, magnesium, zinc), aluminum alkyls and hydrogen. Occassionally, carbon monoxide acts as reducing agent. The mechanism of these reactions involves stepwise reduction of the metal with simultaneous coordination of carbon monoxide. This view is supported by the formation of intermediate, metal-carbonyl halides in the carbonylation of metal halides such as rhodium trichloride, where the products of carbonylation depend upon the conditions:

$$RhCl_3 \cdot 3H_2O \xrightarrow[\text{in methanol}]{60\,°C,\ 40\ \text{bar CO}} Rh_6(CO)_{16}$$

$$RhCl_3 \cdot 3H_2O \xrightarrow{100\,°C,\ \text{CO at atm. pressure}} [Rh(CO)_2Cl]_2$$

$$[Rh(CO)_2Cl]_2 \xrightarrow[H_2O]{\text{CO (atm. pressure)}} Rh_4(CO)_{12}$$

Another example is the carbonylation of iron(II) iodide under mild conditions to give $Fe(CO)_4I_2$.

If the reduction goes too far, anionic metal carbonyls are formed. In this case the anion formed must be re-oxidized, as in the synthesis of vanadium hexacarbonyl:

$$VCl_3 + 6CO + 4Na \xrightarrow[\text{diglyme}]{160\,°C,\ 200\ \text{bar}} [Na(diglyme)_3][V(CO)_6] \xrightarrow[\text{ether}]{HCl} V(CO)_6$$

Of particular value is the matrix-isolation technique, in which a metal vapor reacts with carbon monoxide at low pressure, and the product is trapped in a low-temperature matrix (for example, solidified argon). The preparation and spectral investigation of metal carbonyls which are unstable under ambient conditions is thus possible, for example, $Ti(CO)_6$, $V(CO)_n$ ($n = 1-5$), $V_2(CO)_{10}$ and $V_2(CO)_{12}$, $Cu(CO)_n$ ($n = 1-3$) and $Cu_2(CO)_6$, $Ag(CO)_n$ ($n = 1-3$) and $Ag_2(CO)_6$, $Au(CO)_n$ ($n = 1, 2$) and $Ni(CO)_n$ ($n = 1-3$).

The preparation of selected carbonyls is briefly presented below.

Titanium, Zirconium, Hafnium

Group IV metal carbonyls are unstable; $Ti(CO)_6$ has been obtained by co-condensation of titanium vapor and carbon monoxide in a low-temperature matrix. No zirconium and hafnium carbonyls have been reported so far.

Vanadium, Niobium, Tantalum

In Group V only vanadium forms the binary carbonyl, $V(CO)_6$. The co-condensation of vanadium vapor with carbon monoxide at 10 K affords the matrix isolation of $V(CO)_n$ (n = 2 – 6) species, $V_2(CO)_{10}$ and $V_2(CO)_{12}$, formulated as $(CO)_5V(\mu\text{-}CO)_2V(CO)_5$. Tantalum carbonyls, $Ta(CO)_n$ (n = 1 – 6), are obtained from tantalum atoms in the vapor state with carbon monoxide in a low-temperature matrix.

Chromium, Molybdenum, Tungsten

All three elements of Group VI form metal carbonyls. Chromium hexacarbonyl is prepared by the reactions of anhydrous chromium(III) chloride with carbon monoxide in the presence of aluminum chloride at high temperature and 300–500 bar, in the presence of metallic sodium at 55 bar or with triethylaluminum at high temperature and 30–50 bar. Chromium(III) acetylacetonate is used commercially to form the hexacarbonyl on reduction with zinc or magnesium metal in pyridine at high temperature under 100–300 bar of carbon monoxide.

An active form of chromium metal prepared by reduction of $CrCl_3 \cdot THF$ with potassium metal reacts with carbon monoxide at high temperature and 280 bar to give chromium hexacarbonyl. The yield is increased when the $CrCl_3 \cdot THF$ is reduced with activated magnesium prepared from $MgCl_2$ and potassium metal-potassium iodide.

Molybdenum hexacarbonyl is prepared by the reaction of molybdenum pentachloride with sodium metal and carbon monoxide, or by using triethylaluminum as reducing agent, at high temperature and 30–50 bar carbon monoxide. Tungsten hexacarbonyl is prepared by the reaction of tungsten hexachloride with iron pentacarbonyl under hydrogen atmosphere, transfering the CO ligand from iron to tungsten.

In situ photolysis of hexacarbonyls in a low-temperature matrix produces the unstable lower carbonylated species, $M(CO)_n$ (M = Cr, Mo, W; n = 3, 4, 5).

Manganese, Technetium, Rhenium

These three metals form dinuclear metal carbonyls, $M_2(CO)_{10}$. The manganese derivative is obtained by reducing manganese(II) chloride with sodium-benzophenone ketyl under a carbon monoxide pressure or by the reduction of manganese(II) acetate with triethylaluminum under a carbon-monoxide pressure. At atmospheric pressure, $Mn_2(CO)_{10}$ can be obtained from the reduction of the commercially available

η^5-$CH_3C_5H_4Mn(CO)_3$ with sodium metal under carbon-monoxide pressure followed by treatment with phosphoric acid.

The mononuclear $Mn(CO)_5$ and binuclear $Mn_2(CO)_{10}$ species are produced by co-condensation of manganese vapor with carbon monoxide in a low temperature matrix.

The technetium and rhenium carbonyls, $M_2(CO)_{10}$, are prepared by reduction of the metal heptoxides, M_2O_7, with carbon monoxide at 250 °C under 400 bar carbon monoxide.

Iron, Ruthenium, Osmium

These metals form several carbonyls with mono- and polynuclear structures (see Table 11.1), with osmium forming the largest number of derivatives. Syntheses of iron pentacarbonyl are based upon the reaction of iron metal (powder) with carbon monoxide under pressure at elevated temperatures in the presence of sulfur as catalyst.

Polynuclear iron carbonyls can be obtained by condensation reactions of the mononuclear carbonyl. UV-irradiation of iron pentacarbonyl in glacial acetic acid yields the binuclear compound, $Fe_2(CO)_9$. The trinuclear carbonyl, $Fe_3(CO)_{12}$, can be prepared in two ways, through the intermediate carbonyl anions by treatment with alkalis followed by oxidation with manganese dioxide:

$$3\,Fe(CO)_5 + 6\,OH^- \xrightarrow[-HCO_3^-]{} 3\,HFe(CO)_4^- \xrightarrow[-MnO,\ -OH]{MnO_2} Fe_3(CO)_{12}$$

or, by reaction with triethylamine followed by treatment with hydrochloric acid:

$$Fe(CO)_5 + NEt_3 \xrightarrow[-CO_2,\ -CO,\ -H_2]{+H_2O} [Et_3\overset{+}{N}H][HFe_3(CO)_{11}^-] \xrightarrow[-H_2,\ -FeCl_2,\ -Et_2NH\cdot HCl]{HCl} Fe_3(CO)_{12}$$

The ruthenium carbonyls, $Ru(CO)_5$ and $Ru_3(CO)_{12}$, are obtained from ruthenium acetylacetonate with carbon monoxide and hydrogen at elevated temperature and pressure or by the reduction of ruthenium(III) iodide with silver, in the presence of carbon monoxide; the trinuclear carbonyl can be obtained at low pressure by bubbling carbon monoxide through a solution containing triethylamine and $[Ru_3O(OOCCH_3)_6(H_2O)]\cdot H_2O$. Alternatively, $Ru_3(CO)_{12}$ is prepared by treating $RuCl_3\cdot H_2O$ with carbon monoxide, followed by reduction of the $Ru(CO)_xCl_y$ formed with zinc dust in a carbon monoxide atmosphere.

The osmium carbonyls, $Os(CO)_5$ and $Os_3(CO)_{12}$, have been obtained by the reduction of osmium tetroxide with carbon monoxide.

The less-stable ruthenium and osmium dinuclear carbonyls, $Ru_2(CO)_9$ and $Os_2(CO)_9$, are prepared by UV irradiation of $M(CO)_5$ (M = Ru, Os) at − 40 °C. The products, which possess a dinuclear structure with only one carbonyl bridge:

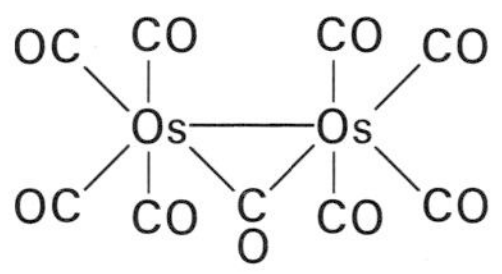

```
OC  CO      CO  CO
  \ |       |  /
   Os-------Os
  / | \   / | \
OC  CO  C   CO  CO
        O
```

decompose to the trinuclear compounds, $M_3(CO)_{12}$ (M = Ru, Os).

Cobalt,,Rhodium, Iridium

Dicobalt octacarbonyl, $Co_2(CO)_8$, is prepared by reduction of cobalt carbonate with hydrogen and carbon monoxide at 240 bar pressure and 150–160 °C. The cobalt(II) or cobalt(III) oxides can also be reduced directly. Iron pentacarbonyl also gives $Co_2(CO)_8$ with cobalt(II) chloride in a ligand-transfer reaction.

The unstable carbonyls, $Co(CO)_4$ and $Co_2(CO)_7$, formed by thermal or photochemical decomposition of $Co_2(CO)_8$, are detectable in a low-temperature matrix. The latter contains no carbonyl bridges. The mononuclear species, $Co(CO)_n$ ($n = 1 - 4$), are obtained from cobalt vapor and carbon monoxide in a low-temperature matrix.

The matrix-isolation technique also shows the successive formation of $Rh(CO)_4$, $Rh_2(CO)_8$, $Rh_4(CO)_{12}$ and $Rh_6(CO)_{16}$ in the reaction of rhodium vapor with carbon monoxide at low temperature. The tetranuclear compound, $Co_4(CO)_{12}$, is formed by gently heating $Co_2(CO)_8$.

Rhodium carbonyls have already been mentioned. Treatment of $[Rh(CO)_2Cl]_2$ in the presence of $NaHCO_3$ with carbon monoxide yields $Rh_4(CO)_{12}$, which precipitates on addition of water. The hexanuclear carbonyl, $Rh_6(CO)_{16}$, is prepared by treating $Rh_2(OOCCH_3)_4$ with acid, and then with carbon monoxide.

The tetranuclear-iridium carbonyl, $Ir_4(CO)_{12}$, is prepared by reducing the complex (*para*- toluidine)$Ir(CO)_2Cl$ with zinc metal under carbon monoxide, or by passing CO over Na_3IrCl_6.

Nickel, Palladium, Platinum

Only nickel forms the defined, stable carbonyl, $Ni(CO)_4$, from the direct reaction of the finely-divided metal with carbon monoxide, or by analogous treatment of nickel salts in the presence of reducing agents. Nickel vapor forms with carbon monoxide all four members of the $Ni(CO)_n$ ($n = 1 - 4$) series in a frozen matrix.

Palladium and platinum carbonyls, $Pd(CO)_n$ ($n = 1 - 4$) and $Pt(CO)_n$ ($n = 1 - 4$), are obtained only as unstable species in low-temperature matrices by co-condensation of the metal vapor with carbon monoxide.

Copper, Silver, Gold

The binary metal carbonyls of the Group IB metals are obtained as matrix-isolated species. Thus, copper vapor with carbon monoxide gives Cu(CO), which forms $Cu(CO)_2$, $Cu(CO)_3$, and $Cu_2(CO)_6$ (believed to have the non-bridged structure $(CO)_3Cu{-}Cu(CO)_3$) on warming. Similarly, silver vapor and carbon monoxide give $Ag(CO)_3$ which dimerizes to $Ag_2(CO)_6$; the species $Ag(CO)_n$ (n = 1 – 2) have also been observed. The gold carbonyls, $Au(CO)_n$ (n = 1 and 2), are obtained by matrix isolation from gold vapor and carbon monoxide.

Heteronuclear Metal Carbonyls

Mixed-metal carbonyls are prepared by coupling reactions of metal carbonyl-halides with the salts of metal-carbonyl anions:

$$Re(CO)_5Cl + NaMn(CO)_5 \longrightarrow (CO)_5ReMn(CO)_5 + NaCl$$
$$Mn(CO)_5Br + NaCo(CO)_4 \longrightarrow (CO)_5MnCo(CO)_4 + NaCl,$$

or by condensation of simple metal carbonyls under ultraviolet irradiation or heating:

$$Mn_2(CO)_{10} + Re_2(CO)_{10} \xrightarrow[\text{or } 220\,°C]{h\nu} (CO)_5MnRe(CO)_5 \xrightarrow[h\nu]{Fe(CO)_5} (CO)_5MnFe(CO)_4Re(CO)_5$$

$$3\,Fe(CO)_5 + Ru_3(CO)_{12} \xrightarrow{110\,°C} Fe_2Ru(CO)_{12} + FeRu_2(CO)_{12} + 3\,CO$$

$$3\,Mn_2(CO)_{10} + Ru_3(CO)_{12} \xrightarrow{205\,°C} 3\,(CO)_5MnRu(CO)_4Mn(CO)_5$$

$$3\,Re_2(CO)_{10} + Os_3(CO)_{12} \xrightarrow{205\,°C} 3\,(CO)_5ReOs(CO)_4Re(CO)_5 .$$

11.1.3. Reactions of Metal Carbonyls

Metal-carbonyl reactions are difficult to systematize since a given reagent does not produce similar reactions with all metal carbonyls. The most common reactions are the following:

-*reduction* to form metal-carbonyl anions;
-*cleavage* by halogens to form metal-carbonyl halides;
-*replacement* of coordinated carbon monoxide by other ligands (organophosphines, arsines, trifluorophosphine, olefins, acetylenes and other unsaturated or aromatic molecules);
-*disproportionation* to form metal-carbonyl anions and cations;
-*polycondensation* to form polynuclear (cluster) compounds, illustrated above;
-*reactions of coordinated carbon monoxide,* for example, to form carbene complexes.

These reactions will be dealt with in the appropriate sections as preparative methods for various classes of compounds, such as metal-carbonyl anions, cations, hydrides, halides or organic derivatives.

11.1.4. Metal-Carbonyl Anions

Metal-carbonyl anions are isoelectronic with other neutral metal carbonyls by formally replacing a carbon-monoxide ligand by an electron pair:

$Cr(CO)_5^{2-}$	$Fe(CO)_4^{2-}$	$Fe_2(CO)_8^{2-}$	$Fe_3(CO)_{11}^{2-}$
$Cr(CO)_6$	$Fe(CO)_5$	$Fe_2(CO)_9$	$Fe_3(CO)_{12}$

Thus, metal-carbonyl anions also satisfy the 18-electron rule, and achieve the electron configuration of the next noble gas.

Tab. 11.3. Binary Metal-Carbonyl Anions.

Group V	Group VI	Group VII	Group VIII		
$V(CO)_6^-$ $V(CO)_5^{3-}$	$Cr(CO)_5^{2-}$ $Cr(CO)_4^{4-}$ $Cr_2(CO)_{10}^{2-}$ $Cr_3(CO)_{14}^{2-}$	$Mn(CO)_5^-$ $Mn(CO)_4^{3-}$ $Mn_2(CO)_9^{2-}$	$Fe(CO)_4^{2-}$ $Fe_2(CO)_8^{2-}$ $Fe_3(CO)_{11}^{2-}$ $Fe_4(CO)_{13}^{2-}$	$Co(CO)_4^-$ $Co(CO)_3^{3-}$ $Co_3(CO)_{10}^-$ $Co_6(CO)_{15}^{2-}$ $Co_6(CO)_{14}^{4-}$	$Ni_2(CO)_6^{2-}$ $Ni_3(CO)_8^{2-}$ $Ni_5(CO)_9^{2-}$ $Ni_5(CO)_{12}^{2-}$ $Ni_6(CO)_{12}^{2-}$ $Ni_9(CO)_{18}^{2-}$
$Nb(CO)_6^-$	$Mo(CO)_5^{2-}$ $Mo(CO)_4^{4-}$ $Mo_2(CO)_{10}^{2-}$ $Mo_3(CO)_{14}^{2-}$	$Te(CO)_5^-$	$Ru(CO)_4^{2-}$ $Ru_3(CO)_{11}^{2-}$ $Ru_4(CO)_{12}^{4-}$ $Ru_4(CO)_{13}^{2-}$ $Ru_6(CO)_{18}^{2-}$	$Rh(CO)_4^-$ $Rh(CO)_3^{3-}$ $Rh_4(CO)_{11}^{2-}$ $Rh_5(CO)_{15}^-$ $Rh_6(CO)_{15}^{2-}$ $Rh_6(CO)_{14}^{4-}$ $Rh_7(CO)_{16}^{3-}$ $Rh_{12}(CO)_{30}^{2-}$ $Rh_{14}(CO)_{25}^{4-}$ $Rh_{15}(CO)_{27}^{3-}$ $Ru_{22}(CO)_{37}^{4-}$	
$Ta(CO)_6^-$	$W(CO)_5^{2-}$ $W(CO)_4^{4-}$ $W_2(CO)_{10}^{2-}$	$Re(CO)_5^-$ $Re(CO)_4^{3-}$ $Re_2(CO)_9^{2-}$ $Re_4(CO)_{16}^{2-}$	$Os(CO)_4^{2-}$ $Os_3(CO)_{11}^{2-}$ $Os_5(CO)_{15}^{2-}$ $Os_6(CO)_{18}^{2-}$ $Os_7(CO)_{20}^{2-}$	$Ir(CO)_3^{3-}$ $Ir_4(CO)_{10}^-$ $Ir_6(CO)_{15}^{2-}$ $Ir_8(CO)_{22}^{2-}$	$[Pt_3(CO)_6^{2-}]_n$ $n = 1, 2, 3, 4, 5,$ 6, 10 $Pt_{19}(CO)_{22}^{4-}$

Table 11.3 lists the metal-carbonyl anions formed by various transition metals. It is interesting to note that some metals which do not form stable, neutral metal carbonyls

(for example, niobium, tantalum and platinum) form the anions. Nickel, which forms only one neutral and mononuclear-metal carbonyl, $Ni(CO)_4$, gives rise to several polynuclear anions.

Heterobimetallic anions, isoelectronic with certain polynuclear metal carbonyls, have also been obtained:

$$MnCr(CO)_{10}^- \quad CoFe(CO)_8^- \quad Fe_2Mn(CO)_{12}^- \quad Rh_5Pt(CO)_{15}^-$$

isoelectronic, respectively, with:

$$Mn_2(CO)_{10} \quad Co_2(CO)_8 \quad Fe_3(CO)_{12} \quad Rh_6(CO)_{16}$$

Formally, these derive by replacement of a metal atom from a neutral carbonyl (bi- or polynuclear) with a metal atom having an atomic number smaller by one unit. In order to keep the noble-gas configuration an additional electron must be added, and this forms the anion. Table 11.4 lists the heteronuclear, metal-carbonyl anions.

Tab. 11.4. Heteronuclear Metal-Carbonyl Anions.

Dinuclear:	$CrMn(CO)_{10}^-$	$CrCo(CO)_9^-$	$MnFe(CO)_9^-$	
	$MoMn(CO)_{10}^-$	$WCo(CO)_9^-$	$ReFe(CO)_{12}^-$	
	$WMn(CO)_{10}^-$			
Trinuclear:	$MnFe_2(CO)_{12}^-$			
	$TcFe_2(CO)_{12}^-$			
	$ReFe_2(CO)_{12}^-$			
Tetranuclear:	$FeCo_3(CO)_{12}^-$	$CoOs_3(CO)_{13}^-$	$CoRu_3(CO)_{13}^-$	
Pentanuclear:	$Cr_2Ni_3(CO)_{16}^{2-}$	$RuIr_4(CO)_{15}^{2-}$	$PtRh_4(CO)_{14}^{2-}$	
	$Mo_2Ni_3(CO)_{16}^{2-}$	$FePt_4(CO)_{16}^{2-}$		
	$W_2Ni_3(CO)_{16}^{2-}$	$FePd_4(CO)_{16}^{2-}$		
Hexanuclear:	$Rh_5Pt(CO)_{15}^-$	$Co_4Ni_2(CO)_{14}^{2-}$	$Mo_2Ni_4(CO)_{14}^{2-}$	$Fe_3Pt_3(CO)_{15}^{2-}$

While the mononuclear species contain coordination polyhedra of expected geometry, the polynuclear anions often contain clusters of unexpected structure as illustrated in Fig. 11.9.

The $Co_2(CO)_8$, $CoFe(CO)_8^-$ and $Fe_2(CO)_8^{2-}$ species are isoelectronic, but not isostructural; $Co_2(CO)_8$ contains two carbonyl bridges, $CoFe(CO)_8$ one bridge and $Fe_2(CO)_8^{2-}$ no carbonyl bridges:

```
          O
OC  CO    C          CO               CO      OC        CO
  \  |  /   \      /                  |         \      /
    Fe--------Co                  OC--Fe-------------Fe--CO
  /  |        |  \                   /   \            |
OC   CO       CO   CO              OC     CO          CO
```

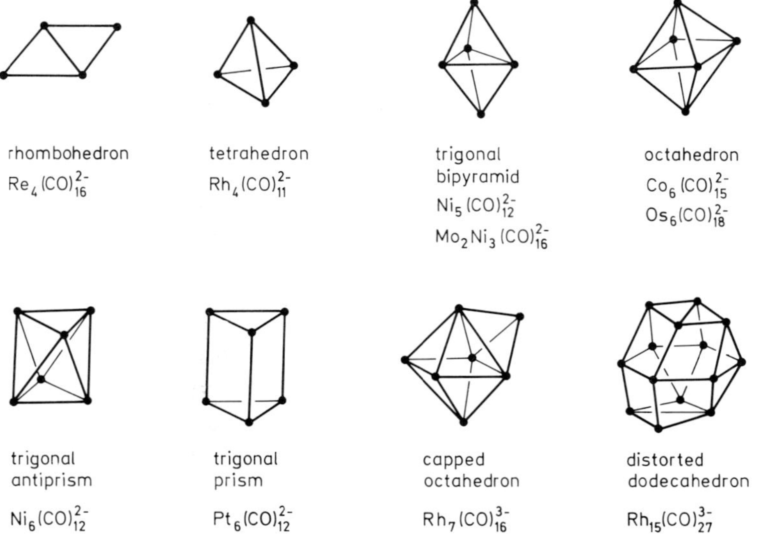

Fig. 11.9. The structures of some polynuclear metal-carbonyl anions (carbonyl groups are omitted for clarity).

Preparation. There is no general procedure for converting metal carbonyls to their isoelectronic anions. However, anions are obtained by the reduction of dinuclear metal carbonyls with alkali metals (or amalgams):

$$Mn_2(CO)_{10} + 2\,Na \longrightarrow 2\,Na^+[Mn(CO)_5]^-$$
$$Re_2(CO)_{10} + 2\,Na \longrightarrow 2\,Na^+[Re(CO)_5]^-$$
$$Co_2(CO)_8 + 2\,Na \longrightarrow 2\,Na^+[Co(CO)_4]^-$$

Potassium-sodium alloy reduces $Mn_2(CO)_{10}$ and $Co_2(CO)_8$ to give the corresponding anions.

The reduction of metal carbonyls with potassium hydride is advantageous, since it produces gaseous hydrogen as the only by-product.

The Group VI metal carbonyls are reduced with sodium to form binuclear anions which can be precipitated with voluminous cations:

$$2\,M(CO)_6 \xrightarrow[Na/NH_3]{Na/Hg} M_2(CO)_{10}^{2-} + 2\,CO$$
$$M = Cr, Mo, W$$

Similarly, reduction of iron pentacarbonyl by sodium amalgam in THF forms a dinuclear anion, but with sodium in liquid ammonia a mononuclear anion results:

$$2\,Fe(CO)_5 \xrightarrow{Na/Hg} Fe_2(CO)_8^{2-} + 2\,CO$$
$$Fe(CO)_5 \xrightarrow[NH_3]{+2\,Na} Fe(CO)_4^{2-} + CO$$

The potassium-graphite intercalation compound, C_8K, readily reduces $M(CO)_6$ to $K_2[M_2(CO)_{10}]$ (M = Cr, Mo, W) and $Fe(CO)_5$ to $K_2[Fe_2(CO)_8]$.

The reduction of nickel tetracarbonyl yields bi- and trinuclear species:

$$2\,Ni(CO)_4 \xrightarrow{Na/NH_3\ liq.} Ni_2(CO)_6^{2-} + 2\,CO$$

$$3\,Ni(CO)_4 \xrightarrow[THF]{Li\ amalgam} Ni_3(CO)_8^{2-} + 4\,CO$$

Reduction of $Ni(CO)_4$ with alkali metals or alkali-metal hydroxides yields $[Ni_5(CO)_{12}]^{2-}$ and $[Ni_6(CO)_{12}]^{2-}$ (identical with a previous reported $[Ni_4(CO)_9]^{2-}$). The anion $[Ni_9(CO)_{18}]^{2-}$ is obtained from $[Ni_6(CO)_{12}]^{2-}$ with $Ni(CO)_4$ or $NiCl_2$.

Highly reduced anions are formed in strongly basic solvents. Thus, $Mn_2(CO)_{10}$ and $Re_2(CO)_{10}$ are reduced to $[M(CO)_4]^{3-}$ (M = Mn, Re) and Group VI metal carbonyls are reduced to $[M(CO)_5]^{2-}$ (M = Cr, Mo, W). Similarly prepared are the trianions $[M(CO)_3]^{3-}$ (M = Co, Rh, Ir) and $[V(CO)_5]^{3-}$. The reduction of the $M(CO)_4$ tetramethylethylenediamine complex with sodium produces the anions $[M(CO)_4]^{4-}$ (M = Cr, Mo, W).

Iron carbonyls react with alcoholic alkalis or organic bases to form anions conserving the cluster size:

$$Fe(CO)_5 + 4\,OH^- \longrightarrow Fe(CO)_4^{2-} + 2\,H_2O + CO_3^{2-}$$

$$Fe_2(CO)_9 + 4\,OH^- \longrightarrow Fe_2(CO)_8^{2-} + 2\,H_2O + CO_3^{2-}$$

$$Fe_3(CO)_{12} + 4\,OH^- \longrightarrow Fe_3(CO)_{11}^{2-} + 2\,H_2O + CO_3^{2-}$$

Analogous reductions of polynuclear osmium carbonyls with bases lead to a variety of anions:

$$Os_5(CO)_{16} \xrightarrow{OH^-} Os_5(CO)_{15}^{2-}$$

$$Os_6(CO)_{18} \xrightarrow{OH^-} Os_5(CO)_{15}^{2-}$$

$$Os_6(CO)_{21} \xrightarrow{OH^-} Os_6(CO)_{18}^{2-}$$

$$Os_8(CO)_{23} \xrightarrow{OH^-} Os_7(CO)_{20}^{2-}$$

Mild reduction of $Os_6(CO)_{18}$ with zinc powder or $NaBH_4$ yields $[Os_6(CO)_{18}]^{2-}$

Rhodium-containing carbonyl anions have been obtained according to the scheme:

$$Rh_4(CO)_{12} \xrightarrow{KOH} Rh_4(CO)_{11}^{2-} \xrightarrow{Rh_4(CO)_{12}} Rh_{12}(CO)_{30}^{2-}$$

$$Rh_4(CO)_{11}^{2-} \xrightarrow{+CO,\ 25\,°C} Rh_6(CO)_{15}^{2-} + Rh(CO)_4^-$$

The reduction of $Ir_4(CO)_{12}$ with KOH or sodium produces $Ir_8(CO)_{20}^{2-}$, $Ir_6(CO)_{15}^{2-}$ and $HIr_4(CO)_{11}^-$.

Reduction of metal carbonyls can proceed in complex reactions, involving condensation followed by fragmentation of large clusters. Thus, the tetrarhodium carbonyl, $Rh_4(CO)_{12}$, reacts with alkali metals:

$$Rh_4(CO)_{12} \longrightarrow [Rh_{12}(CO)_{30}]^{2-} \longrightarrow [Rh_7(CO)_{16}]^{3-} \longrightarrow [Rh_6(CO)_{14}]^{4-} \longrightarrow [Rh(CO)_4]^-$$

The reactions with bases can be complicated by disproportionation and base coordination to the metal:

$$Co_2(CO)_8 + py \longrightarrow [Co(CO)_4py]^+[Co(CO)_4]^-$$
$$Co_2(CO)_8 + 6py \longrightarrow [Co \cdot py_6]^+[Co(CO)_4]^- + 4CO$$
$$5Fe_3(CO)_{12} + 18py \longrightarrow 3[Fe \cdot py_6]^+[Fe_4(CO)_{18}]^- + 21CO.$$

Bases with weak acceptor but strong donor properties (amines) produce disproportionation, with formation of metal carbonyl anions.

The platinum anions, $[Pt_3(CO)_6]_n^{2-}$ (n = 1, 2, 3, 4, 5, 6, 10), are obtained by reducing $Na_2PtCl_6 \cdot 6H_2O$ or $Pt(CO)_2Cl_2$ with alkali metals under CO. Their fluxional structures consist of $Pt_3(\mu\text{-}CO)_3(CO)_3$ triangular units stacked in columns.

Some metal-carbonyl anions can be prepared by direct carbonylation of the metal salts:

$$2CoCl_2 + 12KOH + 11CO \longrightarrow 2K^+[Co(CO)_4]^- + 3K_2CO_3 + 4KCl + 6H_2O$$
$$CoCl_2 \xrightarrow{NH_3} [Co(NH_3)_6]Cl_2 \xrightarrow[120\,°C]{CO,\ 95\ bar,\ H_2O} [Co(CO)_4]^- + NH_4^+ + Cl^- + CO_3^{2-}$$
$$Fe(NH_4)_2(SO_4)_2 \xrightarrow[80\,°C]{CO,\ 115\ bar,\ NH_3} [Fe(CO)_4]^{2-} + NH_4^+ + SO_4^{2-} + CO_3^{2-}$$

Heterobimetallic metal-carbonyl anions can be prepared by reaction of neutral metal carbonyls with anionic carbonyls to eliminate carbon monoxide in photochemical condensation reactions:

$$Fe(CO)_5 + Mn(CO)_5^- \xrightarrow{h\nu} FeMn(CO)_9^- + CO$$
$$2Fe(CO)_5 + Mn(CO)_5^- \longrightarrow MnFe_2(CO)_{12}^{=} + CO$$
$$Fe(CO)_5 + Co(CO)_4^- \xrightarrow{h\nu} FeCo(CO)_8^- + CO$$
$$M(CO)_6 + \underset{(Re)}{Mn(CO)_5^-} \xrightarrow[diglime]{160\,°C} (CO)_5M\underset{(Re)}{Mn}(CO)_5^- + CO$$

M = Cr, Mo, W

11.1.5. Metal-Carbonyl Hydrides and Related Hydrido Anions

Metal-carbonyl hydrides are formally the conjugate acids from which the anions derive by proton displacement, and acidic properties have been established for some hydrides. Thus, $HMn(CO)_5$ is a weak acid, $H_2Fe(CO)_4$ is comparable to acetic acid and $HCo(CO)_4$ is a strong acid. Because of low solubility in water, the estimation of acidic properties is difficult.

Both neutral (Table 11.5) and anionic (Table 11.6) metal-carbonyl hydrides are known.

Tab. 11.5. Metal-Carbonyl Hydrides.

$HV(CO)_6$	$H_2Cr(CO)_5$	$HMn(CO)_5$	$H_2Fe(CO)_4$	$HCo(CO)_4$	$H_2Ni_2(CO)_6$
$H_3V(CO)_3$		$H_2Mn_2(CO)_9$	$H_2Fe_2(CO)_8$	$HCo_3(CO)_9$	$H_2Ni_3(CO)_8$
		$H_3Mn_3(CO)_{12}$	$H_2Fe_3(CO)_{11}$		$H_2Ni_4(CO)_9$
			$H_2Fe_4(CO)_{13}$		
		$HTc(CO)_5$	$H_2Ru(CO)_4$	$HRh(CO)_4$	
			$H_2Ru_3(CO)_{16}$		
			$H_4Ru_4(CO)_{12}$		
			$H_2Ru_4(CO)_{13}$		
			$H_2Ru_6(CO)_{18}$		
		$HRe(CO)_5$	$H_2Os(CO)_4$	$HIr(CO)_4$	
		$H_3Re_3(CO)_{12}$	$H_2Os_2(CO)_8$	$H_2Ir_6(CO)_{15}$	
		$HRe_3(CO)_{14}$	$H_2Os_3(CO)_{10}$		
		$H_4Re_4(CO)_{12}$	$H_2Os_3(CO)_{12}$		
			$H_2Os_4(CO)_{13}$		
			$H_4Os_4(CO)_{12}$		
			$H_2Os_5(CO)_{15}$		
			$H_2Os_5(CO)_{16}$		
			$H_2Os_6(CO)_{18}$		

Tab. 11.6. Hydrido Metal-Carbonyl Anions.

Mononuclear:	$HCr(CO)_5^-$	$HFe(CO)_4^-$		
Binuclear:	$HCr_2(CO)_{10}^-$	$HFe_2(CO)_8^-$		
	$HMo_2(CO)_{10}^-$			
	$HW_2(CO)_{10}^-$			
Trinuclear:	$HRe_3(CO)_{12}^{2-}$	$HFe_3(CO)_{11}^-$	$H_3Rh_3(CO)_{24}^{2-}$	
	$H_2Re_3(CO)_{12}^-$	$HRu_3(CO)_{11}^-$	$H_2Rh_3(CO)_{24}^{3-}$	
	$H_3Re_3(CO)_{10}^{2-}$	$HOs_3(CO)_{11}^-$		
Tetranuclear:	$H_6Re_4(CO)_{12}^{2-}$	$HFe_4(CO)_{13}^-$	$HIr_4(CO)_{11}^-$	
	$H_4Re_4(CO)_{13}^{2-}$	$HRu_4(CO)_{12}^{2-}$		
	$H_4Re_4(CO)_{15}^{2-}$	$H_2Ru_4(CO)_{12}^{2-}$		
		$H_3Ru_4(CO)_{12}^-$		
		$HOs_4(CO)_{13}^-$		
		$H_3Os_4(CO)_{12}^-$		
Pentanuclear:		$HOs_5(CO)_{15}^-$		
Polynuclear:		$HRu_6(CO)_{18}^-$	$HRh_{13}(CO)_{24}^{4-}$	$H_2Ni_8(CO)_{14}^{2-}$
		$HOs_6(CO)_{18}^-$	$HRh_{14}(CO)_{25}^{3-}$	$H_2Ni_{12}(CO)_{21}^{2-}$
				$H_2Ni_{11}(CO)_{20}^{2-}$
				$H_2Ni_{12}(CO)_{21}^{2-}$

Metal-carbonyl hydrides can be prepared directly by reactions of the active metals or salts with carbon monoxide in the presence of water or acids:

$$\mathrm{Co + 4\,CO + 1/2\,H_2O \xrightarrow[250\ bar]{180\,^\circ C} HCo(CO)_4}$$

$$\mathrm{2\,CoI_2 + H_2O + 9\,CO + 4\,Cu \xrightarrow{pressure} 2\,HCo(CO)_4 + CO_2 + 4\,CuI}$$

Carbonyls are treated with amines, alkalis or sodium borohydride, followed by treatment with phosphoric acid to protonate the intermediate anions (not isolated):

$$\mathrm{Os_3(CO)_{12} \xrightarrow[methanol\ \ NaBH_4/THF]{KOH} anions \xrightarrow{H_3PO_4} \begin{cases} H_2Os_3(CO)_{10} + \\ H_2Os_4(CO)_{13} + \\ H_4Os_4(CO)_{12} \end{cases}}$$

$$\mathrm{Re_2(CO)_{10} \xrightarrow{Re(CO)_5^-} anions \xrightarrow{H_3PO_4} H_3Re_3(CO)_{12} \xrightarrow{OH^-} H_2Re_3(CO)_{12}^-}$$

$$\mathrm{anions \xrightarrow[H_3PO_4]{Re(CO)_5^-} HRe_3(CO)_{14}}$$

Iron pentacarbonyl forms various metal-carbonyl hydride anions:

$$\mathrm{Fe(CO)_5 \xrightarrow[{[NEt_4]I}]{NH_3/H_2O} (NEt_4)[HFe(CO)_4]}$$

$$\mathrm{Fe(CO)_5 \xrightarrow{Et_3N/H_2O} [HNEt_3][HFe_3(CO)_{11}]}$$

$$\mathrm{Fe(CO)_5 \xrightarrow[CH_3COOH,\ -50\,^\circ C\ \ [NEt_4]I]{KOH/MeOH} [NEt_4][HFe_2(CO)_8]\ .}$$

The neutral carbonyl, $Ru_3(CO)_{12}$ can likewise be converted to an anion which then gives hydrides by protonation with acids:

$$\mathrm{Ru_3(CO)_{12} \xrightarrow{base} [Ru_6(CO)_{18}]^{2-} \xrightarrow{H^+} [HRu_6(CO)_{18}]^- \xrightarrow{H^+} H_2Ru_6(CO)_{18}}$$

Stepwise deprotonation of $H_4Ru_4(CO)_{12}$ with potassium hydride produces the series of hydrido anions, $[H_3Ru_4(CO)_{12}]^{2-}$, $[H_2Rh_4(CO)_{12}]^{2-}$ and $[HRu_4(CO)_{12}]^{3-}$.

The thermal stability of the metal-carbonyl hydrides is moderate:

	Decomposition temperature
$HCo(CO)_4$	−20 °C
$H_2Fe(CO)_4$	−10 °C
$H_2Ru(CO)_4$	0 °C
$HMn(CO)_5$	80 °C
$H_2Os(CO)_4$	100 °C

Polynuclear hydrides are considerably more stable.

The mononuclear hydrides are sensitive to oxygen, but their substitution products, for example, $HCo(CO)_3PPh_3$, and their polynuclear derivatives are more stable.

The heteronuclear (bi- or polymetallic) carbonyl hydrides, $H_2Re_2Os_3(CO)_{20}$, $HCoRu_3(CO)_{13}$, $H_4FeRu_3(CO)_{12}$, $H_3FeRu_3(CO)_{12}^-$, $H_3CoOs_3(CO)_{12}$, $HFeRu_3(CO)_{13}^-$ and $H_2FeRu_nOs_{3-n}(CO)_{13}$ (n = 0–3) have also been made.

The geometry of the carbonyl-hydride clusters is unrelated to that of their binary carbonyl or carbonyl-anion (hydrogen-free) analogues. Thus, $H_2Os_6(CO)_{18}$ contains an unusual, monocapped square-pyramidal, Os_6, cluster, while $[Os_6(CO)_{18}]^{2-}$ is octahedral. Some unusual structures, $H_2Os_5(CO)_{16}$ (compare with $[H_2Os_5(CO)_{15}]^-$), the anion $[H_4Re_4(CO)_{15}]^{2-}$ (compare with the tetrahedral $[H_4Re_4(CO)_{13}]^{2-}$) $[H_6Re_4(CO)_{12}]^{2-}$, the "butterfly" anion $[HFe_4(CO)_{13}]^-$, the rhomboidal $HOs_3Re(CO)_{15}$ and $H_3Rh_{13}(CO)_{24}$ are illustrated in Fig. 11.10.

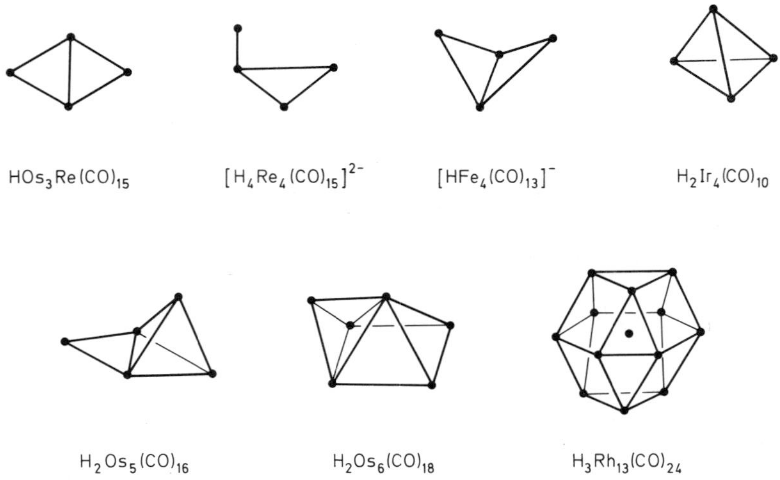

Fig. 11.10. The structure of some polynuclear neutral and anionic metal-carbonyl hydrides. (Carbonyl groups omitted for clarity).

11.1.6. Metal-Carbonyl Cations

Binary metal-carbonyl cations, $[M_x(CO)_y]^+$, are rare, but phosphine derivatives are listed in Table 11.7 for several transition metals.

The cations are prepared from metal-carbonyl halides with Lewis acids:

$$Mn(CO)_5Cl + AlCl_3 + CO \xrightarrow[300\ bar]{100\,°C} [Mn(CO)_6]^+[AlCl_4]^-$$

$$Re(CO)_5Cl + AlCl_3 + CO \xrightarrow[300-350\ bar]{85-95\,°C} [Re(CO)_6]^+[AlCl_4]^-,$$

or by protonation (the formation of $[HFe(CO)_5]^+$, $[HRu_3(CO)_{12}]^+$). Phosphine-substituted cations, $[M(CO)_5PPh_3]^+$ and $[M(CO)_4(PPh_3)_2]^+$ (M = Mn, Re), are prepared similarly.

Tab. 11.7. Metal-Carbonyl Cations and Derivatives ($L = PPh_3$).

Group VII		Group VIII		Group I B
$Mn(CO)_6^+$ $Mn(CO)_5L^+$ $Mn(CO)_4L_2^+$	$HFe(CO)_5^+$ $HFe(CO)_3L_2^+$	$Co(CO)_4L^+$ $Co(CO)_3L_2^+$		$Cu(CO)^+$ $Cu(CO)_3^+$ $Cu(CO)_4^+$
$Tc(CO)_6^+$ $Tc(CO)_4L_2^+$	$HRu_3(CO)_{12}^+$		$Pd(CO)L_2Cl^+$	$Ag(CO)_2^+$
$Re(CO)_6^+$ $Re(CO)_5L^+$ $Re(CO)_4L_2^+$	$HOs_3(CO)_{12}^+$ $HOs(CO)_3L_2^+$ $Os(CO)_4L_2^+$	$Ir(CO)_3L_2^+$ $Ir(CO)_2L_3^+$ $H_2Ir(CO)_2L_2^+$ $H_2Ir(CO)L_3^+$	$Pt(CO)L_2X^+$	

Copper-carbonyl cations are formed when Cu_2O in $HFSO_3$, CF_3SO_3H or $BF_3 \cdot H_2O + HF$ absorb CO to form $[Cu(CO)_4]^+$; addition of sulfuric acid yields unstable $[Cu(CO)_3]^+$ in equilibrium with $[Cu(CO)]^+$. Under similar conditions silver oxide forms only $[Ag(CO)_2]^+$. The copper-carbonyl cation $[Cu(CO)]^+$, isolated as $[Cu(CO)]^+AsF_6$, has been obtained by the reaction of $CuAsF_6$ with carbon monoxide. The explosive salts $[Cu^I(CO)(H_2O)_2]^+ClO_4^-$ and $[Cu_2^I(CO)_2(H_2O)_3]^{2+}(ClO_4)_2$ form from a suspension of copper metal in aqueous copper(II) perchlorate with carbon monoxide.

The halogen-bridged binuclear-carbonyl cations, $[(CO)_5M{-}X{-}M(CO)_5]^+$ (M = Mn, Re; X = Cl, Br, I), are known.

11.1.7. Metal-Carbonyl Halides

Compounds containing carbon monoxide and halogen atoms coordinated to the same metal atom are listed in Table 11.8. Some metals for which no stable binary carbonyls are known are able to form metal-carbonyl halides (for example, gold, copper and palladium).

One method of preparing carbonyl halides is based upon the reactions of metal carbonyls (preferably polynuclear) with halogens:

$$Mn_2(CO)_{10} + X_2 \longrightarrow 2\,Mn(CO)_5X \xrightarrow[-CO]{\Delta} [Mn(CO)_4X]_2.$$

Iron carbonyls, $Fe(CO)_5$ and $Fe_3(CO)_{12}$, react with iodine to form the compounds $Fe(CO)_4I_2$, $Fe_2(CO)_8I$ and $Fe(CO)_4I$. Paramagnetic $Cr(CO)_5I$ is obtained by the oxidation of the $[Cr_2(CO)_{10}]^{2-}$ anion with iodine.

Metal carbonyl hydrides can be halogenated in some cases by carbon tetrahalides (X = Cl, Br; n = 1, 2 and 3):

$$[\{Os(CO)_4\}_n]H_2 + CX_4 \xrightarrow{20°C} [\{Os(CO)_4\}_nX_2] \longrightarrow Os_2(CO)_6X_2$$

Tab. 11.8. Some Typical Metal-Carbonyl Halides.

Group VI	Group VII		Group VIII		Group I B
$Cr(CO)_5I$ $Cr_2(CO)_{10}X$	$Mn(CO)_5X$ $Mn(CO)_4X_2$ $[Mn(CO)_4X]_2$ $Mn(CO)_3F_3$	$Fe(CO)_4X_2$ $[Fe(CO)_3X]_2$ $Fe(CO)_2X_2$ $Fe_2(CO)_8I_2$	$Co(CO)_4I$	$Ni_2(CO)_3Cl_4$	$Cu(CO)X$
$Mo(CO)_4X_2$ $[Mo(CO)_4X_2]_2$ $[Mo(CO)_4I]_2$	$Tc(CO)_5X$	$Ru(CO)_4X_2$ $[Ru(CO)_2X_2]_n$ $[Ru(CO)_3X_2]_2$ $Ru_2(CO)_6X_4$ $Ru_3(CO)_{12}X_6$ $Ru(CO)_3F_3$ $[Ru(CO)_3F_2]_4$	$[Rh(CO)_2X]_2$	$[Pd(CO)X_2]_2$ $[Pd(CO)Cl]_n$	
$W(CO)_4X_2$ $[W(CO)_4X_2]_2$	$Re(CO)_5X$ $[Re(CO)_4X]_2$	$Os(CO)_4X_2$ $[Os(CO)_3X_2]_2$ $[Os(CO)_4X]_2$ $Os_2(CO)_8X_2$ $Os_3(CO)_{10}Cl_2$ $Os_3(CO)_{12}X_2$	$Ir(CO)_3X$ $Ir(CO)_2X_2$ $Ir(CO)_3X_3$ $Ir(CO)_2X_3$	$Pt(CO)_2X_2$ $[Pt(CO)X_2]_2$ $Pt_2(CO)_3X_4$	$Au(CO)Cl$

Carbonylation of the platinum-group halides yields carbonyl halides, for example, platinum(II) chloride forms $Pt(CO)_2Cl_2$, which dimerizes on heating to $[Pt(CO)Cl]_2$. A gold derivative, $Au(CO)Cl$, is formed from $AuCl_3$ and CO in $SOCl_2$.

As seen in Table 11.8, most metal-carbonyl halides are mononuclear, but some are associated through halogen bridging, for example, $[(CO)_4MoI]_2$, $[(CO)_4MnCl]_2$ and $[(CO)_3RuF_2]_4$:

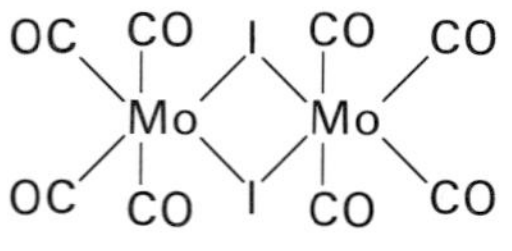

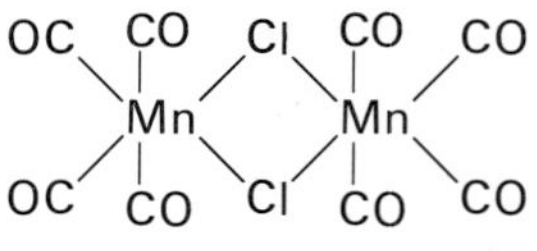

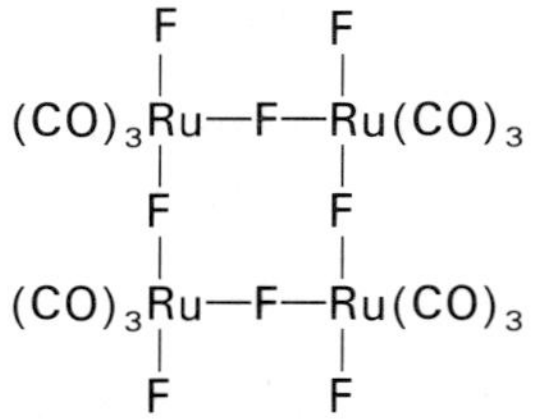

```
OC  CO  I  CO  CO      OC  CO  Cl  CO  CO              F       F
  \ |  / \ |  /          \ |  / \ |  /                 |       |
   Mo      Mo             Mn      Mn         (CO)3Ru—F—Ru(CO)3
  / |  \ / |  \          / |  \ / |  \                 |       |
OC  CO  I  CO  CO      OC  CO  Cl  CO  CO              F       F
                                                       |       |
                                             (CO)3Ru—F—Ru(CO)3
                                                       |       |
                                                       F       F
```

Other metal-carbonyl halides exhibit unexpected structures, like the linear $Os_3(CO)_{12}I_2$ or the metal-metal bonded polymer, $[Ir(CO)_3Cl]_x$:

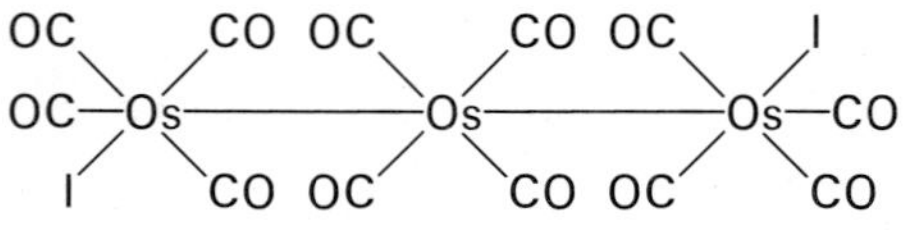

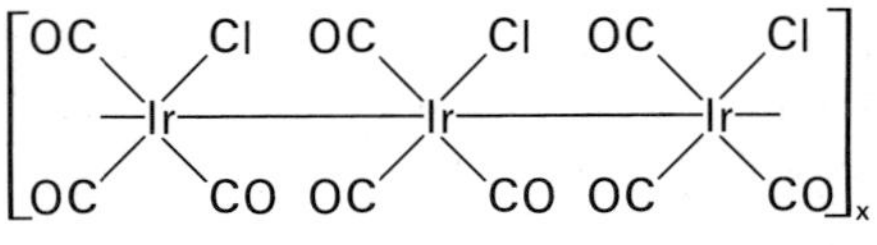

```
OC      CO OC      CO OC      I        [OC     Cl OC     Cl OC     Cl]
   \   /      \   /      \   /          \   /      \   /      \   /
OC—Os————————Os————————Os—CO           —Ir————————Ir————————Ir—
   /   \      /   \      /   \          /   \      /   \      /   \
  I     CO OC      CO OC      CO       [OC     CO OC     CO OC     CO]x
```

Carbonyl halide anions of the type $[M(CO)_5X]^-$, where M = Cr, Mo, W and X = halogen, are obtained by substitution of carbon monoxide in $M(CO)_6$ with halide anions. The anions can be formed by addition reactions accompanied by elimination of carbon monoxide, e. g.:

$$Mn(CO)_5Cl + Cl^- \longrightarrow Mn(CO)_4Cl_2^- + CO$$

$$Cr(CO)_5I^- + W(CO)_6 \xrightarrow{h\nu} CrW(CO)_{10}I^- + CO$$

Metal-carbonyl species with an anionic ligand form when the anion replaces a coordinated CO molecule:

$$M(CO)_6 + [NR_4]^+X^- \longrightarrow [NR_4]^+[M(CO)_5X]^- + CO$$
$$M = Cr, Mo, W \qquad X = Cl, Br, I$$

$$Mn_2(CO)_{10} + 2[NR_4]^+X^- \longrightarrow [NR_4]_2^+[Mn_2(CO)_8X_2]^{2-} + 2CO$$

This reaction can be extended to anions other than halide:

$$Cr(CO)_6 + [PS_2F_2]^- \longrightarrow [Cr(CO)_4PS_2F_2]^- + 2CO$$

and the procedure is used to form $[M(CO)_5(NO_3)]^-$ (M = Cr, W), $[M(CO)_5(RCOO)]^-$ (M = Cr, Mo, W) and $[M(CO)_5SH]^-$ (M = Mo, W).

11.2. Metal Thiocarbonyls

Carbon monosulfide can replace carbon monoxide in metal carbonyls to form thiocarbonyl complexes. The metals which form such compounds are shown in Fig. 11.11.

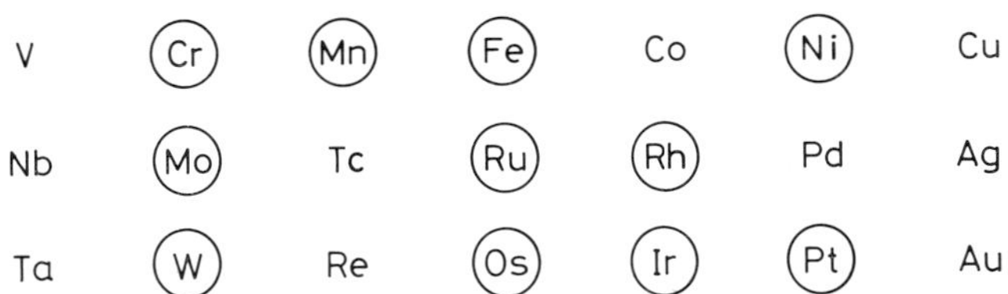

Fig. 11.11. Metals known to form thiocarbonyl complexes.

Only a single binary compound, $Ni(CS)_4$, prepared by co-condensation of nickel atoms with carbon monosulfide has been reported to date. All other derivatives contain carbon monoxide, cyclopentadienyl groups, phosphines, phosphites or a combination of these, as in the carbonyl-thiocarbonyls, $M(CO)_5(CS)$ (M = Cr, Mo, W), $Fe(CO)_4(CS)$ and the cyclopentadienyl derivatives $(\eta^5\text{-}C_5H_5)Mn(CO)_{3-n}(CS)_n$ (n = 1 – 3) and $\eta^5\text{-}C_5H_5Co(CS)_2$.

The thiocarbonyl group acts as a bridging ligand in $[(\eta^5\text{-}C_5H_5)Fe(CO)(CS)]_2$, $[(\eta^5\text{-}C_5H_5)Mn(CS)(NO)]_2$ and $[(\eta^5\text{-}C_5H_5)Ru(CO)(CS)]_2$:

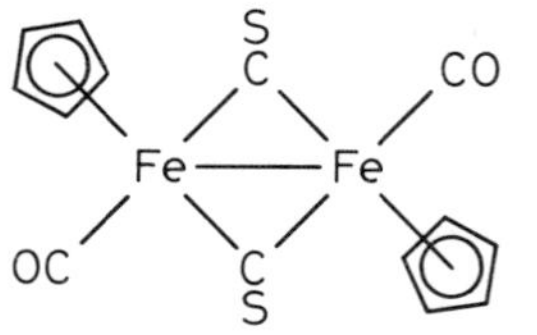

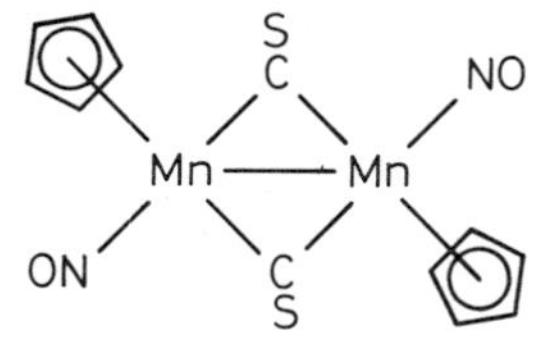

A trimetallic thiocarbonyl bridge is formed in $(\eta^5\text{-}C_5H_5)_3Co_3(CS)S_2$:

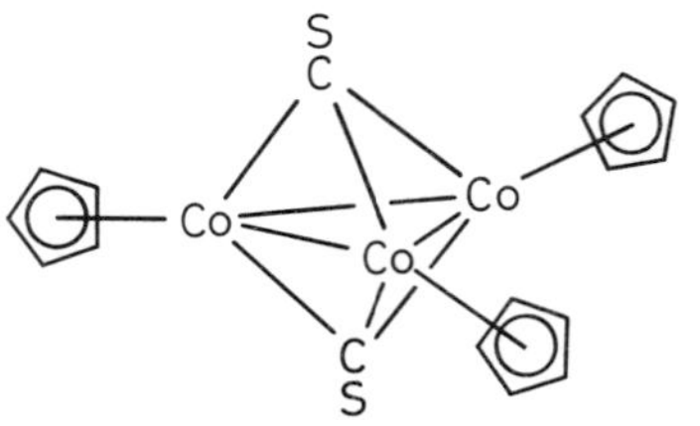

A rare linear bridge, M—C≡S—M′, occurs in $(diphos)_2(CO)W—CS—W(CO)_5$.

Theoretical calculations suggest that CS is both a better σ-donor and π-acceptor than CO, and this is confirmed by the selective replacement of CO rather than CS by phosphines. Also, M—C(S) bonds are shorter than M—C(O) bonds, indicating a higher degree of metal-ligand double bonding.

Metal-carbonyl anions, for example, $[M_2(CO)_{10}]^{2-}$ (M = Cr, W), react with thiophosgene to produce $M(CO)_5(CS)$, and $[Fe(CO)_4]^{2-}$ gives $Fe(CO)_4(CS)$. The CS ligand can also be introduced by carbon disulfide:

$$\underset{M = Mn, Re}{\eta^5\text{-}C_5H_5M(CO)_2THF} + CS_2 + PPh_3 \xrightarrow[-THF]{-SPPh_3} \eta^5\text{-}C_5H_5M(CO)_2(CS)$$

11.3. Metal Selenocarbonyls

The third chalcogenide CSe, can also form transition-metal complexes, as in $(\eta^6\text{-}C_6H_5COOMe)Cr(CO)_2(CSe)$, $RuCl_2(CO)(SCe)(PPh_3)_2$ and $Cr(CO)_5(CSe)$.

The CSe molecule is an even stronger π-acceptor ligand than either CS or CO, and forms shorter metal-carbon bonds in $(\eta\text{-}C_6H_5COOMe)Cr(CO)_2(CSe)$.

11.4. Metal-Isocyanide Complexes

Organic isocyanides :C≡N—R, are formally analogous to carbon monoxide and form similar transition metal complexes. Table 11.8 illustrates the known binary species. The compounds are analogous to metal carbonyls, for example, $M(CNR)_6$ (M = Cr, Mo, W), $Fe(CNR)_5$ or $Ni(CNR)_4$; however (see Tables 11.1 and 11.8) there are important differences:

a) The tendency to form metal-isocyanide cations is stronger;
b) No metal isocyanide anions are known;
c) Few polynuclear, metal-isocyanide complexes are known (only for nickel, while this metal does not form neutral polynuclear carbonyls);
c) Copper, silver and gold, which do not form stable metal carbonyls, coordinate up to four isocyanide molecules in oxidation state +1;
d) Isocyanide complex species are known which have no carbonyl analogues.

Mixed compounds containing carbonyls, halides and cyclopentadienyls are known.

- metal-carbonyl isocyanides:
  $M(CO)_{6-n}(CNR)_n$ M = Cr, Mo, W; n = 1, 2, 3
  $Fe(CO)_{5-n}(CNR)_n$ n = 1, 2
  $Ni(CO)_{4-n}(CNR)_n$ n = 1, 2, 3
- metal-isocyanide halides:
  $Mn(CNR)_5Br$ $Fe(CNR)_4X_2$
  $Co(CNR)_4X_2$ X = Cl, Br, I
  $Pd(CNR)_4X_2$ X = Cl, Br, I
- cyclopentadienylmetal isocyanides:
  η^5-$C_5H_5Mn(CNR)_3$ η^5-$C_5H_5Co(CNR)_2$
  η^5-$C_5H_5Fe(CNR)_2X$ X = Cl, Br, I $[\eta^5$-$C_5H_5Ni(CNR)]_2$
  $[\eta^5$-$C_5H_5Fe(CNR)_2]_2$

Like carbon monoxide, isocyanides can coordinate as bridging groups:

Tab. 11.9. Binary Metal-Isocyanide Complexes.

Group VI	VII		VIII		I B
$Cr(CNR)_6$ $Cr(CNR)_7^{2+}$	$Mn(CNR)_6^{2+}$ $Mn(CNR)_6^+$	$Fe(CNR)_5$ $Fe(CNR)_4^{2+}$ $Fe(CNR)_6^{2+}$	$Co(CNR)_5^+$ $Co(CNR)_5^{2+}$ $Co(CNR)_4^{2+}$ $Co_2(CNR)_8$ $Co_2(CNR)_{10}^{4+}$	$Ni(CNR)_2$ $Ni(CNR)_4$ $Ni_4(CNR)_6$ $Ni_4(CNR)_7$	$Cu(CNR)_4^+$
$Mo(CNR)_6$		$Ru(CNR)_5$ $Ru_2(CNR)_9$	$Rh(CNR)_4^+$	$Pd(CNR)_2$	$Ag(CNR)_2^+$ $Ag(CNR)_4^+$
$W(CNR)_6$ $W(CNR)_7^{2+}$	$Re(CNR)_6^+$		$Ir(CNR)_4^+$	$Pt(CNR)_4^{2+}$ $Pt_3(CNR)_6$	$Au(CNR)_4^+$

The structures of mononuclear isocyanides are analogous to those of the metal carbonyls, for example, in $Fe(CNBu^t)_5$, $Ru(CNBu^t)_5$, $Co_2(CNBu^t)_8$, $Cr(CNC_6H_5)_6$, $[Co_2(CNMe)_{10}]^{4+}$ and $Co(CNMe)_5^+$.

The isocyanides form by replacement of carbon monoxide in metal carbonyls and their derivatives, or by direct reaction of a metal salt with an isocyanide. Complexes with zero-oxidation state of the metal can be obtained from an anhydrous metal halide, the isocyanide and sodium amalgam. An alternative route is the alkylation of metal-cyano complexes, for example, ferricyanide, with ethyl iodide or dimethyl sulfate. The reduction of mononuclear isocyanide cations can lead to dinuclear compounds:

$$[Co(CNR)_5]^+PF_6^- \xrightarrow{K/Hg \text{ in THF}} Co_2(CNR)_8 \qquad R = Bu^t$$

11.5. Metal-Carbene Complexes and Related Compounds

Divalent-carbon compounds (carbenes), $:CR_2$, cannot be isolated, although they are postulated as intermediates in organic reactions. Such species can be stabilized by coordination to a transition metal through an sp^2-hybridized orbital with an electron pair available for donation and a vacant p_z-orbital available for back-donation.

The carbene ligand is a mono*hapto*, two-electron donor; related complexes are known in which the ligand is a cumulated polyene attached to a metal through a terminal carbon atom:

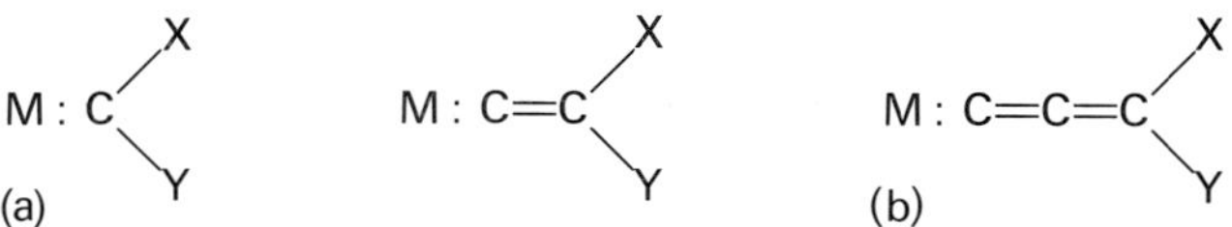

In (a) X and Y can be identical (H, R, OR, NR_2, SR) or different. In (b) frequently $X = Y = CN$ (dicyanovinylidene complexes).

The metals for which carbene complexes have been reported are shown in Fig. 11.12.

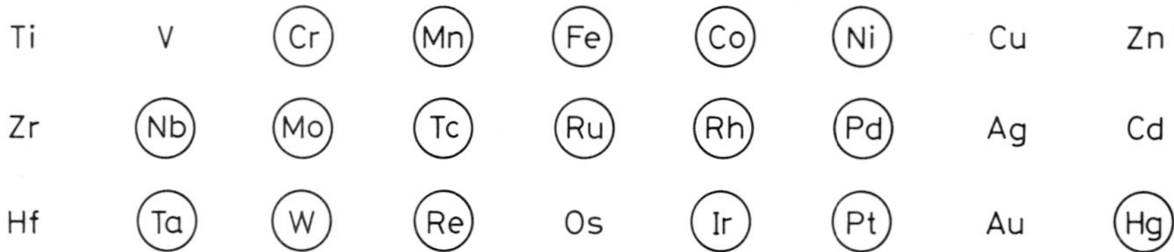

Fig. 11.12. Metals known to form carbene complexes.

The first metal-carbene complex was prepared by treatment of tungsten hexacarbonyl with organolithium reagents, followed by protonation and then reaction with diazomethane, or alkylation with trialkyloxonium tetrafluoroborate:

$$W(CO)_6 + LiR \longrightarrow (CO)_5W{:}C(O^-Li^+)(R) \xrightarrow{H^+} (CO)_5W{:}C(OH)(R)$$

R = Me, Ph

$$(CO)_5W{:}C(OH)(R) \xrightarrow{CH_2N_2} (CO)_5W{:}C(OCH_3)(R)$$

$$(CO)_5W{:}C(O^-Li^+)(R) \xrightarrow{Me_3O^+BF_4^-} (CO)_5W{:}C(OCH_3)(R)$$

This reaction has been extended to Cr, Mo, Mn, Tc, Re, Fe and Ni carbonyls.

A less-common carbene is formed in the reaction of a 1,1-dichlorocyclopropene with sodium carbonyl dichromate:

$$Ph_2C_3Cl_2 + Na_2Cr_2(CO)_{10} \xrightarrow{THF} Ph_2C_3{=}C{:}Cr(CO)_5$$

The addition of alcohols to isocyanide complexes:

$$Et_3P{-}PtCl_2{-}C{\equiv}N{-}Ph + EtOH \longrightarrow Et_3P{-}PtCl_2{:}C(NHPh)(OEt)$$

has been extended to the synthesis of complexes containing four carbene ligands attached to the same metal atom:

$$[M(:CNMe)_4]^{2+} + 4MeNH_2 \longrightarrow \{M[:C(NHMe)_2]_4\}^{2+}$$

There are few complexes with several carbene ligands in the same molecule.

Coordinated carbon disulfide can be converted into a carbene ligand by electrophilic attack with alkyl halides:

$$Os(\pi{-}CS_2)(CO)_2(PPh_3)_2 + 2MeI \longrightarrow OsI_2(CO)_2(PPh_3)_2(:C(SMe)_2)$$

$$Pt(\pi{-}CS_2)(PPh_3)_2 + 2MeI \longrightarrow [(Ph_3P)_2IPt{:}C(SMe)_2]^+ I^-$$

The cleavage of some electron-rich olefins also leads to carbene complexes:

$$[(Et_3P)PtCl_2]_2 + \text{(PhN)}_2C_2H_4C{=}CC_2H_4\text{(NPh)}_2 \xrightarrow{140\,°C} 2\,Et_3P{-}PtCl_2{:}C\text{(NPh)}_2C_2H_4$$

The first difluorocarbene complex resulted by fluorine abstraction from a σ-trifluoromethyl derivative:

$$\eta^5\text{-}C_5H_5(CO)_3Mo{-}CF_3 + SbF_5 \xrightarrow{\text{liq. } SO_2} [\eta^5\text{-}C_5H_5(CO)_3Mo: CF_2]^+SbF_6^-$$

Unsubstituted carbene (methylene) complexes of niobium and tantalum have been prepared by proton abstraction:

$$(\eta^5\text{-}C_5H_5)_2Ta(CH_3)_3 \xrightarrow[-Ph_3CMe]{Ph_3C^+BF_4^-} [(\eta^5\text{-}C_5H_5)_2Ta(CH_3)_2]^+BF_4^- \xrightarrow{\text{base}} (\eta^5\text{-}C_5H_5)_2Ta(CH_3)(=CH_2)$$

The reaction of $(Me_3C{-}CH_2)_3MCl_2$ (M = Nb, Ta) with $LiCH_2CMe_3$ yields carbene complexes of the type:

$$(Me_3C{-}CH_2)_3M: CH{-}CMe_3$$

The trigonal-planar carbene ligands are better σ-donors and weaker π-acceptors than carbon monoxide.

Carbene complexes in which at least one of the carbon substituents is an atom with electron pairs (X = OR, SR, NR_2) are stabilized by (p − p)-π interaction:

$$M \text{---} C(OR)(R')$$

Complexes with $:CR_2$ groups as bridging ligands, (in iron and rhodium compounds) have the structures:

However, these cannot be regarded as carbene complexes, since the carbon atom in the bridge is not trigonal.

Compounds with vinylidene and allylidene ligands, closely related to carbene complexes, are illustrated below:

Addition, substitution, rearrangement, etc., reactions can occur in the carbene ligand without cleavage from the metal, for example, the conversion of a phenylmethoxycarbene into a diphenylcarbene ligand:

$$(CO)_5W{:}C(OMe)(Ph) \xrightarrow{LiPh} Li^+ \, (CO)_5\bar{W}{-}C(Ph)_2{-}OMe \xrightarrow{HCl} (CO)_5W{:}C(Ph)_2$$

Two general features of metal-carbenes should be noted: a) only few compounds with more than one carbene ligand are known, and b) the carbene complexes are usually neutral (seldom cationic or anionic).

11.6. π-Olefinic Complexes

Transition metals are able to form complexes with olefins in which the C=C double bond contributes two electrons to the metal.

The first compound between a transition metal and an olefin, a platinum complex of ethylene, $K[PtCl_3 \cdot C_2H_4] \cdot H_2O$, was obtained in 1827 by Zeise. In 1938 olefinic complexes of palladium from $Pd(PhCN)_2Cl_2$ with olefins were obtained. The nature of these platinum and palladium complexes remained obscure for a long time. It was not until 1951 that the first satisfactory explanation of the metal-olefin bond was made by Dewar and in 1953 the Chatt-Duncanson model was suggested. This model was presented in Section 2.

In Zeise's salt, $K[PtCl_3 \cdot C_2H_4]$, the olefin is situated perpendicular to the plane formed by the central-platinum atom and the chlorine ligands:

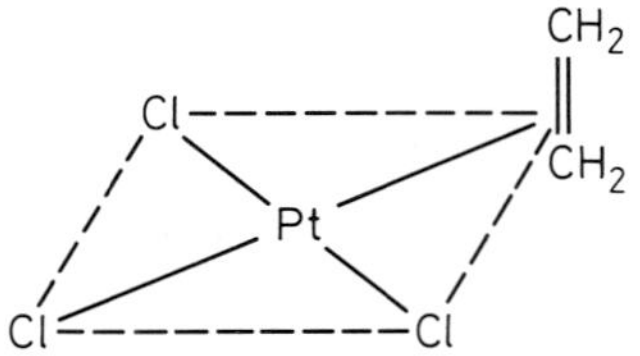

The mono- and polyolefins form complexes with almost all transition metals. Their stability varies with the nature of the metal and olefin, and is strongly influenced by both the metal and the olefinic-carbon atom substituents.

The monoolefins occupy a single coordinative site, as monodentate ligands (a). In polyolefins with non-conjugated (isolated) double bonds, each C=C bond acts as an independent donor, and two situations can arise:

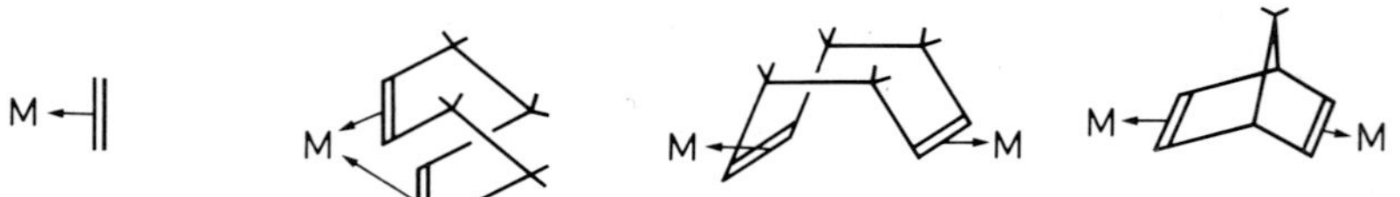

(a) (b) (c)

1. the two isolated C=C bonds of the polyolefin act as a bidentate ligand (b).
2. isolated C=C groups bridge different metal atoms (doubly monodentate ligand) (c)

Even in some conjugated olefins the C=C groups can act independently, [for example, in butadiene (d), cyclohexadiene (e) or fulvenes (f)] to form bridges:

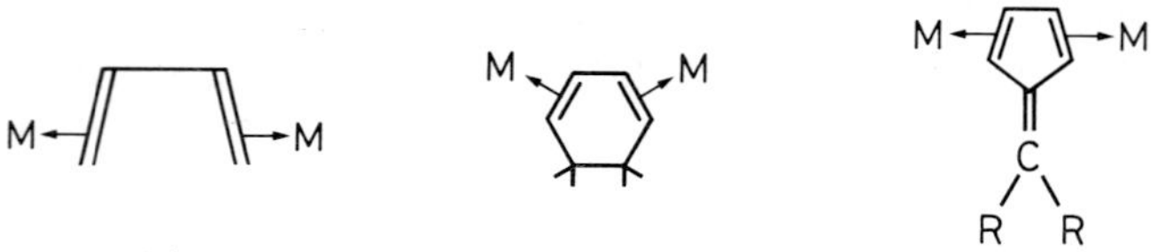

(d) (e) (f)

Molecules with several C=C bonds as in cyclooctatetraene, can form a doubly bidentate bridge:

Examples of tridentate or tetradentate olefins (that is, with 3 or 4 isolated double bonds attached to the same metal atom) are known.

Olefin complexes can be prepared using one of the following procedures:
- addition of the olefin to a metal salt (usually halide):

$$MX_n \ + \ \| \longrightarrow \ \|\!\rightarrow MX_n$$

- substitution of carbon monoxide or other ligands (: CNR, PR_3, etc.):

$$ML_n \ + \ \| \longrightarrow \ \|\!\rightarrow ML_{n-1} \ + \ L$$

- reduction of a metal cation in the presence of the olefin and an additional ligand (usually a phosphine):

$$M^{n+} \ + \ \| \ + \ ne^- \ + \ xL \longrightarrow \ \|\!\rightarrow ML_x$$

- by the gas phase reaction of the olefin with metal-atom vapors:

$$M^0 \ + \ \| \longrightarrow \ M \leftarrow \|$$

- hydride abstraction from σ-alkyl derivatives of the metal by the triphenylmethyl cation:

$$M{-}C_2H_5 \ + \ Ph_3C^+ \longrightarrow \ \begin{matrix} H_2C \\ \|\!\rightarrow M^+ \\ H_2C \end{matrix} \ + \ Ph_3CH$$

11.6.1. Monodentate Olefins

Rare binary complexes containing only monoolefins attached to the metal atom include the ethylene complexes, $M(C_2H_4)_n$ with M = Co (n = 1 and 2), M = Ni (n = 1, 2 and 3), M = Pd (n = 1, 2 and 3), M = Cu (n = 1, 2 and 3), M = Au (n = 1) or $Ni(CF_2{=}CF_2)_n$ (n = 1, 2 and 3), prepared by the metal-vapor synthesis technique, with the complexes isolated in a low-temperature matrix.

The olefin is usually accompanied by additional ligands, as in olefin-metal carbonyl complexes (Fig. 11.13a and 11.13b), cyclopentadienylmetal carbonyl olefin complexes (Fig. 11.13c) or metal-halide olefin complexes (Fig. 11.13d).

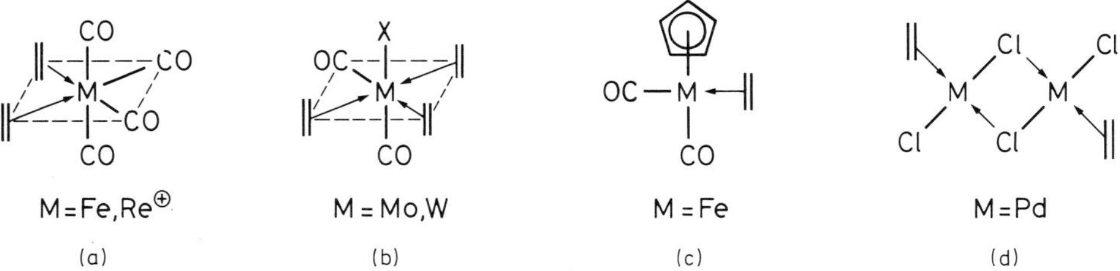

Fig. 11.13. Mixed metal-olefin complexes.

In Fig. 11.14 are shown the transition metals known to form complexes with monolefins.

Sc	Ti	(V)	(Cr)	(Mn)	(Fe)	(Co)	(Ni)	(Cu)	Zn
Y	Zr	Nb	(Mo)	(Tc)	(Ru)	(Rh)	(Pd)	(Ag)	Cd
La	Hf	Ta	(W)	(Re)	(Os)	(Ir)	(Pt)	(Au)	Hg

Fig. 11.14. Metals forming monolefin complexes.

Chromium, Molybdenum, Tungsten

The monolefins form unstable complexes of the types $M(CO)_5$(olefin) and $M(CO)_4$(olefin)$_2$ by irradiation of the molybdenum and tungsten hexacarbonyls. The derivatives of activated olefins like acrolein are prepared by the reaction (M = Mo, W):

$$(MeCN)_3M(CO)_3 + \text{olefin} \longrightarrow M(CO)_{6-n}(\text{olefin})_n$$

Tetracyanoethylene forms complexes by irradiation of the chromium, molybdenum or tungsten hexacarbonyls in the presence of the olefin:

$$M(CO)_6 + (NC)_2C{=}C(CN)_2 \longrightarrow (NC)_2C{=}C(CN)_2 \rightarrow M(CO)_5$$

The cyclopentadienylmetal-carbonyl derivatives of molybdenum and tungsten form ethylene complexes by replacement of carbon monoxide, or by the hydride abstraction of a σ-ethyl derivative:

$$C_5H_5M(CO)_3X + C_2H_4 \xrightarrow[\text{pressure}]{\text{heat}} [C_5H_5M(CO)_3(CH_2{=}CH_2)]^{\oplus}X^{\ominus} \xleftarrow{Ph_3C^{\oplus}X^{\ominus}} C_5H_5M(CO)_3C_2H_5$$

Manganese, Technetium, Rhenium

Ethylene-carbonyl derivatives of manganese form by the reaction of ethylene and manganese-pentacarbonyl chloride, or by hydride abstraction from the σ-ethyl derivative:

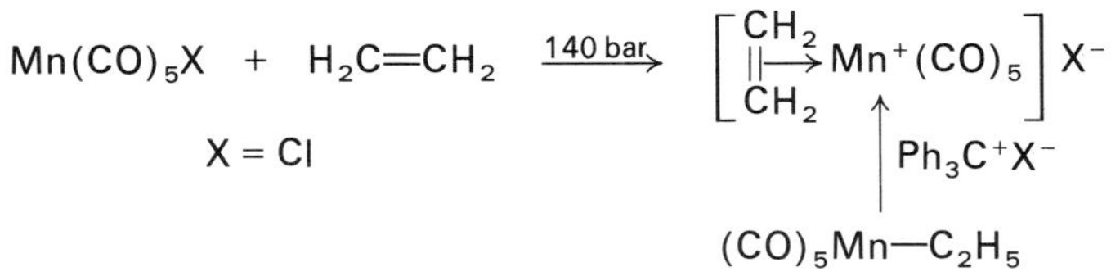

$$Mn(CO)_5X + H_2C{=}CH_2 \xrightarrow{140\ bar} \left[\begin{matrix} CH_2 \\ \| \\ CH_2 \end{matrix} \rightarrow Mn^+(CO)_5\right] X^-$$

X = Cl

$$\uparrow Ph_3C^+X^-$$

$$(CO)_5Mn{-}C_2H_5$$

Photochemical reactions of cyclopentadienylmanganese tricarbonyl with ethylene, butadiene, cyclohexadiene-1,3 and norbornadiene give complexes in which even the dienes act only as monodentate ligands:

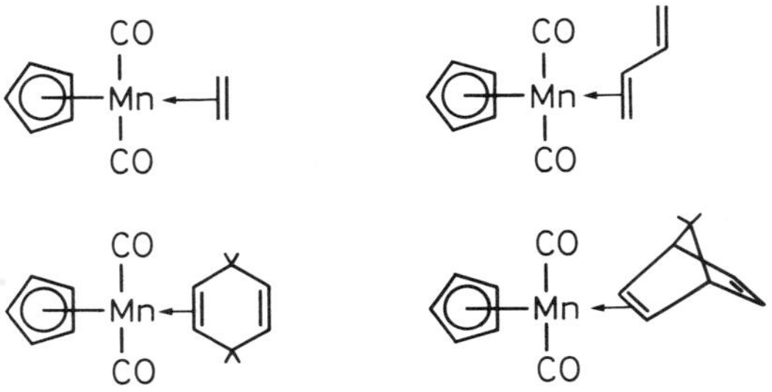

Rhenium-pentacarbonyl chloride reacts with ethylene (250 bar) to form a disubstituted derivative, $[Re(CO)_4(C_2H_4)_2]^+$, with the structure illustrated in Fig. 11.13.a.

The unstable technetium cation, $[Tc(CO)_4(C_2H_4)_2]^+$, is known.

Bis(cyclopentadienyl)rhenium hydride reacts with carbon monoxide to form a complex of monodentate cyclopentadiene which can be further hydrogenated to give the cyclopentene complex:

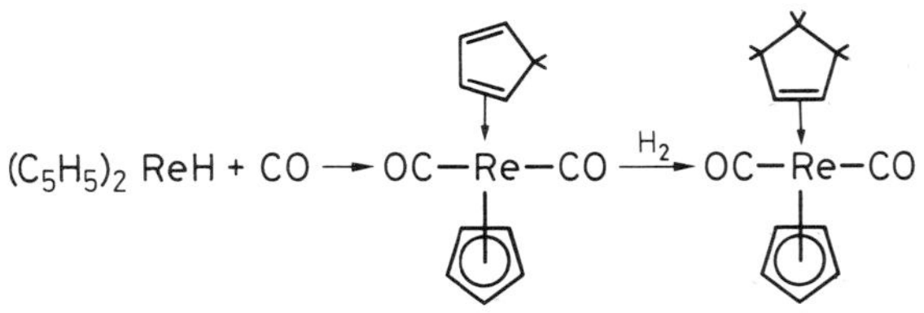

$$(C_5H_5)_2\,ReH + CO \rightarrow OC{-}Re{-}CO \xrightarrow{H_2} OC{-}Re{-}CO$$

Iron, Ruthenium, Osmium

The treatment of the binuclear carbonyl, $Fe_2(CO)_9$, with ethylene gives an unstable complex, $Fe(CO)_4(C_2H_4)$. The perhalo (C_2F_4, C_3F_6, $C_2F_2Cl_2$, cyclo-C_5H_8, etc.) and other substituted olefins ($H_2C{=}CHX$ where X = Me, Ph, OMe, OAc, Cl) form monoolefinic complexes, $Fe(CO)_4$(olefin). Ethylene complexes have also been obtained by hydride abstraction of a σ-ethyl derivative of iron:

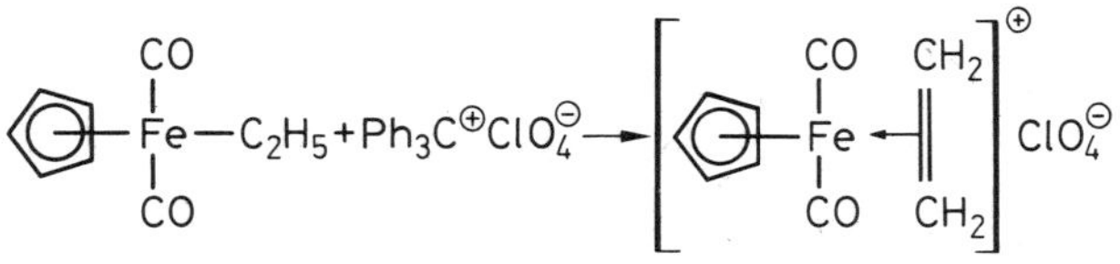

$$C_5H_5Fe(CO)_2{-}C_2H_5 + Ph_3C^{\oplus}ClO_4^{\ominus} \rightarrow \left[C_5H_5Fe(CO)_2 \leftarrow \begin{matrix} CH_2 \\ \| \\ CH_2 \end{matrix}\right]^{\oplus} ClO_4^{\ominus}$$

Such cations can be obtained by replacement of tetrahydrofuran in the cation η^5-$C_5H_5Fe(CO)_2THF^+$.

Ruthenium(II) chloride absorbs ethylene in aqueous hydrochloric solution until a metal : olefin ratio of 1 : 1 is reached, but no complex can be isolated. The ruthenium salts catalyze oligomerizations, polymerizations and dehydrogenations of olefins; these reactions probably involve intermediate formation of olefin complexes.

Ferrocene reacts with lithium metal in the presence of ethylene:

$$Fe(C_5H_5)_2 \xrightarrow[\text{THF, 20°C} \; -C_5H_5Li]{\text{Li, } C_2H_4} (C_2H_4)_2FeLi_2 \xrightarrow[Et_2O]{\text{TMED}} C_2H_4)_2FeLi_2(TMED)_2$$

$$C_2H_4)_2FeLi_2(TMED)_2 \xrightarrow{\text{diphos}} C_2H_4Fe(\text{diphos})_2 \qquad C_2H_4)_2FeLi_2(TMED)_2 \xrightarrow{\text{COD}} (COD)_2FeLi_2$$

COD = cyclooctadiene-1,3

Four ethylene molecules are coordinated to iron in $(C_2H_4)_4FeLi_2(TMED)_2$.

The osmium olefin complex, $[\eta^5\text{-}C_5H_5Os(CO)_2(\eta^2\text{-}C_2H_4)]^+PF_6^-$, is known.

Cobalt, Rhodium, Iridium

In the hexafluorocyclopentadiene-cobalt complex:

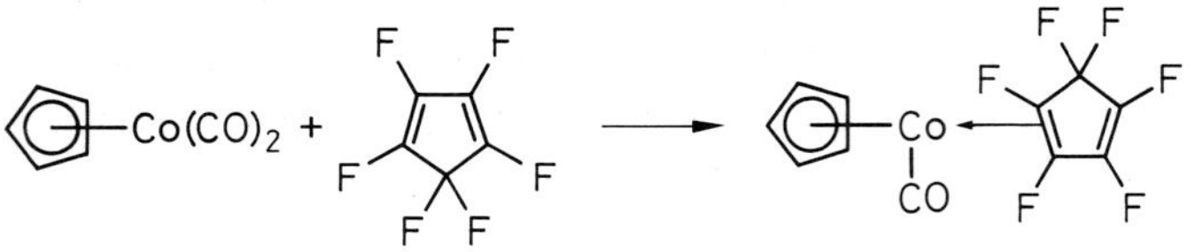

$$(C_5H_5)Co(CO)_2 + C_5F_6 \longrightarrow (C_5H_5)Co(CO)(C_5F_6)$$

the fluoroolefin acts as monodentate ligand.

Cyclopentadienyl ligands can be cleaved from cobaltocene by lithium metal and an olefin such as cyclooctadiene-1,5 (COD) substituted:

$$Co(C_5H_5)_2 \xrightarrow[\text{THF, 0°C} \; -LiC_5H_5]{\text{Li, COD}} (\eta^5\text{-}C_5H_5)CO(\eta^4\text{-}COD) \xrightarrow{\text{Li, COD}} (\eta^4\text{-}COD)_2CoLi(THF)_x$$

to give a compound whose structure is shown below:

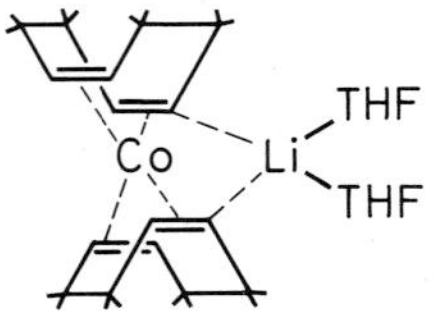

$$(\eta^4\text{-}COD)_2Co\text{-}Li(THF)_2$$

Rhodium salts catalyze the oligomerization and hydrogenation of olefins, via intermediate formation of complexes. The dimers, $[RhCl(\text{olefin})_2]_2$, containing four-coordinated rhodium (with a square-planar geometry) and chlorine bridges:

$$RhCl_3 \cdot aq + \| \longrightarrow [(C_2H_4)_2Rh(\mu\text{-}Cl)_2Rh(C_2H_4)_2]$$

have been prepared from $RhCl_3 \cdot aq$ and olefins.

Ethylene can replace triphenylphosphine in the complex $(Ph_3P)_3RhCl$ to form a *trans*-complex:

$$(Ph_3P)_3RhCl + H_2C{=}CH_2 \longrightarrow trans\text{-}[RhCl(PPh_3)_2(CH_2{=}CH_2)]$$

Salts of the anion, $[(\eta^2\text{-}C_2H_4)_2RhCl_2]^-$, are also known.

Iridium forms an ethylene complex, $IrCl_2(C_2H_4)$, by the action of iridium(III) chloride to dehydrate ethanol. Hydrous hexachloroiridic acid, $H_2[IrCl_6] \cdot 6H_2O$, forms tris(olefin) complexes, $Ir(CO)(olefin)_3Cl$, with cyclooctene and cycloheptene.

Nickel, Palladium, Platinum

Nickel tetracarbonyl reacts with the activated olefins acrylonitrile, acrolein, fumaronitrile, etc., to form $Ni(olefin)_2$. The reduction of nickel(II) acetylacetonate in the presence of ethylene and triphenylphosphine yields a diphosphine-ethylene complex:

$$Ni(acac)_2 + H_2C{=}CH_2 + PPh_3 \xrightarrow{Et_2AlOEt} (Ph_3P)_2Ni \leftarrow \|(CH_2{=}CH_2)$$

The cleavage of cyclopentadienyl ligands from nickelocene by lithium in the presence of an olefin can also lead to olefin complexes:

$$Ni(\eta^5\text{-}C_5H_5)_2 \xrightarrow[THF,\ 0\,°C]{2\,Li,\ COD} Ni(\eta^4\text{-}COD)_2$$

The product can be transformed to the anionic olefin complexes, $[R{-}Ni(C_2H_4)_2]^-$ and $[Ph_2P{-}Ni(C_2H_4)_2]^-$.

Palladium and its salts catalyze the oxidation of olefins to aldehydes and ketones through the intermediacy of π-olefin complexes. Palladium(II) chloride reacts with liquid olefins, for example, ethylene under pressure, to form the dimers, $[PdCl_2(olefin)]_2$, whose square-planar structure is illustrated in Fig. 11.13.d. The starting material of choice for palladium π-olefin complexes is the benzonitrile derivative, $(PhCN)_2PdCl_2$ (Kharasch reagent), in which the weakly bonded nitrile is replaced by olefins. Palladium also forms anionic, $[Pd(olefin)Cl_3]^-$, and neutral, $[Pd(olefin)_2Cl_2]$ and $[Pd(olefin)(PR_3)_2]$, complexes.

The platinum π-olefin complexes, $M[PtX_3(\text{olefin})]$, $[PtX_2(\text{olefin})]_2$ and *trans*-$[PtX_2(\text{olefin})_2]$ were prepared long before the understanding of the nature of metal-olefin bond. The platinum-ethylene complex has also been prepared.

Copper, Silver, Gold

Solid copper(I) halides react with monolefins to form unstable [Cu(olefin)X] complexes. Polyolefins act as monodentate ligands toward Cu(I). Thus, in the cyclooctatetraene complex of copper(I) chloride, $[Cu(C_8H_8)Cl]$, each cyclooctatetraene molecule is coordinated through only one of its double bonds to a copper atom, which in turn, is involved in the formation of a long $(CuCl)_\infty$ chain. The norbornadiene complex is a tetramer, $[Cu(C_7H_8)Cl]_4$, and contains a coordinated ring, Cu_4Cl_4. Each copper atom coordinates a norbornadiene molecule attached through one of its double bonds:

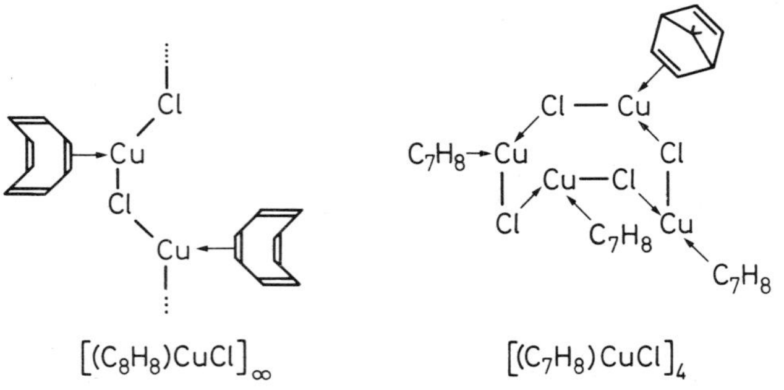

$[(C_8H_8)CuCl]_\infty$ $[(C_7H_8)CuCl]_4$

The stability of the silver-olefin complexes depends upon the nature of the olefin. Each silver ion can bond only one olefin molecule through a C=C double bond. Thus, ethylene forms the nitrate complex, $[Ag(C_2H_4)NO_3]$. Owing to the formation of π-complexes, insoluble olefins can be dissolved in aqueous solutions of silver(I) salts.

Monovalent gold similarly forms olefin complexes with a metal : olefin ration of 1 : 1.

11.6.2. Bidentate Olefins

The polyolefins containing isolated double bonds can act as bidentate ligands. Fig. 11.15 illustrates three types of complexes, obtained with Dewar benzene (a) and

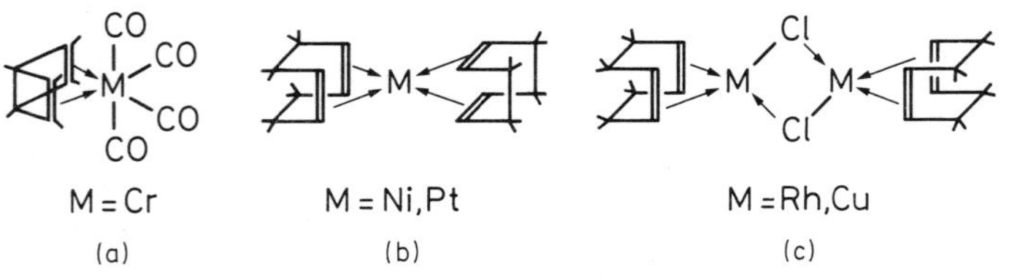

M = Cr (a) M = Ni, Pt (b) M = Rh, Cu (c)

Fig. 11.15. Non-conjugated diolefins as bidentate ligands in metal complexes.

cyclooctadiene-1,5 (b and c). Similar complexes have been obtained with other di- and polyolefins. The metals which form π-complexes with bidentate polyolefins are shown in Fig. 11.16.

Sc	Ti	V	(Cr)	(Mn)	(Fe)	(Co)	(Ni)	(Cu)
Y	Zr	Nb	(Mo)	Tc	(Ru)	(Rh)	(Pd)	(Ag)
La	Hf	Ta	(W)	(Re)	(Os)	(Ir)	(Pt)	(Au)

Fig. 11.16. Metals which form π-complexes with bidentate olefins (cyclooctadiene-1,5, norbornadiene, cyclohexadiene-1,4, etc.).

Chromium, Molybdenum, Tungsten

Hexamethyl Dewar benzene (hexamethylbicyclo[2,2,0]hexa-2,5-diene) reacts with the chromium-acetonitrile carbonyl derivative, $(MeCN)_3Cr(CO)_3$, to form a complex with the structure shown in Fig. 11.15.a. The Group VI metal hexacarbonyls form substitution products with cyclooctadiene-1,5 by replacement of carbon monoxide in which the diene acts as a bidentate ligand:

$$C_8H_{12} + M(CO)_6 \longrightarrow (C_8H_{12})M(CO)_4 + 2CO$$

The cyclooctatriene-1,3,5 complex of molybdenum (in which the triolefin is a six-electron donor) absorbs a mole of carbon monoxide to form a complex in which the olefin becomes bidentate, with one of the double bonds liberated from coordination:

$$(C_8H_{10})Mo(CO)_3 + CO \longrightarrow (C_8H_{10})Mo(CO)_4$$

Iron, Ruthenium, Osmium

Non-conjugated diolefins undergo isomerization in the presence of transition metals to become conjugated (as four-electron donors). However, norbornadiene and cyclooctadiene-1,5 react with the dinuclear iron carbonyl, $Fe_2(CO)_9$, to form complexes of the bidentate diolefins:

$Fe(CO)_3$ $Fe(CO)_3$

Cyclooctadiene-1,5 and norbornadiene also form Ru(diolefin)X_2 compounds with ruthenium(II) halides in which the diolefin is attached as a bidentate ligand.

Cobalt, Rhodium, Iridium

Cyclopentadienylcobalt dicarbonyl takes up cyclooctadiene-1,5 or norbornadiene, and the analagous rhodium compounds with cyclooctatriene-1,3,5 are also known:

Rhodium(III) chloride and the rhodium complex of butadiene react with cyclooctadiene-1,5 to form the dimer, $[RhX(diene)]_2$, whose structure is illustrated in Fig. 11.15.c. The larger ring, cyclodecadiene-1,6 derivative is obtained either directly from the diolefin and rhodium(III) chloride, or by isomerization of cyclodecadiene-1,5:

Cl
Rh Rh
Cl

Norbornadiene and cyclooctadiene-1,5 combine with the salt $Na[IrCl_6] \cdot 6H_2O$ to form $[Ir(diene)Cl]_2$ complexes with structures illustrated in Fig. 11.15.c.

Nickel, Palladium, Platinum

Reduction of nickel(II) acetylacetonate with Et_2AlOEt in the presence of cyclooctadiene-1,5 produces the complex $[Ni(C_8H_{12})_2]$ with the structure shown in Fig. 11.15.b.

Palladium(II) chloride forms the diene complexes, $[Pd(diene)X_2]$, with cyclooctadiene-1,5 and hexadiene-1,5. The Kharasch reagent, $[Pd(PhCN)_2Cl_2]$, reacts with allyl chloride to yield a complex of hexadiene-1,5 formed by coupling of the olefin.

Platinum forms complexes with nonconjugated diolefins in both oxidation states +2 and 0. Thus, $[Pt(diene)X_2]$ complexes have been obtained with cyclooctadiene-1,5 hexadiene-1,5 and cyclooctatetraene. Reduction of the complex [Pt(cyclooctadiene-1,5)Cl_2] with isopropylmagnesium bromide in the presence of additional cyclooctadiene-1,5 gives $[Pt(C_8H_{12})_2]$ with the structure shown in Fig. 11.15.b.

Copper, Silver, Gold

The elements of this triad seldom form bidentate complexes with olefins, but a norbornadiene complex [Cu(diolefin)Cl] whose structure is shown in Fig. 11.15.c., is formed by reduction of copper(II) chloride with sulfur dioxide in the presence of norbornadiene.

11.6.3. Tridentate Olefins

Unconjugated triolefins can act as tridentate ligands, thus, *cis-, cis-, cis*-cyclononatriene-1,4,7 reacts with molybdenum hexacarbonyl to form [$Mo(C_9H_{12})(CO)_3$], with the structure shown in Fig. 11.17.a. The reduction of nickel(II) acetylacetonate with $Et_2Al—OEt$ in the presence of *trans-, trans-, trans*-cyclododecatriene-1,5,9 results in the formation of [$Ni(C_{12}H_{18})$], having the structure illustrated in Fig. 11.17.b. In this compound the nickel atom is coordinatively unsaturated and readily accepts a carbon monoxide molecule, to form [$Ni(C_{12}H_{18})(CO)$].

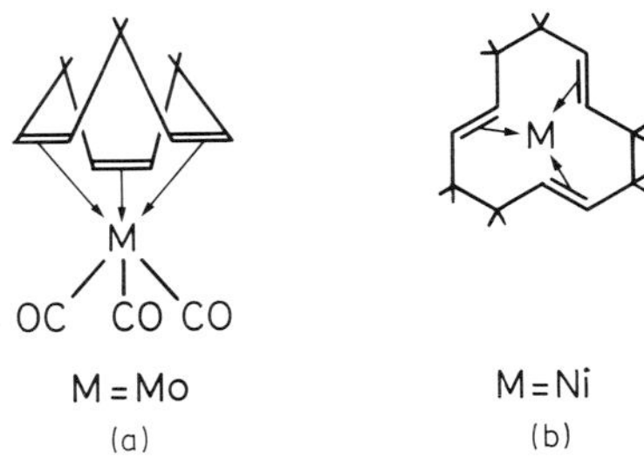

M = Ni
(b)

Fig. 11.17. Metal complexes of triolefins as tridentate ligands.

11.6.4. Bridge-Forming Olefins

In the previous examples both double bonds of a diolefin were attached to the same metal atom, but bridging can also take place, as shown in Fig. 11.18 for butadiene (a) and cyclohexadiene-1,3 (b). Polyolefins can also bridge, as illustrated for cyclooctatetraene in Fig. 11.18.c. Such attachment requires four nonconjugated-double bonds.

Butadiene, cyclohexadiene-1,3, norbornadiene and cyclooctadiene-1,5 all form, apart from the monodentate complexes referred to in Section 11.6.1, bridged bimetallic complexes with cyclopentadienylmanganese tricarbonyl:

OC CO Mn CO CO Mn

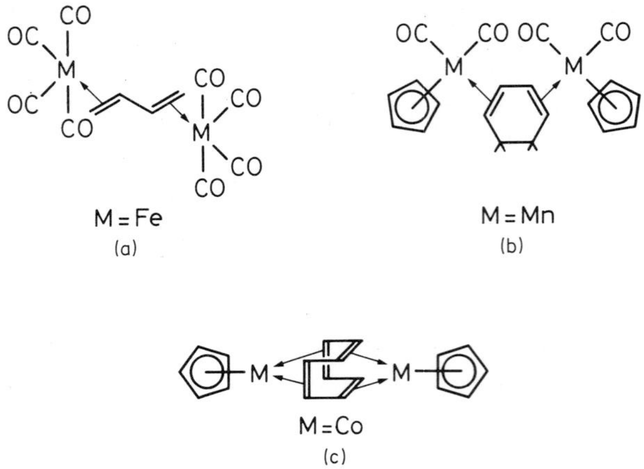

M = Fe
(a)

M = Mn
(b)

M = Co
(c)

Fig. 11.18. Complexes of bridge-forming olefins.

Butadiene forms a diiron derivative, $(CO)_4Fe(H_2C{=}CH{-}CH{=}CH_2)Fe(CO)_4$, from $Fe_2(CO)_9$ with a bridged structure shown in Fig. 11.18.a.

Cyclooctatetraene reacts photochemically with the cyclopentadienyl-metal dicarbonyls, $C_5H_5M(CO)_2$ (M = Co, Rh), to form complexes with the structure shown in Fig. 11.18.c.

The reaction of cyclohexadiene-1,3 with rhodium dicarbonyl chloride produces a complex in which the cyclohexadiene forms an additional bridge:

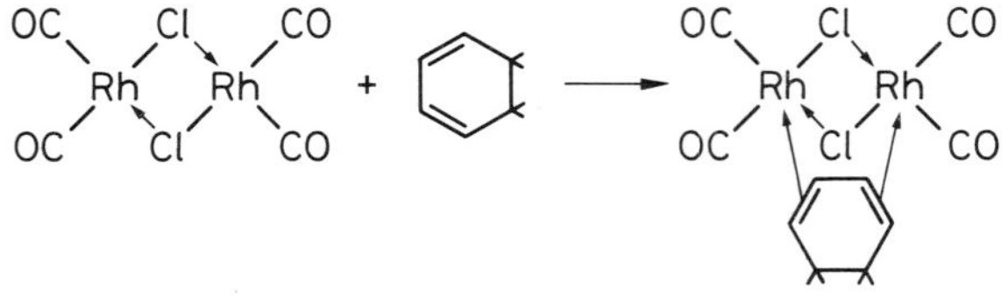

The nickel complex, $[Ni(C_{12}H_{18})]$, cited in Section 11.6.3, reacts with cyclooctatetraene to form polymeric $[Ni(C_8H_8)]_n$, with doubly bidentate-cyclooctatetraene bridges:

Diolefins react with platinum chlorides to form bridged butadiene and cyclohexadiene-1,3 complexes:

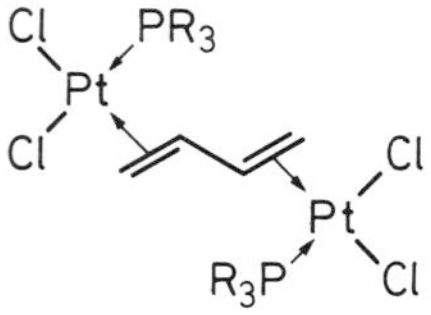

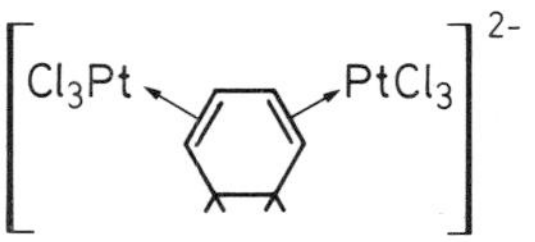

The monovalent metals of the copper-silver-gold triad form bridged complexes, since they can coordinate only one double bond. Thus, copper(I) chloride reacts with butadiene to form $[C_4H_6(CuCl)_2]$, with cyclohexadiene-1,3 to form the unstable $[C_6H_{10}(CuCl)_2]$, and with norbornadiene to form $[C_7H_8(CuBr)_2]$ and $[C_7H_8(CuCl)_2]$:

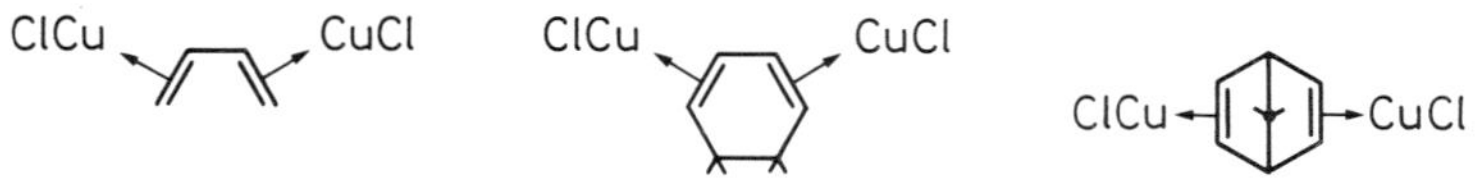

ClCu CuCl ClCu CuCl ClCu CuCl

Silver ion forms bridged-butadiene complexes in 1 : 1 and 2 : 1 ratios by absorption of butadiene by aqueous silver nitrate:

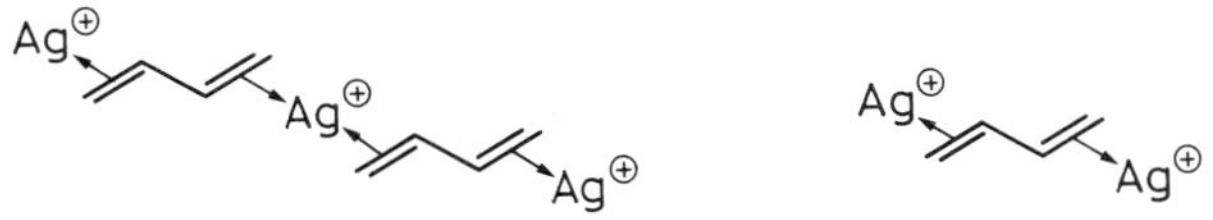

$Ag^{\oplus}$ $Ag^{\oplus}$ $Ag^{\oplus}$ $Ag^{\oplus}$ $Ag^{\oplus}$

Gold forms the bridges $[C_8H_{12}(AuCl)_2]$ on ultraviolet irradiation of cyclooctadiene-1,5 with tetrachloroauric acid, $H[AuCl_4]$.

12. Compounds with Three-Electron Ligands

12.1. π-Allylic Complexes and Related Compounds

The most important three-electron ligand is the π-allyl group (bonded trihapto), which has an open chain of three sp^2-hybridized carbon atoms, each having a π-electron available for metal-ligand bond formation:

$$H_2\dot{C}\!-\!\underset{H}{\dot{C}}\!-\!\dot{C}H_2$$

The allylic group can occur either as an isolated group or as a fragment of a larger chain or ring (π-enyl). Figure 12.1 illustrates η^3-allylic complexes with isolated allyl groups.

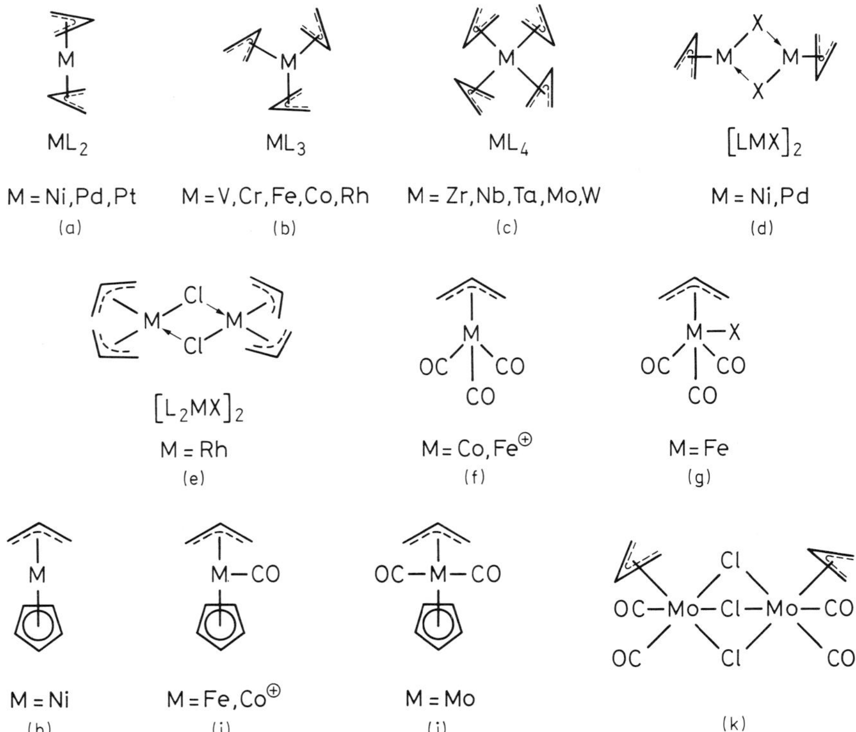

Fig. 12.1. Structural types of η^3-allylic complexes.

The transition metals known to form η-allylic complexes are shown in Fig. 12.2.

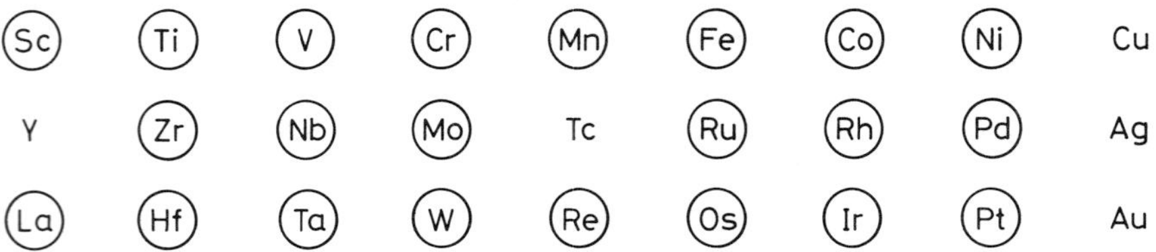

Fig. 12.2. Transition metals known to form η^3-allylic complexes.

Bridging allylic ligands are present in some palladium complexes and in dichromium and dimolybdenum tetraallyls:

L— Pd — Pd —L

M —— M

M = Cr,Mo

Most of the binary complexes, $M_m(C_3H_5)_n$ (Table 12.1), are mononuclear. No cluster compounds containing only π-allylic ligands are known.

Tab. 12.1. Binary π-Allyl Metal Complexes*

Group IV	Group V	Group VI	Group VII	Group VIII		
**	$V(C_3H_5)_3$	$Cr(C_3H_5)_3$ $Cr_2(C_3H_5)_8$	–	$Fe(C_3H_5)_3$	$Co(C_3H_5)_3$	$Ni(C_3H_5)_2$
$Zr(C_3H_5)_4$	$Nb(C_3H_5)_3$	$Mo(C_3H_5)_4$ $Mo_2(C_3H_5)_8$	–	–	$Rh(C_3H_5)_3$	$Pd(C_3H_5)_2$
$Hf(C_3H_5)_4$	$Ta(C_3H_5)_4$	$W(C_3H_5)_4$	$Re_2(C_3H_5)_4$	–	$Ir(C_3H_5)_3$	$Pt(C_3H_5)_2$

* $Th(C_3H_5)_4$ and $U(C_3H_5)_4$ have also been reported.
** A methyl-substituted derivative of titanium, $Ti(C_3H_4Me)_4$, is known.

η^3-Benzyl, η^3-cycloheptatrienyl, η^3-cycloheptadienyl, η^3-cyclohexenyl, η^3-cyclopentenyl, or even η^3-cyclopentadienyl groups can act as π-allylic ligands as illustrated in Fig. 12.3.

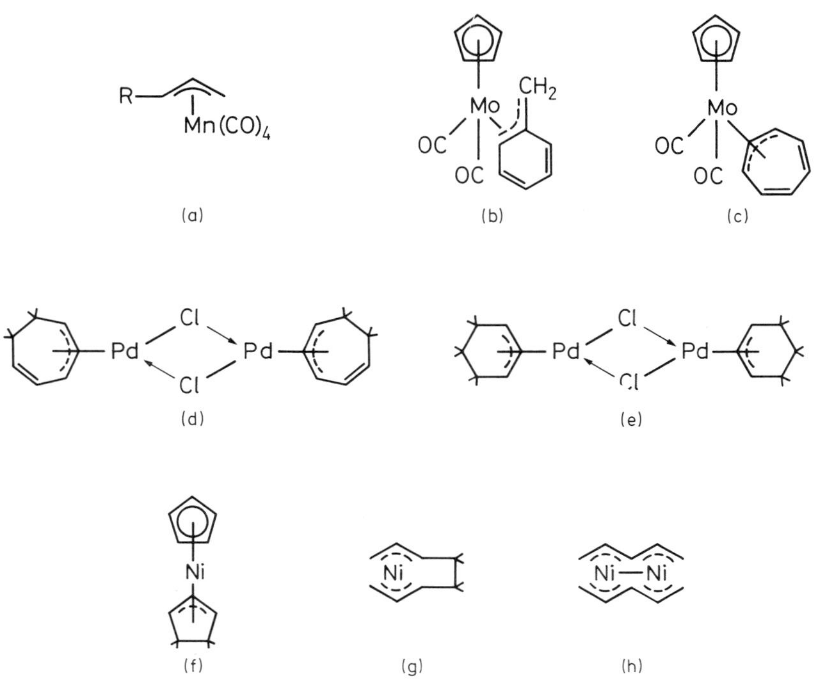

Fig. 12.3. Metal complexes in which the π-allyl group is a fragment of a larger ligand.

In addition to the η^3-bonding type (discussed in Section 2.6), allylic groups occur as σ-allyl ligands or as σ,η^2-ligands:

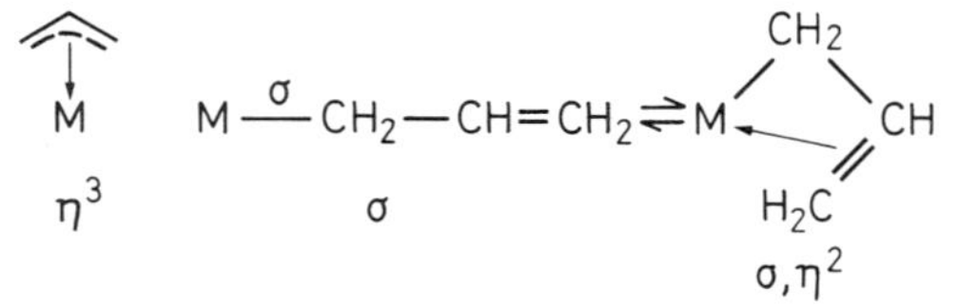

Therefore, caution must be exercised in writing the structures of allylic derivatives. The 18-electron rule is useful in discriminating between η^3- and η^1-allyl types, but the third type (σ, η^2) can be identified only by structure determination.

The η^3-allylic group is attached symmetrically so that the plane of the group is perpendicular to the metal-allyl bonds. σ,η^2-Allyl complexes are fluxional, with the metal changing positions from one side to another of the plane comprising the three allylic-carbon atoms:

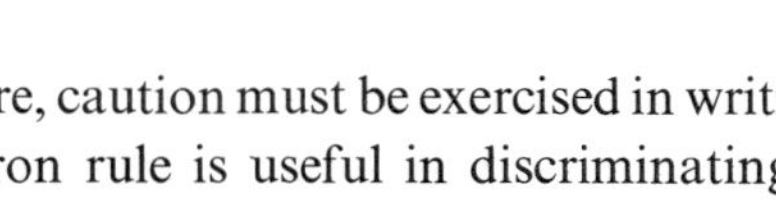

The preparative methods for π-allylic complexes are:

a) Alkali metal salts of a metal-carbonyl anion react with an allyl halide to give a σ-bonded derivative followed by UV-irradiation to promote σ-π-rearrangement:

$$[M(CO)_n]^{\ominus} + XCH_2CH{=}CH_2 \xrightarrow{-X^{\ominus}} (CO)_nM{-}CH_2CH{=}CH_2 \xrightarrow[-CO]{h\nu} (CO)_{n-1}M{-}(\eta^3\text{-}C_3H_5)$$

b) Allyl halides or alcohols react with metal halides or metal carbonyls;

c) Allyl Grignard reagents react with metal halides:

$$MX + CH_2{=}CH{-}CH_2MgX \longrightarrow M(\eta^3\text{-}C_3H_5) + MgX_2$$

d) Metal hydrides add to dienes:

$$MH + \text{diene} \longrightarrow M(\eta^3\text{-allyl})\text{-}CH_3$$

e) Olefins or allenes react with metal salts:

$$\text{olefin} + MX \longrightarrow (\eta^3\text{-allyl})MX$$

Scandium, Lanthanides, Actinides

The only known scandium derivative is the mixed ligand complex, $(\eta^5\text{-}C_5H_5)_2Sc(\eta^3\text{-}C_3H_5)$, prepared from $(\eta^5\text{-}C_5H_5)_2ScCl$ and C_3H_5MgCl. The lanthanide derivatives include $(\eta^5\text{-}C_5H_5)_3Ln(\eta^3\text{-}C_5H_5)$ (Ln = Sm, Ho, Er) and are prepared from $(\eta^5\text{-}C_5H_5)_3LnCl$ and C_3H_5MgCl.

Thorium and uranium form binary tetraallyl derivatives, $M(\eta^3\text{-}C_3H_5)_4$, from Grignard reactions.

Titanium, Zirconium, Hafnium

The dichloride, $(\eta^5\text{-}C_5H_5)_2TiCl_2$, reacts with allylmagnesium bromide to give $(\eta^5\text{-}C_5H_5)_2Ti(\eta^3\text{-}C_3H_5)$. The analogous zirconium compound yields a diallyl derivative, $(\eta^5\text{-}C_5H_5)_2Zr(\eta^3\text{-}C_3H_5)_2$.

Vanadium, Niobium, Tantalum

η^3-Allylvanadium pentacarbonyl, $\eta^3\text{-}C_3H_5V(CO)_5$, is obtained from allyl chloride and sodium hexacarbonylvanadate. Niobium forms a mixed ligand complex, $(\eta^5\text{-}C_5H_5)_2Nb(\eta^3\text{-}C_8H_9)$, in which a cyclooctatrienyl group is bonded tri*hapto* to satisfy the 18-electron rule:

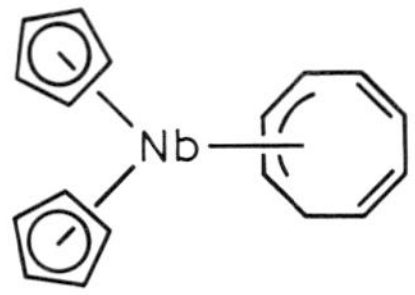

Chromium, Molybdenum, Tungsten

The dimer, $[\eta^5\text{-}C_5H_5Cr(CO)_3]_2$, irradiated with UV in the presence of butadiene forms a methylallyl complex:

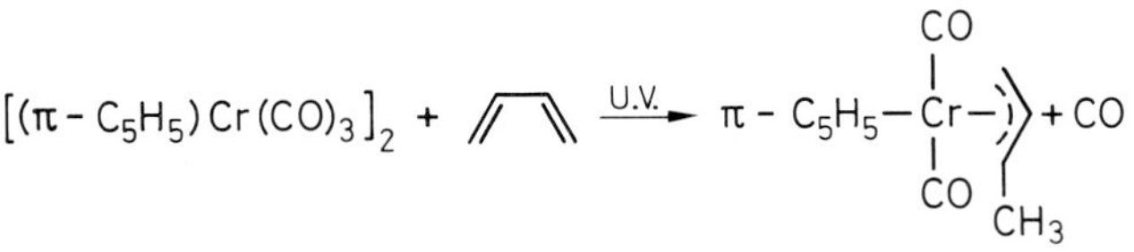

Chromocene is reduced by a mixture of hydrogen and carbon monoxide to a complex in which one of the five-membered rings becomes a η^3-ligand:

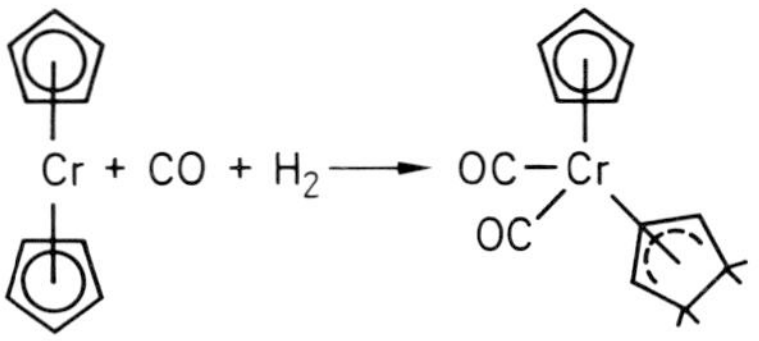

The sodium salt of the anion, $[C_5H_5Mo(CO)_3]^-$, reacts with allyl chloride to form a η^1-allyl derivative which eliminates a molecule of carbon monoxide under ultraviolet irradiation to form a η^3-allylic complex with the structure shown in Fig. 12.1.j. The hepta*hapto* cycloheptatrienyl ligand in $\eta^7\text{-}C_7H_7Mo(CO)_2I$ is converted into a tri*hapto* ligand by reaction with sodium cyclopentadienide, to form $\eta^5\text{-}C_5H_5Mo(CO)_2(\eta^3\text{-}C_7H_7)$, whose structure is shown in Fig. 12.3.c. The $C_5H_5Mo(CO)$ group requires only three electrons to achieve a noble gas configuration.

Molybdenum hexacarbonyl reacts with cyclohexa-1,3-diene to form a complex in which one of the rings is a five-electron and the other is a three-electron donor:

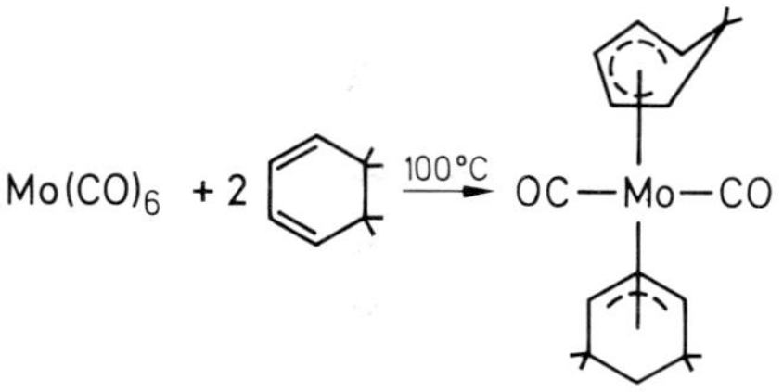

σ-Benzyl derivatives are converted to fluxional, η^3-benzylic compounds under UV irradiation with the resulting perturbation of the aromatic conjugation in the phenyl group:

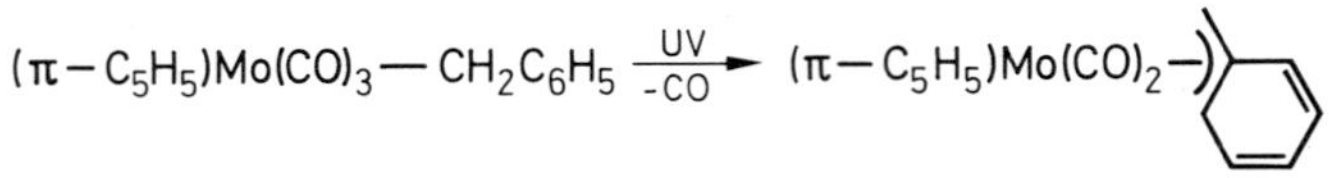

$$(\pi-C_5H_5)Mo(CO)_3-CH_2C_6H_5 \xrightarrow[-CO]{UV} (\pi-C_5H_5)Mo(CO)_2-$$

A thiophene derivative undergoes similar transformation:

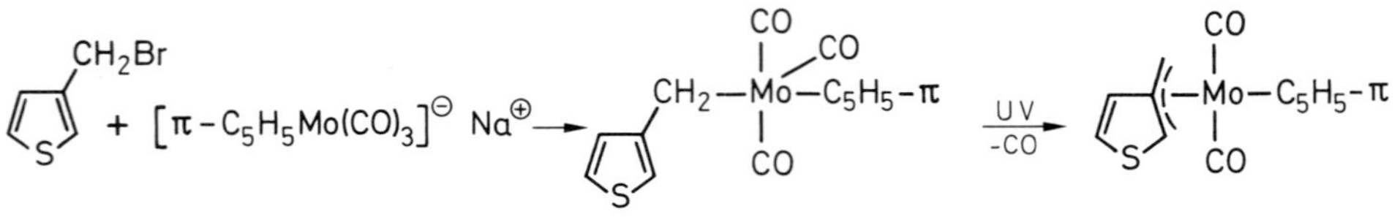

$$\text{(3-thienyl)}CH_2Br + [\pi-C_5H_5Mo(CO)_3]^{\ominus}Na^{\oplus} \longrightarrow \text{(3-thienyl)}CH_2-Mo(CO)_3-C_5H_5-\pi \xrightarrow[-CO]{UV} \text{(thienyl)}-Mo(CO)_2-C_5H_5-\pi$$

The tungsten complex, $(\eta^5\text{-}C_5H_5)_2W(CO)_2$ contains a bent tri*hapto*cyclopentadienyl ligand (the 18-electron rule requires one of the C_5H_5 groups to be bonded as an η^3-allylic fragment):

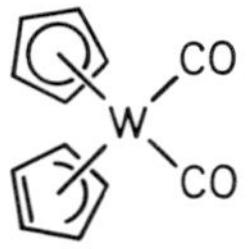

$$(C_5H_5)_2W(CO)_2$$

Manganese, Technetium, Rhenium

Manganese pentacarbonyl hydride adds to butadiene to form a η^3-allylic complex:

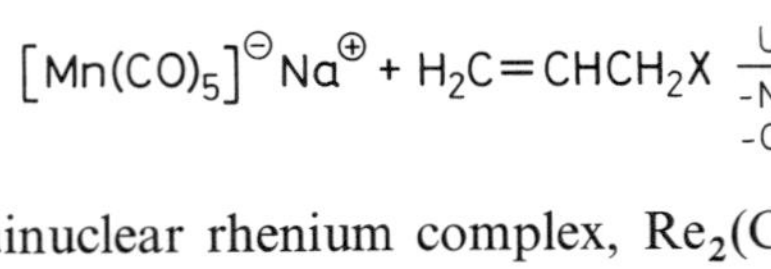

$$(CO)_5MnH + C_4H_6 \longrightarrow (CH_3\text{-}C_3H_4)-Mn(CO)_4$$

and sodium pentacarbonylmanganate reacts with allyl chloride under UV irradiation to form η^3-allymanganese tetracarbonyl:

$$[Mn(CO)_5]^{\ominus}Na^{\oplus} + H_2C{=}CHCH_2X \xrightarrow[-NaX,\ -CO]{U.V.} (C_3H_5)-Mn(CO)_4$$

The dinuclear rhenium complex, $Re_2(C_3H_5)_4$, contains no allylic bridge, unlike other $M_2(C_3H_5)_4$ compounds known (M = Cr,Mo):

$$(C_3H_5)_2Re-Re(C_3H_5)_2$$

Iron, Ruthenium, Osmium

Numerous π-allyl derivatives are known for the metals of this triad. Iron pentacarbonyl forms $(\eta^3\text{-}C_3H_5)Fe(CO)_3X$ (X = halogen) with allyl halides (Fig. 12.1.g). The dimer $[\eta^3\text{-}C_3H_5Fe(CO)_3]_2$:

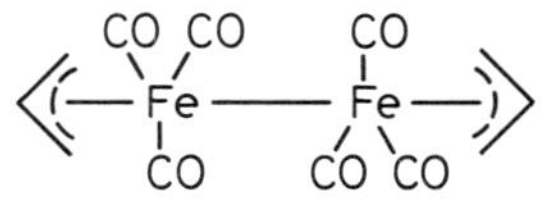

CO CO CO
Fe —— Fe
CO CO CO

is formed by elimination of iodine from $(\eta^3\text{-}C_3H_5)Fe(CO)_3I$ during chromatography over alumina, or in the reaction:

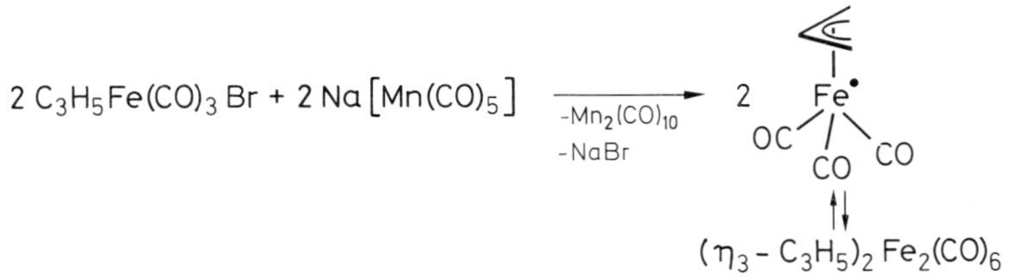

$$2\,C_3H_5Fe(CO)_3Br + 2\,Na[Mn(CO)_5] \xrightarrow[-NaBr]{-Mn_2(CO)_{10}} 2\ (\eta^3\text{-}C_3H_5)Fe^{\bullet}(CO)_3 \rightleftarrows (\eta_3\text{-}C_3H_5)_2Fe_2(CO)_6$$

The tris(η^3-allyl) iron complex $Fe(\eta^3\text{-}C_3H_5)_3$ is formed in a Grignard reaction with iron(III) chloride:

$$FeCl_3 \;+\; 3\,C_3H_5MgCl \xrightarrow[-3\,MgCl_2]{-78\,°C} Fe(\eta^3\text{-}C_3H_5)_3$$

Iron pentacarbonyl readily forms $(\eta^3\text{-}C_3H_5)Fe(CO)_3X$ with allyl halides. Allyl chloride reacts with $Na^+\,[(\eta^5\text{-}C_5H_5)Fe(CO)_2]^-$ to form a η^1-allyl derivative, convertible to a η^3-allylic compound by UV irradiation:

$$[\eta^5\text{-}C_5H_5Fe(CO)_2]^-Na^+ + H_2C{=}CH_2CH_2Cl \longrightarrow \eta^5\text{-}C_5H_5Fe(CO)_2 \overset{\sigma}{—} CH_2CH{=}CH_2$$

$$\downarrow \text{UV, } -CO$$

$$\eta^5\text{-}C_5H_5Fe(CO)(\pi\text{-}C_3H_5)$$

Because the $(\eta^5\text{-}C_5H_5)Fe(CO)$ group requires only a three-electron ligand, a cycloheptatrienyl group attaches as a tri*hapto* ligand:

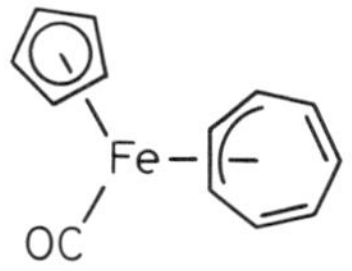

Fe —
OC

The η^4-butadiene complex of iron can be converted to a η^3-methylallyl derivative by protonation:

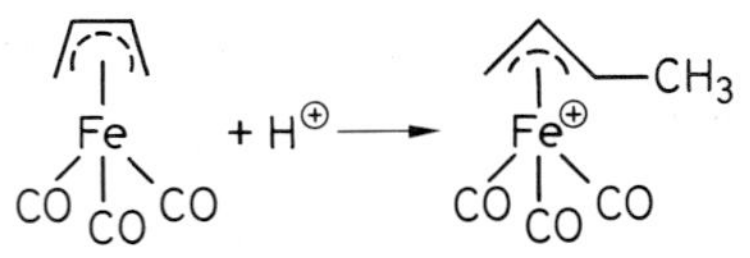

The dinuclear iron carbonyl, $Fe_2(CO)_9$, reacts with cycloheptatriene-1,3,5 to yield a product in which the dimetallic, $Fe_2(CO)_6$ unit is attached to the ring through two η^3-allylic fragments:

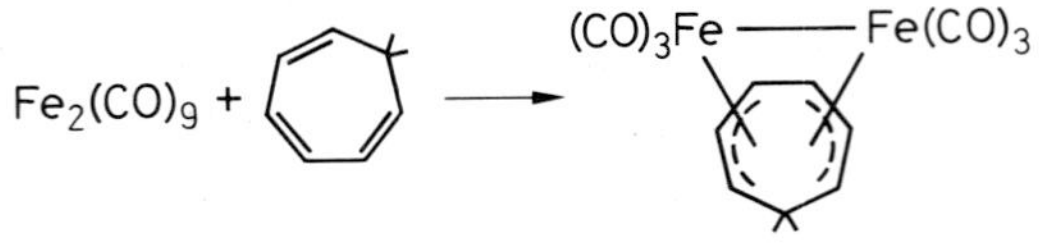

Allene reacts with $Fe_3(CO)_{12}$ to form a diallene (tetramethyleneethane) derivative, in which the ligand is again attached through two η^3-allylic bonds; the same product is obtained by reduction of an iron chloride complex by zinc metal:

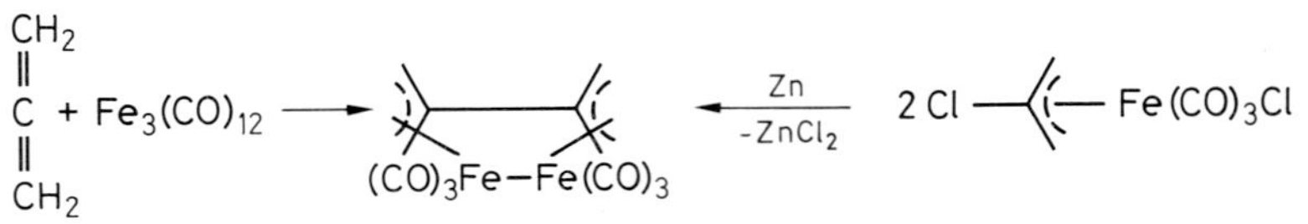

Related compounds are formed by intramolecular cyclizations of macrocyclic diallenes with iron carbonyls:

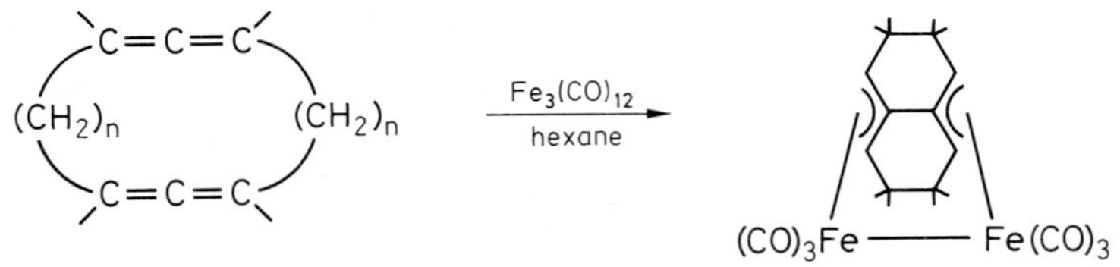

Phenylallene forms two diallylic isomers:

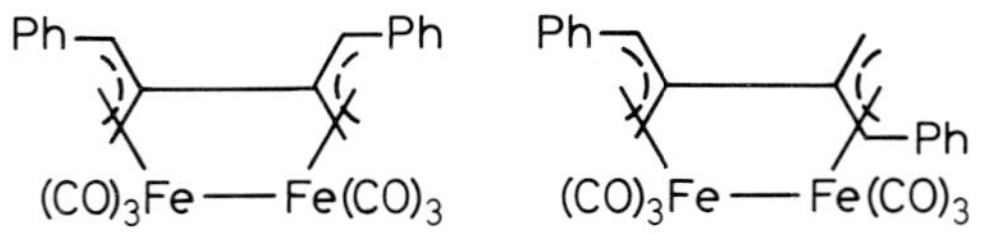

Tetracene forms a product with $Fe_3(CO)_{12}$ in which the ligand is also bonded through a pair of η^3-allylic bonds:

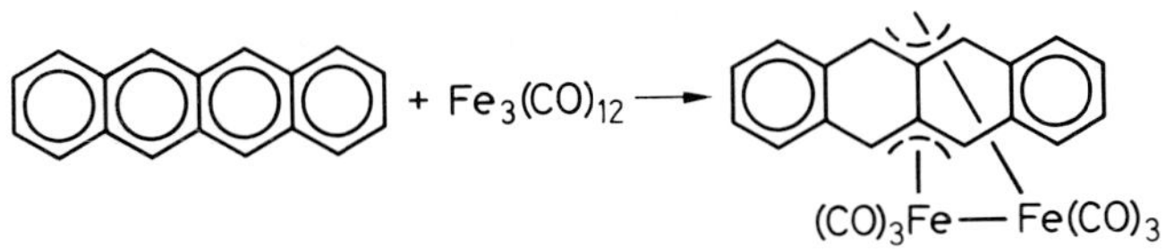

A η^3-1-silapropenyliron complex, containing a silicon atom in the allylic fragment, participating in the tri*hapto* bond, has been claimed from the following route, but this was later shown to be incorrect:

$$Fe_2(CO)_9 + RSiMe_2{-}SiMe_2{-}CH{=}CH_2 \xrightarrow[\text{- Fe(CO)}_5,\ \text{- CO}]{\text{benzene}} (\eta^3\text{-}Me_2SiCHCH_2){-}Fe(CO)_3SiMe_2R$$

$$R = Me \text{ or } CH{=}CH_2$$

The ruthenium η^3-allyl derivatives, η^3-$Ru(CO)_3X$, form from allyl halides with $Ru_3(CO)_{12}$. Ruthenium compounds tend to promote olefin isomerization with coordination of the product. Thus, ruthenium(III) chloride forms $RuCl_2(C_{12}H_{18})$ with butadiene which is a complex of 2,6,10-dodecatriene attached to the metal through two η^3-allylic bonds and a η^2-olefinic bond:

[Structure: $RuCl_2(C_{12}H_{18})$ — Ru bonded to two Cl and the dodecatriene ligand]

Cobalt, Rhodium, Iridium

The reaction of $Na^+[Co(CO)_4]^-$ with allyl halides produces an unstable η^1-allyl derivative which readily isomerizes to the η^3-allylic $(\eta^3\text{-}C_3H_5)Co(CO)_3$, whose structure is shown in Fig. 12.1.f. The η^3-methylallyl, $(\eta^3\text{-}MeC_3H_4)Co(CO)_3$, is formed by addition of the hydride $HCo(CO)_4$ to butadiene.

The cation, $[(\eta^5\text{-}C_5H_5)Co(CO)(\eta^3\text{-}C_3H_5)]^+$ (Fig. 12.1.i), is formed by reacting $(\eta^5\text{-}C_5H_5)Co(CO)_2$ and an allyl halide. Since the $Co(CO)_3$ group requires only three electrons to acquire a noble gas configuration in the cycloheptatriene complex, $C_7H_7Co(CO)_3$, prepared from $Co_2(CO)_8$ and cycloheptatriene under UV irradiation, the cyclic ligand must be bonded as a η^3-allylic fragment:

$$Co_2(CO)_8 + C_7H_8 \longrightarrow (\eta^3\text{-}C_7H_7)Co(CO)_3$$

leaving a butadiene fragment in the ring not involved in coordination.

The rhodium dimer, $[(\eta^3\text{-}C_3H_5)_2RhCl]_2$ (Fig. 12.1.e), has been obtained from the carbonyl chloride, $[Rh(CO)_2Cl]_2$, with allyl chloride. The allylic dimer reacts with allylmagnesium chloride to form tris(η^3-allyl)rhodium, $Rh(\eta^3\text{-}C_3H_5)_3$ (Fig. 12.1.b):

$$[Rh(CO)_2Cl]_2 + C_3H_5Cl \xrightarrow{H_2O} [(\eta^3\text{-}C_3H_5)_2RhCl]_2$$

$$[(\eta^3\text{-}C_3H_5)_2RhCl]_2 \xrightarrow{+ C_3H_5MgCl} Rh(\eta^3\text{-}C_3H_5)_3$$

Allyl chloride forms a σ,η^2-allyl product with tris(triphenylphosphine)rhodium chloride:

$$(PPh_3)_3RhCl + H_2C{=}CHCH_2Cl \longrightarrow (Ph_3P)_2Rh(Cl)(Cl)(CH_2{-}CH{=}CH_2)$$

Nickel, Palladium, Platinum

The air-sensitive, bis(allyl) derivative, $Ni(\eta^3\text{-}C_3H_5)_2$ (Fig. 12.1.a), is obtained from nickel bromide with allylmagnesium chloride, by disproportionation of the halides $[(\eta^3\text{-}C_3H_5)NiX]_2$ or by the reaction of nickel vapor with tetraallyltin. The compound $Ni(\eta^3\text{-}C_3H_5)_2$ has two electrons fewer than the noble gas configuration.

Dimeric $[\eta^3\text{-}C_3H_5NiX]_2$ (X = halogen) (Fig. 12.1.d) forms on heating nickel tetracarbonyl with allyl halides which replace completely the carbon monoxide ligands. Highly active forms of nickel and palladium, prepared by reduction of the anhydrous dihalides with potassium, react with allyl halides to produce $[(\eta^3\text{-}C_3H_5)MX]_2$ (M = Ni, Pd).

The 18-electron compound, $Ni(\eta^5\text{-}C_5H_5)(\eta^3\text{-}C_3H_5)$ (Fig. 12.1.h), can be obtained by treatment of nickelocene with allylmagnesium chloride, treatment of $[\eta^3\text{-}C_3H_5NiBr]_2$ with sodium cyclopentadiene, or by treatment of nickel(II) chloride with allylmagnesium chloride and lithium cyclopentadienide:

$$\left.\begin{array}{l} Ni(\eta^5\text{-}C_5H_5)_2 + C_3H_5MgCl \\ [\eta^3\text{-}C_3H_5NiBr]_2 + NaC_5H_5 \\ NiCl_2 + C_3H_5MgCl + LiC_5H_5 \end{array}\right\} \longrightarrow (\eta^5\text{-}C_5H_5)Ni(\eta^3\text{-}C_3H_5)$$

The $\eta^5\text{-}C_5H_5Ni$ group requires only three electrons to achieve a noble-gas configuration, and tends to form η^3-allylic complexes, as in the η^5-cyclopentadienyl-η^3-cyclopentenylnickel complex, $Ni(\eta^5\text{-}C_5H_5)(\eta^3\text{-}C_5H_7)$ (Fig. 12.3.f):

$$\left.\begin{array}{l} Ni(CO)_4 + C_5H_6 \\ Ni(\pi\text{-}C_5H_5)_2 + Na/Hg/EtOH \\ NiBr_2 + NaC_5H_5 + C_5H_7MgBr \end{array}\right\} \longrightarrow (\pi\text{-}C_5H_5)Ni(C_5H_7)$$

In some nickel-promoted, olefin-oligomerization reactions, the oligomerized olefin remains attached to the metal through η^3-allylic bonds. Thus, nickel(II) acetylacetonate reduced in the presence of butadiene forms a bidentate bis(η^3-allylic) complex (Fig. 12.3.g) derived from octadiene-1,6 which has been isolated as a trialkylphosphite adduct:

$$Ni(acac)_2 + C_4H_6 + P(OR)_3 \xrightarrow{\text{reduction}} (RO)_3P{-}Ni(C_8H_{12})$$

Nickel cyclododecatriene, $Ni(C_{12}H_{18})$, polymerizes butadiene with formation of a nickel(0) complex of an octatriene, attached to the metal through two η^3-allylic fragments and a η^2-olefinic bond:

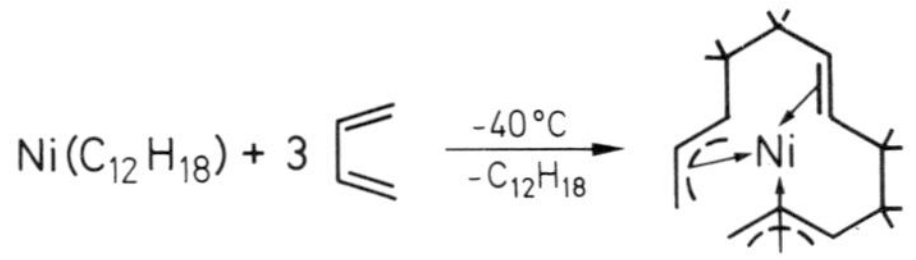

$$Ni(C_{12}H_{18}) + 3 \; C_4H_6 \xrightarrow[-C_{12}H_{18}]{-40\,°C}$$

Pentadiene-1,4 forms a bis(η^3-allylic) compound (Fig. 12.3.h) with nickel chloride by reduction with triethylaluminum.

Cyclooctatetraene forms the dimeric $[Ni(C_8H_8)]_2$ in which each eight-membered ring is bonded to two nickel atoms through η^3-allylic fragments:

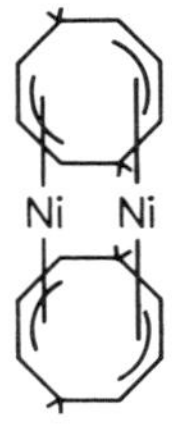

Ni Ni

Cyclobutadiene derivatives of nickel and palladium can be reduced to their η^3-allylic analogues. Thus, treatment of bis(tetramethylcyclobutadiene)nickel chloride with sodium cyclopentadienide results in the attachment of the four-membered ring through a η^3-allylic fragment:

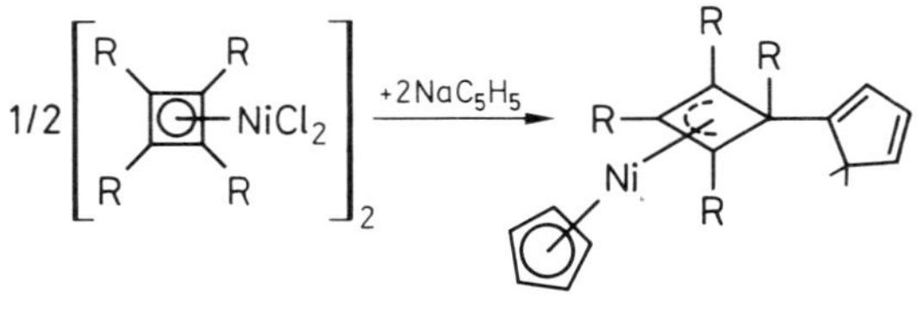

$$1/2\,[R_4C_4NiCl_2]_2 \xrightarrow{+2NaC_5H_5}$$

The stronger bonding of a η^5-C_5H_5 group to nickel forces the four-membered ring to become tri*hapto*, since the η^5-C_5H_5Ni group requires only three electrons to achieve a noble-gas configuration.

Palladium(II) chloride forms the allylic dimer, $[(\eta^3\text{-}C_3H_5)PdCl]_2$ (Fig. 12.1.d), with allyl chloride in acetic acid or allyl alcohol. Monoolefins react with palladium(II) chloride to form the η^3-allylic dimers, $[(\eta^3\text{-}C_3H_4R)PdCl]_2$. The dimeric compound $[(\eta^3\text{-}C_6H_9)PdCl]_2$ (Fig. 12.3.e) is also formed in the reaction of cyclohexene with $PdCl_2$ in acetic acid or from cyclohexadiene-1,3 with $[Pd(CO)Cl]_2$. A compound first described as $[Pd(butadiene)Cl_2]_2$ has been identified as the η^3-allylic derivative, $[(\eta^3\text{-}C_3H_4CH_2Cl)PdCl]_2$.

The strong tendency of palladium to form η^3-allyl complexes is shown by the following "disproportionation" of cyclohexadiene-1,3:

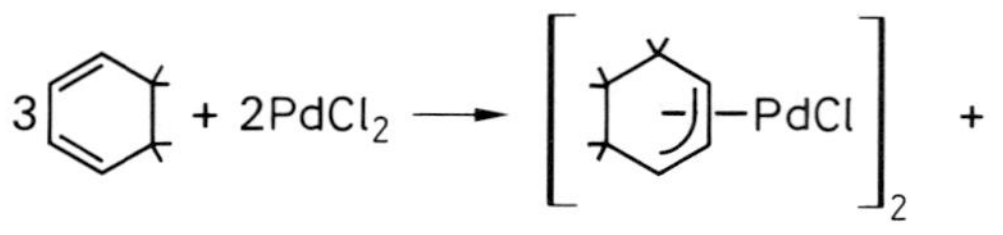

The σ-benzyl derivative, $[C_6H_5CH_2PdCl(PPh_3)]_2$, treated with silver tetrafluoroborate gives a complex in which the benzyl group is attached as a tri*hapto*-ligand as for molybdenum:

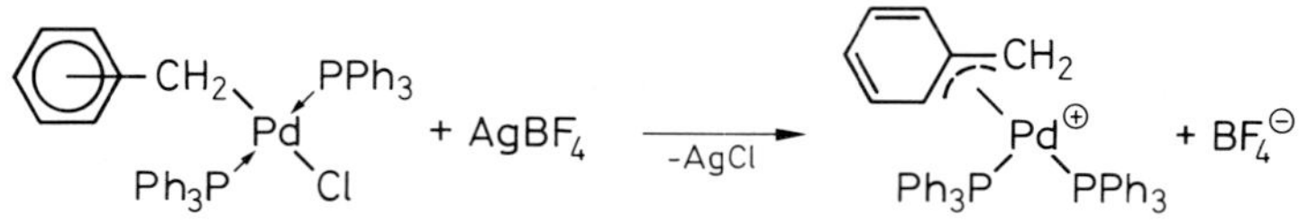

Cyclopentadienyl groups can be attached to a dipalladium unit as bridging η^3-allylic ligands as in μ-(η^3-C_5H_5)(μ—X)Pd_2L_2, prepared by reducing η^5-C_5H_5PdLX with Mg metal, Na amalgam, $LiAlH_4$ or $NaBH_4$ ($L = PR_3$):

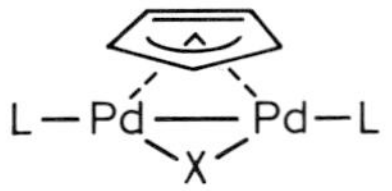

The platinum η^5-cyclopentadienyl-η^3-allyl derivative (Fig. 12.1.h) has been prepared by treatment of [Pt(propylene)Cl_2] with allylmagnesium chloride and sodium cyclopentadienide.

12.2. Cyclopropenyl Complexes

The unsaturated three-membered cyclopropenyl ring can act as a three-electron ligand:

in transition metal complexes. The starting materials are the trialkyl(aryl)-cyclopropenium salts, $C_3R_3^+X^-$. The dimeric nickel compound satisfies the 18-electron rule:

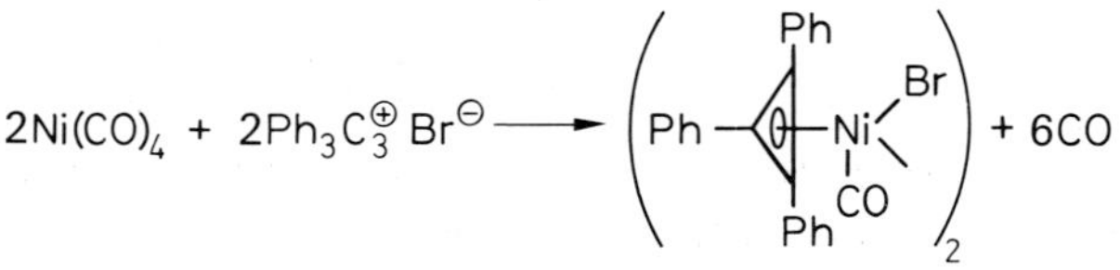

The tri*hapto*-bonding is found in $[Ni(C_3Ph_3\text{-}\eta^3)Cl(py)_2] \cdot py$ and $(\eta^5\text{-}C_5H_5)Ni(\eta^3\text{-}C_3Ph_3)$. The latter is prepared as follows:

$$Ni(C_3Ph_3)Br(py)_2 \cdot py + TlC_5H_5 \xrightarrow{C_6H_6} (\eta^5\text{-}C_5H_5)Ni(\eta^3\text{-}C_3Ph_3)$$

Nickel derivatives containing the trialkylcyclopropenyl ligand are similarly prepared:

$$Ni(CO)_4 + C_3R_3^{\oplus}BF_4^{\ominus} + NaBr \longrightarrow Ni(CO)Br(\eta^3\text{-}C_3R_3)$$

$R = Bu^t$

$$Ni(CO)Br(\eta^3\text{-}C_3R_3) \xrightarrow{CO} Ni(CO)_2Br(\eta^3\text{-}C_3R_3)$$

$$Ni(CO)Br(\eta^3\text{-}C_3R_3) \xrightarrow{80^\circ,\ vacuum} [Ni(C_3R_3)Br]_2$$

$$Ni(CO)Br(\eta^3\text{-}C_3R_3) \xrightarrow{NaC_5H_5} (\eta^5\text{-}C_5H_5)Ni(\eta^3\text{-}C_3R_3)$$

A molybdenum-carbonyl derivative containing the triphenylcyclopropenyl ligand has been obtained from the chloride:

$$Mo(CO)_3(MeCN)_3 + C_3Ph_3^+Cl^- \longrightarrow MoCl(MeCN)_2(CO)_2(\eta^3\text{-}C_3Ph_3)$$

$$\downarrow +LiC_5H_5$$

$$(\eta^5\text{-}C_5H_5)Mo(CO)_2(\eta^3\text{-}C_3Ph_3)$$

Dicobalt octacarbonyl also reacts with a triphenylcyclopropenium salt:

$$C_3Ph_3^{\oplus}BF_4^{\ominus} + Co_2(CO)_8 \longrightarrow Co(CO)_3(\eta^3\text{-}C_3Ph_3)$$

Metal carbonyl or other anionic complexes react with cyclopropenium salts to give products other than η^3-cyclopropenyl complexes. Thus, triphenylcyclopropenium cation form a salt with the hexacarbonylvanadate anion, $C_3Ph_3^+V(CO)_6^-$, which is converted by UV irradiation to a cyclopropenium complex:

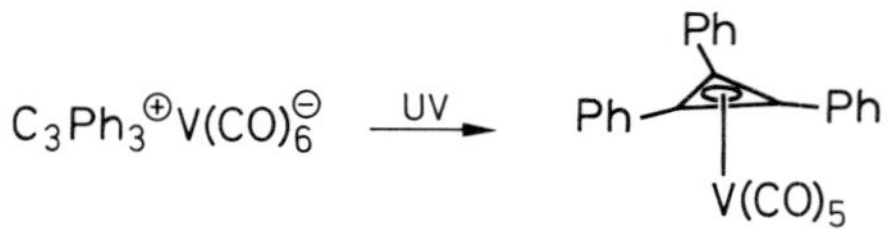

$$C_3Ph_3^{\oplus}V(CO)_6^{\ominus} \xrightarrow{UV} (\eta^3\text{-}C_3Ph_3)V(CO)_5$$

Triphenylcyclopropenium bromide, $C_3Ph_3^+Br^-$, gives an oxocyclobutenyl tricarbonylcobalt complex with tetracarbonylcobaltate anion, and $C_3Ph_3^+Cl^-$ with *trans*-$Rh(CO)Cl(PMe_2Ph)_2$ or $Ir(CO)Cl[P(OMe)_3]_3$ gives metallocycles:

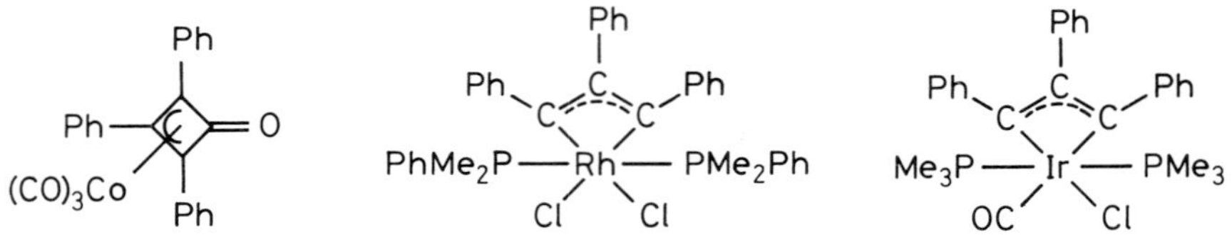

Triarylcyclopropenium salts behave similarly with zero-valent platinum and palladium compounds.

The formation of such compounds results from the strain in the three-membered ring. Caution should thus be exercised when assigning a tri*hapto* structure to a compound containing a C_3R_3 unit.

12.3. Metal-Carbyne Complexes

Carbynes contribute three electrons to the metal:

$$M \vdots C{-}R$$

and pairing with the metal electrons leads to the formation of a metal-carbon triple bond. Alternatively, the carbyne complexes can be interpreted as resulting from σ-bond, forward-donation from carbon to the metal with back-donation from the metal into π-orbitals of carbon.

The metal-carbyne complexes were discovered in an attempt to exchange halogens in methoxy-carbene complexes by reacting them with boron trihalides:

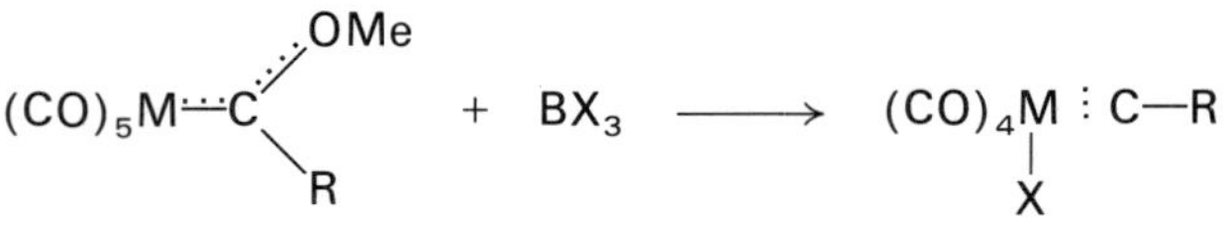

$$(CO)_5M{=}C(OMe)R + BX_3 \longrightarrow X(CO)_4M \vdots C{-}R$$

M = Cr, Mo, W
R = Me, Et, Ph
X = Cl, Br, I

The linear arrangement of the atoms in the M—C—R fragment is expected for carbyne complexes.

The above reaction has been extended to other functional carbene complexes:

$$(CO)_5Cr{\cdots}C(OH)(Ph) + BBr_3 \xrightarrow{-40\,^\circ C} Br{-}Cr(CO)_4{\vdots}C{-}Ph$$

$$(CO)_5W{\cdots}C(NH{-}CH_2COOMe)(Ph) + BBr_3 \xrightarrow{-25\,^\circ C} Br{-}W(CO)_4{\vdots}C{-}Ph$$

to other Lewis acids besides boron trihalides and to phosphine-, arsine- and stibine-substituted metalcarbonyl-carbene complexes.

Carbyne complexes can also be prepared from dihalophosphoranes:

$$(CO)_5W{\cdots}C(OLi)(Ph) + Br_2PPh_3 \xrightarrow[-OPPh_3,\ -CO,\ -LiBr]{-50\,^\circ C} Br{-}W(CO)_4{\vdots}C{-}Ph$$

N-Alkylation of isocyanide complexes with triethyloxonium tetrafluoroborate, dimethylsulfate or methylfluorosulfonate yields carbyne complexes:

$$trans\text{-}[M(CNMe)_2(dppe)_2] \xrightarrow{Et_3O^+BF_4^-} trans\text{-}[M({\vdots}C{-}NMeR)_2(dppe)_2]^+X$$

The carbyne complexes decompose thermally to form disubstituted acetylenes:

$$2\,Br(CO)_4Cr{\vdots}C{-}R \xrightarrow{30\,^\circ C} R{-}C{\equiv}C{-}R + \text{other products}$$

The metal-halides can couple to metal-metal bonded compounds without cleavage of the carbyne ligand:

$$NaRe(CO)_5 + Br(CO)_4Mo{\vdots}C{-}Ph \xrightarrow{THF} (CO)_5Re{-}Mo(CO)_4{\vdots}C{-}Ph$$

Ligand-transfer reactions of carbyne complexes with dicobalt octacarbonyl and nickelocene lead to alkylidyne and μ-acetylene complexes:

$$Br(CO)_4Cr{\vdots}C{-}R + Co_2(CO)_8 \longrightarrow Co_3(CO)_9CR + Co_2(CO)_6(RCCR)$$

$$Br(CO)_4Cr{\vdots}C{-}R + Ni(C_5H_5\text{-}\eta^5)_2 \longrightarrow (\eta^5\text{-}C_5H_5)_2Ni_2(RCCR)$$

Bi- and trimetallic-bridged carbyne complexes are known. Thus, the reaction of $MoCl_5$ and WCl_6 with $LiCH_2CMe_3$ yields the dimeric $[M(CH_2CMe_3)_3({\vdots}C\text{-}CMe_3)]_2$

with bridging-carbyne ligands. The dimer dissociates into monomeric carbyne complexes:

$$MCl_n + LiCH_2CMe_3 \xrightarrow{-78°C} (Me_3C{-}CH_2)_2M(\mu\text{-}CCMe_3)_2M(CH_2{-}CMe_3)_2 \rightleftharpoons 2\ Me_3C{-}CH_2{-}M(CH_2{-}CMe_3)_2{\equiv}C{-}CMe_3$$

M = Mo, n = 5

M = W , n = 6

Unexpected formation of carbyne complexes involves the cleavage of acetylenes in a reaction with a triple bonded W≡W derivative ($R = CMe_3$):

$$(RO)_3W{\equiv}W(OR)_3 + R'C{\equiv}CR' \longrightarrow 2\ (RO)_3W{\equiv}CR' \qquad (R' = Me, Et, Pr)$$

The tungsten derivative reacts with tetramethyldiphosphinoethane to give product containing both a carbene and a carbyne ligand:

$$(Me_2PCH_2CH_2PMe_2)W({\equiv}CCMe_3)({=}CHCMe_3)(CH_2CMe_3)$$

A bridged carbyne, tetrakis(trimethylsilylmethyl)bis(μ-trimethylsilylmethylidine) ditungsten, is obtained in the reaction of WCl_4 with $Hg(CH_2SiMe_3)_2$ or Me_3SiCH_2MgX:

$$(Me_3SiCH_2)_2W(\mu\text{-}CSiMe_3)_2W(CH_2SiMe_3)_2$$

Trimetallic carbyne (alkylidyne)-bridging ligands are found in the alkylidyne tricobalt compounds obtained from cobalt-carbonyl acetylene complexes,

$Co_2(CO)_6(C_2R_2)$, with acids, or from $Co_2(CO)_8$ and RCX_3 in compounds, $(C_5H_5)_3(CR)(CR')$, obtained by thermolysis of $C_5H_5Co(RC{\equiv}CR')(PPh_3)$ in the nickel compounds, $(C_5H_5Ni)_3CR$, or in the ruthenium and osmium derivatives, $H_3M_3(CO)_9CR$.

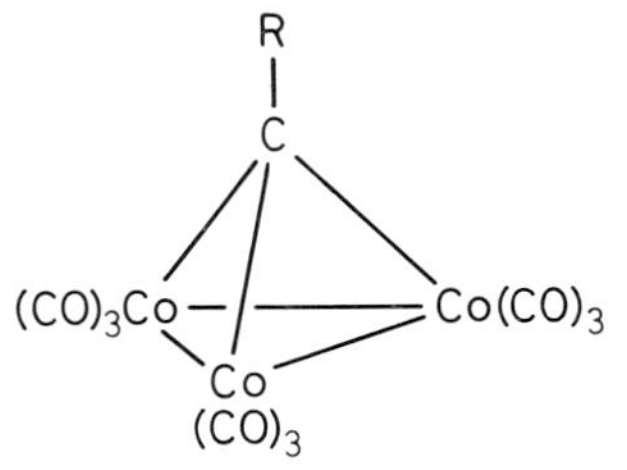

R C (CO)$_3$Co Co(CO)$_3$ Co (CO)$_3$

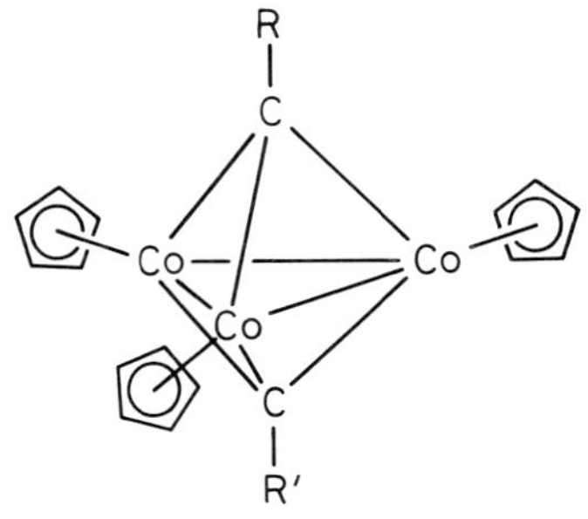

R C Co Co Co C R'

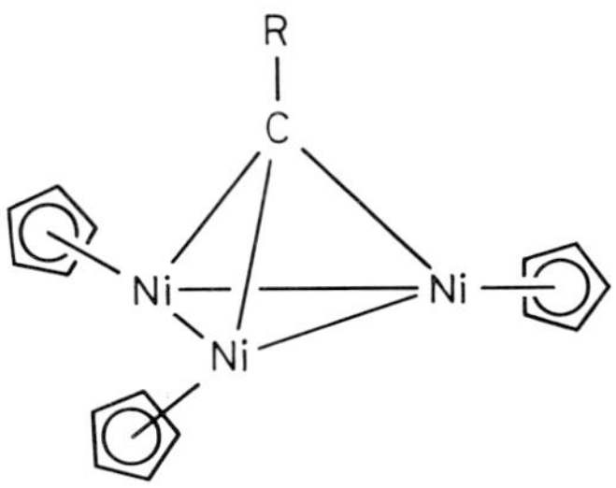

R C Ni Ni Ni

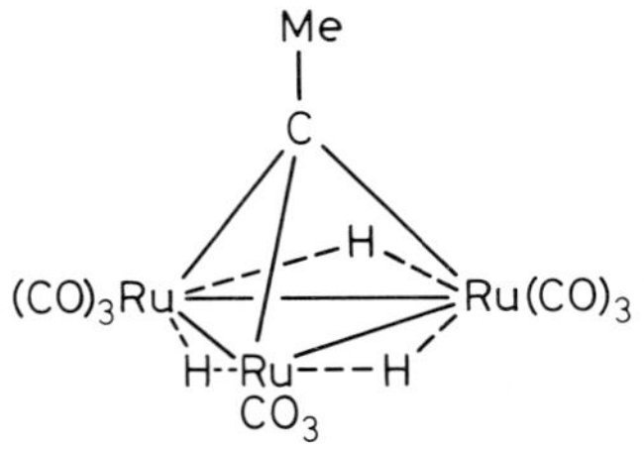

Me C H (CO)$_3$Ru Ru(CO)$_3$ H Ru H CO$_3$

13. Compounds With Four-Electron Ligands

The diolefins can contribute four electrons to a metal. Conjugated diolefins can be linear (as in butadiene-1,3 and its derivatives) or cyclic (as in cyclobutadiene, cyclopentadiene, cyclohexadiene-1,3, cycloheptadiene-1,3, fulvenes, cyclopentanone, etc.). Cyclic polyolefins having more than two double bonds, but using only a butadiene fragment, can also participate in tetra*hapto*-bonding (as in cycloheptatriene-1,3,5 or cyclooctatetraene).

The four carbon atoms of an sp^2-hybridized butadiene fragment must be coplanar. In Fig. 13.1 are shown the most important four electron ligands.

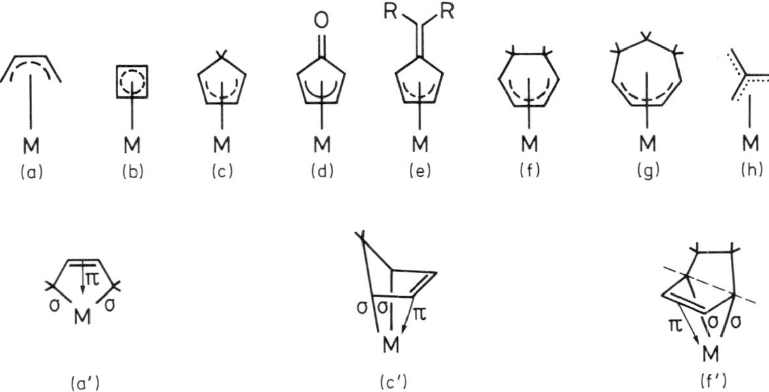

Fig. 13.1. Some typical four-electron ligands and their attachment to the metal atom.

In the trimethylenemethyl group, the four sp^2-carbon atoms are in a branched arrangement (Fig. 13.1.h).

An alternative mode of bonding, involving a η^2-olefin-metal bond and two σ-metal-carbon bonds, is possible as illustrated in Fig. 13.1.a′ for butadiene, in 13.1.c′ for cyclopentadiene, and in 13.1.f′ for cyclohexadiene. The non-equivalence of the three carbon-carbon bonds involved in coordination (especially in fluorinated derivatives) is the main argument in favor of this alternative bonding type. However, we will assume in this chapter η^4-donation in all cases.

The metals which form conjugated diene and cyclobutadiene complexes are illustrated in Fig. 13.2 and 13.3, respectively.

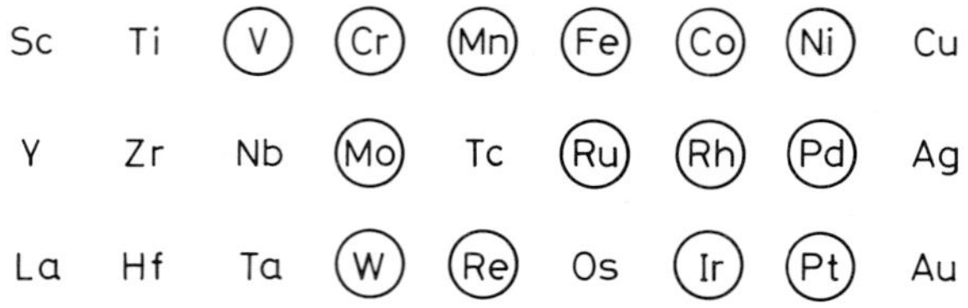

Sc Ti V Cr Mn Fe Co Ni Cu
Y Zr Nb Mo Tc Ru Rh Pd Ag
La Hf Ta W Re Os Ir Pt Au

Fig. 13.2. Metals known to form diene complexes (excluding cyclobutadiene).

Sc Ti V Cr Mn Fe Co Ni Cu
Y Zr Nb Mo Tc Ru Rh Pd Ag
La Hf Ta W Re Os Ir Pt Au

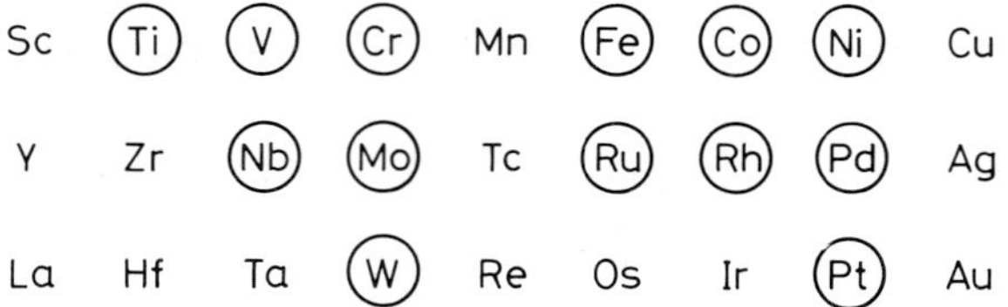

Fig. 13.3. Metals forming cyclobutadiene complexes.

13.1. Butadiene Complexes and Related Compounds

The first transition metal complex of butadiene was $C_4H_6Fe(CO)_3$, prepared in 1930 by the reaction of butadiene and iron pentacarbonyl:

Fe
OC
CO
CO

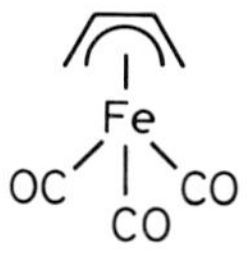

The butadiene ligand is planar, suggesting that the π-electron pairs are delocalized over the C_4 skeleton.

Vanadium forms a butadiene complex photochemically from cyclopentadienylvanadium tetracarbonyl, $(\eta^5\text{-}C_5H_5)V(CO)_4$, and butadiene by replacement of two carbonyl molecules to yield $(\eta^5\text{-}C_5H_5)V(CO)_2(\eta^4\text{-}C_4H_6)$. Molybdenum hexacarbonyl reacts similarly to form $(\eta^4\text{-}C_4H_6)Mo(CO)_4$ and $(\eta^4\text{-}C_4H_6)_2Mo(CO)_2$:

OC—V—CO
Mo
OC
CO
CO CO
Mo
CO
CO

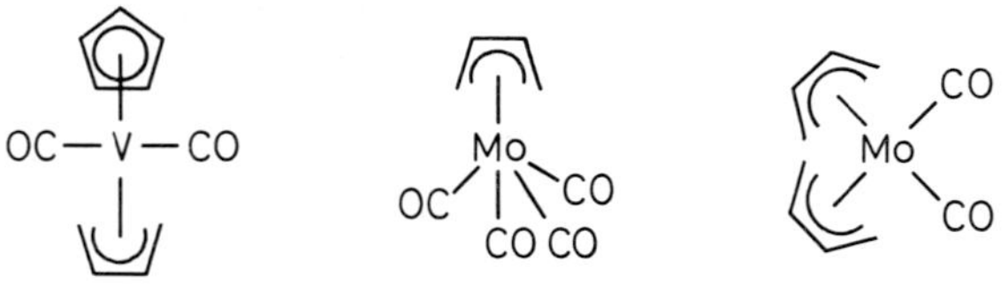

The unstable chromium complex, $(\eta^4\text{-}C_4H_6)Cr(CO)_4$, is obtained by reacting chromium vapor with butadiene and carbon monoxide:

$$Cr_{at} + C_4H_6 + CO \longrightarrow (C_4H_6)Cr(CO)_4$$

The metal-atom synthesis affords tris(η^4-butadiene) complexes of molybdenum and tungsten:

$$M_{at} + 3\,C_4H_6 \longrightarrow M(\eta^4\text{-}C_4H_6)_3$$

M = Mo,W

Cyclopentadienylmanganese tricarbonyl reacts with butadiene photochemically to replace two carbon monoxide groups and form $(\eta^5\text{-}C_5H_5)Mn(\eta^4\text{-}C_4H_6)CO$:

$$(C_5H_5)Mn(CO)_3 + C_4H_6 \xrightarrow{h\nu} (C_5H_5)Mn(C_4H_6)\text{—}CO$$

The electronic requirements of the $Fe(CO)_3$ group favor butadiene complexes, as in butadieneiron tricarbonyl. Irradiation of $Fe(CO)_5$ and butadiene yields a bis(η^4-butadiene) complex:

$$Fe(CO)_5 + C_4H_6 \xrightarrow{UV} (C_4H_6)_2Fe\text{—}CO + 4CO$$

while the reaction of iron vapor with butadiene yields an unstable tris-complex (not isolated) which can be readily converted into stable bis(butadiene) derivatives:

$$Fe_{at} + C_4H_6 \longrightarrow (C_4H_6)_2Fe(C_4H_6) \xrightarrow{L} (C_4H_6)_2Fe\text{—}L$$

L = CO,PR$_3$

Cyclopentadienyliron dicarbonyl bromide, $(\eta^5\text{-}C_5H_5)Fe(CO)_2Br$ reacts with butadiene in the presence of aluminum bromide to form a butadiene-containing cation, $[(\eta^5\text{-}C_5H_5)Fe(CO)_2(\eta^4\text{-}C_4H_6)]^+$

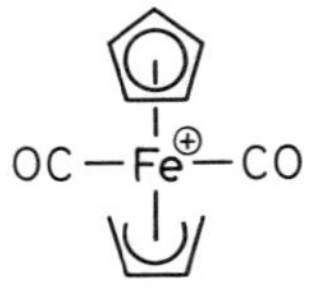

Organic molecules which initially do not contain a butadiene fragment can react with iron carbonyls to form butadiene complexes as a result of rearrangement. Thus, tetramethylallene reacts with $Fe_2(CO)_9$ to form a trimethylbutadiene complex in addition to the expected tetramethylallene derivative:

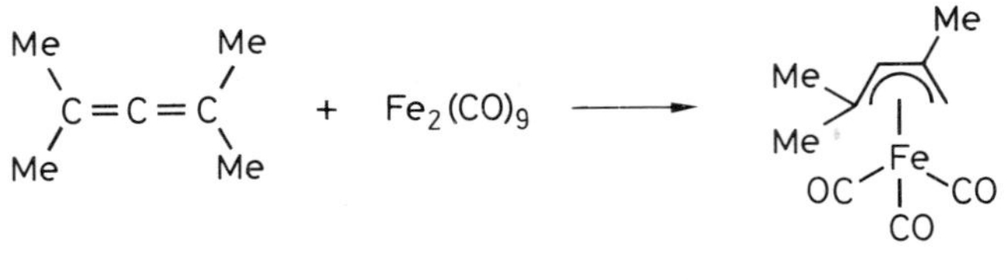

α,α′-Dibromoxylene reacts with $Fe_2(CO)_9$ to form a butadiene fragment following bromide elimination and π-electron redistribution:

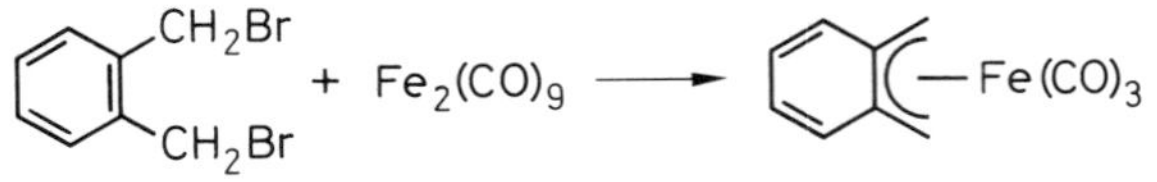

Irradiation of the product in the presence of iron pentacarbonyl results in complex formation with the second butadiene fragment of the molecule to yield two isomers:

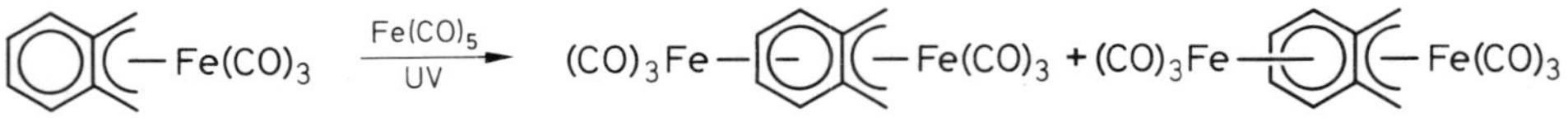

Vinylbenzene and *para*-divinylbenzene form butadiene complexes with iron carbonyls by perturbing the aromatic conjugation in the six-membered ring on association of a ring double bond with the exocyclic vinyl group:

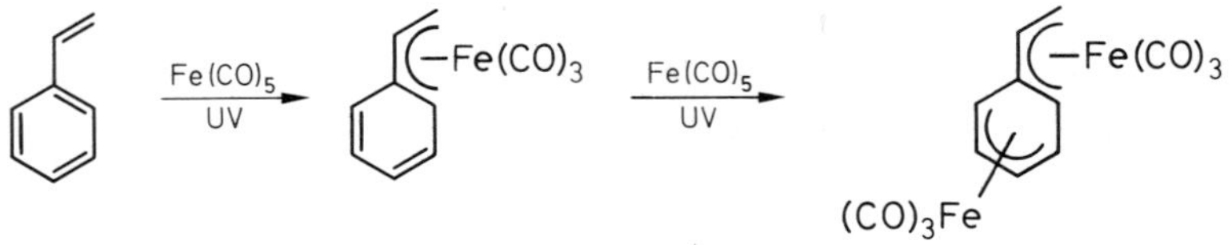

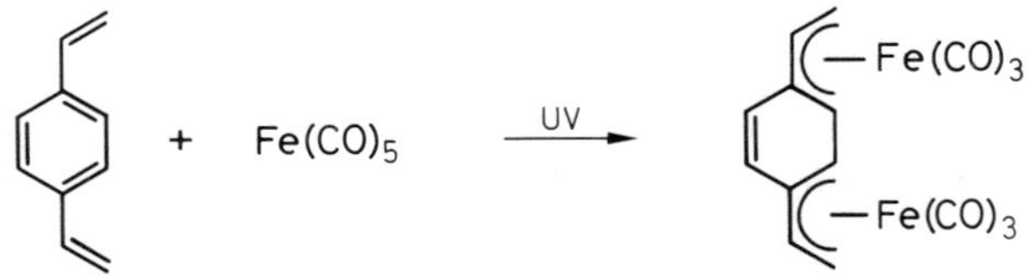

Cobaltocene reacts with butadiene to form the complex $(\eta^5\text{-}C_5H_5)Co\text{-}(\eta^4C_4H_6)$, and dicobalt octacarbonyl forms the bimetallic complexes $Co_2(CO)_6(\eta^4\text{-}C_4H_6)$ and $Co_2(CO)_4(\eta^4\text{-}C_4H_6)_2$:

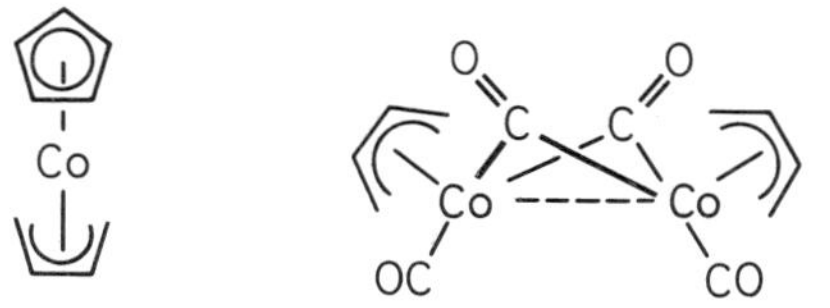

The reaction of rhodium(III) chloride with butadiene yields the compound $(\eta^4\text{-}C_4H_6)_2RhCl$, and an analogous iridium complex is formed in the reaction of Na_2IrCl_4 with butadiene. The compound $(\eta^5\text{-}C_5H_5)Ir(\eta^4\text{-}C_4H_6)$ has been prepared by treating $(\eta^4\text{-}C_4H_6)_2IrCl$ with TlC_5H_5:

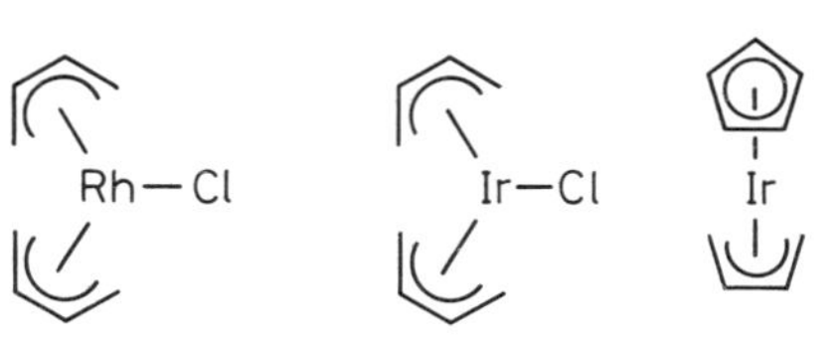

13.2. Cyclobutadiene Complexes

Cyclobutadiene itself cannot be isolated, but coordination to a transition metal stabilizes many derivatives. Because of the non-existence of the free ligand, reactions in which the cyclobutadiene ligand is formed simultaneously with the complex must be used for the synthesis of cyclobutadiene complexes. The precursors are usually compounds containing a four-membered ring (for example, 1,2-dichlorocyclobutene, photo-α-pyrone, etc.). The most convenient procedures start from acetylenes, since these readily available reagents can often be converted into cyclobutadiene complexes by reaction with transition-metal carbonyls or halides.

The first synthesis of a cyclobutadiene complex was accomplished in 1959 by reacting 1,2-dichlorotetramethylcyclobutene with nickel tetracarbonyl to give the complex $[(\eta^4\text{-}C_4Me_4)NiCl_2]_2$:

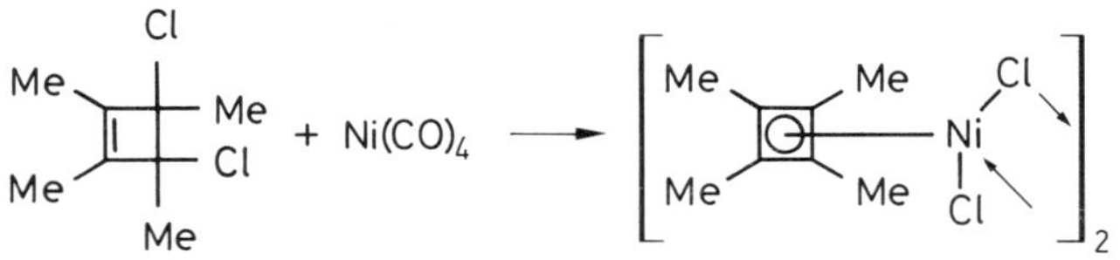

$$\text{Me}_4\text{C}_4\text{Cl}_2 + \text{Ni(CO)}_4 \longrightarrow [(\eta^4\text{-C}_4\text{Me}_4)\text{NiCl}_2]_2$$

The first complex of unsubstituted cyclobutadiene was (η^4-C_4H_4)$Fe(CO)_3$ prepared in 1965 by the action of 1,2-dichlorocyclobutene-3,4 on $Fe_2(CO)_9$:

$$C_4H_4Cl_2 + Fe_2(CO)_9 \longrightarrow (\eta^4\text{-}C_4H_4)Fe(CO)_3$$

in which the planar four-membered ring is coordinated to iron. Many other cyclobutadiene complexes have now been prepared.

Of the Group IV element triad, only the titanium cyclobutadiene complex is known. Titanium(III) chloride with cyclooctatetraene and diphenylacetylene in the presence of iso-PrMgCl gives a mixed ligand complex containing tetraphenylcyclobutadiene:

$$TiCl_3 + C_8H_8 + C_2R_2 \xrightarrow{\text{iso-PrMgCl}} (\eta^4\text{-}C_4R_4)Ti(C_8H_8)$$

R = Ph

Vanadium and niobium also form tetraphenylcyclobutadiene complexes starting from diphenylacetylene (R = Ph):

$$\eta^5\text{-}C_5H_5V(CO)_4 \xrightarrow[UV]{C_2R_2} (\eta^5\text{-}C_5H_5)V(CO)(VO)(R{-}{\equiv}{-}R) \xrightarrow[110°C]{C_2R_2} (\eta^5\text{-}C_5H_5)V(CO)_2(\eta^4\text{-}C_4R_4)$$

$$\eta^5\text{-}C_5H_5Nb(CO)_4 \xrightarrow[UV]{C_2R_2} (\eta^5\text{-}C_5H_5)Nb(CO)_2(R{-}{\equiv}{-}R) \xrightarrow[UV]{C_2R_2} (\eta^5\text{-}C_5H_5)Nb(CO)(R{-}{\equiv}{-}R)_2 \xrightarrow[80°C]{C_2R_2}$$

$$\longrightarrow (\eta^5\text{-}C_5H_5)Nb(CO)(R{-}{\equiv}{-}R)(\eta^4\text{-}C_4R_4)$$

Chromium, molybdenum and tungsten form the cyclobutadiene complexes, $(\eta^4\text{-}C_4R_4)M(CO)_4$, by reaction of 1,2-dichlorocyclobutenes with the metal hexacarbonyls in the presence of sodium amalgam. A bis(η^4-tetraphenylcyclobutadiene) complex of molybdenum has been obtained from $Mo(CO)_6$ and diphenylacetylene:

$$Mo(CO)_6 + 4\,C_2R_2 \longrightarrow (\eta^4\text{-}C_4R_4)_2Mo(CO)_2 \qquad R = Ph$$

Iron forms many cyclobutadiene complexes, and $(\eta^4\text{-}C_4H_4)Fe(CO)_3$ has already been cited. This compound can also be prepared via photochemical transformation of α-pyrone, followed by coordination to iron and elimination of carbon dioxide. The procedure has been extended to vanadium-, cobalt- and rhodium-cyclobutadiene complexes (Fig. 13.4):

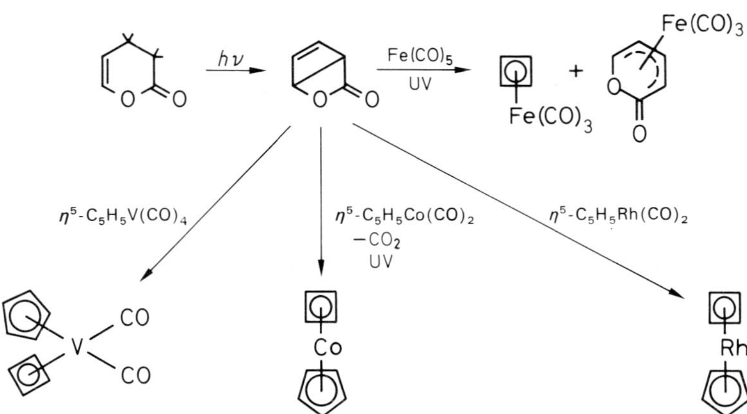

Fig. 13.4. Formation of cyclobutadiene complexes from α-pyrone.

Friedel-Crafts acetylation, aminomethylation, mercuration and other metallations, etc., of $C_4H_4Fe(CO)_3$ reflect the aromatic character of this complex. In Fig. 13.5 some aromatic substitution reactions of the iron-coordinated cyclobutadiene ring are shown.

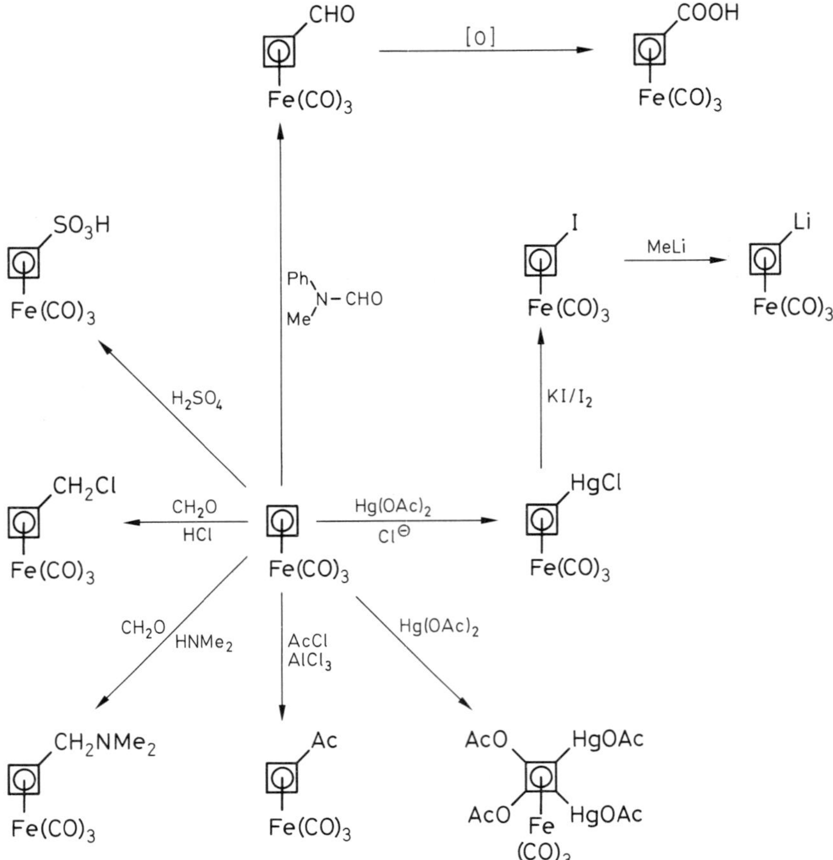

Fig. 13.5. Some aromatic-type reactions of $(\eta^4\text{-}C_4H_4)Fe(CO)_3$ and its derivatives.

Oxidation of $(\eta^4\text{-}C_4H_4)Fe(CO)_3$ with Fe^{3+}, Ce^{4+} or Ag^+ generates transient, free cyclobutadiene which can be trapped with various organic compounds. This is cleverly exploited for the preparation of many unusual compounds which are not available by other methods.

Fused cyclobutadiene complexes are formed in reactions similar to those mentioned above:

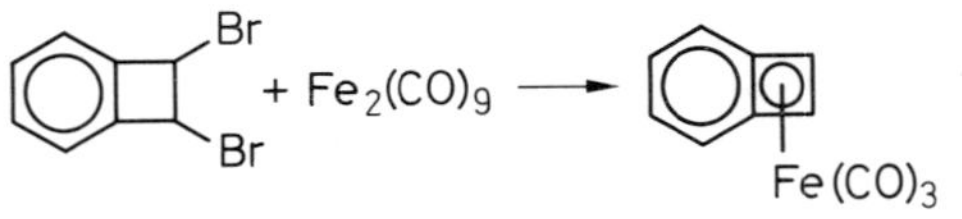

Br, Br + $Fe_2(CO)_9$ ⟶ $Fe(CO)_3$

Diphenylacetylene acts on iron pentacarbonyl at 240 °C to yield a tetraphenylcyclobutadiene complex, $(\eta^4\text{-}C_4Ph_4)Fe(CO)_3$ in addition to a tetraphenylcyclopentadienone complex:

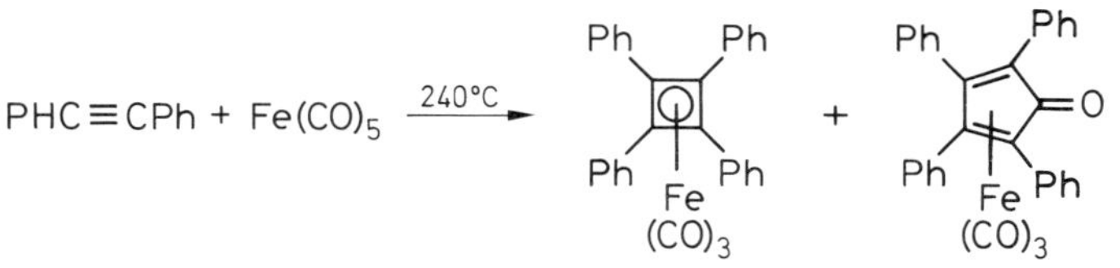

Macrocyclic diacetylenes also react with iron carbonyls to form cyclobutadiene complexes, among other products (See Chapter 17).

Cobalt forms cyclobutadiene complexes, especially when the metal atom is part of a η^5-C_5H_5Co fragment which requires four electrons to fulfill a noble-gas configuration. Thus, 1,2-dichlorocyclobutene forms a cyclobutadiene complex by reaction with $NaCo(CO)_4$; the primary product can be converted to a cyclopentadienylcobalt derivative:

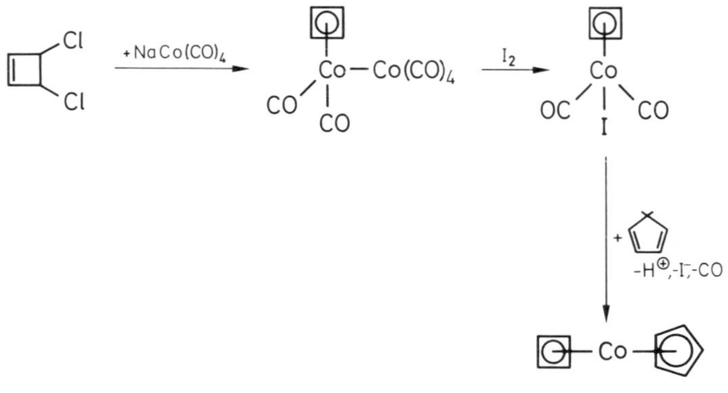

Disubstituted acetylenes and macrocyclic diacetylenes react with cyclopentadienylcobalt dicarbonyl to form cyclobutadiene complexes:

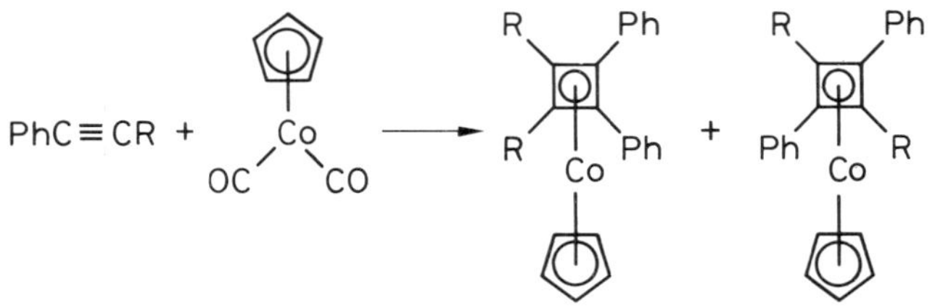

A rhodium complex, $(\eta^5$-$C_5H_5)Rh(\eta^4$-$C_4Ph_4)$, has been prepared similarly.

Bis(tetraphenylcyclobutadiene)nickel is obtained from $(\eta^4$-$C_4Ph_4)NiBr_2$ and dilithiotetraphenylbutadiene:

R R R R NiBr$_2$ + R R R R Li Li $\xrightarrow{-2LiR}$ R R R R Ni R R R R

R = Ph

Disubstituted acetylenes form the important complexes $[(\eta^4\text{-}C_4R_4)PdBr_2]_2$ by reacting with the Kharasch reagent which can transfer the cyclobutadiene ligand to other metals (shown in Fig. 13.6).

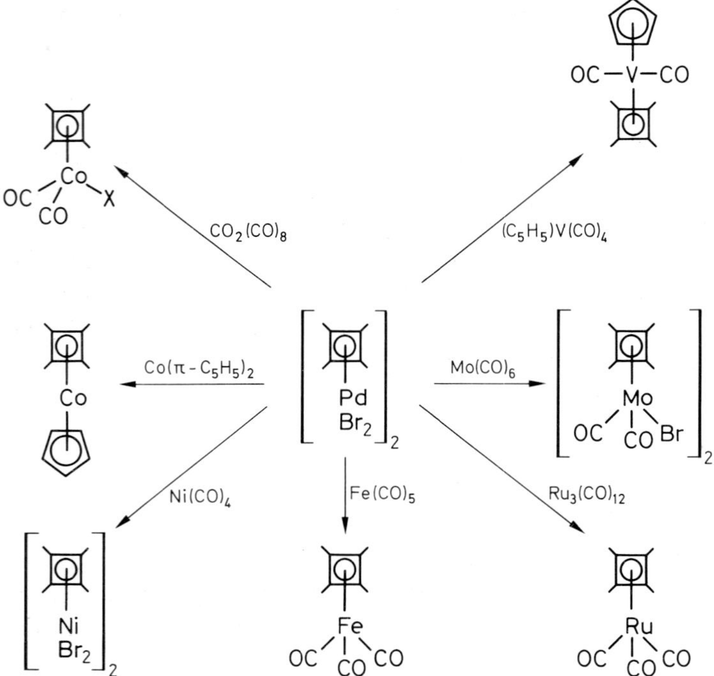

Fig. 13.6. Cyclobutadiene ligand transfer from palladium.

The palladium complexes undergo nucleophilic addition to the four-membered ring (for example, with ethanol) to form the trihapto-complexes:

[Ph Ph Ph Ph PdCl$_2$]$_2$ $\xrightarrow{EtOH}$ [Ph Ph PdCl EtO Ph Ph]$_2$

The chemistry of the cyclobutadiene complexes of transition metals is a beautiful illustration of the role which complexation by transition metals can play in stabilizing organic molecules incapable of independent existence, and in modifying the chemical reactivity of a coordinated organic group.

13.3. η^4-Complexes of Cyclopentadiene, Cyclopentadienone, Fulvene and Heterocycles Derived from Cyclopentadiene

The butadiene fragment incorporated in a title cyclic system can act as a four-electron donor to form tetra*hapto*-complexes. Typical examples are illustrated in Fig. 13.1.

Cyclopentadiene

Heated with $Fe(CO)_5$, cyclopentadiene forms the cyclopentadienyliron dimer, $[(\eta^5\text{-}C_5H_5)Fe(CO)_2]_2$, in which the ring is a five-electron donor. It is, however, possible to isolate a cyclopentadiene complex in which the hydrocarbon acts as a four-electron donor, attached to iron in a tetra*hapto*-fashion, as in Fig. 13.1c. This is believed to be an intermediate in the formation of the cyclopentadienyl dimer cited. Thus, if cyclopentadiene reacts with the more reactive iron carbonyl, $Fe_2(CO)_9$, at only 40 °C, the compound $(\eta^4\text{-}C_5H_6)Fe(CO)_3$ can be isolated:

+ $Fe_2(CO)_9$ —40 °C→ $Fe(CO)_3$

Heating to 140 °C leads to $[(\eta^5\text{-}C_5H_5)Fe(CO)_2]_2$.

Spirocyclopentadienes, which cannot be converted to penta*hapto*-cyclopentadienyls, form complexes with $Fe_2(CO)_9$ containing the $Fe(CO)_3$ fragment coordinated to the butadiene part of the ring:

$(CH_2)_n$ + $Fe_2(CO)_9$ → $Fe(CO)_3$ and $Fe(CO)_3$

n = 4 and 2

η^4-Cyclopentadiene complexes can also be obtained by addition of hydride ion to η^5-cyclopentadienyl complexes. Thus, the compound $[(\eta^5\text{-}C_5H_5)Fe(CO)_2PPh_3]^+$ is reduced with sodium borohydride to $(\eta^4\text{-}C_5H_6)Fe(CO)_2PPh_3$. The cation $[(\eta^5\text{-}C_5H_6)Fe(\eta^6\text{-}C_6H_6)]^+$ also undergoes a reduction of the five-membered ring with $LiAlH_4$ to give neutral $(\eta^4\text{-}C_5H_6)Fe(\eta^6\text{-}C_6H_6)$:

The latter is isoelectronic with ferrocene, and all four compounds shown in these reactions have noble-gas configurations.

Similar compounds are derived from cobalt, rhodium and iridium. The metal-hydride reduction of cobalticenium chloride, $[Co(\eta^5\text{-}C_5H_5)_2]^+Cl^-$, yields $(\eta^5\text{-}C_5H_5)Co(\eta^4\text{-}C_5H_6)$. The reaction of iridium(III) chloride with potassium cyclopentadienide and cyclopentadiene leads to $(\eta^5\text{-}C_5H_5)Ir(\eta^4\text{-}C_5H_6)$:

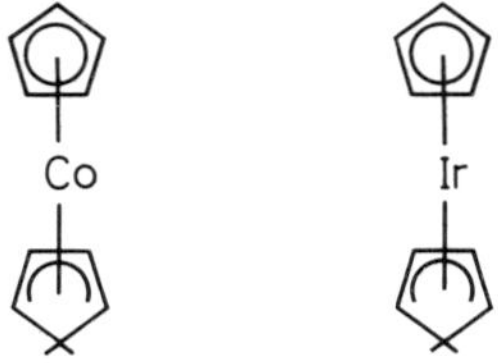

Sometimes the bonding of an unsaturated C_4 fragment to a transition metal can be better represented by a $2\sigma + \eta^2$ model. This is suggested by the structure of $C_5H_5ReMe_2(C_5H_5Me\text{-}\pi)$, which is prepared from $(LiC_5H_5)_2ReH$ and methyl iodide:

The $2\sigma + \eta^2$ model may sometimes be the true picture of bonding which in most cases is an η^4-type.

Cyclopentadienones

These also behave as four-electron donors forming η^4-complexes, especially with iron. Such compounds can be obtained directly from cyclopentadienones, or are sometimes formed by reactions of disubstituted acetylenes with iron carbonyls:

R R R R O + $Fe(CO)_5$ → R R R R O $Fe(CO)_3$ ← RC≡CR + $Fe(CO)_5$ $R = CF_3$

The structure of the trifluoromethyl derivative ($R = CF_3$) has been confirmed.

Fulvenes

These form transition-metal complexes in which they act as four-electron donors, but an alternative formulation, in which a π-electron redistribution occurs with formation of a η^5-cyclopentadienyl ligand, must also be considered:

R R M R ⊖ R M

An iron tricarbonyl derivative is obtained by the reaction of diphenylfulvene with $Fe_2(CO)_9$, and a cobalt derivative by the reaction of diphenylfulvene, cobalt(II) chloride and isopropyl magnesium chloride:

Ph Ph $Fe(CO)_3$ ← $Fe_2(CO)_9$ Ph Ph → $CoCl_2$ / iso-PrMgBr Ph Ph $Co^{\oplus}$ Ph Ph

Heterocycles

Several η^4-complexes derive from five-membered heterocyclic ligands containing a butadiene fragment. Typical is a group of silacyclopentadiene complexes of iron and cobalt:

Si R_2 → $+Fe(CO)_5$ / 150-200° OC CO C O Fe Si R_2 ← $Fe_2(CO)_9$ / 50-55° Br Si Br R_2

Co$(CO)_2$ + Si R_2 → -CO Co SiR$_2$

Thiophene dioxide forms an η^4-complex with iron on UV irradiation with iron carbonyl:

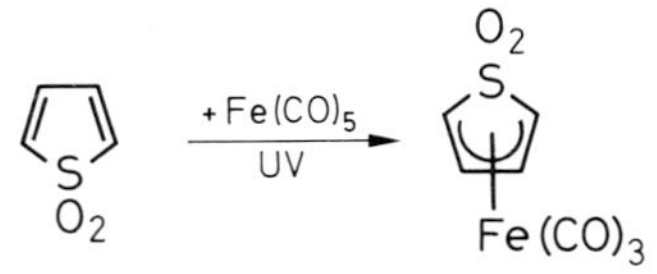

O_2 S + $Fe(CO)_5$ / UV → O_2 S $Fe(CO)_3$

In the related compound derived from pentaphenylphosphole, the tertiary phosphorus atom is not involved in electron donation:

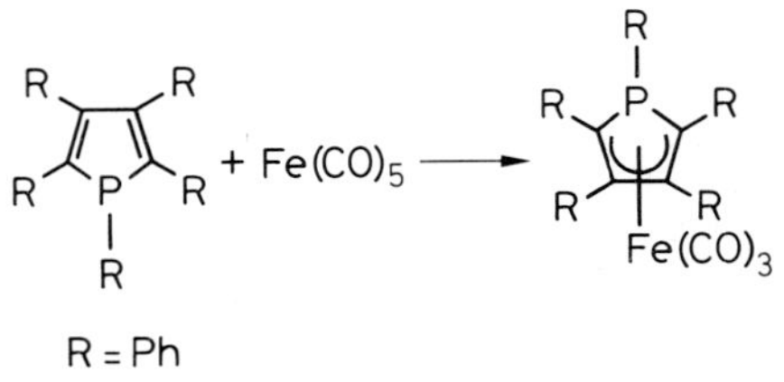

R R R P R R + $Fe(CO)_5$ → R P R R R R $Fe(CO)_3$

R = Ph

and thus seems to be a weaker donor than the butadiene fragment.

13.4. Complexes with Other Cyclic Dienes and Polyenes

Virtually all conjugated cyclic dienes are able to form η^4-complexes, contributing the four π-electrons of their butadiene fragment. Some seven- and eight-membered polyenes can behave in a similar manner when the metal atom requires four electrons (or a multiple) to achieve a noble-gas configuration.

Cyclohexadiene-1,3

This diene forms many η^4-complexes of the type shown in Fig. 13.1.f. Irradiation of cyclopentadienylvanadium tetracarbonyl with cyclohexadiene-1,3 yields $(\eta^5\text{-}C_5H_5)V(CO)_2(\eta^4\text{-}C_6H_8)$. This general method has been used to yield other metal carbonyl derivatives of cyclohexadiene-1,3, for example, $(\eta^4\text{-}C_6H_8)_2W(CO)_2$ [from $(CH_3CN)_3W(CO)_3$]:

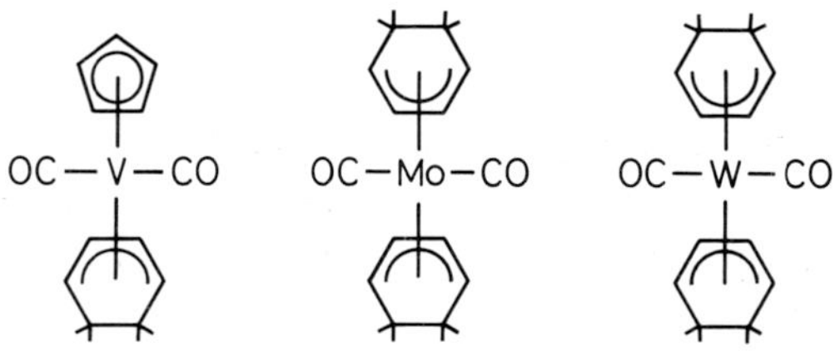

OC—V—CO OC—Mo—CO OC—W—CO

The coordinated η^6-benzene ring can be reduced with metal hydrides to form η^4-cyclohexadiene complexes. This procedure has been used for the synthesis of $(\eta^5\text{-}C_5H_5)W(H)(CO)(\eta^4\text{-}C_6H_8)$ from $[(\eta^5\text{-}C_5H_5)W(CO)(\eta^6\text{-}C_6H_6)]^+$ and of $(\eta^4\text{-}C_6H_8)Mn(H)(CO)_3$ from $[(\eta^6\text{-}C_6H_6)Mn(CO)_3]^+$:

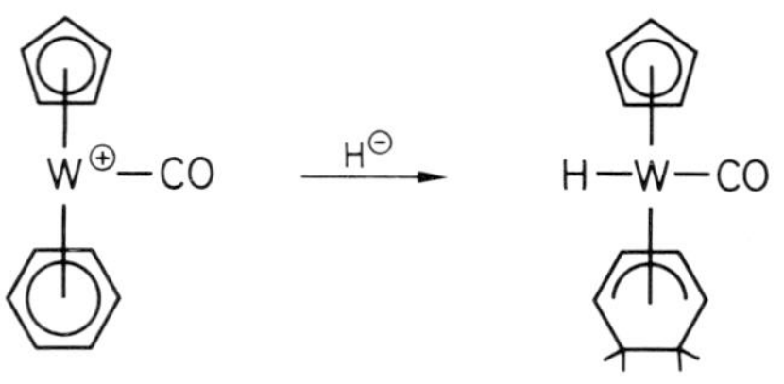

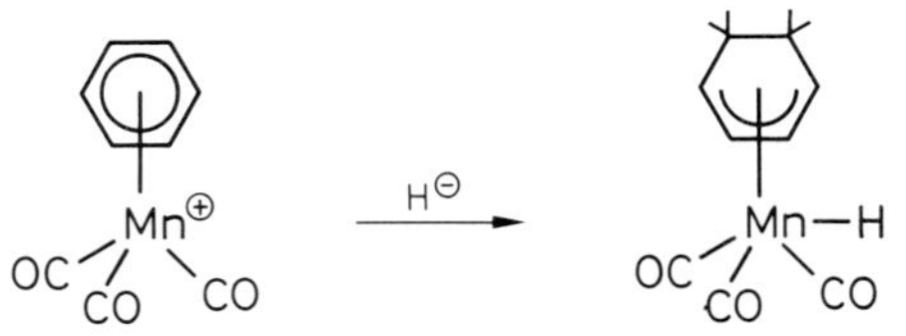

Since the $Fe(CO)_3$ group requires only four electrons, it readily forms η^4-complexes with cyclohexadienes. Thus, heating iron pentacarbonyl with cyclohexadiene-1,3 yields $\eta^4\text{-}C_6H_8Fe(CO)_3$. The same compound is formed even if a non-conjugated cyclohexadiene such as the 1,4-isomer is used. By heating with $Fe(CO)_5$, cyclohexadiene-1,4 undergoes isomerization, and the complex of the conjugated diene is formed. Thus, complex formation can be a driving force strong enough to produce isomerization of a cyclic diolefin:

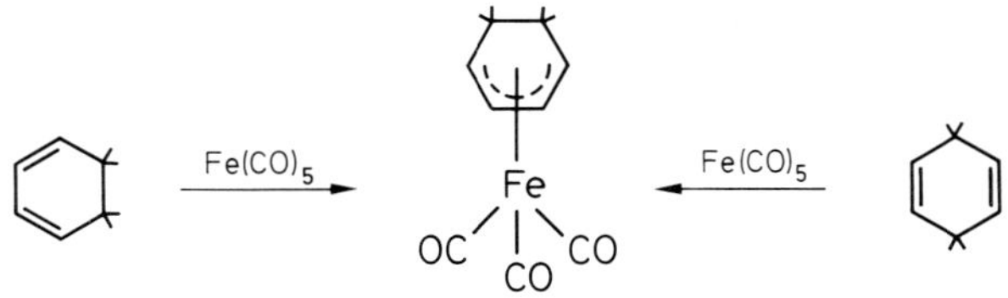

The bis(η^4-cyclohexadiene) complex, $(\eta^4\text{-}C_6H_8)_2FeCO$, is also known:

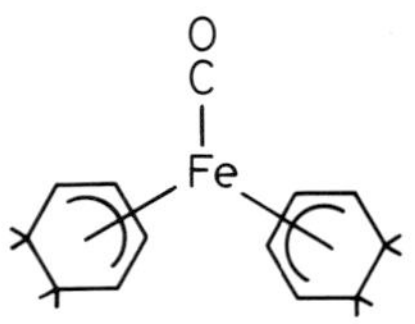

The fluorinated dienes tend to form $2\sigma + \eta^2$ type complexes (*vide supra*), and this happens also with fluorinated cyclohexadienes. Thus, octafluorocyclohexadiene-1,3 forms the compound $(C_6F_8)Fe(CO)_3$ with $Fe_3(CO)_{12}$, for which the structure suggests $2\sigma + \eta^2$ bonding (as shown in Fig. 13.1.f′).

The $(\eta^6\text{-}C_6H_6)M(\eta^4\text{-}C_6H_8)$ (M = Fe, Ru, Os) complexes have been prepared by reduction of $[M(\eta^6\text{-}C_6H_6)_2]^{2+}$ cations with metal hydrides or by direct treatment of the metal(III) halide with cyclohexadiene-1,3 and iso-PrMgCl in ether:

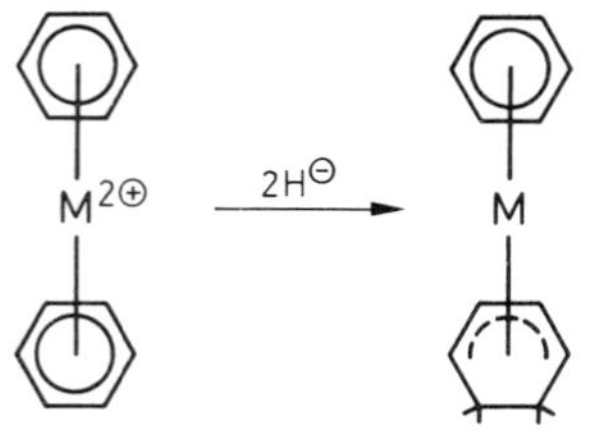

An iron derivative is formed in the reaction of iron vapor with benzene:

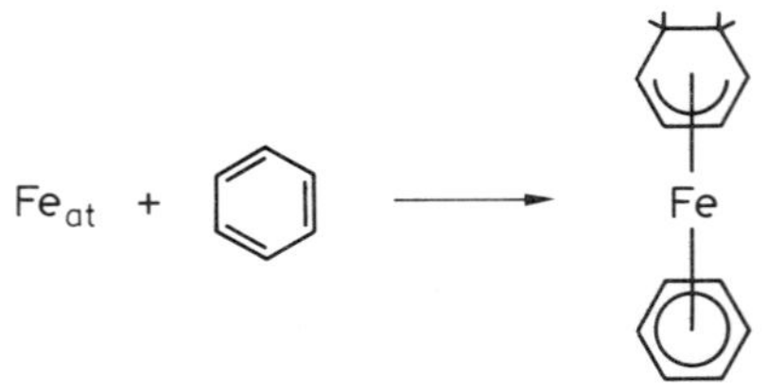

Another group which requires four electrons is $\eta^5\text{-}C_5H_5Co$, and it readily forms a cyclohexadiene-1,3 complex by reaction of the diolefin with $(\eta^5\text{-}C_5H_5)Co(CO)_2$ on heating:

The presence of a heteroatom in a cyclohexadiene ring does not influence the ability to form η^4-complexes; thus, dimethylsilacyclohexadiene reacts with iron pentacarbonyl to form the expected tetra*hapto* complex:

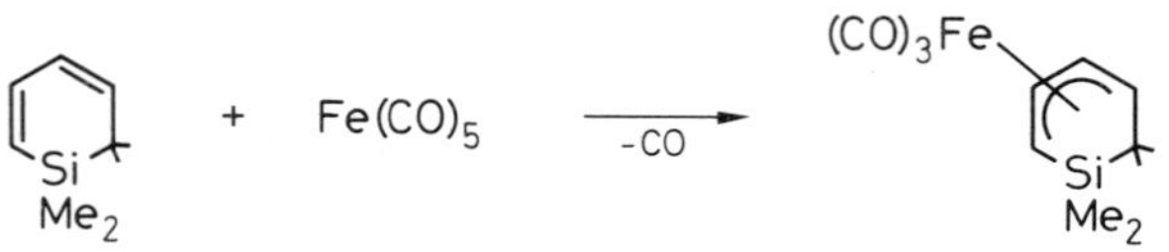

Cycloheptadiene-1,3

This diolefin can form the same type of complexes as cyclohexadiene-1,3, but few examples are known. Thus, chromium(III) chloride reacts with cycloheptatriene in the presence of cyclopentadienylmagnesium bromide to form $(\eta^7\text{-}C_7H_7)Cr(\eta^4\text{-}C_7H_{10})$, containing a tropylium cation and a cycloheptadiene molecule coordinated to chromium. With iron pentacarbonyl, cycloheptadiene-1,3 forms the expected $(\eta^4\text{-}C_7H_{10})Fe(CO)_3$:

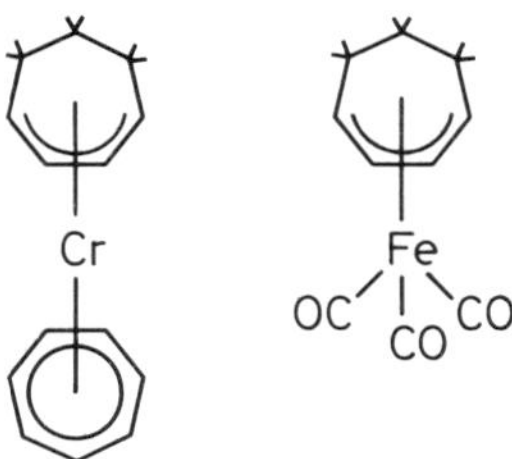

The simultaneous reaction of cobalt(II) chloride with cyclopentadiene and cycloheptadiene in the presence of iso-PrMgCl in ether yields a η^4-cyclopentadiene derivative:

$CoCl_2$ + + —iso-PrMgCl→ Co

Cycloheptatriene-1,3,5

This diene uses only two of its three double bonds to coordinate a $Fe(CO)_3$ fragment. The product can be hydrogenated in the presence of Raney nickel to form a cycloheptadiene-1,3 complex:

+ $Fe(CO)_5$ → $Fe(CO)_3$ —H_2/Ni→ $Fe(CO)_3$

Cyclooctatetrane

This tetraene can also use butadiene fragments in coordination to transition metals. The reaction between cyclooctatetraene and iron pentacarbonyl yields several compounds, depending upon the conditions, two of which are of the η^4-diene type. In $C_8H_8Fe(CO)_3$ two double bonds act as a conjugated-diene system, leaving the other two double bonds free. The complex $C_8H_8Fe_2(CO)_6$, obtained from cyclooctatetraene and $Fe_2(CO)_9$, contains two $Fe(CO)_3$ units attached to the two butadiene fragments:

$Fe(CO)_3$ $Fe(CO)_3$ $Fe(CO)_3$

In the chromium, molybdenum and tungsten compounds of the type $M_2(C_8H_8)_3$, the cyclooctatetraene ligands are attached through butadiene fragments to the dimetal unit:

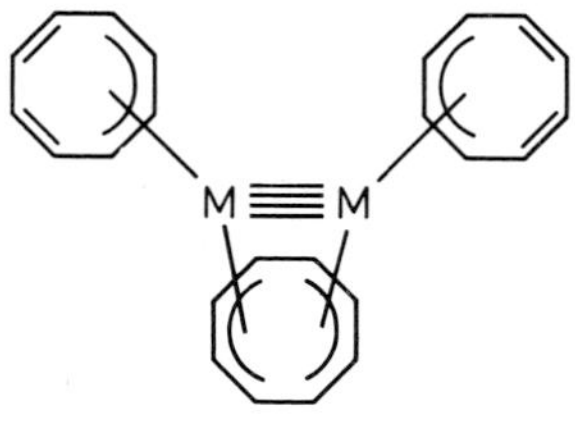

M≡M

The ruthenium complex of cyclooctatetraene, $(C_8H_8)_2Ru_3(CO)_4$, contains an Ru_3 triangle and two C_8H_8 rings attached to it through butadiene fragments of the ring:

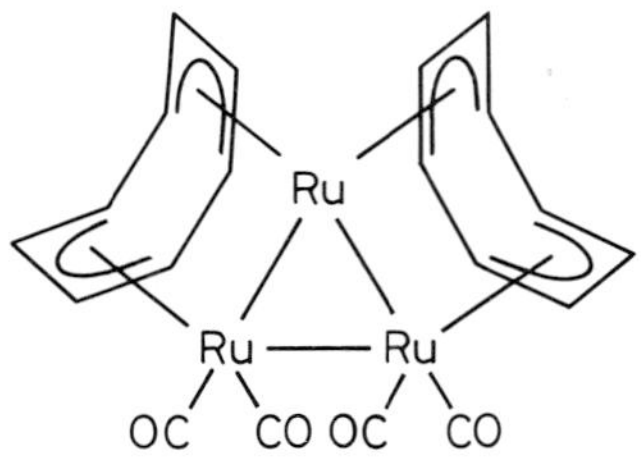

Ru

Ru — Ru

OC CO OC CO

The compound has been prepared by the reaction of cyclooctatetraene with $Ru_3(CO)_{12}$. Highly fluxional complexes, $Ru(\eta^4\text{-}C_8H_8)(\eta^6\text{-arene})$, have also been reported.

13.5. Trimethylenemethyl Complexes

The ability of transition metals to stabilize organic molecules incapable of independent existence through complexation has been mentioned for cyclobutadiene. A more unusual case is the trimethylenemethyl free radical in which each carbon atom is sp^2-hybridized and has a π-electron, available for donation:

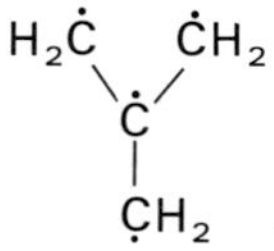

$H_2\dot{C}$ $\dot{C}H_2$

$\dot{C}$

$\dot{C}H_2$

Such a species cannot exist free, but by coordination to a transition metal it can act as a four-electron donor and can be stabilized. An iron derivative has been prepared by reacting 1-chloro-2-chloromethylpropene-2 with $Fe_2(CO)_9$:

$$H_2C{=}C(CH_2Cl)_2 + Fe(CO)_9 \longrightarrow (\text{trimethylenemethyl})Fe(CO)_3$$

The four carbon atoms are nearly coplanar and all three C—C bonds are equivalent.

Trimethylenemethyl complexes are also formed by ring opening of methylene-cyclopropane derivatives by $Fe_2(CO)_9$:

$$\text{RR}'\text{C}_3\text{H}_2{=}CH_2 + Fe_2(CO)_9 \longrightarrow H_2C{=}C(CRR')(CH_2)Fe(CO)_3$$

Chromium- and molybdenum-carbonyl derivatives are formed from 1-chloro-2-chloromethylpropene-2:

$$(\text{C}(CH_2)_3)Cr(CO)_4 \xleftarrow{Cr(CO)_6} H_2C{=}C(CH_2Cl)_2 \xrightarrow{Na_2Mo(CO)_5} (\text{C}(CH_2)_3)Mo(CO)_4$$

Compounds with a more complicated structure containing the trimethylenemethyl group as a fragment of a larger molecule have been obtained by the reaction of $Fe_2(CO)_9$ with chloromethylcycloheptatriene and bromomethylnaphthalene:

$$Fe(CO)_3 \qquad C_{10}H_7{-}CH_2 \quad Fe(CO)_3$$

The isolation of these compounds suggests that many interesting reactions may occur with haloalkylene derivatives and metal carbonyls.

13.6. Boron-Containing Four-Electron Ligands

Some divinylboranes, boron-containing heterocycles and carboranes can also act as four-electron ligands. Thus, alkoxydivinylboranes react with $Fe_2(CO)_9$ to form iron-tricarbonyl complexes on UV irradiation:

$$RO{-}B(CH{=}CH_2)_2 \xrightarrow{Fe_2(CO)_9} RO{-}B(CH{=}CH_2)_2Fe(CO)_4 \xrightarrow{UV} RO{-}B(CH{=}CH_2)_2Fe(CO)_3$$

The boron atom participates in the π-electron conjugation, although it contributes no electrons.

The five-membered heterocycle, thiadiborolene, also behaves as a four-electron ligand and forms an iron-tricarbonyl complex. Two electrons come from the olefinic double bond and another two from the sulfur atom, while the boron atoms having vacant p_z-orbitals seem to permit delocalization in the ring:

$$Et_2C_2(BMe)_2S + Fe(CO)_5 \longrightarrow (Et_2C_2(BMe)_2S)Fe(CO)_3$$

Some carborane complexes containing the four-electron accepting group, $Fe(CO)_3$, are illustrated in Fig. 13.7.

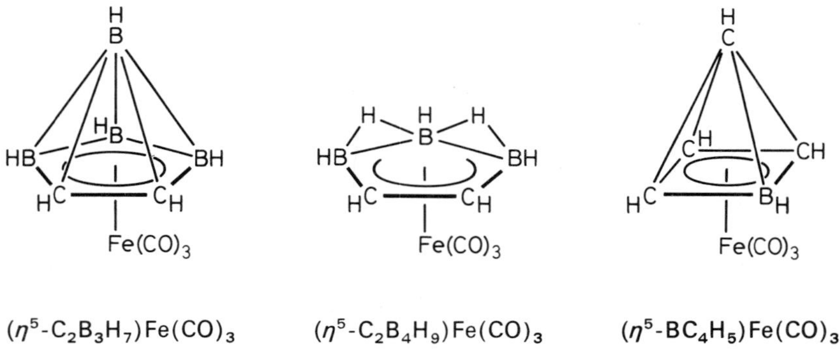

Fig. 13.7. Some carborane-iron complexes.

The compounds $(C_2B_3H_7)Fe(CO)_3$ and $(C_2B_4H_9)Fe(CO)_3$ are obtained by heating $C_2B_4H_8$ with iron pentacarbonyl, while $(BC_4H_5)Fe(CO)_3$ is formed in a photochemical reaction of $(\eta^4\text{-}C_4H_4)Fe(CO)_3$ with pentaborane:

$$(\eta^4\text{-}C_4H_4)Fe(CO)_3 + B_5H_9 \xrightarrow{UV} (\eta^4\text{-}BC_4H_5)Fe(CO)_3$$

14. Compounds with Five-Electron Ligands

The best known five-electron ligand is the cyclopentadienyl group, C_5H_5, bonded in penta*hapto*-fashion. Ferrocene, or bis(cyclopentadienyl)iron, $Fe(\eta^5\text{-}C_5H_5)_2$, the compound whose accidental discovery in 1951 marked the beginning of a new era in transition metal organometallic chemistry, is a very stable derivative of this ligand. The two cyclopentadienyl ligands (considered as C_5H_5 radicals, with 5π-electrons each distributed over a planar network of five sp^2-hybridized carbon atoms) complete a noble-gas configuration at iron. The strong tendency of metals to form complexes with this ligand is demonstrated by the formation of many other "metallocenes", even when the 18-electron rule is not obeyed.

In addition to cyclopentadienyl (a), the acyclic pentadienyl (b), cyclohexadienyl (c), cycloheptadienyl (d), and other groups act as five-electron ligands:

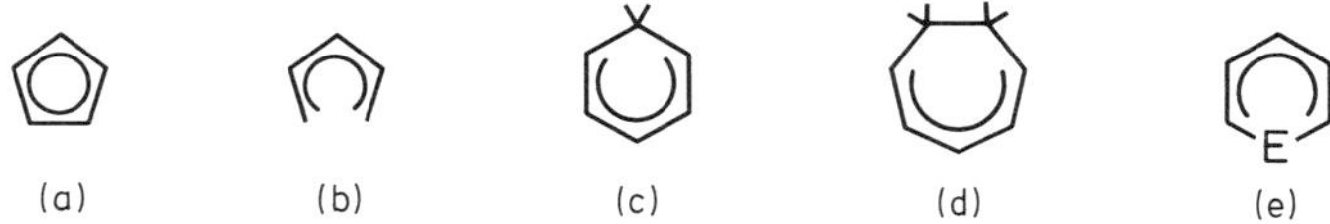

The open-chain pentadienyl ligand (b) is not a favored one, and much better ligands are obtained if the five-atom chain is held in a cyclic conformation by a CH_2 or CH_2CH_2 group [as in (c) or in (d), or by a heteroatom closing a six-membered ring (e) where E = B, Si, P, etc.] without contributing electrons.

14.1. Transition-Metal, η^5-Cyclopentadienyl Complexes

The η^5-cyclopentadienyl group, alone or in association with other ligands, forms a large number of compounds with a variety of structures. Depending upon electronic requirements, a metal atom may bond one, two, or sometimes more η^5-cyclopentadienyl groups. The most common are the "metallocenes" or bis(η^5-cyclopentadienyl) metals, $M(\eta^5\text{-}C_5H_5)_2$, in which the two C_5H_5 rings are in parallel planes, in either eclipsed or staggered orientation. Thus, solid $V(C_5H_5)_2$, $Cr(C_5H_5)_2$, $Ru(C_5H_5)_2$ are eclipsed and $Fe(C_5H_5)_2$ and $Co(C_5H_5)_2$ are staggered.

Binary cyclopentadienyl-metal compounds, for example, derivatives containing only C_5H_5 groups and metal atoms in the molecule, are known for almost all transition metals (See Table 14.1). When several C_5H_5 groups are present in a molecule, they may

Tab. 14.1. Binary metal-cyclopentadienyl compounds (neutral and ionic species); $Cp = C_5H_5$.

Group III	Group IV	Group V	Group VI	Group VII	Group VIII			Group IB
$ScCp_3$	$(TiCp_2)_2$ $TiCp_3$ $TiCp_4$	VCp_2 VCp_2^+	$CrCp_2$ $CrCp_2^+$	$MnCp_2$	$FeCp_2$ $FeCp_2^+$	$CoCp_2$ $CoCp_2^+$	$NiCp_2$ $NiCp_2^+$ $Ni_2Cp_3^+$ $[Ni_6Cp_6]^{0,+}$	$CuCp \cdot L$
YCp_3	$(ZrCp_2)_2$ $ZrCp_4$	$NbCp_4$	$MoCp_4$ $(MoCp_2)_2$	$(TcCp_2)_2$	$RuCp_2$ $RuCp_2^+$	$RhCp_2$ $RhCp_2^+$		$AgCp \cdot L$
$LrCp_3$ Ln = La, Ge, Pr, Nd, Sm, Gd, Dy, Er, Yb $(NdCp_3)_4$	$HfCp_4$	$TaCp_4$	$(WCp_2)_2$		$OsCp_2$ $OsCp_2^+$	$IrCp_2$ $IrCp_2^+$	Pt_2Cp_4	$AuCp \cdot L$
$(ThCp_3)_2$ $ThCp_4$ UCp_3 UCp_4								

Tab. 14.2. Cyclopentadienyl-metal Carbonyls (Neutral and Ionic Species)

Group IV	Group V	Group VI	Group VII	Group VIII			Group IB
$Cp_2Ti(CO)_2$	$CpV(CO)_4$ $CpV(CO)_3^{2-}$ $CpV(CO)_2^+$ $[CpV(CO)_2]_3$ $CpV(CO)_4$ $Cp_2V(CO)_2^+$ $Cp_2V_2(CO)_5$	$CpCr(CO)_4^+$ $[CpCr(CO)_3]_2$ $CpCr(CO)_3^-$ $[CpCr(CO)_2]_2$	$CpMn(CO)_3$	$CpFe(CO)_3^+$ $[CpFe(CO)_2]_2$ $CpFe(CO)_2^-$ $[CpFe(CO)]_4$	$CpCo(CO)_2$ $CpCo(CO)_3$ $Cp_2Co_2(CO)_3$ $Cp_2Co_2(CO)_2$	$CpNi(CO)_2$ $Cp_3Ni_3(CO)_2$	CpCuCO
$Cp_2Zr(CO)_2$	$CpNb(CO)_4$ $Cp_3Nb_3(CO)_7$	$CpMo(CO)_4^+$ $[CpMo(CO)_3]_2$ $CpMo(CO)_3^-$ $[CpMo(CO)_2]_2$	$CpTc(CO)_3$	$[CpRu(CO)_2]_2$	$[CpRh(CO)_2]_3$ $CpRh(CO)_3$ $Cp_2Rh_2(CO)_3$ $Cp_3Rh_3(CO)_4^-$		
$Cp_2Hf(CO)_2$	$CpTa(CO)_4$	$CpW(CO)_4^+$ $[CpW(CO)_3]_2$ $CpW(CO)_3^-$ $[CpW(CO)_2]_2$	$CpRe(CO)_3$ $Cp_2Re_2(CO)_5$	$[CpOs(CO)_2]_2$	$CpIr(CO)_2$ $Cp_3Ir_3(CO)_2$	$CpPt(CO)_2$	

be bonded in different ways (for example, penta*hapto* and mono*hapto*). Simple metallocenes of early transition metals are difficult to obtain, and for some (for example, "titanocene") the structure was found to be more complicated than believed initially.

Many mixed-ligand complexes have also been prepared, and these include cyclopentadienylmetal carbonyls (Table 14.2), cyclopentadienylmetal halides (Table 14.3), cyclopentadienyl metal sulfides, and many others. In these compounds the η^5-C_5H_5 rings are usually in bent rather than in parallel planes.

Tab. 14.3. Some Cyclopentadienyl-Metal Halides; Cp = C_5H_5, X = Halogen

Cp_2ScX_2	Cp_2TiX_2	Cp_2VX_2	Cp_2CrX	Cp_2FeX
X = Cl	X = F, Cl, Br, I	X = Cl, Br	X = I	X = F
$(CpScX)_2$	(Cp_2TiX)	Cp_2VX	$CpCrX_2$	
X = Cl	X = Cl	X = Cl, Br, I	X = Cl, Br, I	
	$CpTiX_3$	$CpVX_3$		
	X = Cl, Br, I			
Cp_2YX	Cp_2ZrX_2	Cp_2NbX_3	Cp_2MoX_2	$(CpRhX_2)_n$
X = Cl	X = Cl, Br	X = Cl, Br	X = Cl, Br, I	X = Cl, Br
$CpYX_2$		X = Cl, Br		
X = Cl		Cp_2NbCl_2	$CpMoX_4$	
		Cp_2NbCl	X = Cl	
*	Cp_2HfX_2	Cp_2TaX_3	Cp_2WX_2	$(CpPdX)_2$
	X = Cl	X = Cl, Br	X = Cl, Br, I	X = Cl
		$Cp_4Ta_2Cl_3$		

* Note: Several cyclopentadienyl-lanthanide halides are known, e.g.: Cp_2MCl (M = Sm, Gd, Tb, Dy, Ho, Er, Tm, Yb, Lu), and $CpMCl_2$ (M = Sm, Eu, Gd, Dy, Ho, Er, Yb, Lu) as THF adducts, $CpMCl_2 \cdot 3THF$.

Another source of structural diversity is the possibility of forming di- and polynuclear species with different metal atoms in the same molecule. The structural possibilities are further expanded by the ability of cyclopentadienyl groups to bond as a penta*hapto*-ligand to one atom and simultaneously as a mono*hapto*-ligand to another atom in the same molecule, as in the following two examples of iron and thorium derivatives:

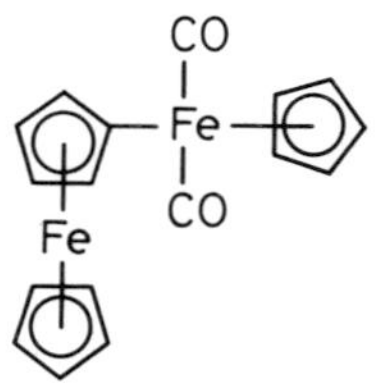

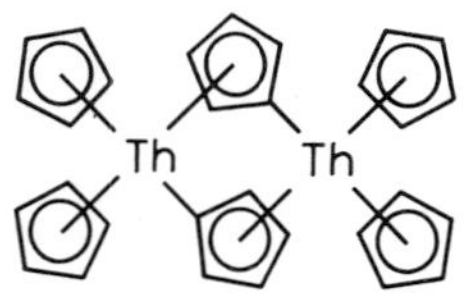

There are few heterobimetallic compounds, but a mixed niobium-molybdenum derivative illustrates this possibility:

An additional feature of interest in such compounds is the possible occurence of semi-bridging carbonyl groups, as in the example cited.

In Fig. 14.1. are illustrated some basic types of mononuclear cyclopentadienyl derivatives, and Fig. 14.2 shows some examples of binuclear and polynuclear derivatives. As shown in Fig. 14.3, nearly all transition metals, including some lanthanides and actinides (even transuranium elements), form cyclopentadienyl-metal complexes.

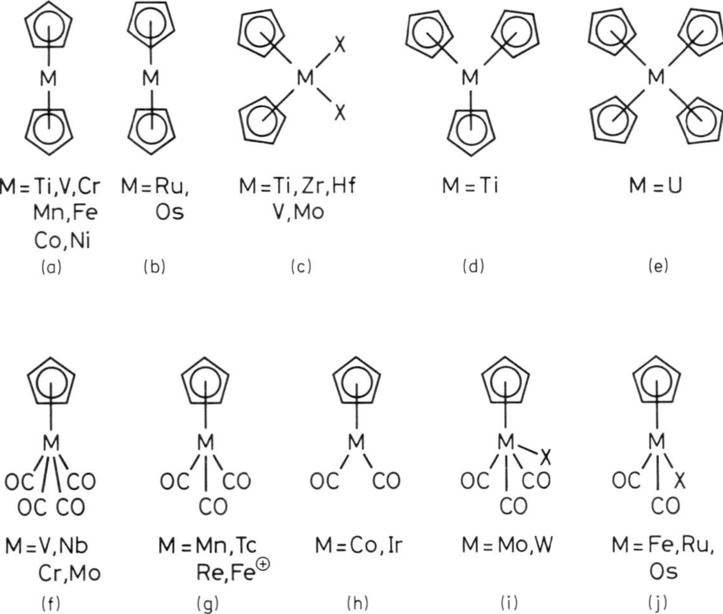

Fig. 14.1. Some basic types of monuclear cyclopentadienyl metal complexes.

The η^5-cyclopentadienyl derivatives of most transition metal are stable thermally, but their oxidative stability varies from metal to metal. Many binary derivatives, $M(\eta^5\text{-}C_5H_5)_2$, are paramagnetic and do not obey the effective atomic number rule, having either an electron deficit or an electron excess. However, the compounds which have 18-electrons in their valence shell, for example, $Fe(C_5H_5)_2$, $Ru(C_5H_5)_2$, and $Os(C_5H_5)_2$ among binary compounds, or $\eta^5\text{-}C_5H_5Mn(CO)_3$ and $\eta^5\text{-}C_5H_5Re(CO)_3$ are the most stable and exhibit aromatic character, obvious in numerous aromatic substitution reactions.

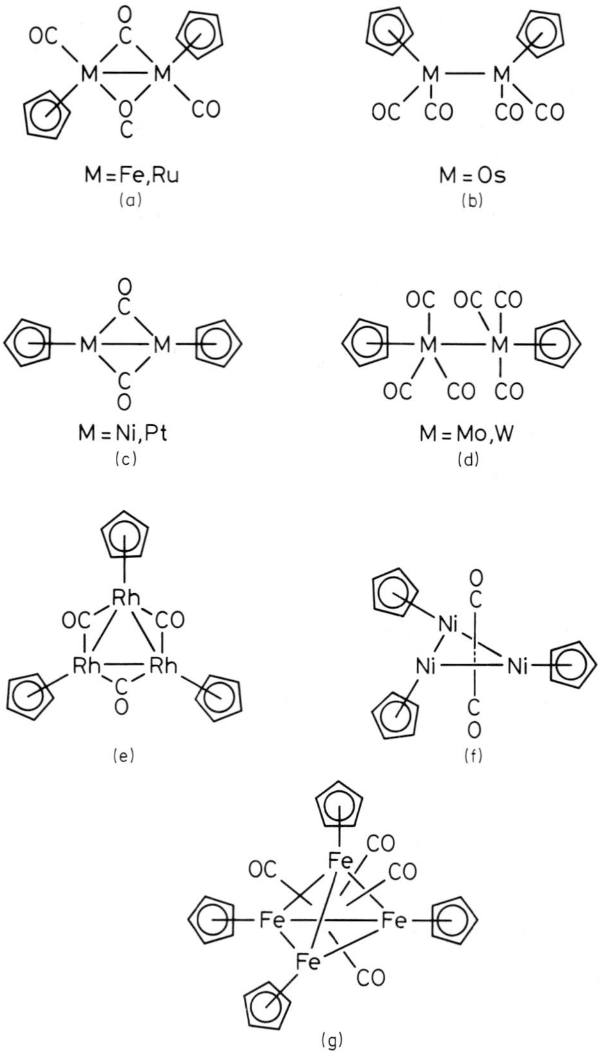

Fig. 14.2. Some typical bi- and polynuclear metal-cyclopentadienyl derivatives.

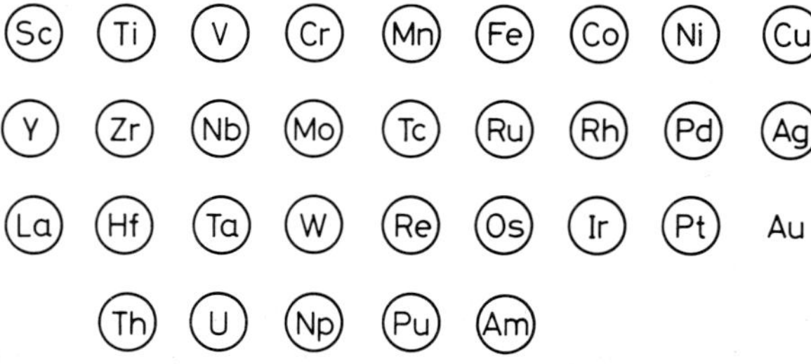

Fig. 14.3. Metals forming cyclopentadienyl complexes.

The preparative methods for π-cyclopentadienyl-metal derivatives are rather general and can be applied to different metals. The main reaction types used for this purpose are the following:

a) Sodium cyclopentadienide and cyclopentadienylmagnesium halides with anhydrous transition-metal halides:

$$C_5H_5^{\ominus}Na^{\oplus} + MX \longrightarrow (C_5H_5)\text{—}M + NaX$$

b) Cyclopentadiene with anhydrous metal halide in the presence of a base:

$$C_5H_6 \quad + \quad MX \quad \xrightarrow[\text{base HX}]{\text{base}} \quad (\pi\text{-}C_5H_5)M$$

c) Cyclopentadiene with the free metal (active or activated):

$$C_5H_6 \quad + \quad M \quad \longrightarrow \quad (\pi\text{-}C_5H_5)M \quad + \quad 1/2\,H_2$$

d) Cyclopentadiene and metal carbonyls:

$$C_5H_6 \quad + \quad M(CO)_n \quad \xrightarrow[-CO]{} \quad (\pi\text{-}C_5H_5)M(CO)_{n-1} \quad + \quad {}^1\!/_2\,H_2$$

e) Ligand-transfer reactions (from one metal to another):

$$(\pi\text{-}C_5H_5)M \quad + \quad M'X \quad \longrightarrow \quad (\pi\text{-}C_5H_5)M' \quad + \quad MX$$

Further examples will be given for each group of elements.

Scandium, Yttrium, Lanthanides and Actinides. (Group III Elements)

Cyclopentadienyl derivatives of Group III elements, prepared by the reaction of scandium, yttrium, and lanthanide halides with sodium cyclopentadienide are known. The nature of the bonding between the metal and the cyclopentadienyl group has been unclear for some time, their properties suggesting an ionic structure. In $[(C_5H_5)_2ScCl]_2$ the cyclopentadienyl groups are bonded in a penta*hapto*-fashion to the metal:

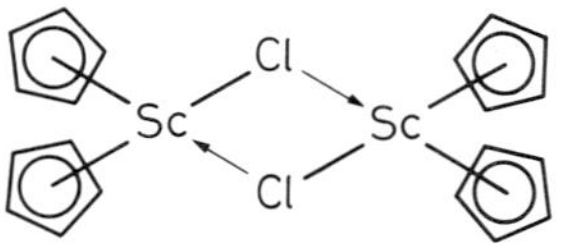

Cl Cl Sc Sc

The lanthanides form air-sensitive, but thermally stable tris(cyclopentadienyl) derivatives, $M(C_5H_5)_3$. Europium and ytterbium also form bis(cyclopentadienyl) complexes, $M(C_5H_5)_2$, by the reaction of cyclopentadiene with the metals.

Cyclopentadienyl derivatives of actinides have also been prepared. Thorium tetrachloride forms with sodium cyclopentadienide the tetrakis derivative, $Th(C_5H_5)_4$, and with thallium cyclopentadienide the compound $(C_5H_5)_3ThCl$.
The uranium compounds $U(C_5H_5)_4$ and $U(C_5H_5)_3$ have similarly been prepared. The bis(cyclopentadienyl) derivatives $(\eta^5\text{-}C_5Me_5)_2MCl_2$ (M = Th, U) are obtained from $Na^+C_5H_5^-$ with MCl_5. The cation $U(C_5H_5)_3^+$ is formed by treatment of $U(C_5H_5)_3Cl$ with potassium tetracyanonickelate and -platinate.

For the synthesis of transuranium cyclopentadienyls such as $(C_5H_5)_3NpCl$, $Pu(C_5H_5)_4$, $Am(C_5H_5)_3$, $Cm(C_5H_5)_3$ and $Bk(C_5H_5)_3$, the reaction between bis(cyclopentadienyl)beryllium and metal trichlorides has been used (sometimes on a microgram scale). For the preparation of $Cm(C_5H_5)_3$ the reaction of $CmCl_3$ with bis(cyclopentadienyl)magnesium can also be used.

Titanium, Zirconium, Hafnium

Titanium tetrachloride forms with sodium cyclopentadienide an air-stable dichloride, $(\eta^5\text{-}C_5H_5)_2TiCl_2$ with a bent sandwich structure (Fig. 14.1.c) in which the chlorines can be replaced by other groups (Me, Ph, OR, SR, OCOR, NCS, etc.). The reduction of $(C_5H_5)_2TiCl_2$ with zinc metal, or the reaction of titanium trichloride with bis(cyclopentadienyl) magnesium yields the dimer, chlorine-bridged, $[(C_5H_5)_2TiCl]_2$:

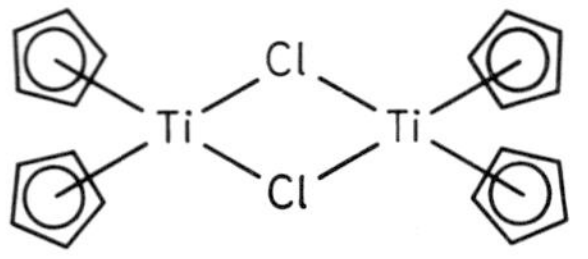

Cl
Ti Ti
Cl

The monosubstituted compound $C_5H_5TiCl_3$ has been synthesized by ligand transfer between $[C_5H_5Fe(CO)_2]_2$ and titanium tetrachloride or by ligand redistribution between $(C_5H_5)_2TiCl_2$ and titanium tetrachloride.

If the reaction between titanium tetrachloride and sodium cyclopentadienide is carried out under carbon monoxide the product is $(\eta^5\text{-}C_5H_5)_2Ti(CO)_2$; the same bent sandwich compound is obtained by reducing $(C_5H_5)_2TiCl_2$ with aluminum powder in THF, in a carbon monoxide atmosphere:

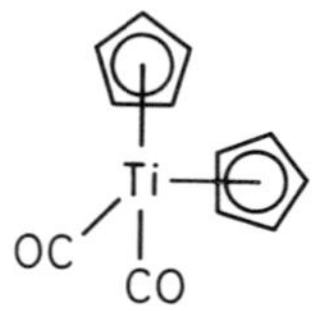

Ti
OC CO

With excess sodium cyclopentadienide, $(C_5H_5)_2TiCl_2$ yields the tricyclopentadienyl derivative, $Ti(C_5H_5)_3$, which contains only two cyclopentadienyl rings bound in a penta*hapto*-fashion, while the third is di*hapto*:

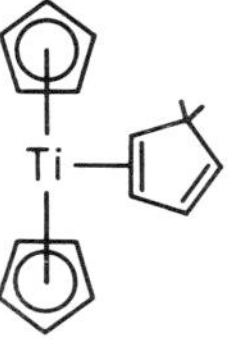

The compound first reported as "titanocene", $Ti(C_5H_5)_2$, prepared from $TiCl_2$ and sodium cyclopentadienide, by reducing the resulting $(\eta^5\text{-}C_5H_5)_2TiCl_2$, exists in several forms described as green, black and metastable titanocenes. Two of these are now known to be dimers. The green form is μ-(η^5 : η^5-fulvalene)di-μ-hydrido-bis(η^5-cyclopentadienyl)dititanium while the black form contains a η^1 : η^5-bridge:

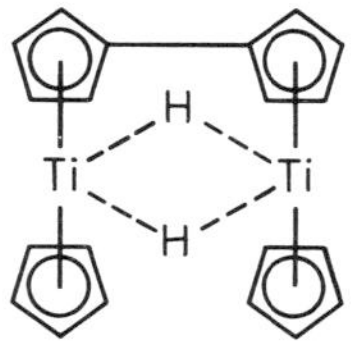

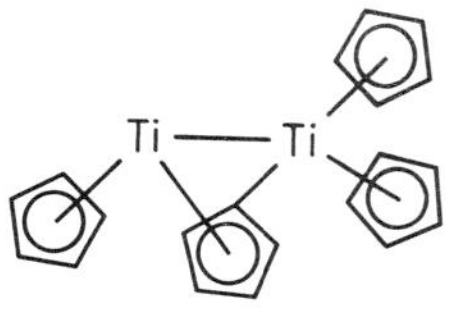

Zirconium and hafnium tetrachlorides form bis(cyclopentadienyl) metal dichlorides, $(\eta^5\text{-}C_5H_5)_2MCl_2$ (M = Zr, Hf), which can be converted to tetrasubstituted derivatives, $M(C_5H_5)_4$, by reaction with excess NaC_5H_5. In $Zr(C_5H_5)_4$ only three cyclopentadienyls are penta*hapto* while the fourth is mono*hapto*:

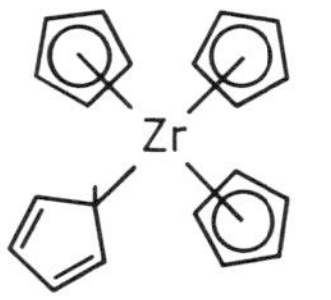

In solution the compound is fluxional and exhibits only a single ^{1}H-NMR-signal.

Zirconium and hafnium tetrachlorides form monocyclopentadienyl trichlorides, $C_5H_5ZrCl_3$ and $C_5H_5HfCl_3$, in reactions with bis(cyclopentadienyl)magnesium.

Cyclopentadienylmetal carbonyls, $(\eta^5\text{-}C_5H_5)_2M(CO)_2$ (M = Zr, Hf), are among the few carbonyl derivatives of these two elements.

Vanadium, Niobium, Tantalum

Vanadium tetrachloride and sodium cyclopentadienide form $(C_5H_5)_2VCl_2$, and with excess reagent, $V(C_5H_5)_2$ (vanadocene). The latter can be conveniently prepared from $VCl_2 \cdot 2THF$ with NaC_5H_5.

Bis(cyclopentadienyl)vanadium monochloride, $(C_5H_5)_2VCl$, is formed by redistribution between $V(C_5H_5)_2$ and $(C_5H_5)_2VCl_2$.

Stable cyclopentadienyl vanadium tetracarbonyl, $C_5H_5V(CO)_4$, with a noble-gas configuration, is formed in the reaction between sodium hexacarbonyl vanadate and cyclopentadienylmercury chloride.

$$Na[V(CO)_6] + \eta^1\text{-}C_5H_5HgCl \longrightarrow \eta^5\text{-}C_5H_5V(CO)_4 + Hg + 2CO + NaCl$$

The structure of this compound is shown in Fig. 14.1.f. Irradiation or heating of $C_5H_5V(CO)_4$ yields $(C_5H_5)_2V_2(CO)_5$, which is a prototypal example of a compound with semibridging carbonyl groups:

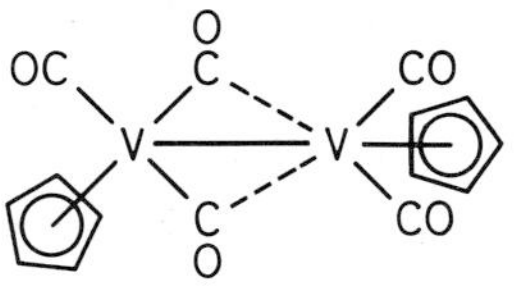

Refluxing in THF leads to a tetranuclear compound $[\eta^5\text{-}C_5H_5V(CO)]_4$.
The anion, $[(\eta^5\text{-}C_5H_5)V(CO)_3]^{2-}$ is formed by the sodium amalgam reduction of $C_5H_5V(CO)_4$, and the cation, $[(\eta^5\text{-}C_5H_5)_2V(CO)_2]^+$, is obtained by the reaction between vanadocene and molybdenum hexacarbonyl under carbon monoxide or in the following sequence of reactions:

$$(C_5H_5)_2VCl \xrightarrow[BPh_4^-]{H_2O/\text{acetone}} [V(C_5H_5)_2]^+BPh_4^- \xrightarrow{CO} [(C_5H_5)_2V(CO)_2]^+BPh_4^-$$

Cyclopentadienylniobium tetracarbonyl, $(\eta^5\text{-}C_5H_5)Nb(CO)_4$, whose structure is shown in Fig. 14.1.f, is obtained by the reaction of NaC_5H_5 with the $[Nb(CO)_6]^-$ anion, or by reduction of $(C_5H_5)_2NbCl_2$ with an Na/Cu/Al alloy under carbon monoxide. Irradiation of $C_5H_5Nb(CO)_4$ by UV yields a trinuclear compound, $(\eta^5\text{-}C_5H_5)_3Nb_3(CO)_7$, which contains a novel carbonyl bridge:

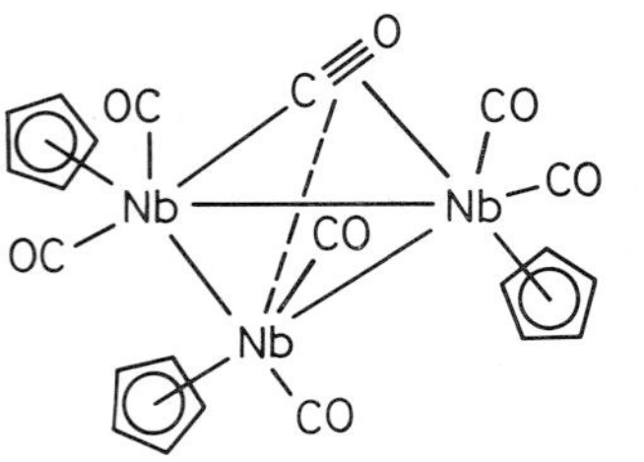

Cyclopentadienyl halides of niobium and tantalum, $C_5H_5MX_4$ and $C_5H_5NbBr_3$, have been prepared by reacting MX_5 (M = Nb, Ta) with bis(cyclopentadienyl)-magnesium or $C_5H_5SnMe_3$.

Niobocene, $(\eta^5\text{-}C_5H_5)_2Nb$ can be obtained by reducing $(\eta^5\text{-}C_5H_5)_2NbCl_2$ but exists only in solution. A solid described earlier as "niobocene" was found to be a dimeric hydride with $(\eta^1:\eta^5\text{-}C_5H_4)$ bridges:

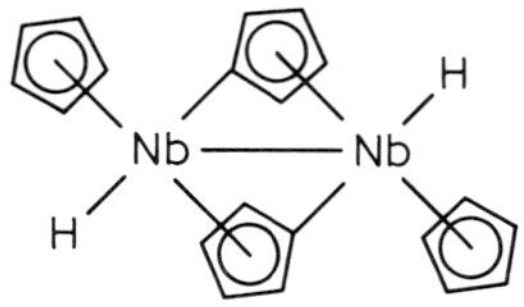

Chromium, Molybdenum, Tungsten

Chromocene, $Cr(\eta^5\text{-}C_5H_5)_2$, is prepared by the reaction of anhydrous chromium(III) chloride with sodium cyclopentadienide, or better from $CrCl_2 \cdot THF$. With only 16 electrons in its valence shell, this compound is unstable and air-sensitive. Molybdenum and tungsten analogues could not be prepared, the hydrides, $(\eta^5\text{-}C_5H_5)_2MH_2$, forming instead. Chromocene is also obtained from chromium hexacarbonyl and cyclopentadiene at 280–350 °C, an unusually high temperature for organometallic compounds! Chromium and molybdenum hexacarbonyl, heated with cyclopentadiene yields dimers, $[(\eta^5\text{-}C_5H_5)M(CO)_3]_2$, containing metal-metal bonds and no carbonyl bridges:

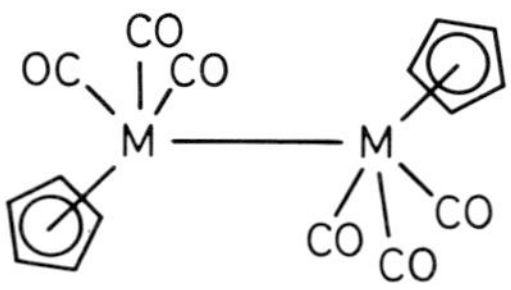

Their solutions are mixtures of *anti-* and *gauche*-rotamers with barriers to rotation around the metal-metal bonds of *ca.* 50–63 kJ/mole.
These dimers are the starting materials for many syntheses. With sodium amalgam they are reduced to the anion, $[(\eta^5\text{-}C_5H_5)M(CO)_3]^-$. They can be oxidized to $(\eta^5\text{-}C_5H_5)MoO_2Cl$ and $[(\eta^5\text{-}C_5H_5)MoO_2]_2$. Hydrochloric acid converts the dioxochloride to $(\eta^5\text{-}C_5H_5)MoCl_4$, which hydrolyzes to $[(\eta^5\text{-}C_5H_5)MoO]_4$.

Unlike cyclopentadiene, which forms the Mo-Mo bonded dimer, pentamethylcyclopentadiene forms with molybdenum hexacarbonyl a dimer which contains a triple Mo≡Mo bond with two fewer carbonyl groups:

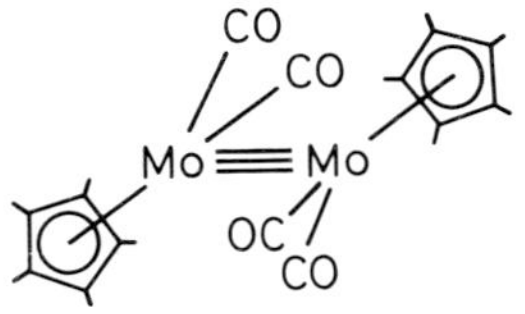

The chromium analogue, $[(\eta^5\text{-}C_5Me_5)Cr(CO)_2]_2$, containing a triple Cr≡Cr bond has the hydrocarbon ligands in a *trans*-position:

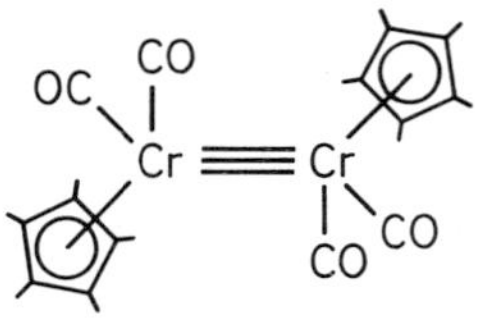

The metal-metal triple-bonded compounds, $[\eta^5\text{-}C_5H_5M(CO)_2]_2$, (M = Cr, Mo, W), are formed on refluxing the metal-metal single-bonded dimers, $[(\eta^5\text{-}C_5H_5)M(CO)_3]_2$, in toluene. Both the chromium and molybdenum compounds contain metal-metal triple bonds, but the molecular geometry is different: the chromium derivative is a *trans*-isomer, while the molybdenum compound contains a linear $C_5H_5\text{-}Mo{\equiv}Mo\text{-}C_5H_5$ fragment:

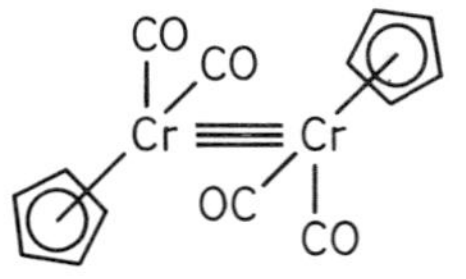

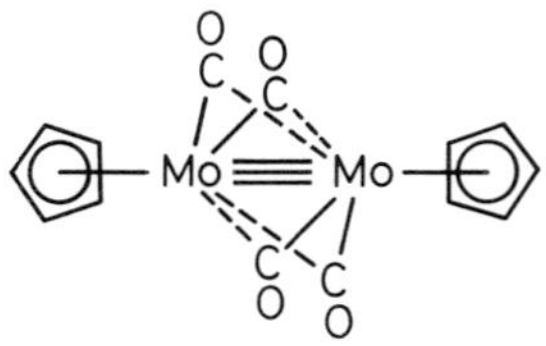

Ultraviolet irradiation of the dicyclopentadienylmolybdenum dihydride, $(\eta^5\text{-}C_5H_5)_2MoH_2$, whose neutron diffraction structure is available yields a dimeric "dehydromolybdocene". Two isomeric hydrido molybdocenes are also known:

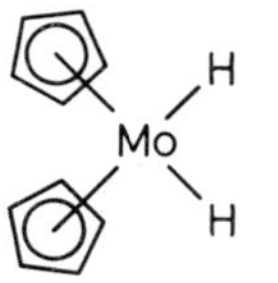

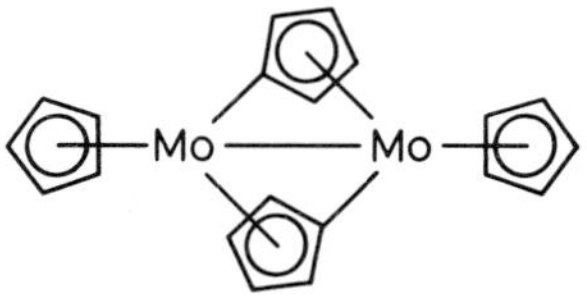

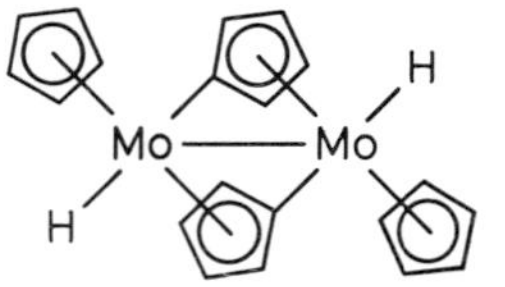

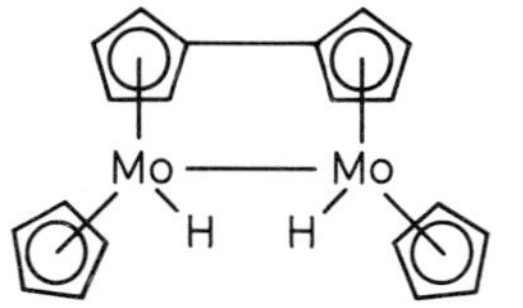

The anion $[(\eta^5\text{-}C_5H_5)W(CO)_3]^-$ is, obtained from the corresponding dimer with sodium amalgam. All three anions, $[\eta^5\text{-}C_5H_5M(CO)_3]^-$ (M = Cr, Mo, W), have only moderate nucleophilic character.

Tungstenocene is dimeric and has a *trans*-η^1,η^5-bridged cyclopentadienyl structure:

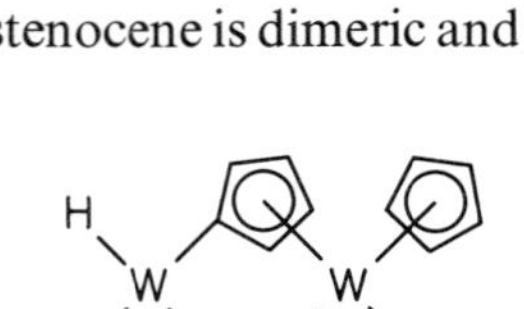

Manganese, Technetium, Rhenium

Manganocene, $Mn(\eta^5\text{-}C_5H_5)_2$, from anhydrous manganese(II) chloride and sodium cyclopentadienide, is an ionic, polymeric solid formed of alternating Mn^{2+} cations and $C_5H_5^-$ anions in a chain. This is the only metallocene of a first row transition element having an ionic structure.

In the vapor phase 1,1 -dimethylmanganocene contains two molecular forms: an ionic high-spin form (with a $Mn—C_5H_5$ distance of 243.3 pm≡2.433 Å) and low-spin η^5-complex with a $Mn—C_5H_5$ distance of 214.4 pm (2.144 Å).

Carbon monoxide converts manganocene to η^5-cyclopentadienylmanganese tricarbonyl, which can also be obtained from sodium cyclopentadienide and $Mn_2(CO)_{10}$, or $Mn(py)_2Cl_2$ with cyclopentadiene, magnesium metal and carbon monoxide. The

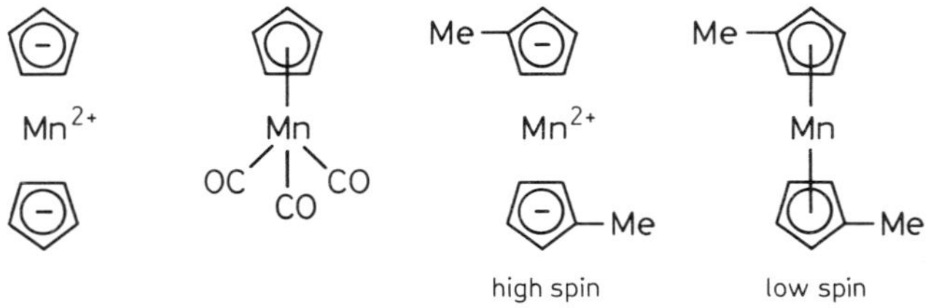

Mn²⁺ OC CO CO Mn Me Mn²⁺ Me high spin Me Mn Me low spin

product has a noble gas configuration for the metal, and undergoes aromatic substitution reactions.

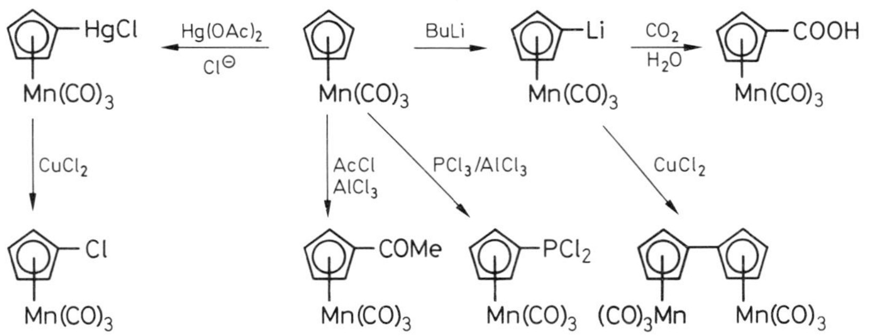

HgCl Mn(CO)₃ ← Hg(OAc)₂ / Cl⊖ — Mn(CO)₃ — BuLi → Li Mn(CO)₃ — CO₂ / H₂O → COOH Mn(CO)₃; CuCl₂ → Cl Mn(CO)₃; AcCl AlCl₃ → COMe Mn(CO)₃; PCl₃/AlCl₃ → PCl₂ Mn(CO)₃; CuCl₂ → (CO)₃Mn–Mn(CO)₃ bis(cyclopentadienyl)

Technetium and rhenium derivatives of the type $C_5H_5M(CO)_3$, prepared by coupling the pentacarbonylmetal chlorides with sodium cyclopentadienide:

$$M(CO)_5Cl + NaC_5H_5 \xrightarrow[-2CO]{-NaCl} \eta^5\text{-}C_5H_5M(CO)_3$$

are similar to their manganese analogue.

Iron, Ruthenium, Osmium

Bis(cyclopentadienyl)iron or ferrocene, $Fe(\eta^5\text{-}C_5H_5)_2$, discovered independently in two laboratories in 1951 is the subject of a huge volume of literature systematized in several reviews and monographs (see, for example, the appropriate volumes of the Gmelin Handbook). It was the discovery of ferrocene that initiated an explosive growth in organic transition metal chemistry after the peculiar type of bonding in this compound was recognized.

Almost all the ferrocene preparations have in common the conversion of cyclopentadiene to the $C_5H_5^-$ anion by base (dimethylamine, sodium metal, Grignard reagents, etc.), followed by the reaction between the anion and an iron(II) salt, usually the chloride:

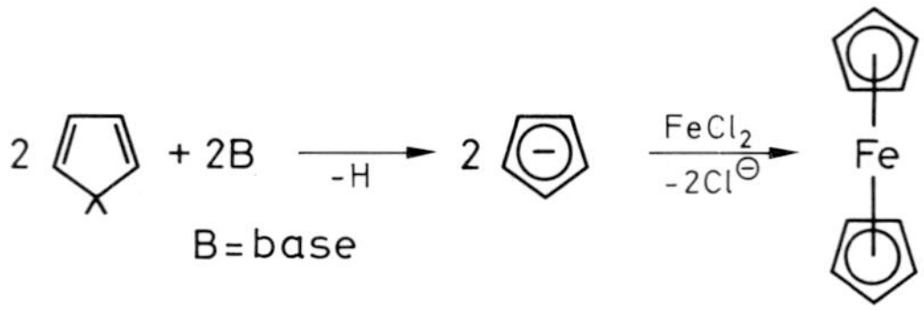

2 C_5H_6 + 2B $\xrightarrow{-H}$ 2 $C_5H_5^-$ $\xrightarrow[-2Cl^-]{FeCl_2}$ $Fe(C_5H_5)_2$

B = base

A simple preparation involves treatment of cyclopentadiene with KOH and $FeCl_2$ in tetrahydrofuran in the presence of [18]-crown-6 ether; the method can also be applied to substituted derivatives.

Ferrocene is prepared industrially from cyclopentadiene with iron oxides at elevated temperatures.

Substituted ferrocenes are obtained either by substitution reactions on the ferrocene molecule, or by starting from substituted cyclopentadienes with iron carbonyls or iron(II) chloride.

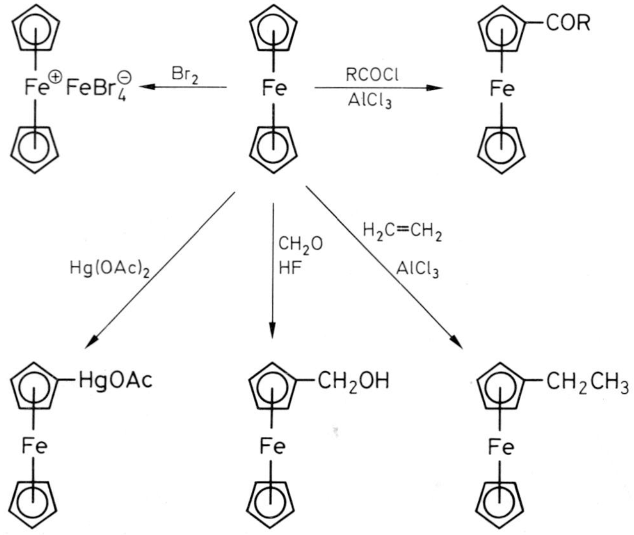

Fig. 14.4. Aromatic substitution reactions of ferrocene.

Because of its aromatic character, observed shortly after its discovery, ferrocene is an extremely interesting compound. Ferrocene can be acylated in the presence of aluminum chloride, can be mercurated, sulfonated, and by indirect methods nitrated or halogenated. These aromatic substitution reactions have been extensively investigated, and Fig. 14.4 illustrates some of the most typical.

Ferrocenyllithium is the starting material for the preparation of many derivatives as shown in Fig. 14.5.

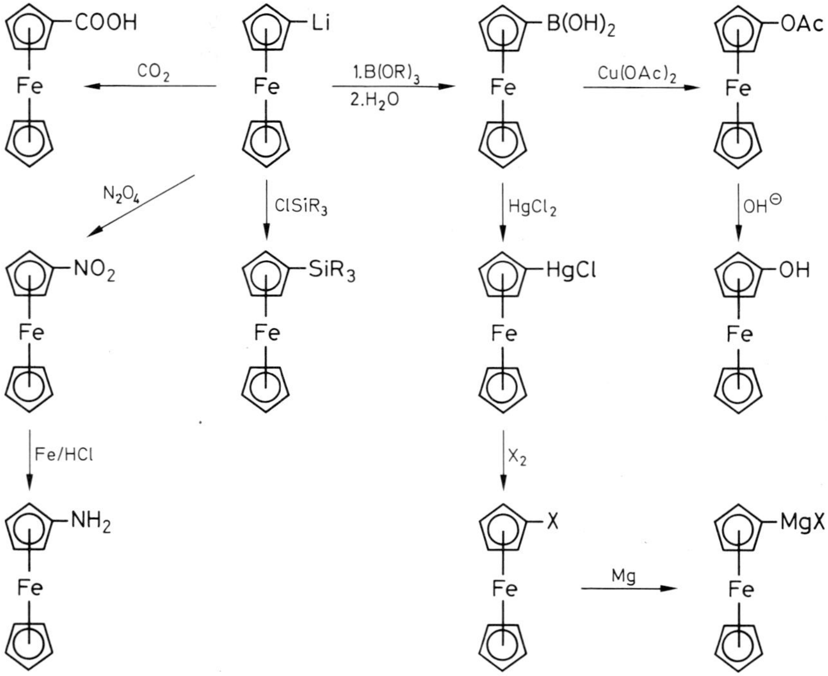

Fig. 14.5. Ferrocenyllithium as a starting material for functional ferrocene derivatives.

The ferrocene molecule has an antiprismatic structure with parallel, staggered C_5H_5 rings (Fig. 14.1.a) with all carbon atoms equidistant from metal. In the ferrocenium salt $[Fe(\eta^5\text{-}C_5H_5)_2]^+BiCl_4^-$, the rings are eclipsed. The cyclopentadienyl rings can be forced into a bent orientation by a diphenylsilicon bridge:

Fe Si(Ph)(Ph)

The electron density is higher than in benzene, and thus ferrocenylamine is a stronger base than aniline, and the ferrocenylcarboxylic acid is weaker than benzoic acid.

The dimer, $[(\eta^5\text{-}C_5H_5)Fe(CO)_2]_2$, from the stepwise reaction between iron pentacarbonyl and cyclopentadiene:

+ Fe(CO)5 -CO → Fe OC CO CO -CO → Fe—H OC CO -H [(π—C5H5)Fe(CO)2]2

contains carbonyl bridges with a *trans*-conformation in the solid, but an equilibrium of *cis*- and *trans*-isomers in solution. The dimer is fluxional with bridging and terminal carbonyl groups interchanging through a bridge-free form:

trans ⇌ ⇌ cis

The dimer is reduced by sodium amalgam to give a strongly nucleophilic anion, $(\eta^5\text{-}C_5H_5)Fe(CO)_2^-$, which is a useful starting material:

$[\eta^5\text{-}C_5H_5Fe(CO)_2]_2$ Na/Hg THF

+RCOCl → $\eta^5\text{-}C_5H_5Fe(CO)_2COR$; +RX → $\eta^5\text{-}C_5H_5Fe(CO)_2R$; $+C_6F_6$ → $\eta^5\text{-}C_5H_5Fe(CO)_2C_6F_5$; $+R_3ECl$ → $\eta^5\text{-}C_5H_5Fe(CO)_2ER_3$

E = Si, Ge, Sn

Halogens cleave the metal-metal bond, to form cyclopentadienyliron dicarbonyl halides, $C_5H_5Fe(CO)_2X$, which are also useful starting materials:

$[\eta^5\text{-}C_5H_5Fe(CO)_2]_2$

X_2

$[\eta^5\text{-}C_5H_5Fe(CO)_2L]^{\oplus}X^{\ominus}$

L = CO, PPh_3, py

+L

$\eta^5\text{-}C_5H_5Fe(CO)_2X$

+RMgX

$\eta^5\text{-}C_5H_5Fe\text{-}R$

$+NaM'(CO)_m$

$\eta^5\text{-}C_5H_5Fe(CO)_2\text{—}M'(CO)_n$

M′ = CO; n = 4

M′ = Mn, Re; n = 5

$+Na[\eta^5\text{-}C_5H_5Mo(CO)_3]^{\ominus}$

$\eta^5\text{-}C_5H_5Fe(CO)_2\text{—}Mo(CO)_3C_5H_5\text{-}\eta^5$

Carbon monoxide is lost on refluxing of the dimer in benzene to form the tetramer, $[(\eta^5\text{-}C_5H_5)Fe(CO)]_4$, whose structure is shown in Fig. 14.2.g. Carbon monoxide is lost from the solid on heating:

$$[\eta^5\text{-}C_5H_5Fe(CO)_2]_2 \xrightarrow[-CO]{165\,°C} [\eta^5\text{-}C_5H_5Fe(CO)]_4 \xrightarrow[-CO]{210\,°C} Fe(\eta^5\text{-}C_5H_5)_2$$

The tetramer, which also forms on refluxing $[\eta^5\text{-}C_5H_5Fe(CO)_2]_2$ with triphenylphosphine can be oxidized electrochemically to form the cations $[(\eta^5\text{-}C_5H_5)Fe(CO)]_4^{n+}$ (n = 1 or 2), or can be reduced electrochemically to the anion, $[(\eta^5\text{-}C_5H_5)Fe(CO)]_4^-$.

Compounds containing both iron and nickel are prepared from iron pentacarbonyl and the very reactive nickelocene via cyclopentadienyl ligand transfer:

$$Ni(\pi\text{-}C_5H_5)_2 + Fe(CO)_5 \xrightarrow[-2CO]{} (C_5H_5)Ni(\mu\text{-}CO)_2Fe(CO)(C_5H_5)$$

Ruthenocene, $Ru(\eta^5\text{-}C_5H_5)_2$, which is prepared from ruthenium(III) chloride and sodium cyclopentadienide and has prismatic structure (Fig. 14.1.b) with eclipsed C_5H_5 rings, is even more stable thermally than ferrocene, but its aromatic-substitution reactions are more difficult. Lithiation is possible with BuLi·TMED, and the product can be further converted to iodoruthenocene:

$$(\eta^5\text{-}C_5H_5)_2Ru + BuLi.TMED \longrightarrow (\eta^5\text{-}C_5H_5)Ru(\eta^5\text{-}C_5H_4Li) \xrightarrow{I_2} (\eta^5\text{-}C_5H_5)Ru(\eta^5\text{-}C_5H_4I)$$

A dimer analogous to that of iron, $[(\eta^5\text{-}C_5H_5)Ru(CO)_2]_2$, has been prepared from the dimeric carbonyl chloride, $[Ru(CO)_2Cl_2]_2$, and sodium cyclopentadienide.

Osmocene, $Os(\eta^5\text{-}C_5H_5)_2$, prepared form osmium tetrachloride and sodium cyclopentadienide, can be acetylated in a Friedel-Crafts reaction to form a monosubstituted derivative. Its prismatic, eclipsed structure (Fig. 14.1.b) in the solid is similar to that of the ruthenium analogue. The metal-metal bonded dimer, $[(\eta^5\text{-}C_5H_5)Os(CO)_2]_2$, prepared form $Os(CO)_3Cl_2$ and sodium cyclopentadienide, contains no carbonyl bridges (Fig. 14.2.b). Thus there is a remarkable difference between the iron and osmium compounds.

Cobalt, Rhodium, Iridium

Cobaltocene, $Co(\eta^5\text{-}C_5H_5)_2$, is prepared form cobalt(II) chloride and sodium cyclopentadienide. With 19 electrons in the valence shell, for example, one electron more than the noble-gas configuration, this compound is readily oxidized to the 18-electron cation, $[Co(\eta^5\text{-}C_5H_5)_2]^+$. This tendency to achieve the 18-electron configuration is also manifest in the addition of carbon tetrachloride, reduction with lithium alanate, and addition of alkyl halides, in which one of the cyclopentadienyl rings in the products becomes attached to the metal as a four-electron donor:

$$(\eta^5\text{-}C_5H_5)Co(\eta^4\text{-}C_5H_5HR) \xleftarrow{RX} Co(\eta^5\text{-}C_5H_5)_2 \xrightarrow{CCl_4} (\eta^5\text{-}C_5H_5)Co(\eta^4\text{-}C_5H_5CCl_3) + [Co(\eta^5\text{-}C_5H_5)_2]^{\oplus}Cl^{\ominus}$$

$$Co(\eta^5\text{-}C_5H_5)_2 \xrightarrow{+LiAlH_4} Co(\pi\text{-}C_5H_5)_2^{\oplus}$$

Cyclopentadiene reacts with dicobalt octacarbonyl to form $(\eta^5\text{-}C_5H_5)Co(CO)_2$, which on refluxing in hexane yields the trimer, $[(\eta^5\text{-}C_5H_5)Co(CO)]_3$ (Fig. 14.2.e). The photochemical reaction of $\eta^5\text{-}C_5H_5Co(CO)_2$ leads to unstable $\eta^5\text{-}C_5H_5Co(CO)$, which either dimerizes or reacts with the parent compound, followed by further transformation:

$$C_5H_5Co(CO)_2 \xrightarrow[-CO]{h\nu} C_5H_5Co(CO) \xrightarrow{+C_5H_5Co(CO)_2} (C_5H_5Co)_2(\mu\text{-CO})(CO)_2 \longrightarrow (C_5H_5Co)_2(\mu\text{-CO})_2$$

$$C_5H_5Co(CO) \xrightarrow{\text{dimerization}} (C_5H_5Co)_2(\mu\text{-CO})_2 \longrightarrow (C_5H_5)_3Co_3(CO)_3$$

Reduction of the analogous complex containing η^5-pentamethylcyclopentadienyl groups forms a dinuclear radical anion, $[(\eta^5\text{-}C_5Me_5)_2Co_2(CO)_2]^-$ which can be oxidized to a neutral species:

$$(\eta^5\text{-}C_5Me_5)Co(CO)_2 \xrightarrow{Na/THF} [(\eta^5\text{-}C_5Me_5)_2Co_2(CO)_2]^- \xrightarrow{FeCl_3} (\eta^5\text{-}C_5Me_5)_2Co_2(CO)_2\,.$$

The neutral compound contains a Co-Co double bond (233 pm) and the radical-anion a bond of 1.5 order (237 pm $\equiv$ 2.37 Å).

Ring expansion with formation of a coordinated boron heterocycle, occurs on treating cobaltocene with phenylboron dichloride:

$$(C_5H_5)_2Co \xrightarrow{PhBX_2} (\eta^5\text{-}C_5H_5)Co(\eta^4\text{-}C_5H_5\text{-}BX\text{-}Ph) \longrightarrow [(\eta^5\text{-}C_5H_5)Co(\eta^6\text{-}C_5H_5B\text{-}Ph)]^{\oplus}X^{\ominus}$$

Rhodium and iridium behave otherwise. Thus, rhodium(III) and iridium(III) chlorides form compounds with sodium cyclopentadienide in which only one of the two rings is a five-electron donor:

$$MCl_3 + NaC_5H_5 \longrightarrow (\eta^5\text{-}C_5H_5)M(\eta^4\text{-}C_5H_6)$$

These compounds have a noble-gas configuration.

The dimeric rhodium-carbonyl chloride forms with sodium cyclopentadienide $(\eta^5\text{-}C_5H_5)Rh(CO)_2$, which on storage undergoes self-condensation with elimination of carbon monoxide to form $(\eta^5\text{-}C_5H_5)_2Rh_2(CO)_3$. Ultraviolet irradiation of this binuclear compound yields the trimer, $[(\eta^5\text{-}C_5H_5)Rh(CO)]_3$. The binuclear compound contains a carbonyl bridge and a metal-metal bond. The trimer can exist in two isomeric forms, one of which is shown in Fig. 14.2.e, while the second is less symmetrical:

Both are sterochemically non-rigid in solution owing to carbonyl-group scrambling, but are not interconvertible.

Reduction of $\eta^5\text{-}C_5H_5Rh(CO)_2$ with sodium amalgam yields the anion $[(\eta^5\text{-}C_5H_5)_2Rh_3(CO)_4]^-$, which contains an unusual unsymmetrical trimetallic carbonyl bridge:

Iridium also forms a cyclopentadienyl dicarbonyl derivative, $\eta^5\text{-}C_5H_5Ir(CO)_2$, from $Ir(CO)_3Cl$ and sodium cyclopentadienide.

Nickel, Palladium, Platinum

Nickelocene, $Ni(\eta^5\text{-}C_5H_5)_2$, made from anhydrous nickel bromide, cyclopentadiene and diethylamine, or from the complex $Ni(NH_3)_4Cl_2$ and sodium cyclopentadienide, has 20 electrons which makes it sensitive to oxidation to the cation, $Ni(\eta^5\text{-}C_5H_5)_2^+$, but removal of a second electron leads to decomposition, rather than to formation of an 18-electron dication.

A triple-decker sandwich cation, $[Ni_2(C_5H_5)_3]^+$, contains three parallel-C_5H_5 rings with the nickel atoms sandwiched between them:

Cyclopentadiene and nickel tetracarbonyl produce a dicyclopentadienyl derivative in which the second ring is bonded through an η^3-allylic fragment, since the η^5-C_5H_5Ni group requires only three electrons to achieve a noble-gas configuration. Nickelocene itself can be reduced with sodium amalgam to form the same compound:

$$2\,C_5H_6 + Ni(CO)_4 \longrightarrow (\eta^5\text{-}C_5H_5)Ni(\eta^3\text{-}C_5H_7) \xleftarrow{Na/Hg} (\eta^5\text{-}C_5H_5)_2Ni$$

Nickelocene reacts with nickel carbonyl to form the dimer, $[(\eta^5\text{-}C_5H_5)Ni(CO)]_2$, which is converted by sodium amalgam to a trinuclear compound, $(\eta^5\text{-}C_5H_5)_3Ni_3(CO)_2$, whose structure is shown in Fig. 14.2.f. A tetranuclear nickel hydride, $(\eta^5\text{-}C_5H_5)_4Ni_4H_3$, is also known.

The dimer, $[(\eta^5\text{-}C_5H_5)Ni(CO)]_2$, reacts with cobalt, iron and manganese dinuclear carbonyls to form $(\eta^5\text{-}C_5H_5)NiCo_3(CO)_9$, $(\eta^5\text{-}C_5H_5)_2Ni_2Fe(CO)_5$ and $[(\eta^5\text{-}C_5H_5)_2Ni_2Mn(CO)_5]^-$, respectively.

Because the η^5-C_5H_5M fragment (M = Ni, Pd, Pt) requires only three electrons to complete a noble-gas configuration, and the nitrosyl group is a three-electron donor, complexes of the type $(\eta^5\text{-}C_5H_5)M(NO)$ form. The nickel compound, for example, can be obtained from nickelocene and nitrogen monoxide. Palladium(II) chloride reacts with cyclopentadiene in water to give $[\eta^5\text{-}C_5H_5PdCl]_2$.

Platinum forms a cyclopentadienyl derivative, $(\eta^5\text{-}C_5H_5)_2Pt_2(CO)_2$, obtained from $Pt(CO)_2Cl_2$ and sodium cyclopentadiene; $Pt_2(C_5H_5)_4$, prepared from platinum(II) chloride and $Na^+C_5H_5^-$, has only one cyclopentadienyl ring attached to each atom as a five-electron donor:

$$(\eta^5\text{-}C_5H_5)Pt\text{—}Pt(\eta^5\text{-}C_5H_5)\ \ (\mu\text{-}C_5H_5\text{-}C_5H_5)$$

Copper, Silver, Gold

The reaction of copper(I) oxide, triphenylphosphine and cyclopentadiene, or treatment of $R_3P \cdot CuCl$ with TlC_5H_5 yields $(\eta^5\text{-}C_5H_5)Cu \cdot PR_3$; a related carbon monoxide derivative, η^5-C_5H_5CuCO, is formed when CuCl reacts with TlC_5H_5 in the presence of carbon monoxide.

The related silver compound, η^5-$C_5H_5Ag \cdot PPh_3$, obtained from $AgCF_3SO_3$ with $Na^+C_5H_5^-$ and triphenylphosphine, decomposes slowly at room temperature:

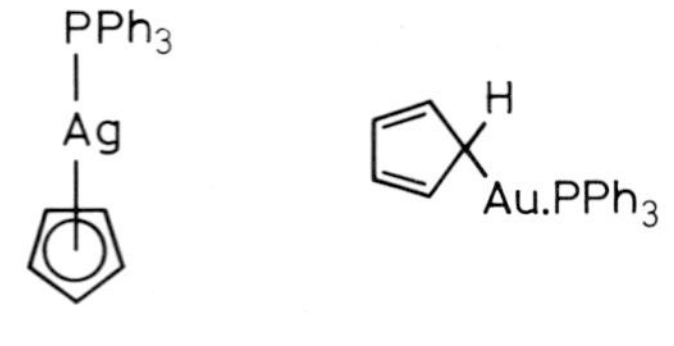

The gold compound, $C_5H_5Au \cdot PPh_3$, unlike the copper and silver analogues, contains σ-bonded η^1-cyclopentadienyl group.

14.2. Acyclic Pentadienyl η^5-Complexes

Iron butadiene complexes are converted to pentadienyl complexes containing an open-chain, five-electron donor either by protonation of a hydroxymethylbutadiene, or by hydride abstraction of a methylbutadiene, complex:

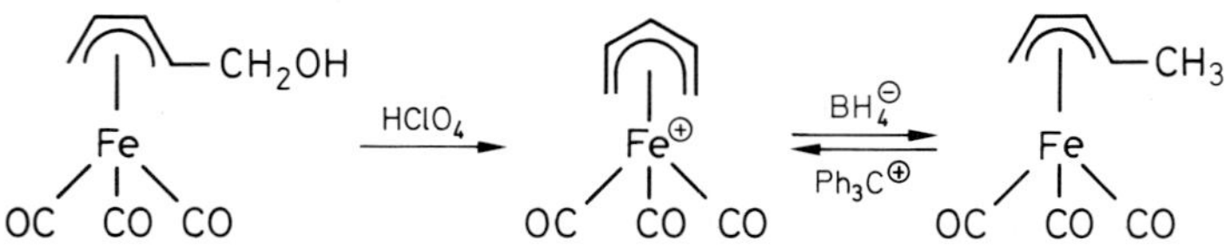

Friedel-Crafts acylation of the iron tricarbonyl methylbutadiene also produces a pentadienyl derivative:

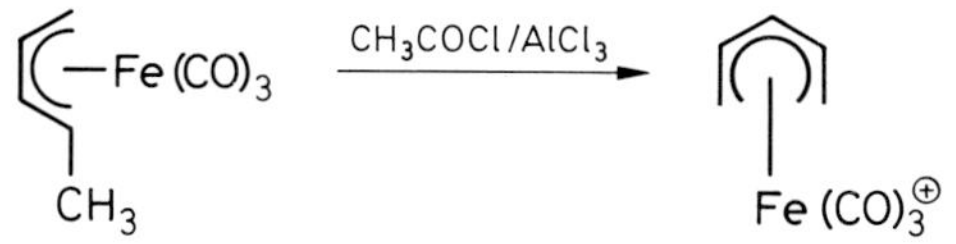

The manganese complex, η^5-$C_5H_7Mn(CO)_3$, is also known.

14.3. η^5-Complexes of Some Heterocyclic Ligands

Fixing the pentadienyl fragment in a ring by closure with a CH_2 unit (see cyclohexadienyl complexes below), or by a heteroatom, improves its coordination ability.

Silacyclohexadienyl complexes are prepared by hydride abstraction from a cyclohexadiene complex:

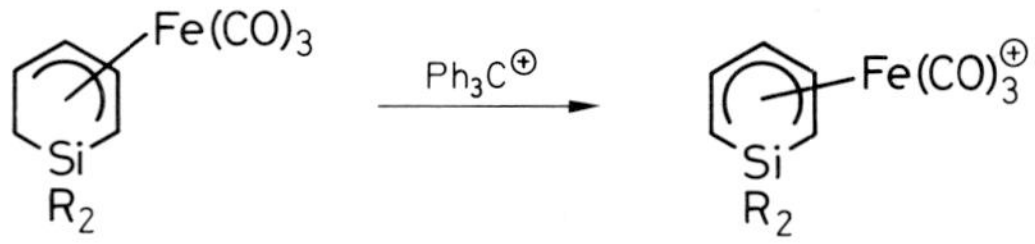

Borabenzene ligands form several transition metal complexes:

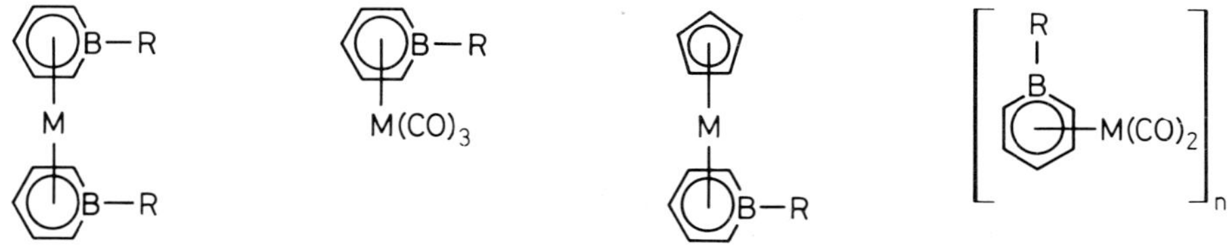

M = Fe, Os, Co, Ru   M = Mn, Re   M = Fe, Co   M = Fe, n = 2; M = Co, n = 1

A typical preparation is shown:

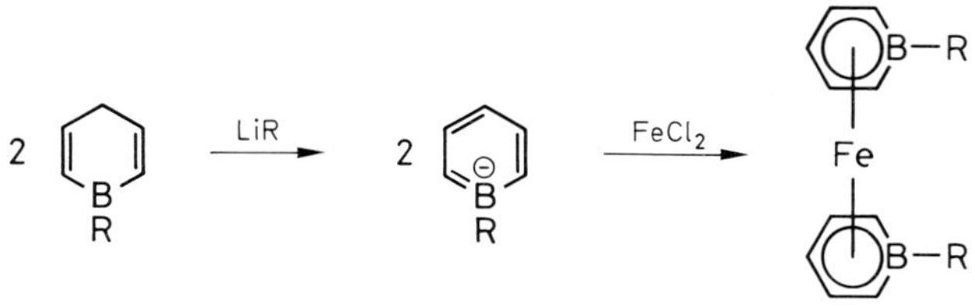

In these ligands the boron atom contributes no electrons, but its vacant p_z-orbital permits cyclic conjugation in the ring.

The pyrrole ring can form an azaferrocene:

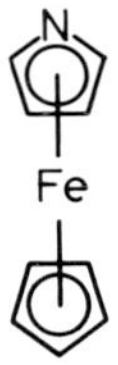

P-Phenylphospholes react with $Mn_2(CO)_{10}$ with P—Ph bond cleavage to give η^5-phosphole complexes:

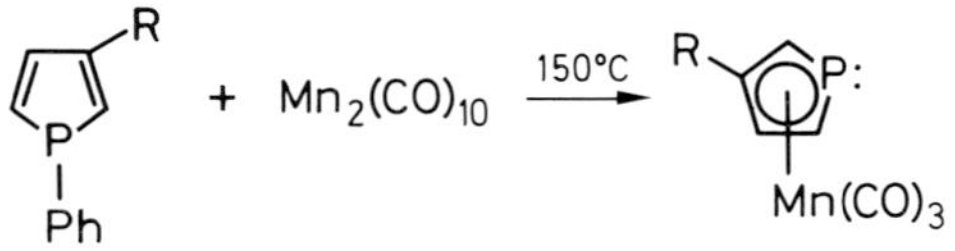

$$\text{(R-substituted P-Ph phosphole)} + Mn_2(CO)_{10} \xrightarrow{150^\circ C} \text{R-}C_4H_3P\text{:}\,Mn(CO)_3$$

and phosphaferrocenes have been obtained similarly:

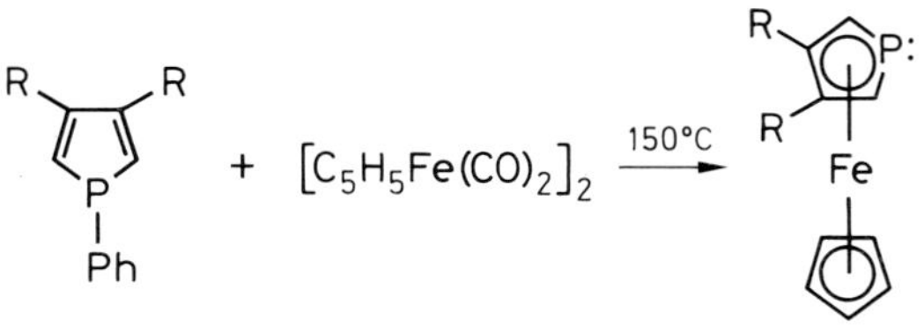

$$\text{(R,R-substituted P-Ph phosphole)} + [C_5H_5Fe(CO)_2]_2 \xrightarrow{150^\circ C} \text{R}_2C_4H_2P\text{:}\,Fe\,C_5H_5$$

Diphosphaferrocenes have been obtained with the aid of lithiophospholes:

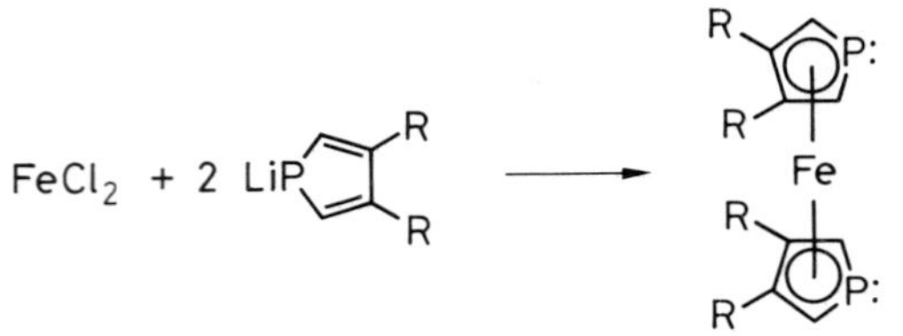

$FeCl_2$ + 2 LiP(C_4H_2R_2) ⟶ (R_2C_4H_2P)_2Fe

Manganese and iron complexes of arsoles are also known:

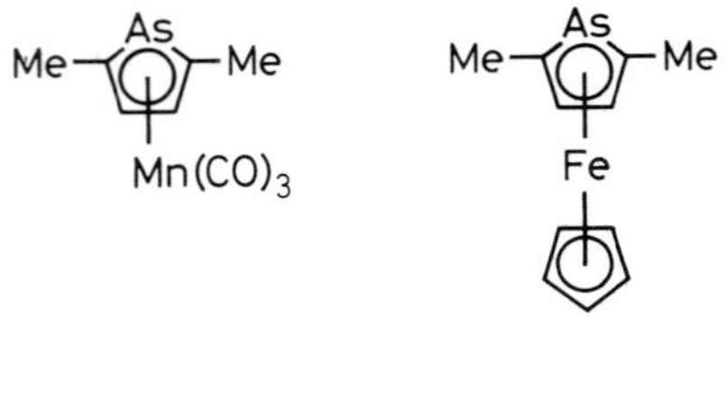

14.4. η^5-Cyclohexadienyl Complexes

The six-electron donor aromatic groups in η^6-arene complexes (described in the next section) can be reduced with hydride ion to five-electron η^5-cyclohexadienyl ligands. Thus, η^6-arene vanadium carbonyl complexes are reduced with borohydride to form η^5-cyclohexadienyl derivatives:

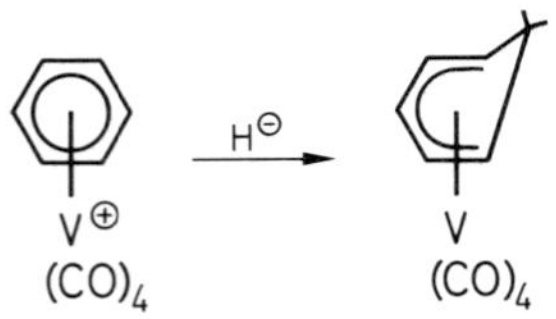

and the manganese complex, $(\eta^6\text{-}C_6H_6)Mn(CO)_3$, is converted to $(\eta^5\text{-}C_6H_7)Mn(CO)_3$. The six-membered ring loses its aromaticity on reduction and the sixth, sp^3-hybridized carbon atom lies outside the plane of the other five carbons.

The rhenium complex of hexamethylbenzene is also reduced with $LiAlH_4$ to a η^5-cyclohexadienyl complex.

The cation, $[(\eta^5\text{-}C_5H_5)Fe(\eta^6\text{-}C_6H_6)]^+$, is reduced by $LiAlH_4$ to the neutral η^5-cyclopentadienyl-η^5-cyclohexadienyl-iron:

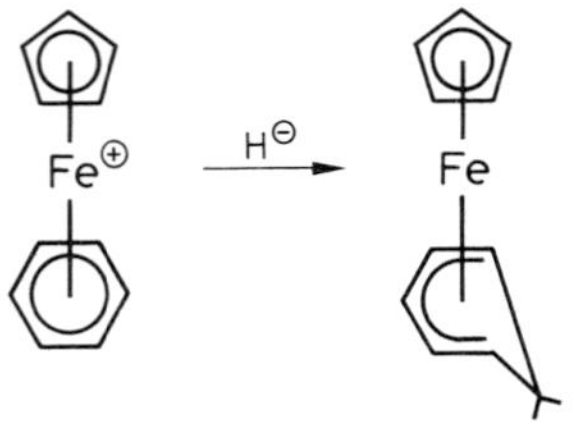

Another route to iron η^5-cyclohexadienyl complexes involves hydride abstraction of the four-electron ligand, η^4-cyclohexadiene:

$$\text{Fe(CO)}_3 \xrightarrow{Ph_3C^{\oplus}BF_4^{\ominus}} \text{Fe}^{\oplus}\text{(CO)}_3$$

The ruthenium complex, $[Ru(\eta^6\text{-}C_6H_6)_2]^+$, has been reduced with sodium to $Ru(\eta^5\text{-}C_6H_7)_2$.

14.5. η^5-Cycloheptadienyl Complexes

Even seven-membered rings can act as five-electron donors. Thus, a complex of cycloheptatriene containing the $Fe(CO)_3$ group attached to a butadiene fragment of the ring can be protonated in acid to form a η^5-cycloheptadienyl complex, $[(\eta^5\text{-}C_7H_7)Fe(CO)_3]^+$, and the cycloheptadiene complex $(\eta^4\text{-}C_7H_{10})Fe(CO)_3$ can be converted to the same compound:

$$\eta^4\text{-}C_7H_8Fe(CO)_3 \xrightarrow{H^{\oplus}} \eta^5\text{-}C_7H_9Fe(CO)_3^{\oplus} \xleftarrow{H^{\ominus}} \eta^4\text{-}C_7H_{10}Fe(CO)_3$$

The iron atom in each case has a noble-gas configuration.

The seven-membered ring in azulene can also act as a five-electron donor. Thus, the product from azulene and $Mn_2(CO)_{10}$ has two $Mn(CO)_3$ fragments coordinated as five-electron acceptors:

$$\text{azulene} + Mn_2(CO)_{10} \xrightarrow{100°C} \text{Mn(CO)}_3 \ldots \text{Mn(CO)}_3$$

The related iron complex of benzocycloheptatriene is hydride abstracted to give a compound in which the seven-membered ring becomes a five-electron donor:

$$\text{Fe(CO)}_3 \xrightarrow[-Ph_3CH]{+Ph_3C^{\oplus}BF_4^{\ominus}} \text{Fe(CO)}_3^{\oplus}\ BF_4^{\ominus}$$

The ruthenium complex, $(\eta^4\text{-}C_7H_8)RuCl_2$, reacts with cycloheptatriene and iso-PrMgBr to give the unusual η^5-cycloheptadienyl-η^5-cycloheptatrienylruthenium derivative which contains two penta*hapto*-bonded seven-membered rings:

$$(C_7H_8)RuCl_2 + C_7H_8 \xrightarrow{\text{iso-PrMgBr}} (C_7H_9)Ru(C_7H_7)$$

14.6. Complexes of Carborane Ligands

Typical examples of these interesting five-electron donor complexes are illustrated in Figs. 14.6 to 14.11, and are now classic compounds.

The behavior of the $[C_2B_9H_{11}]^{2-}$ anion, called carbollyl, is like that of the C_5H_5 group in forming with iron(II) a η^5-complex, $[Fe(1,2\text{-}C_2B_9H_{11})_2]^-$, shown in Fig. 14.6, which is similar to ferrocene.

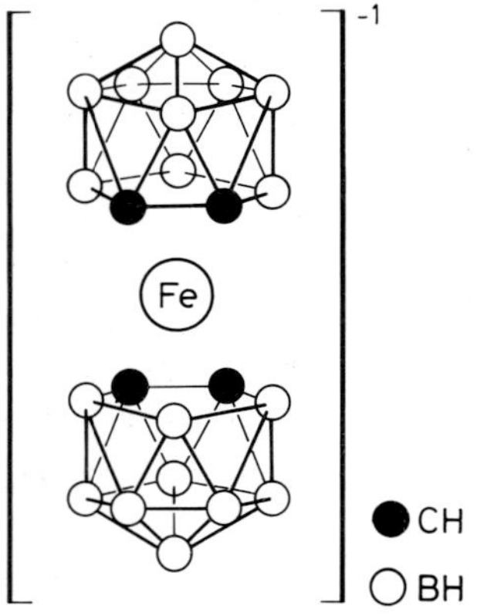

Fig. 14.6. The structure of $[Fe(1,2\text{-}C_2B_9H_{11})]^-$.

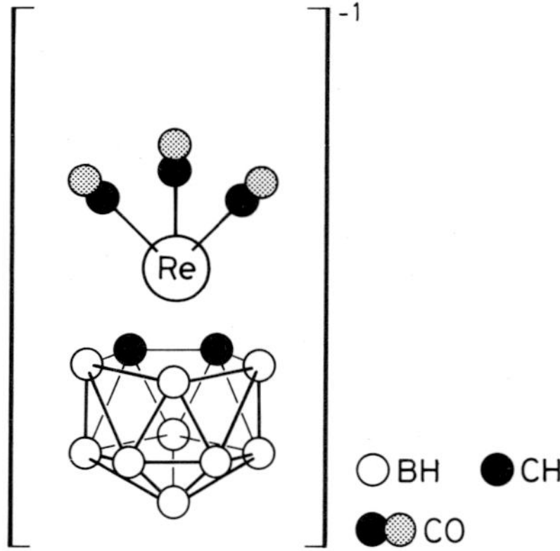

Fig. 14.7. The structure of $[(1,2\text{-}C_2B_9H_{11})Re(CO)_3]^-$.

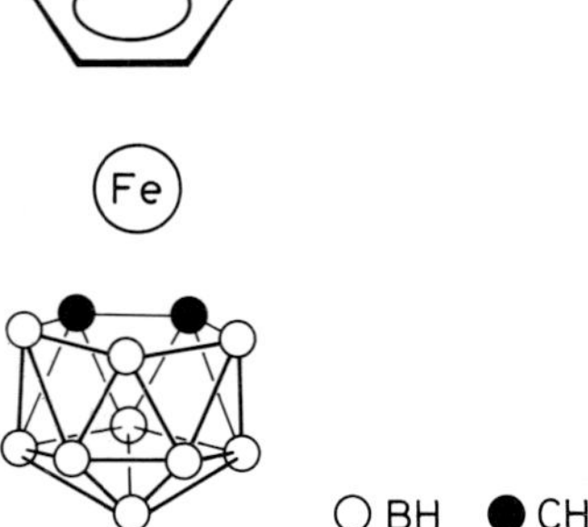

Fig. 14.8. The structure of $(C_5H_5)Fe(1,2\text{-}C_2B_9H_{11})$.

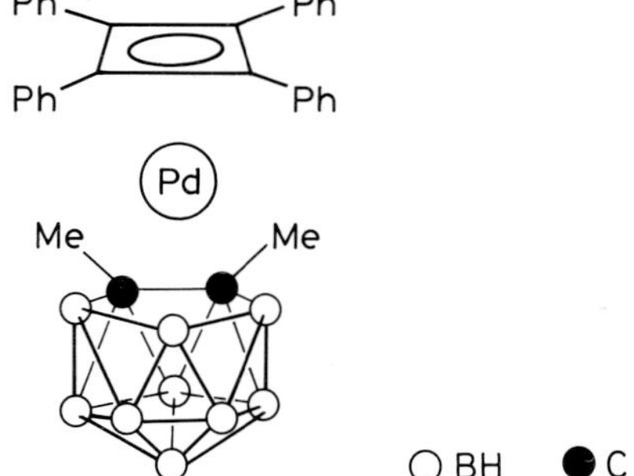

Fig. 14.9. The structure of $(C_4Ph_4)Pd(1,2\text{-}Me_2C_2B_9H_9)$.

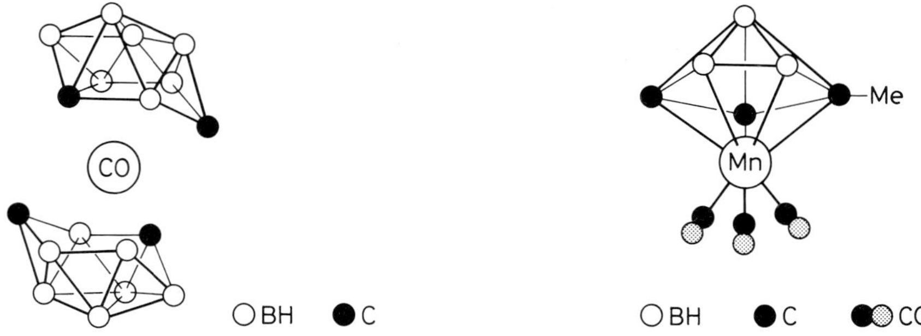

Fig. 14.10. The structure of $Co(1,6\text{-}C_2B_7H_9)_2$.

Fig. 14.11. The structure of $(MeC_3B_3H_5)Mn(CO)_3$.

A manganese complex is formed by the reaction of the anion with the metal carbonyl bromide, $Mn(CO)_5Br$.

The mixed iron complex, $(\eta^5\text{-}C_5H_5)Fe(\eta^5\text{-}1,2\text{-}C_2B_9H_{11})$, obtained by simultaneous action of the two anionic ligands upon iron(II) chloride, contains both a η^5-cyclopentadienyl ligand and a carbollyl ligand as shown in Fig. 14.8.

Another mixed compound, $(\eta^4\text{-}C_4Ph_4)Pd(\eta^5\text{-}1,2\text{-}Me_2C_2B_9H_{11})$, prepared by the reaction of the carbollyl anion with the complex $[(\eta^4\text{-}C_4Ph_4)PdCl_2)]_2$, contains a cyclobutadiene and a carbollyl ligand as shown in Fig. 14.9.

The lower carboranes are also utilized. Thus, the $[C_2B_7H_{11}]^{2-}$ anion forms with cobalt(II) $Co(1,6\text{-}C_2B_7H_9)_2$, as shown in Fig. 14.10. The methyl derivative, $MeC_3B_3H_6$, reacts with $Mn_2(CO)_{10}$ to form $(MeC_3B_3H_5)Mn(CO)_3$, shown in Fig. 14.11. This ligand also has a pentagonal face, similar to the cyclopendienyl group.

Some interesting sandwich complexes, containing $C_2B_3H_5$ (in a triple-decker structure), $C_2B_3H_7$ and $C_2B_4H_6$ ligands, have been obtained from $2,3\text{-}C_2B_4H_8$:

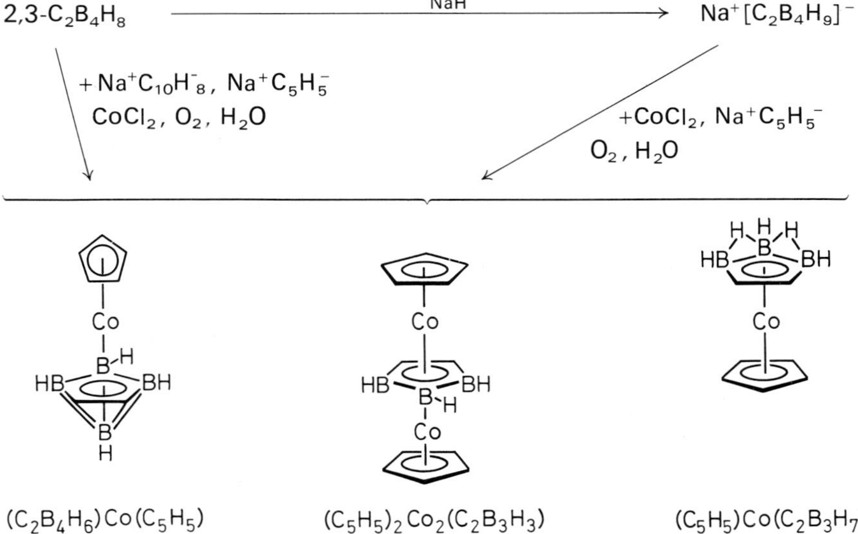

15. Compounds with Six-Electron Ligands

The most important six-electron ligand is benzene, and together with other arenes it forms the complexes illustrated in Fig. 15.1. Polyphenyls and condensed polyarenes can use one or more of their aromatic rings in metal bonding. In addition, non-aromatic cyclic trienes like cycloheptatriene and cyclooctatriene or even cyclooctatetraene can behave as six-electron donors (Fig. 15.2).

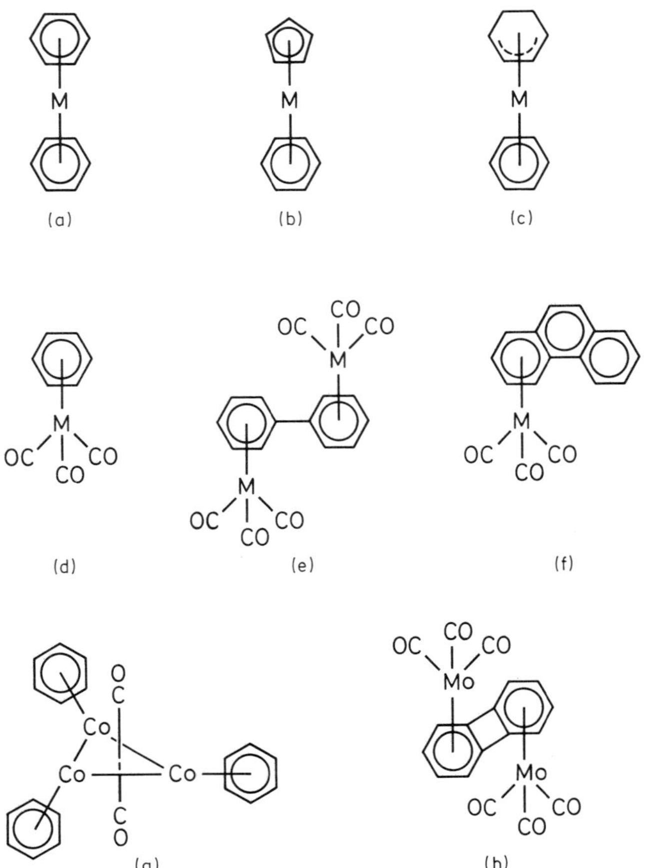

Fig. 15.1. Some types of benzene complexes and other arene-metal compounds.

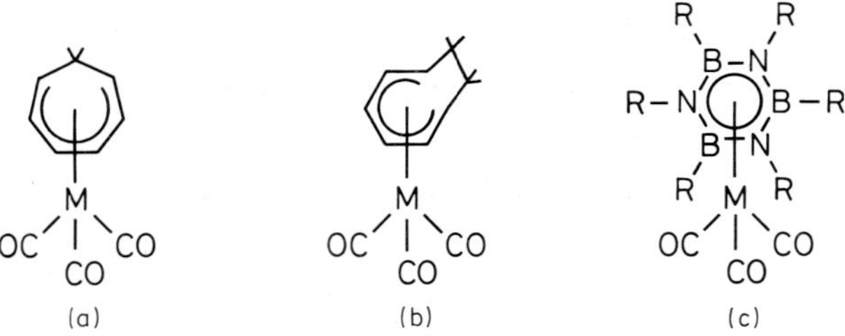

Fig. 15.2. Some six-electron ligands other than arenes.

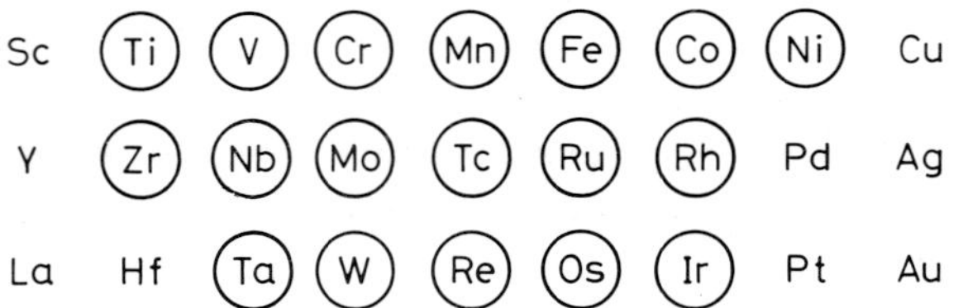

Fig. 15.3. Metals known to form complexes with six-electron ligands.

The metals for which six-electron ligand complexes are known are shown in Fig. 15.3.

The graphite intercalation compounds of some transition metals (for example, Cr, Mo, W, Mn, Co, Ni, etc.) are complexes of η^6-type in which the metal is attached to six-electron fragments of the graphite layer.

15.1. η^6-Complexes of Benzene and its Derivatives

Benzene forms bis(arene)metal, arene-metal carbonyls, arene-cyclopentadienyl-metal and other types of compounds.

Titanium, Zirconium, Hafnium

Bis(benzene)titanium is formed in the reaction of the metal vapor with benzene at 77 K, and the reaction can be extended to other arenes (toluene, mesitylene, etc.). The reaction of titanium tetrachloride with hexamethylbenzene and Et_2AlCl gives the cations $[(\eta^6\text{-}C_6Me_6)TiCl_2]^+$ and $[(\eta^6\text{-}C_6Me_6)_3Ti_3Cl_6]^+$, and another η^6-hexamethylbenzene complex was found to have the structure:

Ti
Cl Cl
Cl—Al— Cl Cl—Al— Cl
Cl Cl

Vanadium, Niobium, Tantalum

Paramagnetic bis(benzene)vanadium, $V(\eta^6\text{-}C_6H_6)_2$, obtained by the reaction of vanadium tetrachloride with benzene, in the presence of aluminum chloride and aluminum powder, followed by alkaline hydrolysis, has 17 electrons and can be readily reduced with alkali metals to the 18-electron anion, $[V(\eta^6\text{-}C_6H_6)_2]^-$. Bis(benzene)vanadium can be metallated with n-BuLi, to give $V(\eta^6\text{-}C_6H_5Li)_2$.

Vanadium hexacarbonyl reacts with benzene and its substituted derivatives to form $[(arene)V(CO)_4]^+[V(CO)_6]^-$ complexes, in which both vanadium atoms have noble-gas electronic configuration. Metathesis reactions lead to hexafluoro-salts of this cation.

Paramagnetic hexaphenylbenzene vanadium tricarbonyl, $(\eta^6\text{-}C_6Ph_6)V(CO)_3$, forms in the reaction of $V(CO)_6$ with diphenylacetylene.

Niobium vapor and benzene, toluene or mesitylene give bis(arene)niobium derivatives.

The polynuclear complexes of hexamethylbenzene, $(\eta^6\text{-}C_6Me_6)_2M_2Cl_4$ and $[(\eta^6\text{-}C_6Me_6)_3M_3Cl_6]^+Cl^-$, are formed in the reaction of MCl_5 ($M = Nb$, Ta) with hexamethylbenzene.

Chromium, Molybdenum, Tungsten

The group VI elements require 12 electrons, and can achieve a noble-gas configuration by coordinating two benzene molecules. Thus, bis(benzene)chromium, $Cr(\eta^6\text{-}C_6H_6)_2$, is obtained from a reductive Friedel-Crafts reaction with chromium chloride and benzene in the presence of aluminum chloride and aluminum powder, followed by reduction with sodium dithionite:

$$3\,CrCl_3 + 2\,Al + AlCl_3 + 6\,C_6H_6 \longrightarrow 3\,[(\eta^6\text{-}C_6H_6)_2Cr]^+AlCl_4^-$$

$$\xrightarrow[OH^-]{Na_2S_2O_4} Cr(\eta^6\text{-}C_6H_6)_2$$

Alkylbenzenes react similarly, but not benzenes which contain lone pairs of electrons in the substituent. Chromium vapor affords bis(benzene)chromium with benzene at low temperatures.

In bis(benzene)chromium the two rings are parallel (Fig. 15.1.a). Substitution reactions are difficult, but metallation is possible, and further reactions of the metallated derivatives can lead to various products.

Chromium hexacarbonyl forms compounds of the type $Cr(arene)(CO)_3$ with aromatic compounds. Some aromatic derivatives do not react with chromium hexacarbonyl.

The coordinated-benzene molecule retains its aromaticity as shown by substitution reactions known for free benzene. Reactivity is, however, modified by coordination. Thus, benzenechromium tricarbonyl is mercurated and lithiated, and the organometallic derivatives obtained can be converted to other compounds:

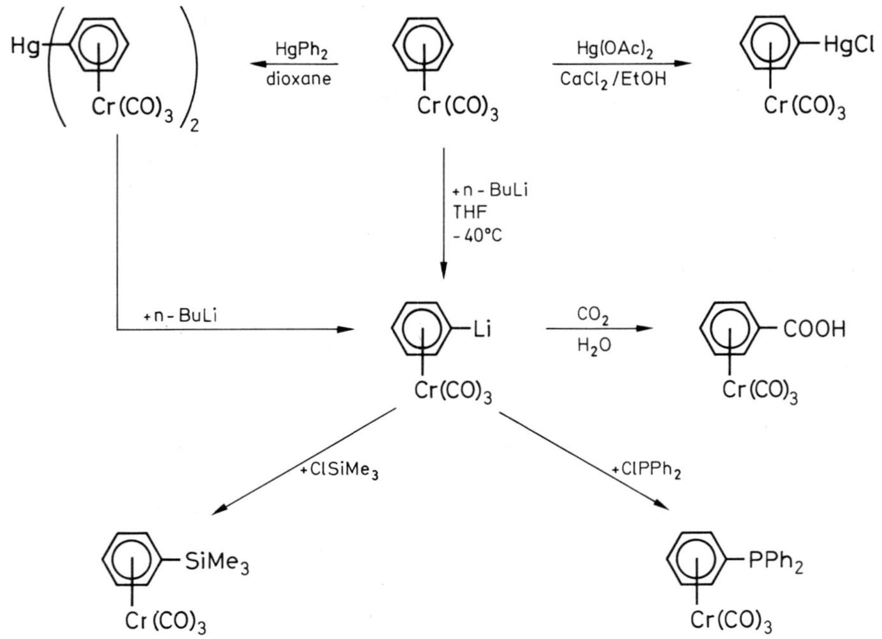

Bis(benzene) chromium can also be metalated with the n-BuLi·TMED complex:

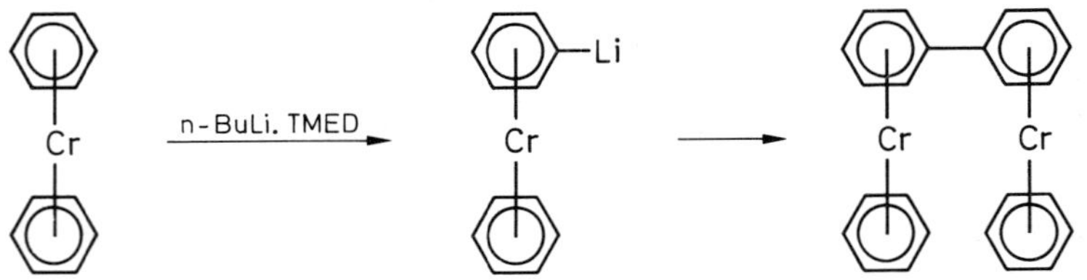

Paramagnetic benzenechromium cyclopentadienyl, $(\eta^6\text{-}C_6H_6)Cr(\eta^5\text{-}C_5H_5)$, with 17 electrons, formed in the reaction of chromium(III) chloride with sodium cyclopentadienide and phenylmagnesium bromide is shown in Fig. 15.1.b.

The less-stable bis(benzene) derivatives of molybdenum and tungsten, $M(\eta^6\text{-}C_6H_6)_2$ (Fig. 15.1.a), are obtained by reductive Friedel-Crafts reactions or from metal atom-vapors and benzene. Benzenemolybdenum cyclopentadienyl (Fig. 15.1.b) is obtained by reducing the complex $[(\eta^5\text{-}C_5H_5)(\eta^6\text{-}C_6H_6)Mo(CO)]^+$, with $LiAlH_4$.

Manganese, Technetium, Rhenium

Manganese(II) chloride reacts with sodium cyclopentadienide and phenylmagnesium bromide to give a mixed derivative (Fig. 15.1.b), along with a bimetallic compound derived from biphenyl:

$$MnCl_2 + NaC_5H_5 + PhMgBr \longrightarrow (\eta^5\text{-}C_5H_5)Mn(\eta^6\text{-}C_6H_6) + [(\eta^5\text{-}C_5H_5)Mn]_2(\eta^6,\eta^6\text{-}C_6H_5\text{-}C_6H_5)$$

Similarly, $(\eta^5\text{-}C_5H_5)Re(\eta^6\text{-}C_6H_6)$, forms from rhenium(V) chloride, C_5H_5MgBr and cyclohexadiene, under UV irradiation.

Iron, Ruthenium, Osmium

Iron, ruthenium and osmium chlorides heated with aromatic hydrocarbons in the presence of aluminum chlorides and aluminum powder, after hydrolysis, yield $[M(\eta^6\text{-}arene)_2]^{2+}$ cations, which can be precipitated as hexafluorophosphates. The hexamethylbenzene derivative can be reduced to a monopositive cation, $[Fe(\eta^6\text{-}arene)_2]^{+}$, and to the unstable, neutral $[Fe(\eta^6\text{-}arene)_2]$. Similarly, the complex salt, $[Fe(\eta^6\text{-}C_6H_6)_2]^{2+}(PF_6)_2$ is reduced with $NaBH_4$ to a cyclohexadienyl complex, and with sodium dithionite to a monopositive cation:

$$(\eta^6\text{-}C_6H_6)Fe^{\oplus}(\eta^5\text{-}C_6H_7) \xleftarrow{NaBH_4} [(\eta^6\text{-}C_6H_6)_2Fe]^{2\oplus} \xrightarrow{Na_2S_2O_4} [(\eta^6\text{-}C_6H_6)_2Fe]^{\oplus}$$

The cation, $[(\eta^5\text{-}C_5H_5)Fe(\eta^6\text{-}C_6H_6)]^+$, (Fig. 15.1.b) ist obtained from $(\eta^5\text{-}C_5H_5Fe(CO)_2Cl$ and benzene with aluminum chloride, or from ferrocene and benzene in the presence of aluminum chloride and aluminum metal powder:

$$(\eta^5\text{-}C_5H_5)_2Fe + C_6H_6 \xrightarrow{Al + AlCl_3} [(\eta^6\text{-}C_6H_6)Fe(\eta^5\text{-}C_5H_5)]^{\oplus}Cl^{\ominus}$$

Ultraviolet irradiation of iron(III) chloride with cyclohexadiene-1,3 and isopropylmagnesium bromide forms a complex containing both benzene and cyclohexadiene coordinated to iron (Fig. 15.1.c) with a noble gas configuration.

Cyclohexadiene-1,3 undergoes dehydrogenation with ruthenium(III) chloride to give a η^6-benzene complex, $[(\eta^6\text{-}C_6H_6)RuCl]_x$. Benzene complexes are also formed by irradiation of cyclohexadiene with iron or osmium halides.

Bis(hexamethylbenzene)ruthenium has an 18-electron structure, achieved by disturbing the aromatic conjugation in the second ring and bonding the ruthenium atom only to a butadiene fragment; as a result this six-membered ring is not planar:

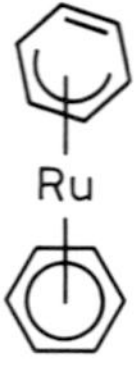

Like in ferrocene, one of the cyclopentadienyl rings of ruthenocene can be replaced by an aromatic ring (mesitylene, hexamethylbenzene, etc.) by heating with an aluminum chloride-aluminum metal powder mixture:

Ru + mesitylene $\xrightarrow[Al + AlCl_3]{165\,°C}$ $Ru^{\oplus}$

Benzeneruthenium-chloro complexes, for example, dimeric $[(\eta^6\text{-}C_6H_6)RuCl_2]_2$, and monomeric $[(\eta^6\text{-}C_6H_6)RuCl_3]^-$ and $[(\eta^6\text{-}C_6H_6)_2Ru(\mu\text{-}Cl)_3]^+$ are also known.

Cobalt, Rhodium, Iridium

A cation of the type $[(\eta^5\text{-}C_5H_5)Co(\eta^6\text{-}C_6H_6)]^{2+}$, shown in Fig. 15.1.b, can be obtained by hydride abstraction of the η^5-cyclopentadienyl-cobalt cyclohexadiene complex with trityl tetrafluoroborate:

$Co + 2Ph_3C^{\oplus}BF_4^{\ominus} \longrightarrow Co^{2\oplus}$

15.2. η^6-Cycloheptatriene Complexes

Cycloheptatriene-1,3,5 can act as a six-electron ligand, and form complexes with Group VI elements.

Chromium hexacarbonyl or $py_3Cr(CO)_3$ forms $(\eta^6\text{-}C_7H_8)Cr(CO)_3$ with BF_3:

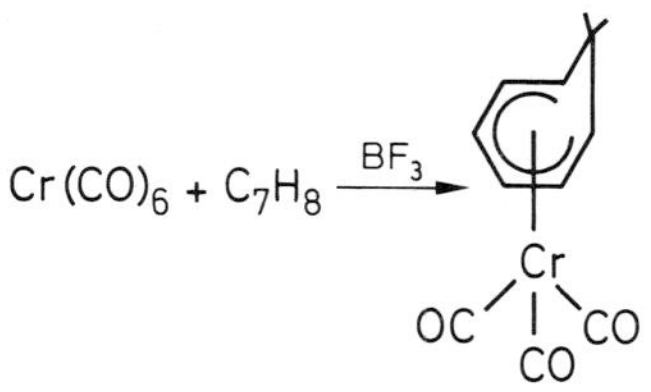

$Cr(CO)_6 + C_7H_8 \xrightarrow{BF_3}$ Cr OC CO CO

and the tropyllium-chromium tricarbonyl complex can add hydride or R^- anions to form a cycloheptatriene complex:

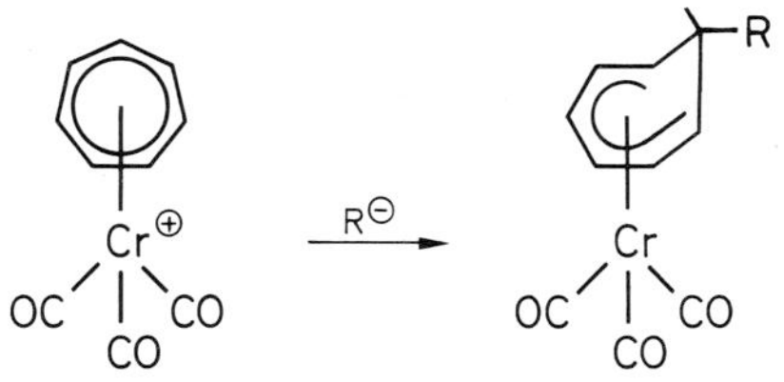

Cr⊕ OC CO CO $\xrightarrow{R^\ominus}$ R Cr OC CO CO

The reaction of metal vapors with cycloheptatriene produces η^6-complexes, as in the vanadium and chromium derivatives:

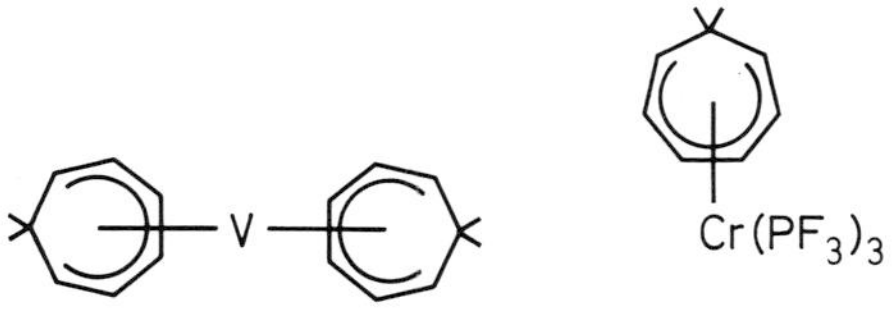

V $Cr(PF_3)_3$

Molybdenum hexacarbonyl reacts with cycloheptatriene to form a complex, $(\eta^6\text{-}C_7H_8)Mo(CO)_3$, similar to the chromium analogue.

15.3. Cyclooctatriene and Cyclooctatetraene as Six-Electron Ligands

Cyclooctatriene-1,3,5 forms with chromium, molybdenum and tungsten hexacarbonyls $(\eta^6\text{-}C_8H_{10})M(CO)_4$ in which the metal is bonded to a hexatriene fragment which is planar, while the two sp^3-hybridized carbon atoms of the ring are out of the plane:

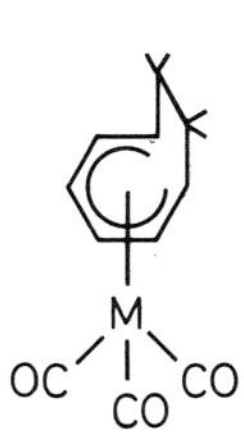

M OC CO CO

Cyclooctatetraene reacts with chromium, molybdenum and tungsten hexacarbonyls to form η^6-complexes in which only three of the four double bonds of the ring are coordinated to the metal as a hexatriene six-electron donor fragment. The fourth double bond is not involved in conjugation and is located out of the plane:

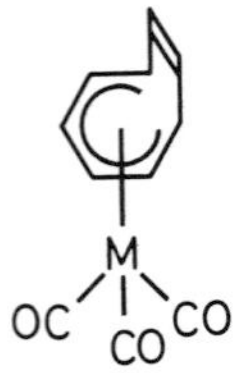

M
OC CO CO

The chromium compound, $(\eta^6\text{-}C_8H_8)Cr(CO)_3$, exhibits fluxional behavior.

15.4. Some η^6-Complexes with Heterocyclic Ligands

Heterocyclic systems containing nitrogen, phosphorus(III), sulfur or other heteroatoms tend to coordinate to transition metals as σ-donors by donation of the electron pair of the heteroatom. However, it is possible to obtain π-complexes with such heterocycles as ligands. Thus, pyridine can form both σ-donor and hexa*hapto* π-complexes with molybdenum:

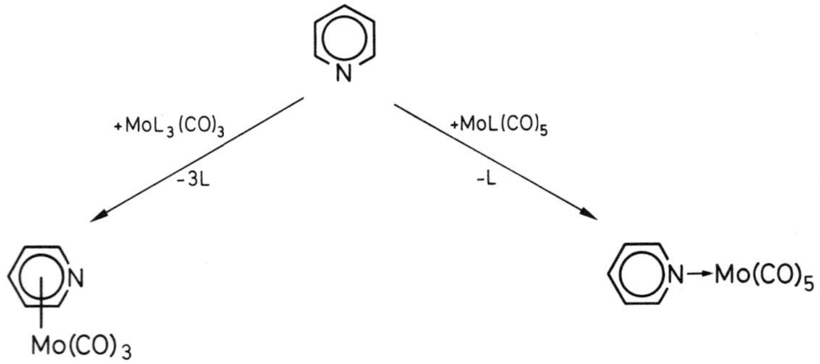

N

$+MoL_3(CO)_3$ / $-3L$ → $Mo(CO)_3$ (N)

$+MoL(CO)_5$ / $-L$ → $N \rightarrow Mo(CO)_5$

Similar η^6-complexes have been obtained with phosphabenzene, arsabenzene and stibabenzene.

Bis(2,6-dimethylpyridine) chromium has been obtained by co-condensation of chromium vapor with 2,6-dimethylpyridine at 77 K:

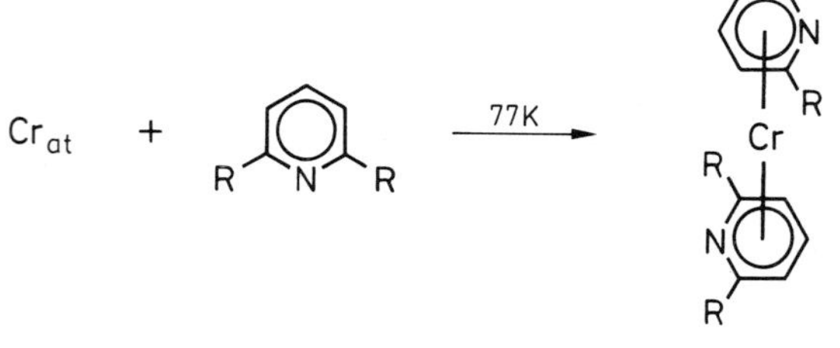

Cr_{at} + (R, N, R) $\xrightarrow{77K}$ Cr (R, N, R; R, N, R)

Tricarbonyl-chromium, -molybdenum and -tungsten complexes of λ^5-phosphorines have been prepared; in these compounds the phosphorus atom has no lone pair, and, therefore, does not compete for metal coordination:

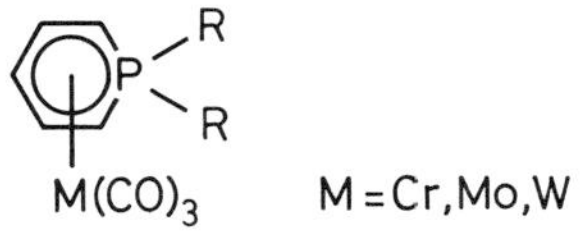

$M(CO)_3$ M = Cr, Mo, W

Thiophene behaves like benzene in its reaction with ferrocene and an aluminum metal/aluminum bromide mixture, to give a six-electron, η^5-thiophene complex:

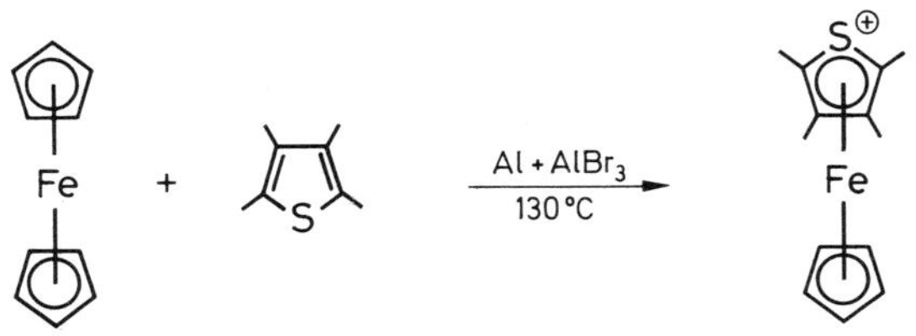

Al + $AlBr_3$, 130 °C

The six-electron heterocycle, borazine, which is isoelectronic with benzene, can act as a similar ligand with Group VI elements. Thus, hexamethylborazine, $(MeBNMe)_3$, reacts with chromium hexacarbonyl to form a complex in which the heterocycle behaves as a six-electron ligand:

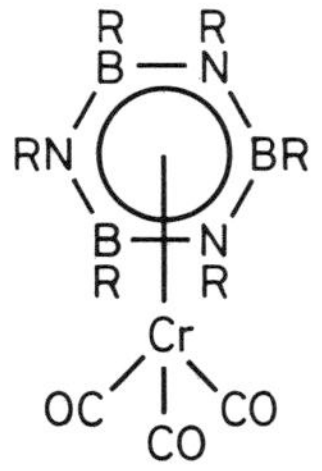

Similar complexes with other substituted borazines and $Cr(CO)_3$ and $Mo(CO)_3$ fragments have been obtained.

16. Compounds with Seven- and Eight-Electron Ligands

The cycloheptatrienyl, C_7H_7, and cyclooctatrienyl groups, C_8H_9, can act as seven-electron donors, and the most common eight-electron donor is cyclooctatetraene, C_8H_8.

16.1. Seven-Electron Ligands

Cycloheptatriene reacts with vanadium hexacarbonyl to form two products, both containing the cycloheptatrienyl groups coordinated to the metal as seven-electron ligands:

+ V(CO)6 —50°C→ V(CO)3 + V⊕

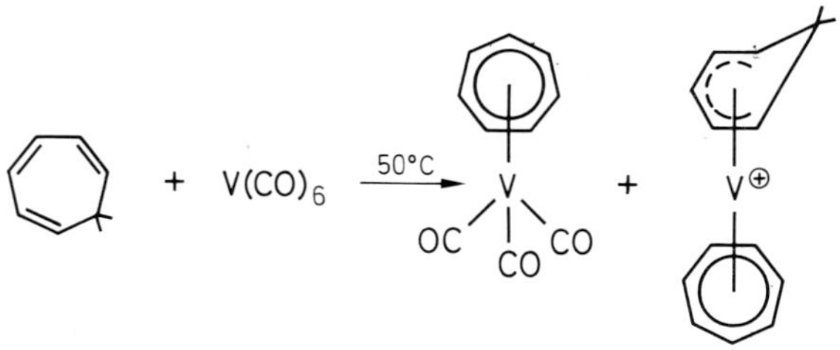

The analogous reaction with cyclopentadienyl vanadium tetracarbonyl yields the η^7-complex, $V(\eta^7\text{-}C_7H_7)(\eta^5\text{-}C_5H_5)$:

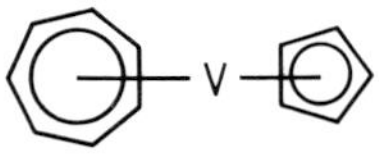

η^6-Cycloheptatriene-metal tricarbonyl complexes of chromium, molybdenum and tungsten, $(\eta^6\text{-}C_7H_8)M(CO)_3$, are readily hydride abstracted to form η^7-cycloheptatrienyl complexes:

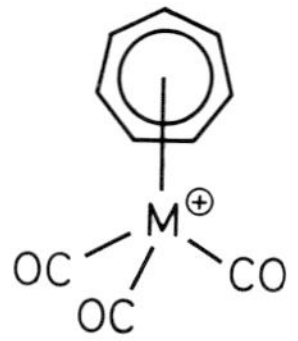

M⊕ OC OC CO

The mixed η^5-cyclopentadienyl- η^7-cycloheptatrienyl complex of chromium:

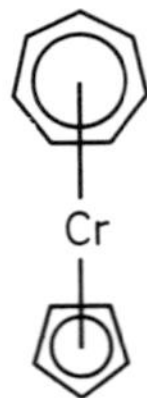

has been prepared by the reaction of anhydrous chromium(III) chloride with cyclopentadiene, cycloheptatriene and iso-PrMgBr, treatment of $(\eta^5\text{-}C_5H_5)CrCl_2$ with cycloheptatriene and iso-PrMgBr or by treatment of $(\eta^5\text{-}C_5H_5)Cr(\eta^6\text{-}C_6H_6)$ with a cycloheptatrienyl salt, followed by reduction with an alkali metal dithionite.

The protonation of the η^6-cyclooctatetraene-molybdenum tricarbonyl complex yields a η^7-derivative:

$Mo(CO)_3$ $\xrightarrow{+H_2SO_4}$ $Mo(CO)_3^{\oplus}$

Related η^7-cycloheptatrienyl-molybdenum complexes with a η^6-ligand as second substituent, must also be in cationic form in order to preserve the 18-electron configuration:

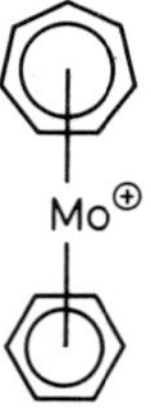

Cycloheptatriene reacts with several vaporized metals (Ti, V, Fe, Co) to yield η^7-cycloheptatrienyl complexes with titanium, vanadium and chromium while iron gives $Fe(\eta^5\text{-}C_7H_7)(\eta^5\text{-}C_7H_9)$:

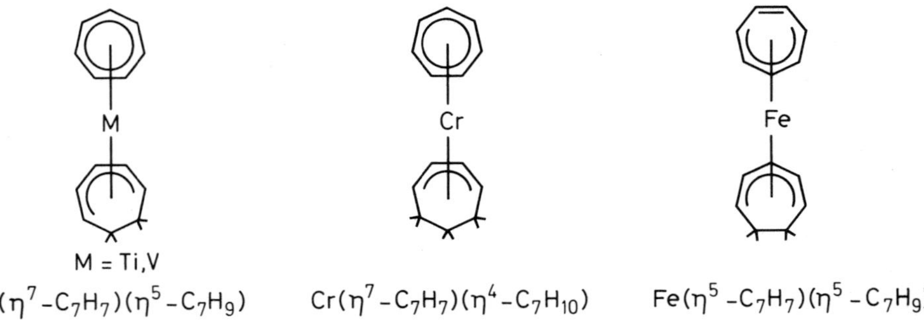

16.2. Eight-Electron Ligands

Cyclooctatetraene forms η^8-complexes only with the early transition metals which require a large number of electrons to achieve a noble-gas electronic configuration. No η^8-cyclooctatetraene complex of a transition metal beyond Group VI is known. The lanthanides and actinides form η^8-cyclooctatetraene complexes, since the η^8-C_8H_8 ligand has molecular orbitals able to interact with the f-orbitals of the metals.

Scandium derivatives of cyclooctatetraene can be prepared according to the sequence:

$$ScCl_3 \cdot 3THF \xrightarrow{K_2C_8H_8} \eta^8\text{-}C_8H_8ScCl \cdot THF \begin{cases} \xrightarrow{K_2C_8H_8} K^+[Sc(\eta^8\text{-}C_8H_8)_2]^- \\ \xrightarrow[NaC_5H_5]{} (\eta^8\text{-}C_8H_8)Sc(\eta^5\text{-}C_5H_5) \end{cases}$$

Cyclooctatetraene, tetrabutyltitanate and triethylaluminum react to yield $Ti_2(C_8H_8)_3$, which contains two terminal cyclooctatetraenes as eight-electron donors with the third molecule forming a bridge:

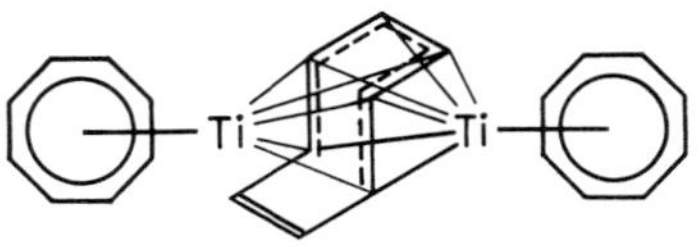

Ti Ti

The product is reduced with potassium to a triple-decker sandwich anion with three planar C_8H_8 rings:

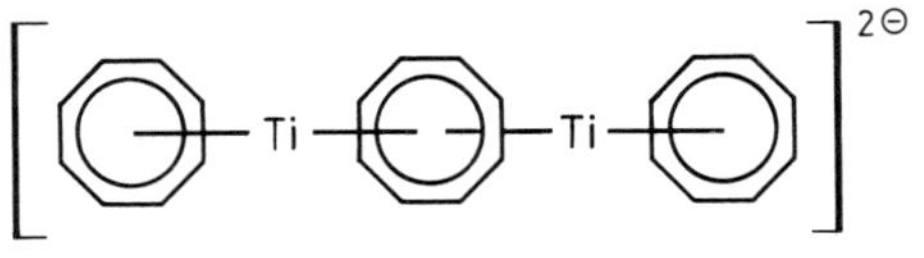

$$[(C_8H_8)Ti(C_8H_8)Ti(C_8H_8)]^{2\ominus}$$

One η^8-C_8H_8 ring in $Ti(C_8H_8)_2$ is planar, while the second adopts a boat conformation and is coordinated only by half of the molecule. Rapid interconversion of the two eight-membered rings occurs in solution:

Ti ⇌ Ti

In the derivative $(\eta^5\text{-}C_5H_5)Ti(\eta^8\text{-}C_8H_8)$ the cyclooctatetraene ring is bonded in octa*hapto*-fashion and both rings are coplanar:

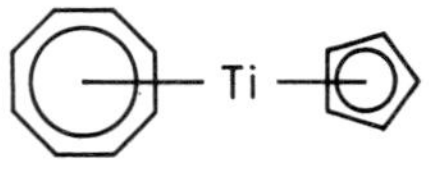

Ti

The η^8-cyclooctatetraene titanium halides, $[C_8H_8TiCl]_4$ and $[C_8H_8TiCl \cdot THF]_2$, are also known; the dimer reacts with sodium cyclopentadienide to form $(\eta^8\text{-}C_8H_8)Ti(\eta^5\text{-}C_5H_5)$ and with allylmagnesium halides, RC_3H_4MgX, to give $(\eta^8\text{-}C_8H_8)Ti(\eta^3\text{-}C_3H_4R)$.

Niobium(V) chloride forms $C_8H_8NbCl_2 \cdot THF$ with $K_2C_8H_8$; with NaC_5H_5 this is converted to $(\eta^5\text{-}C_5H_5)_2Nb(\eta^2\text{-}C_8H_8)$, in which cyclooctatraene is bonded as a simple olefin.

Cyclopentadienylchromium cyclooctatetraene exists in two equilibrium forms, and is readily converted to a cation:

$$C_5H_5CrCl_2 + C_8H_8 \xrightarrow[THF]{i\text{-}PrMgBr} Cr \rightleftharpoons Cr \rightleftharpoons Cr^{\oplus}$$

The pyrophoric bis(cyclooctatetraene) complex, $U(\eta^8\text{-}C_8H_8)_2$, is obtained from uranium(IV) chloride and $K_2C_8H_8$. Both rings are coplanar and act as 8-electron donors; f-orbitals participate in the bonding (Fig. 16.1).

Fig. 16.1. Interaction of f-orbitals with a molecular orbital of cyclooctatetraene.

The tetraphenylcyclooctatetraene complex, $U(\eta^8\text{-}C_8H_4Ph_4\text{-}1,3,5,7)_2$, prepared from UCl_4 and $K_2C_8H_4Ph_4$ is air-stable and sublimes at 400 °C (!) at $1.3 \cdot 10^{-5}$ mbar; it has a nearly eclipsed configuration with the phenyl groups tilted away from the C_8-plane.

The analogous lanthanides, thorium, protactinium, neptunium and plutonium demonstrate the ability of the f-block elements to form η^8-cyclooctatetraene complexes.

The uranium derivative ("uranocene") can be anodically oxidized to an air-stable cation $[U(C_8H_8)_2]^+$, but anions like $[Np(\eta^8\text{-}C_8H_8)_2]^-$, $[Pu(\eta^8\text{-}C_8H_8)_2]^-$, $[Am(\eta^8\text{-}C_8H_8)_2]^-$, $[Ln(\eta^8\text{-}C_8H_8)_2]^-$ (Ln = La, Ce, Nd, Er) have also been prepared.

17. Organometallic Compounds Derived from Acetylenes

Acetylenes react with the metal-carbonyl derivatives to form three possible products: 1) the acetylene molecule becomes coordinated to the transition metal as a di*hapto*(terminal)-ligand (Fig. 17.1.a–c); 2) the acetylene molecule forms a bridge between two (Fig. 17.1.d, e) or more metal atoms (Fig. 17.1.f, g) with the formation of polynuclear (cluster) compounds; and 3) during the reaction with the transition metal, acetylene undergoes cyclooligomerization with the product acting as a ligand. Sometimes acetylenes undergo organic reactions catalyzed by the transition metal (trimerization, etc.) with formation of metal-free end products.

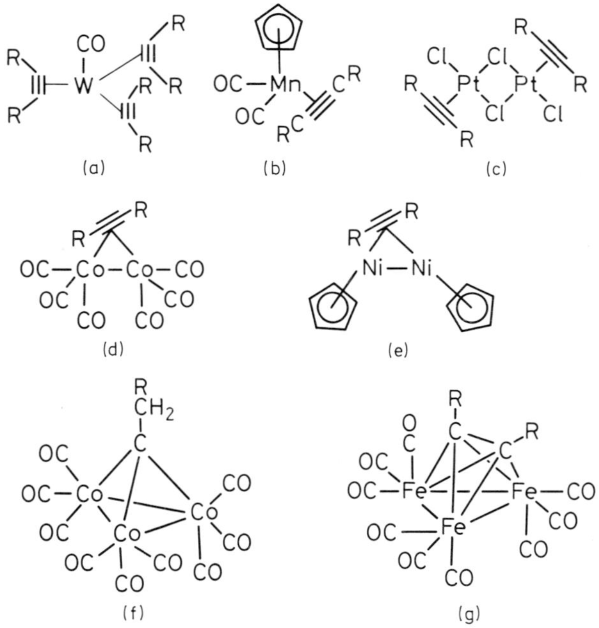

Fig. 17.1. Various complexes containing acetylene ligands.

The organic ligands formed from acetylenes include cyclobutadienes, metallocycles, cyclopentadienones and quinones, as illustrated in Fig. 17.2. Such reactions have synthetic utility.

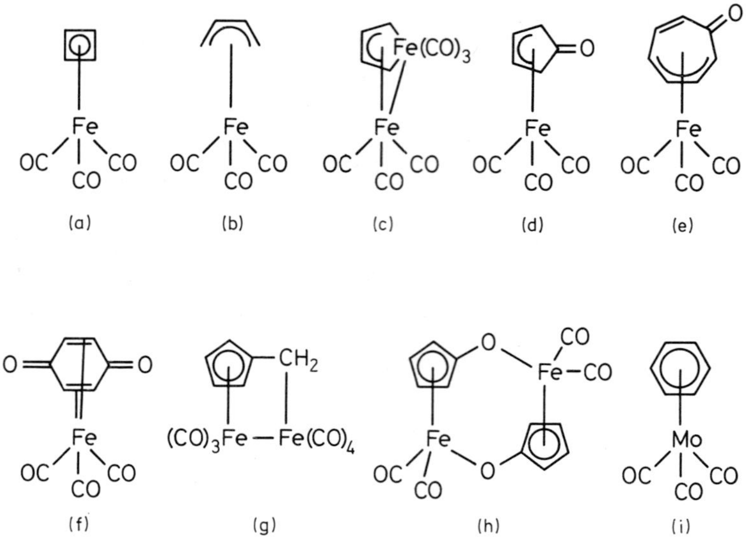

Fig. 17.2. Complexes of ligands formed by transformations of acetylenes during the reaction with the transition metal.

17.1. Complexes of Untransformed Acetylenes as Terminal (Dihapto) Ligands

Simple acetylene complexes are rare, since they undergo further reactions of the coordinated acetylene.

There are two ways in which an acetylene molecule can coordinate to a metal: (a) as a two-electron donor, participating with only one of its π-systems (Fig. 17.3.a), or (b) as a four-electron donor, participating with both its π-systems (Fig. 17.3.b).

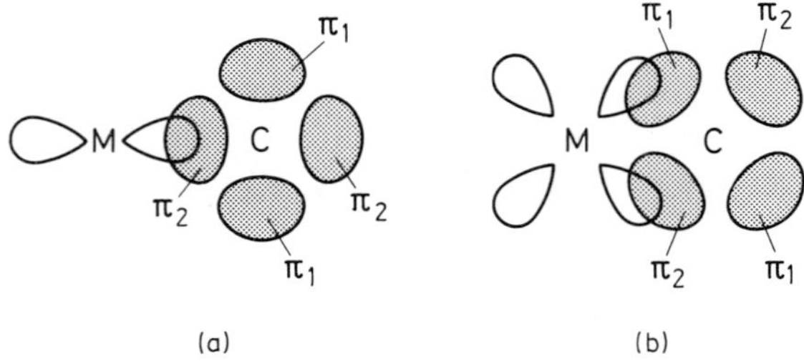

Fig. 17.3. Two ways of coordinating an acetylene molecule to a transition metal.

Titanium, Zirconium, Hafnium

Only $(\eta^5\text{-}C_5H_5)_2Ti(PhC{\equiv}CPh)$ has been isolated from the reaction of $(\eta^5\text{-}C_5H_5)_2Ti(CO)_2$ with diphenylacetylene.

Vanadium, Niobium, Tantalum

Irradiation of (η^5-C_5H_5)V(CO)$_4$ with disubstituted acetylenes yields the complex (η^5-C_5H_5)V(CO)$_2$(RC≡CR). Since a single acetylene replaces two carbon monoxide ligands, both π-systems of the acetylene must participate in bonding, and the ligand is thus a four-electron donor. In the complex (η^5-C_5H_5)$_2$V(CF_3C≡CCF_3) the acetylene ligand acts as a two-electron donor.

Niobium and tantalum complexes of diphenylacetylene are implicated as intermediates in the formation of cyclobutadiene complexes (Chapter 13.2). Reactive bis(cyclopentadienyl)niobium and tantalum complexes are obtained from the corresponding hydrides with acetylenes:

(η^5-C_5H_5)$_2$MH_3 + RC≡CR $\xrightarrow{-H_2}$ (η^5-C_5H_5)$_2$M(H)(RC≡CR)

M = Nb, Ta

Chromium, Molybdenum, Tungsten

Chromium complexes are difficult to obtain because of the catalyzed cyclotrimerization which tends to convert the acetylenes to substituted benzenes, but (η^6-C_6Me_6)Cr(CO)$_2$(PhC≡CPh) has been isolated. A bis(trifluoromethyl) acetylene complex, (η^5-C_5H_5)$_2$Cr(CF_3C≡CCF_3), can be prepared from chromocene and the acetylenic ligand.

Irradiation of molybdenum and tungsten carbonyls with acetylenes yields the Mo(CO)$_5$(RC≡CR′) and W(CO)$_5$(RC≡CR′) in which acetylene replaces only one carbon monoxide ligand, behaving as a two-electron donor. In the dithiophosphinato- and dithiocarbamato derivatives of molybdenum and tungsten, Mo(CO)(C_2H_2)(S_2PR_2) and W(CO)(C_2H_2)(S_2C-NEt$_2$)$_2$, the acetylene ligand must act as a four-electron donor in order to ensure an 18-electron configuration in the metal.

Tungsten forms triacetylene complexes (RC≡CR)$_3$W(CO) (Fig. 17.1.a) from tris(acetonitrile) tungsten tricarbonyl, W(CO)$_3$(MeCN)$_3$.

Bis(cyclopentadienyl)metal-acetylene complexes are obtained from the dihydride or by photolysis of (η^5-C_5H_5)$_2$Mo(CO):

(η^5-C_5H_5)$_2$MH_2 $\xrightarrow[-\text{PhCH=CHPh}]{2\,\text{PhC≡CPh}}$ (η^5-C_5H_5)$_2$M(RC≡CR) $\xleftarrow[h\nu]{\text{RC≡CR}}$ (η^5-C_5H_5)$_2$M–CO

M = Mo, W; M = Mo

Replacement of carbon monoxide in (η^5-C_5H_5)M(CO)$_3$X (where M = Mo, W; X = halogen) produces mono- and diacetylene complexes:

$$(C_5H_5)M(CO)_3X \xrightarrow[-2CO]{RC\equiv CR} (C_5H_5)M(X)(CO)(RC\equiv CR) \xrightarrow[-CO]{RC\equiv CR} (C_5H_5)M(X)(RC\equiv CR)_2$$

Manganese, Technetium, Rhenium

Manganese forms the acetylene complexes, $(\eta^5\text{-}C_5H_5)Mn(CO)_2(RC{\equiv}CR)$ (with the structure illustrated in Fig. 17.1.b) by irradiation of $\eta^5\text{-}C_5H_5Mn(CO)_3$. Here acetylene replaces a single carbon monoxide ligand and acts as a two-electron donor.

Rhenium(III) chloride forms the dimeric $[ReCl(PhC_2H)_2]_2$ containing two phenylacetylene molecules coordinated to each metal atom.

Iron, Ruthenium, Osmium

Iron forms the mono-substituted $Fe(CO)_4(RC{\equiv}CR)$ only with R = *tert*-Bu or $SiMe_3$ and other bulky acetylenes. Otherwise, the acetylenes undergo transformation, and the reaction products contain these transformed acetylenes. The rare ruthenium and osmium complexes are illustrated by $Ru(PPh_3)_2(NO)Cl(CF_3C{\equiv}CCF_3)$ and $Os(CO)_2(PPh_3)_2(CF_3C{\equiv}CCF_3)$.

Cobalt, Rhodium, Iridium

Cobalt complexes with terminal acetylene ligands include the unstable $(\eta^5\text{-}C_5H_5)Co(PPh_3)(PhC{\equiv}CPh)$. Rhodium complexes such as $Rh(PPh_3)_2Cl(RC{\equiv}CR)$ ($R = CF_3$, Ph) are also difficult to isolate. Heating $Rh_4(CO)_{12}$ with diphenylacetylene and PPh_3 yields the binuclear rhodium complex:

$$(PhC{\equiv}CPh)(CO)_2Rh{-}Rh(CO)(PPh_3)(PhC{\equiv}CPh)$$

Iridium forms a mononuclear compound with bis(trifluoromethyl) acetylene, $IrCl(PPh_3)_2(CF_3C{\equiv}CCF_3)$.

Nickel, Palladium, Platinum

These metals catalyze oligomerization of acetylenes and, therefore, acetylene complexes such as $Ni(CO)_2(CF_3C{\equiv}CCF_3)$ in which the acetylenic ligand behaves as a four-electron donor are rare.

Palladium complexes are obtained from reactions of $Pd(PPh_3)_4$ with acetylenes.

The dimeric $[Pt(RC{\equiv}CR)Cl_2]_2$ with the structure shown in Fig. 17.1.c is obtained by treatment of Na_2PtCl_4 with tert-butylacetylene. The compounds $Pt(PPh_3)_2(C_2R_2)$ are also known. Diphenylacetylene replaces 1,5-cyclooctadiene to give a diacetyleneplatinum complex which reacts with trimethylphosphine to produce a dinuclear platinum complex containing a terminal and a bridging acetylene ligand:

Pt(1,5 - COD)$_2$ $\xrightarrow{C_2Ph_2}$ Pt(RC≡CR)$_2$ $\xrightarrow{PMe_3}$ (RC≡CR)Pt(μ-RC≡CR)Pt(PMe$_3$)$_2$

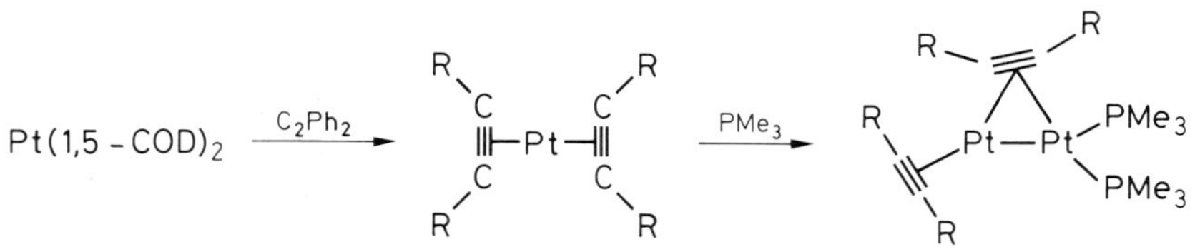

Only the bulky acetylene derivatives of platinum can be isolated.

17.2. Binuclear Complexes with Bridging Acetylene Ligands

Acetylenes react with dicobalt octacarbonyl to yield derivatives in which the acetylene molecule forms a bimetallic bridge (Fig. 17.1.d). Analogous nickel compounds (Fig. 17.1.e) are prepared from $[(\eta^5\text{-}C_5H_5)Ni(CO)]_2$ and acetylenes. Dicobalt octacarbonyl reacts with perfluorocyclohexadiene-1,3 to yield a product containing a fluorinated derivative of benzyne as a ligand:

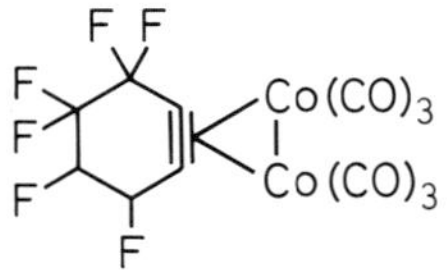

C_6F_6 (benzyne ligand) bridging Co(CO)$_3$–Co(CO)$_3$

This is another example of the stabilization of an organic molecule incapable of independent existence via complexation with transition metals.

Macrocyclic diynes can attach both their triple bonds to the bimetallic fragments, $Co_2(CO)_6$ or $Ni_2(C_5H_5)_2$, to form tetranuclear compounds:

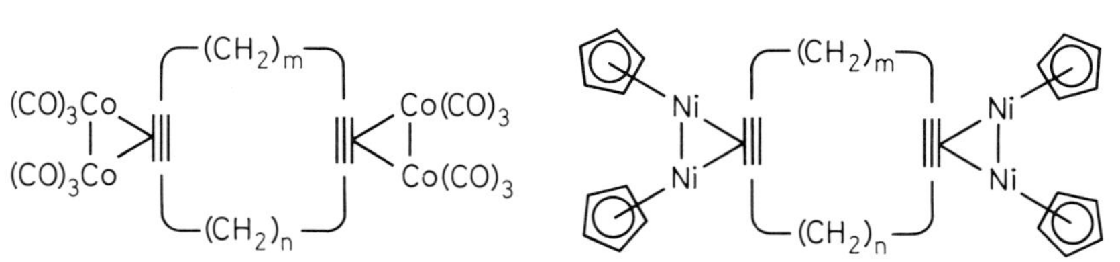

(CO)$_3$Co–Co(CO)$_3$ and Ni–Ni (C$_5$H$_5$) units bridged by macrocyclic diyne with (CH$_2$)$_m$ and (CH$_2$)$_n$ chains

The dinuclear compound $[(\eta^5\text{-}C_5H_5)Mo(CO)_2]_2$ reacts with acetylenes to form $(\eta^5\text{-}C_5H_5)_2Mo_2(CO)_4(\mu\text{-}RC{\equiv}CR')$ which contains metal-metal bonds and bridging acetylenes.

The analogous tungsten derivative, $(\eta^5\text{-}C_5H_5)_2W_2(CO)_4(\mu\text{-}HC{\equiv}CH)$, forms from $(\eta^5\text{-}C_5H_5)_2W_2(CO)_4$ and acetylenes.

The reaction of $Fe_3(CO)_{12}$ with bis(tert-butyl)acetylene gives a compound containing two bridging acetylenes and an iron-iron double bond, while $Fe_2(CO)_9$ with the same acetylene produces a singly-bridged derivative:

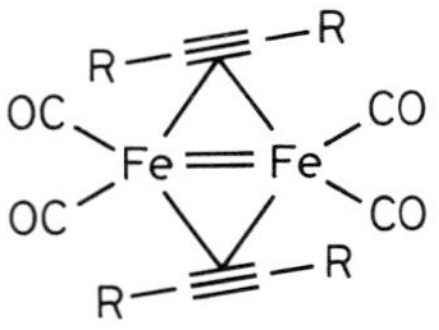

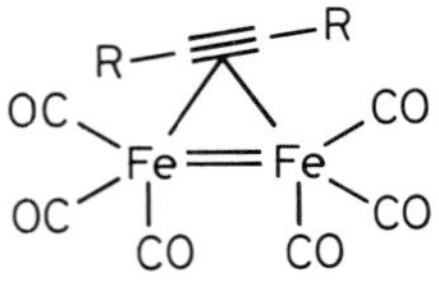

Palladium complexes with bridging acetylenes are also known:

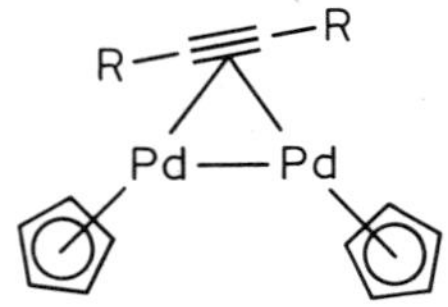

The formation of a binuclear η^5-pentaphenylcyclopentadienylpalladium derivative, $(\eta^5\text{-}C_5Ph_5)_2Pd_2(\mu\text{-}PhC{\equiv}CPh)$, in a reaction involving partial loss of PhC units as $PhC(OR)_3$ occurs on heating diphenylacetylene with Pd(II) acetate:

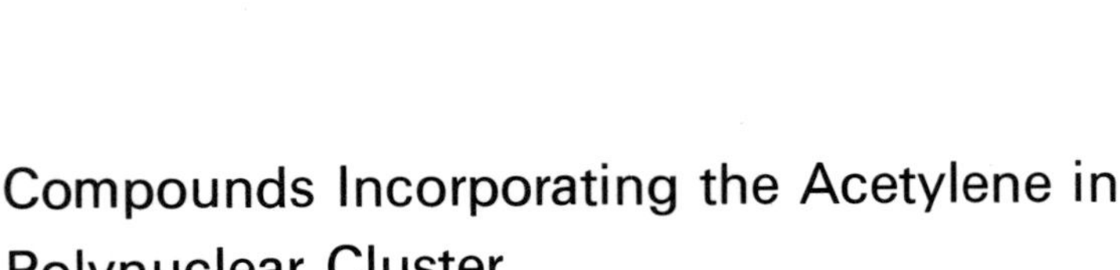

$$PhC{\equiv}CPh + Pd(OAc)_2 \xrightarrow[-PhC(OR)_3]{ROH}$$

17.3. Compounds Incorporating the Acetylene in a Polynuclear Cluster

Reactions of metal carbonyls with acetylenes often form polynuclear metal clusters. Thus, the cobalt carbonyls, $Co_2(CO_6)(RC{\equiv}CR)$, are converted by strong acids to the polynuclear-cluster compounds, $RCH_2CCo_3(CO)_9$, with the structure shown in Fig. 17.1.f. These complexes contain carbyne units forming a trimetallic bridge (see Section 12.3). The polynuclear $Co_4(CO)_{10}(RC{\equiv}CR)$, formed from acetylenes with $Co_2(CO)_8$, have open Co_4 skeletons with the acetylene ligand sitting above it:

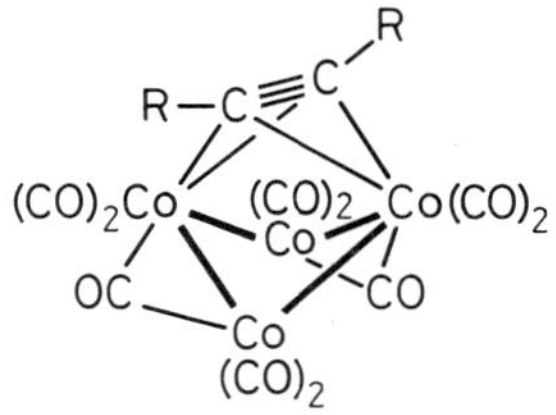

The iron compound, $Fe_3(CO)_{12}$, is degraded by acetylenes to give a large variety of compounds. Diphenylacetylene yields with $Fe_2(CO)_9$ the polynuclear compound $Fe_3(CO)_9(PhC{\equiv}CPh)$ (structure shown in Fig. 17.1.g). Ruthenium and osmium carbonyls preserve the parent cluster in reactions with acetylenes. Thus, $Ru_3(CO)_{12}$ forms $Ru_3(CO)_{12}(RC{\equiv}CR)$ and $Ru_3(CO)_8(RC{\equiv}CR)_2$:

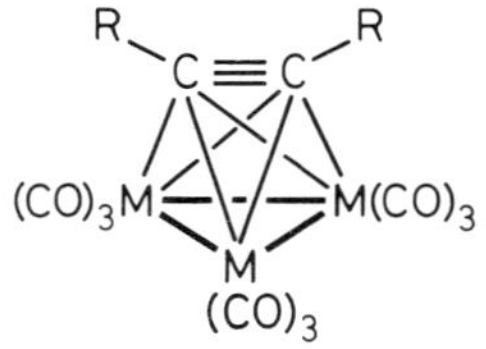

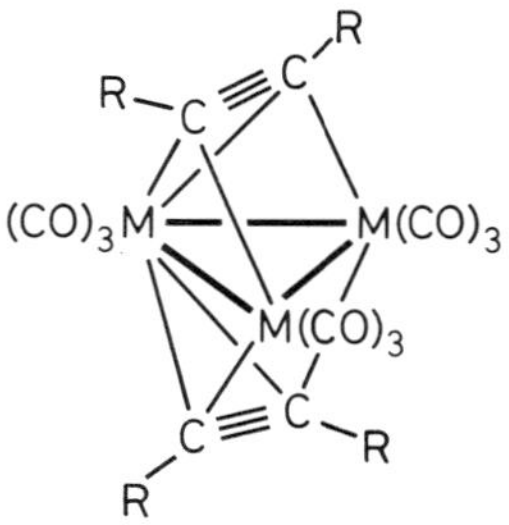

Diphenylacetylene forms the tetranuclear, $Ru_4(CO)_{12}(PhC{\equiv}CPh)$:

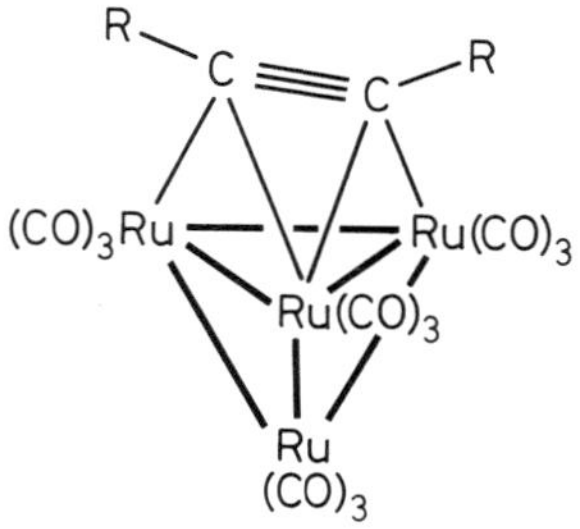

Similar $Os_3(CO)_{10}(PhC{\equiv}CPh)$ and $H_2Os_3(CO)_9C_2H_2$ compounds are known.

Heterobimetallic clusters, incorporating acetylene fragments are also known:

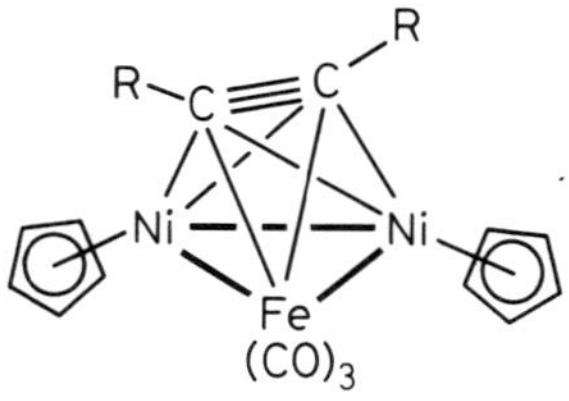

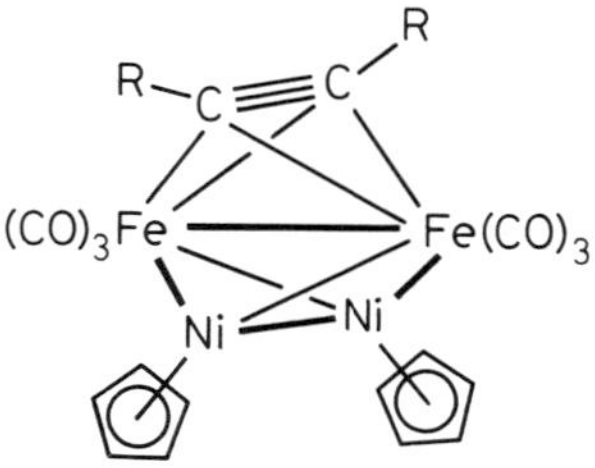

17.4. Complexes Formed by Chemical Transformations of Acetylenes

Acetylenes are transformed by metal carbonyls, especially those of iron. The product from an acetylene and iron carbonyl depends upon the iron carbonyl used, the reaction conditions and the nature of the acetylene. The temperatures at which iron carbonyls react with acetylenes are in the following ranges: $Fe_3(CO)_{12}$, 60–100 °C; $Fe_2(CO)_9$ 25–30 °C; and $Fe(CO)_5$, 150 °C or higher; UV irradiation is also useful in promoting these reactions. Several compounds are often formed. Thus, the reaction of $Fe_3(CO)_{12}$ with diphenylacetylene produces:

$(PhC_2Ph)Fe_2(CO)_6$ $(PhC_2Ph)_2Fe(CO)_3$
$(PhC_2Ph)_2Fe(CO)_6$ $(PhC_2Ph)_2COFe(CO)_3$
$OC(PhC_2Ph)_2Fe_2(CO)_6$ $(PhC_2Ph)_2Fe_3(CO)_8$.

The formation of complex molecules from acetylenes and iron carbonyls probably occurs through a multicenter mechanism involving a succession of organometallic compounds.

One compound, $Fe(CO)_3(PhC{\equiv}CPh)_2$, contains a tetraphenylcyclobutadiene ligand (Fig. 17.2.a). Other cyclobutadiene complexes form in the reaction of iron carbonyls with macrocyclic diacetylenes (Fig. 17.4):

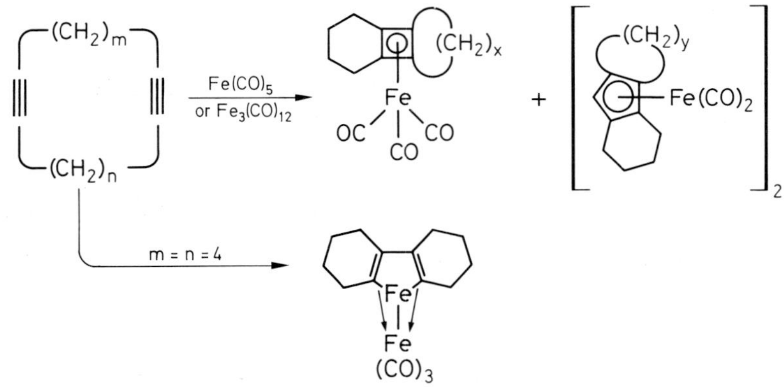

Fig. 17.4. Compounds formed in reactions of macrocyclic diynes with iron carbonyls.

The ferracyclopentadiene (ferrole) complexes, $Fe_2(CO)_6(RC{\equiv}CR)_2$, shown in Fig. 17.2.c, form from acetylenes with iron carbonyls:

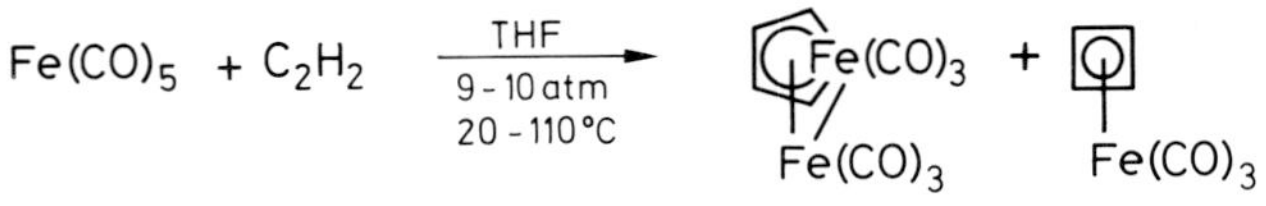

The compounds $Fe_3(CO)_8(RC{\equiv}CR)_2$ also contain a ferrole system, as does the product from unsubstituted acetylene along with the cyclobutadiene complex.

Acetylenes with iron carbonyls can incorporate carbon monoxide units to give complexes of cyclopentadienones (Fig. 17.2.d), which can also be obtained directly from iron carbonyls and cyclopentadienones. Phenylacetylene with $Fe_3(CO)_{12}$ (Fig. 17.2.e) yields a complexed tropone ring (Fig. 17.2.e).

The stability of the cyclopentadienyl iron group can be the driving force that causes this unit to form in reactions of acetylenes with iron carbonyls. Thus, the unsubstituted acetylene forms with $Fe_3(CO)_{12}$ the compound shown in Fig. 17.2.g. The compound shown in Fig. 17.2.h was isolated from acetylene with $Fe(CO)_5$. Cyclopentadienyl ligands are also formed in the reactions of macrocyclic diacetylenes (diynes) with iron carbonyls (Fig. 17.4).

Vanadium hexacarbonyl also reacts with diphenylacetylene to form η^6-hexaphenylbenzenevanadium tricarbonyl:

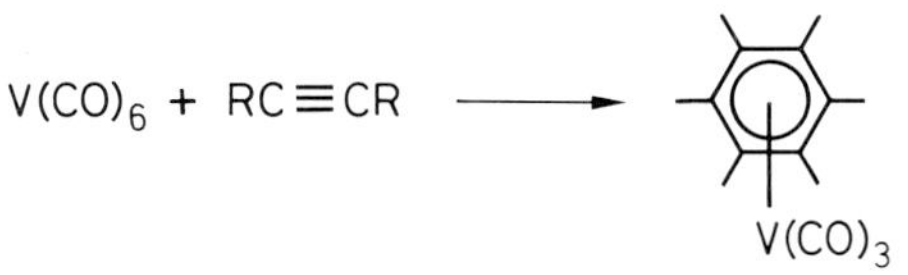

$$V(CO)_6 + RC\equiv CR \longrightarrow \eta^6\text{-}C_6R_6V(CO)_3$$

Manganese carbonyl, $Mn_2(CO)_{10}$, and acetylene react to form a manganese complex of dihydropentalene:

$Mn(CO)_3$

Acetylenes react with η^5-$C_5H_5Co(CO)_2$ to form cyclopentadienone, cyclobutadiene and other complexes:

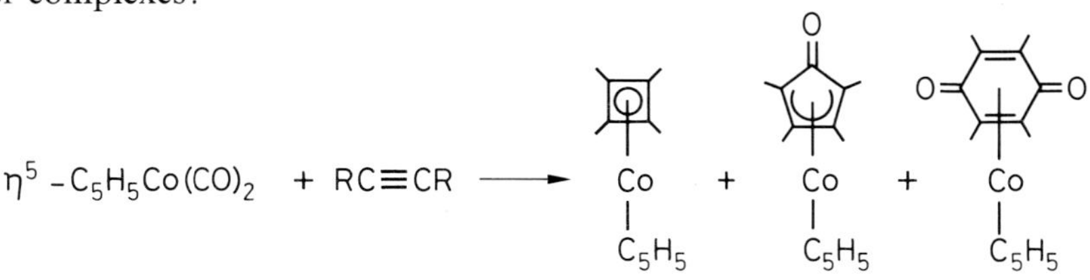

$$\eta^5\text{-}C_5H_5Co(CO)_2 + RC\equiv CR \longrightarrow \text{Co–}C_5H_5 + \text{Co–}C_5H_5 + \text{Co–}C_5H_5$$

With macrocyclic diynes, transannular ring closure occurs to form cyclobutadiene derivatives:

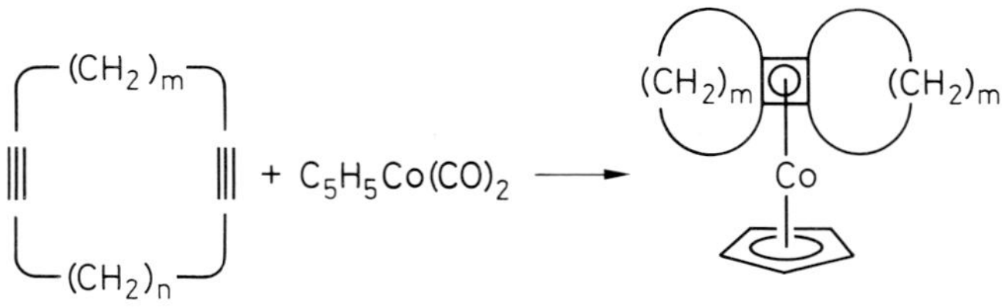

$$(CH_2)_m,\ (CH_2)_n\text{ diyne} + C_5H_5Co(CO)_2 \longrightarrow (CH_2)_m \ (CH_2)_m \ \text{Co}$$

Similar transformations occur in the reactions of $C_5H_5Rh(CO)_2$ with acetylenes and macrocyclic diynes.

The mechanism of the cyclizations is suggested by the isolation of intermediates whose structures demonstrate the formation of increasingly long carbon-carbon chains (M = Cr or Mo) (Fig. 17.5).

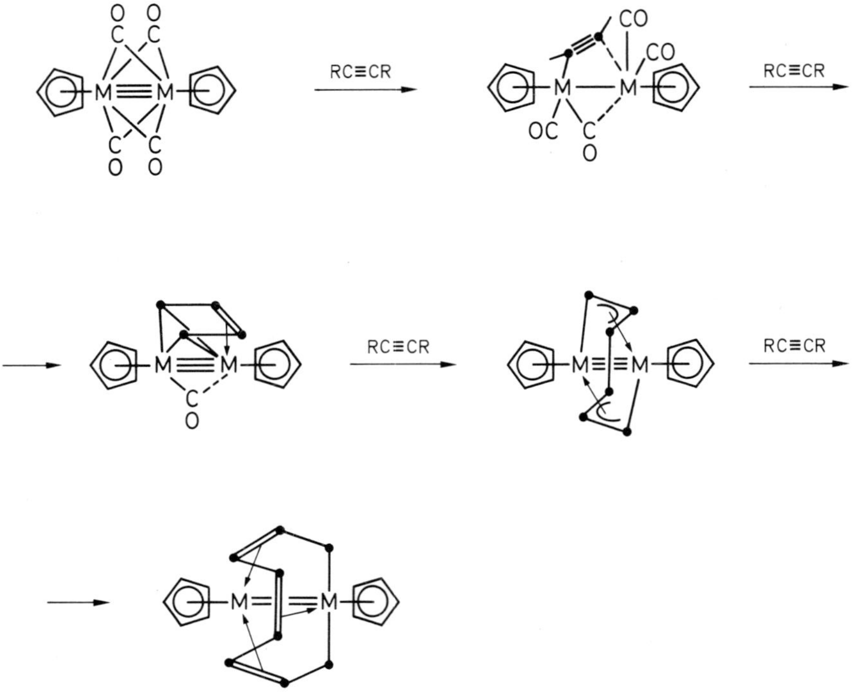

Fig. 17.5. The increase of the C-C chain in reactions of acetylenes with transition metal complexes.

18. Organometallic Compounds with σ-Metal-Carbon Bonds

Compounds with σ-metal-carbon bonds are typical for the main group elements, but until recently there was general belief that transition metal derivatives would be unstable. However, many σ-bond derivatives have now been isolated, and the reason for kinetical instability of simple transition-metal alkyls or aryls is now seen to be the incomplete occupancy of the valence-shell orbitals.

The stability of σ-bonded transition-metal organic derivatives is increased when stabilizing factors are involved:

a) *Coordination of certain ligands* to the transition metals. Particularly good are π-acceptors, like CO, PR_3 or η^5-C_5H_5, but donors like bipyridyl or phenanthroline are also suitable;

b) *Avoiding β-elimination* (*vide infra*) by using organic groups of appropriate structure;

c) *Steric protection* with the aid of bulky substituents;

d) *Chelate ring formation.*

π-Acceptor ligands facilitate use of both occupied and vacant metal *d*-orbitals to achieve noble gas configurations (see Section 2.3). Thus, metal carbonyl or cyclopentadienylmetal fragments lacking only one electron form σ-bonded organic derivatives, but the role of these ligands is only secondary in determining the structure and properties.

Alkyl groups σ-bonded to transition metals tend to eliminate olefin (see Section 2.3) with the migration of the hydrogen atom from β-position to the metal to form a hydride (β-elimination):*

$$\text{M–C–C–H} \longrightarrow \text{M–H} + \text{C=C}$$

Organic groups such as $-CH_2SiR_3$, $-CH(SiR_3)_2$, $-CH_2Ph$, $-CH_2CMe_3$, etc., lacking a β-hydrogen, form more stable σ-derivatives than normal alkyl groups. Such groups are also bulky, and therefore, exhibit a salutary steric influence as well. Stable

* Remember that α-elimination leads to carbene complexes (see Section 11.5).

homoleptic** derivatives, MR_n, are obtained almost exclusivley with such substituents. This is illustrated by Table 18.1 which lists neutral and anionic homoleptic compounds, and by Table 18.2, which lists alkylmetal halides.

Tab. 18.1. Representative Homoleptic Molecular and Anionic Derivatives of Transition Metals, MR_n.

	R						
M	CH_2SiMe_3	$CH(SiMe_3)_2$	Ph	C_6F_5	CH_2Ph	CH_3	Other
Sc		ScR_3	ScR_3				$Sc(C{\equiv}CPh)_3$
Y		YR_3	YR_3				
Lan-tani-des	ErR_3 MR_4^- (Y, Er, Yb, Lu)	YbR_3	LaR_4^- PrR_4^-	YbR_2 4THF		ErR_6^{3-} LuR_6^{3-}	$Eu(C{\equiv}CR)_3$
Acti-nides	UR_6^{2-} UR_8^{3-}		UR_6^{2-}		ThR_4	UR_6^{2-} UR_8^{3-}	$U(CH_2Bu^t)_6^{2-}$ $U(CH_2Bu^t)_8^{3-}$
Ti	TiR_3 TiR_4	TiR_3	TiR_2 TiR_4		TiR_2 TiR_3 TiR_4	TiR_4 TiR_5^-	$Ti(CH_2Bu^t)_4$
Zr	ZrR_4		ZrR_4	ZrR_4	ZrR_4	ZrR_4 ZrR_6^{2-}	$Zr(CH_2Bu^t)_4$
Hf	HfR_4				HfR_4		$Hf(CH_2Bu^t)_4$
V	VR_3 VR_4	VR_3	VR_3 VR_4 VR_6^{4-}		VR_4		$VR_{3\text{ and }4}$ (R = mesityl)
Nb			NbR_6^{4-} NbR_7^{3-}	NbR_4		NbR_5	
Ta			TaR_6^-		TaR_5	TaR_5	
Cr	CrR_4 CrR_4 CrR_4^-	CrR_3	CrR_3 3THF CrR_4^- CrR_5^{2-} CrR_6^{3-} $Cr_2R_6^{2-}$		CrR_3	CrR_4 CrR_4^{2-} CrR_6^{3-} $Cr_2R_8^{4-}$	$CrPr^i_4$ $CrBu^t_4$
Mo	Mo_2R_6		MoR_6^{3-}		MoR_4 Mo_2R_6	$Mo_2R_8^{4-}$	MoR_4 (mesityl)
W	W_2R_6		WR_6^{2-}	WR_5 WR_5^-	WR_4 W_2R_6	WR_6 WR_8^{2-} $W_2R_8^{2-}$	

** *Homoleptic* compounds are derivatives, MR_n, in which all substituents are the same. A pair of compounds is called *isoleptic* when they contain identical substituents, for example, $PbMe_4$ and $TiMe_4$; when a compound contains two kinds of substituents, it is called heteroleptic, for example, R_mMCl_n.

Tab. 18.1. Continued

	R						
M	CH_2SiMe_3	$CH(SiMe_3)_2$	Ph	C_6F_5	CH_2Ph	CH_3	Other
Mn	MnR_2	MnR_2			MnR_2	MnR_3^- MnR_4^{2-}	$Mn(C{\equiv}CR)_2^t$ $Mn(CH_2Bu^t)_2$
Tc							
Re	Re_3R_{12}					ReR_6 $Re_2R_8^{2-}$	
Fe			FeR_4^{2-}			FeR_4^{2-}	FeR_2 (mesityl) $Fe(C{\equiv}CR)_6^{2-}$
Ru							FeR_4
Os							(norbornyl)
Co	CoR_4^{2-}	CoR_4^{2-}		CoR_2 CoR_4^{2-}		CoR_4^{2-}	$Co(C{\equiv}CR)_6^{3-}$ $Co(C{\equiv}CR)_6^{4-}$ CoR_4
Rh							(norbornyl)
Ir							
Ni	NiR_2.bipy		NiR_4^{2-}	NiR_2 NiR_4^{2-}		NiR_4^{2-}	$NiBu_2^t$ Ni_2R_4 ($R = CH_2CH_2PMe_2$)
Pd				PdR_3^{2-} PdR_4^{2-}			
Pt				PtR_2 PtR_3^{2-}		PtR_5^- PtR_6^{2-} Pt_2R_6	
Cu	$(CuR)_4$		CuR CuR_4^{3-}	$(CuR)_4$		CuR_2^-	CuR_2 ($R = CH_2CH_2PMe_2$) $Cu(C{\equiv}CR)_2^-$
Ag			AgR AgR_2^-	AgR AgR_2^-			$Ag(C{\equiv}CR)_2^-$
Au		AuR·L		AuR_2^- AuR_3		AuR_2^- AuR_4^-	$Au(C{\equiv}CR)_2^-$ $Au(C{\equiv}CR)$

Tab. 18.2. Transition-Metal Halides with σ-M—C Bonds.

M	Compounds
Lanthanides	$[(Me_3SiCH_2)_3MCl]^-$ M = Er, Yb $[((Me_3Si)_2CH)_3MCl]^-$ M = Er, Yb
Ti	$Me_3SiCH_2TiCl_3$ $MeTiCl_3$ $(Me_3SiCH_2)_2TiCl_2$ Me_2TiCl_2
Zr	$[(Me_3Si)_2CH]_3ZrCl$
Hf	$[(Me_3Si)_2CH]_3HfCl$
V	$(Me_3C_6H_2)VCl_2 \cdot 2THF$
Nb	Me_2NbCl_3 Me_3NbCl_2
Ta	Me_3TaCl_2
W	$PhWCl_3$ $PhWCl_5$
Re	$Re_3Cl_3(CH_2SiMe_3)_6$
Pd	$PhCH_2PdCl$
Pt	$(Me_3PtX)_4$
Au	$(PhAuCl_2)_2$ $[C_6F_5AuX]^-$ $(Ph_2AuCl)_2$ $[(C_6F_5)_3AuX]^-$ $[PhAuCl_3]^-$ $[C_6F_5AuBr_3]^-$

Chelate ring formation is usually beneficial to the stability of the σ-derivatives, and σ-bonded organic groups, containing donor substituents in positions favorable for chelate ring formation, form more stable compounds as illustrated by the chromium compound:

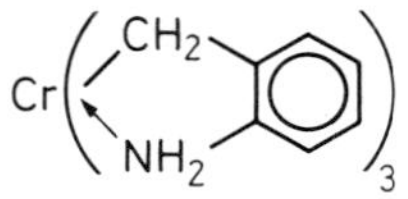

Cr(CH2–C6H4–NH2)3

Some phosphorus ylids are particularly favorable ligands:

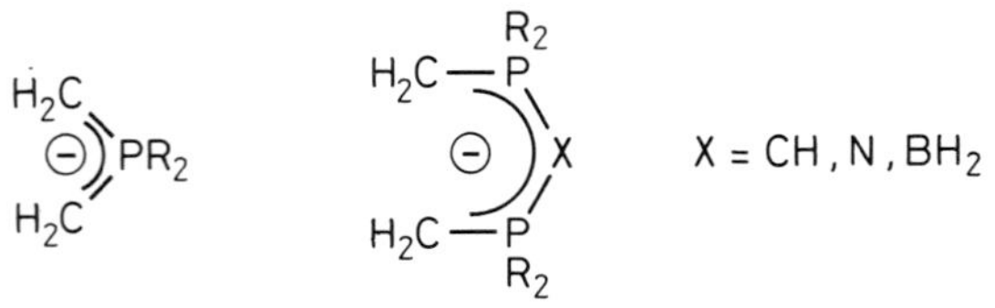

H2C ⊖ PR2 H2C   H2C—P(R2) ⊖ X H2C—P(R2)   X = CH, N, BH2

and form stable cyclic compounds of the type:

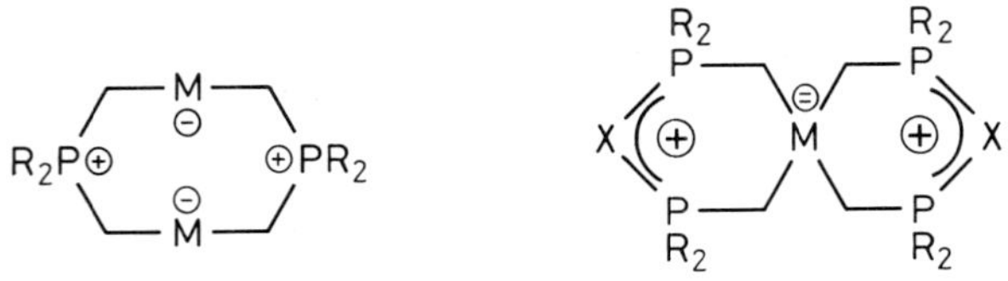

R2P⊕ ⊖M ⊖M ⊕PR2   R2P X ⊕ P R2 ⊖M R2P ⊕ X P R2

In these derivatives the metal forms only σ-carbon bonds.

The participation of the metal as a heteroatom in a metallocycle also leads to stable compounds.

From these stabilizing factors the following types of compounds with σ-metal-carbon bonds will be encountered:

1. *Compounds containing σ-carbon bonds exclusively:*

a) homoleptic alkyl and aryl derivatives: neutral MR_n and anionic MR_n^- (few cations are known).

2. *Compounds containing σ-carbon bonds and additional ligands:*

a) adducts of homoleptic derivatives with donors, such as neutral $CrPh_3 \cdot 3THF$, cationic $[CrPh_2(bipy)_2]^+$, or anionic $[R_3AuX]^-$;

b) compounds with monofunctional ligands, for example, halide, $R{-}MX_n$, alkoxy, $R{-}M(OR\,)_n$, amino, $R{-}M(NR_2')_n$ and mercapto derivatives, $R{-}M(SR')_n$;

c) compounds with π-acceptor ligands (usually n = 1):
- metal carbonyl derivatives, $(CO)_mMR_n$,
- cyclopentadienylmetal derivatives, $(\eta^5\text{-}C_5H_5)_mMR_n$;
- cyclopentadienylmetal carbonyl derivatives, $(\eta^5\text{-}C_5H_5)M(CO)_mR_n$;

d) chelate rings and metallocycles with σ-carbon and M—X (X = O, N, S, P, As, etc.) bonds.

The σ-bonded group, R, can be aliphatic, aromatic, olefinic (σ-vinyl, σ-allyl), perfluorinated or perchlorinated (CF_3, C_3F_7, C_6F_5, C_6Cl_5, etc.), acyl (—CO—R), or an alkynyl group, —C≡C—R (in acetylides). The most favored are those incapable of β-elimination. Electronegative character of the organic group also increases the stability of the σ-bonded compounds. Thus, aromatic derivatives and polyfluorinated or polychlorinated groups yield more stable compounds.

The syntheses of σ-bonded derivatives are no different from those used in main-group organometallic chemistry. Among the most important are:

a) a metal halide or halogeno complex with organolithium, organomagnesium or other organometallic reagents able to transfer an organic group:

$$M{-}X + M'R \longrightarrow M{-}R + M'X$$

M' = Li, MgX, Al, Hg, Tl, etc.

b) a metal hydride with an olefin (addition):

$$M{-}H + H_2C{=}CHR \longrightarrow M{-}CH_2CH_2R$$

This reaction is the reverse of β-elimination.

c) a metal hydride with diazomethane to give σ-methyl derivatives:

$$M{-}H + CH_2N_2 \longrightarrow M{-}CH_3 + N_2$$

d) an anionic metal complex (metal carbonyl or cyclopentadienylmetal carbonyl anion) with a halogenated organic compound:

$$L_xM^- + RX \longrightarrow L_xM{-}R + X^-$$

e) metal vapor with organic halides:

$$M + RX \longrightarrow R{-}MX \text{ or } M{-}R + X$$

In the following pages some of the most representative σ-derivatives will be briefly introduced, illustrating the preparative methods used and the types of compounds mentioned above.

Scandium, Yttrium, Lanthanides and Actinides (Group III Elements)

Trisubstituted derivatives, MR_3, where M = Sc and Y, R = $CH(SiMe_3)_2$ and M = Er, R = CH_2SiMe_3, as well as anionic, $[Lu(CH_2SiMe_3)_4]^-$, are prepared from organolithium derivatives. Scandium and yttrium tris-alkynyls, $M(C{\equiv}CPh)_3$, are also known. The disubstituted $Yb(C_6F_5)_2 \cdot 4THF$ is obtained from ytterbium metal and $Hg(C_6F_5)_2$ in THF.

The cyclopentadienylmetal derivatives, $(\eta^5\text{-}C_5H_5)_2M{-}R$ (M = Y, Dy, Ho, Er, Yb, Gd, Tm; R = Me, Ph, C≡C—Ph), are prepared from the corresponding halides and LiR, while the dimeric halides, $[(\eta^5\text{-}C_5H_5)_2MCl]_2$, react with $Li[AlR_4]$ or $Mg[AlR_4]_2$ to give the alkyl-bridged compounds, $(\eta^5\text{-}C_5H_5)_2M(\mu\text{-}R)_2AlR_2$ with M = Sc, Y, Dy, Ho, Er, Tm, Yb; R = Me, and M = Sc, Y, Ho, R = Et:

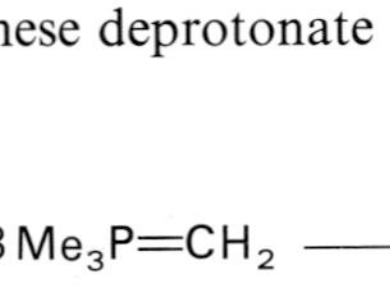

In the rare earth cationic complexes, $[M(CH_2PMe_3)_3]^{3+}3Cl^-$ (M = La, Pr, Nd, Sm, Gd, Ho, Er, Lu), the positive charge is localized at phosphorus rather than at the metal; these deprotonate to give neutral compounds containing chelate rings:

$$MCl_3 + 3Me_3P{=}CH_2 \longrightarrow [M(CH_2\overset{+}{P}Me_3)_3]3Cl^- \xrightarrow{LiBu} M^{3-}\left(\begin{matrix} CH_2 & & Me \\ & \overset{+}{P} & \\ CH_2 & & Me \end{matrix}\right)_3$$

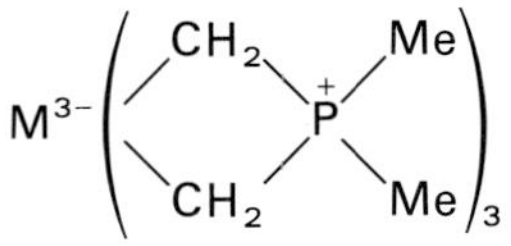

Other anions containing six M—C bonds are $[MR_6]^{3-}$ (M = Y, La, Pr, Nd, Sm, Gd, Tb, Dy, Ho, Er, Tm, Yb, Lu; R = Me).

Uranium forms the σ-derivatives, $(\eta^5\text{-}C_5H_5)_3U\text{-}R$ (R = C≡CH, etc.). A high degree of σ-substitution is reached in the homoleptic anions, $[UR_6]^{2-}$ (R = CH_2SiMe_3,

Me, Ph, and *ortho*-$Me_2C_6H_4$) and $[UR_8]^{3-}$ (R = Me, CH_2Bu^t, CH_2SiMe_3), isolated as the solvated lithium salts. Tetrasubstituted thorium derivatives include $Th(CH_2Ph)_4$.

Titanium, Zirconium, Hafnium

The stable bis(cyclopentadienyl) titanium derivatives, $(\eta^5\text{-}C_5H_5)_2TiR_2$, are prepared from the corresponding dichloride with organolithium reagents. The less-stable homoleptic TiR_4 are prepared from titanium tetrachloride and alkyllithiums. The unstable tetramethyltitanium can be stabilized in the *ortho*-phenanthroline or bipyridyl complexes.

Tetrasubstituted derivatives, $M(CH_2SiMe_3)_4$ (M = Ti, Zr, Hf) and $Ti(CH_2Ph)_4$, are stable, and the trisubstituted complexes, $Ti(CH_2SiMe_3)_3$ and $Ti[CH(SiMe_3)_2]_3$, are also known. The compound, $Ti[CH(SiMe_3)_2]_3$ forms from $TiCl_4$ and $LiCH(SiMe_3)_2$, but zirconium and hafnium tetrachlorides yield the triorganometal chlorides, $[(Me_3Si)_2CH_2]_3MCl$.

Titanium tetrachloride forms organotitanium halides, $MeTiCl_3$ and Me_2TiCl_2, in the reaction with organoaluminum and organolead reagents. The former can be stabilized as a tertiary phosphine complex. The fully substituted $TiMe_4$ can be further converted to $Li[TiMe_5]$. Titanium tetrachloride and phenyllithium form tetraphenyltitanium, which polymerizes to give $(TiPh_2)_x$ and reacts with cyclopentadiene to give $(\eta^5\text{-}C_5H_5)_2TiPh_2$.

Tetrakis(pentafluorophenyl)zirconium, $Zr(C_6F_5)_4$, and tetrabenzylzirconium, $Zr(CH_2Ph)_4$, are known, but tetraphenylzirconium is unstable. Six Zr—C bonds form in the anion, $[ZrMe_6]^{2-}$

Ylid derivatives are known for titanium and zirconium:

$$MCl_4 + nMe_3P{=}CH_2 \longrightarrow Cl_4M(CH_2PMe_3)_n$$

M = Ti, n = 2 and 3
M = Zr, n = 2, 3 and 4

$$(\eta^5\text{-}C_5H_5)_2MCl_2 + 2\,Me_3P{=}CH_2 \longrightarrow [(\eta^5\text{-}C_5H_5)_2M(\text{—}CH_2\text{—}PMe_3)_2]^{2+}2\,Cl^-$$

M = Ti, Zr

$$(\eta^5\text{-}C_5H_5)_2TiCl_2 + \underset{CH_3}{Me_2P}{=}N\text{—}\underset{CH_2}{\overset{\|}{PMe_2}} \xrightarrow[-LiCl,\ -RH]{LiR} (\eta^5\text{-}C_5H_5)_2Ti\langle CH_2\text{—}PMe_2\text{—}N\text{—}PMe_2\text{—}CH_2\rangle$$

Metallocycles containing titanium heteroatoms are prepared from diphenylacetylene and organodilithium compounds:

$$(\eta^5\text{-}C_5H_5)_2Ti(CO)_2 + PhC{\equiv}CPh \longrightarrow (\eta^5\text{-}C_5H_5)_2Ti(CO)(C_2Ph_2) \xrightarrow{30\,°C}$$

Ti

$$(\eta^5\text{-}C_5H_5)_2TiCl_2 + Li(CH_2)_4Li \xrightarrow[-78\,°C]{Et_2O}$$

Ti

Vanadium, Niobium, Tantalum

Vanadium trichloride forms the anionic, hexasubstituted $Li_4[VPh_6]$, with phenyllithium and bis(cyclopentadienyl)vanadium chloride yields $(C_5H_5)_2V{-}Ph$. Triphenylvanadium is obtained from $VCl_3 \cdot 3THF$ and phenyllithium in THF. Both trimethylsilylmethyl derivatives, $V(CH_2SiMe_3)_n$ ($n = 3$ and 4), are known. 2,4,6-Trimethylphenyl derivatives of the neutral MR_4 and anionic $[MR_4]^-$ are also known.

Niobium and tantalum pentachlorides react with dimethylzinc to form the unstable trimethyl dichlorides, Me_3MCl_2. The pentamethylniobium and -tantalum obtained from the corresponding methylmetal chlorides and methyllithium decompose by α-hydrogen abstraction. The hexamethyl derivative, $TaMe_6$, explodes even *in vacuo*. Highly-substituted phenyl anions, $[NbPh_6]^{4-}$, $[TaPh_6]^-$ and $[NbPh_7]^{3-}$ are also known.

Chromium, Molybdenum, Tungsten

Chromium(III) chloride gives the trisubstituted $CrPh_3 \cdot 3THF$ from phenylmagnesium bromide in THF, which is readily converted to the η^6-complexes of benzene and biphenyl. Excess phenyllithium gives the hexasubstituted anion, $Li_3[CrPh_6] \cdot nEt_2O$, which reacts with cyclopentadiene to give the complex $Li[(\eta^5\text{-}C_5H_5)CrPh_3]$. The homoleptic $CrPh_3$ is unsaturated and incapable of independent existence. Disproportionation of $Li_3[CrPh_6] \cdot nEt_2O$ with $CrCl_3$ leads to $LiCrPh_3 \cdot nEt_2O$ or $LiCrPh_4$. The pentaphenylchromium derivative, $Na_2CrPh_5(Et_2O)_3(THF)$, contains a trigonal-bipyramidal $CrPh_5$ unit with interactions between sodium and the phenyl substituents:

Et_2O, Et_2O — Na ---- Ph; Ph — Cr (Ph, Ph, Ph, Ph); Ph --- Na (OEt_2, THF)

Di- and tetraphenyl derivatives of chromium, $CrPh_2(PEt_3)_2$ and $Li_2[CrPh_4] \cdot 4THF$, are also known.

Benzylmagnesium bromide forms with $CrCl_3$, a trisubstituted derivative, $Cr(CH_2Ph)_3$, which decomposes to form η^6-arene complexes. The cationic benzyl derivative, $[PhCH_2Cr(H_2O)_5]^{2+}$, forms in the reaction of chromium(III) chloride with benzyl chloride in aqueous acid. Anionic species, $[R{-}M(CO)_5]^-$ (R = Me, Et, $PhCH_2$), are also formed in the reaction of $Na_2[Cr(CO)_5]$ with the corresponding organic halides.

A binuclear compound containing a chromium-chromium triple bond is obtained from $CrCl_2$ and methyllithium:

$$CrCl_2 + nLiMe \longrightarrow 2\,Li^+ \left[Me_2Cr{\equiv}CrMe_2 \right]^{2-}$$

Multiple metal-metal bonding is also found in $Cr_2(CH_2SiMe_3)_4(PMe_3)_2$, which contains both bridging and terminal $-CH_2SiMe_3$ groups:

$$(Me_3SiCH_2)(Me_3P)Cr(\mu\text{-}CH_2SiMe_3)_2Cr(CH_2SiMe_3)(PMe_3)$$

Chromium-carbon σ-bonds can be part of a metallocycle in the following two structures:

$$[(C_4H_8)_2Cr{-}Cr(C_4H_8)_2] \qquad [Cr(C_5H_{10})_3]^{3\ominus}$$

Chromium compounds containing four-membered chelate rings derived from phosphorus ylids, are obtained in the reaction:

$$[P(CH_3)_4]Cl + Li_3[CrPh_6] \longrightarrow Me_2P(CH_2)_2Cr(Ph)_2(CH_2)_2PMe_2 \longrightarrow Cr((CH_2)_2PMe_2)_3$$

The cyclopentadienylmetal tricarbonyl alkyls of molybdenum and tungsten form from the corresponding anions and alkyl halides:

$$[(\eta^5\text{-}C_5H_5)M(CO)_3]_2 \xrightarrow{Na/THF} [(\eta^5\text{-}C_5H_5)M(CO)_3]^- \xrightarrow{RX} (\eta^5\text{-}C_5H_5)M(CO)_3\text{—}R$$

M = Mo, W

σ-Allyl and σ-benzyl derivatives of molybdenum prepared by this procedure undergo photochemical σ-π rearrangement to give η^3-allylic complexes.

Six Mo—C σ-bonds are found in the anion, $[MoPh_6]^{3-}$, while tungsten forms $[WMe_8]^{2-}$ anions. The reaction of WCl_6 with methyllithium or trimethylaluminum gives WMe_6 which is explosive. Excess methyllithium forms the $[WMe_8]^{2-}$ anion. The adduct $WMe_6 \cdot PMe_3$ obtained from the components decomposes thermally or on photolysis to give *trans*-$WMe_2 \cdot 4PMe_3$:

Me
Me₃P PMe₃
W
Me₃P PMe₃
Me

Molybdenum forms the dinuclear, metal-metal triple-bonded compounds, $Mo_2(CH_2SiMe_3)_6$, from $MoCl_6$ and $Me_3Si\text{-}CH_2MgCl$:

$$R_3Mo{\equiv}MoR_3 \qquad R = CH_2SiMe_3$$

A quadruple Mo-Mo bond is found in $Li_2Mo_2Me_8 \cdot 4THF$. Related tungsten compounds are also known, including $Li_2[W_2Me_8] \cdot 4Et_2O$ and $W_2(CH_2SiMe_3)_6$, which contain multiple metal-metal bonds.

Manganese, Technetium, Rhenium

For Group VII metals the pentacarbonyl alkyls are typical σ-derivatives. The sodium salt of the anion, $[Mn(CO)_5]^-$, forms the σ-carbon $(CO)_5Mn\text{—}CH_3$ with methyl iodide. This absorbs carbon monoxide reversibly, to form an acetyl derivative, also available by an alternative route:

$$Na[Mn(CO)_5] + MeI \longrightarrow (CO)_5Mn\text{—}Me \underset{80\,°C}{\overset{CO}{\rightleftarrows}} (CO)_5Mn\text{—}CO\text{—}Me$$

$$Na[Mn(CO)_5] \xrightarrow{+ClCO\text{—}Me} (CO)_5Mn\text{—}CO\text{—}Me$$

Other σ-alkyls, $(CO)_5M\text{—}R$ (M = Mn, Re), form by the decarbonylation of acyl derivatives, $(CO)_5M\text{—}CO\text{—}R$ (R = Ph, perfluoroalkyl).

Hydrometallation, in the addition of perfluoroethylene to a carbonyl hydride, can also be used:

$$(CO)_5Mn{-}H + F_2C{=}CF_2 \longrightarrow (CO)_5Mn{-}CF_2{-}CF_2H$$

The thermally stable manganese(II) derivatives, MnR_2 (R = CH_2SiMe_3, CH_2Bu^t, CH_2CMe_2Ph), are prepared using alkylmetal intermediates. The anionic species, $[MnR_3]^-$ and $[MnR_4]^{2-}$, with high numbers of metal-carbon bonds are obtained with R = Me, and C≡C—R′ (R′ = H, Me, Ph).

The rhenium anion, $Li_2[Re_2Me_8]$, is prepared from $ReCl_5$ and LiMe and contains a quadruple Re-Re bond; and $Re_3(CH_2SiMe_3)_6(\mu\text{-}Cl)_3$ is based on a Re_3 cluster:

$$[Me_4Re{\equiv}ReMe_4]^{2\ominus}$$

$R = CH_2SiMe_3$

Iron, Ruthenium, Osmium

Tetracarbonyliron diiodide, $Fe(CO)_4I_2$, reacts with pentafluorophenyllithium to give $C_6H_5{-}Fe(CO)_4I$, and analogous compounds are formed from iron pentacarbonyl with perfluoroalkyl iodides:

$$Fe(CO)_5 + R_FI \xrightarrow{70\,°C} R_F{-}Fe(CO)_4I + CO$$

Cyclopentadienyliron dicarbonyl derivatives are obtained by the reaction of the nucleophilic anion, $[(\eta^5\text{-}C_5H_5)Fe(CO)_2]^-$, with alkyl halides, hexafluorobenzene or substituted perfluorobenzenes:

$$(\eta^5\text{-}C_5H_5)Fe(CO)_2^{\ominus} + RX \longrightarrow (\eta^5\text{-}C_5H_5)Fe(CO)_2{-}R + X^{\ominus}$$

This anion also reacts with acyl halides, to form the acyliron derivatives, $(\eta^5\text{-}C_5H_5)Fe(CO)_2{-}CO{-}R$, which can be decarbonylated to $(\eta^5\text{-}C_5H_5)Fe(CO)_2{-}R$ (R = perfluoroalkyl, Ph, etc.).

The fluxional, mixed $\eta^5\text{-}C_5H_5{-}\sigma\text{-}C_5H_5$ derivative is prepared from the corresponding halide and sodium cyclopentadienide:

$$(\eta^5\text{-}C_5H_5)Fe(CO)_2{-}Cl + Na^{\oplus}C_5H_5^{\ominus} \xrightarrow[-CO]{} (\eta^5\text{-}C_5H_5)Fe(CO)_2(\sigma\text{-}C_5H_5)$$

Metal-hydride addition to olefins leads to σ-derivatives of iron:

$$(\eta^5\text{-}C_5H_5)Fe(CO)_2H + CH_2{=}CH{-}CH_3 \longrightarrow (\eta^5\text{-}C_5H_5)Fe(CO)_2{-}CH_2CH{=}CH_2$$

A rare homoleptic dimesityliron derivative is prepared from $FeCl_2$ by a Grignard route, and lithium tetrasubstituted ferrates and $Li_2[FeR_4]$ (R = Me, Ph) can be isolated as the dioxane adducts. Six σ-carbon bonds are found in the alkynyl-iron anions, $[Fe(C{\equiv}CR)_6]^{2-}$ (R = H, Me, Ph).

Cyclic σ-carbon derivatives of ruthenium and osmium are known:

$$(Ph_3P)_4M\langle(CH_2)_2\rangle SiR_2$$

Cobalt, Rhodium, Iridium

The tetracarbonylcobaltate anion reacts with methyl iodide to give the σ-carbon derivative, $(CO)_4Co{-}Me$, and with acetyl chloride to yield $(CO)_4Co{-}COMe$. Stable phosphino-derivatives, $(R'_3P)_2CoR_2$, form when the phenyl group attached to cobalt is *ortho*-substituted. Stable CoR_2(bipy) is prepared from cobalt(II) acetylacetonate, bipyridine and aluminum trialkyls:

$$(R'_3P)_2CoR_2 \qquad (bipy)CoR_2$$

The coordinatively unsaturated anion, $[Co(CN)_5]^-$, reacts with alkyl halides to form six-coordinated complexes, containing a σ-carbon bond, $[R{-}Co(CN)_5]^{3-}$.

Dimethylglyoximato chelates of cobalt, $Co(D_2H_2)LR$ (L = pyridine, H_2, etc.), have been investigated as B_{12}-vitamin models, since the coenzyme of this vitamin also contains a Co—C bond in a similar coordinative environment:

$$R{-}Co(D_2H_2){-}L$$

Tetrasubstituted cobalt anions, $[CoR_4]^{2-}$ with R = CH_2SiMe_3, Ph, C_6F_5, C_6Cl_5, Me, etc., are known. Hexasubstitution is achieved in the ethynyl derivatives, $[Co(C{\equiv}CR)_6]^{3-}$ and $[Co(C{\equiv}CR)_6]^{4-}$.

Cobalt vapor and bromopentafluorobenzene afford $Co(C_6F_5)_2$, which reacts further with toluene:

$$Co_{at} + C_6F_5Br \longrightarrow Co(C_6F_5)_2 \xrightarrow{PhMe} (\eta^6\text{-}MeC_6H_5)Co(C_6F_5)_2$$

The treatment of a phosphine rhodium or iridium halide with Grignard reagents gives σ-substituted derivatives:

$$(PR'_3)_3MBr_3 + 2\,RMgBr \longrightarrow (PR'_3)_3MR_2Br$$

M = Rh, Ir

Other Rh—C bonded derivatives include metallocycles derived from phosphorus ylides and polymethylene di-Grignard reagents:

$$LRh\text{-}[CH_2\text{-}PMe_2\text{=}N^{+}\text{=}PMe_2\text{-}CH_2] \qquad (\eta^5\text{-}C_5Me_5)Rh[(CH_2)_n](PPh_3)$$

L = 1,5 cyclooctadiene   n = 1, 2, 3

Nickel, Palladium, Platinum

Stable bis(phosphine)metal dialkyls, $(PR'_3)_2MR_2$, can be isolated for all three elements from the corresponding dihalides with organolithium reagents. The pentafluorophenyl derivative, $(PPh_3)_2Ni(C_6F_5)_2$, is more stable.

A bipyridyl-stabilized adduct has been prepared with the aid of organoaluminum reagents:

$$Ni(acac)_2 + Al(CH_2SiMe_3)_3 \xrightarrow{bipy} bipy \cdot Ni(CH_2SiMe_3)_2$$

The tetrasubstituted anions, $[NiR_4]^{2-}$ (with R = Me, Ph), and halogen-bridged dinuclear anions, $[(C_6F_5)_2NiX_2Ni(C_6F_5)_2]^{2-}$ (X = Cl, Br, also CN), are also known.

Cyclopentadienylmetal derivatives are obtained from the dimer with perfluoroalkyl iodides:

$$[(\eta^5\text{-}C_5H_5)Ni(CO)]_2 + R_FI \longrightarrow (\eta^5\text{-}C_5H_5)Ni(CO)\text{—}R_F + (C_5H_5)Ni(CO)I$$

Nickel vapor reacts with bromopentafluorobenzene to form unstable C_6F_5NiBr, which reacts further with toluene:

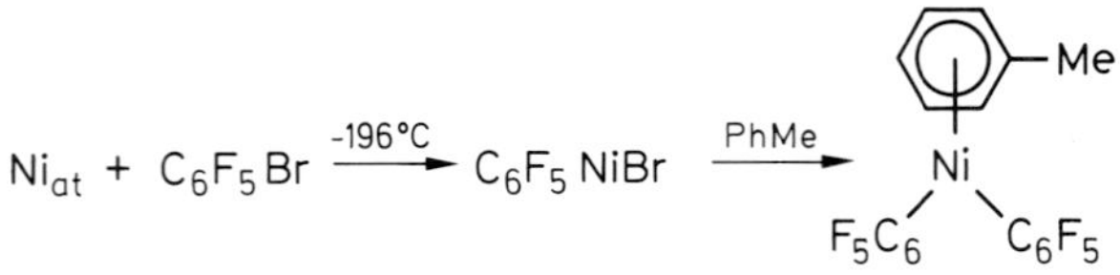

$$Ni_{at} + C_6F_5Br \xrightarrow{-196\,°C} C_6F_5NiBr \xrightarrow{PhMe} (\eta\text{-}C_6H_5Me)Ni(C_6F_5)_2$$

Nickel-containing metallocycles and spirocycles have been obtained from appropriate organolithium reagents:

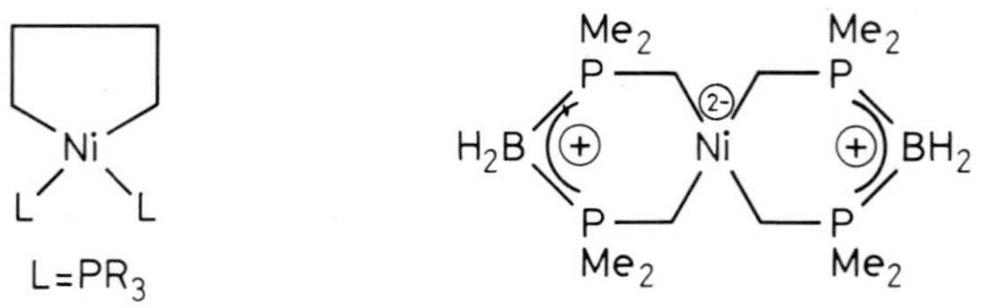

With the phosphorus ylid, $[Me_2P(CH_2)_2]^-$, the two $Ni_2[Me_2P(CH_2)_2]_4$ isomers have been prepared:

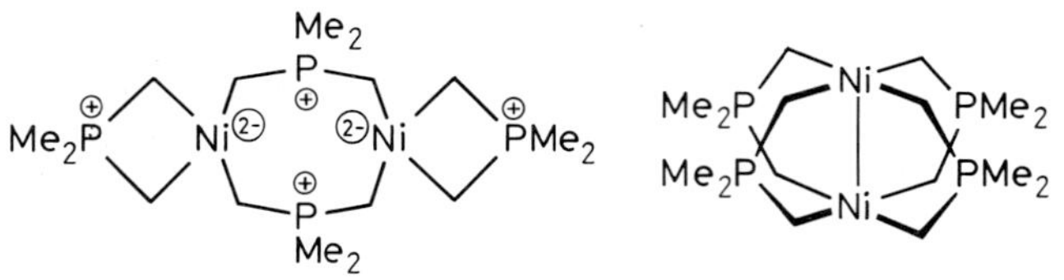

The square-planar palladium compounds, $(RP'_3)_2PdR_2$ and $(PR'_3)_2PdRX$, are prepared from the dihalides with Grignard reagents as both the *cis*- and *trans*-isomers.

Palladium forms σ-carbon bonds in *σ-ortho*-substituted derivatives by *ortho*-metalation or cyclometalation as in the complexation of palladium(II) chloride with azobenzene:

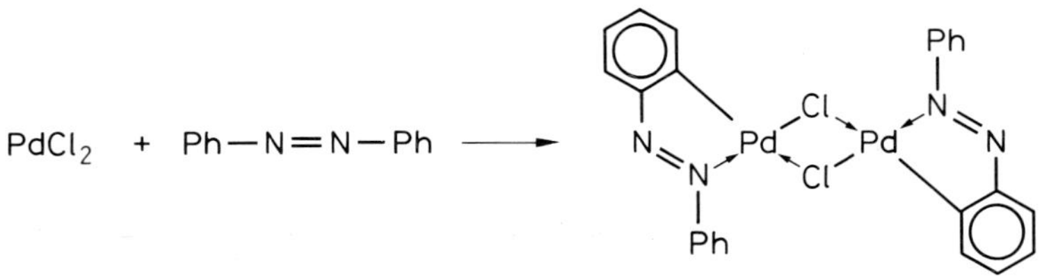

$$PdCl_2 + Ph{-}N{=}N{-}Ph \longrightarrow$$

Platinum forms cubic, tetrameric $(Me_3PtX)_4$ (X = Cl, I, etc.) σ-derivatives from the reaction of platinum(II) chloride with Grignard reagents:

Six σ-alkyls are found in the hexamethylplatinate anion, $[PtMe_6]^{2-}$, which is obtained from $(Me_3PtI)_4$ or $(NR_4)_2[PtCl_6]$ with methyllithium.

Double ylids of phosphorus react with metal halides to form nickel, palladium and platinum spirocyclic compounds:

$$Me_3P{=}C{=}PMe_3 + MCl_2 \longrightarrow$$

M = Ni, Pd, Pt

$$Me_3P{=}N{-}PMe_2{=}CH_2 + (Me_3P)_2MCl_2 \longrightarrow$$

M = Ni, Pt

Chelating ligands react with platinum and palladium halides to form complexes with σ-carbon bonds:

$$2\ o\text{-}(R_2PCH_2)C_6H_4Li + MCl_2 \longrightarrow$$

M = Pd, Pt

Copper, Silver, Gold

Polymeric monovalent RCu compounds are obtained from copper(I) halides with organolithium, organozinc or Grignard reagents.

The pentafluorophenyl derivative, $(C_6F_5Cu)_4$, is more stable than the alkyls. The trimethylsilylmethyl derivative is a tetramer, and contains alkyl bridges and weak Cu-Cu bonding interactions:

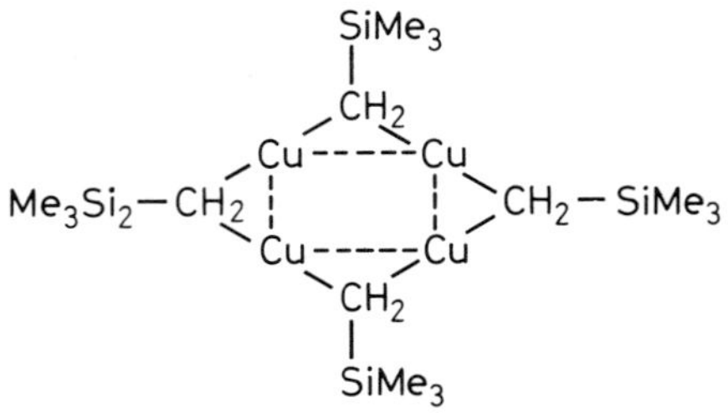

$$\begin{array}{c} SiMe_3 \\ | \\ CH_2 \\ Me_3Si_2-CH_2 \;\; [Cu_4] \;\; CH_2-SiMe_3 \\ CH_2 \\ | \\ SiMe_3 \end{array}$$

With excess phenyllithium, copper(I) iodide forms the unstable tetrasubstituted $Li[CuPh_4] \cdot nEt_2O$. The disubstituted anionic complexes, $[CuR_2]^-$, serve as reagents in synthetic organic chemistry. Anionic alkynyl derivatives, $[Cu(C{\equiv}CR)_2]^-$ (R = Me, Ph) and $[Cu(C{\equiv}CR)_3]^{2-}$ (R = H, Me, Ph), and their silver and gold analogues are also known in addition to neutral compounds, $(Cu{-}C{\equiv}CR)_n$.

Phosphorus and arsenic ylids form metallocycles containing two metal atoms:

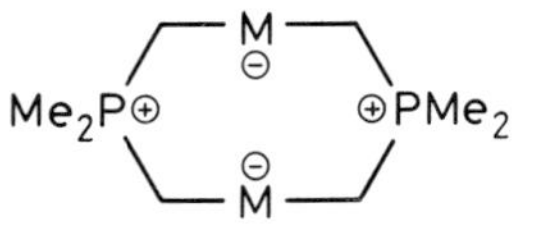

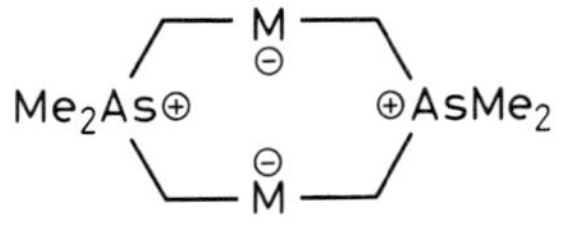

$$Me_2P^{\oplus}(CH_2)_2M^{\ominus}(CH_2)_2{}^{\oplus}PMe_2 \;(\text{cyclic, two } M^{\ominus}) \qquad Me_2As^{\oplus}(CH_2)_2M^{\ominus}(CH_2)_2{}^{\oplus}AsMe_2$$

M = Cu, Ag

Organolead, tin and bismuth compounds form with silver nitrate the σ-carbon, $(RAg)_2 \cdot AgNO_3$. The polymeric phenylsilver and other σ-aryl derivatives, $(AgR)_n$, are prepared from silver(I) salts and organozinc reagents. The dimeric, disubstituted silver derivatives,which contain aryl groups bridging silver and lithium atoms, are obtained from lithium aryls.

In the cationic silver derivatives, $[AgR_2]^+$ ($R = CH_2PPh_3$), the positive charge is localized at phosphorus and not at the metal:

$$Ph_3\overset{+}{P}{-}CH_2{-}\overset{-}{Ag}{-}CH_2{-}\overset{+}{P}Ph_3$$

Trimethylgold, prepared from gold(III) bromide and methyllithium, is stabilized by complexation with amines or tertiary phosphines, in $Me_3Au \cdot L$. The dimeric compounds, $(R_2AuX)_2$, are prepared from $[Au(py)Cl_3]$ and methylmagnesium iodide.

Gold(I) derivatives, stabilized by complexation with tertiary phosphines, R_3P AuR, are obtained from the halides, $R_3P \cdot AuX$, with organolithium or Grignard reagents. Examples are $(Me_3Si)_2CH{-}Au \cdot L$ ($L = PPh_3$, $AsPh_3$). The triorganogold complexes, $R_3Au \cdot L$, undergo reductive elimination, to form monovalent gold compounds, $RAu \cdot L$.

The di- and tetramethylaurate anions, $[AuMe_2]^-$ and $[AuMe_4]^-$, are prepared from organolithium reagents:

$$Me_3Au \cdot PPh_3 + LiMe \longrightarrow Li^+[AuMe_4]^-$$
$$MeAu \cdot PPh_3 + LiMe \longrightarrow Li^+[AuMe_2]^-$$

The former is linear, while the latter is square-planar. The anions are more thermally stable than neutral species, but less stable to oxygen.

The pentafluorophenyl derivatives of gold include the neutral species, $(C_6F_5)_nAu$ PR_3 (n = 1 or 3), the anionic species $[Au(C_6F_5)_n]^-$ (n = 2 or 4), $[(C_6F_5)_nAuX]^-$ (n = 1 or 3), or the cationic, $[(PPh_3)(C_6F_5)_2Au{-}X{-}Au(C_6F_5)_2(PPh_3)]^+$.

Gold compounds derived from phosphorus ylides include $Me_3Au{-}CH_2PMe_3$, $[Me_2Au(CH_2PMe_3)_2]Br$, $[Au(CH_2PMe_3)_2]^+X^-$ and several cyclic compounds:

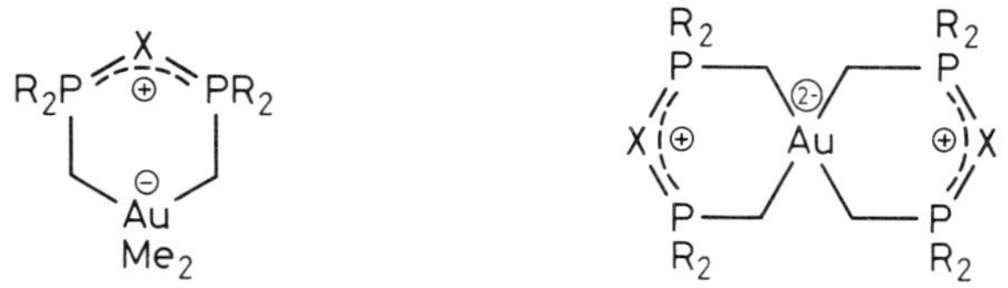

X = CH, N

X = CH,N

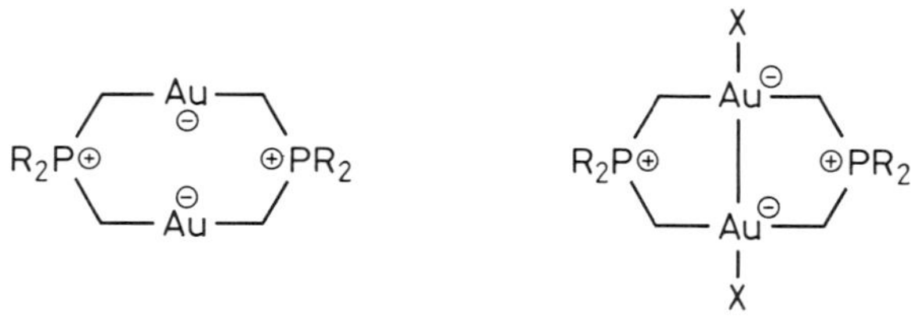

Heterocyclic gold compounds can be obtained by replacement of tin from a tetraphenylstannole with $AuCl_3$:

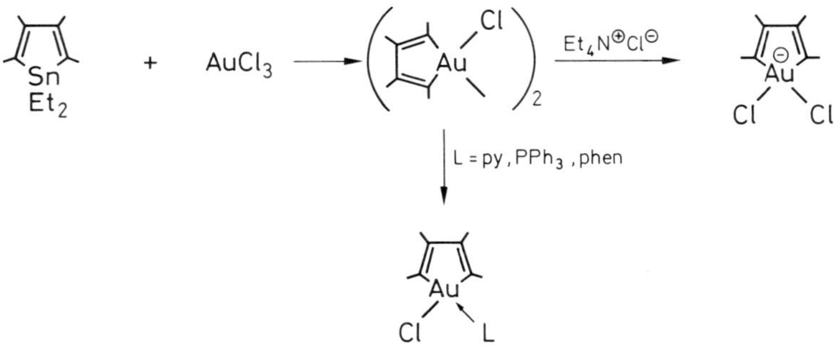

L = py, PPh₃, phen

The auration of benzene occurs on treatment with gold(III) chloride to give $(PhAuCl_2)_2$. Ferrocene and other η^5-cyclopentadienylmetal derivatives form compounds in which a ring carbon atom is bonded to two gold atoms. Their relation to σ-bonded compounds is illustrated by the interconversions:

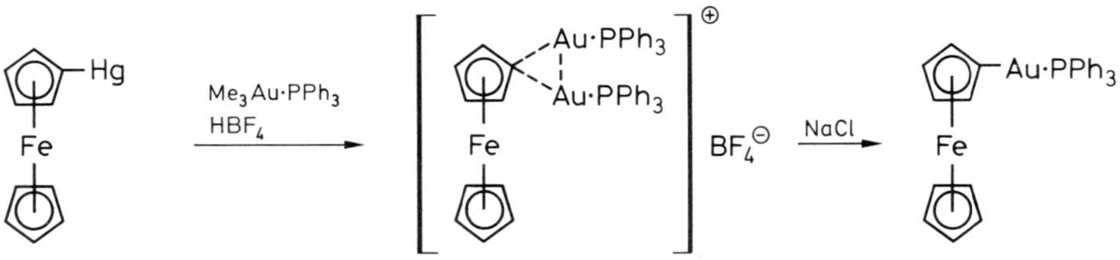

$$(C_5H_4AuPPh_3)Mn(CO)_3 \xrightarrow{HBF_4} [(C_5H_4(AuPPh_3)_2)Mn(CO)_3]^{\oplus} BF_4^{\ominus}$$

These structures involve polycenter —Au—Au—C-bonds.

Bibliography

Each chapter bibliography subdivision begins by listing books and monographs in chronological order of publication date with the most recent first. Review articles are then listed in the same order, with the two lists separated by a white line.

Chapter 1

a) Scope

J. Buckingham, *Dictionary of Organometallic Compounds*, 3 Vols., Chapman and Hall, London (1984).

A. Wojcicki, ed., *Organometallic Chemistry (Pure Appl. Chem.* Vol. *56*(1)), Pergamon Press, Oxford (1984) (Lectures from the XIth International Conference, Calloway Gardens, GA).

M. H. Chisholm, ed., *Inorganic Chemistry: Toward the 21st Century*, ACS Symposium Series, No. 211, American Chemical Society, Washington, DC (1983).

M. Tsutsui, Y. Ishii, H. Yaozeng, eds., *Fundamental Research in Organometallic Chemistry*, Van Nostrand Reinhold, New York (1982).

F. R. Hartley, S. Patai, eds., *The Chemistry of the Metal-Carbon Bond*, Vol. 1, *The Strucure, Preparation, Thermochemistry and Characterization of Organometallic Compounds*, Wiley, New York (1982).

F. R. Hartley, *Elements of Organometallic Chemistry*, Royal Society of Chemistry, London (1981).

R. B. King, ed., *Inorganic Compounds with Unusual Properties–II*, Advances in Chemistry Series, No. 173, American Chemical Society, Washington, DC (1980).

C. Eaborn, ed., *Further Perspectives in Organometallic Chemistry, J. Organomet. Chem.,200*, (1980) (Retrospectives and perspectives by 22 chemists).

R. B. King, ed., *Inorganic Compounds with Unusual Properties*, Advances in Chemistry Series, No. 150, American Chemical Society, Washington, DC (1976).

C. Eaborn, ed., *Perspectives in Organometallic Chemistry, J. Organomet. Chem., 100* (1975) (Retrospectives and new perspectives by 20 distinguished organometallic chemists).

G. E. Coates, M. L. H. Green, P. Powell, K. Wade, *Principles of Organometallic Chemistry*, Chapman and Hall, London (1968).

M. N. Hughes, *Mech. Inorg. Organomet. React., 65* (1983) 324 (Non-metallic reactions).

J. C. Martin, *Science, 221* (1983) 509 (Organo-nonmetallic chemistry).

G. A. Olah, G. K. S. Prakash, *Chem. Br., 19* (1983) 916 (Hypercarbon and hypercoordinated carbon compounds).

R. Hoffmann, *Science, 211* (1982) 995 (Theoretical organometallic chemistry).

J. P. Collman, *J. Organomet. Chem., 200* (1980) 79 (Correlations between coordination, organometallic and bioinorganic chemistry).

O. A. Reutov, *Tetrahedron, 34* (1978) 2827 (Non-transition metal organometallics).

J. B. Cence, *Chemistry, 46* (1973) 6 (Covalent carbon-metalloid compounds).

I. Haiduc, *J. Chem. Doc., 12* (1972) 175 (The literature of organometallic chemistry).

F. G. A. Stone, *Nature*, *232* (1971) 534 (Perspectives in organometallic chemistry).
M. F. Lappert, *Chem. Br.*, *5* (1969) 342 (Organometallic chemistry).
W. P. Neumann, *Naturwissenschaften*, *15* (1968) 553 (The increasing significance of organometallic chemistry).

b) History

J. J. Eisch, *J. Chem. Educ.*, *60* (1983) 1009 (Karl Ziegler biography).
E. C. Ashby, *J. Organomet. Chem.*, *200* (1980) 1 (A personal account of the early days of main-group hydrides).
H. Behrens, *Adv. Organomet. Chem.*, *18* (1980) 1 (A chemical autobiography – metal carbonyl chemistry in liquid ammonia).
R. Calas, *J. Organomet. Chem.*, *200* (1980) 11 (Thirty years in organosilanes).
H. C. Clark, *J. Organomet. Chem.*, *200* (1980) 63 (A personal account of a quarter century of research).
J. P. Collman, *J. Organomet. Chem.*, *200* (1980) 79 (Reminiscences of a career in coordination, organometallic and bioinorganic chemistry).
M. L. H. Green, *J. Organomet. Chem.*, *200* (1980) 119 (Transition-metal arene developments).
J. Halpern, *J. Organomet. Chem.*, *200* (1980) 133 (A homogeneous catalytic-hydrogenation retrospective).
H. D. Kaesz, *J. Organomet. Chem.*, *200* (1980) 145 (The author's research into transition-metal hydride complexes and clusters).
P. M. Maitlis, *J. Organomet. Chem.*, *200* (1980) 161 (A personal account of palladium-acetylene and cyclobutadiene research).
H. Sakurai, *J. Organomet. Chem.*, *200* (1980) 261 (The author's research on organopolysilanes).
H. Werner, *J. Organomet. Chem.*, *200* (1980) 335 (Discovery of the triple-decker sandwich compounds).
G. Wilke, *J. Organomet. Chem.*, *200* (1980) 349 (Contributions to homogeneous catalysis, 1955–1980).
H. C. Brown, *Chem. Eng. News*, *59* (1980) 24 (Priestly Medal address; adventures in research).
H. C. Brown, K. Krishnamurthy, *Tetrahedron*, *35* (1979) 567 (Forty years of hydride reductions).
P. L. Pauson, *Pure Appl. Chem.*, *49* (1977) 839 (A quarter-century of aromatic transition-metal complexes).
K. P. Callahan, M. F. Hawthorne, *Adv. Organomet. Chem.*, *14* (1976) 145 (Ten years of metallocarboranes).
J. A. Zubieta, J. J. Zuckerman, *Chem. Eng. News*, American Chemical Society Centennial Issue, 6 April, 1976 (Inorganic chemistry: the past 100 years).
J. J. Zuckerman, in *A Century of Chemistry*, H. Skolnik, K. M. Reese, eds., American Chemical Society, Washington, DC (1976) (Inorganic chemistry).
J. Chatt, *Adv. Organomet. Chem.*, *15* (1974) 1 (Organotransition-metal complexes).
E. O. Fischer, *Adv. Organomet. Chem.*, *14* (1976) 1 (The discovery of carbene and carbyne complexes).
F. A. Cotton, *J. Organomet. Chem.*, *100* (1975) 29 (Fluxionality discovered).
D. Seyferth, *J. Organomet. Chem.*, *100* (1975) 237 (The preparation of the elusive silacyclopropanes).
F. G. A. Stone, *J. Organomet. Chem.*, *100* (1975) 257 (Synthetic applications of d^{10}-metal complexes).

G. Wilkinson, *J. Organomet. Chem.*, *100* (1975) 275 (A recollection of the first four months after the discovery of ferrocene).
J.S. Thayer, *Adv. Organomet. Chem.*, *13* (1974) 1 (A very readable presentation of the history of organometallic chemistry).
H.C. Brown, *Adv. Organomet. Chem.*, *11* (1973) 1; *J. Organomet. Chem.*, *100* (1975) 3 (Boranes in organic synthesis).
A.N. Nesmeyanov, *Adv. Organomet. Chem.*, *10* (1972) 1 (An account of his many contributions).
W. Hieber, *Adv. Organomet. Chem.*, *8* (1970) 1 (Early investigations in metal-carbonyl chemistry).
E.G. Rochow, *Adv. Organomet. Chem.*, *9* (1970) 1 (An historical account of some fundamental developments).
J.S. Thayer, *J. Chem. Educ.*, *46* (1969) 764 (Historical origins of organometallic chemistry).
H. Gilman, *Adv. Organomet. Chem.*, *7* (1968) 1 (Some personal notes on organometallic chemistry; a charming account).
O. Yu. Okhlobystin, *J. Gen. Chem. USSR (Engl. Transl.)*, *37* (1967) 2261 (Development of organometallic chemistry in the Soviet Union).
K. Ziegler, *Adv. Organomet. Chem.*, *6* (1967) 1 (Personal experiences in organometallic chemistry).
M.D. Rausch, in *Werner Centennial*, G.B. Kauffman, ed., Advances in Chemistry Series, No. 62, American Chemical Society, Washington, DC, (1966) p. 486 (The history and development of organotransition-metal chemistry).

c) Bio-Organometallic Chemistry

J.S. Thayer, *Organometallic Compounds and Living Organisms*, Academic Press, New York (1984).

A.J. Crowe, *Chem. Ind. (London)*, (1983) 304 (Organometallics in medicine).
E.A. Koerner von Gustorf, L.H.G. Leendos, I. Fischler, R.N. Perutz, *Adv. Inorg. Chem. Radiochem.*, *19* (1976) 65 (Biological implications of transition-metal photochemistry).
D.D. Perrin, *Top. Curr. Chem.*, *64* (1976) 181 (Inorganic medicinal chemistry).
R.A.D. Wentworth, *Coord. Chem. Rev.*, *18* (1976) 1 (Mechanisms of Mo reactions in enzymes).
W.G. Zumft, *Struct. Bonding (Berlin)*, *29* (1976) 1 (Biological dinitrogen fixation).
J.S. Thayer, *J. Organomet. Chem.*, *76* (1974) 265 (Organometallic compounds and living organisms).
G.N. Schrauzer, *Angew. Chem., Int. Ed. Engl.*, *14* (1975) 514 (Non-enzymatic simulation of nitrogenase reactions).
J.S. Thayer, *J. Chem. Educ.*, *48* (1971) 806 (Biological aspects of organometallic chemistry).
J.M. Barnes, L. Magos, *Organomet. Chem. Rev., A*. *3* (1968) 137 (Organometallic toxicology).

d) Environmental-Organometallic Chemistry

M. Webb, ed., *The Chemistry, Biochemistry and Biology of Cadmium*, Vol. 2, Elsevier, Amsterdam (1979).

M. Webb, ed., *The Biogeochemistry of Mercury in the Environment*, Elsevier, Amsterdam (1979).
F. E. Brinkman, J. M. Bellama, eds., *Organometals and Organometalloids. Occurrence and Fate in the Environment*, ACS Symposium Series, No. 82, American Chemical Society, Washington, DC (1978).
J. O. Nriagu, ed., *Topics in Environmental Health*, Vol. 1, *The Biogeochemistry of Lead*, Parts A and B, Ann Arbor Sci. Publ., Ann Arbor, MI (1978).

P. J. Craig, in *Comprehensive Organometallic Chemistry*, Vol. 2, G. Wilkinson, F. G. A. Stone, E. W. Abel, eds., Pergamon Press, Oxford (1982), p. 979 (Environmental aspects).
J. S. Thayer, F. E. Brinckman, *Adv. Organomet. Chem.*, *20* (1982) 313 (Biomethylation of metals and metalloids).
J. S. Thayer, *J. Chem. Educ.*, *58* (1981) 764 (Controlled-release apllications of organometallics).
W. H. Zoller, *Science*, *208* (1980) 500 (Methyl metals in nature).
J. M. Barnes, L. Magos, *Organomet. Chem. Rev.*, *3* (1968) 137 (Toxicology of organometallic compounds).

e) Organometallic Radiochemistry

D. R. Wiles, *Adv. Organomet. Chem.*, *11* (1973) 207 (Organometallic radiochemistry).

f) Organometallic Polymers

B. M. Culbertson, C. U. Pittmann, Jr., *New Monomers and Polymers*, Plenum Press, New York (1984).
A. H. Cowley, ed., *Rings, Clusters and Polymers of the Main-Group Elements*, ACS Symposium Series, No. 232, American Chemical Society, Washington, DC (1983).
C. E. Carraher, Jr., *Advances in Organometallic and Inorganic Polymer Science*, M. Dekker, New York (1982).
E. Neuse, H. Rosenberg, *Metallocene Polymers*, Marcel Dekker, New York (1979).
C. E. Carraher, Jr., J. Sheats, C. U. Pittman, Jr., eds., *Organometallic Polymers*, Academic Press, New York (1978).
A. L. Rheingold, ed., *Homoatomic Rings, Chains and Macromolecules of the Main-Group Elements*, Elsevier, Amsterdam (1977).
K. A. Andrianov, *Metallorganic Polymers*, Wiley, New York (1964).

C. E. Carraher, Jr., *J. Chem. Educ.*, *58* (1981) 921 (Organometallic polymers).
I. Hagihara, Sonagashira, Takahashi, *Adv. Polm. Sci.*, *41* (1981) 149 (Linear organotransition-metal polymers).
D. W. Slocum, M. Hodgman, K. Kuchel, M. Moronski, R. Noble, K. Webber, S. Duraj, A. Siegel, D. A. Owen, *J. Macromol. Sci., Chem.*, *16* (1981) 357 (Polymers with organometallic residues).
C. Pittman, in *Organometallic Reactions*, eds. E. Becker, M. Tsutsui, Vol. 6, Plenum Press, New York (1977) (Organometallic polymers).
C. U. Pittman, Jr., *Organomet. React. Synth.*, *6* (1977) 1 (Vinyl polymers containing transition metals).

g) Mechanisms

J. D. Atwood, *Mechanisms of Inorganic and Organometallic Reactions*, Wadsworth, Belmont, CA (1984).

M. V. Twigg, ed., *Mechanisms of Inorganic and Organometallic Reactions*, Vol. 1, Plenum Press, New York (1983); Vol. 2 (1984).

A. G. Sykes, ed., *Advances in Inorganic and Bioinorganic Reaction Mechanisms*, Vol. 1, Academic Press, New York (1982).

J. H. Brewster, ed., *Aspects of Mechanisms in Organometallic Chemistry*, Plenum Press, New York (1978).

J. K. Kochi, *Organometallic Mechanisms and Catalysis*, Academic Press, New York (1978).

R. F. Heck, *Organotransition-Metal Chemistry: A Mechanistic Approach*, Academic Press, New York (1974).

D. S. Matteson, *Organometallic Reaction Mechanisms of the Non-Transition Elements*, Academic Press, New York (1974).

C. H. Bamford, C. F. H. Tipper, eds., *Comprehensive Chemical Kinetics*, Vol. 4, *Decomposition of Inorganic and Organometallic Compounds*, Elsevier, Amsterdam (1972); Vol. 7, *Reactions of Metal Salts and Complexes and Organometallic Compounds* (1972).

J. J. Eisch, *Pure Appl. Chem., 56* (1984) 35 (Mechanisms of organometallic oxidative additions).

R. J. Klingler, S. Fukuzumi, J. K. Kochi, in *Inorganic Chemistry Toward the 21st Century*, M. H. Chisholm, ed., ACS Symposium Series, No. 211, American Chemical Society, Washington, DC (1983), p. 117 (Electron-transfer mechanisms).

A. R. Manning, *Coord. Chem. Rev., 51* (1983) 41 (Mechanisms of electrophilic attack on polynuclear transition-metal carbonyls).

A. Poë, *Chem. Br., 19* (1983) 997 (Kinetic behavior of metal-metal bonded carbonyls).

B. A. Dolgoplosk, *Russ. Chem. Rev. (Engl. Transl.), 52* (1983) 1086 (Organotransition metal decomposition mechanisms).

O. A. Reutov, *J. Organomet. Chem., 250* (1983) 145 (Ion pairs in organometallic S_E-Reactions).

J. Halpern, *Science, 217* (1982) 401 (Asymmetric hydrogenation mechanisms).

P. Biloen, W. M. H. Sachtler, *Adv. Catal., 30* (1981) 165 (Fischer-Tropsch catalysis mechanisms).

D. Forster, A. Herschman, D. E. Morris, *Catal. Rev., 23* (1981) 89 (Mechanism of olefin hydrocarboxylation catalysis by Co).

F. C. Gault, *Adv. Catal., 30* (1981) 1 (Mechanisms of skeletal isomerization of hydrocarbons on metals).

C. K. Rofer-DePoorter, *Chem. Rev., 81* (1981) 447 (Fischer-Tropsch mechanism).

G. Henrici-Olivé, S. Olivé, *ChemTech. 11* (1981) 746 (Mechanism of Ziegler-Natta polymerization).

V. A. Zakharov, G. D. Bukatov, Yu. I. Ermakov, *Russ. Chem. Rev. (Engl. Transl.) 49* (1981) 1097 (Mechanism of Ziegler-Natta polymerization).

E. L. Muetterties, *Inorg. Chim. Acta, 50* (1981) 1 (Mechanisms of homogeneous catalytic hydrogenation).

I. Beletskaya, *Sov. Sci. Rev., Chem. Rev., B, 1* (1981) 1 (Organometallic reaction mechanisms).

J. Halpern, *Inorg. Chim. Acta, 50* (1981) 11 (Mechanisms of homogeneous catalytic hydrogenation).

G. K. Anderson, R. J. Cross, *Chem. Soc. Rev., 9* (1980) 185 (Isomerization mechanisms of square-planar complexes).

J. K. Kochi, *Pure Appl. Chem., 52* (1980) 571 (Electron- and charge-transfer in organometallic chemistsry).

V. A. Zakharov, G. D. Bukatov, Yu. I. Ermakov, *Russ. Chem. Rev. (Engl. Transl.)*, *49* (1980) 1097 (Mechanism of catalytic polymerization of olefins).
I. P. Beletskaya, *Sov. Sci. Rev., Sect. B, Chem. Rev.*, *1* (1979) (Reaction mechanisms of organometallic compounds).
J. Halpern, *Pure Appl. Chem.*, *51* (1979) 2171 (Free-radical mechanisms in coordination chemistry).
E. L. Muetterties, J. Stein, *Chem. Rev.*, *79* (1979) 479 (Mechanisms of catalytic CO hydrogenation).
O. A. Reutov, *Pure Appl. Chem.*, *50* (1978) 717; *J. Organomet. Chem.*, *100* (1975) 219 (Mechanism of substitution of non-transition-metal organometallics).
V. Gutmann, *Coord. Chem. Rev.*, *18* (1976) 225 (Solvent effects on organometallic reactivity).
R. A. D. Wentworth, *Coord. Chem. Rev.*, *18* (1976) 1 (Mechanism of Mo reactions in enzymes).
J. K. Kochi, *Acc. Chem. Res.*, *7* (1974) 351 (Electron transfer in catalysis).
M. Orchin, W. Rupilius, *Catal. Rev.*, *6* (1972) 85 (Oxo-reaction mechanisms).
I. P. Beletskaya, K. P. Butin, O. A. Reutov, *Organomet. Chem. Rev.*, *A*, *7* (1971) 51 ($S_E1(N)$ mechanism in organometallic chemistry).
R. E. Dessy, W. Kitching, *Adv. Organomet. Chem.*, *4* (1966) 267 (Organometallic reaction mechanisms).

h) Catalysis

M. Graziani, G. M. Giongo, eds., *Fundamental Research in Homogeneous Catalysis*, Vol. 4, Plenum Press, New York (1984).
J. D. Morrison, ed., *Asymmetric Synthesis*, 2 Vols., Academic Press, New York (1984).
W. Keim, ed., *Catalysis in C_1 Chemistry*, D. Reidel, Dordrecht (1983).
J. H. Sinfelt, *Bimetallic Catalysis: Discoveries, Concepts and Applications*, Wiley, New York (1983).
E. Dehmlow, S. Dehmlow, *Phase-Transfer Catalysis*, 2nd. ed., Verlag Chemie, Weinheim (1983).
L. M. Pignolet, ed., *Homogeneous Catalysis with Metal Phosphine Complexes*, Plenum Press, New York (1983).
R. P. Quirk, H. L. Hsieh, G. B. Klingensmith, P. J. T. Tait, *Transition-Metal Catalyzed Polymerizations: Alkenes and Dienes*, Harwood, Chur, Switzerland (1983).
E. C. Alyea, D. W. Meck, eds., *Catalytic Aspects of Metal Phosphine Complexes*, Advances in Chemistry Series, No. 196, American Chemical Society, Washington, DC (1982).
J. S. Miller, ed., *Chemically-Modified Surfaces in Catalysis and Electrocatalysis*, ACS Symposium Series, No. 192, American Chemical Society, Washington, DC (1982).
C. Masters, *Homogeneous Transition-Metal Catalysis – A Gentle Art*, Chapman and Hall, London (1981).
R. Ugo, ed., *Aspects of Homogeneous Catalysis*, Vol. 4, Reidel, Dordrecht (1981).
R. A. Sheldon, J. K. Kochi, *Metal-Catalyzed Oxidations of Organic Compounds*, Academic Press, New York (1981).
T. Seiyma, K. Tanabe, eds., *New Horizons in Catalysis*, 2 Vols., Elsevier, Amsterdam (1981).
Y. I. Yermakov, B. N. Kuznetsov, V. A. Zakharov, *Catalysis by Supported Complexes*, Elsevier, Amsterdam (1981).
P. C. Ford, ed., *Catalytic Activation of Carbon Monoxide*, ACS Symposium Series, No. 152, American Chemical Society, Washington, DC (1981).
H. Pines, *The Chemistry of Catalytic Hydrocarbon Conversions*, Academic Press, New York (1981).

A. Nakamura, M. Tsutsui, *Principles and Applications of Homogeneous Catalysis*, John Wiley, New York (1980).
G. W. Parshall, *Homogeneous Catalysis. The Applications and Chemistry of Catalysis by Soluble Transition-Metal Complexes*, Wiley-Interscience, New York (1980).
A. Nakamara, M. Tsutsui, *Principles and Applications of Homogeneous Catalysis*, Wiley, New York (1980).
J. Falbe, *New Syntheses with Carbon Monoxide*, Springer Verlag, Berlin (1980).
M. Tsutsui, ed., *Fundamental Research in Homogeneous Catalysis*, Vol. 3, Plenum Press, New York (1979).
B. C. Gates, J. R. Katzer, G. C. A. Schuit, *Chemistry of Catalytic Processes*, McGraw-Hill, New York (1979).
E. L. Kugler, F. W. Steffgen, eds., *Hydrocarbon Synthesis from Carbon Monoxide and Hydrogen*, Advances in Chemistry Series, No. 178, American Chemical Society, Washington, DC (1979).
S. D. Christian, J. J. Zuckerman, L. J. Guggenberger, eds., *Structural Aspects of Homogeneous, Heterogeneous and Biological Catalysis, Trans. Am. Crystallog. Assoc.*, Vol. 14 (1978).
Y. Ishii, M. Tsutsui, eds., *Fundamental Research in Homogeneous Catalysis*, Plenum Press, New York, Vol. 2 (1978).
J. K. Kochi, *Organometallic Mechanisms and Catalysis*, Academic Press, New York (1978).
M. Freifelder, *Catalytic Hydrogenation in Organic Synthesis: Procedures and Commentary*, Wiley, New York (1978).
G. V. Smith, ed., *Catalysis in Organic Chemistry*, Academic Press, New York (1977–1978).
G. Henrici-Olivé, S. Olivé, *Coordination and Catalysis*, Verlag Chemie, Weinheim (1977).
R. Ugo, ed., *Aspects of Homogeneous Catalysis*, Vol. 3, E. Reidel, Dordrecht (1977).
F. J. McQuillan, *Homogeneous Hydrogenation in Organic Chemistry*, Reidel, Dordrecht (1976).
B. L. Shaw, N. I. Tucker, *Organotransition-Metal Compounds and Related Aspects of Homogeneous Catalysis*, Pergamon Press, Oxford (1975).
B. Delmon, G. Jannes, eds., *Catalysis, Homogeneous and Heterogeneous*, Elsevier, Amsterdam (1975).
M. M. T. Khan, A. E. Martell, *Homogeneous Catalysis by Metal Complexes*, 2 Vols., Academic Press, New York (1974).
B. R. James, *Homogeneous Hydrogenation*, Wiley, New York (1973).
P. N. Rylander, *Organic Synthesis with Noble-Metal Catalysts*, Academic Press, New York (1973).
G. N. Schrauzer, ed., *Transition Metals in Homogeneous Catalysis*, Marcel Dekker, New York (1971).
R. Ugo, ed., *Aspects of Homogeneous Catalysis*, Vol. 1, Manfredi, Milan (1970).
R. L. Augustine, *Catalytic Hydrogenation*, M. Dekker, New York (1965).

Yu. I. Ermakov, *J. Mol. Catal., 21* (1983) 35 (Organometallics for supported catalysts).
J. M. Bassett, A. Choplin, *J. Mol. Catal., 21* (1983) 95 (Surface organometallic heterogeneous catalysis).
Z. Paal, P. Z. Menon, *Catal. Rev., 25* (1983) 229 (Hydrogen effects in metal catalysis).
L. C. Costa, *Catal. Rev., 25* (1983) 325 (Product selectivity in the homogeneous hydrogenation of CO to oxygenates).
E. Furimsky, *Catal. Rev., 25* (1983) 421 (Catalytic hydrodeoxygenation).
T. H. Maugh, II, *Science, 219* (1983) 474, 944, 1413; *220* (1983) 592, 1032, 1261 (A series of research news articles on catalysis).
T. Bartik, P. Heimbach, H. Schenkluhn, *Kontakte*, (1983) 16 (Controlling metal-catalyzed organic synthesis).
H. S. Mosher, J. D. Morrison, *Science, 221* (1983) 1013 (Asymmetric synthesis).

R. R. Schrock, *Science*, *219* (1983) 13; in *Inorganic Chemistry Toward the 21st Century*, M. H. Chisholm, ed., ACS Symposium Series, No. 211, American Chemical Society, Washington, DC (1983), p. 369 (Multiple metal-carbon bonds in catalysis).

E. L. Muetterties, *Pure Appl. Chem.*, *54* (1982) 83 (Organometallic Chemistry of metal surfaces).

M. Chanon, M. L. Tobe, *Angew. Chem., Int. Ed. Engl.*, 21 (1982) 1 (Electron-transfer catalysis).

W. A. Herrmann, *Angew. Chem., Int. Ed. Engl.*, *21* (1982) 117 (Fischer-Tropsch synthesis).

R. J. Puddephatt, *Comments Inorg. Chem.*, *2* (1982) 69 (Metallocycloalkanes as catalytic intermediates).

D. R. Armstrong, P. G. Perkins, *Coord. Chem. Rev.*, *38* (1981) 139 (Electronic structure and homogeneous catalysis by main-group compounds).

P. Heimbach, H. Schenkluhn, K. Wisseroth, *Pure Appl. Chem.*, *53* (1981) 2419 (Control of catalyzed organic reactions).

P. J. Brothers, *Prog. Inorg. Chem.*, *28* (1981) 1 (Heterolytic activation of hydrogen by complexes).

I. P. Beletskaya, D. I. Makhon'kov, *Russ. Chem. Rev. (Engl. Transl.)*, *50* (1981) 534 (Oxidation of alkyl-substituted aromatics by transition-metal salts).

G. H. Posner, *Pure Appl. Chem.*, *53* (1981) 2307 (Asymmetric synthesis using organometallic reagents).

R. Noyori, *Pure Appl. Chem.*, *53* (1981) 2315 (Asymmetric synthesis via axially dissymmetric molecules).

B. M. Trost, *Pure Appl. Chem.*, *53* (1981) 2357 (Transition-metal templates for selectivity in organic synthesis).

H. Mimoun, *Pure Appl. Chem.*, *53* (1981) 2389 (Selective oxidation of olefins).

H. Alper, *Adv. Organomet. Chem.*, *19* (1981) 183 (Phase-transfer catalysis).

V. Caplar, G. Comisso, V. Sunjic, *Synthesis*, (1981) 85 (Homogeneous asymmetric hydrogenation).

D. C. Bailey, S. H. Langer, *Chem. Rev.*, *81* (1981) 109 (Supported metal-carbonyl catalysts).

C. K. Rofer-De Poorter, *Chem. Rev.*, *81* (1981) 447 (Fischer-Tropsch synthesis).

J. Evans, *Chem. Soc. Rev.*, *10* (1981) 159 (Supported metal-carbonyl catalysts).

S. D. Jackson, P. B. Wells, R. Whyman, P. Worthington, *Catal.*, *4* (1981) 75 (Metal-cluster catalysts).

E. Negishi, *Pure Appl. Chem.*, *53* (1981) 2333 (Catalyzed cross-coupling reactions).

E. L. Muetterties, *Catal. Rev.*, *23* (1981) 69 (Metal-cluster catalysts).

D. Foster, A. Hershman, D. E. Morris, *Catal. Rev.*, *23* (1981) 89 (Mechanism of olefin hydrocarboxylation by Co, Rh and Ir complexes).

G. W. Parshall, *Catal. Rev.*, *23* (1981) 107 (Industrial Co and Rh catalysis).

G. Henrici-Olivé, S. Olivé, *ChemTech.*, *11* (1981) 746 (Ziegler-Natta polymerization).

R. E. Merrill, *ChemTech.*, *11* (1981) 118 (Asymmetric catalysis).

B. Bosnich, M. D. Fryzuk, *Top. Stereochem.*, *12* (1981) 119 (Asymmetric catalysis).

U. Matteoli, P. Frediani, M. Bianohi, C. Botteghi, S. Gladiali, *J. Mol. Catal.*, *12* (1981) 265 (Asymmetric catalysis).

D. W. Slocum, M. Hodgman, K. Kuchel, M. Moronski, R. Noble, K. Webber, S. Duraj, A. Siegel, D. A. Owen, *J. Macromol. Sci., Chem.*, *16* (1981) 357 (Catalytic polymers containing organometallic residues).

G. A. Somarjai, *Catal. Rev.*, *23* (1981) 189 (Fischer-Tropsch catalysis).

D. L. King, J. A. Cusumano, R. L. Garten, *Catal. Rev.*, *23* (1981) 233 (Fischer-Tropsch catalysis).

J. A. Davies, F. R. Hartley, *Chem. Rev.*, *81* (1981) 79 (Pt-metal catalysis with weak-donor ligands).

H. Wynberg, *Recl. J. R. Neth. Chem. Soc.*, *100* (1981) 393 (Asymmetric catalysis).

P.C. Ford, *Acc. Chem. Res., 14* (1981) 31 (Water-gas shift reaction catalysis).
L. Moggi, A. Juris, D. Sandrini, M.F. Manfrin, *Rev. Chem. Intermed., 4* (1981) 171 (Photocatalysis with transition-metal complexes).
J.A. Osborn, *Strem Chem., 9* (1981) 1 (Homo- and heterogeneous catalysts).
A.L. Lapidus, Y.Y. Ping, *Russ. Chem. Rev. (Engl. Transl.), 50* (1981) 63 (CO_2-fixation by transition-metal complexes).
P. Heimbach, H. Schenkluhn, *Top. Curr. Chem., 92* (1980) 45 (Control of metal-catalyzed organic reactions).
J. Furukawa, *Acc. Chem. Res., 13* (1980) 1 (Stereoregular and sequence-regular polymerization of butadiene).
P. Pino, R. Mulhaupt, *Angew. Chem., Int. Ed. Engl., 19* (1980) 857 (Stereospecific polymerization of propylene – a quarter century).
C.P. Casey, S.M. Neumann, M.A. Andrews, D.R. McAlister, *Pure Appl. Chem., 52* (1980) 625 (Metal-catalyzed CO reduction).
A.J. Deeming, I.P. Rothwell, *Pure Appl. Chem., 52* (1980) 649 (C-H bond activation in transition-metal compounds).
Yu. I. Yermakov, *Pure Appl. Chem., 52* (1980) 2075 (Anchored complex catalysts).
J. Halpern, *J. Organomet. Chem., 200* (1980) 133 (Homogeneous catalytic hydrogenation).
K.I. Zamaraev, *Kinet. Catal. (Engl. Transl.), 21* (1980) 223 (ESR and NMR of paramagnetic complexes in homogeneous catalysis).
R.L. Pruett, *Adv. Organomet. Chem., 17* (1979) 1 (Hydroformylation).
C. Masters, *Adv. Organomet. Chem., 17* (1979) 61 (The Fischer-Tropsch reaction).
N. Calderon, J.P. Lawrence, E.A. Ofstead, *Adv. Organomet. Chem., 17* (1979) 449 (Olefin metathesis).
E.L. Muetterties, J.R. Bleeke, *Acc. Chem. Res., 12* (1979) 324 (Catalytic hydrogenation of aromatic hydrocarbons).
V.A. Zakharov, Yu. I. Yermakov, *Catal. Rev., 19* (1969) 67 (Supported organometallic catalysts for olefin polymerization).
C.A. Tolman, *Chem. Rev., 77* (1977) 313 (Phosphorus ligands in homogeneous catalysis).
V.A. Sergeev, V.K. Shitikov, V.A. Pankratov, *Russ. Chem. Rev. (Engl. Transl.), 48* (1979) 79 (Polycyclotrimerization).
S.F. Martin, *Synthesis*, (1979) 633 (Aldehyde, ketone and carboxylic acid synthesis from lower carbonyls by C-C coupling).
J.W. Apsimon, R.P. Seguin, *Tetrahedron, 35* (1979) 2797 (Asymmetric synthesis).
G. Manecke, W. Storck, *Angew. Chem., Int. Ed. Engl., 17* (1978) 657 (Polymeric catalysts).
A.J. Birch, D.H. Williamson, *Org. React., 24* (1978) 1 (Homogeneous hydrogenation catalysts).
R.H. Grubbs, *Prog. Inorg. Chem., 24* (1978) 1 (Olefin metathesis).
A. Ya. Yuffa, G.V. Lisichkin, *Russ. Chem. Rev. (Engl. Transl.), 47* (1978) 751 (Heterogeneous metal-complex catalysts).
D. Valentine, J.W. Scott, *Synthesis*, (1978) 329 (Asymmetric synthesis).
H.B. Kagan, J.C. Fiand, *Top. Stereochem., 10* (1978) 175 (Asymmetric synthesis).
A.E. Shilov, *Pure Appl. Chem., 50* (1978) 725 (Alkane activation by transition-metal complexes).
T.J. Katz, *Adv. Organomet. Chem., 16* (1977) 283 (Olefin metathesis).
R.F. Heck, *Adv. Catal., 26* (1977) 323 (Organic-halide reactions with CO, olelfins and acetylenes).
K.P.C. Vollhardt, *Acc. Chem. Res., 10* (1977) 1 (Acetylene cyclizations).
C.A. Tolman, *Chem. Rev., 77* (1977) 313 (Phosphorus ligands in homogeneous catalysis).
A.E. Shilov, A.A. Shteinman, *Coord. Chem. Rev., 24* (1977) 97 (Activation of hydrocarbons in solution).

G. W. Daub, *Prog. Inorg. Chem.*, *22* (1977) 409 (Oxidative cleavage of transition metal-carbon bonds).
D. E. Webster, *Adv. Organomet. Chem.*, *15* (1976) 147 (Alkane activation).
N. Calderon, E. A. Ofstead, W. A. Judy, *Angew. Chem., Int. Ed. Engl.*, *15* (1976) 401 (Olefin metathesis mechanisms).
I. Wender, *Catal. Rev.*, *14* (1976) 97 (Catalytic synthesis from coal).
S. M. Atlan, H. F. Mark, *Catal. Rev.*, *13* (1976) 1 (Polymerization catalysis).
Yu. I. Yermakov, *Catal. Rev.*, *13* (1976) 77 (Oxide-supported catalysts).
K. C. Bishop, III, *Chem. Rev.*, *76* (1976) 461 (Catalyzed rearrangements of small ring organics).
R. H. Grubbs, *J. Organomet. Chem. Libr.*,*1* (1976) 423 (Olefin metathesis).
B. A. Krentsel, *Russ. Chem. Rev. (Engl. Transl.)*, *45* (1976) 738 (Hydrodehydropolymerization and isomerization polymerization of unsaturated hydrocarbons).
A. P. Kozikowski, H. F. Wetter, *Synthesis* (1976) 561 (Transition metals in organic synthesis).
G. Wilke, *Pure Appl. Chem.*, *50* (1978) 677 (Organotransition metals as homogeneous catalytic intermediates).
J. C. Mol, J. A. Moulijn, *Adv. Catal.*, *24* (1975) 131 (Metathesis of unsaturated hydrocarbons).
Yu. I. Yermakov, V. Zakharov, *Adv. Catal.*, *24* (1975) 173 (One-component catalyst for olefin polymerization).
G. W. Parshall, *Acc. Chem. Res.*, *8* (1975) 113 (C-H Bond activation).
R. Ugo, *Catal. Rev.*, *11* (1975) 225 (Understanding surface reactions).
R. J. Haines, G. J. Leigh, *Chem. Soc. Rev.*, *4* (1975) 155 (Olefin metathesis).
F. D. Mango, *Coord. Chem. Rev.*, *15* (1975) 109 (Pericyclic reactions).
L. G. Volkova, I. Ya. Levitin, H. E. Vol'pin, *Russ. Chem. Rev. (Engl. Transl.)*, *44* (1975) 552 (Alkylation and arylation of unsaturated compounds via transitionmetal complexes).
F. D. Mango, *Top. Curr. Chem.*, *45* (1974) 39 (Removal of orbital-symmetry restrictions to organic reactions).
J. C. Bailar, *Catal. Rev.*, *10* (1974) 17 (Heterogenizing homogeneous catalysis).
G. Brieger, T. J. Nestrick, *Chem. Rev.*, *74* (1974) 567 (Catalytic transferhydrogenation).
G. Dolcetti, N. W. Hoffmann, *Inorg. Chim. Acta*, *9* (1974) 269 (Homogeneous hydrogenation).
V. I. Labunskaya, A. D. Shebaldova, M. L. Khidekel', *Russ. Chem. Rev. (Engl. Transl.)*, *43* (1974) 1 (Catalysis of symmetry-disallowed reactions).
M. E. Vol'pin, *Russ. Chem. Rev. (Engl. Transl.)*, *43* (1974) 399 (Hydrogen transfer from organic compounds catalyzed by transition-metal complexes).
L. J. Kricka, A. Ledworth, *Synthesis*, (1974) 539 (Cyclobutanes from photochemical metal-catalyzed and cation-radical induced monoolefin dimerization).
E. Müller, *Synthesis*, (1974) 761 (Transiton-metal 1,4-, 1,5-, 1,6- and 1,7- diyne reactions).
B. Bogdanovic, *Angew. Chem., Int. Ed.Engl.*, *12* (1973) 954 (Homogeneous catalysis of asymmetric syntheses).
P. Heimbach, *Angew. Chem., Int. Ed. Engl.*, *12* (1973) 975 (Transition-metal catalyzed cyclooligomerization).
G. P. Chiusoli, *Acc. Chem. Res.*, *6* (1973) 422 (Olefin and CO insertions).
L. Marko, Bilteil, *Catal. Rev.*, *8* (1973) 269 (Asymmetric homogeneous hydrogenation).
R. E. Harmon, S. K. Gupta, D. J. Brown, *Chem. Rev.*, *73* (1973) 21 (Homogeneous catalysis of hydrogenation).
P. J. Staples, *Coord. Chem. Rev.*, *11* (1973) 277 (Acid-catalyzed reactions of transition-metal complexes).
Yu. A. Alexandrov, *J. Organomet. Chem.*, *55* (1973) 1 (Liquid-phase autooxidation of non-transition metal organometallics).
E. B. Milowskaya, *Russ. Chem. Rev. (Engl. Transl.)*, *42* (1973) 384 (Homolytic substitution at the metal atom in organometallics).

L. Cassar, G.B. Chiusoli, F. Guerri, *Synthesis*, (1973) 509 (Atmospheric pressure carbonylations yielding carboxylic acids and others).
C.A. Tolman, *Chem. Soc. Rev.*, *1* (1972) 337 (The 16- and 18-electron rule in homogeneous catalysis).
F.E. Paulik, *Catal. Rev.*, *6* (1972) 49 (Hydroformylation).
G. Henrici-Olivé, S. Olivé, *Angew. Chem., Int. Ed. Engl.*, *10* (1971) 105 (Influence of ligands on soluble catalyst activity and specificity).
J.R. Garnett, *Catal. Rev.*, *6* (1971) 229 (Homo- and heterogeneous catalytic exchange of hydrocarbons with metals).
H. Arzonmanian, J. Metzger, *Synthesis*, (1971) 527 (Olefin oxidations by Hg salts).
A.J. Hubert, H. Reimlinger, *Synthesis*, (1970) 405 (Thermal and catalytic isomerization of olefins).
G.C. Bailey, *Catal. Rev.*, *3* (1969) 37 (Olefin disproportionation).

i) Organometallics in Organic Synthesis

W. Bartmann, B.M. Trost, eds., *Selectivity – A Goal for Synthetic Efficiency*, Verlag Chemie, Weinheim (1984).
H.M. Colquhoun, J. Holton, D.J. Thompson, M.V. Twigg, *New Pathways for Organic Synthesis: Practical Applications of Transition Metals*, Plenum Press, New York (1983).
R. Scheffold, ed., *Transition Metals in Organic Chemistry, Modern Synthetic Methods*, Verlag Sauerlaender, Aarau, Switzerland (1983).
S.G. Davies, *Organotransition Metal Chemistry: Applications to Organic Synthesis*, Pergamon Press, Oxford (1982).
P. Houghton, *Metal Complexes in Organic Chemistry*, Cambridge University Press, Cambridge (1979).
H. Alper, ed., *Transition-Metal Organometallics in Organic Synthesis*, Vol. II, Academic Press, New York (1978).
D.W. Slocum, ed., *The Place of Transition Metals in Organic Synthesis*, New York Academy of Science, New York (1977).
D. Seyferth, ed., *New Applications of Organometallic Reagents in Organic Synthesis*, Elsevier, Amsterdam (1976).
H. Alper, ed., *Transition-Metal Organometallics in Organic Synthesis*, Vol. I, Academic Press, New York (1976).
I. Wender, P. Pino, *Organic Synthesis via Metal Carbonyls*, Vol. 2, Wiley New York (1976).
J. Tsuji, *Organic Synthesis by Means of Transition-Metal Complexes*, Vol. 1, *Reactivity and Structure*, Springer Verlag, Berlin (1975).
J.M. Swan, D.St.C. Black, *Organometallics in Organic Synthesis*, Chapman and Hall, London (1974).
E.L. Augustine, ed., *Oxidation Techniques in Organic Synthesis*, M. Dekker, New York (1969).
I. Wender, P. Pino, *Organic Synthesis via Metal Carbonyls*, Vol. 1, Wiley, New York (1968).
C.W. Bird, *Transition-Metal Intermediates in Organic Synthesis*, Academic Press, New York (1967).

B.M. Trost, *Chem. Br.*, *20* (1984) 315 (Selectivity in organic synthesis).
J.K. Stille, *Mod. Synth. Methods*, *3* (1983) 1 (Transition metals in organic synthesis).
L.S. Hegedus, *Mod. Synth. Methods*, *3* (1983) 61 (Group-VIII metals in organic synthesis).
S.V. Ley, R.A. Porter, P.F. Gordon, A.J. Nelson, *Gen. Synth. Methods*, *6* (1983) 218 (Organometallics in synthesis).

N. Nilsson, *Kem.-Kemi, 10* (1983) 133 (Organotransition metals in organic synthesis).
J.V.N. Vara Prasad, C.N. Pillai, *J. Organomet. Chem., 259* (1982) 1 (Carbometallation).
T. Kauffmann, *Angew. Chem., Int. Ed. Engl., 21* (1982) 401 (Heavy main-group elements in organic synthesis).
L.S. Hegedus, *J. Organomet. Chem., 237* (1982) 231 (Transition metals in organic synthesis).
J.F. Normant, A. Alexakis, *Synthesis*, (1981) 841 (Carbometallation of alkynes: stereospecific synthesis of alkenes).
A.P. Kozikowski, H.F. Wetter, *Synthesis*, (1976) 561 (Transition metals in organic synthesis).

Chapter 2

a) General

J.K. Burdett, *Molecular Shapes*, Wiley, New York (1980).

A.H. Cowley, *Polyhedron, 3*(1984) 389 (Double bonding between the heavier main-group elements).
R. Hoffmann, *Angew. Chem., Int. Ed. Engl., 21* (1982) 711 (Nobel lecture: building bridges between inorganic and organic chemistry).
M.E. O'Neill, K. Wade, in *Comprehensive Organometallic Chemistry*, G. Wilkinson, F.G.A. Stone, E.W. Abel, eds., Pergamon Press, Oxford, (1982), Vol. 1, p. 1. (Structure and bonding relationships among main-group organometallics).
T.A. Albright, *Tetrahedron, 38* (1982), 1339 (Structure and reactivity by MO theory).
E.L. Muetterties, *Pure Appl. Chem., 54* (1982) 83 (A coordination chemist's view of surface science).
D.R. Armstrong, P.G. Perkins, *Coord. Chem. Rev., 38* (1981) 139 (Electronic structures of main-group organometallics).
D.W. Clack, K.D. Warren, *Struct. Bonding (Berlin)*, 39 (1980) 1 (Bonding in 3d-sandwich compounds).
P.S. Braterman, *Topp. Curr. Chem., 92* (1980) 149 (Orbital correlation in the making and breaking of transition metal-carbon bonds).
J. Chatt, G.J. Leigh, *Angew. Chem., Int. Ed. Engl., 17* (1978) 400 (Distribution of charge in complexes).
E.L. Muetterties, *Angew. Chem., Int. Ed. Engl., 17* (1978) 545 (A coordination chemist's view of surface science).
J.K. Burdett, *Adv. Inorg. Chem. Radiochem., 21* (1978) 113 (A new look at transition-metal complexes).
J.K. Burdett, *Chem. Soc. Rev., 7* (1978) 507 (Molecular shapes).
D.M.P. Mingos, *Adv. Organomet. Chem., 15* (1977) 1 (Recent developments in theoretical organometallic chemistry).
K.D. Warren, *Struct. Bonding (Berlin), 27* (1976) 45 (Ligand-field theory of sandwich compounds).
R.F. Fenske, *Prog. Inorg. Chem., 21* (1976) 179 (MO theory and photoelectron spectroscopy).
P.S. Bratterman, R.J. Cross, *Chem. Soc. Rev., 2* (1973) 271 (Organotransition-metal complexes: stability, reactivity, orbital correlations).
R.G. Pearson, *Top. Curr. Chem., 41* (1973) 75 (Orbital-symmetry rules for inorganic reactions).

G. Wilkinson, *Pure Appl. Chem.*, *30* (1972) 627 (The transition-metal sigma-bond).
R. Mason, *Chem. Soc. Rev.*, *1* (1972) 431 (Valence in transition-metal complexes).
D.A. Brown, W.J. Chambers, N.J. Fitzpatrick, *Inorg. Chim. Acta Rev.*, *6* (1972) 7 (MO theory of transition-metal complexes).
D.R. Davies, G.A. Webb, *Coord. Chem. Rev.*, *6* (1971) 95 (MO calculations on transition-metal complexes).
I.I. Kritskaya, *Russ. Chem. Rev. (Engl. Transl.)*, *35* (1969) 177 (Transition metal-carbon sigma-bonds: conditions of formation and stabilization).
D.A. Brown, *Rec. Chem. Prog.*, *30* (1969) 211 (Bond theory of organometallic compounds).

b) Bond Strength, Thermochemistry

J.J. Christensen, *Handbook of Metal-Ligand Heats and Related Thermodynamic Quantities*, 3rd ed., M. Dekker, New York (1983).
J.D. Cox, G. Pilcher, *Thermochemistry of Organic and Organometallic Compounds*, Academic Press, New York (1970).
C.T. Mortimer, *Reaction Heats and Bond Strengths*, Pergamon Press, Oxford (1962) (Discussion of M-C bond strengths).

L.N. Sakharov, G.A. Domrachev, Yu.T. Struchkov, *J. Struct. Chem. (Engl. Transl.)*, *24* (1983) 75 (Structural aspects of organometallic thermal stability).
T. Bartik, P. Heimbach, H. Schenkluhn, *Kontakte*, (1983) 16 (Organometallic free-energy relationships).
J. Halpern, *Acc. Chem. Res.*, *15* (1982) 238 (Transition metal-alkyl bond-dissociation energies).
P.B. Armentrout, L.F. Halle, J.F. Beauchamp, *J. Am. Chem. Soc.*, *103* (1981) 6501 (Metal-H, C and O dissociation energies).
V.I. Tel'noi, I.B. Rabinovich, *Russ. Chem. Rev. (Engl. Transl.)*, *49* (1980) 603 (Thermochemistry of non-transition-metal organic compounds).
V.I. Tel'noi, I.B. Rabinovich, *Russ. Chem. Rev. (Engl. Transl.)*., *46* (1977) 689 (Thermochemistry of transition-metal organic compounds).
J.A. Connor, *Top. Curr. Chem.*, *71* (1977) 71 (Thermochemistry of organotransition-metal carbonyls).
R.H.T. Bleyerveld, T. Höhle, K. Vrieze, *J. Organomet. Chem.*, *94* (1975) 281 (Thermochemistry of metal carbonyls).
F.R. Hartley, *Chem. Rev.*, *73* (1973) 163 (Thermodynamics of olefin and acetylene transition-metal complexes).
A.G. Lee, *Organomet. Chem. Rev.*, *A*, *6* (1970) 139 (Thermodynamics of redistribution).
M.A. Skinner, *Adv. Organomet. Chem.*, *2* (1965) 49 (The strength of metal-carbon bonds: thermochemistry).

c) Electron-Deficient Bonds

K. Wade, *Electron Deficient Compounds*, Nelson, London (1973).
K. Wade, *Electron-Deficient Compounds*, Appleton-Century-Crofts, New York (1971).

J.P. Oliver, *Adv. Organomet. Chem.*, *15* (1977) 235 (Structures of organometallic compounds containing electron-deficient bridge bonds).

d) Delocalized Bonds, Polynuclear Systems

B. F. G. Johnson, ed., *Transition-Metal Clusters*, Wiley, Chichester, England (1980).
R. N. Grimes, *Carboranes*, Academic Press, New York (1970)

H. B. v. Schnering, *Angew. Chem., Int. Ed. Engl.*, *20* (1981) 33 (Homoatomic bonding of the main-group elements).
M. C. Manning, W. C. Trogler, *Coord. Chem. Rev.*, *39* (1981) 89 (Electronic structures of transition-metal clusters).
B. F. G. Johnson, R. E. Benfield, *Top. Stereochem.*, *12* (1981) 253 (Stereochemistry of transition-metal clusters).
D. M. P. Mingos, *Pure Appl. Chem.*, *52* (1980) 705 (Theory of clusters).
K. Wade, *Adv. Inorg. Chem. Radiochem.*, *18* (1976) 1 (Bonding and structure of carboranes, clusters and related systems).
R. E. Williams, *Adv. Inorg. Chem. Radiochem.*, *18* (1976) 67 (Coordination number pattern-recognition theory of carborane structures).
R. W. Rudolph, *Acc. Chem. Res.*, *9* (1976) 446 (Boranes and heteroboranes as paradigms for clusters).

e) Non-Rigidity, Fluxional Molecules, Rearrangements and Exchanges

L. M. Jackman, F. A. Cotton, eds., *Dynamic Nuclear Magnentic Resonance Spectroscopy*, Academic Press, New York (1975) (Contains several chapters dealing with stereochemically nonrigid molecules).

A. J. Deeming, *Mech. Inorg. Organomet. React.*, (1983) 249, 343 (Rearrangements, intramolecular exchanges, and isomerizations of organometallics).
T. A. Albright, *Acc. Chem. Res.*, *15* (1982) 149 (Rotational barriers and conformations in transition-metal chemistry).
L. A. Fedorov, D. N. Kravtsov, A. S. Peregudov, *Russ. Chem. Rev. (Engl. Transl.)*, *50* (1981) 682 (Metallotropic tautomerism of the σ,π-type).
K. Tatsumi, M. Tsutsui, *J. Mol. Catal.*, *13* (1981) 117 (σ-π Rearrangements).
F. A. Cotton, B. E. Hanson, in *Rearrangements in Ground and Excited States*, ed. by P. de Mayo, Vol. 2, Academic Press, New York (1980), p. 379 (Fluxional metal carbonyls).
B. F. G. Johnson, R. E. Benfield, in *Transition-Metal Clusters*, B. F. G. Johnson, ed., J. Wiley, Chichester, England (1980), p. 471 (Ligand mobility).
G. L. Geoffroy, *Acc. Chem. Res.*, *13* (1980) 469 (Fluxionality of mixed-metal clusters).
G. B. Shul'pin, *Russ. Chem. Rev. (Engl. Transl.)*, *49* (1980) 645 (Diasterotropy in transition-metal complexes).
L. A. Federov, A. S. Peregudov, D. N. Kravtsov, *Russ. Chem. Rev. (Engl. Transl.)*, *48* (1979) 840 (Equilibrium exchange of univalent heavy non-metal organometallics).
E. Band, E. L. Muetterties, *Chem. Rev.*, *78* (1978) 639 (Metal-cluster rearrangements).
B. Gorewit, M. Tsutsui, *Adv. Catal.*, *12* (1978) 277 (σ-π Transitions).
M. R. Reetz, *Adv. Organomet. Chem.*, *16* (1977) 33 (Dyotropic rearrangements and related σ-π exchanges).

M. Tsutsui, A. Courtney, *Adv. Organomet. Chem.*, *16* (1977) 241 (σ-π Rearrangements).
R. B. King, *Inorg. Chem.*, *16* (1977) 1822 (A graph-theoretical interpretation of stereochemically nonrigid structures).
J. W. Faller, *Adv. Organomet. Chem.*, *16* (1977) 211 (Fluxionality of organotransition-metal π-complexes).
J. Evans, *Adv. Organomet. Chem.*, *16* (1977) 319 (Fluxionality of polynuclear transition-metal complexes).
F. A. Cotton, *J. Organomet. Chem.*, *100* (1975) 29 (Fluxionality of organometallic compounds and metal carbonyls).
M. Hancock, M. N. Levy, M Tsutsui, *Organomet. React.*, *4* (1974) 1 (σ-π Rearrangements).
H. L. Clarke, *J. Organomet. Chem.*, *80* (1974) 155 (Fluxional π-allyls).
R. B. Larrabee, *J. Organomet. Chem.*, *74* (1974) 313 (Fluxional main-group IV organometallics).
J. R. Shapley, J. A. Osborn, *Acc. Chem. Res.*, *6* (1973) 305 (Rearrangements in pentacoordinated transition-metal compounds).
H. Beal, C. H. Bushweller, *Chem. Rev.*, *73* (1973) 465 (Dynamic processes in boranes, borane complexes and carboranes).
L. A. Fedorov, *Russ. Chem. Rev. (Engl. Transl.)*, *42* (1973) 678 (^{1}H-NMR studies of fluxionality).
A. Efraty, *J. Organomet. Chem.*, *57* (1973) 1 (π-Ligand transfer).
A. Z. Rubezhov, S. P. Gubin, *Adv. Organomet. Chem.*, *10* (1972) 347 (Ligand substitution in π-organotransition complexes).
K. Vrieze, P. W. N. M. Van Leuwen, *Prog. Inorg. Chem.*, *14* (1971) 1 (Dynamic organometallic structures).
G. W. Parshall, *Acc. Chem. Res.*, *3* (1970) 139 (Intramolecular aromatic substitution in transition-metal complexes).
F. A. Cotton, *Chem. Br.*, *4* (1968) 345 (Fluxional organometallic compounds).
F. A. Cotton, *Acc. Chem. Res.*, *1* (1968) 257 (Fluxional organometallic compounds).
S. J. Lippard, *Trans. N. Y. Acad. Sci.*, *29* (1967) 917 (Stereochemically dynamic organometallic compounds).

f) Dative Bonds (Metal Carbonyls, Olefin and Polyene Complexes)

G. Deganello, *Transition-Metal Complexes of Cyclic Polyolefins*, Academic Press, New York (1979).
M. Herberhold, *Metal π-Complexes*. Vol. II, *Complexes with Mono-Olefinic Ligands*, Part 2, *Specific Aspects*, Elsevier, Amsterdam (1974).
M. Herberhold, *Metal π-Complexes*, Vol. II, *Complexes with Mono-Olefinic Ligands*, Part 1, *General Survey*, Elsevier, Amsterdam (1972).

F. G. A. Stone, *Angew. Chem., Int. Ed. Engl.*, *23* (1984) 89 (Isolobal principles in transition-metal complexes).
S. Y. Chu, R. Hoffmann, *J. Phys. Chem.*, *86* (1982) 1289 (Cyclopentadiene complexes).
O. Eisenstein, R. Hoffmann, A. R. Rossi, *J. Am. Chem. Soc.*, *103* (1981) 5582 (Olefin metathesis intermediates).
C. V. Senoff, *Coord. Chem. Rev.*, *32* (1980) 111 (Substituent effects in complexes with non-fused phenyls).

D. W. Clark, K. D. Warren, *Structure Bonding (Berlin)*, *39* (1980) 1 (MO Calculations on sandwich compounds).

R. Hoffmann, B. E. R. Schilling, R. Bau, H. D. Kaesz, D. M. P. Mingos, *J. Am. Chem. Soc.*, *100* (1979) 6088 (Electronic structure of $M_4(CO)_{12}H_n$ and $M_4Cp_4H_n$).

T. A. Albright, R. Hoffmann, J. C. Thibeault, D. L. Thorn, *J. Am. Chem. Soc.*, *101* (1979) 3801 (Ethylene complexes. Bonding rotational barriers, conformational preferences).

T. A. Albright, R. Hoffmann, Y. C. Tse, T. D'Ottario, *J. Am. Chem. Soc.*, *101* (1979) 3812 (Polyene-ML_2 and -ML_4 complexes. Conformational preferences and barriers of rotation).

R. H. Summerville, R. Hoffmann, *J. Am. Chem. Soc.*, *101* (1979) 3821 (Bonding and structure of M_2L_9 complexes).

B. E. R. Schilling, R. Hoffmann, D. L. Lichtenberger, *J. Am. Chem. Soc.*, *101* (1979) 585 (Bonding and structure of $CpM(CO)_2$ (ligand) complexes).

B. E. R. Schilling, R. Hoffmann, *J. Am. Chem. Soc.*, *101* (1979) 3456 (Bonding and structure of M_2L_9(ligand) complexes).

B. E. R. Schilling, R. Hoffmann, J. W. Faller, *J. Am. Chem. Soc.*, *101* (1979) 592 (Structure and reactivity of CpMLL′(allyl) and π-(ethylene) complexes).

A. R. Pinhas, R. Hoffmann, *Inorg. Chem.*, *18* (1979) 654 (Ring-puckering and metal-metal separations in dibridged binuclear compounds).

A. A. Bagatur'yants, O. V. Gritsenko, I. I. Moiseev, *Koord. Khim.*, *4* (1978) 1779 (Theoretical aspects of the coordination of olefins to transition metals).

D. L. Thorn, R. Hoffmann, *Inorg. Chem.*, *17* (1978) 126 (Bonding and structure of $M_2(CO)_6$(ligand) complexes).

J. W. Lauher, *J. Am. Chem. Soc.*, *100* (1978) 5305 (The bonding capabilities of transition-metal clusters; stoichiometry and geometry of polynuclear metal carbonyls).

E. Maslowsky, *J. Chem. Educ.*, *55* (1978) 276 (Structures of metal-cyclopentadienyl derivatives).

R. Hoffmann, T. A. Albright, D. L. Thorn, *Pure Appl. Chem.*, *50* (1978) 1 (Theoretical aspects of the coordination of unsaturated molecules to transition-metal centers: olefins, acetylenes, cyclolyenes).

K. D. Warren, *Struct. Bonding, (Berlin)*, *33* (1977) 97 (Ligand-field theory of f-orbital sandwich complexes).

R. B. King, *Isr. J. Chem.*, *15* (1977) 181 (Systematization of metal-polyene structure types; the "elplacarnet tree").

T. A. Albright, P. Hofmann, R. Hoffmann, *J. Am. Chem. Soc.*, *99* (1977) 7546 (Polyene-ML_3 complexes. conformational preferences and rotational barriers).

D. M. P. Mingos, *J. Chem. Soc., Dalton Trans.*, (1977) 20, 26, 31 (A topological Hückel model for organometallics).

S. D. Itell, J. A. Ibers, *Adv. Organomet. Chem.*, *14* (1976) 33 (Coordination of unsaturated molecules to transition metals).

P. Chini, G. Longoni, V. G. Albano, *Adv. Organomet. Chem.*, *14* (1976) 285 (High-nuclearity metal-carbonyl clusters).

E. C. Baker, G. W. Halstead, K. N. Raymond, *Struct. Bonding (Berlin)*, *26* (1976) 23 (Structure and bonding of 4f- and 5f-series organometallic compounds).

M. Elian, M. M. L. Chen, D. M. P. Mingos, R. Hoffmann, *Inorg. Chem.*, *15* (1976) 1148 (Comparative study of bonding in conical fragments, for example, $E(CO)_n$, $M(CH)_n$, etc.).

J. W. Lauher, R. Hoffmann, *J. Am. Chem. Soc.*, *98* (1976) 1729 (Bonding in bis(cyclopentadienyl)metal complexes; bent $M(C_5H_5)_2$ derivatives).

J. W. Lauher, M. Elian, R. H. Summerville, R. Hoffmann, *J. Am. Chem. Soc.*, *98* (1976) 3219 (Bonding and structure of triple-decker sandwich compounds).

D. A. Bochvar, N. O. Gambaryan, in *Methody Elementoorganicheskoi Khimii. Tipy Metalloorganicheskikh Soedinenii Perekhodnykh Metallov* (Methods of organoelement chemistry. Ty-

pes of organometallic compounds of transition metals), Nauka, Moscow (1975), p. 9 (in Russian) (Chemical bonding in organic complexes of transition metals).
M. Elian, R. Hoffmann, *Inorg. Chem.*, *14* (1975) 1058 (Bonding capabilities of metal-carbonyl fragments).
A. N. Nesmeyanov, *J. Organomet. Chem.*, *100* (1975) 161 (Metallotropy and dual reactivity).
N. Rösch, R. Hoffmann, *Inorg. Chem.*, *13* (1974) 2656 (Geometry of transition-metal complexes with ethylene and allyl groups as the only ligands).
P. S. Braterman, *Struct. Bonding (Berlin)*, *10* (1972) 57 (Bonding in metal carbonyls).
C. Tolman, *Chem. Soc. Rev.*, *1* (1972) 327 (The 16- and 18-electron rule in organometallic chemistry and homogeneous catalysis).
F. R. Hartley, *Angew. Chem., Int. Ed. Engl.*, *11* (1972) 596 (Metal-olefin and metal-acetylene bonding).
G. Hafelinger, *Top. Curr. Chem.*, *28* (1972) 1 (Theoretical considerations for cyclic systems).
L. D. Petit, *Quart. Rev.*, *25* (1971) 1 (Multiple bonding and back coordination).
E. W. Abel, F. G. A. Stone, *Quart. Rev.*, *23* (1969) 325 (Structure and bonding in metal carbonyls).
R. B. King, *J. Am. Chem. Soc.*, *91* (1969) 7217 (Topology of metal complexes of planar unsaturated molecules).
D. A. Brown, in *Transition-Metal Chemistry*, R. L. Carlin, ed., M. Dekker, New York (1966), Vol. 3, p. 2 (Electronic structures of organometallic compounds).

Chapter 3

a) Laboratory Manipulations

A. Weissberger, ed., *Techniques in Chemistry*, Wiley-Interscience, New York (1970 – continuing).
J. J. Eisch, *Organometallic Syntheses*, Vol. 2, *Nontransition-Metal Compounds*, Academic Press, New York (1981).
E. Meyer, *Chemistry of Hazardous Materials*, Prentice-Hall, Englewood Cliffs, NJ (1977).
R. J. Angelici, *Synthesis and Technique in Inorganic Chemistry*, 2nd ed., Saunders, Philadelphia (1977).
D. F. Shriver, *The Manipulation of Air-Sensitive Compounds*, McGraw-Hill, New York (1969).
J. T. Clerc, E. Pretsch, J. Seibel, *Structural Analysis of Organic Compounds by Combined Application of Spectroscopic Methods*, Elsevier, Amsterdam (1981).
R. S. Drago, *Physical Methods in Chemistry*, Interscience, New York (1977).
F. C. Nachod, J. J. Zuckerman, eds., *Determination of Organic Structures by Physical Methods*, Vols. 1–6, Academic Press, New York (1955–1976).
J. Yarwood, ed., *Spectroscopy and Structure of Molecular Complexes*, Plenum Press, New York (1973).
M. Tsutsui, ed., *Characterization of Organometallic Compounds*, 2 Vols., Wiley-Interscience, New York (1969, 1971).
W. O. George, ed., *Spectroscopic Methods in Organometallic Chemistry*, Butterworths, London (1970).
R. B. King, *Organometallic Syntheses*, Vol. 1, *Transition-Metal Compounds*, Academic Press, New York (1965).

A. Krause, E. von Grosse, *Die Chemie der Metall-Organischen Verbindungen*, Bornträger Verlag, Berlin (1937), Chapter XII, p. 802.
A.C. Bond, in *Techniques and Methods of Organic and Organometallic Chemistry*, D.E. Denny, ed., Vol. 1, M. Dekker, New York (1969), p. 32 (Chapter on high-vacuum techniques).
S. Herzog, H. Oberender, R. Berger, *Z. Chem.*, *5* (1965) 372 (Liquid-ammonia methods).
S. Herzog, J. Dehnert, *Z. Chem.*, *4* (1964) 1 (Anaerobic methods).
S.B. Mirviss, A.J. Rutowski, C.W. Seelbach, H.T. Oakley, *Ind. Eng. Chem.*, *53* (1961) 58 A (Handling of pyrophoric organometallics).

b) Direct Synthesis

Y.-H. Lai, *Synthesis*, (1981) 585 (Grignard reagents from activated Mg).
R.D. Rieke, *Acc. Chem. Res.*, *10* (1977) 301; *Top. Curr. Chem.*, *59* (1975) 1 (Reactive metal powders).
V. Bazant, J. Joklik, J. Rathousky, *Angew. Chem., Int. Ed. Engl.*, *7* (1968) 112 (Direct synthesis of organohalosilanes).
E.G. Rochow, *J. Chem. Educ.*, *43* (1966) 58 (The direct synthesis).
J.J. Zuckerman, *Adv. Inorg. Chem. Radiochem.*, *6* (1964) 383 (Group-IV organometallic direct syntheses).

c) Metal-Atom Syntheses

K.J. Klabunde, *Chemistry of Free Atoms and Particles*, Academic Press, New York (1980).
J.R. Blackbrow, D. Young, *Metal Vapor Synthesis in Organometalic Chemistry*, Springer Verlag, Berlin (1979).
M.J. McGlinchey, P.A. Skell, B.M. Moskovits, G.A. Ozin, eds., *Cryochemistry*, J. Wiley, New York (1976), pp. 137, 167 (Organometallic syntheses using vapors).
S. Craddock, A.J. Hinchcliffe, *Matrix Isolation*, Cambridge University Press, Cambridge (1975).

G.A. Ozin, J.G. McCaffey, D.F. McIntosh, *Pure Appl. Chem.*, *56* (1984) 111 (Methane and H_2 reactions with Mn and Fe atoms).
G.A. Ozin, *Coord. Chem. Rev.*, *48* (1983) 203 (Taking metal-atom chemistry out of the cold).
G.A. Ozin, S.A. Mitchell, *Angew. Chem., Int. Ed. Engl.*, *2* (1983) 674 (Ligand-free metal clusters).
S.C. Davis, K.J. Klabunde, *Chem. Rev.*, *82* (1982) 153 (Unsupported small metal particles).
K.J. Klabunde, in *Reactive Intermediates*, Vol. 1, R.A. Abromovich, ed., Plenum Press, New York (1980), p. 37 (Metal atoms).
M.L.H. Green, *J. Organomet. Chem.*, *200* (1980) 119 (Group IV, V and VI transition-metal atoms in the synthesis of arenes).
R.J. Lagow, J.A. Morrison, *Adv. Inorg. Chem. Radiochem.*, *23* (1980) (Trifluoromethyl organometallics by plasma-generated radicals and metal-vapor cocondensations).
G.A. Ozin, W.J. Power, *Adv. Inorg. Chem. Radiochem.*, *23* (1980) (Metal-vapor cryochemistry).
R.G. Gastinger, J.K. Klabunde, *Trans. Met. Chem.*, *4* (1979) (π-Arene complexes of the group-VIII metals by metal-vapor synthesis).

G. A. Domrachev, V. D. Zinov'ev, *Russ. Chem. Rev. (Engl. Transl.)*, *47* (1978) 354 (Reactions of transition-metal atoms with organic compounds).
K. J. Klabunde, *Ann. N. Y. Acad. Sci.*, *295* (1977) 59 (Metal-vapor synthesis of π-complexes).
P. L. Timms, T. W. Turney, *Adv. Organomet. Chem.*, *15* (1977) 53 (Metal-atom synthesis of organometallic compounds).
G. A. Ozin, *Acc. Chem. Res.*, *10* (1977) 21 (Metal-atom matrix chemistry).
G. A. Ozin, *Catal. Rev.*, *16* (1977) 191 (Very small metallic and bimetallic clusters).
G. A. Ozin, A. Van der Voet, *Prog. Inorg. Chem.*, *10* (1975) 105 (Metal-gas reactions studied by vibrational spectroscopy).
P. S. Skell, M. J. McGlinchey. *Angew. Chem., Int. Ed. Engl.*, *14* (1975) 195 (Reactions of transition-metal atoms with organic substrates).
E. A. Koerner von Gustorf, O. Jaenicke, O. Wolfbeis, C. R. Eady, *Angew. Chem., Int. Ed. Engl.*, *14* (1975) 278 (Laser evaporation of metals and its use in the synthesis of organometallic compounds).
P. L. Timms, *Angew. Chem., Int. Ed. Engl.*, *14* (1975) 273 (Metal-atom syntheses).
K. J. Klabunde, *Angew. Chem., Int. Ed. Engl.*, *14* (1975) 287 (Metal atoms and fluorocarbons).
E. D. Kundig, M. Moskovits, G. A. Ozin, *Angew. Chem., Int. Ed. Engl.*, *14* (1975) 292 (Metal-atom syntheses of binuclear complexes).
J. J. Turner, *Angew. Chem., Int. Ed. Engl.*, *14* (1975) 304 (Photochemical generation of metal atoms in matrices).
G. A. Ozin, *Coord. Chem. Rev.*, *28* (1974) 117 (Olefin chemisorption on Co, Ni, and Cu naked-metal clusters).
P. L. Timms, *Adv. Inorg. Chem. Radiochem.*, *14* (1972) 121 (Metal-atom synthesis).

d) Photochemistry

J. J. Zuckerman, ed., *Inorganic Reactions and Methods*, Vol. 15, Verlag Chemie, Weinheim (1985) (Photochemical and other energized reactions).
M. S. Wrighton, ed., *Inorganic and Organometallic Photochemistry*, Advances in Chemistry Series, No. 168, American Chemical Society, Washington, DC (1979).
G. L. Geoffroy, M. S. Wrighton, *Organometallic Photochemistry*, Academic Press, New York (1979).

G. L. Geoffroy, *J. Chem. Educ.*, *60* (1983) 861 (Organometallic photo-chemistry).
J. M. Kelly, C. Long, *Photochemistry*, *13* (1982) 196; *12* (1982) 155; *11* (1981) 259 (Photochemistry of organotransition metals).
A. W. Maverick, H. B. Gray, *Pure Appl. Chem.*, *52* (1980) 2339 (Solar-energy storage involving metal complexes).
G. L. Geoffroy, *Prog. Inorg. Chem.*, *27* (1980) 123 (Photochemistry of transition-metal hydrides).
J. J. Turner, *Angew. Chem., Int. Ed. Engl.*, *14* (1975) 304 (Photochemistry in matrices).

e) Electrochemical Syntheses

J. J. Zuckerman, ed., *Inorganic Reactions and Methods* Vol. 15, Verlag Chemie, Weinheim (1985) (Electron transfer and electrochemical reactions).

D. G. Tuck, *Pure Appl. Chem.*, *51* (1979) 2005 (Electrochemical synthesis of organometallics).
L. I. Denisovich, S. P. Gubin, *Russ. Chem. Rev. (Engl. Transl.)*, *46* (1977) 27 (Electrochemistry of π-organotransition-metal compounds).
S. G. Mairanovski, *Russ. Chem. Rev. (Engl. Transl.)*, *45* (1976) 298 (Polarography of non-transition metal organometallics).
H. Lehmkuhl, *Synthesis*, (1973) 377 (Preparative organometallic chemistry).
G. A. Teodoradze, *J. Organomet. Chem.*, *88* (1975) 1 (Electrochemical synthesis of organometallic compounds).
W. J. Settineri, L. D. McKeever, *Techn. Chem. (New York)*, *5* (1975) Pt. 5, p. 397 (Electrolytic synthesis and reactions of organometallic compounds).
M. D. Morris, *Electroanal. Chem.*, *7* (1974) 79 (Organometallic electrochemistry).
R. E. Dessy, L. A. Bares, *Acc. Chem. Res.*, *5* (1972) 415 (Organometallic electrochemistry).
B. L. Laube, C. D. Schmulbach, *Prog. Inorg. Chem.*, *14* (1971) 65 (Electrosynthesis in non-aqueous solvents).
S. P. Gubin, *Pure Appl. Chem.*, *23* (1970) 463 (Electrochemical methods in organometallic chemistry).

f) Diffraction

O. Kennard, D. G. Watson, F. H. Allen, S. M. Weeds, *Molecular Structure and Dimensions*, Vols. 1–12 (1935–1980), D. Reidel, Dordrecht (1977)-continuing.
M. Hargittai, T. Hargittai, *The Molecular Geometries of Coordination Compounds in the Vapor Phase*, Akademiai Kiado, Budapest (1977) (Electron diffraction of organometallic compounds).
O. Kennard, D. G. Watson, F. H. Allen, N. W. Isaacs, W. D. S. Motherwell, R. C. Petterson, W. G. Town, eds., *Interatomic Distances 1960–1965*, Vol. Al: *Organic and Organometallic Crystal Structures*, D. Reidel, Dordrecht (1972).

I. Hargittai, *Top. Curr. Chem.*, *96* (1981) (Electron diffraction of metallocenes and metal borohydrides).
R. G. Teller, R. Bau, *Struct. Bonding (Berlin)*, *44* (1981) 1 (Transition-metal hydride structures).
S. D. Ittel, J. A. Ibers, *Adv. Organomet. Chem.*, *14* (1976) 33 (Coordination of unsaturated molecules to transition metals).
A. Haaland, *Top. Curr. Chem.*, *53* (1975) 1 (Organometallic electron diffraction).
C. Krüger, *Angew. Chem., Int. Ed. Engl.*, *11* (1972) 387 (Automated X-ray diffraction and structure determination of organometallic compounds).
W. C. Hamilton, *Science*, *169* (1970) 133 (The revolution in crystallography).
N. C. Baenziger, in *Characterization of Organometallic Compounds*, M. Tsutsui, ed., Wiley-Interscience, New York (1969), p. 213 (Determination of organometallic structures by X-ray diffraction).

g) Mass Spectrometry

J. Charalambous, ed., *Mass Spectrometry of Metal Compounds*, Butterworth, London (1975) (Contains chapters on fundamental aspects, fragmentation of metal-containing ions, metal carbonyls and transition-metal hydrocarbon compounds).

M.R. Litzow, T.R. Spalding, *Mass Spectrometry of Inorganic and Organometallic Compounds*, Elsevier, Amsterdam (1973).
D.H. Williams, ed., *Mass Spectrometry* (Specialist Periodical Reports Series), Royal Society of Chemistry, London, (Vol. 1, 1968-continuing).

I.K. Gregor, M. Guilhaus, *Mass Spectrom. Rev., 3* (1984) 39 (Metalorganic negative ions).
J.M. Miller, *J. Organomet. Chem., 249* (1983) 299 (Fast-atom bombardment (FAB) for organometallics).
J.M. Miller, G.L. Wilson, *Adv. Inorg. Chem. Radiochem., 18* (1976) 229 (Mass spectrometry applied to inorganic and organometallic compounds).
P.E. Gaivoronskii, N.V. Lamin, *Russ. Chem. Rev. (Engl. Transl.), 43* (1974) 466 (Mass spectroscopy of transition-metal π-complexes).
R.H. Cragg, A.F. Weston, *J. Organomet. Chem., 67* (1974) 161 (Mass spectra of boron compounds).
V. Yu. Orlov, *Russ. Chem. Rev. (Engl. Transl.), 42* (1973) 529 (Mass spectra of organo-group IVB compounds).
J. Müller, *Angew. Chem., Int. Ed. Engl., 11* (1972) 653 (Decomposition of organometallic compounds in the mass spectrometer).
R.W. Kiser, in *Characterization of Organometallic Compounds*, M. Tsutsui, ed., Wiley-Interscience, New York (1971), Vol. 2, p. 137.
M. Cais, M.S. Lupin, *Adv. Organomet. Chem., 8* (1970) 211 (Mass spectra of metallocenes).
R.B. King, *Top. Curr. Chem., 14* (1970) 92 (Mass spectra of transition-metal derivatives).
T.R. Spalding, in *Spectroscopic Methods in Organometallic Chemistry*, W.O. George, ed., Butterworths, London (1970), p. 95 (Organometallic mass spectroscopy).
M.I. Bruce, *Adv. Organomet. Chem., 6* (1968) 273 (Mass spectra of organometallics).

h) Electronic Spectroscopy

P.S. Braterman, *Metal-Carbonyl Spectra*, Academic Press, New York (1975).
B.G. Ramsay, *Electronic Transitions in Organometalloids*, Academic Press, New York (1969).

M.C. Manning, W.C. Trogler, *Coord. Chem. Rev., 39* (1981) 89 (Electronic structures of transition-metal cluster compounds).

i) Photoelectron Spectroscopy

P.S. Braterman, *Metal-Carbonyl Spectra*, Academic Press, New York (1975).

J.C. Green, *Struct. Bonding (Berlin), 43* (1981) 37 (Gas-phase photoelctron spectra of d- and f-block organometallics).
I.D. Hiller, *Pure Appl. Chem., 51* (1979) 2183 (Photoelectron spectroscopy of transition-metal complexes).
A.H. Cowley, *Prog. Inorg. Chem., 26* (1979) (UV photoelectron spectroscopy of transition-metal compounds).

C. Furlani, C. Cauletti, *Struct. Bonding (Berlin)*, *35* (1978) 119 (He(I) photoelectron spectra of transition-metal compounds).
K. Siegbahn, *Pure Appl. Chem.*, *48* (1976) 77 (Photoelectron spectroscopy and molecular structure).
R.F. Fenske, *Prog. Inorg. Chem.*, *21* (1976) 179 (Photoelectron spectra of transition-metal complexes).
A.F. Orchard, in *Electronic States of Inorganic Compounds: New Experimental Techniques*, NATO Advanced Studies Institute, Dodrecht, (1975).
W.L. Jolly, *Coord. Chem. Rev.*, *13* (1974) 47; *Top. Curr. Chem.*, *71* (1977) 149 (X-ray photoelectron spectroscopy in inorganic chemistry).

j) Vibrational Spectra (Infrared, Raman)

K. Nakamoto, *Infrared and Raman Spectra of Inorganic and Coordination Compounds*, 3rd ed., Wiley, New York (1978).
K. Leicht, P. Reich, *Literature Data for Infrared, Raman and NMR Spectroscopy of Si, Ge, Sn, and Pb-Organic Compounds*, VEB Deutscher Verlag der Wissenschaften, Berlin (1977).
E. Maslowsky, Jr., *Vibrational Spectra of Oganometallic Compounds*, Wiley-Interscience, New York (1977).
V.T. Aleksanyan, B.V. Lokshin, *Kolebatel'nye Spektry, π-Kompleksov Perekhodnykh Elementov* (Vibrational Spectra of π-Complexes of Transition Metals, Itogi Nauki, VINITI, Akademiya Nauk, Moscow (1976), (In Russian).
J.R. Ferraro, *Low-Frequency Vibrations of Inorganic and Coordination Compounds*, Plenum Press, New York (1971).
N.N. Greenwood, B.J.F. Ross, B.P. Straughan, *Index of Vibrational Spectra of Inorganic and Organometallic Compounds* (1935–1960), Butterworth, London (1972).

E. Maslowsky, Jr., *Chem. Soc. Rev.*, *9* (1980) 25 (Methyl-metal vibrations).
S.F. Kettle, *Top. Curr. Chem.*, *71* (1977) 111 (Metal-carbonyl spectra).
P.S. Braterman, *Struct. Bonding (Berlin)*, *26* (1976) 1 (Metal-carbonyl spectra).
G.A. Ozin, A. Vandervoet, *Prog. Inorg. Chem.*, *19* (1975) 105 (Metal-gas reactions studied by vibrational spectroscopy).
G. Davidson, *Organomet. Chem. Rev.*, *A*, *8* (1972) 303 (Vibrational spectra of π-organotransition-metal complexes).
S.F.A. Kettle, *Adv. Organomet. Chem.*, *10* (1972) 199 (Intensities of metal-carbonyl stretching vibrations).
J.R.Hall, in *Essays in Structural Chemistry*, D.A. Long, A.J. Downs, L.A.K. Stavely, eds., Macmillan, New York (1971), p. 433 (Infrared and Raman spectra of organometallic compounds).
E. Maslowsky, Jr., *Chem. Rev.*, *71* (1971) 507 (Vibrations of intra- and inter-metal and semimetal bonds).
K. Nakamoto, in *Characterization of Organometallic Compounds*, M. Tsutsui, ed., Vol. 1, Wiley-Interscience, New York (1969), p. 73 (Vibrational spectra).
L.M. Haines, M.H.B. Stiddard, *Adv. Inorg. Chem. Radiochem.*, *12* (1969) 53 (Vibrational spectra of transitional-metal carbonyl systems).
H.P. Fritz, *Adv. Organomet. Chem.*, *1* (1965) 239 (Vibrational spectra of π-complexes).
D.R. Huggins, H.D. Kaesz, *Prog. Solid-State Chem.*, *1* (1964) 417 (IR and Raman spectroscopy in the study of organometallic compounds).

k) NMR Spectroscopy

P. Laszlo, *NMR of Newly Accessible Nuclei*, 2 Vols., Academic Press, New York (1984).
J. W. Akitt, *NMR and Chemistry: An Introduction to the Fourier-Transform Multinuclear Era*, 2nd ed., Chapman and Hall (Methuen), London (1983).
R. K. Harris, *NMR Spectroscopy*, Pitman, London (1983).
J. B. Lambert, F. G. Riddell, *The Multinuclear Approach to NMR Spectroscopy*, D. Reidel, Dordrecht (1983).
B. E. Mann, B. Taylor, *^{13}C NMR Data for Organometallic Compounds*, Academic Press, New York (1981).
R. K. Harris, B. E. Mann, eds., *NMR and the Periodic Table*, Academic Press, New York (1979).
A. Steigel, *Dynamic NMR Spectroscopy*, Springer Verlag, Berlin (1978).

P. W. Jolly, R. Mynott, *Adv. Organomet. Chem.*, *19* (1981) 257 (Carbon-13 NMR in transition-metal complexes).
K. I. Zamaraev, *Kinet. Catal. (Engl. Transl.)*, *21* (1980) 223 (NMR of paramagnetic complexes in homogeneous catalysis).
H. Marsmann, in *NMR: Basic Principles and Progress*, P. Diehl, E. Fluck, R. Kosfeld, eds., Vol. 17, Springer Verlag, Berlin (1981) p. 65 (^{29}Si NMR).
P. W. Hickmott, M. Cais, A. Modiano, *Ann. Rev. NMR Spectrosc. 6C* (1977) 1 (Data on organotransition-metal carbonyls).
J. Evans, *Adv. Organomet. Chem.*, *16* (1977) 319 (Rearrangements in polynuclear transition-metal complexes).
R. K. Harris, *Chem. Soc. Rev.*, *5* (1976) 1 (NMR and the periodic table).
M. H. Chisholm, S. Godleski, *Prog. Inorg. Chem.*, *20* (1976) 299 (^{13}C NMR in inorganic and organometallic chemistry).
J. W. Faller, in *Determination of Organic Structure by Physical Methods*, Vol. 5, F. C. Nachod, J. J. Zuckerman, eds., Academic Press, New York, 1975, p. 75 (Spin-saturation labeling).
B. E. Mann, *Adv. Organomet. Chem.*, *12* (1974) 135 (^{13}C NMR of organometallic compounds).
L. A. Fedorov, *Russ. Chem. Rev. (Engl. Transl.)*, *42* (1973) 678 (Fluxional organotransition-metal compounds).
G. Mavel, *Annu. Rev. NMR Spectrosc.*, *B5* (1973) 1 (^{31}P NMR).
W. McFarlane, *Annu. Rev. NMR Spectrosc.*, *A*, *5* (1972) 353 (Heteronuclear double resonance).
K. Vrieze, P. N. M. Van Leuwen, *Prog. Inorg. Chem.*, *14* (1971) 1 (NMR spectroscopy and dynamic sterochemistry of organometallic compounds).
R. G. Kidd, in *Characterization of Organometallic Compounds*, M. Tsutsui, ed., Vol. 2, Wiley-Interscience, New York (1971), p. 373 (Organometallic NMR).
M. L. Maddox, H. D. Kaesz, *Adv. Organomet. Chem.*, *3* (1965) 1 (NMR spectroscopy in organometallic chemistry).

l) Electron-Spin Resonance

A. Schweiger, ed., *Electron-Nuclear Double Resonance of Transition-Metal Complexes with Organic Ligands (Struct. Bonding (Berlin)*, Vol. 51) Springer Verlag, Berlin (1982).

M. C. R. Symons, *Electron-Spin Reson.*, *7* (1982) 127 (Inorganic and organometallic radicals).
E. R. Milaeva, H. Z. Rubezhov, A. L. Prokofev, O. Yu. Okhlobystin, *Russ. Chem. Rev. (Engl. Transl.)*, *51* (1982) 1638 (Paramagnetic transition-metal complexes).

S. P. Solodovnikov, *Russ. Chem. Rev. (Engl. Transl.)*, *51* (1982) 1674 (Paramagnetic transition-metal metallocenes and diarenes).
A. G. Milaev, O. Yu,. Okhlobystin, *Russ. Chem. Rev. (Engl. Transl.)*, *49* (1980) 873 (Organo-non-transition-metal free radicals).
K. I. Zamaraev, *Kinet. Catal. (Engl. Transl.)*, *21* (1980) 223 (Paramagnetic complexes in homogeneous catalysis).
P. R. Jones, *Adv. Organomet. Chem.*, *15* (1977) 273 (Organometallic radical anions).
M. F. Lappert, P. W. Lednor, *Adv. Organomet. Chem.*, *14* (1976) 345 (Organometallic free radicals).
F. J. Smentowski, in *Characterization of Organometallic Compounds*, M Tsutsui, ed., Vol. 2, Wiley-Interscience, New York (1971), p. 481 (ESR).
B. A. Goodman, J. B. Raynor, *Adv. Inorg. Chem. Radiochem.*, *13* (1970) 136 (ESR of transition-metal complexes).
G. Urry, in *Radical Ions*, E. T. Kaiser, L. Kevan, eds., Wiley-Interscience, New York (1968), p. 275 (Group-IV organometallic radicals and radical ions).

m) Mössbauer Spectroscopy

R. H. Herber, ed., *Chemical Mössbauer Spectroscopy*, Plenum Press, New York (1985).
B. V. Thosar, P. K. Iyengar, J. K. Srivastava, S. C. Bhargava, eds., *Advances in Mössbauer Spectroscopy. Applications to Physics, Chemistry and Biology*, Elsevier, Amsterdam (1983).
J. G. Stevens, G. K. Shenoy, eds., *Mössbauer Spectroscopy and Its Chemical Applications*, Advances in Chemistry Series, No. 194. American Chemical Society, Washington, DC (1981).
A. Vertes, K. Korecz, K. Burger, *Mössbauer Spectroscopy*, Elsevier, Amsterdam (1979).
P. Gütlich, *Mössbauer Spectroscopy and Transition-Metal Chemistry*, Springer Verlag, Berlin (1978).
G. K. Shenoy, F. E. Wagner, eds., *Mössbauer Isomer Shifts*, North Holland, Amsterdam (1977).
T. C. Gibb, *Principles of Mössbauer Spectroscopy*, Methuen, New York (1976).
V. I. Goldanskii, R. H. Herber, eds., *Chemical Applications of Mössbauer Spectroscopy*, Academic Press, New York (1968).
J. J. Zuckerman, in *Chemical Mössbauer Spectroscopy*, R. H. Herber, ed., Plenum Press, New York (1985) (Organotin-119m Mössbauer spectroscopy).
W. H. Armstrong, E. E. Dorflinger, O. T. Anderson, B. R. Willeford, *J. Chem. Educ.*, *58* (1981) 515 (General review of Mössbauer spectroscopy).
R. V. Parish, in *The Organic Chemistry of Iron*, E. A. Koerner von Gustorf, F. W. Grevels, I. Fischler, eds., Academic Press, New York (1978), p. 175 (Iron-57 Mössbauer spectroscopy).
J. N. R. Ruddick, *Rev. Si, Ge, Sn, Pb Compnds.*, *2* (1976) 115 (Tin-119m Mössbauer data).
N. W. G. Debye, J. J. Zuckerman, in *Determination of Organic Structure by Physical Methods*, Vol. 5, F. C. Nachod, J. J. Zuckerman, eds., Academic Press, New York (1973), p. 235 (Organometallic Mössbauer spectroscopy).
W. M. Reiff, *Coord. Chem. Rev.*, *10* (1973) 37 (Magnetically perturbed Fe-57 and Sn-119m spectra).
K. Burger, *Inorg. Chim. Acta Rev.*, *6* (1972) 31 (Mössbauer spectra of mixed-ligand complexes).
G. M. Bancroft, R. H. Platt, *Adv. Inorg. Chem. Radiochem.*, *15* (1972) 59 (Bonding and structure from Mössbauer spectra).
R. V. Parish, *Prog. Inorg. Chem.*, *15* (1972) 101 (Interpretation of tin-Mössbauer spectra).
R. H. Herber, in *Characterization of Organometallic Compounds*, M. Tsutsui, ed., Vol. 2, Wiley-Interscience, New York, (1971) p. 315.

T.C. Gibb, in *Spectroscopic Methods in Organometallic Chemistry*, W.O. George, ed., Butterworths, London (1970), p. 33 (Mössbauer applications).
J.J. Zuckerman, *Adv. Organomet. Chem.*, *9* (1970) 22 (Applications of ^{119m}Sn Mössbauer spectroscopy to the study of organotin compounds).
P.J. Smith, *Organomet. Chem. Rev.*, *A*, *5* (1970) 373 (Organotin Mössbauer).
V.I. Goldanski, V.V. Khrapov, R.A. Stukin, *Organomet. Chem. Rev.*, *A*, *4* (1969) 225 (Organotin Mössbauer).
R.H. Herber, *Prog. Inorg. Chem.*, *8* (1967) 1 (Chemical applications of Mössbauer spectroscopy).
E. Fluck, *Top. Curr. Chem.*, *5* (1966) 395 (Mössbauer spectroscopy).

n) Nuclear-Quadrupole Resonance

E.A.C. Lucken, *Nuclear Quadrupole Coupling Constants*, Academic Press, New York (1969).

M.G. Voronkov, V.P. Feshkin, *Org. Magn. Reson.*, *9* (1977) 665 (NQR of haloorganometallics).
L. Ramakrishnan, S. Soundararajan, V.S.S. Sastry, J. Ramakrishna, *Coord. Chem. Rev.*, *22* (1977) 123 (NQR of coordination compounds).
M.G. Voronkov, V.P. Feshkin, in *Determination of Organic Structures by Physical Methods*, Vol. 5, F.C. Nachod, J.J. Zuckerman, eds., Academic Press, New York (1973) p. 169 (NQR of organometallics).

o) Optical Activity

B.E. Douglas, Y. Saito, eds., *Stereochemistry of Optically Active Transition-Metal Compounds*, ACS Symposium Series, No. 119, American Chemical Society, Washington, DC (1980).

G.B. Shul'pin, *Russ. Chem. Rev. (Engl. Tansl.)*, *49* (1980) 645 (Diastereotropy in transition-metal complexes).
H.P. Jensen, F. Woldbye, *Coord. Chem. Rev.*, *29* (1979) 213 (Optically active coordination compounds).
K. Bernauer, *Top. Curr. Chem.*, *65* (1976) 1 (Diastereoisomerism and diastereoselectivity in transition-metal complexes).
H. Brunner, *Top. Curr. Chem.*, *56* (1975) 67 (Reaction stereochemistry of optically active organotransition-metal compounds).
H. Brunner, *Angew. Chem., Int. Ed. Engl.*, *10* (1971) 249 (Asymmetric transition-metal atoms).

p) Dipole Moments

S. Sorriso, *Chem. Rev.*, *80* (1980) 313 (Dipole moments of inorganic complexes).

q) Microwave

R. Varma, in *Characterization of Organometallic Compounds*, M. Tsutsui, ed., Vol. 1, Wiley-Interscience, New York (1969), p. 277 (Microwave of metalloid and organometallics).

r) Magnetic Susceptibility

L.N. Mulay, J.T. Dehn, in *Characterization of Organometallic Compounds*, M. Tsutsui, ed., Vol. 2, Wiley-Interscience, New York (1971), p. 439 (Magnetic susceptibility of organometallics).

s) Chemical-Vapor Deposition

C.H. Bamford, C.F.H. Tipper, eds., *Comprehensive Chemical Kinetics*, Vol. 4, *Decomposition of Inorganic and Organometalic Compounds*, Elsevier, Amsterdam (1972).

G.A. Domrachev, *Vestn. Akad. Nauk SSSR*, (1983) 17 (Materials and coatings by decomposition of organometallics).

G.A. Domrachev, O.N. Suvorova, *Russ. Chem. Rev. (Engl. Tranl.)*, *49* (1980) 810 (Formation of inorganic coatings by decomposition of organometallics).

B.G. Gribov, *Russ. Chem. Rev. (Engl. Transl.)*, *42* (1973) 678 (Superpure materials from organometallics).

H.E. Podall, M.M. Mitchel, *Ann. NY Acad. Sci.*, *125* (1965) 218 (Organometallics in chemical-vapor deposition).

t) Gas Chromatography

T.R. Crompton, *Gas Chromatography of Organometallic Compounds*, Plenum Press, New York (1982).

H. Veening, B.R. Willeford, *Rev. Inorg. Chem.*, *1* (1979) 281 (High-performance liquid chromatography (HPLC) for organometallics).

N.T. Ivanova, L.A. Frangulyan, *Russ. Chem. Rev. (Engl. Transl.)*, *46* (1977) 171 (GLC of unstable inorganic and organometallic compounds).

V.A. Chernoplekova, V.M. Sakharov, K.I. Sakodynskii, *Russ. Chem. Rev. (Engl. Transl.)*, *42* (1973) 1063 (GLC of organo-group I–IV compounds).

H. Russell, G. Tölg, *Top. Curr. Chem.*, *33* (1972) 74 (GLC for separation and determination of organometallics).

u) Elemental Analysis

T.R. Compton, *Chemical Analysis of Organometallic Compounds*, Academic Press, Vol. 1 (1973); Vol. 2 (1974); Vol. 3 (1974); Vol. 4 (1975); Vol. 5 (1977).

A. L. Smith, ed., *Analysis of Silicones*, Wiley, New York (1974).
J. G. A. Luijten, *A Bibliography of Organotin Analysis*, 2nd ed., Tin Research Institute, Greenford, Middlesex (1970).
T. R. Crompton, *Chemical Analysis of Organoaluminum and Organozinc Compounds*, Pergamon, Oxford (1968).

J. M. McCall, D. E. Leyden, C. W. Blount,, *Anal. Chem.*, *43* (1971) 1324 (X-Ray fluorescence determination of heavy elements in organometallics).
O. Schwarzkopf, F. Schwarzkopf, in *Characterization of Organometallic Compounds*, M. Tsutsui, ed., Vol. 1, Wiley-Interscience, New York (1969), p. 35 (Microanalysis of organometallics).

Note: Specific applications of various physical methods to individual elements or groups of compounds are cited in the bibliographies of the appropriate chapters.

Chapter 5

a) Group IA (Alkali Metals): Li, Na, K, Rb, Cs

G. E. Coates, K. Wade, *Organometallic Compounds*, 4th ed., Vol. 1, Part 1: *Groups I–III*, Chapman and Hall, London (1981).
Gmelin Handbuch der Anorganischen Chemie. Perfluoroorgano Compounds of Main-Group Elements, Part 4, *Compounds with Elements of Main Groups 1 to 4* (New Supplement Series, Vol. 25) (1975).
A. W. Langer, ed., *Polyamine-Chelated Alkali Metal Compounds*, Advances in Chemistry Series, No. 130, American Chemical Society, Washington, DC (1974).
M. Schlosser, *Struktur und Reaktivität polarer Organometalle. Eine Einführung in die Chemie organischer Alkali- und Erdalkalimetall-Verbindungen*, Springer Verlag, Berlin (1972).
T. V. Talalaeva, K. A. Kocheshkov, *Methody Elementoorganischeskoi Khimii* (Methods of Organoelement Chemistry); *Li, Na, K, Rb, Cs*, 2 Vols, Nauka, Moscow (1971) (in Russian).
E. Müller, O. Bayer, H. Meerwein, K. Ziegler, eds., *Methoden der Organischen Chemie (Houben-Weyl)*, Band XIII/1, *CH-Acidität, Metallorganische Verbindungen der I. Gruppe des Periodensystems: Li, Na, K, Rb, Cs, Cu, Ag, Au*, by G. Bahr, P. Burba, F. H. Ebel, A. Luttringhaus, U. Scholkopf, G. Thieme Verlag, Stuttgart (1970).
A. A. Morton, *Solid Organoalkali-Metal Reagents*, Gordon & Breach, New York (1964).

K. Jonas, *Adv. Organomet. Chem.*, *19* (1981) 97 (Alkali metal-transition-metal complexes).
J.-P. Quintard, M. Pereyra, *Rev. Si, Ge, Sn, Pb Compnds.*, *4* (1980) 151 (Organotin-alkali reagents).
K. Jonas, C. Krüger, *Angew. Chem., Int. Ed. Engl.*, *19* (1980) 520 (Alkali metal-transition metal π-complexes).
B. J. Wakefield, in *Comprehensive Organic Chemistry.*, D. Barton, W. D. Ollis, eds., Vol. 3, D. N. Jones, ed., Pergamon Press, Oxford (1979), p. 943 (Group-I organic compounds).
E. Grovenstein, A. A. Morton, *Angew. Chem., Int. Ed. Engl.*, *17* (1978) 313 (Structure rearrangements of organoalkali-metal compounds).

J. D. Petty, P. L. Knutson, *Synthesis*, (1977) 509 (Di- and polyalkali-metal derivatives of heterofunctionally substituted organic molecules).
E. Grovenstein, *Adv. Organomet. Chem.*, *16* (1977) 167 (Aryl migrations in organoalkalis).
A. A. Morton, *Chem. Rev*, *75* (1975) 767 (Displacement of alkali metal by Hg and the dissociation of ion pairs).
H. Gilman, *J. Organomet. Chem.*, *100* (1975) 83 (Perfluoro derivatives).
H. Normant, *J. Organomet. Chem.*, *100* (1975) 189 (α-Halo-enolates).
M.E. Volpin, I.S. Kolomnikov, *Organomet. React.*, *5* (1975) 313 (Reactions with CO_2).
J.P. Oliver, *MPT Int. Rev. Sci., Inorg. Chem., Ser. Two*, *4* (1975) 1 (Organoalkali-metal compounds).
E.M. Kaiser, *J. Organomet. Chem.*, *103* (1975) 1 (Heavier organoalkalis).
N.S. Wyazankin, G.A. Razuvaev, O.A. Kruglaya, *Organomet. React.*, *5* (1975) 101 (Metal-alkali compounds).
V. Kalyanaraman, M.V. George, *J. Organomet. Chem.*, *47* (1973) 225 (Alkali-metalorganic compounds, alkali-metal addition to unsaturated systems).
V.A. Chernoplekova, V.M. Sakharov, K.I. Sakodynskii, *Russ. Chem. Rev. (Engl. Trans.)*, *42* (1973) 1063 (GLC of group-I organometallics).
D.V. Ioffe, M.I. Mostova, *Russ. Chem. Rev. (Engl. Transl.)*, *42* (1973) 56 (Dual reactivity of arylmethylmetal derivatives).
D.J. Peterson, *Organomet. Chem. Rev., A*, *7* (1972) 295 (α-Neutral heteroatom-substituted organometallics).
P. West, *MTP Int. Rev. Sci., Inorg. Chem., Ser. One*, *4* (1972) 1 (Organoalkali-metal compounds).
T. Chivers, *Organomet. Chem. Rev., A*, *6* (1970) 1(Chloro- and bromocarbon derivatives).
J.P. Oliver, *Adv. Organomet. Chem.*, *8* (1970) 167 (Fast exchanges of group I–III organometallics).
E. de Boer, *Adv. Organomet. Chem.*, *2* (1964) 115 (Electronic structure of alkali-metal adducts of aromatic hydrocarbons).

b) Lithium

B.J. Wakefield, *The Chemistry of Organolithium Compounds*, Pergamon Press, London (1974).

P.v.R. Schleyer, *Pure Appl. Chem.*, *56* (1984) 151 (Remarkable organolithium structures).
H.J. Siegel, *Top. Curr. Chem.*, *106* (1982) 55 (Li halocarbenoid carbanions).
W.E. Parham, C.K. Bradsher, *Acc. Chem. Res.*, *15* (1982) 300 (Phenyllithiums bearing electrophilic groups).
H.W. Gschwend, H.R. Rodriguez, *Org. React.*, *26* (1980) 1 (Heteroatom-facilitated lithiations).
A.F. Halasa, D.N. Schule, D.P. Tate, V.D. Mochel, *Adv. Organomet. Chem.*, *18* (1980) 49 (Organolithium catalysis of olefin and diene polymerization).
T. Kauffmann, *Top. Curr. Chem.*, *92* (1980) 109 (New reagents for organic synthesis).
J.P. Oliver, *Adv. Organomet. Chem.*, *15* (1977) 235, 255 (Structure of organolithium compounds).
D. Seebach, K.-H. Geiss, *J. Organomet. Chem. Libr.*, *1* (1976) 1 (Organolithiums in organic synthesis).
H.O.Howe, *Acc. Chem. Res.*, *9* (1976) 59 (Li organocuprate additions).

H. Gilman, *J. Organomet. Chem.*, *100* (1975) 83 (Perfluorinated organolithium compounds).
D. Ivanov, G. Vassilev, I. Panayotov, *Synthesis*, (1975) 83 (Organolithiums from weakly acidic C-H compounds).
T. Kauffmann, *Angew. Chem., Int. Ed. Engl.*, *13* (1974) 627 (1,3-Anionic cyclo-additions of organolithium compounds).
W. N. Smith, in *Polyamine-Chelated Alkali Metal Compounds*, A. W. Langer, ed., Advances in Chemistry Series, No. 130, American Chemical Society, Washington, DC (1974), p. 23 (Tertiary amine complexes of organolithium compounds, synthetic aspects).
D. W. Slocum, D. I. Sugarman, in *Polyamine-Chelated Alkali Metal Compounds,* A. W. Langer, ed., Advances in Chemistry Series, No. 130, American Chemical Society, Washington, DC (1974), p. 222 (Directed metallation with organolithium compounds).
R. West, in *Polyamine-Chelated Alkali Metal Compounds,* A. W. Langer, ed., Advances in Chemistry Series, No. 130, American Chemical Society, Washington, DC (1974), p. 211 (Polylithiation of hydrocarbons).
T. L. Brown, *Pure Appl. Chem.*, *23* (1970) 447 (Organolithium structures).
M. J. Jorgenson, *Org. React.*, *18* (1970) 1 (Ketones from organolithiums with carboxylic acids).
J. M. Mallan, R. L. Bebb, *Chem. Rev.*, *69* (1969) 693 (Metallation with organolithium compounds).
M. Schlosser, in *Newer Methods of Preparative Organic Chemistry*, W. Foerst, ed., Academic Press, New York (1968), Vol. 5, p. 238 (Organosodium and organopotassium compounds).
G. Köbrich, *Angew. Chem., Int. Ed. Engl.*, *6* (1967) 41 (α-Halogenated organolithium derivatives).
H. Heaney, *Organomet. Chem. Rev.*, *1* (1966) 27 (Organolithiums derived from polyhalogenated compounds).
T. L. Brown, *Adv. Organomet. Chem.*, *3* (1965) 365 (Organolithium structures).
I. T. Millar, H. Heaney, *Quart. Rev.*, *1* (1957) 109 (Organolithiums derived from organic halides).
E. A. Braude, *Prog. Org. Chem.*, *3* (1955) 172 (Preparation and reactions of organolithiums).
H. Gilman, J. W. Morton, *Org. React.*, *8* (1954) 258 (Metallation with organolithiums).
R. G. Jones, H. Gilman, *Org. React.*, *6* (1951) 339 (Lithium-halogen exchange reactions).

c) Sodium

J. F. Nobis, L. F. Moormeir, R. E. Robinson, in *Metal-Organic Compounds*, M. Sittig, ed., Advances in Chemistry Series, No. 23, American Chemical Society, Washington, DC (1959), p. 63 (Organosodium compounds in the preparation of other metal-carbon bonds).
R. A. Benkeser, D. L. Forster, D. M. Saure, J. F. Nobis, *Chem. Rev.*, *57* (1957) 867 (Metallation with organosodiums).

d) Sodium and Potassium

M. Schlosser, in *New Methods of Preparative Organic Chemistry*, Vol. 5, W. Foerst, ed., Academic Press, New York (1968), p. 238 (Organosodiums and organopotassiums).
M. Schlosser, *Angew. Chem., Int. Ed. Engl.*, *3* (1964) 287, 362 (Organosodiums and organopotassiums).

e) Rubidium and Cesium

F.M. Perel'man, *Rubidium and Cesium*, Pergamon Press, Oxford (1965).

Chapter 6

a) Group IIA: Be, Mg, Ca, Sr, Ba

G.E. Coates, K. Wade, *Organometallic Compounds*, 4th ed., Vol. 1, Part 1, *Groups I–III*, Chapman and Hall, London (1981).

Gmelin Handbuch der Anorganischen Chemie. Perfluorohalogenoorgano Compounds of Main-Groups 1 to 4 (New Supplement Series, Vol. 25), Springer Verlag, Berlin (1975) (Beryllium, calcium and magnesium derivatives).

E. Müller, O. Bayer, H. Meerwein, K. Ziegler, eds., *Methoden der Organischen Chemie (Houben-Weyl)*, Band XIII/2a. *Metallorganische Verbindungen: Be, Mg, Ca, Sr, Ba, Zn, Cd*, by K. Nützel, G. Bähr, H. Gilman, H.O. Kalinowski, G.F. Wright, G. Thieme Verlag, Stuttgart (1973).

M. Schlosser, *Struktur und Reaktivität polarer Organometalle. Eine Einführung in die Chemie organischer Alkali- und Erdalkalimetall-Verbindungen*, Springer Verlag, Berlin (1972).

S.T. Yoffe, A.N. Nesmeyanov, *The Organic Chemistry of Magnesium, Calcium, Strontium and Barium*, North-Holland, Amsterdam (1967); English transl. of. S.T. Yoffe, A.N. Nesmeyanov, *Methods Elementoorganicheskoi Khimii* (Methods of Organoelement Chemistry): *Mg, Be, Ca, Sr, Ba*, Nauka, Moscow (1963).

R.D. Chambers, T. Chivers, *Organomet. Chem. Rev.*, *1* (1966) 279 (Pentafluorophenyl derivatives of B, Al, Ga, In, Tl).

B.J. Wakefield, in *Comprehensive Organic Chemistry*, D. Barton, W.D. Ollis, eds., Vol. 3, D.N. Jones, ed., Pergamon Press, Oxford (1979), p. 969 (Group II organic compounds).

B.G. Gowenlock, W.E. Lindsell, *J. Organomet. Chem. Libr.*, *3* (1977) 1 (Organometallic compounds of the alkaline earths).

Yu.A. Alekandrov, V.P. Maslennikov, *J. Organomet. Chem. Libr.*, *3* (1977) 75 (Organic peroxides of main-group II elements).

M.A. Coles, *MTP Int. Rev. Sci., Inorg. Chem., Ser. Two*, *4* (1975) 359 (Organoderivatives of Be, Ca, Sr. Ba).

V.A. Chernoplekova, V.M. Sakharov, K.I. Sakodyneskii, *Russ. Chem. Rev. (Engl. Transl.)*, *42* (1973) 1063 (GLC of group II organometallics).

J.P. Oliver, *Adv. Organomet. Chem.*, *8* (1970) 167 (Fast exchanges of group I–III organometallics).

C.A. Balueva, S.T. Yoffe, *Russ. Chem. Rev., (Engl. Transl.)*, *31* (1962) 439 (Organometallic compounds of Be, Ca, Sr, Ba).

b) Beryllium

D.A. Everest, *The Chemistry of Beryllium*, Elsevier, Amsterdam (1965), Ch. 7 (Organometallic derivatives).

J.P. Oliver, *Adv. Organomet. Chem.*, *15* (1977) 235, 252 (Structures of organoberylliums).
F. Bertin, G. Thomas, *Bull. Soc. Chim. Fr.*, (1971) 3951 (Organic compounds of beryllium and their complexes).
G.E. Coates, G.L. Morgan, *Adv. Organomet. Chem.*, *9* (1970) 1095 (Organoberylliums).
N.R. Fetter, *Organomet. Chem. Rev.*, *A*, *3* (1968) 1 (Organoberylliums).

c) Magnesium

H. Felkin, C. Swierczewski, *Activation of Grignard Reagents by Transition Metal Compounds*, Pergamon Press, Oxford (1976).
A.N. Nesmeyanov, S.T. Yoffe, *Handbook of Magnesium-Organic Compounds*, Vols. 1–3, Pergamon Press, Oxford (1957).
M.S. Kharasch, O. Reinmuth, *Grignard Reactions of Nonmetallic Substances*, Prentice-Hall, Englewood Cliffs, NJ (1954).

T. Holm, *Acta Chem. Scand., Ser. B. 37* (1983) 567 (Electron-transfer mechanisms from alkylmagnesiums).
M. Dagonneau, *Bull. Soc. Chim. Fr.*, (1982) 269 (Radical Grignard reactions).
Y.-H. Lai, *Synthesis*, (1981) 585 (Grignard reagents from activated Mg).
J.-P. Quintard, M. Pereyre, *Rev. Si. Ge, Sn, Pb Compnds.*, *4* (1980) 151 (Organotin-magnesium reagents).
A.G. Pinkus, *J. Chem. Educ.*, *55* (1978) 704 (Magnesocene, structure and bonding).
A.G. Pinkus, *Coord. Chem. Rev.*, *25* (1978) 173 (Three-coordinated Mg).
P. Sobota, B. Jezowska-Trzebiatowska, *Coord. Chem. Rev.*, *26* (1978) 71 (Transition-metal compound-Mg systems for fixing and activating N_2, CO, H_2, etc.).
C. Blomberg, F.A. Hartog, *Systhesis*, (1977) 18 (The Barbier reaction – one-step synthesis of organomagnesiums).
N.A. Bell, *MTP Int. Rev. Sci., Inorg. Chem., Ser. Two*, *4* (1975) 381 (Organic derivatives of magnesium).
E.A. Hill, *J. Organomet. Chem.*, *91* (1975) 123 (Rearrangements in organomagnesium chemistry).
E.C. Ashby, *Chem. Rev.*, *75* (1975) 521; *Acc. Chem. Res.*, *7* (1974) 272 (Mechanism of Grignard addition to ketones).
S.S. Dua, R.D. Howells, H. Gilman, *J. Fluorine Chem.*, *4* (1974) 409 (Perfluoroalkyl-Grignard reagents).
F. Bickelhaupt, *Angew. Chem., Int. Ed. Engl.*, *13* (1974) 419 (Free radicals in Grignard reactions).
G. Courtois, L. Miginiac, *J. Organomet. Chem.*, *69* (1974) 1 (Allyl derivatives of Mg, Zn and Cd).
N.A. Bell, *Educ. Chem.*, *10* (1973) 143 (Constitution of Grignard reagents).
H. Normant, *Pure Appl. Chem.*, *30* (1972) 463 (Chemistry of organomagnesiums after V. Grignard).
L.F. Elsom, J.D. Hunt, A. McKillop, *Organomet. Chem. Rev.*, *A*, *8* (1972) 135 (Co(II) chloride catalysis in Grignard reactions).
J. Villieras, *Organomet. Chem. Rev.*, *A*, *7* (1971) 81 (α-Haloalkyl Grignard reagents).
R.A. Benkeser, *Synthesis*, (1971) 347 (Allyl- and crotyl-Grignard reagents).
E.C. Ashby, *Quart. Rev.*, *21* (1967) 259 (The structure of Grignard reagents).

H. Heaney, *Organomet. Chem. Rev.*, *1* (1966) 27 (Grignard reagents from polyhalogenated compounds).
B. J. Wakefield, *Organomet. Chem.*, *1* (1966) 131 (Stucture of Grignard reagents).
H. Normant, *Adv. Org. Chem., Methods Results*, *2* (1960) 1 (Alkenylmagnesium halides).
D. A. Shirley, *Org. React.*, *8* (1954) 28 (Synthesis of ketones from acyl halides and organometallic compounds of Mg, Zn and Cd).
C. Courtot, in *Traite de Chimie Organique*, V. Grignard, G. Dupont, E. Locquin, eds., Masson, Paris (1937), Vol. 5 (Extensive review – 415 pages – of earlier literature on Grignard reagents).

d) Group IIB: Zn, Cd, Hg

N. I. Sheverdina, K. A. Kocheshkov, *The Organic Compounds of Zinc and Cadmium*, North-Holland, Amsterdam (1967); Engl. transl. of N. I. Sheverdina, K. A. Kocheshkov, *Methods Elementoorganicheskoi Khimii* (Methods of Organoelement Chemistry): *Zn, Cd*, Nauka, Moscow (1964).

L. G. Kuzmina, N. G. Bokii, Yu. T. Struchkov, *Russ. Chem. Rev., (Engl. Transl.)*, *44* (1975) 73 (Structural chemistry of organic compounds of Hg, Zn, Cd.).
K. C. Bass, *MTP Int. Rev. Sci., Inorg. Chem., Ser. One*, *4* (1972) 41 (Zn-, Cd-, Hg-organic compounds).
K.-H. Thiele, P. Zdunek, *Organomet. Chem. Rev.*, *1* (1966) 331 (Donor-acceptor complexes of organometallic compounds of Zn, Cd, Hg).

e) Zinc

J. Boersman, J. G. Noltes, *Organozinc Coordination Chemistry,* ILZRO, New York (1968).

M. W. Rathke, *Org. React.*, *22* (1975) 423 (The Reformatsky reaction).
J. Furukawa, N. Kawabata, *Adv. Organomet. Chem.*, *12* (1974) 83 (Organozinc compounds in synthesis).
H. E. Simmons, T. L. Cairns, S. A. Vladuchick, C. M. Hoiness, *Org. React.*, *20* (1973) 1 (Halomethylzinc-organic compounds in the synthesis of cyclopropanes).
G. J. M. Van der Kerk, *Russ. Chem. Rev. (Engl. Transl.)*, *30* (1972) 389 (Coordination chemistry and catalytic action of zinc-organic compounds).

f) Cadmium

P. R. Jones, P. J. Desio, *Chem. Rev.*, *78* (1978) 491 (Less-familiar reactions of organocadmiums).
J. Casson, *Chem. Rev.*, *40* (1940) 15 (Organocadmiums in the preparation of ketones).

g) Mercury

C.A. McAuliffe, ed., *The Chemistry of Mercury*, Macmillan, London (1977).

E. Müller, O. Bayer, H. Meerwein, K. Ziegler, eds., *Methoden der Organichen Chemie (Houben-Weyl). Band XIII/2b. Quecksilber-Organische Verbindungen: Hg*, by H. Straub, K.P. Zeller, H. Leditschke, G. Thieme Verlag, Stuttgart (1974).

M.W. Miller, T.W. Clarkson, eds., *Mercury, Mercurials and Mercaptans*, Thomas Books, New York (1973).

L.T. Friberg, J.J. Vostal, *Mercury in the Environment*, CRC Press, Cleveland, OH (1972).

F.R. Jensen, B. Rickborn, *Electrophilic Substitution of Organomercurials*, McGraw-Hill, New York (1968).

O.A. Reutov, I.P. Beletskaya, *Reactions of Organometallic Compounds*, North Holland, Amsterdam (1968) (Mainly Hg compounds, despite title).

L.G. Makarova, A.N. Nesmeyanov, *The Organic Compounds of Mercury*, North Holland, Amsterdam (1967); English transl. of L.G. Makarova, A.N. Nesmeyanov, *Methody Elementoorganicheskoi Khimii* (Methods of Organo-element Chemistry): *Hg*, Nauka, Moscow (1965).

I.P. Beletskaya, *J. Organomet. Chem.*, *250* (1983) 551 (Organomercury cross-coupling with organic halides catalyzed by Pd).

R.C. Larock, *Tetrahedron*, *38* (1982) 1713 (Organomercurials in organic synthesis).

J.A. Mangravite, *J. Organomet. Chem. Libr.*, *7* (1979) 45 (Allylmercurials).

W.P. Neumann, K. Reuter, *J. Organomet. Chem. Libr.*, *7* (1979) 229 (Silylmercurials in organic synthesis).

D.L. Rabenstein, *Acc. Chem. Res.*, *11* (1978) 101; *J. Chem. Educ.*, *55* (1978) 292 (The chemistry of methylmercury toxicology).

R.C. Larock, *Angew. Chem., Int. Ed. Engl.*, *17* (1978) 27 (Organomercury compounds in organic synthesis).

R.C. Larock, in *Aspects of Mechanism in Organometallic Chemistry*, J.H. Brewster, ed., Plenum Press, New York (1978), p. 251 (Organomercurials in organic synthesis).

M.G. Voronkov, N.F. Chernov, *Russ. Chem. Rev. (Engl. Transl.)*, *48* (1978) 964 (Silicon-containing organomercurials).

V.I. Sokolov, O.A. Reutov, *Coord. Chem. Rev.*, *27* (1978) 89 (Organomercurials with Pd(0) and Pt(0) complexes).

K.H. Falchuk, L.J. Goldwater, B.L. Vallee, in *The Chemistry of Mercury*, C.A. McAuliffe, ed., MacMillan, New York (1977), p. 259 (Biochemistry and toxicology of Hg).

A.J. Bloodworth, in *The Chemistry of Mercury*, C.A. McAuliffe, ed., MacMillan, New York (1977), p. 137 (Review of organomercurials with 500 references).

V.S. Petrosyan, O.A. Reutov, *J. Organomet. Chem.*, *76* (1976) 123 (NMR spectra and structure of organomercury compounds).

R.C. Larock, *J. Organomet. Chem. Libr.*, *1* (1976) 257 (Organomercury compounds in organic synthesis).

O.A. Reutov, K.P. Butin, *J. Organomet. Chem.*, *99* (1975) 171 (Organic calomels; transmetalations with mercury metal).

A.A. Morton, *Chem. Rev.*, *75* (1975) 767 (Displacement of alkali metals by Hg and the dissociation of ion pairs).

S.F. Zhil'tsov, O.N. Druzhkob, *Tr. Khim Khim. Tekhnol.* (In Russian), (1974), No. 4, 13 (Liquid-phase homolytic reactions of organomercury compounds).

P. Beletskaya, K.P. Butin, A.N. Ryabtsev, O.A. Reutov, *J. Organomet. Chem*, *59* (1973) 1 (Stability of organomercury complexes with anionic and neutral ligands).

D. Seyferth, *Acc. Chem. Res.*, *5* (1972) 65 (Phenyl(trihalomethyl)mercurials as dihalocarbene precursors).
E. G. Makarova, *Organomet. React.*, 2 (1971) 335 (Reactions of organomercury compounds).
H. Arzonmanian, J. Metagar, *Synthesis*, (1971) 527 (Olefin oxidation by Hg salts).
E. G. Makarova, *Organomet. React.*, *1* (1970) 119 (Reactions of organomercury compounds).
D. E. Matteson, *Organomet. Chem. Rev.*, *A*, *4* (1969) 263 (Electrophilic displacements).
J. J. Lagowski, in *New Pathways in Inorganic Chemistry*, E. A. V. Ebsworth, A. G. Maddock, A. G. Sharpe, eds., Cambridge Univ. Press, Cambridge (1968) p. 137 (Fluoroalkylmercurials).
W. Kitching, *Organomet. Chem. Rev.*, *A3* (1968) 35 (Mercuration reactions).
O. A. Reutov, *Top. Curr. Chem.*, *8* (1967) 61 (Electrophilic substitution in organomercury compounds).
R. E. Dessy, W. Kitching, *Adv. Organomet. Chem.*, *4* (1966) 268 (Mechanisms of organomercury reactions).
D. Grdenic, *Quart. Rev.*, *19* (1965) 303 (Structural chemistry of mercury, including organomercury compounds).

Chapter 7

a) Group-IIIA: B, Al, Ga, In, Tl

Gmelin Handbuch der Anorganischen Chemie, Perfluorohalogeno Organo Compounds of Main-Group I to IV (New Supplement Series, Vol. 25), Springer Verlag, Berlin (1975).
E. Müller, O. Bayer, H. Meerwein, K. Ziegler, eds., *Methoden der Organischen Chemie (Houben-Weyl)*, 4th ed., Band XIII/4, *Metallorganische Verbindungen: Al, Ga, In, Tl*, by G. Bähr, P. Burba, H. Lehmkuhl, K. Ziegler, G. Thieme Verlag, Stuttgart (1970).
A. N. Nesmeyanov, R. A. Sokolik, *Methody Elementoorganicheskoi Khimii: B, Al, Ga, In, Tl*, (Methods of Organoelement Chemistry), Nauka, Moscow (1964) (In Russian); Engl. transl.: A. N. Nesmeyanov, R. A. Sokolik, *The Organic Compounds of Boron, Aluminum, Gallium, Indium and Thallium*, North Holland, Amsterdam (1969).

B. Singaram, G. G. Pai, *Heterocycles*, *18* (1982) 387 (Borane and alane addition compounds).
G. Zweifel, in *Comprehensive Organic Chemistry*, D. Barton, W. D. Ollis, eds., Vol. 3, D. N. Jones, ed., Pergamon Press, Oxford (1979), 1013 (Group-III organic compounds).
J. J. Eisch, *Adv. Organomet. Chem.*, *16* (1977) 67 (Rearrangements of unsaturated organic B and Al compounds).
Yu. A. Aleksandrov, V. P. Maslennikov, *J. Organomet. Chem. Libr.*, *3* (1977) 103 (Organic peroxides of group IIIA elements).
E. Negishi, *J. Organomet. Chem. Libr.*, *1* (1976) 93 (Organoboranes and alanes as nucleophiles in organic syntheses).
J. Weidlein, *J. Organomet. Chem.*, *49* (1973) 257 (Eight-membered organoaluminum, gallium, indium and thallium heterocycles).
V. A. Chernoplekova, V. M. Sakharov, K. I. Sakodynskiii, *Russ. Chem. Rev. (Engl. Transl.)*, *42* (1973) 1063 (GLC of group-IIIA organometallics)

J. B. Farmer, K. Wade, *MTP Int. Rev. Sci., Inorg. Chem., Ser. One, 4* (1972) 105 (Organometallic compounds of Al, Ga, In, Tl).
J. P. Oliver, *Adv. Organomet. Chem., 8* (1970) 167 (Fast exchange reactions of group-III organometallics).
H. Schmidbaur, *Adv. Organomet. Chem., 9* (1970) 259 (Isoelectronic organo-P, Si and Al series).
K. Yasuda, R. Okawara, *Organomet. Chem. Rev., 2* (1967) 255 (Organometallic derivatives of Ga, In, Tl).
W. E. Davidsohn, M. C. Henry, *Chem. Rev., 67* (1967) 73 (Group-III acetylenes).

b) Boron

i) Organoboron Compounds: General

E. Müller, O. Bayer, H. Meerwein, K. Ziegler, eds., *Methoden der Organischen Chemie (Houben-Weyl)*. Band XIII/3. *Bor-Organische Verbindungen, I*, by R. Köster, W. Siebert, G. Thieme Verlag, Stuttgart (1982); *II*, by R. Köster, G. Schmid (1983).
A. Pelter, ed., *IMEBORON V.* (in *Pure Appl. Chem., 55* (1983) (Lectures at 5th International Symposium on Boron Chemistry, 1983).
P. Mellor, ed., *Mellor's Modern Treatise on Inorganic and Theoretical Chemistry*, Part 5, *Boron*, Longmans Green, London (1980), R. Thompson ed., Supplement (1982).
Gmelin Handbuch der Anorganischen Chemie. Boron Compounds, 1st Supplement, Vol. 2 (1980) (Organoboron halides, borazines, other heterocycles). Vol. 3 (1981) (Carboranes, organothioboranes). 2nd Supplement, Vol. 1 (1982) (Organoboron-oxygen and nitrogen compounds). Vol. 2 (1982) (Organoboron-halogen, oxygen and sulfur compounds; carboranes). (see page 39, Ch. 4 for a complete list of volume titles).
R. W. Parry, G. Kodama, *Boron Chemistry*, Pegamon Press, Oxford (1980).
D. N. Jones, ed., *Sulphur, Selenium, Silicon, Boron Organometallic Compounds*, Vol. 3, *Comprehensive Organic Chemistry*, D. Barton, W. D. Ollis, eds., Pergamon Press, Oxford (1974).
H. Nöth, H. Wrackmeyer, *Nuclear Magnetic Resonance of Boron Compounds*, Springer Verlag, Berlin (1978).
T. P. Onak, *Organoborane Chemistry*, Academic Press, New York (1975).
E. L. Muetterties, ed., *Boron Hydride Chemistry*, Academic Press, New York (1975).
I. Haiduc, *The Chemistry of Inorganic Ring Systems*, J. Wiley & Sons, New York (1970), Part I. Chapter III (Organoderivatives of borazines, boroxines and borthianes).
R. J. Brotherton, H. Steinberg, eds., *Progress in Boron Chemistry*, Pergamon Press, Oxford, Vol. 1 (1964), Vol. 2 (1966), Vol. 3 (1970).
R. T. Holzmann, ed., *Production of Boranes and Related Research*, Academic Press, New York (1967).
E. L. Muetterties, ed., *The Chemistry of Boron and Its Compounds*, Wiley, New York (1967) (Contains several chapters dealing with organoboron compounds).
W. Gerrard, *The Organic Chemistry of Boron*, Academic Press, New York (1961).

I. Ander, in *Comprehensive Heterocyclic Chemistry*, A. Katritzsky, C. W. Rees, eds., Vol. 1, O. Meth-Cohn, ed., Pergamon Press, Oxford (1984) (Boronheterocycles).
B. M. Mikhailov, *Izv. Akad. Nauk. SSSR, Ser. Khim.*, (1984) 225 (1-Boraadamantane).
B. M. Mikhailov, *Pure Appl. Chem., 55* (1983) 1439 (1-Boraadamantane).
H. Nöth, S. Weber, B. Rasthofer, C. Narula, A. Konstantinov, *Pure Appl. Chem., 55* (1983) 1453 (Di- and tricoordinated boron cations).

R.F. Porter, L.J. Turbini, *Top. Curr. Chem.*, *96* (1981) (Photochemistry of organoboranes, carboranes and metallocarboranes).

L.H. Long, in *Mellor's Comprehensive Treatise on Inorganic and Theoretical Chemistry*, Supplement, Vol. V, *Boron*, Part BI, *Boron-Hydrogen Compounds*, Longmans, London (1981), p. 52 (Diboranes); p. 399 (Borane(3) derivatives).

R.W. Jotham, L.H. Long, in *Mellor's Comprehensive Treatise on Inorganic and Theoretical Chemistry*, Supplement, Vol. V, *Boron*, Part BI, *Boron-Hydrogen Compounds*, Longman, London (1981), p. 235 (Higher boranes).

H.C. Brown, *Chem. Eng. News, 59* (1981) 24 (A personal survey of research).

H.C. Brown, J.B. Campbell, Jr., *Aldrichim. Acta, 14* (1981) 3 (Vinylboranes).

B.M. Mikhailov, *Sov. Sci. Rev., Sect. B, 2* (1980) 283 (Boraheterocycles from allylboranes).

H. Schmidbaur, *J. Organomet. Chem.*, *200* (1980) 287 (Organophosphane-borane chemistry).

W. Siebert, *Adv. Organomet. Chem.*, *18* (1980) 301 (Boron heterocycles as transition-metal ligands).

A. Pelter, in *Rearrangements in Ground and Excited States*, P. de Mayo, ed., Vol. 2, Academic Press, New York (1980), p. 95 (Rearrangements involving boron).

A. Pelter, K. Smith, in *Comprehensive Organic Chemistry*, D. Barton, W.D. Ollis, eds., Vol. 3, D.N. Jones, ed., Pergamon Press, Oxford (1979), p. 687 (General review).

C.W. Allen, D.E. Palmer, *J. Chem. Educ.*, *55* (1978) 497 (Borabenzene anions).

B.M. Mikhailov, *Pure. Appl. Chem.*, *49* (1977) 749 (Cyclic coordination of boron compounds).

R. Köster, *Pure Appl. Chem.*, *49* (1977) 765 (Organoboranes in synthesis and analysis).

B.M. Mikhailov, *Russ. Chem. Rev. (Engl. Transl.)*, *45* (1976) 557 (Allylboranes).

E.I. Neghishi, *J. Organomet. Chem.*, *108* (1976) 281 (Organoborates).

S. Gronowitz, *J. Heterocycl. Chem.*, *3* (1976) 17 (Aromatic boron-containing heterocycles).

D.F. Gaines, M.B. Fischer, S.J. Hildebrandt, J.A. Ulman, J.W. Lott, in *Inorganic Compounds with Unusual Properties*, R.B. King, ed., Advances in Chemistry Series, No. 150, American Chemical Society, Washington, DC (1976), p. 311 (Borane-anion ligands).

H.C. Brown, *Pure Appl. Chem.*, *47* (1976) 49 (Organoboranes).

A.T.T. Hsieh, *Inorg. Chim. Acta*, *14* (1975) 87 (Organometallics containing bonds between metals and boron).

K. Smith, *Chem. Soc. Rev.*, *3* (1974) 443 (Organoborane preparations).

R.H. Cragg, A.F. Weston, *J. Organomet. Chem.*, *67* (1974) 161 (Mass spectra of organoboron compounds).

T.D. Coyle, J.J. Ritter, *Adv. Organomet. Chem.*, *10* (1972) 237 (Organometallic diboron chemistry).

K. Niedenzu, *MTP Int. Rev. Sci., Inorg. Chem., Ser. One*, *4* (1972) 3 (Organoboron compounds, general).

B.M. Mikhailov, *Organomet. Chem. Rev.*, *A*, *8* (1972) 1 (Allylboranes).

W. Kliegel, *Organomet. Chem. Rev.*, *A*, *8* (1972) 153 (Organoboron betaines).

G. Schmid, *Angew. Chem., Int. Ed. Engl.*, *9* (1970) 819 (Metal-boranes).

K. Niedenzu, *Organomet. Chem. Rev.*, *1* (1966) 305 (Organohaloboranes).

M.J.S. Dewar, *Prog. Boron Chem.*, *1* (1964) 235 (Heteroaromatic boron compounds).

R. Köster, *Adv. Organomet. Chem.*, *2* (1964) 257 (Cyclic organoboranes).

P.M. Treichel, F.G.A. Stone, *Adv. Organomet. Chem.*, *1* (1964) 143 (Fluoroorganoboranes).

H.D. Kaesz, F.G.A. Stone, in *Organometallic Chemistry*, H. Zeiss, ed., Reinhold, New York (1960), Ch. 4 (Vinylboranes).

H.C. Brown, in *Organometallic Chemistry,* H. Zeiss, ed., Reinhold, New York (1960), Ch. 4, p. 150 (Organoboranes).

B.M. Mikhailov, *Russ. Chem. Rev., (Engl. Transl.)*, *28* (1959) 1450(Organoboron compounds; much Russian literature cited).

A. J. Barnard, *Chem. Anal.*, *44* (1955) 104; *48* (1959) 44 (Analytical applications of $NaBPh_4$).
M. F. Lappert, *Chem. Rev.*, *56* (1956) 959 (Early review of organoboron compounds).

ii) Hydroboration and Application to Organic Synthesis

Gmelin Handbuch der Anorganischen Chemie, 8th ed., *New Supplement Series*, Vol. 45, *Boron Compounds*, Part 14, *Boron-Hydrogen Compounds*, *1*, Springer Verlag, Berlin (1977).
H. C. Brown, *Organic Synthesis via Boranes*, J. Wiley New York (1975).
G. M. L. Cragg, *Organoboranes in Organic Synthesis*, M. Dekker, New York (1973).
H. C. Brown, *Boranes in Organic Chemistry*, Cornell Univ. Press, Ithaca, New York (1972).
H. C. Brown, *Hydroboration*, W. A. Benjamin, New York (1962).

H. C. Brown, J. Chandrasekharan, K. K. Wang, *Pure Appl. Chem.*, *55* (1983) 1387 (Hydroboration – kinetics and mechanism).
H. C. Brown, B. Singaram, S. Singaram, *J. Organomet. Chem.*, *239* (1982) 43 (Alkylborohydrides).
H. C. Brown, P. K. Jadhav, A. K. Mandel, *Tetrahedron*, *37* (1981) 3547 (Asymmetric synthesis via chiral organoboranes).
A. Pelter, *Chem. Soc. Rev.*, *11* (1980) 191 (C-C bond formation by boron reagents).
K. Avasthi, D. Devaprabhakara, A. Suzuki, *J. Organomet. Chem. Libr.*, *7* (1979) 1 (Non-catalytic hydrogenation via organoboranes).
H. C. Brown, S. Krishnamurthy, *Tetrahedron*, *35* (1979) 567 (Hydride reductions).
K. Avasthi, D. Devaprabhakara, A. Suzuki, *J. Organomet. Chem. Libr.*, *7* (1979) 1 (Hydrogenation of carbon-carbon multiple bonds by boron hydrides).
G. M. L. Cragg, K. R. Koch, *Chem. Soc. Rev.*, *6* (1977) 393 (Alkenyl, alkynyl and cyanoborates as synthetic intermediates).
H. C. Brown, E. Negishi, *Tetrahedron*, *33* (1977) 2331 (Boraheterocycles via cyclic hydroboration).
C. F. Lane, in *Synthetic Reagents*, J. S. Pitzey, ed., Vol. 3, Ellis Harwood, Ltd., London (1977), p. 1 (Diborane and its Lewis adducts as synthetic reagents in organic chemistry).
J. Weill-Raynal, *Synthesis*, (1976) 633 (Formation of C-C bonds by organoboranes).
E. Neghishi, S. U. Kulkarni, H. C. Brown, *Heterocycles*, *5* (1976) 883 (Organoborane heterocycles via hydroboration of dienes).
G. W. Kabalka, J. D. Baker, G. W. Neal, *J. Chem. Educ.*, *53* (1976) 549 (Organoboranes as alkylating agents).
G. M. L. Cragg, *Chemsa (South Africa)*, *2* (1976) 216 (Synthesis of alcohols via organoboranes).
H. C. Brown, *J. Organomet. Chem.*, *100* (1975) 3 (Hydroboration, historical notes).
H. L. Long, *Adv. Inorg. Chem. Radiochem.*, *16* (1974) 201 (Diborane reactions).
E. Negishi, H. C. Brown, *Synthesis*, (1974) 77 (Thexylborane – a versatile hydroborating agent).
H. C. Brown, *Adv. Organomet. Chem.*, *11* (1973) (Boranes in organic chemistry).
H. C. Brown, M. M. Midland, *Angew. Chem., Int. Ed. Engl.*, *11* (1972) 692 (Free-radical displacement reactions of organoboranes).
J. A. Marshall, *Synthesis*, (1971) 229 (Diene synthesis via boronate fragmentation).
H. C. Brown, *Acc. Chem. Res.*, *2* (1969) 65 (Hydroboration).
G. Zweifel, H. C. Brown, *Org. React.*, *13* (1963) Ch. 1 (Hydroboration).

iii) Organoboron-Nitrogen Compounds

Gmelin Handbuch der Anorganischen Chemie, 8th ed., *New Supplement Series*, Vol. 13, *Boron Compounds*, Part 1, *Binary Boron-Nitrogen Compounds, B-N-C Heterocycles, Polymeric Boron-Nitrogen Compounds*, Springer Verlag, Berlin (1974). Vol. 46, Part 15, *Amine-Boranes* (1977). Vol. 51, Part 17. *Borazine and its Derivatives* (1978).

J.J. Lagowski, *Coord. Chem. Rev.*, *22* (1977) 185 (Borazines as ligands).
K. Niedenzu, *Chem.-Ztg.*, *98* (1974) 487 (Organoboron compounds containing B-N bonds).
W. Kliegel, *Organomet. Chem. Rev.*, *A*, *8* (1972) 153 (In German) (Boron-nitrogen betaines).
S. Trofimenko, *Chem. Rev.*, *72* (1972) 497 (Polypyrazolylborate ligands).
A. Meller, *Top. Curr. Chem.*, *26* (1972) 37 (Iminoboranes).
S. Trofimenko, *Acc. Chem. Res.*, *4* (1971) 17 (Polypyrazolylborate ligands).
I.B. Atkins, B.R. Currell, *Inorg. Macromol. Rev.*, *1* (1971) 203 (Organoboron-nitrogen polymers).
A. Meller, *Top. Curr. Chem.*, *15* (1970) 145 (Organoboron-nitrogen chemistry, preparative aspects).
H. Nöth, *Prog. Boron Chem.*, *3* (1970) 211 (Boron-nitrogen compounds).
A. Finch, J.B. Leach, J.H. Morris, *Organomet. Chem. Rev.*, *A*, *4* (1969) 1 (Cyclic B-N compounds).
J.J. Lagowski, E.K. Mellon, *Adv. Inorg. Chem. Radiochem.*, *5* (1963) 295 (Borazines).
J.C. Sheldon, B.C. Smith, *Quart. Rev.*, *14* (1960) 200 (Early review on borazines).

iv) Organoboron-Oxygen Compounds

Gmelin Handbuch der Anorganischen Chemie, 8th ed., *New Supplement Series*, Vol. 44, *Boron Compounds*. Part 13. *Boron-Oxygen Compounds*, *1*, Springer Verlag, Berlin (1977). 1st Supplement, Vol. 1 (1980) (Organoboron-oxygen compounds).

A. Suzuki, *Acc. Chem. Res.*, *15* (1982) 178 (Organoborates in synthesis).
P.L. Strong, in *Mellor's Comprehensive Treatise on Inorganic and Theoretical Chemistry*, Supplement, Vol. V, *Boron*, Part A, *Boron-Oxygen Compounds*, Longmans, London (1980), p. 765 (Boronic and borinic acids).
K. Torsell, in *Progress in Boron Chemistry*, R.J. Brotherton, H. Steinberg, eds., Vol. 1, Pergamon Press, Oxford (1964), p. 369 (Boronic and borinic acids).

v) Organoboron-Halides

Gmelin Handbuch der Anorganischen Chemie, 8th ed., *New Supplement Series*, Vol. 34, *Boron Compounds*, Part 9, *Boron-Halogen Compounds*, *1*, Springer Verlag, Berlin (1976). Part 19. *Boron-Halogen Compounds*, *2* (1979).

H.C. Brown, S.U. Kulkarni, *J. Organomet. Chem.*, *239* (1982) 23 (Alkylhaloboranes).
K. Niedenzu, *Organomet. Chem. Rev.*, *1* (1966) 305 (Organohaloboranes).

vi) Organoboron-Sulfur Compounds

Gmelin Handbuch der Anorganischen Chemie, 8th ed., *New Supplement Series*, Vol. 19, *Boron Compounds*, Part 3, *Compounds of Boron with the Non-metals S, Se, Te, P, As, Sb and Si and with Metals*, Springer Verlag, Berlin (1975).

W. Siebert, *Chem.-Ztg.*, *98* (1974) 479 (Organoboron compounds containing S, Se, Te).
R. H. Cragg, M. F. Lappert, *Organomet. Chem. Rev.*, *1* (1966) 43 (Sulfur-containing organoboranes).

vii) Carboranes and Metallocarboranes

R. N. Grimes, ed., *Metal Interaction with Boron Clusters*, Plenum Press, New York (1982).
Gmelin Handbuch der Anorganischen Chemie, 8th ed., *New Supplement Series*, Vol. 15, *Boron Compounds*, Part 2, *Carboranes, 1*, Springer Verlag, Berlin (1974). Vol. 27, Part 6, *Carboranes, 2* (1975), Vol. 42, Part II, *Carboranes, 3*. Vol. 43, Part 12, *Carboranes, 4*.
R. N. Grimes, *Carboranes*, Academic Press, New York (1970).
E. L. Muetterties, W. H. Knoth, *Polyhedral Boranes*, M. Dekker, New York (1968) (Chapter on carboranes).

N. N. Greenwood, *Pure Appl. Chem.*, *55* (1983) 1415 (Metalloboranes).
R. N. Grimes, *Adv. Inorg. Chem. Radiochem.*, *26* (1983) 55 (Carbon-rich carboranes and their metal derivatives).
R. N. Grimes, *Acc. Chem. Res.*, *16* (1983) 22 (Metals in boron clusters).
R. N. Grimes, *Pure Appl. Chem.*, *54* (1982) 43 (Metalocarboranes in organic synthesis).
C. E. Housecroft, T. P. Fehlner, *Adv. Organomet. Chem.*, *21* (1982) 57 (Metalloborane relations to metal hydrocarbon complexes and clusters).
R. W. Jotham, in *Mellor's Comprehensive Treatise on Inorganic and Theoretical Chemistry*, Supplement, Vol. V, *Boron*, Part BI, *Boron-Hydrogen Compounds*, Longmans, London (1981), p. 450 (Carboranes).
M. C. Favas, D. L. Kepert, *Prog. Inorg. Chem.*, *28* (1981) 309 (Stereochemistry of 9-, 10- and 12-coordination).
V. V. Korshak, S. A. Pavlova, P. N. Gribkova, T. N. Balykova, *Acta Polym.*, *32* (1981) 61 (Degradation of carborane-polyamides).
V. S. Mastryukov, O. V. Dorofeeva, L. V. Vitkov, *Russ. Chem. Rev. (Engl. Transl.)*, *49* (1980) 2377 (Carbaborane structural data).
B. M. Mikhailov, *Pure Appl. Chem.*, *52* (1980) 691 (Boron-cage compounds).
E. V. Leonova, *Russ. Chem. Rev. (Engl. Transl.)*, *49* (1980) 147 (Cobalt carbaborane complexes).
V. N. Kalinin, *Russ. Chem. Rev. (Engl. Transl.)*, *49* (1980) 1084 (Boron-substituted carbaboranes (12)).
R. N. Grimes, *Coord. Chem. Rev.*, *28* (1979) 47 (Sandwich complexes of cyclic, planar and pyramidal boron-containing ligands).
R. N. Grimes, *Acc. Chem. Res.*, *11* (1978) 420 (Metalloborane cages).
R. N. Grimes, *Rev. Si, Ge, Sn, Pb Compnds.*, (1977) 223 (Carboranes containing group-IV elements).
R. N. Grimes, *Organomet. React. Synth.*, *6* (1977) 63 (Reactions of metallocarboranes).

A. Shaver, *J. Organomet. Chem. Libr.*, *3* (1977) 157 (Metal complexes of polyprazolylborates).
K.P. Callahan, M.F. Hawthorne, *Adv. Organomet. Chem.*, *14* (1976) 145 (Metallocarboranes).
R.E. Williams, *Adv. Inorg. Chem. Radiochem.*, *18* (1976) 67 (Bonding in carboranes).
K. Wade, *Adv. Inorg. Chem. Radiochem.*, *18* (1976) 1 (Bonding in carboranes).
V.I. Stanko, V.A. Brattsev, S.P. Knyazev, *Russ. Chem. Rev., (Engl. Transl.)*, *45* (1976) 643 (Structure and reactivity of *closo*- and *nido*-carboranes).
R. Snaith, K. Wade, *MTP Int. Rev. Sci., Inorg. Chem., Ser. Two*, *1* (1975) 95 (Carboranes and metallocarboranes).
M.F. Hawthorne, *J. Organomet. Chem.*, *100* (1975) 97 (Metallocarboranes).
K. Wade, *Chem. Ber.*, *11* (1975) 177 (Carborane and other cluster compounds: a structural analogy).
T.E. Paxton, K.P. Callahan, E.L. Hoel, M.F. Hawthorne, in *Organotransition-Metal Chemistry*, Y. Ishii, M. Tsutsui, eds., Plenum Press, New York (1975), p. 1 (Metallocarboranes).
K.P. Callahan, W.J. Evans, M.F. Hawthorne, *Ann. N.Y. Acad. Sci.*, *239* (1974) 88 (Carboranes).
K.P. Callahan, M.F. Hawthorne, *Pure Appl. Chem.*, *39* (1974) 475 (Metallocarboranes and metalloboranes).
J. Plesek, *Pure Appl. Chem.*, *39* (1974) 431 (Intermediate *nido*-carboranes).
R.N. Grimes, *Pure Appl. Chem.*, *39* (1974) 455 (Metalloboron-cage compounds derived from small carboranes).
L.I. Zakharkin, V.N. Kalinin, *Russ. Chem. Rev. (Engl. Transl.)*, *43* (1974) 551 (Metallocarboranes).
R.N. Grimes, *Ann. N.Y. Acad. Sci.*, *239* (1974) 180 (Small metalloboron-cage compounds as analogues of metal clusters; unifying concepts of bonding).
N.N. Greenwood, I.M. Ward, *Chem. Soc. Rev.*, *3* (1974) 231 (Metalloboranes and metal-boron bonding).
G.B. Dunks, M.F. Hawthorne, *Acc. Chem. Res.*, *6* (1973) 124 (Non-icosahedral carboranes).
R. Snaith, K. Wade, *MTP Int. Rev. Sci., Inorg. Chem., Ser. One*, *1* (1972) 139 (Carboranes and metallocarboranes).
L.J. Todd, *Adv. Organomet. Chem.*, *8* (1970) 87 (Transition-metal carborane complexes).
H.A. Schroeder, *Inorg. Macromol. Rev.*, *1* (1970) 45 (Carborane polymers).
V.I. Bregadze, O.Yu. Okhlobystin, *Organomet. Chem. Rev., A*, *4* (1969) 345 (Organoelement derivatives of carboranes).
R. Köster, M.A. Grassberger, *Angew. Chem., Int. Ed. Engl.*, *6* (1967) 218 (Carboranes).
T.P. Onak, *Adv. Organomet. Chem.*, *3* (1965) 2631 (Carboranes).
I. Haiduc, *Stud. Cercet. Chim.*, *13* (1964) 783 (In Romanian) (Carboranes).

c) Aluminum

N.N. Korneev, *Khimiya i Tekhnologiya Alyuminiiorganicheskikh Soedinenii*, Khimiya, Moscow (1979) (Chemistry and technology of organoaluminum compounds).
J. Boor, *Ziegler-Natta Catalysts and Polymerizations*, Academic Press, New York (1978).
T. Mole, E.A. Jeffrey, *Organoaluminum Compounds*, Elsevier, Amsterdam (1972).
Gmelin Handbuch der Anorganischen Chemie. System Nummer 35: *Aluminum*, Teil A, Abt. 1, Kapitel: *Metallorganische Verbindungen*. Verlag Chemie, Berlin (1934/35), p. 426.

A. Sporzynski, K.B. Starowieyski, *J. Organomet. Chem. Libr.*, *9* (1980) 19 (Organoaluminum complexes of metal carbonyls).

H. Sinn, W. Kaminsky, *Adv. Organomet. Chem.*, *18* (1980) 99 (Ziegler-Natta catalysis).
G. Zweifel, in *Comprehensive Organic Chemistry*, D. Barton, W.D. Ollis, eds., Vol. 3, D.N. Jones, ed., Pergamon Press, Oxford (1979), p. 1013 (General review).
V.P. Mardykin, P.N. Gaponik, A.F. Popov, *Russ. Chem. Rev. (Engl. Transl.)*, *48* (1979) 487 (Organoaluminum complexes of ethers).
G. Zweifel, in *Aspects of Mechanisms and Organometallic Chemistry*, J.H. Brewster, ed., Plenum Press, New York (1978), p. 229 (Use of organoaluminum compounds in organic synthesis).
H. Yamamoto, H. Nozaki, *Angew. Chem., Int. Ed. Engl.*, *17* (1978) 169 (Reactions with organoaluminums).
J.P. Oliver, *Adv. Organomet. Chem.*, *15* (1977) 235 (Structures of organoaluminums).
J.P. Oliver, *Adv. Organomet. Chem.*, *16* (1977) 111 (Exchange reactions of organoaluminiums).
J.J. Eisch, *Adv. Organomet. Chem.*, *16* (1977) 67 (Rearrangements of unsaturated organoboron and organoaluminum compounds).
J.J. Ligi, D.B. Malpass, in *Encyclopedia of Chemical Processes*, J.J. McKetta, W.A. Cunningham, eds., Dekker, New York (1977), Vol. 3, p. 1 (Aluminum alkyls).
C.F. Lane, in *Synthetic Reagents*, J.S. Pitzey, ed., Vol. 3, Ellis Harwood, Ltd., London (1977), p. 1 (Diborane and its Lewis adducts as synthetic reagents in organic chemistry).
K.L. Henold, J.P. Oliver, *Org. React.*, *5* (1975) 387 (Unsaturated organoaluminum compounds).
E. Winterfeldt, *Synthesis*, (1975) 617 (Diisobutylaluminum hydride and triisobutylaluminum as reducing agents in organic synthesis).
G. Courtois, L. Miginiac, *J. Organomet. Chem.*, *69* (1974) 1 (Allyl derivatives of aluminum).
S. Pasynkiewicz, *Pure Appl. Chem.*, *30* (1972) 509 (Reactions of organoaluminums with electron donors).
J.W. Akitt, *Annu. Rev. NMR Spectrosc.*, *5A* (1972) 465 (NMR of Al and Ga compounds).
T. Mole, *Organomet. React.*, *1* (1970) 1 (Organoalane redistributions).
H. Schmidbaur, *Adv. Organomet. Chem.*, *9* (1970) 259 (Isoelectronic organo-P, Si and Al compounds).
R. Köster, P. Binger, *Adv. Inorg. Chem. Radiochem.*, *7* (1965) 263 (Organoaluminum compounds).
H. Jenkner, *Chem.-Ztg.*, *86* (1962) 527, 563 (Alkylations with organoaluminum compounds).

d) Gallium

I.A. Sheka, I.S.Chares, T.T. Mityueva, *The Chemistry of Gallium*, Elsevier, Amsterdam (1966).

e) Thallium

A.G. Lee, *The Chemistry of Thallium*, Elsevier, Amsterdam (1971).

M. Veith, *Nachr. Chem. Tech. Lab.*, *30* (1982) 940 (Organothallium(I) compounds).
A.G. Lee, *Organomet. React.*, *5* (1975) 1 (Reactions of organothallium compounds).
B. Walther, H. Albert, *Z. Chem.*, *15* (1975) 293 (Chemistry of thallium-organic compounds).

A. Banerji, R. Das, *J. Sci. Ind. Res.*, *33* (1974) 510 (Thallium in organic synthesis).
A. McKillop, E.C. Taylor, *Chem. Br.*, *9* (1973) 4 (Organothallium compounds in organic synthesis).
A. McKillop, E.C. Taylor, *Adv. Organomet. Chem.*, *11* (1973) 3 (Organothallium chemistry).
E.C. Taylor, A. McKillop, *Acc. Chem. Res.*, *3* (1970) 338 (Thallium in organic synthesis).
H. Kurosawa, R. Okawara, *Organomet. Chem. Rev.*, *A*, *6* (1970) 65 (Organothallium chemistry).
A.G. Lee, *Quart. Rev.*, *24* (1970) 310 (Organothallium chemistry).

Chapter 8

a) Group-IVA: Si, Ge, Sn, Pb

Gmelin Handbuch der Anorganischen Chemie. Perfluorohalogenoorgano Compounds of Main-Group Elements. Part 4. *Compounds with Elements of Main-Groups I to IV* (New Supplement Series,Vol. 25) (1975) Perfluoroorgano derivatives of Si, Ge, Sn, Pb). Vol. Bl, *Compounds with Hydrogen*, Springer Verlag, Berlin (1982).
B.J. Aylett, *Organometallic Compounds*, 4th ed., Vol. 1, *The Main-Group Elements*, Part 2, Chapman and Hall, London (1979) (Groups IV and V).
A.G. MacDiarmid, ed., *Organometallic Compounds of the Group-IV Elements*, M. Dekker, Inc., New York, Vol. 1 (Parts 1 and 2) (1968); Vol. 2 (Parts 1 and 2) (1972).
K. Licht, P. Reich, *Literature Data for IR, Raman and NMR Spectroscopy of Si, Ge, Sn and Pb-Organic Compounds*, VEB Deutscher Verlag der Wissenschaften, Berlin, DDR (1971).
I. Haiduc, *The Chemistry of Inorganic Ring Systems*, Part I, Chapter 4, p. 365, Wiley, New York (1970) (Organoderivatives of inorganic cyclic systems of Si, Ge, Sn, Pb).
E. Ya. Lukevits, M.G. Voronkov, *Organic Insertion Reactions of Group-IV Elements*, Consultants Bureau, New York (1966) (Engl. transl., *Gidrosililirovanie, Gidrogermilirovanie, i Gidrostannilirovanie*, Riga, 1964) (In Russian).

D.A. Armitage, in *Comprehensive Heterocyclic Chemistry*, A.R. Katritzky, C.W. Rees, eds., Vol. 1, O. Meth-Cohn, ed., Pergamon Press, Oxford (1984) p. 573 (Group-IV heterocycles).
L. Fábry, *Rev. Si, Ge, Sn, Pb Compnds.*, *7* (1983) (Photochemistry of unsaturated group-IV derivatives).
P.G. Harrison, *Coord. Chem. Rev.*, *49* (1983) 193 (Group-IV review).
Yu. A. Alexandrov, N.B. Yablokova, *Rev. Si, Ge, Sn, Pb Compnds.*, *6* (1982) 1 (Organo-group IV peroxide reactions with electron donors).
A. Shanzer, E. Schwartz, J. Libman, *Rev. Si, Ge, Sn, Pb Compnds.*, *6* (1982) 149 (Si and Sn templates in organic synthesis).
E. Lukevics, A.E. Skorova, O.A. Pudova, *Sulfur Rep.*, *2* (1982) 177 (Group-IV thiophenes).
A.S. Gordetsov, V.P. Kozyukov, I.A. Vostokov, S.V. Sheludyakova, Yu.L. Dergunov, V.F. Mironov, *Russ. Chem. Rev. (Engl. Transl.)*, *51* (1982) 848 (Group-IV carbodiimides and cyanamides).
W. Neumann, *Nachr. Chem. Tech. Lab.*, *30* (1982) 190, 192 (Silylenes, germylenes and stannylenes).

J. Dubac, P. Mazerolles, *J. Organomet. Chem. Libr., 12* (1981) 149 (Group-IV metallocycloalkanes).

A.G. Davies, *J. Organomet. Chem. Libr., 12* (1981) 181 (Group-IV radicals).

V.P. Feshin, L.S. Romanenko, M.G. Voronkov, *Russ. Chem. Rev. (Engl. Transl.), 50* (1981) 248 (The α-effect in organo-group-IV derivatives).

V.F. Mironov, T.K. Gar, N.S. Fedotov, G.E. Evert, *Russ. Chem. Rev. (Engl. Transl.), 50* (1981) 262 (Adamantane structures of Si, Ge and Sn).

R.Kh. Freidlina, A.B. Terent'ev, *Adv. Free Radical Chem., 6* (1980) (R_3Si and R_3Ge radicals).

M. Ishikawa, M. Kumada, *Rev. Si, Ge, Sn, Pb Compnds., 4* (1979) 7 (Photolysis of catenated group-IV derivatives).

R.C. Poller, in *Comprehensive Organic Chemistry*, D. Barton, W.D. Ollis, eds., Vol. 3, D.N. Jones, ed., Pergamon Press, Oxford (1979), p. 1111 (Group-IV derivatives).

J.A. Mangravite, *J. Organomet. Chem. Libr., 7* (1979) 45 (Group-IV allyls).

T.-L. Ho, *Synthesis*, (1979) 1 (Organic reductions by low-valent group-IV species).

N.V. Fomina, N.I. Sheverdina, K.A. Kocheshkov, *Russ. Chem. Rev. (Engl. Transl.), 47* (1978) 238 (Radiation effects in group-IV chemistry).

H.E. Henderson, J.E. Drake, *Rev. Si, Ge, Sn, Pb Compnds., 3* (1978) 145 (Synthesis and reactivity of organometallic compounds containing group IV-group V bonds).

A. Bonny, *Coord. Chem. Rev., 25* (1978) 229 (Group IV-Fe triad carbonyls).

R. West, *Adv. Organomet. Chem., 16* (1977) 1 (1,2-Anionic rearrangements of organosilanes and germanes).

R.N. Grimes, *Rev. Si, Ge Compnds., 2* (1977) 223 (Carborane derivatives of Si, Ge, Sn, Pb).

Yu.A. Aleksandrov, B.I. Tarunin, *Russ. Chem. Rev. (Engl. Transl.), 46* (1977) 905 (Oxidation of organometallic compounds of the silicon subgroup with ozone).

R. Livingstone, in *Rodd's Chemistry of Carbon Compounds*, 2nd ed., S. Coffey, M.F. Ansell, eds., Elsevier, Amsterdam, Vol. 4 (1977), p. 347 (Six-membered group-IV heterocycles).

O.N. Nefedov, S.P. Kolesnikov, I.A. Ioffe, *J. Organomet. Chem. Libr., 5* (1977) 181 (Group-IV carbene analogues).

M. Devaud, *Rev. Si, Ge, Sn, Pb Compnds., 2* (1976) 87 (Organo-group IV electrochemistry).

M.F. Lappert, P.W. Lednor, *Adv. Organomet. Chem., 14* (1976) 345 (Free radicals of group-IV elements).

R.C. Mehrotra, V.D. Gupta, G. Srivastava, *Rev. Si, Ge, Sn, Pb Compnds., 1* (1975) 299 (Alkoxides and alkylalkoxides of Si, Ge, Sn).

B. Majee, *Rev. Si, Ge, Sn, Pb Compnds., 2* (1975) 6 (Properties of organo-Si, Ge, Sn, Pb derivatives interpreted by the del-re method).

R.B. Larrabee, *J. Organomet. Chem., 74* (1974) 313 (Fluxional main group IV-organometallic compounds: implications for orbital-symmetry rules).

E.W. Abel, *Ann. N.Y. Acad. Sci., 29* (1974) 306 (Applicatins of organo-Si and -Sn compounds as synthetic intermediates in organometallic chemistry).

D. Brandes, *J. Organomet. Chem., 78* (1974) 1 (Organic peroxides of group-IV elements).

V.Yu. Orlov, *Russ. Chem. Rev. (Engl. Transl.), 42* (1973) 529 (Mass-spectra of organo-group IV compounds).

C. Glidewell, *Inorg. Chim. Acta Rev., 7* (1973) 69 (Unusual stereochemistry of Si and Ge-organic compounds).

E.W. Abel, M.O. Dunster, A. Waters, *J. Organomet. Chem., 49* (1973) 287 (Cyclopentadienyl compounds of Si, Ge, Sn, Pb).

H. Sakurai, in *Free Radicals*, 2 Vols., J.K. Kochi, ed., Wiley, New York (1973), p. 591 (Organo-group-IV radicals).

I. Omae, *Rev. Si, Ge, Sn, Pb Compnds., 1* (1972) 59 (Organometallic intramolecular-coordination compounds, containing group-IV elements).

M. Gielen, C. Dehouck, H. Mokhtar-Jamai, J. Topart, *Rev. Si, Ge, Sn, Pb Compnds., 1* (1972) 9 (Stereochemistry of 4-, 5- and 6-coordinated group-IV complexes).

A. N. Egorochkin, N. S. Vyazankin, S. Ya. Rhorshev, *Russ. Chem. Rev. (Engl. Transl.), 41* (1972) 425 ((p-d)-π Bonding in group IV).

R. Damreau, *Organomet. Chem. Rev., 8* (1972) 67 (Cyclobutanes with Si and Ge heteroatoms).

C. H. Yoder, J. J. Zuckerman, *Prep. Inorg. React., 6* (1971) 81 (Heterocyclic compounds of the group-IV elements).

K. Moedritzer, *Organomet. React., 2* (1971) 1 (Redistribution reactions).

H. Schumann, H. Benda, *Chem. Ber., 104* (1971) 333 (Cyclic derivatives with Si-P, Ge-P and Sn-P bonds).

E. W. Abel, S. M. Illingworth, *Organomet. Chem. Rev., A, 5* (1970) 143 (Phosphines, arsines, stibines and bismuthines containing Si, Ge, Sn, Pb).

S. C. Cohen, A. G. Massey, *Adv. Fluorine Chem., 6* (1970) 18 (Polyfluoroaromatic derivatives of Si, Ge, Sn, Pb).

Yu. I. Baukov, I. F. Lutsenko, *Organomet. Chem. Rev., A, 6* (1970) 355 (Organo-Si, Ge, Sn, Pb derivatives of keto-enols).

C. J. Attridge, *Organomet. Chem. Rev., A, 5* (1970) 323 (π-Bonding in group IV).

U. Belluco, G. Deganello, R. Pietropaulo, P. Uguagliati, *Inorg. Chim. Acta Rev., 4* (1970) 7 (Group-IV donor complexes with Pt(II)).

C. F. Shaw, A. L. Allred, *Organomet. Chem. Rev., A, 5* (1970) 95 (Nonbonded interactions in organometallic compounds of group IV).

R. A. Jackson, in *Essays on Free.Radical Chemistry*, Chemical Society Special Publication No. 24 (1970), Chemical Society, London, p. 295 (Organo-Si, -Ge, -Sn, -Pb radicals).

E. H. Brooks, R. J. Cross, *Organomet. Chem. Revs., A, 6* (1970) 227 (Group IV-metal derivatives of transition elements).

D. D. Davis, C. E. Gray, *Organomet. Chem. Revs., A, 6* (1970) 283 (Alkali metal and Mg derivatives of organo-Si, -Ge, -Sn, -Pb compounds).

Yu. A. Aleksandrov, *Organomet. Chem. Rev., A, 6* (1970) 209 (Oxidation of group-IV non-transition elements by ozone).

W. P. Neumann, *Pure Appl. Chem., 23* (1970) 433 (Organoderivatives of the group-IV Elements).

H. Schmidbaur, *Adv. Organomet. Chem., 9* (1970) 259 (Isoelectronic organo-P, Si and Al series).

R. A. Jackson, *Adv. Free-Radical Chem., 3* (1969) 231 (Group-IV radical reactions).

N. E. Kolobova, A. B. Antonova, K. N. Anismov, *Russ. Chem. Rev. (Engl. Transl.), 38* (1969) 822 (Derivatives of metal carbonyls containing a bond between transition metal atoms and group-IV elements).

W. P. Neumann, *Ann. N.Y. Acad. Sci., 159* (1969) 56 (Substituent exchange equilibria on Ge, Sn, Pb).

B. J. Aylett, *Prog. Stereochem., 4* (1969) 213 (Stereochemistry of Si and Ge-organic compounds).

K. M. McKay, R. Watt, *Organomet. Chem. Rev., A, 6* (1969) 137 (Catenated organoderivatives of Si, Ge, Sn, Pb).

A. G. Brook, *Adv. Organomet. Chem., 7* (1968) 96 (Keto derivatives of group-IV organometalloids).

K. Moedritzer, *Organomet. Chem. Rev., 1* (1966) 179; *A, 3* (1968) 135 (Redistribution reactions of organometallic compounds of Si, Ge, Sn, Pb).

N. G. Bokii, Yu. T. Struchkov, *J. Struct. Chem. (Engl. Transl.), 9* (1968) 633 (Structures of group-IV organometallics).

O. J. Scherer, *Organomet. Chem. Rev., 3* (1968) 281 (Cleavage of Si-N, Ge-N, Sn-N bonds in organometallic derivatives by halides of main-group elements).

W.E. Davidsohn, M.C. Henry, *Chem. Rev.*, *67* (1967) 73 (Acetylenic derivatives of Si, Ge, Sn, Pb).

E.W. Abel, D.A. Armitage, *Adv. Organomet. Chem.*, *5* (1967) 2 (Organometallic compounds with Si-S, Ge-S, Sn-S and Pb-S bonds).

M.F. Lappert, H. Pyszora, *Adv. Inorg. Chem. Radiochem.*, *9* (1966) 133 (Organosilicon, -germanium, -tin and -lead pseudohalides).

H. Gilman, W.H. Atwell, F.K. Cartledge, *Adv. Organomet. Chem.*, *4* (1966) 1 (Catenated organic compounds of Si, Ge, Sn, Pb).

K. Moedritzer, *Organomet. Chem. Rev.*, *1* (1966) 179 (Redistribution reactions of organometallic compounds of Si, Ge, Sn, Pb).

M. Gielen, N. Sprecher, *Organomet. Chem. Rev.*, *1* (1966) 455 (Influence of *d*-orbitals on coordination at group-IV metals).

M. Schmidt, *Pure Appl. Chem.*, *13* (1966) 15 (Analogies and differences among organometallic compounds of Si, Ge and Sn).

R.S. Drago, *Rec. Chem. Prog.*, *26* (1965) 157 (Understanding group-IV reactivity trends).

J.C. Lockhart, *Chem. Rev.*, *65* (1965) 131 (Redistribution reactions of organometallic compounds of group IV).

K.A. Andrianov, I. Haiduc, L.M. Khananashvili, *Russ. Chem. Rev. (Engl. Transl.)*, *34* (1965) 13 (Ability of elements to form polymers with inorganic chains and rings; dealing mainly with group-IV elements).

J.J. Zuckerman, *Adv. Inorg. Chem. Radiochem.*, *5* (1964) 383 (Direct synthesis of organometallic compounds).

N.A. Chumaevskii, *Russ. Chem. Rev. (Engl. Transl.)*, *32* (1963) 509 (Vibrationa spectra of group-IV organometallic compounds).

b) Germanium, Tin, Lead

M. Gielen, P.G. Harrison, eds., *Organometallic and Coordination Chemistry of Ge, Sn and Pb*, Georgi, Zürich (1979) (Plenary lectures from a conference of the same name held in 1977).

E. Müller, O. Bayer, H. Meerwein, K. Ziegler, eds., *Methoden der Organischen Chemie (Houben-Weyl)*, 4 Aufl., Band XIII/6: *Germanium- und Zinnorganische Verbindungen, Ge, Sn*, by G. Bähr, H.-O. Kalinowsky, S. Pawlenko, G. Thieme Verlag, Stuttgart (1978).

M. Dub, R. Weiss, *Organometallic Compounds.*, Vol. 2. *Compounds of Ge, Sn, Pb*, Springer Verlag, Berlin (1967). First Supplement, 1973.

D.A. Kochkin, I.N. Azerbaev, *Olovo-i Svinets-Organicheskie Monomery i Polymery*, Nauka, Alma-Ata (1968) (Tin- and lead-organic monomers and polymers).

W.P. Neumann, *Naturwissenschaften, 68* (1981) 354 (Organo-Ge, Sn and Pb).

M. Gielen, *Top. Stereochem.*, *12* (1981) 218 (Stereochemistry of Ge and Sn compounds).

G. Mengoli, *Rev. Si, Ge, Sn, Pb Compnds.*, *4* (1979) 59 (Electrochemical synthesis of alkyl-Sn and Pb derivatives).

W.N. Aldridge, *Rev. Si, Ge, Sn, Pb Compnds.*, *4* (1978) 9 (Biological properties of organo-Ge, Sn and Pb compounds).

P.G. Harrison, *Coord. Chem. Rev.*, *20* (1976) 1 (Structural chemistry of bivalent Ge, Sn, Pb-organic compounds).

P.G. Harrison, *MTP Int. Rev. Sci., Inorg. Chem., Series Two*, *4* (1975) 81 (Organic derivatives of Sn and Pb).

B.C. Pant, *J. Organomet. Chem.*, *66* (1974) 321 (Cycloalkanes containing Ge, Sn and Pb).

T. Chivers, B.J. Wakefield, in *Polychloroaromatic Compounds*, H. Suschitzky, ed., Plenum Press, New York (1974), p. 365 (Polychloroaryl derivatives of metals and metalloids, including group-IV elements).

I.P. Beletskaya, K.P. Butin, A.N. Ryabtsev, O.A. Reutov, *J. Organomet. Chem.*, *59* (1973) 1 (Stability of organotin and lead complexes).

A.J. Bloodworth, *MTP Int. Rev. Sci., Inorg. Chem., Ser. One*, *4* (1972) 275 (Organo-Sn and -Pb compounds).

T. Tanaka, *Organomet. Chem. Rev.*, *A*, *5* (1970) 1 (Vibrational spectra of organo-Sn and Pb-compounds).

I. Ruidisch, H. Schmidbaur, H. Schumann, in *Halogen Chemistry*, V. Gutmann, ed., Vol. 2, Academic Press, New York (1967), p. 233 (Organohalides of Ge, Sn, Pb).

c) Silicon

i) Organosilanes: General

V. Chvalovský, J.M. Bellama, *Carbon-Functional Organosilicon Compounds*, Plenum Press, New York (1984).

E. Müller, O. Bayer, H. Meerwein, K. Ziegler, eds., *Methoden der Organischen Chemie (Houben Weyl)*, 4th ed., Band XII/5, *Silicium-Organische Verbindungen, Si*, by S. Pawlenko, G. Thieme Verlag, Stuttgart (1980).

V. Bazant, J. Hetflejs, V. Chvalovsky, J. Jolik, O. Kruchna, J. Ratousky, J. Schraml, *Handbook of Organosilicon Compounds, Advances Since 1961*, Vols. 5–6, M. Dekker, New York (1980).

D.N. Jones, ed., *Sulphur, Selenium, Silicon, Boron Organometallic Compounds*, Vol. 3, *Comprehensive Organic Chemistry*, D. Barton, D.W. Ollis, eds., Pergamon Press, Oxford (1979).

V. Bazant, J. Hetflejs, V. Chvalovsky, J. Joklik, O. Kruchna, J. Ratousky, J. Schraml, *Handbook of Organosilicon Compounds, Advances Since 1961*, Vols. 1–4, M. Dekker, New York (1975).

F. Boschke, ed., *Silicon Chemistry*, 2 Vols. (*Top. Curr. Chem.* Nos. 51, 52), Springer Verlag, Berlin (1974).

S.N. Borisov, M.G. Voronkov, E.Ya. Lukevits, *Organosilicon Heteropolymers and Heterocompounds*, Plenum Press, New York (1970).

R.J.H. Voorhoeve, *Organohalosilanes. Precursors to Silicones*, Elsevier, Amsterdam (1967).

S.N. Borisov, M.G. Voronkov, E.Ya. Lukevits, *Kremne-elementoorganicheskie Soedineniya. Proizvodnye Neorganogenov*, Khimiya, Moscow (1966) (Silico-elemento-organic compounds).

K.A. Andrianov, *Metalorganic Polymers*, Interscience, New York (1965) (Almost exclusively organosilicon polymers).

V. Bazant, V. Chvalovsky, J. Rathousky, *Organosilicon Compounds*, Vols. 1, 2/1, 2/2. Czech. Acad. Sci. Publ. House, Prague, and Academic Press, New York (1965).

L.H. Sommer, *Stereochemistry, Mechanism and Silicon*, McGraw-Hill, New York (1965).

A.D. Petrov, V.F. Mironov, V.A. Ponomarenko, E.A. Chernyshev, *Synthesis of Organosilicon Monomers*, Heywood, London (1964).

C. Eaborn, *Organosilicon Compounds*, Butterworths, London (1960).

Gmelin Handbuch der Anorganischen Chemie, System No. 15, *Silicium*, Teil C. *Organische Silicium-Verbindungen*, Springer Verlag, Berlin (1958).

K.A. Andrianov, *Kremniiorganicheskie Soedineniya*, Goskhimizdat, Moscow (1955) (Organosilicon compounds).

R. J. P. Corriu, C. Guerin, J. J. E. Moreau, *Top. Stereochem., 15* (1984) 43 (Stereochemistry at silicon).

J. Pola, in *Carbon-Functional Organosilicon Compounds*, V. Chvalovsky, J. M. Bellama, eds., Plenum Press, New York (1984), p. 35 (Intramolecular interactions).

R. Ponec, in *Carbon-Functional Organosilicon Compounds*, V. Chvalovsky, J. M. Bellama, eds., Plenum Press, New York (1984), p. 233 (Organosilicon bonding theory).

C. Biran, *Ann. Chim. (Paris), 8* (1983) 151 (Organosilicon chemistry, new trends).

M. Weidenbruch, A. Schäfer, *Rev. Si, Ge, Sn, Pb Compnds., 7* (1983) (Steric effects in organosilanes).

I. S. Akhrem, N. M. Chistovalova, M. E. Volpin, *Russ. Chem. Rev. (Engl. Transl.), 52* (1983) 953 (Si-C bond breaking by Pd and Pt compounds).

C. Eaborn, *J. Organomet. Chem., 239* (1982) 93 (Hindered organosilanes).

H. F. Schaefer, III, *Acc. Chem. Res., 15* (1982) 283 (The Si-C double bond).

H. Sakurai, *Pure Appl. Chem., 54* (1982) 1 (Allylsilanes in organic synthesis).

R. J. P. Corriu, C. Guerin, *Adv. Organomet. Chem., 20* (1982) 265 (Nucleophilic displacement at Si).

M. G. Voronkov, V. B. Pukhnarevich, *Izv. Akad. Nauk SSSR, Ser. Khim.*, (1982) 1056 (Si-H Bonds in organosilanes).

D. J. Ager, *Chem. Soc. Rev., 11* (1982) 493 (Si-containing carbonyl derivatives).

I. Ojima, T. Kogure, *Rev. Si, Ge, Sn, Pb Compnds., 5* (1981) 7 (Hydrosilation).

R. Walsh, *Acc. Chem. Res., 14* (1981) 246 (Silicon bond energies).

V. P. Mileshkevich, N. F. Novikova, *Russ. Chem. Rev. (Engl. Transl.), 50* (1981) 49 (Correlation of organosilane properties by the Taft relation).

R. Calas, *J. Organomet. Chem., 200* (1980) 11 (Thirty years in organosilicon chemistry).

S. Murai, N. Sonoda, *Angew. Chem., Int. Ed. Engl., 19* (1980) 837 (Catalytic reactions of hydrosilanes with carbon monoxide).

N. T. Anh, *Top. Curr. Chem., 88* (1980) (Regio- and stereoselectivities in nucleophilic reactions at silicon).

S. Blechert, *Nachr. Chem. Tech. Lab., 28* (1980) 801 (Vinyl and allylsilanes).

A. G. Brook, A. R. Bassindale, in *Rearrangements in Ground and Excited States*, P. de Mayo, ed., Vol. 2, Academic Press, New York (1980), p. 149 (Rearrangements of organosilicon compounds).

V. F. Mironov, V. D. Sheludyakov, V. P. Kozyukov, *Russ. Chem. Rev. (Engl. Transl.), 48* (1979) 473 (Phosgene in organosilane chemistry).

K. A. Andrianov, J. Soucek, L. M. Khananashvili, *Russ. Chem. Rev. (Engl. Transl.), 48* (1979) 657 (Hydride addition of organohydridosiloxanes to multiple C-C bond).

D. Häblich, F. Effenbacher, *Synthesis* (1979) 841 (Preparation of aryl- und heteroarylsilanes).

J. L. Speier, *Adv. Organomet. Chem., 17* (1979) 407 (Homogeneous catalysis of hydrosilation by transition-metal compounds).

I. Fleming, in *Comprehensive Organic Chemistry*, D. Barton, W. D. Ollis, eds., Vol. 3, D. N. Jones, ed., Pergamon Press, Oxford (1979), p. 541 (General review).

M. S. Wrighton, in *Inorganic and Organometallic Photochemistry*, Advances in Chemistry Series, No. 168, M. S. Wrighton, ed., American Chemical Society, Washington, DC (1978), p. 189 (Photocatalyzed reactions of alkenes with silanes using trinuclear metal-carbonyl cluster precursors).

S. Murai, N. Sonoda, *Angew. Chem., Int. Ed. Engl., 17* (1978) 169 (Catalytic reactions of hydrosilanes with CO).

E. Lukevics, Z. V. Belyakova, M. G. Pomerantseva, M. G. Voronkov, *J. Organomet. Chem. Libr., 5* (1977) 1 (Hydrosilylation).

J. F. Harrod, A. J. Chalk, in *Organic Syntheses via Metal Carbonyls*, I. Wender, P. Pino, ed.,

J. Wiley & Sons, New York (1977) Vol. 2, p. 673 (Hydrosilylation with group-VIII metal catalysts).
B. Bøe, *J. Organomet. Chem.*, *107* (1976) 139 (Solvolysis mechanisms of Si-O, Si-N and Si-C compounds).
V. Sheludyakov, V. P. Kozukov, V. F. Mironov, *Russ. Chem. Rev. (Engl. Transl.)*, *45* (1976) 227 (Organosilylcarbamates – silylurethanes).
V. Chvalovsky, *MTP Int. Rev. Sci. Inorg. Chem., Series Two*, *4* (1975) 141 (Organosilicon compounds).
H. Bürger, *MTP Int. Rev. Sci., Inorg. Chem., Series Two*, *4* (1975) 195 (Organosilicon compounds).
C. Eaborn, *J. Organomet. Chem.*, *100* (1975) 43 (Cleavage of arylsilicon and related bonds by electrophiles).
G. Fritz, *Chem.-Ztg.*, *97* (1973) 111 (Carbosilanes).
R. J. P. Corriu, M. Henner, *J. Organomet. Chem.*, *74* (1974) 1 (The siliconium ion problem).
A. G. Brook, *Acc. Chem. Res.*, *7* (1974) 77 (Molecular rearrangements of organosilicon compounds).
G. Fritz, *Top. Curr. Chem.*, *50* (1975) 43 (Organometallic synthesis of carbosilanes).
H. Bürger, R. Eujen, *Top. Curr. Chem.*, *50* (1974) (Low-valent Si).
E. Ya. Lukevits, *Russ. Chem. Rev. (Engl. Transl.)*, *42* (1973) 662 (Hydrosilylation).
L. V. Nozdrina, Ya. I. Mindlin, K. A. Andrianov, *Russ. Chem. Rev. (Engl. Transl.)*, *42* (1973) 509 (Organosilylepoxides).
P.Boudjouk, R. West, *Intra-Sci. Chem. Rept.*, *7* (1973) 65 (Organosilyl hydroxylamines).
V. P. Kozyukov, V. D. Sheludyakov, V. F. Mironov, *Russ. Chem. Rev. (Engl. Transl.)*, *42* (1973) 662 (Si isocyanates).
M. Weidenbruch, *Chem.-Ztg.*, *97* (1973) 116 (Polyfluoroaromatic organosilicon derivatives).
V. F. Mironov, *Khim. Teknol. Elementoorg. Soedin.*, (1972) 63 (Carbofunctional organosilicon compounds).
E. Lukevits, A. E. Pestunovich, *Russ. Chem. Rev. (Engl. Transl.)*, *41* (1972) 938 (Organosilicon derivatives of mononitrogen heterocycles).
H. Bürger, *MTP. Int. Rev. Sci., Inorg. Chem., Series One*, *4* (1972) 205 (Organosilicon compounds with Si-N, Si-P, Si-O, Si-S, Si-Se, Si-Te bonds).
V. Chvalovsky, *Organomet. React.*, *3* (1972) 191 (Cleavage reactions of the carbon-silicon bond).
G. V. Motsarev, K. A. Andrianov, V. I. Zetkin, *Russ. Chem. Rev. (Engl. Transl.)*, *40* (1971) 485 (Halogenation of organosilicon compounds).
R. A. Benkeser, *Acc. Chem. Res.*, *4* (1971) 94 (The chemistry of $HSiCl_3$ – tertiary amine combinations).
D. H. O'Brien, T. J. Hairston, *Organomet. Chem. Rev., A*, *7* (1971) 95 (Reactions of organosilanes with Lewis acids).
W. R. Dunnavant, *Inorg. Macromol. Rev.*, *1* (1970) 33 (Silylarene polymers).
H. Schmidbaur, *Adv. Organomet. Chem.*, *9* (1970) 259 (Isoelectronic organo-P, Si and Al compounds).
J. F. Klebe, *Acc. Chem. Res.*, *3* (1970) 299 (Silyl-proton exchange).
N. V. Komarov, V. K. Roman, *Russ. Chem. Rev. (Engl. Transl.)*., *39* (1970) 578 (Organosilicon ketones).
D. F. Ballard, K. Shiina, T. Brennan, F. W. G. Fearon, I. Haiduc, H. Gilman, *Pure Appl. Chem.*, *19* (1969) 449 (Silylation of polyhaloaromatic compounds).
D. R. Weyenberg, L. G. Mahone, W. H. Atwell, *Ann. N. Y. Acad. Sci.*, *159* (1969) 38 (Redistribution reactions in organosilicon chemistry).
V. Bazant, J. Joklik, J. Ratousky, *Angew. Chem., Int. Ed. Engl.*, *7* (1968) 112 (Direct synthesis of organohalosilanes).

B. J. Aylett, *Adv. Inorg. Chem. Radiochem.*, *11* (1968) 249 (Organosilicon hydrides).
R. N. Haszeldine, in *New Pathways in Inorganic Chemistry*, E. A. V. Ebsworth, A. G. Maddock, A. G. Sharpe, eds., Cambridge University Press, Cambridge (1968), p. 115 (Perfluoroalkylsilanes).
C. Eaborn, R. W. Bott, in *Organometallic Compounds of the Group-IV Elements*, A. G. MacDiarmid, ed., Vol. 1, Part 1, M. Dekker, New York (1968), p. 90 (The synthesis and reactions of the silicon-carbon bond).
G. Fritz, *Angew. Chem., Int. Ed. Enl.*, *6* (1967) 677 (Carbosilanes).
K. A. Andrianov, A. I. Petrashko, *Organomet. Chem. Rev.*, *2* (1967) 383 (Halogeno derivatives of alkyl, aryl-Halogenosilanes).
R. Müller, *Organomet. Chem. Rev.*, *1* (1966) 357 (Organofluorosilicates).
G. Fritz, J. Grobe, D. Kummer, *Adv. Inorg. Chem Radiochem.*, 7 (1965) 349 (Carbosilanes).
J. J. Zuckerman, *Adv. Inorg. Chem. Radiochem.*, *6* (1964) 383 (Direct synthesis of organosilicon compounds).
R. G. Neville, *J. Chem. Educ.*, *39* (1962) 276 (Functionally substituted aromatic silanes).
V. M. Vdovin, A. D. Petrov, *Russ. Chem. Rev. (Engl. Transl.)*, *31* (1962) 393 (Organosilicon compounds containign cyano groups).
A. G. MacDiarmid, *Adv. Inorg. Chem radiochem.*, *3* (1961) 207 (Silanes and their derivatives).
G. V. Odabashyan, V. A. Ponomarenko, A. d. Petrov, *Russ. Chem. Rev. (Engl. Transl.)*, *30* (1961) 407 (Fluoroorganosilicon compounds).

ii) Organosilicon-Oxygen Compounds

H. A. Liebhafsky, *Silicones under the Monogram: A Story of Industrial Research*, J. Wiley, New York (1978).
M. G. Voronkov, V. P. Mileshkevich, Yu. A. Yuzhelevskii, *The Siloxane Bond: Physical Properties and Chemical Transformations*, J. Livak, transl., Plenum Press, New York (1978).
A. L. Smith, ed., *Analysis of Silicones*, Wiley, New York (1974).
K. A. Andrianov, L. M. Khananashkvili, *Tekhnologiya Elementoorganicheskikh Monomerov i Polimerov*, Khimiya, Moscow (1973) (Technology of organometallic monomers and polymers – industrial manufacture of organosilicon compounds).
W. Noll, *Chemie und Technologie der Silicone*, Verlag Chemie, Weinheim (1968); translated as *Chemistry and Technology of Silicones*, Academic Press, New York (1968).
R. J. H. Voorhoeve, *Organohalosilanes: Precursor to Silicones*, Elsevier, Amsterdam (1967).
R. N. Meals, *Silicones*, Rheinhold, New York (1959).
N. N. Sokolov, *Metody Sinteza Polyorganosiloxanov*, Gosenergoizdat, Moscow (1959) (Methods of polyorganosiloxane synthesis).
R. R. McGregor, *Silicones and Their Uses*, McGraw-Hill, New York (1954).
E. G. Rochow, *An Introduction to the Chemistry of Silicones*, 2nd Ed., J. Wiley, New York (1951).

R. H. Cragg, R. D. Lane, *J. Organomet. Chem.*, *267* (1984) 1 (Organo-1,3,2-dioxasilaheterocycles).
A. M. Ton'shin, B. A. Kamaritskii, V. N. Spektor, *Russ. Chem. Rev. (Engl. Transl.)*, *52* (1983) 1365 (Silicon dielectrics: polyorganosilasesquioxanes).
Yu. A. Aleksandrov, *J. Organomet. Chem.*, *238* (1982) 1 (Organosilicon peroxides).
V. D. Sheludyakov, V. F. Mironov, *Russ. Chem. Rev. (Engl. Transl.)*, *46* (1977) 1167 (Organosilicon carbonates and chlorocarbonates).
G. Srivastava, R. C. Mehrotra, *Rev. Si, Ge, Sn, Pb Compnds.*, *2* (1977) 307 (Si β-diketonates).

M. G. Voronkov, V. P. Mileshkevich, Yu. A. Yuzhelerskii, *Russ. Chem. Rev. (Engl. Transl.)*, *45* (1976) 1167 (Organosiloxane complexes).
E. Ya. Lukevits, L. Liberts, M. G. Voronkov, *Russ. Chem. Rev. (Engl. Transl.)*, *39* (1970) 953 (Organosilicon derivatives of aminoalcohols).
K. A. Andrianov, *Inorg. Macromol. Rev.*, *1* (1970) 33 (Polyelementosiloxanes).
H. Schmidbaur, *Top. Curr. Chem.*, *13* (1969) 167 (Siloxanes and silazanes).
A. J. Barry, H. N. Beck, in *Inorganic Polymers*, F. G. A. Stone, W. A. G. Graham, eds., Academic Press, New York (1962) (Silicone polymers).
J. S. Hughes, in *Developments in Inorganic Polymer Chemistry*, M. F. Lappert, G. J. Leigh, eds., Elsevier, Amsterdam (1962), p. 138 (Silicone polymers).
Yu. K. Yuriev, S. V. Belyakova, *Russ. Chem. Rev. (Engl. Transl.)*, *29* (1960) 383 (Acyloxysilanes).
P. D. George, M. Prober, J. R. Elliot, *Chem. Rev.*, *56* (1956) 1065 (Carbon-functional silicones).

iii) Organosilicon Halides

R. J. H. Voorhoeve, *Organohalosilanes: Precursors to Silicones*, Elsevier, Amsterdam (1967).

M. Kumada, K. Tamao J. Yoshida, *J. Organomet. Chem.*, *239* (1982) 115 (Organopentafluorosilicates).
R. M. Pike, M. F. Mangano, *J. Organomet. Chem. Libr.*, *12* (1981) 53 (Organopseudohalosilanes).
A. H. Schmidt, *Chem.-Ztg.*, *104* (1980) 253 (Organosilyl bromides and iodides).
E. F. Hengge, *Rev. Inorg. Chem.*, *2* (1980) 139 (Chlorooligosilanes).
G. V. Motsarev, K. A. Andrianov, V. I. Zetkin, *Russ. Chem. Rev. (Engl. Transl.)*, *40* (1971) 485 (Halogenation of organosilicon compounds).
B. Bazant, J. Joklik, J. Ratousky, *Angew. Chem., Int. Ed. Engl.*, *7* (1968) 112 (Direct synthesis of organohalosilanes).
K. A. Andrianov, A. I. Petrashko, *Organomet. Chem. Rev.*, *2* (1967) 383 (Halogeno derivatives of alkyl, aryl-halogenosilanes).
R. Müller, *Organomet. Chem. Rev.*, *1* (1966) 357 (Organofluorosilicates).
J. J. Zuckerman, *Adv. Inorg. Chem. Radiochem.*, *6* (1964) 383 (Direct synthesis of organosilicon compounds).
G. V. Odabashyan, V. A. Ponomarenko, A. D. Petrov, *Russ. Chem. Rev. (Engl. Transl.)*, *30* (1961) 407 (Fluoroorganosilicon compounds).

iv) Organosilicon-Nitrogen Compounds

V. F. Mironov, *J. Organomet. Chem.*, (1984) (Organosilicon synthesis of isocyanates).
R. M. Pike, N. Sobinski, P. J. McManus, *J. Organomet. Chem.*, *253* (1983) 183 (Azidosilanes).
Yu. M. Varezhkin, D. Ya. Zhinkin, M. M. Morgunova, *Russ. Chem. Rev. (Engl. Transl.)*, *50* (1981) 2212 (Organocyclodisilazanes).
D. Ya. Zhinkin, Yu. M. Varezhkin, M. M. Morgunowa, *Russ. Chem. Rev. (Engl. Transl.)*, *49* (1980) 1149 (Cyclodisilazanes).
D. H. Harris, M. F. Lappert, *J. Organomet. Chem. Libr.*, *2* (1976) 13 (Metal and metalloid dialkylamides containing the $(Me_3Si)_2N$- or $Me_3C(Me_3Si)N$-ligand).
W. R. Peterson, *Rev. Si, Ge, Sn, Pb Compnds.*, *1* (1974) 193 (Chemistry of organoazidosilanes).

N. Wiberg, *Angew. Chem., Int. Ed. Engl., 10* (1971) 374 (Bis-trimethylsilyl-diimine).
U. Wannagat, *Top. Curr. Chem., 9* (1967) 102 (Silicon-nitrogen compounds).
B.J. Aylett, *Prep. Inorg. React., 2* (1965) 93 (Silicon-nitrogen polymers).
U. Wannagat, *Adv. Inorg. Chem. Radiochem. 6* (1964) 225 (Nitrogen-containing organosilicon compounds).
R.Fessenden, J.S. Fessenden, *Chem. Rev., 61* (1961) 36 (Organosilylamines, silazanes).

v) Organosilicon-Sulfur and Phosphorus Compounds

S.N. Borisov, M.G. Voronkov, E. Ya. Lukevits, *Organosilicon Derivatives of Phosphorus and Sulfur*, Plenum Press, New York (1971).

A. Haas, R. Hitze, in *Sulfur in Organic and Inorganic Chemistry*, A. Senning, ed., Vol. 4. M. Dekker, New York (1982), p. 1 (The Si-S bond).
E.W. Abel, S.A. Muckeljohn, *Phosphorus Sulfur*, 9 (1981) 235 (Phosphinimes).
D. Brandes, *J. Organomet. Chem. Libr., 7* (1979) 257 (Organosilicon dervatives of S, Se, Te).
B.M. Glawicevski, J. E. Drake, *Rev. Si, Ge, Sn, Pb Compnds., 3* (1978) 279 (Organosilicon compounds containing bonds to group-VI elements).
P.J. Moehs, G.E. Legrow, *Rev. Si, Ge, Sn, Pb Compnds., 1* (1974) 155 (Disilthianes).
E.A. Chernyshev, E.F. Bugerenko, *Organomet. Chem. Rev., A, 3* (1968) 469 (Organosilicon compounds containing phosphorus).
G. Fritz, *Angew. Chem., Int. Ed. Engl., 5* (1966) 53 (Organosilicon-phosphorus compounds).
A. Haas, *Angew. Chem., Int. Ed. Engl., 4* (1965) 1014 (Organo-silicon-sulfur compounds).

vi) Rings Containing Organosilicon Groups

I. Haiduc, *The Chemistry of Inorganic Ring Systems*, J. Wiley, New York (1970), Part I (Cyclosiloxanes, cyclosilazanes, cyclosilthianes and other inorganic rings).

R.H. Cragg, R.D. Lane, *J. Organomet. Chem., 267* (1984) 1 (Organo-1,3,2-dioxasilaheterocycles).
G. Märkl, H. Höllriegel, W. Schlosser, *J. Organomet. Chem., 260* (1984) 129 (Silacyclohexadienyl anions).
O.A. Dyachenko, L.O. Atovmyan, *J. Struct. Chem. (Engl. Transl.)* , *24* (1983) 144 (Heterocyclic organosilicon structures).
R.J. McMahon, *Coord. Chem. Rev., 47* (1982) 1 (π-Complexes of silacycles).
L.A. Paquette, *Isr. J. Chem., 21* (1981) 128 (Silacyclopropanes).
T.J. Barton, *Pure Appl. Chem., 52* (1980) 615 (Reactive intermediates from organosilacycles).
D.Ya. Zhinkin, Yu.M. Varezhkin, M.M. Morgunova, *Russ. Chem. Rev. (Engl. Transl.), 49* (1980) 1149 (Cyclodisilazanes).
K.A. Andrianov, V.N. Emelyanov, *Russ. Chem. Rev. (Engl. Transl.), 46* (1977) 1092 (Organocyclosilazoxanes).
U. Wannagat, M. Schlingmann, H. Autzen, *Z. Naturforsch., Teil B, 31* (1976) 621 (Inorganic five-membered rings containing silicon, nitrogen and a heteroelement).
M.V. George, R. Balasubramanian, *J. Organomet. Chem. Libr., 2* (1976) 103 (Organosilacyclenes).
D. Seyferth, *J. Organomet. Chem., 100* (1975) 237 (Silacyclopropanes).

E. Ya. Lukevits, A. E. Pestunovich, *Russ. Chem. Rev. (Engl. Transl.)*, *41* (1972) 929 (Organosilicon nitrogen-containing heterocycles).
N. S. Nametkin, I. Kh. Islamov, L. E. Guselnikov, V. M. Vdovin, *Russ. Chem. Rev. (Engl. Transl.)*, *41* (1972) 111 (Cyclocarbosilanes).
R. Damrauer, *Organomet. Chem. Rev., A*, *8* (1972) 67 (Si and Ge cyclobutanes).
O. K. Johannson, C. Lee, in *Cyclic Monomers*, K. C. Frisch, ed., Wiley, New York (1972) (Cyclic siloxanes and silazanes).
K. A. Andrianov, L. M. Khananashvili, *Organomet. Chem. Rev.*, *2* (1967) 141 (Cyclic organosilicon compounds).
W. Fink, *Angew. Chem., Int. Ed. Engl.*, *5* (1966) 53 (Cyclosilazanes).
H. Gilman, G. L. Schwebke, *Adv. Organomet. Chem.*, *1* (1964) 90 (Organocyclosilanes).
K. A. Andrianov, I. Haiduc, L. M. Khananashvili, *Russ. Chem. Rev. (Engl. Transl.)*, *32* (1963) 243 (Cyclic organosilicon compounds).
G. Fritz, *Top. Curr. Chem.*, *4* (1963) 459 (Pyrolysis of organosilanes: formation of cyclocarbosilanes).

vii) Polysilanes and Related Species

E. A. Chernyshev, N. G. Komalenkova, S. A. Bashkirova, *J. Organomet. Chem.*, (1984) (Gas-phase reactions of dichlorosilylene).
R. West, *Pure Appl. Chem.*, *56* (1984) 163 (The disilenes: Si=Si double bond).
R. West, *Pure Appl. Chem.*, *54* (1982) 1041 (Cyclic polysilanes).
R. Calas, J. Dunogues, G. Deleris, N. Duffaut, *J. Organomet. Chem.*, *225* (1982) 117 (The disilane residue from the direct synthesis of methylchlorosilanes).
P. P. Gaspar, in *Reactive Intermediates*, vol. 2, M. Jones, Jr., R. A. Moss, eds., Wiley, New York (1981), p. 335 (Silylenes).
G. Bertrand, G. Trinquier, P. Mazerolles, *J. Organomet. Chem. Libr.*, *12* (1981) (The C-Si double bond).
M. Ishikawa, M. Kumada, *Adv. Organomet. Chem.*, *19* (1981) 51 (Photochemistry of organopolysilanes).
B. Coleman, M. Jones, Jr., *Rev. Chem. Intermed.*, *4* (1980) 291 (Silenes).
H. Sakurai, *J. Organomet. Chem.*, *200* (1980) 261 (Organopolysilanes).
L. E. Gusel'nikov, N. S. Nametkin, *Chem. Rev.*, *79* (1979) 529 ((p-p)-π group-IV intermediates).
M. Ishikawa, *Pure Appl. Chem.*, *50* (1978) 11 (Photolysis of organopolysilanes; Si = C intermediates).
E. Hengge, in *Homoatomic Rings, Chains, Macromolecules of Main-Group Elements*, A. L. Rheingold, ed., Academic Press, New York (1977), p. 235 (Cyclopolysilanes und linear polysilanes).
O. M. Nefedov, S. P. Kolesnikov, A. I. Ioffe, *J. Organomet. Chem. Libr.*, *5* (1977) 181 (Group-IV carbene analogues).
R. E. Ballard, P. J. Wheatley, *J. Organomet. Chem. Libr.*, *2* (1976) 1 (Carbon-silicon double bonds).
E. A. Chernyshev, N. G. Komalenkova, S. A. Bashkirova, *Russ. Chem. Rev. (Engl. Transl.)*, *45* (1976) 913 (Silicon analogues of carbenes).
M. Kumada, *J. Organomet. Chem.*, *100* (1975) 127 (Skeletal transformations of organopolysilanes).
R. West, E. Carberry, *Science*, *189* (1975) 179 (Permethylpolysilanes, silicon analogues of hydrocarbons).

L. E. Gusel'nikov, N. S. Nametkin, V. M. Vdovin, *Acc. Chem. Res.*, *8* (1975) 18 (Unstable silicon analogues of unsaturated compounds).
L. E. Gusel'nikov, N. S. Nametkin, V. M. Vdovin, *Russ. Chem. Rev. (Engl. Transl.)*, *43* (1974) 620 (Unstable Si analogues of olefins and ketones).
R. West, *Ann. N.Y. Acad. Sci.*, *239* (1974) 1 (Aromatic properties of cyclopolysilanes).
E. Hengge, *Top. Curr. Chem.*, *51* (1974) 1 (Organosilicon compounds with Si-Si bonds).
M. Kumada, *Intra-Sci. Chem. Rept.*, *7* (1973) 121 (Skeletal transformations of organopolysilanes).
E. Hengge, *Angew. Chem., Int. Ed. Engl.*, *8* (1969) 901 (Properties of the Si-Si bond, cyclosilanes).
W. H. Atwell, D. R. Weyenberg, *Angew. Chem., Int. Ed. Engl.*, *8* (1969) 469 (Silylenes).
M. Kumada, K. Tamao, *Adv. Organomet. Chem.*, *6* (1967) 19 (Aliphatic organopolysilanes).
G. Schott, *Top. Curr. Chem.*, *9* (1967) 60 (Oligo-and polysilanes).
H. Gilman, G. L. Schwebke, *Adv. Organomet. Chem.*, *1* (1964) 90 (Cyclic and linear organopolysilanes).

viii) Metal-Containing Organosilanes

R. J. P. Corriu, E. Colmer, *Ann. Chim. (Paris)*, *8* (1983) 121 (Si-transition-metal bonds from optically active organosilanes).
R. J. McMahon, *Coord. Chem. Rev.*, *47* (1982) 1 (π-Complexes of silacycles).
B. J. Aylett, *Adv. Inorg. Chem. Radiochem.*, *25* (1982) 1 (Si-transition-metal chemistry).
E. Colomer, R. J. P. Corriu, *Top. Curr. Chem.*, *96* (1981) 79 (Si- and Ge-transition-metal compounds).
M. D. Curtis, P. S. Epstein, *Adv. Organomet. Chem.*, *19* (1981) 213 (Silicon redistributions catalyzed by transition metals).
M. G. Voronkov, N. F. Chernov, *Russ. Chem. Rev. (Engl. Transl.)*, *48* (1979) 964 (Mixed organosilyl mercurials).
I. Haiduc, V. Popa, *Adv. Organomet. Chem.*, *15* (1977) 113 (π-Complexes of transition metals with organosilicon ligands).
M. F. Lappert, in *Inorganic Compounds with Unusual Properties*, R. B. King, ed., Advances in Chemistry Series, No. 150, American Chemical Society, Washington, DC (1976), p. 256 (Bulky $(Me_3Si)_3CH$-stabilized transition-metal and Sn derivatives).
F. Höfler, *Top. Curr. Chem.*, *50* (1974) 129 (Si-Transition metal compounds).
C. S. Cundy, B. M. Kingston, M. F. Lappert, *Adv. Organomet. Chem.*, *11* (1973) 253 (Organometallic complexes with silicon-transition metal or Si-C-metal bonds).
H. G. Ang, P. T. Lau, *Organomet. Chem. Rev.*, *A*, *8* (1972) 235 (Compounds containing silicon-transition metal bonds).
N. S. Vyazankin, G. A. Razuvaev, O. A. Kruglaya, *Organomet. Chem. Synth.*, *1* (1971) 205 (Organosilyl-mercury compounds).
K. A. Andrianov, *Inorg. Macromol. Rev.*, *1* (1970) 33 (Polyelemento-siloxanes).
N. S. Vyzankin, G. A. Razuvaev, O. A. Kruglaya, *Organomet. Chem. Rev.*, *A*, *3* (1968) 323 (Organosilicon compounds with silicon-metal bonds).
J. F. Young, *Adv. Inorg. Chem. Radiochem.*, *11* (1968) 92 (Organosilicon derivatives of transition metals).
F. Schindler, H. Schmidbaur, *Angew. Chem., Int. Ed. Engl.*, *6* (1967) 683 (Siloxane compounds of the transition metals).
E. Wiberg, O. Stecher, H. J. Andrascheck, L. Kreutzbichler, E. Straude, *Angew. Chem., Int. Ed. Engl.*, *2* (1963) 507 (Metal silyls of the type $M(SiR_3)_n$).

J. Idris-Jones, in *Developments in Inorganic Polymer Chemistry*, M. F. Lappert, G. J. Leigh, eds., Elsevier, Amsterdam (1962), p. 162 (Polymetallosiloxanes).

H. Gilman, H. J. Winkler, in *Organometallic Chemistry*, H. Zeiss, ed., Reinhold, New York (1960), p. 270 (Organosilyl-metal derivatives).

D. Wittenberg, H. Gilman, *Quart. Rev.*, *13* (1959) 116 (Organosilyl-metal compounds).

ix) Applications of Organosilanes to Organic Synthesis

W. P. Weber, *Silicon Reagents for Organic Synthesis*, Springer Verlag, Berlin (1983).

H. Reich, ed., *Recent Developments in the Use of Silicon in Organic Synthesis*, *Tetrahedron*, Vol. *39* (1983).

E. Colvin, *Silicon in Organic Synthesis*, Butterworths, London (1981).

H. Brunner, *Angew. Chem., Int. Ed. Engl., 22* (1983) 897 (Enantioselective hydrosilylation).

P. Brownbridge, *Synthesis, 1* (1983) 85 (Silyl enol ethers in synthesis).

S. Murai, I. Ryu, N. Sonoda, *J. Organomet. Chem., 250* (1983) 121 (Siloxycyclopropane synthetic intermediates).

J. Donogues, *Ann. Chim. (Paris), 8* (1983) 135 (Silylation-desilylation in organic synthesis).

H. Sakurai, *Pure Appl. Chem., 54* (1982) 1 (Allylsilanes applied to organic synthesis).

P. Magnus, *Rev. Si, Ge, Sn, Pb Compnds., 6* (1982) 37 (Organosilane reagents for C-C bond formation).

A. K. Banerjee, *J. Sci. Ind. Res., 41* (1982) 699 (Trimethyliodosilane in organic synthesis).

L. A. Paquette, *Science, 217* (1982) 793 (Silicon-mediated organic synthesis).

H. Emde, D. Domsch, H. Feger, U. Frick, A. Goeta, H. H. Hergott, K. Hofmann, W. Kober, K. Kraegeloh, *Synthesis*, (1982) 1 (Trialkylsilyl perfluoroalkane sulfonates in organic synthesis).

I. Ojima, T. Kogure, *Rev. Si, Ge, Sn, Pb Compnds., 5* (1981) 8 (Hydrosilation).

S. Danishevsky, *Acc. Chem. Res., 14* (1981) 400 (Siloxydienes in total synthesis).

I. Fleming, *Chem. Soc. Rev., 10* (1981) 83 (Silicon compounds in organic synthesis).

P. Magnus, *Aldrichim. Acta, 13* (1980) 43 (Silicon in organic synthesis).

L. Birkofer, O. Stuhl, *Top. Curr. Chem., 88* (1980) 33 (Silylated synthons in organic synthesis).

I. Fleming, *Chimia, 34* (1980) 265 (Silicon in organic synthesis).

W. C. Groutas, D. Felker, *Synthesis*, (1980) 861 (Synthetic applications of Me_3Si-CN, -I, $-N_3$ and -SMe).

S. Blechert, *Nachr. Chem. Techn. Lab., 28* (1980) 801 (Vinyl- and allylsilanes in synthesis).

A. H. Schmidt, *Chem. Ztg., 104* (1980) 253 (Bromo- and iodosilanes in synthesis).

W. P. Neumann, K. Reuter, *J. Organomet. Chem. Libr., 7* (1979) 229 (Silyl-mercurials in organic synthesis).

T. H. Chan, I. Fleming, *Synthesis*, (1979) 761 (Electrophilic substitution of organosilicon compounds – applications to organic synthesis).

E. C. Colvin, *Chem. Soc. Rev., 7* (1978) 15 (Silicon in organic synthesis).

B. E. Cooper, *Chem. Ind. (London)*, (1978) 794 (Silylation as a protective method in organic synthesis).

T. H. Chan, *Acc. Chem. Res., 10* (1977) 442 (Alkene synthesis via β-functionalized organosilicon compounds).

R. Calas, J. Donogues, *J. Organomet. Chem. Libr., 2* (1976) 277 (Novel applications of chlorosilane/Mg or Li-donor solvent systems in synthesis).

P. F. Hudrlik, *J. Organomet. Chem. Libr.*, *1* (1976) 127 (Organosilanes in organic synthesis).
I. Fleming, *Chem. Ind. (London)*, (1975) 449 (Bond formation controlled by silicon – applications to organic synthesis).
M. V. Kashutina, S. L. Ioffe, V. A. Tartakovskii, *Russ. Chem. Rev. (Engl. Transl.)*, *44* (1975) 733 (Silylation of organic compounds).
J. F. Klebe, *Adv. Org. Chem.*, *8* (1972) 97 (Silylation in organic synthesis).
K. Rühlmann, *Synthesis*, (1971) 236 (Use of Me_3SiCl in organic synthesis).
D. Ballard, K. Shiina, T. Brennan, F. W. G. Fearon, I. Haiduc, H. Gilman, *Pure Appl. Chem.*, *19* (1969) 449 (Silylation of polyhaloaromatic compounds).
L. Birkofer, A. Ritter, *Angew. Chem. Int. Ed. Engl.*, *4* (1965) 417 (Use of silylation in organic synthesis).
K. Rühlmann, *Z. Chem.*, (1965) 130 (Use of organosilicon compounds in organic synthesis).

x) Miscellaneous Topics

E. P. Pluddemann, *Silane Coupling Agents*, Plenum Press, New York (1982).
D. Leyden, W. Collins, *Silylated Surfaces*, Gordon and Breach, New York (1980).
F. L. Boschke, ed., *Bioactive Organo-Silicon Compounds* (*Top. Curr. Chem.*, Vol. 84), Springer Verlag, Berlin (1979).
M. G. Voronkov, G. I. Zelchan, E. Ya. Lukevits, *Kremnii i Zhizn* (Silicon and Life), 2nd ed., Zinatne, Riga, (1978) (In Russian) (Biological activity of organosilicon compounds).
H. Kwart, K. King, *d-Orbital Involvement in the Organic Chemistry of Silicon, Phosphorus and Sulfur*, Springer Verlag, Berlin (1977).

J. Schraml, in *Carbon-Functional Organosilicon Compounds*, V. Chvalovsky, J. M. Bellama, eds., Plenum Press, New York (1984), p. 121 (NMR applications).
J. W. Witt, in *Reactive Intermediates*, Vol. 3, R. A. Abromovitch, ed., Plenum Press, New York (1983), p. 113 (Silane-radical reactions).
E. M. Genies, F. El Omar, *Electrochim. Acta*, *28* (1983) 541 (Organosilicon electrochemistry).
Y.-N. Tang, in *Reactive Intermediates*, Vol. 2, R. A. Aromovitch, ed., Plenum Press, New York (1982), p. 297 (Si-atom and silylene reactions).
H. Bock, W. Kaim, *Acc. Chem. Res.*, *15* (1982) 9 (Organosilicon radical cations).
M. G. Voronkov, V. M. D'yakov, S. V. Kirpichenko, *J. Organomet. Chem.*, *233* (1981) 1 (Silatranes).
R. Walsh, *Acc. Chem. Res.*, *14* (1981) 246 (Bond dissociation values).
M. D. Curtis, P. S. Epstein, *Adv. Organomet. Chem.*, *19* (1981) 188 (Redistributions on silicon catalyzed by transition-metal complexes).
R. J. Fessenden, J. S. Fessenden, *Inorg. Chim. Acta Rev.*, *4* (1979) 97 (Diamagnetic behavior and structure of silanes).
M. G. Voronkov, *Top. Curr. Chem.*, *88* (1979) 77 (Bioactive organosilatranes).
R. Take, U. Wannagat, *Top. Curr. Chem.*, *88* (1979) 1 (Synthesis of bioactive organosilicon compounds).
E. A. Williams, J. D. Cargioli, *Annu. Rev. NMR Spectrosc.*, *91* (1979) 221 (Si-29 NMR).
R. K. Harris, B. J. Kimber, *Appl. Spectrosc. Rev.*, *10* (1975) 117 (Si-29 NMR as a tool for studying silicones).
H. Bürger, *Angew. Chem., Int. Ed. Engl.*, *12* (1973) 474 (Anomalies in the structural chemistry of silicon).
M. G. Voronkov, *Chem. Br.*, *9* (1973) 411 (Bio-organosilicon chemistry).

I. M. T. Davidson, *Quart. Rev.*, *25* (1971) 111 (Silicon-radical chemistry).
Y. Nagai, *Intra-Sci. Chem. Rept.*, *4* (1970) 115 (Chemistry of organosilicon free radicals).
R. L. Mital, R. R. Gupta, *Inorg. Chim. Acta Rev.*, *4* (1970) 97 (Diamagnetic behavior and structure of silanes).
M. G. Voronkov, E. Ya. Lukevits, *Russ. Chem. Rev. (Engl. Transl.)*, *38* (1969) 975 (Biologically active organosilanes).

d) Germanium

M. Lesbré, P. Mazerolles, J. Satgé, *The Organic Chemistry of Germanium*, Wiley-Interscience, New York (1971).
F. Glockling, *The Chemistry of Germanium*, Academic Press, New York (1969) (Contains 117 pages on organogermanium compounds).
V. F. Mironov, T. K. Gar, *Organicheskie Soedineniya Germaniya* (Organic Compounds of Germanium), Nauka, Moscow (1967).
V. I. Davydov, *Germanium*, Gordon and Breach, New York (1966).
F. Rijkens, G. J. M. Van der Kerk, *Organogermanium Chemistry*, Germanium Research Committee, Utrecht (1964).

J. Satgé, *Pure Appl. Chem.*, *56* (1984) 137 (Organogermane intermediates).
R. J. P. Corriu, E. Colomer, *Ann. Chim. (Paris)*, *8* (1983) 121 (Ge-transitionmetal bonds from optically active organogermanes).
K. C. Molloy, J. J. Zuckerman, *Adv. Inorg. Chem. Radiochem.*, *27* (1983) 113 (Organogermanium structures).
M. Dräger, L. Ross, D. Simon, *Rev. Si, Ge, Sn, Pb Compnds.*, *7* (1983) (Polygermanes and heterocycles).
J. Satgé, *Bull. Soc. Chim. Belg.*, *91* (1982) 1019 (Ge heterocycles).
J. Satgé, *Adv. Organomet. Chem.*, *21* (1982) 241 (Multiply bonded Ge).
G. A. Razuvaev, M. N. Bochkarev, *J. Organomet. Chem. Libr.*, *12* (1981) 241 (Polygermanes).
H. Sakurai, *J. Organomet. Chem. Libr.*, *12* (1981) 267 (Organogermyl radicals).
E. Colomer, R. J. P. Corriu, *Top. Curr. Chem.*, *96* (1981) 79 (Ge-transitionmetal compounds).
R. C. Poller, in *Comprehensive Organic Chemistry*, D. Barton, W. D. Ollis, eds., Vol. 3, D. N. Jones, ed., Pergamon Press, Oxford (1979), p. 1111 (General review).
B. M. Glavincevski, J. E. Drake, *Rev. Si, Ge, Sn, Pb Compnds.*, *3* (1978) 333 (Organogermanium compounds containing Ge-group VI element bonds).
T. Gar, V. F. Mironov, *Elementoorganicheskie Soedineniya*, *3* (1976) 83 (Direct synthesis of organogermanium compounds).
E. J. Bulten, *MTP Int. Rev. Sci., Inorg. Chem., Ser. Two*, *4* (1975) 246 (Organogermanes).
J. Satgé, M. Massol, P. Riviere, *J. Organomet. Chem.*, *56* (1973) 1 (Divalent germanium species as starting materials and intermediates in organogermanium chemistry).
J. J. Zuckerman, in *Organometallic Compounds of the Group IV Elements*, A. G. MacDiarmid, ed., Vol. 2, Part 2, M. Dekker, New York (1972), p. 1 (Organogermanium-halide synthesis).
E. J. Bulten, *MTP Int. Rev. Sci., Inorg. Chem., Ser. One*, *4* (1972) 247 (Organogermanium chemistry).
V. F. Mironov, T. Gar, (Organomet. Chem. Rev., *A*, *3* (1968) 311 (Trichlorogermane chemistry, organogermanium derivatives).
K. A. Hooton, *Prep. Inorg. React.*, *4* (1968) 85 (Organogermanium compounds).

F. Glockling, K.A. Hooton, in *Organometallic Compounds of the Group IV Elements*, A.G. MacDiarmid, ed., Vol. 1, Part 2, M. Dekker, New York (1968) (Formation and reactions of Ge-C bonds).
I. Ruidisch, H. Schmidbaur, H. Schumann, in *Halogen Chemistry*, V. Gutman, ed., Academic Press, New York (1967), p. 133 (Organogermanium halides).
F. Glockling, *Quart. Rev., 20* (1966) 45 (Organogermanium introductory review).
D. Quane, R.S. Bottei, *Chem. Rev., 63* (1963) 406 (Organogermanes).
E. Gastinger, *Top. Curr. Chem., 3* (1955) 603 (Research developments in organogermanium chemistry, early results).

e) Tin

i) Organotins: General Review

P.G. Harrison, *The Chemistry of Tin*, Elsevier, Amsterdam (1984).
Gmelin Handbook of Inorganic Chemistry, Organotin Compounds (by H. Schumann, I. Schumann), Springer Verlag, Berlin: Part 11, *Trimethyl- and Trimethyltin-Oxygen Compounds* (1984); Part 10, *Mono- and Diorganotin-Sulfur Compounds, Organotin Selenium and Organotin-Tellurium Compounds* (1983); Part 9, *Triorganotin Sulfur Compounds* (1982).
F.L. Boschke, ed., *Organotin Compounds* (*Top. Curr. Chem.*, Vol. 104) Springer Verlag, Berlin (1982).
Gmelin Handbook of Inorganic Chemistry, Organotin Compounds (by H. Schumann, I. Schumann), Springer Verlag, Berlin: Part 8, *Organotin Iodides, Organotin Pseudohalides* (1981); Part 7, *Organotin Bromides* (1980); Part: 6, *Diorganotin Dichlorides, Organotin Trichlorides* (1978); Part 5, *Organotin Fluorides, Triorganotin Chlorides* (1978); Part 4, *Organotin Hydrides* (1976); Part: 3, *Tin Tetraorganyls, $R_2SnR'_2$, Heterocycles and Spiranes* (1976).
J.J. Zuckerman, ed., *Organotin Compounds: New Chemistry and Applications*, Advances in Chemistry Series, No. 157, American Chemical Society, Washington, DC (1976).
Gmelin Handbook of Inorganic Chemistry, Organotin Compounds (by H. Schumann, I. Schumann), Springer Verlag, Berlin: Part 2, *Tin Tetraorganyls, R_3SnR'* (1975); Part 1, *Tin Tetraorganyls, SnR_4* (1975).
J.G.A. Luijten, *Toxicity of Organotin Compounds. A Bibliography*, Tin Research Institute, Greenford (1972).
A.K. Sawyer, ed., *Organotin Compounds*, 3 Vols., M. Dekker, New York (1970–1972).
J.G.A. Luijten, *A. Bibliography of Organotin Analysis*, Tin Research Institute, Greenford (1970).
R.C. Poller, *The Chemistry of Organotin Compounds*, Academic Press, New York (1970).
W.P. Neumann, *The Organic Chemistry of Tin*, J. Wiley & Sons, New York (1970); English transl. of *Die Organische Chemie des Zinns*, F. Enke, Stuttgart (1967).
H.M.J.C. Creemers, *Hydrostannolysis. A General Method for Establishing Tin-Metal Bonds*, Institute of Organic Chemistry, TNO, Utrecht (1967).
A.J. Leusink, *Hydrostannation*, Organic Chemical Institute, TNO, Utrecht (1966).
J.G. Noltes, G.J.M. Van der Kerk, *Functionally Substituted Organotin Compounds*, Tin Research Institute, Greenford (1958).

J. G. A. Luijten, G. J. M. Van der Kerk, *Investigations in the Field of Organotin Chemistry*, Tin Research Institute, Greenford (1955).

R. F. Bennett, *Ind. Chem. Bull., 2* (1983) 171 (Commercial organotins).

P. A. Cusack, P. J. Smith, *Rev. Si, Ge, Sn, Pb Compnds., 7* (1983) 1 (Organotin fire retardants).

C. J. Evans, R. Hill, *Rev. Si, Ge, Sn, Pb Compnds., 7* (1983) 57 (Organotin antifoulants).

V. S. Petrosyan, *J. Organomet. Chem., 250* (1983) 157 (Solvent effects on alkyltin exchange).

J. M. Fukuto, R. R. Jensen, *Acc. Chem. Res., 16* (1983) 177 (Mechanisms of organotin S_E2 reactions).

V. I. Shiryaev, V. F. Mironov, *Russ. Chem. Rev. (Engl. Transl.)., 52* (1983) 321 (Bivalent tin carbene analogues).

K. C. Molloy, J. J. Zuckerman, *Acc. Chem. Res., 16* (1983) 386 (Organotin oxy and thio phosphorus acid structures).

I. P. Beletskaya, *J. Organomet. Chem., 250* (1983) 551 (Organotin halide cross-coupling catalyzed by Pd).

A. Tzschach, K. Jurkschat, *Comments Inorg. Chem., 3* (1983) 35 (Pentacoordinated tins).

M. Veith, *Top. Curr. Chem., 104* (1982) 1 (Molecular tin(II) monomers).

R. C. Poller, *J. Organomet. Chem., 239* (1982) 189 (Reviews author's work).

M. Gielen, *Top. Curr. Chem., 104* (1982) 57 (Organotin chirality, static and dynamic stereochemistry).

R. Hani, R. Geanangel, *Coord. Chem. Rev., 44* (1982) 229 (Sn-119 NMR).

M. Pereyre, J.-P. Quintard, A. Rahm, *Pure Appl. Chem., 54* (1982) 29 (Sn-119 NMR).

Z. M. O. Rzaev, *Top. Curr. Chem., 104* (1982) 107 (Coordination effects in formation and crosslinking of organotin macromolecules).

J. W. Nicholson, *Coord. Chem. Rev., 47* (1982) 263 (Organostannate(IV) complexes).

B. Wrackmeyer, *Rev. Si, Ge, Sn, Pb Compnds., 6* (1982) 75 (Alkynyltins form C-C bonds in organoborations).

R. C. Mehrotra, G. Srivastava, B. S. Saraswat, *Rev. Si, Ge, Sn, Pb Compnds., 6* (1982) 171 (Organotin Schiff-base complexes).

M. Veith, *Nachr. Chem. Tech. Lab., 30* (1982) 940 (Organotin(II)).

M. Gielen, B. dePoorter, *Rev. Si, Ge, Sn, Pb Compnds., 5* (1981) 7 (Synthesis of tetraorganotin derivatives).

J. W. Connolly, C. H. Hoff, *Adv. Organomet. Chem., 19* (1981) 110 (Organotin(II) compounds).

P. J. Smith, *J. Organomet. Chem. Libr., 12* (1981) 97 (Organotin X-ray structures).

M. Gielen, I. V. Eynde, *J. Organomet. Chem. Libr., 12* (1981) 193 (Organotin chirality and dynamic stereochemistry).

M. Pereyre, J. P. Quintard, A. Rahm, *J. Organomet. Chem. Libr., 12* (1981) 213 (Tin-carbon bond chemistry).

A. Tzschach, H. Weichmann, K. Jurkschat, *J. Organomet. Chem. Libr., 12* (1981) 293 (Pentacoordinated cyclic organotins).

M. Veith, *J. Organomet. Chem. Libr., 12* (1981) 319 (Tin(II) in rings and cages).

F. E. Brinckman, *J. Organomet. Chem. Libr., 12* (1981) 319 (Environmental organotin chemistry).

M. Gielen, *Pure Appl. Chem., 52* (1980) 657 (Stereoselective substitution at optically active tin centers).

A. G. Davies, P. J. Smith, *Adv. Inorg. Chem. Radiochem., 23* (1980) 1 (General review).

J.-P. Quintard, M. Pereyre, *Rev. Si, Ge, Sn, Pb Compnds., 4* (1980) 151 (Organotin alkali-metal and magnesium reagents).

J. E. Drake, L. N. Khasrou, *Rev. Si, Ge, Sn, Pb Compnds., 4* (1980) 271 (Tin-group VI compounds).

J.J. Zuckerman, R.P. Reisdorf, H.V. Ellis, R.R. Wilkinson, in *Organometals and Organometal Occurrence and Fate in the Environment*, F.E. Brinckman, J.M. Bellama, eds., ACS Symposium Series, No. 82, American Chemical Society, Washington, DC (1978), p. 388 (Organotins in biology and environment).

J.A. Zubieta, J.J. Zuckerman, *Prog. Inorg. Chem.*, *24* (1978) 251 (Structural tin chemistry).

W.L. Yeager, V.J. Castelli, in *Organometallic Polymers*, C.E. Carraher, Jr., J.E. Sheats, C.U. Pittman, Jr., eds., Academic Press, New York (1978), p. 175 (Organotin antifouling polymers).

P.J. Smith, A.P. Tupciauskas, *Annu. Rev. NMR Spectrosc.*, *8* (1978) 292 (Sn-119 NMR).

R.C. Poller, *Rev. Si, Ge, Sn, Pb Compnds.*, *3* (1978) 243 (Organotin free radicals).

V.S. Petrosyan, *Prog. NMR Spectrosc.*, *11* (1977) 115 (Organotin NMR).

P.A. Flinn, in *Mössbauer Isomer Shifts*, G.K. Shenoy, F.E. Wagner, eds., North Holland, Amsterdam (1977), p. 593 (Tin-119m isomer shifts).

U. Kunze, *Rev. Si, Ge, Sn, Pb Compnds.*, *2* (1977) 251 (Organotins in liquid SO_2).

V.S. Petrosyan, *Prog. NMR Spectrosc.*, *11* (1977) Part 2, p. 115 (NMR spectra and structures of organotin compounds).

M. Gielen, B. dePoorter, *Rev. Si, Ge, Sn, Pb Compnds.*, *3* (1977) 9 (Synthesis of tetraorganotin derivatives).

Yu.I. Dergunov, V.F. Gerega, O.S. Dyachkovskaya, *Russ. Chem. Rev. (Engl. Transl.)*, *46* (1977) 1132 (Synthesis, reactivity and properties of Sn-N compounds).

W.P. Neumann, in *Homoatomic Rings, Chains and Macromolecules of Main-Group Elements*, A.L. Rheingold, ed., Elsevier (1977), p. 277 (Di-, poly- and cyclostannanes).

V.S. Petrosyan, N.S. Yashina, O.A. Reutov, *Adv. Organomet. Chem.*, *14* (1976) 63 (Methyltin halides and their molecular complexes).

M.F. Lappert, in *Organotin Compounds: New Chemistry and Applications*, J.J. Zuckerman, ed., Advances in Chemistry Series, No. 157, American Chemical Society, Washington, DC (1976), p. 256 (Use of bulky alkyls to stabilize stannylenes).

P.J. Smith, *Chem. Ind. (London)*, (1976) 1025 (Organotin intermediates in organic synthesis).

J.N.R. Ruddick, *Rev. Si, Ge, Sn, Pb Compnds.*, *2* (1976) 115 (Tin-119m Mössbauer data).

V.I. Shiryaev, E.M. Stepina, V.F. Mironov, *Elementoorg. Soedinen.*, *3* (1976) 8 (Direct synthesis of organotin compounds).

P.J. Smith, L. Smith, *Chem. Br.*, *11* (1975) 208 (Organotin applications).

J.G. Noltes, *J. Organomet. Chem.*, *100* (1975) 177 (Organotin hydrides).

M.F. Lappert, *J. Organomet. Chem.*, *100* (1975) 139 (Dialkylstannylenes and their complexes).

R. Barbieri, L. Pellerito, G.C. Stocco, *Inorg. Chim. Acta*, *11* (1974) 173 (Mössbauer spectra of monoorganotin(IV) derivatives).

U. Kunze, J.D. Koola, *J. Organomet. Chem.*, *80* (1974) 281 (Mechanism of SO_2 insertion into Sn-C bonds).

J.D. Kennedy, W. McFarlane, *Rev. Si, Ge, Sn, Pb Compnds*, *1* (1974) 235 (Tin-119 NMR).

M. Gielen, S. Boué, M. DeClerq, B. DePoorter, *Rev. Si, Ge, Sn, Pb Compnds.*, *1* (1974) 97 (Organotin physical properties).

E.W. Abel, in *Comprehensive Inorganic Chemistry*, J.C. Bailar, H.J. Emeléus, R. Nyholm, A.F. Trotman-Dickenson, eds., Pergamon Press, Oxford (1973), Vol. 2, p. 43 (Tin review).

B.Y.K. Ho, J.J. Zuckerman, *J. Organomet. Chem.*, *49* (1973) 1 (Structural organotin chemistry).

P.J. Smith, L. Smith, *Inorg. Chim. Acta Rev.*, *7* (1973) 11 (Applications of ^{119}Sn NMR chemical shifts to structural tin chemistry).

J. Nasielski, *Pure Appl. Chem.*, *23* (1973) 449 (Five-coordinated organotin compounds).

M. Gielen, *Acc. Chem. Res.*, *6* (1973) 198 (Optically active organotin compounds).

M. Gielen, *Ind. Chim. Belg.*, *38* (1973) 20, 138 (Synthesis and properties of tetraorganotins).

R.V. Parish, *Prog. Inorg. Chem.*, *15* (1972) 101 (Interpretation of ^{119m}Sn Mössbauer spectra).

C. R. Dillard, in *Organotin Compounds*, A. K. Sawyer, ed., Vol. 3, M. Dekker, New York (1972), p. 997 (Organotin microanalysis).
J. G. A. Luijten, in *Organotin Compounds*, A. K. Sawyer, ed., Vol. 3, M. Dekker, New York (1972), p. 931 (Biological effects of organotins).
H. C. Clark, R. J. Puddephatt, in *Organometallic Compounds of the Group-IV Elements*, A. G. MacDiarmid, ed., M. Dekker, New York (1972), Vol. 2, Part 2, p. 71 (Organotin syntheses).
E. Lindner, U. Kunze, *Rev. Si, Ge, Sn, Pb Compnds.*, *1* (1972) 35 (Sulfinatotins).
E. J. Kupchik, in *Organotin Compounds*, A. K. Sawyer, ed., Vol. 1, M. Dekker, New York (1971), p. 7 (Organotin hydrides).
G. P. van der Kelen, E. V. van den Berghe, L. Verdonck, in *Organotin Compounds*, Vol. 1, A. K. Sawyer, ed., M. Dekker, New York (1971), p. 81 (Organotin halides).
J. J. Zuckerman, *Adv. Organomet. Chem.*, *9* (1970) 21 (Applications of Mössbauer spectroscopy to organotin compounds).
P. J. Smith, *Organomet. Chem. Rev.*, *A*, *5* (1970) 373 (Mössbauer parameters of organotin compounds).
G. J. M. van der Kerk, *Chem. Ind. (London)*, (1970) 644 (Prospectives in organotin chemistry).
A. Bokranz, H. Plum, *Top. Curr. Chem.*, *16* (1970) 366 (Manufacture and uses of organotin compounds).
C. K. Banks, in *Kirk-Othmer Encyclopedia of Chemical Technology*, R. E. Kirk, D. F. Othmer, eds., 2nd ed., Vol. 20, McGraw-Hill, New York (1969), p. 304 (Tin compounds).
A. G. Davies, *Chem. Br.*, *4* (1968) 403 (Organotin chemistry).
J. G. A. Luijten, G. J. M. Van der Kerk, in *Organometallic Compounds of the Group-IV Elements,* A. G. MacDiarmid, ed., Vol. 1, Part I, M. Dekker, New York (1968), p. 91 (Organotin syntheses).
H. G. Kuivila, *Acc. Chem. Res.*, *1* (1968) 299 (Addition of organotin hydrides to olefins).
R. Okawara, M. Wada, *Adv. Organomet. Chem.*, *5* (1967) 137 (Structural aspects of organotin chemistry).
G. Tagliavini, P. Zanella, M. Fioriani, *Coord. Chem. Rev.*, *1* (1966) 249 (Pentacoordination of organotin compounds in nonaqueous solvents).
A. Ross, *Ann. N.Y. Acad. Sci.*, *125* (1965) 107 (Industrial applications of organotin compounds).
H. G. Kuivila, *Adv. Organomet. Chem.*, *1* (1964) 47 (Organotin hydrides, reactions with organic compounds).
W. P. Newmann, *Angew. Chem., Int. Ed. Engl.*, *2* (1963) 165 (Developments in organotin chemistry).
R. K. Ingham, S. D. Rosenberg, H. Gilman, *Chem. Rev.*, *60* (1960) 459 (Organotin compounds, early chemistry).

ii) Applications of Organotins to Organic Synthesis

P. J. Smith, D. V. Sanghani, K. D. Bos, J. D. Donaldson, *Chem. Ind. (London)*, (1984) 167 (Tin(II) reducing agents).
M. Pereyre, J.-P. Quintard, *Pure Appl. Chem.*, *53* (1981) 2401 (Organotin reagents in organic synthesis).
M. Pereyre, J.C. Pommier, *J. Organomet. Chem. Libr.*, *1* (1976) 661 (Organotins in organic synthesis).
P. J. Smith, *Chem. Ind. (London)*, (1976) 1025 (Organic synthesis via organotin intermediates).

A.J. Bloodworth, A.G. Davies, *Chem. Ind. (London)* (1972) 490 (Organotins in organic synthesis).
J.G.A. Luijten, *Chem. Ind. (London)*, (1972) 103 (Organotin compounds in organic synthesis).
H.G. Kuivila, *Synthesis*, (1970) 499 (Reduction of organics with organotin hydrides).
A.G. Davies, *Synthesis*, (1969) 56 (Organotin oxides and alkoxides in organic synthesis).
H.G. Kuivila, *Adv. Organomet. Chem.*, *1* (1964) 47 (Organotin hydrides with organics).

f) Lead

J.M. Ratcliffe, *Lead in Man and the Environment*, Halsted Press, New York (1981).
J.O. Nriagu, ed., *Topics in Environmental Health*, Vol. 1, *The Biogeochemistry of Lead*, Parts A and B, Ann Arbor Sci. Publ., Ann Arbor, MI (1978).
E. Müller, O. Bayer, H. Meerwein, K. Ziegler, eds., *Methoden der Organischen Chemie (Houben Weyl)*. Band XIII/7. *Metalloorganische Verbindungen von Blei sowie den Metallen der IV.–VI. Nebengruppe des Periodensystems: Pb, Ti, Zr, Hf, V, Nb, Ta, Cr, Mo, W*, by G. Bähr, E. Langer, A. Segnitz, G. Thieme Verlag, Stuttgart (1975).
H. Shapiro, F.W. Frey, *The Organic Compounds of Lead*, Interscience, New York (1968).
L.C. Willemsens, G.J.M. van der Kerk, *Investigations in the Field of Organolead Chemistry*, ILZRP, New York (1965).
L.C. Willemsens, *Organolead Chemistry*, ILZRO, New York (1964).

D. DeVos, J. Wolters, *Rev. Si, Ge, Sn, Pb Compnds.*, *4* (1980) 209 (Monoorganolead(IV)s).
W.P. Neumann, K. Kühlein, *Adv. Organomet. Chem.*, *7* (1968) 242 (Preparations and reactions of compounds with Pb-C, Pb-H, Pb-N and Pb-O bonds).
L.C. Willemsens, G.J.M. Van der Kerk, in *Organometallic Compounds of the Group-IV Elements*, A.G. MacDiarmid, ed., Vol. 1, Part II, M. Dekker, New York (1968), p. 1911 (Organolead compounds).
R.W. Leeper, L. Summers, H. Gilman, *Chem. Rev.*, *54* (1954) 101 (Early literature on organo-Pb compounds).

Chapter 9

a) Group-VA: As, Sb, Bi

B.J. Aylett, *Organometallic Compounds*, 4th ed., Vol. 1, *The Main-Group Elements*, Part 2, Groups IV and V, Chapman and Hall, London (1979).
C.A. McAuliffe, W. Levason, *Phosphine, Arsine and Stibine Complexes of Transition Elements*, Elsevier, Amsterdam (1979).
E. Müller, S. Bayer, H. Meerwein, K. Ziegler, eds., *Methoden der Organischen Chemie (Hou-*

ben-Weyl), Band XIII/8, *Organo-Arsen-, Antimon-, Wismuth-Verbindungen, As, Sb, Bi*, by S. Samaan, G. Thieme Verlag, Stuttgart (1978).

Gmelin Handbuch der Anorganischen Chemie. Perfluoroorgano Compounds of Main-Group Elements. Part 3. *Compounds of Phosphorus, Arsenic, Antimony and Bismuth* (New Supplement Series, Vol. 24), Springer Verlag, Berlin (1975).

C. A. McAuliffe, ed., *Transition-Metal Complexes of Phosphorus, Arsenic and Antimony Ligands*, MacMillan, New York (1973).

R. Luckenbach, *Dynamic Stereochemistry of Pentacoordinated Phosphorus and Related Elements*, G. Thieme Verlag, Stuttgart (1973).

I. Haiduc, *The Chemistry of Inorganic Ring Systems*, Part II, J. Wiley, New York (1970), Chapter 5, p. 876 (Organoelement derivatives containing inorganic rings of As, Sb).

G. O. Doak, L. D. Freedman, *Organometallic Compounds of Arsenic, Antimony and Bismuth*, J. Wiley, New York (1970).

F. G. Mann, *The Heterocyclic Derivatives of Phosphorus, Arsenic, Antimony and Bismuth*, 2nd ed., J. Wiley, New York (1970).

M. Dub, *Organometallic Compounds*. Vol. III *Compounds of Arsenic, Antimony and Bismuth*, 2nd ed., Springer Verlag, Berlin (1968); First Supplement (1972).

A. W. Johnson, *Ylid Chemistry*, Academic Press, New York (1966) (Contains a chapter on As and Sb ylides).

G. T. Morgan, *Organic Compounds of Arsenic and Antimony*, Longmans & Green, London (1918) (Early literature).

R. E. Atkinson, in *Comprehensive Heterocyclic Chemistry*, A. R. Katritzky, C. W. Rees, eds., Vol. 1, O. Meth-Cohn, ed., Pergamon Press, Oxford (1984) p. 539 (Heterocycles of As, Sb, Bi).

D. Hellwinkel, *Top. Curr. Chem., 109* (1983) 1 (Penta- and hexaorgano group-V derivatives).

B. Arbuzov, R. P. Arshinova, N. A. Polezhaeva, *Izv. Akad. Nauk SSSR, Ser. Khim.*, (1983) 2507 (Group-V and -VI heterocycles).

A. J. Ashe, III, *Top. Curr. Chem., 105* (1982) 125 (Group-V heterobenzenes).

W. R. Callen, J. D. Woollins, *Coord. Chem. Rev., 39* (1981) 1 (Ferrocenylphosphines and arsines).

A. J. Ashe, III, *Acc. Chem. Res., 11* (1978) 153 (Group-V heterobenzenes).

R. E. Atkinson, in *Rodd's Chemistry of Carbon Compounds*, 2nd ed., Elsevier (1978), Vol. 4G, p. 83 (Six-membered heterocycles with P, A, Sb, Bi heteroatoms).

F. Bickelhaupt, H. Vermeer, *Method. Chim., 78* (Part B) (1978) 549 (Organoarsenic, antimony and bismuth compounds).

Yu. A. Aleksandrov, V. P. Maslennikov, V. P. Sergeyeva, *J. Organomet Chem. Libr., 5* (1977) 219 (Organic peroxides of the group-V elements).

A. L. Rheingold, in *Homoatomic Rings, Chains and Macromolecules of Main-Group Elements*, Elsevier, Amsterdam (1977), p. 385 (Structures of catenated homonuclear compounds of the group-V elements).

N. G. Bokii, Yu. T. Struckhov, A. E. Kalinin, V. G. Andrianov, T. N. Salnikova, *Itogi Nauki, Kristalokhim., 12* (1977) 56 (Structures).

W. Levason, C. A. McAuliffe, *Coord. Chem. Rev., 19* (1976) 173 (Phosphine, arsine and stibine complexes of main-group elements).

H. Schmidbaur, *Adv. Organomet. Chem., 14* (1976) 205 (Pentaalkyls and alkylidene trialkyls of group-V elements).

G. O. Doak, L. D. Freedman, *Synthesis*, (1974) 328 (Syntheses based on arsenides, stibides and bismuthides).

R. J. Cross, *MTP Int. Rev. Sci., Inorg. Chem., Ser. Two, 5* (1974) 147 (Arsine, stibine and bismuthine complexes).

E. Maslowsky, Jr., *J. Organomet. Chem.*, *70* (1974) 153 (Vibrational spectra of organo-group V derivatives).

C.S. Kraihanzel, *J. Oganomet. Chem.*, *73* (1974) 137 (Reactions of coordinated pnictogen ligands).

D.I. Hall, J.H. Ling, R.S. Nyholm, *Struct. Bonding (Berlin)*, *15* (1973) 3 (Metal complexes of chelating olefin-group V ligands).

H. Hilmer, in *Arzneimittel: Entwicklung, Wirkung, Darstellung*, 2nd ed., Vol. 4, (1972) p. 57 (Organoarsenic and -antimony compounds as therapeutic agents).

O.A. Reutov, O.A. Ptitsina, *Organomet. React.*, *4* (1972) 73 (Preparation of organo-As, Sb, Bi compounds via onium compounds of N, I, Br, Cl).

J.P. Crow, W.R Cullen, *MTP Int. Rev. Sci., Inorg. Chem., Ser. One*, *4* (1972) 355 (Organo-As, Sb, Bi compounds).

K. Moedritzer, *Organomet. React.*, *2* (1971) 1 (Redistribution reactions).

E.W. Abel, S.M. Ilingworth, *Organomet. Chem. Rev.*, *A*, *5* (1970) 143 (Arsines, stibines and bismuthines containing Si, Ge, Sn and Pb).

S.C. Cohen, A.G. Massey, *Adv. Fluorine Chem.*, *6* (1970) 185 (Polyfluoroaromatic derivatives of metals and metalloids).

D.D. Davis, C.E. Gray, *Organomet. Chem. Rev.*, *A*, *6* (1970)283 (Alkali-metal and magnesium derivatives of P, As, Sb and Bi compounds).

M. Field, O. Glemser, *Fluorine Chem. Rev.*, *3* (1969) 129 (Pentafluorophenyl derivatives of P, As, Sb).

J.K. Ruff, *Ann. N.Y. Acad. Sci.*, *159* (1969) 234 (Reorganization reactions of As-, S-, and Bi-organic compounds).

K. Moedritzer, *Adv. Organomet. Chem.*, *6* (1967) 171 (Redistribution equilibria of organometallic compounds).

J.C. Lockhard, *Chem. Rev.*, *65* (1965) 131 (Redistribution and exchange reactions in groups IIB–VIIB).

L. Kolditz, *Adv. Inorg. Chem. Radiochem*, *7* (1965) 1 (Organohalo derivatives of As, Sb, Bi).

G. Booth, *Adv. Inorg. Chem. Radiochem.*, *6* (1964) 1 (Complexes of transition metals with phosphines, arsines and stibines).

H.L. Yale, in *Pyridine and its Derivatives*, F. Klingsberg, ed., Interscience, New York (1964) (As and Sb-analogues of pyridine).

F.G. Mann, *Prog. Stereochem.*, *2* (1958) 196 (Stereochemistry of the group-V elements).

A.E. Goddard, in *Textbook of Inorganic Chemistry*, J.N. Friend, ed., Vol. XI, Part III, Griffin, London (1936) (Derivatives of phosphorus, antimony and bismuth).

b) Arsenic

W.H. Lederer, R.J. Fensterheim, eds., *Arsenic: Industrial, Biomedical, Environmental Perspectives*, Von Nostrand-Reinhold, New York (1983).

A. Tzschach, J. Heinicke, *Arsenheterocyclen*, VEB Deutscher Verlag für Grundstoffindustrien, Leipzig (1978).

L.K. Krannich, *Compounds Containing As-N Bonds*, Wiley, New York (1976).

E.A. Woolson, ed., *Arsenical Pesticides*, ACS Symposium Series, No. 7, American Chemical Society, Washington, DC (1975).

G.W. Raizis, J.L. Gavron, *Organic Arsenical Compounds*, Chemical Catalog Co., New York (1923).

A. Bertheim, *Handbuch der Organischen Arsenverbindungen*, F. Enke Verlag, Stuttgart (1913).

I. Omae, *Coord. Chem. Rev.*, *42* (1982) 245 (Intramolecular As coordination).
F. Kober, *Synthesis*, (1982) 173 (Aminoarsine preparative reagents).
Y. Huang, Y. Shen, *Adv. Organomet. Chem.*, *20* (1982) 115 (Arsonium ylides).
F. Kober, *Chem.-Ztg.*, *105* (1981) 199 (The arsenic-arsenic bond).
F.D. Yambushev, V.I. Savin, *Russ. Chem. Rev. (Engl. Transl.)*, *48* (1979) 582 (Organoarsenic stereochemistry).
S.S. Sandhu, M. Arshad, S. Baweja, S.S. Parmar, *J. Chem. Sci.*, *4* (1978) 76 (Cyclic arsines).
L.C. Duncan, *Annu. Rep. Inorg. Gen. Synth.*, *5* (1977) 111 (Synthetic aspects).
A.S. Levinson, *J. Chem. Educ.*, *54* (1977) 98 (Salvarsan: structure and the As=As double bond; textbook errors column).
B.O. West, in *Homoatomic Rings, Chains and Macromolecules of the Main-Group Elements*, A.L. Rheingold, ed., Elsevier, Amsterdam (1977).
F. Kober, *Chem.-Ztg.*, *100* (1976) 313 (Chemistry of aminoarsines).
L.R. Smith, J.L. Mills, *J. Organomet. Chem.*, *84* (1975) 1 (Cyclopolyarsines).
I.N. Azerbaev, Z.A. Abramova, Yu.G. Bosyakov, *Russ. Chem. Rev. (Engl. Transl.)*, *43* (1974) 657 (Acetylenic organoarsenic compounds).
S.C. Cohen, A.G. Massey, *Adv. Fluorine Chem.*, *6* (1970) 185 (Polyfluoraromatic derivatives of metals and metalloids).
P.S. Elmes, B.O. West, *Coord. Chem. Rev.*, *3* (1968) 279 (Coordinating properties of $(MeAs)_5$).
W.R. Cullen, *Adv. Organomet. Chem.*, *4* (1966) 145 (Organoarsenic compounds).
C.S. Hamilton, J.F. Morgan, *Org. React.*, *2* (1944) 415 (Aromatic arsonic and arsinic acids).

c) Antimony

Gmelin Handbuch der Anorganischen Chemie, New Supplement Series, *Organoantimony Compounds*, Part 1, *Compounds of Trivalent Antimony with Three Sb-C Bonds*, Springer Verlag, Berlin (1981). Part 3 (1983).
W.G. Christiansen, *Organic Derivatives of Antimony*, Chemical Catalog Co., New York (1925).

V.K. Jain, R. Bohra, R.C. Mehrotra, *Struct. Bonding (Berlin)*, *52* (1982) 147 (Organoantimony(V)).
R.C. Poller, in *Comprehensive Organic Chemistry*, D. Barton, W.D. Ollis, eds., Vol. 3, D.N. Jones, ed., Pergamon Press, Oxford (1979), p. 1111 (General review).
W. Levason, C.A. McAuliffe, *Acc. Chem. Res.*, *11* (1978) 363 (Coordination chemistry of organostibines).
R. Okawara, Y. Matsumura, *Adv. Organomet. Chem.*, *14* (1976) 187 (Organoantimony chemistry).
J.G. Noltes, H.A. Meinema, *Ann. N.Y. Acad. Sci.*, *239* (1974) 278 (Organoantimony coordination chemistry).
R.A. Zingaro, *Ann. N.Y. Acad. Sci.*, *192* (1972) 72. (Trialkylstibine oxides, sulfides, selenides).
M. Ida, Z. Toyoshima, T. Nakamura, *J. Pharm. Chem.*, *19* (1947) 158 (Chemistry of stibonic acids).

d) Bismuth

Gmelin Handbuch der Anorganischen Chemie, Bismuth-Organische Verbindungen, by M. Wieber, Springer Verlag, Berlin (1977).

L. D. Freedman, G. O. Doak, *Chem. Rev.*, *82* (1982) 15 (Organobismuth compounds).
K. C. Moss, M. A. R. Smith, *MTP Int. Rev. Sci., Inorg. Chem., Ser. Two*, *2* (1975) 287 (Compounds of Bi(III) and Bi(V)).
P. G. Harrison, *Organomet. Chem. Rev.*, *A*, *5* (1970) 183 (Organobismuth chemistry).

Chapter 10

a) General

R. Scheffold, ed., *Modern Synthetic Methods*, Vol. 3, *Transition Metals in Organic Synthesis*, Wiley, New York (1983) (Reports from the triennial Interlaken Seminar of 1983).
B. L. Shapiro, ed., *Organometallic Compounds: Synthesis, Structure and Theory*, Texas A&M Univ. Press, College Station, TX (1983).
M. Tsutsui, Y. Ishii, eds., *Fundamental Research in Organometallic Chemistry*, Van Nostrand-Reinhold, New York (1982).
S. G. Davies, *Organotransition-Metal Chemistry: Applications to Organic Synthesis*, Pergamon Press, Oxford (1982).
A. Müllen, E. Diemann, *Transition Metal Chemistry*, Verlag Chemie, Weinheim (1981).
C. Masters, *Homogeneous Transition-Metal Catalysis--A Gentle Art*, Chapman and Hall, London (1981).
B. E. Douglas, Y. Saito, eds., *Stereochemistry of Optically-Active Transition-Metal Complexes*, ACS Symposium Series, No. 119, American Chemical Society, Washington, DC (1980).
G. W. Parshall, *Homogeneous Catalysis: the Applications and Chemistry of Catalysis by Soluble Transition-Metal Complexes*, Wiley-Interscience, New York (1980).
J. P. Collman, L. S. Hegedus, *Principles and Applications of Organotransition-Metal Chemistry*, University Science Books, Ann Arbor, MI (1980).
G. Deganello, *Transition-Metal Complexes of Cyclic Polyolefins*, Academic Press, New York (1980).
R. P. Houghton, *Metal Complexes in Organic Chemistry*, Cambridge Univ. Press, Cambridge (1979).
D. St. C. Black, W. R. Jackson, J. M. Swan, in *Comprehensive Organic Chemistry*, D. Barton, W. D. Ollis, eds., Vol. 3, D. N. Jones, ed., Pergamon Press, Oxford (1979), p. 1127 (General review).
A. N. Nesmeyanov, K. A. Kocheshkov, eds., *Methody Elementoorganicheskoi Khimii. Kobal't, Nikel'. Platinovye Metaly* (Methods of Organoelement Chemistry, Cobalt, Nickel, Platinum Metals), Nauka, Moscow (1978) (In Russian).
J. K. Kochi, *Organometallic Mechanisms and Catalysis*, Academic Press, New York (1978).
H. Alper, ed., *Transition-Metal Organometallics in Organic Synthesis*, Academic Press, New York (1976).
D. Seyferth, ed., *New Applications of Organometallic Reagents in Organic Synthesis*, Elsevier, Amsterdam (1976).
J. Tsuji, *Organic Synthesis by Means of Transition-Metal Complexes*, Springer Verlag, Berlin (1975).
Y. Ishii, M. Tsutsui, eds., *Organotransition-Metal Chemistry*, Plenum Press, New York (1975).

A. N. Nesmeyanov, K. a. Kocheshkov, eds., *Methody Elementoorganicheskoi Khimii. Podgrupy Medi, Skandiya, Titana, Vanadiya, Khroma i Margantsa, Lantanoidy i Aktinoidy* (Methods of Organoelement Chemistry. Copper, Scandium, Titanium, Vanadium, Chromium and Manganese Subgroups. Lanthanides and Actinides), Nauka, Moscow (1974) (2 Vols, in Russian).

R. F. Heck, *Organotransition-Metal Chemistry: A Mechanistic Approach*, Academic Press, New York (1974).

L. Malatesta, S. Cemni, *Zerovalent Compounds of Metals*, Academic Press, New York (1974).

R. B. King, *Transition-Metal Organometallic Chemistry. An Introduction*, Academic Press, New York (1969) (A textbook in which organometallic compounds are arranged element-by-element).

C. W. Bird, *Transition-Metal Intermediates in Organic Synthesis*, Logos, London (1967).

M. Dub, ed., *Organometallic Compounds.*, Vol. 1, *Compounds of Transition Metals*, Springer Verlag, New York (1966).

R. J. Cross, *Mech. Inorg. Organomet. React., 2* (1984) 105 (Substitutions at inert metal complexes.

D. A. Sweigart, *Mech. Inorg. Organomet. React., 2* (1984) 237 (Organometallic substitutions and insertions).

L. A. P. Kane-McGuire, *Mech. Inorg. Organomet. React., 2* (1984) 301 (Reactivity of coordinated hydrocarbons).

A. J. Deeming, *Mech. Inorg. Organomet. React., 2* (1984) 319 (Organometallic rearrangements, intramolecular exchanges and isomerizations).

W. E. Watts, in *Comprehensive Heterocyclic Chemistry*, A. R. Katritzky, C. W. Rees, eds., Vol. 1, O. Meth-Cohn, ed., Pergamon Press, Oxford (1984) p. 665 (Transition-metal heterocycles).

M. L. H. Green, *Pure Appl. Chem., 56* (1984) 47 (Group IV–VII transition-metal chemistry).

M. Pascal, C. Lapinte, D. Astruc, *Ann. N. Y. Acad. Sci., 415* (1983) 97 (Organometallic electron, hydrogen-atom and hydride reservoirs).

J. K. Stille, *Mod. Synth. Methods, 3* (1983) 1 (Transition metals in organic synthesis).

Yu. I. Ermakov, *J. Mol. Catal., 21* (1983) 35 (Organometallics for supported catalysts).

J. M. Bassett, A. Choplin, *J. Mol. Catal., 21* (1983) 95 (Surface organometallic heterogeneous catalysis).

S. V. Ley, H. A. Porter, *Gen. Synth. Methods, 6* (1983) 218 (Organometallics in synthesis).

L. S. Hegedus, *J. Organomet. Chem., 237* (1982) 231 (Transition metals in organic synthesis).

S. D. Chapell, D. J. Cole-Hamilton, *Polyhedron, 1* (1982) 739 (Metalacyclic transition-metal derivatives).

J. Halpern, *J. Organomet. Chem., 200* (1980) 133 (Homogeneous catalytic hydrogenation).

P. L. Pauson, *J. Organomet. Chem., 200* (1980) 207 (Nucleophilic addition to transiton-metal complexes).

H. Werner, *J. Organomet. Chem., 200* (1980) 335 (Novel sandwich compounds).

G. Wilke, *J. Organomet. Chem., 200* (1980) 349 (Homogeneous catalysis).

H. Brunner, *Adv. Organomet. Chem., 18* (1980) 132 (Chiral metal atoms in optically-active organotransition-metal compounds).

B. R. James, *Adv. Organomet. Chem., 17* (1979) 319 (Hydrogenations catalyzed by transition-metal complexes).

H. Brunner, *Acc. Chem. Res., 12* (1979) 250 (Optical induction in organotransition metal compounds and asymmetric catalysis).

B. Gorewit, M. Tsutsui, *Adv. Catal., 27* (1978) 227 (σ-π Rearrangements in catalysis).

F. R. Hartley, P. N. Vezey, *Adv. Organomet. Chem., 15* (1976) 189 (Supported transition-metal complexes).

J. D. Morrison, W. F. Master, M. K. Newberg, *Adv. Catal., 25* (1976) 81 (Asymmetric homogeneous hydrogenation).

b) Group IIIB: Scandium, Yttrium, Lanthanides, Actinides

Gmelin Handbook of Inorganic Chemistry, Sc, Y, La-Lu, Rare-Earth Elements. Part D 6. *Organometallic Compounds*, Springer Verlag, Berlin (1983).

Müller, O. Bayer, H. Meerwein, K. Ziegler, eds., *Methoden der Organischen Chemie (Houben-Weyl)*, Vol. XIII/4, *Metallorganische Verbindungen der III. Gruppe des Periodensystems (ausser Bor): Al, Ga, In, Tl, Sc, Y, La*, by G. Bähr, P. Burba, H. Lehmkuhl, K. Ziegler, G. Thieme Verlag, Stuttgart (1970).

i) Scandium

M.E. Thompson, J.E. Bercaw, *Pure Appl. Chem., 56* (1984) 1 (Permethylscandocene derivatives).

T.V. Nikitina, in *Methody Elementoorganischeskoi Khimii. Podgrupy Medi, Skandiya, Titana, Vanadiya, Khroma, i Margantsa, Lantanoidy i Atkinoidy*, A.N. Nesmeyanov, K.A. Kocheskov, eds., Nauka, Moscow (1974), p. 138 (In Russian) (Organoscandium compounds).

G.A. Melson, R.W. Stotz, *Coord. Chem. Rev., 7* (1971) 133 (Coordination chemistry of Sc).

ii) Yttrium

T.V. Nikitana, in *Methody Elementoorganicheskoi Khimii. Podgrupy Medi, Skandiya, Titana, Vanadiya, Khroma i Margantsa, Lantanoidy i Aktinoidy*, A.N. Nesmeyanov, K.A. Kocheshkov, eds., Nauka, Moscow (1974), p. 143 (In Russian) (Organoyttrium compounds).

iii) Lanthanides and Actinides

T.J. Marks, R.D. Fischer, eds., *Organometallics of the f-Elements*, D. Reidel, Dordrecht (1979).

K.N. Raymond, C.W. Eigenbrett, Jr., *Acc. Chem. Res., 13* (1980) 276 (Structural criteria for bonding).

T.J. Marks, *Prog. Inorg. Chem., 25* (1979) 223 (f-Element organometallics).

S.A. Cotton, *J. Organomet. Chem. Libr., 3* (1977) 89 (Organometallic chemistry of lanthanides and actinides).

K.D. Warren, *Struct. Bonding (Berlin), 33* (1977) 97 (Ligand field theory of f-orbital sandwich compounds).

E.C. Baker, G.W. Halstead, K.N. Raymond, *Struct. Bonding (Berlin), 25* (1976) 23 (4f and 5f Organometallics).

M. Tsutsui, N. Ely, R. Dubois, *Acc. Chem. Res., 9* (1976) 217(σ-Organic f-element derivatives).

R.K. Sheline, J.L. Slater, *Angew. Chem., Int. Ed. Engl., 14* (1975) 309 (Spectral evidence for lanthanide and actinide carbonyls).

M. Tsutsui, C. Hyde, A. Gebala, N. Ely, in *Organotransition-Metal Chemistry*, Y. Ishii, M. Tsutsui, eds., *Proc. Jpn.-Am. Semin.* (1974), Plenum Press, New York (1975), p. 93. (Organolanthanides and actinides).

R.G. Hayes, J.L. Thomas, *Organomet. Chem. Rev., A, 7* (1971) 1 (Organometallic compounds of lanthanides and actinides).

H. Gysling, M. Tsutsui, *Adv. Organomet. Chem., 9* (1970) 361 (Organometallic compounds of lanthanides and actinides).

iv) Lanthanides

H. Schumann, W. Genthe, in *Handbook on the Physics and Chemistry of Rare Earths*, K.A. Gschneidner, L.R. Eyring, eds., North Holland, Amsterdam (1984) (Organolanthanides).

H. Schumann, *Comments Inorg. Chem., 2* (1983) 247 (Homoleptic rare earths); *Angew. Chem., Int. Ed. Engl.*, 23 (1984) 474 (Organolanthanides).

W.J. Evans, *J. Organomet. Chem., 250* (1983) 217 (Organolanthanides).

M.F. Lappert, A. Singh, *J. Organomet. Chem., 239* (1982) 133 (Bis-η^5-cyclopentadienyllanthanoid(III) chlorides).

H. Schumann, *Nachr. Chem. Tech. Lab., 27* (1979) L389 (Rare-earth organometallic chemistry).

T.J. Marks, *Prog. Inorg. Chem., 24* (1978) 51 (Chemistry and spectroscopy of lanthanide organometallics).

N.S. Vyazankin, R.N. Shehelokov, O.A. Kruglaya, in *Methody Elementoorganicheskoi Khimii. Podgrupi Medi, Skandiya, Titana, Vanadiya, Khroma i Margantsa, Lantanoidy i Aktinoidy*, A.N. Nesmeyanov, K.A. Kocheskov, eds., Nauka, Moscow (1974), p. 909 (Organometallic compounds of lanthanides).

K.O. Hodgson, F. Mares, D.F. Starks, A. Streitwieser, Jr., *J. Am. Chem. Soc., 95* (1973) 8650 (Lanthanide complexes with the cyclooctatetraene dianion).

v) Actinides

Gmelin Handbuch der Anorganischen Chemie, Uranium. Supplement Vol. E2, *Coordination Compounds (Including Organouranium Compounds)*, Springer Verlag, Berlin (1980), Vol. 4, C12 *Uranium* (1983)1, C13 *Uranium and Carbon* (1983). *Transuranium Elements.* Part C. *The Compounds* (1972) (Includes organometallics).

C. Keller, in *The Chemistry of Transuranium Elements*, Verlag Chemie, Weinheim 1971, Chapter 8, p. 187 (Organometallic compounds).

T.J. Marks, *Science, 217* (1982) 989 (Organoactinides).

P.J. Fagan, E.A. Maatta, J.M. Manriquez, K.G. Moloy, A.M. Seyam, T.J. Marks, in *Actinide Prospectives*, N.M. Edelstein, ed., Pergamon Press, Oxford (1982), p. 433 (Organoactinides).

B. Kanellakopulos, in *Organometallics of the f Elements*, T.J. Marks, R.D. Fischer, eds., D. Reidel, Doedrecht (1979), p. 1 (Cyclopentadienyl compounds of the actinides).

P.J. Fagon, J.M. Manriquez, T.J. Marks, in *Organometallics of the f Elements*, T.J. Marks, R.D. Fischer, eds., D. Riedel, Doedrecht (1979), p. 113 (Properties of actinide-to-carbon-σ-bonds).

T.J. Marks, *Prog. Inorg. Chem., 25* (1979) 224 (Actinide organometallics).

T.J. Marks, *Acc. Chem. Res.*, *9* (1976) 223 (Actinide organometallic chemistry: importance of 5f electrons in bonding).
T.J. Marks, in *Inorganic Compounds with Unusual Properties*, R.B. King, ed., Advances in Chemistry Series, No. 150, American Chemical Society, Washington, DC (1976), p. 232 (Organoactinide chemistry: coordination patterns).
T.J. Marks, in *Organotransition-Metal Chemistry*, Y. Ishii, M. Tsutsui, eds., *Proc. Jpn.-Am. Semin.* (1974), Plenum Press, New York (1973), p. 81 (Organoactinides: coordination patterns and chemical reactivity).
R.N. Shchelokov, O.T. Bolotova, N.S. Vyazankin, in *Methody Elementoorganicheskoi Khimii. Podgrupi Medi, Skandiya, Titana, Vanadiya, Khroma i Margantsa, Lantanoidy i Aktinoidy*, A.N. Nesmeyanov, K.A. Kocheshkov, eds., Nauka, Moscow (1974) (In Russian) (Organometallic compounds of actinides).
E. Cernia, A. Mazzei, *Inorg. Chim. Acta Rev.*, *10* (1974) 239 (Organometallic chemistry of uranium).
A. Streitwieser, Jr., in *Topics in Nonbenzoid Aromatic Chemistry*, Vol. 1 (1973) 221 (Uranocene: f-orbital aromatic system derived from cyclooctatetraene).
G.T. Seaborg, *Pure Appl. Chem.*, *30* (1972) 539 (Organometallic compounds of actinides).

c) Group IVB: Titanium, Zirconium, Hafnium

E. Müller, O. Bayer, H. Meerwein, K. Ziegler, eds., *Methoden der Organischen Chemie (Houben-Weyl)*, Band XIII/7. *Metallorganische Verbindungen: Ti, Zr, Hf*, by G. Bähr, E. Langer, A. Segnitz, Thieme Verlag, Stuttgart (1975).
P.C. Wailes, R.S.P. Coutts, H. Weigold, *Organometallic Chemistry of Titanium, Zirconium and Hafnium*, Academic Press, New York (1974).

B. Weidmann, D. Seebach, *Angew. Chem., Int. Ed. Engl.*, *22* (1983) 12 (Organo-Ti and Zr).
G.E. Toogood, M.G.H. Wallbridge, *Adv. Inorg. Chem. Radiochem.*, *25* (1982) 267 (Ti group-metal hydrides).
E. Negishi, *Pure Appl. Chem.*, *53* (1981) 2333 (Bimetallic catalysts containing Ti and Zr applied to selective organic synthesis).
G.P. Pez, J.N. Armor, *Adv. Organomet. Chem.*, *19* (1981) 1 (Titanocene and zirconocene).
M.L.H. Green, *J. Organomet. Chem.*, *200* (1980) 119 (Group-IVB zerovalent arenes).
G.L. Soloveichik, B.M. Bulychev, *Russ. Chem. Rev. (Engl. Transl).*, *51* (1982) 507 (Mono-transition-metal hydrides).
A. Müller, R. Josters, F.A. Cotton, *Angew. Chem., Int. Ed. Engl.*, *19* (1980) 875 (Trinuclear clusters).
M.D. Rausch, W.H. Boon, H.G. Att, *Ann. N.Y. Acad. Sci.*, *295* (1977) 103 (Photochemical studies on organic derivatives of Ti, Zr, Hf).
R.R. Schrock, G.W. Parshall, *Chem. Rev.*, *76* (1975) 243 (σ-Alkyl and aryl complexes).
E.M. Larsen, *Adv. Inorg. Chem. Radiochem.*, *13* (1970) 1 (Zr and Hf chemistry; Compounds containing M-C bonds).

i) Titanium

Gmelin Handbuch der Anorganischen Chemie. Organotitanium Compounds. Part 3, *Mononuclear Compounds* 3 (1983); Part 4, *Mononuclear Compounds 4* Springer Verlag, Berlin (1984).

Gmelin Handbook der Anorganischen Chemie. Organotitanium Compounds. Part 2, *Mononuclear Compounds* 2, Springer Verlag, Berlin (1980).
Gmelin Handbuch der Anorganischen Chemie. Organotitanium Compounds. Part 1, *Mononuclear Compounds* 1, Springer Verlag, Berlin (1977).
A.N. Nesmeyanov, K.A. Kocheshkov, eds., *Methody Elementoorganicheskoi Khimii. Podgrupy Medi, Skandiya, Titana, Vanadiya, Khroma i Margantsa, Lantanoidy i Aktinoidy*, Nauka, Moscow (1974) (In Russian) (Organotitanium compounds).
R.J.H. Clark, *The Chemistry of Titanium and Vanadium*, Elsevier, Amsterdam (1968).
R. Feld, P.L. Cowe, *Organic Chemistry of Titanium*, Plenum Press, New York (1965).

C.S. Rondestvedt, in *Kirk-Othmer Encyclopedia of Chemical Technology*, 3rd ed., M. Grayson, D. Eckroth, eds., Wiley, New York (1983), Vol. 23, p. 176.
M.T. Reetz, *Top. Curr. Chem.*, *106* (1982) 1 (Organotitanium reagents for organic synthesis).
G. Pez, J.N. Armor, *Adv. Organomet. Chem.*, *19* (1981) 2 (Titanocene).
R.J.H. Clark, S. Moorhouse, J.A. Stockwell, *J. Organomet. Chem. Libr.*, *3* (1977) 223 (Organometallic chemistry of Ti).
R.S.P. Coutts, P.C. Wailes, *Adv. Organomet. Chem.*, *9* (1970) 135 (Organic compounds of lower-valent Ti).
G.A. Razuvaev, V.N. Latyaeva, *Organomet. Chem. Rev.*, *2* (1967) 349 (Covalent organic compounds of Ti).
I. Shiihara, W.T. Schwartz, H.W. Post, *Chem. Rev.*, *61* (1961) 1 (Organic chemistry of Ti).
A.L. Suvorov, S.S. Spaskii, *Russ. Chem. Rev. (Engl. Transl.)*, *28* (1959) 1267 (Organic compounds of Ti).

ii) Zirconium

Gmelin Handbuch der Anorganischen Chemie, Organozirconium Compounds (New Supplement Series, Vol. 10), Springer Verlag, Berlin (1973).

P.T. Wolczanski, J.E. Bercaw, *Acc. Chem. Res.*, *13* (1980) 121 (Carbon-monoxide reduction with Zr hydrides).
J. Schwartz, *Pure Appl. Chem.*, *52* (1980) 733 (Organozirconium reagents).
J.R. Bercaw, in *Transition-Metal Hydrides*, R. Bau, ed., Advances in Chemistry Series, No. 167, American Chemical Society, Washington, DC (1978), p. 136 ($(\eta^5\text{-}C_5Me_5)_2ZrH_2$).
D.B. Carr, M. Yashifuji, L.I. Shoer, K.I. Gell, J. Schwartz, *Ann. N.Y. Acad. Sci.*, *295* (1977) 127 (Organozirconium reagents in organometallic synthesis).
J. Schwartz, J.A. Labinger, *Angew. Chem., Int. Ed. Engl.*, *15* (1976) 333 (Hydrozirconation: organic syntheses with a new organometallic reagent).
E.M. Brainina, in *Methody Elementoorganicheskoi Khimii, Podgrupy Medi, Skandiya, Titana, Vanadiya, Khroma i Margantsa, Lantanoidy i Atkinoidy*, A.N. Nesmeyanov, K.A. Kocheskov, eds., Nauka, Moscow (1974), p. 320 (In Russian) (Organozirconium compounds).

iii) Hafnium

Gmelin Handbuch der Anorganischen Chemie. Organohafnium Compounds (New Supplement Series, Vol. 11), Springer Verlag, Berlin (1973).

E. M. Brainina, in *Methody Elementoorganicheskoi Khimii. Podgrupy Medi, Skandiya, Titana, Vanadiya, Khroma i Margantsa, Lantanoidy i Aktinoidy*, A. N. Nesmeyanov, K. A. Kocheshkov, eds., Nauka, Moscow (1974), p. 373 (In Russian) (Organohafnium compounds).

d) Group VB: Vanadium, Niobium, Tantalum

E. Müller, O. Bayer, H. Meerwein, K. Ziegler, eds., *Methoden der Organischen Chemie (Houben-Weyl)*, Band XIII/7, *Metallorganische Verbindungen: V, Nb, Ta*, by G. Bähr, E. Langer, A. Stegnitz, Thieme Verlag, Stuttgart (1975).
F. Fairbrother, *Chemistry of Niobium and Tantalum*, Elsevier, Amsterdam (1967).

G. E. Toogood, M. G. H. Wallbridge, *Adv. Inorg. Chem. Radiochem.*, *25* (1982) 267 (Group-V metal hydrides).
A. Müller, R. Josters, F. A. Cotton, *Angew. Chem., Int. Ed. Engl.*, *19* (1980) 875 (Trinuclear clusters).
R. R. Schrock, *Acc. Chem. Res.*, *12* (1979) 98 (Alkylidene complexes of Nb and Ta).
R. R. Schrock, G. W. Parshall, *Chem. Rev.*, *76* (1976) 243 (σ-Alkyl and aryl complexes).
R. C. Mehrotra, A. K. Rai, P. N. Kapoor, R. Bohra, *Inorg. Chim. Acta*, *16* (1976) 237 (Organic derivatives of Nb and Ta).

i) Vanadium

Gmelin Handbuch der Anorganischen Chemie, Organovanadium Compounds (New Supplement Series, Vol. 2), Springer Verlag, Berlin (1971).
R. J. H. Clark, *The Chemistry of Titanium and Vanadium*, Elsevier, Amsterdam (1968).

G. Erker, *Pure Appl. Chem.*, *55* (1983) 103 (Carbonylation of zirconocenes).
G. Pez, J. N. Armor, *Adv. Organomet. Chem.*, *19* (1981) 2 (Vanadocene).
A. A. Pasynskii, in *Methody Elementoorganicheskoi Khimii. Podgrupy Medi, Skandiya, Titana, Vanadiya, Khroma i Margantsa, Lantanoidy i Aktinoidy*, A. N. Nesmeyanov, K. A. Kocheshkov, eds., Nauka, Moscow (1974), p. 389 (In Russian) (Organovanadium compounds).

ii) Niobium

Gmelin Handbuch der Anorganischen Chemie. Niobium. Part B. Section 4. *Carbon Compounds of Niobium*, Springer Verlag, Berlin (1973).

A. A. Pasynskii, in *Methody Elementoorganicheskoi Khimii. Podgrupy Medi, Skandiya, Titana, Vanadiya, Khroma i Margantsa, Lantanoidy i Aktinoidy*, A. N. Nesmeyanov, K. A. Kocheshkov, eds., Nauka, Moscow (1974), p. 434 (In Russian) (Organoniobium compounds).

iii) Tantalum

Gmelin Handbuch der Anorganischen Chemie. Tantalum. Part B. Section 2. *Complex Compounds of Tantalum*, Springer Verlag, Berlin (1971).

R. R. Schrock, S. J. McLain, J. Saneho, *Pure Appl. Chem.*, *52* (1980) 729 (Tantalacyclopentanes in the catalytic dimerization of olefins).
A. A. Pasynskii, in *Methody Elementoorganicheskoi Khimii. Podgrupy Medi, Skandiya, Titana, Vandiya, Khroma i Margantsa, Lantanoidy i Aktinoidy*, A. N. Nesmeyanov, K. A. Kocheshkov, eds., Nauka, Moscow (1974), p. 453 (In Russian) (Organotantalum compounds).

e) Group VIB: Chromium, Molybdenum, Tungsten

E. Müller, O. Bayer, H. Meerwein, K. Ziegler, eds., *Methoden der Organischen Chemie (Houben-Weyl)*, Band XIII/7. *Metallorganische Verbindungen: Cr, Mo, W*, by G. Bähr, E. Langer, A. Stegnitz, Thieme Verlag, Stuttgart (1975).

M. H. Chisholm, *Polyhedron, 2* (1983) 681 (Organomolybdenum and tungsten alkoxides).
M. H. L. Green, *J. Organomet. Chem.*, *200* (1980) 119 (Group-VI zerovalent arenes).
A. Müller, R. Josters, F. A. Cotton, *Angew. Chem., Int. Ed. Engl.*, *19* (1980) 875 (Trinuclear clusters).
M. H. Chisholm, in *Inorganic Compounds with Unusual Properties-II*, R. B. King, ed., Advances in Chemistry Series, No. 173, American Chemical Society, Washington, DC (1979), p. 396 (Metal-metal triple bonds in Mo and W chemistry).
M. H. Chisholm, F. A. Cotton, *Acc. Chem. Res.*, *11* (1978) 356 (Metal-metal triple bonds between Mo and W).
M. L. H. Green, *Pure Appl. Chem.*, *50* (1978) 27 (Organomolybdenum and tungsten synthesis, mechanism and reactivity).
M. F. Farona, *Organomet. React. Synth.*, *6* (1977) 223 (Homogeneous catalysis by arene-group VI tricarbonyls).
R. R. Schrock, G. W. Parshall, *Chem. Rev.*, *76* (1976) 243 (σ-Alkyl and aryl complexes).
K. W. Barnett, D. W. Slocum, *J. Organomet. Chem.*, *44* (1972) 1 (Cyclopentadienyl compounds of Cr, Mo and W).
R. Colton, *Coord. Chem. Rev.*, *6* (1971) 269 (Steric effects in Mo halocarbonyls).

i) Chromium

Gmelin Handbuch der Anorganischen Chemie. Organochromium Compounds (New Supplement Series, Vol. 3), Springer Verlag, Berlin (1973).

J. H. Espenson, *Prog. Inorg. Chem.*, *30* (1983) 189 (Homolytic and radical pathways for organochromium reactions).
J. Kalousova, J. Holecek, J. Votinsky, L.Benes, *Z. Chem.*, *23* (1983) 327 (Chromocene reactions).
J. H. Espensen, *Adv. Inorg. Bioinorg. Mech.*, *1* (1982) 1 (Organochromium(III) reactions and mechanisms).

D.J. Darensbourg, M.Y. Darensbourg, R.R. Buroh, Jr., J.A. Froelich, M.J. Incorvia, in *Inorganic Compounds with Unusual Properties-II*, R.B. King, ed., Advances in Chemistry Series, No 173, American Chemical Society, Washington, DC (1980), p. 106 (Oxygen exchage and ligand substitution in Cr carbonyls).
A.N. Artemov, *Khim. Elementoorg. Soedin.*, (1977) 17 (Organometallic derivatives of benzenechromium tricarbonyl).
G. Jaouen, *Ann. N.Y. Acad. Sci.*, *295* (1977) 58 (Use of arene chromium tricarbonyl complexes in organic synthesis).
G.K. Magomedov, in *Methody Elementoorganicheskoi Khimii. Podgrupy Medi,Skandiya, Titana, Vanadiya, Khroma i Margantsa, Lantanoidy i Actinoidy*, A.N. Nesmeyanov, K.A. Kocheshkov, eds., Nauka, Moscow (1974), p. 477 (In Russian) (Organochromium compounds).
M.A. Bennett, in *Rodd's Chemistry of Carbon Compounds*, 2 nd ed., Vol. 13, Part B, S. Coffey, ed., Elsevier, Amsterdam (1974), p. 357 (Aromatic compounds of transistion metals).
H.H. Zeiss, R.P.A. Sneeden, *Angew. Chem., Int. Ed. Engl.*, *6* (1967) 435 (σ-Bonded organochromium compounds).
M. Tsutsui, *Z. Chem.*, *3* (1963) 215 (Bis-arene chromiums).

ii) Molybdenum

D. Coucouvanis, *Acc. Chem. Res.*, *14* (1981) 201 (The Mo site in nitrogenase).
M.L.H. Green, *J. Less-Common Met.*, *54* (1977) 159 (New synthetic routes in organomolybdenum chemistry).
J.W. Faller, A.M. Rosan, *Ann. N.Y. Acad. Sci.*, *295* (1977) 186 (Cationic η^4-diene and η^3-allyl complexes of Mo).
E.I. Steifel, *Prog. Inorg. Chem.*, *22* (1977) 1 (Coordination and bioinorganic chemistry of Mo).
S.J. Lippard, *Prog. Inorg. Chem.*, *21* (1976) 91 (Seven- and eightcoordinated Mo complexes and Mo(IV) oxocomplexes with cyanide and isocyanide ligands).
E.F. Ashworth, M.L.H. Green, J. Knight, *J. Less-Common Met.*, *36* (1974) 213 (Cycloheptatrienyl-Mo chemistry).
G.K. Magomedov, in *Methody Elementoorganicheskoi Khimii. Podgrupy Medi, Skandiya, Titana, Vanadiya, Khroma i Margantsa, Lantanoidy i Atkinoidy*, A.N. Nesmeyanov, K.A. Kocheshkov, eds., Nauka, Moscow (1974), p. 577 (In Russian) (Organomolybdenum compounds).

iii) Tungsten

J.C. Hayes, P. Jernakoff, G.A. Miller, J.N. Cooper, *Pure Appl. Chem.*, *56* (1984) 25 (Cationic W alkylidenes and alkyls).
F.G.A. Stone, in *Inorganic Chemistry Toward the 21st Century*, M.H. Chisholm, ed., ACS Symposium Series, No. 211, American Chemical Society, Washington, DC (1983), p. 383 (The alkylidyne reagent η^5-$C_5H_5(OC)_2WCC_6H_4CH_3$-4 for synthesizing W bonds to other metals).
Z. Dori, *Prog. Inorg. Chem.*, *28* (1981) 234 (Coordination chemistry of W).
G.K. Magomedov, in *Methody Elementoorganicheskoi Khimii. Podgrupy Medi, Skandiya, Titana, Vanadiya, Khroma i Margantsa, Lantanoidy i Actinoidy*, A.N. Nesmeyanov, K.A. Kocheshkov, eds., Nauka, Moscow (1974) (In Russian) (Organotungsten compounds).

f) Group VIIB: Manganese, Technetium, Rhenium

E. Müller, O. Bayer, H. Meerwein, K. Ziegler, eds., *Methoden der Organischen Chemie (Houben-Weyl)*, Vol. XIII/9, *Metalloorganische Verbindungen der VII. und VIII. Nebengruppe des Periodensystems: Mn, Tc, Re*, by H.F. Klein, E. Langer, H. Segnitz, K. von Werner, G. Thieme Verlag, Stuttgart (1985).

R.D. Peacock, *Chemistry of Technetium and Rhenium*, Elsevier, Amsterdam (1965).

K.G. Caulton, *Coord. Chem. Rev.*, *38* (1981) 1 (Coordination chemistry of the fragment η^5-$C_5H_5M(CO)_2$ (M = Mn, Re)).

R.R. Schrock, G.W. Parshall, *Chem. Rev.*, *76* (1976) 243 (σ-Alkyl and aryl complexes of group VII).

i) Manganese

K.G. Caulton, *Coord. Chem. Rev.*, *38* (1981) 1 (η^5-$CpMn(CO)_2$ coordination).

K.N. Anisimov, A.A. Joganson, N.E. Kolobova, *Russ. Chem. Rev. (Engl. Transl.)*, *37* (1978) 184 (Organomanganese compounds).

A.A. Joganson, K.N. Anisimov, N.E. Kolobova, in *Methody Elementoorganicheskoi Khimi. Podgrupy Medi, Skandiya, Titana, Vanadiya, Khroma i Margantsa, Lantanoidy i Aktinoidy*, A.N. Nesmeyanov, K.A. Kocheshkov, eds., Nauka, Moscow (1974); p. 692 (Organomanganese compounds). K.N. Anisimov, Z.P.Valueva, p. 781 (In Russian) (Cyclopentadienyltricarbonylmanganese, cymantrene).

ii) Technetium

K.N. Anisimov, N.E. Kolobova, in *Methody Elementoorganicheskoi Khimii. Podgrupy Medi, Skandiya, Titana, Vanadiya, Khroma i Margantsa, Lantanoidy i Aktinoidy,* A.N. Nesmeyanov, K.A. Kocheshkov, eds., Nauka, Noscow (1974), p. 851 (Organotechnetium compounds; in Russian).

iii) Rhenium

K.G. Caulton, *Coord. Chem. Rev.*, *38* (1981) 1 (η^5-$CpRh(CO)_2$ coordination).

A.A. Joganson, K.N. Anisimov, N.E. Kolobova, in *Methody Elementoorganicheskoi Khimii. Podgrupy Medi, Skandiya, Titana, Vanadiya, Khroma i Margantsa, Lantanoidy i Actinoidy*, A.N. Nesmeyanov, K.A. Kocheshkov, eds., Nauka, Moscow (1974), p. 858 (Organorhenium compounds; in Russian).

G. Rouschias, *Chem. Rev.*, *74* (1974) 531 (Recent advances in Re chemistry, including organometallic compounds).

J.E. Fergusson, *Coord. Chem. Rev.*, *1* (1966) 459 (Coordination chemistry of Re).

g) Group VIII: Iron, Ruthenium, Osmium

E. Müller, O. Bayer, H. Meerwein, K. Ziegler, eds., *Methoden der Organischen Chemie (Houben-Weyl)*, Vol. XIII/9 *Metallorganische Verbindungen der VII. und VIII. Nebengruppe des Periodensystems: Fe, Ru, Os* by H.F. Klein, E. Langer, H. Stegnitz, G. Thieme Verlag, Stuttgart (1985).

W.P. Griffith, *The Chemistry of the Rarer Platinum Metals, (Os, Ru, Ir and Rh)*, Interscience, New York (1967) (Including organometallic compounds).

M. Rosenblum, *Chemistry of the Iron-Group Metallocenes: Ferrocene, Ruthenocene, Osmocene*, Wiley, New York (1965).

B.F.G. Johnson, J. Lewis, W.J.H. Nelson, J.N. Nicholls, M.D. Vargas, *J. Organomet. Chem., 249* (1983) 255 (Re- and Os-carbido complexes).

E. Sappa, A. Tiripicchio, P. Braunstein, *Chem. Rev., 83* (1983) 203 (Fe triad alkyne-carbonyl clusters).

L.S. Hegedus, *Mod. Synth. Methods, 3* (1983) 61 (Group-VIII metals in organic synthesis).

A.J. Carty, in *Catalytic Aspects of Metal-Phosphine Complexes*, E.C. Alyea, D.W. Meek, eds., Advances in Chemistry Series, No. 196, American Chemical Society, Washington, DC (1982), p. 163 (Phosphido-bridged iron-group clusters).

J. Lewis, B.F.G. Johnson, *Pure Appl. Chem., 54* (1982) 97 (Carbonyl clusters of Ru and Os).

P.S. Pregosin, *Annu. Rev. NMR Spectrosc., 11 A* (1981) 227 (Carbon-13 NMR of group-VIII complexes).

G.P. Chiusoli, *Pure Appl. Chem., 52* (1980) 635 (Organic syntheses catalyzed by group-VIII complexes).

A. Bonny, *Coord. Chem. Rev., 25* (1978) 229 (Group-IV derivatives of the iron-triad carbonyls).

J.K. Stille, K.S.Y. Lau, *Acc. Chem. Res., 10* (1977) 434 (Oxidative-addition mechanisms of organic halides to group-VIII complexes).

G. Deganello, P. Uguagliati, L. Calligano, P.L. Sandrin, F. Zingales, *Inorg. Chim. Acta, 13* (1975) 247 (Polyolefin carbonyls of Fe, Ru, Os).

S.C. Tripathi, S.C. Srivastava, R.P. Mani, A.K. Shrimal., *Inorg. Chim. Acta, 15* (1975) 249 (Ru and Os carbonyls).

i) Iron

Gmelin Handbuch der Anorganischen Chemie. Vol 14. *Organoiron Compounds, Part A: Ferrocene and Its Derivatives,* Section 1: *Ferrocene, 1* (1974); Vol. 49, Section 2: *Ferrocene, 2* (1977); Vol. 50, Section 3: *Ferrocene, 3* (1978); Vol. 41, Section 6: *Ferrocene, 6* (1977); Section 4: *Ferrocene, 4* (1980); Section 5: *Ferrocene, 5* (1981); Section 7: *Ferrocene 7* (1980). Vol. 36, *Ferrocene, 5* (1981); *Part B: Mononuclear Compounds* (Excluding Ferrocene) Section 1: Part 1 (1976); Section 2: Part 2 (1979); Section 2: Part 3 (1978); Section 3: Part 3 (1979); Section 4: Part 4 (1978); Section 5: Part 5 (1978); Section 6: Part 6 (1981); Section 7: Part 7 (1981) Section 11: Part 11 (1983). *Part C: Binuclear Compounds,* Section 1 (1979): Section 2 (1979); Section 3 (1980); Section 4 (1981); Section 11 (1983).

E.A. Koerner von Gustorf, F.W. Grevels, I. Fischer, eds., *The Organic Chemistry of Iron*, 2 Vols., Academic Press, New York (1978, 1981).

M. Rosenblum, *Chemistry of Iron-Group Metallocenes. Ferrocene, Ruthenocene, Osmocene*, Interscience, New York (1965).

S. B. Fergusson, L. J. Sanderson, T. A. Shackelton, M. C. Baird, *Inorg. Chim. Acta, 83* (1984) L45 (Reassessment of η^5-$C_5H_5Fe(CO)_2H$).

D. Astruc, *Tetrahedron, 39* (1983) 4027 (Applications of aromatic iron compounds in synthesis).

R. Desiderato, Jr., G. R. Dobson, *J. Chem. Educ., 59* (1982) 752 (The $Fe_3(CO)_{12}$ structure).

D. Hoppe, *Nachr. Chem. Tech. Lab., 30* (1982) 706, 711 (1,3-Dienetricarbonyliron).

M. I. Rybinskaya, *Pure Appl. Chem., 54* (1982) 145 (α-Carbenium stabilization in olefin-Fe complexes).

W. R. Cullen, J. D. Woollins, *Coord. Chem. Rev., 39* (1981) 1 (Ferrocene-containing phosphine and arsine metal complexes).

A. J. Pearson, *Trans. Met. Chem., 6* (1981) 67; *Acc. Chem. Res., 13* (1980) 463 (Tricarbonyl (diene)iron complexes in synthesis).

H.-F. Klein, *Angew. Chem., Int. Ed. Engl., 19* (1980) 362 (Trimethylphosphine complexes – models for homogeneous catalysis).

D. Mansuy, *Pure Appl. Chem., 52* (1980) 681 (Phosphyrines with an Fe-C bond).

D. Ballivet-Tkatchenko, in *Fundamental Research in Homogeneous Catalysis*, Vol. 3., M. Tsutsui, ed., Plenum Press, New York (1979), p. 257 (Zeolite-bound organoiron Fischer-Tropsch reactivity).

S. D. Ittel, in *Fundamental Research in Homogeneous Catalysis*, Vol. 3, M. Tsutsui, ed., Plenum Press, New York (1979), p. 305 (C-H activation by low-valent Fe complexes).

S. D. Ittel, C. A. Tolman, A. D. Englisch, J. P. Jesson, in *Inorganic Compounds with Unusual Properties-II*, R. B. King, ed., Advances in Chemistry Series, No. 173, American Chemical Society, Washington, DC (1979), p. 67 (Activation of C–H bonds by bidentate phosphine Fe complexes).

C. U. Pittman, Jr., W. D. Honnick, M. S. Wrighton, R. D. Sanner, R. G. Austin, in *Fundamental Research in Homogeneous Catalysis*, Vol. 3, M. Tsutsui, ed., Plenum Press, New York (1979), p. 603 (Photogeneration of polymer-anchored catalysts from iron carbonyls).

W. E. Watts, *J. Oranomet. Chem. Libr., 7* (1979) 401 (Ferrocenyl carbocations).

A. N. Nesmeyanov, M. I. Rybinskaya, L. V. Rybin, *Russ. Chem. Rev. (Engl. Transl.), 48* (1979) 213 (Dinuclear Fe carbonyls with nitrogen-containing bridges).

M. Poliakoff, *Chem. Soc. Rev., 7* (1978) 527 ($Fe(CO)_4$).

T. Hayashi, M. Kumada, in *Fundamental Research in Homogeneous Catalysis*, Vol. 2, Y. Ishii, M. Tsutsui, eds., Plenum Press, New York (1978), p. 159 (Chiral ferrocenyl phosphines in asymmetric catalysis).

R. B. King, in *The Organic Chemistry of Iron*, E. A. Koerner von Gustorf, F. W. Grevels, I. Fischler, eds., Academic Press, New York (1978), p. 463 (Fe Allyls).

R. G. Sutherland, *J. Organomet. Chem. Libr., 3* (1977) 311 (π-Arene-π-cyclopentadienyliron cations).

R. Noyori, *Ann. N. Y. Acad. Sci., 295* (1977) 225 (Iron carbonyls in organic synthesis).

M. Brookhart, C. R. Graham, G. O. Nelson, G. Scholes, *Ann. N. Y. Acad. Sci., 295* (1977) 254 (Chemistry of diene and enone iron-tricarbonyl complexes).

E. Weissburger, D. Laszlo, *Acc. Chem. Res., 9* (1976) 209 (Stereospecific cyclic ketone formation by interligand reaction with Fe(0).

C. U. Pittman, Jr., B. Surynarayanan, Y. Sasaki, in *Inorganic Compounds with Unusual Properties*, R. B. King, ed., *Advances in Chemistry Series,* No. 150, American Chemical Society, Washington, DC (1976), p. 46 (Semiconducting ferrocene polymers).

K. Schlögl, H. Falk, in *Methodicum Chimicum*, Vol. 8, *Preparation of Transition Metal Derivatives*, K. Niedenzu, H. Zimmer, eds., G. Thieme Verlag, Stuttgart (1976), p. 469 (Ferrocene); H. Rosenberg, p. 500 (Sandwich compounds, metallocenes).

J. P. Collman, *Acc. Chem. Res., 8* (1975) 342 (Disodium tetracarbonylferrate – a transition-metal Grignard).

R. Petit, *J. Organomet. Chem., 100* (1975) 205 (Cyclobutadienyliron tricarbonyl).
G. Wilkinson, *J. Organomet. Chem., 100* (1975) 273 (Fe sandwich compounds – the first four months).
E.A. Sullivan, T.F. Jula, *Int. Lab.,* (1974) 51 (Disodium tetracarbonylferrate as reagent in organic synthesis).
A.N. Nesmeyanov, N.S. Kochetkova, *Russ. Chem. Rev. (Engl. Transl.), 43* (1974) 710 (Practical applications of ferrocenes).
G.B. Shul'pin, M.I. Rybinskaya, *Russ. Chem. Rev. (Engl. Transl.), 43* (1974) 716 (Ferrocenophanes).
M. Rosenblum, *Acc. Chem. Res., 7* (1974) 122 (Organoiron complexes as reagents in organic synthesis).
R. Morassi, I. Bertini, L. Sacconi, *Coord. Chem. Rev., 11* (1973) 343 (Fe(II) five-coordinated complexes).
G. Davidson, *Organomet. Chem. Rev., A, 8* (1972) 303 (Vibrational spectra of π-organotransition-metal complexes).
P. Chini, A. Cavalieri, S. Martinengo, *Coord. Chem. Rev., 8* (1972) 3 (Mixed cluster-carbonyl anions containing Ni and Co).
M.D. Rausch, *Pure Appl. Chem., 30* (1972) 523 (Metal-cyclopentadienyl complexes).
D.E. Bublitz, K.E. Rinehart, *Org. React., 17* (1969) 1 (Synthesis of substituted ferrocenes and related metallocenes).
A.N. Nesmeyanov, *Pure Appl. Chem., 17* (1968) 211 (Organic chemistry of ferrocenes).
E.W. Neuse, *Adv. Macromol. Chem., 1* (1968) 1 (Ferrocene polymers).
F.G. Emerson, K. Ehrlich, W.P. Giering, D. Ehntholt, *Trans. N.Y. Acad. Sci., 30* (1968) 1001 (The chemistry of iron-olefin complexes).
H.J. Lorkowski, *Top. Curr. Chem., 9* (1967) 207 (Ferrocene as building unit in polymers).
W.E. Watts, *Organomet. Chem. Rev., 2* (1967) 231 (Ferrocenophane systems).
A.N. Nesmeyanov, E.G. Perevalova, *Ann. N.Y. Acad. Sci., 125* (1965) 67 (Organic chemistry of ferrocene).
R. Pettit, *Ann. N.Y. Acad. Sci., 125* (1965) 89 (Chemical properties of olefin π-complexes of iron).
. Pettit, G.F. Emerson, *Adv. Organomet. Chem., 1* (1964) 1 (Diene-iron-carbonyl complexes and related species).
R. Pettit, G.F. Emerson, J. Mahler, *J. Chem. Educ., 40* (1963) 175 (Chemistry of diene-iron tricarbonyl complexes).
L. Plesske, *Angew. Chem., Int. Ed. Engl., 1* (1962) 301 (Ring substitutions in ferrocene).

ii) Ruthenium

K.R. Seddon, E.A. Seddon, *The Chemistry of Ruthenium,* Elsevier, Amsterdam (1984).

S.A.R. Knox, *Pure Appl. Chem., 56* (1984) 81 (Di- and triruthenium complexes).
F.H. Jardine, *Prog. Inorg. Chem., 31* (1984) 265 ($Cl_2[(C_6H_5)_3P]_3Ru(II)$ and its derivatives).
P.C. Ford, *Acc. Chem. Res., 14* (1981) 31 (Water-gas shift reaction catalysis).
B.R. James, in *Transition-Metal Hydrides,* R. Bau, ed., Advances in Chemistry Series, No. 167, American Chemical Society, Washington, DC (1978), p. 122 (Ru-hydrides with chiral ligands in asymmetric hydrogenation).
A.A. Koridze, in *Methody Elementoorganicheskoi Khimii. Kobal't, Nikel', Platinovye Metaly,* A.N. Nesmeyanov, K.A. Kocheshkov, eds., Nauka, Moscow (1978), p. 246 (Organoruthenium compounds).

B. R. James, in *Fundamental Research in Homogeneous Catalysis,* Vol. 2, Y. Ishii, M. Tsutsui, eds., Plenum Press, New York (1978), p. 35 (Activation of hydrocarbons by Ru(I)).

J. Catterick, P. Thornton, *Adv. Inorg. Chem. Radiochem., 20* (1977) 189 (Hydride complexes of Ru, Rh and Ir).

B. R. James, *Inorg. Chim. Acta Rev., 4* (1970) 73 (Organoruthenium compounds in catalysis).

M. I. Bruce, F. G. A. Stone, *Angew. Chem., Int. Ed. Engl., 7* (1968) 427 (Dodecacarbonyltriruthenium).

iii) Osmium

D. S. Moore, *Coord. Chem. Rev., 44* (1982) 127 (Alkylosmiums).

J. R. Norton, *Acc. Chem. Res., 12* (1979) 139 (Elimination mechanisms of organoosmium compounds).

A. A. Koridze, in *Methody Elementoorganicheskoi Khimii. Kobal't, Nikel, Platinovye Metaly,* A. N. Nesmeyanov, K. A. Kocheshkov, eds., Nauka, Moscow (1978), p. 315 (In Russian) (Organoosmium compounds).

J. R. Norton, in *Transition-Metal Hydrides,* R. Bau, ed., Advances in Chemistry Series, No. 167, American Chemical Society, Washington, DC (1978), p. 170 (Os hydridoalkyls and their elimination mechanisms).

h) Group VIII: Cobalt, Rhodium, Iridium

E. Müller, O. Bayer, H. Meerwein, K. Ziegler, eds., *Methoden der Organischen Chemie (Houben-Weyl)*, Vol. XIII/9, *Metallorganische Verbindungen der VII. und VIII. Nebengruppe des Periodensystems: Co, Rh, Ir* (by H. F. Klein, E. Langer, H. Stegnitz, K. von Werner), G. Thieme Verlag, Stuttgart (1985).

L. S. Hegedus, *Mod. Synth. Methods, 3* (1983) 61 (Group-VIII metals in organic synthesis).

E. Sappa, A. Tiripicchio, P. Braunstein, *Chem. Rev., 83* (1983) 203 (Cobalt triad alkyne-carbonyl clusters).

D. Forster, A. Hershman, D. E. Morris, *Catal. Rev., 23* (1981) 89 (Mechanisms in Rh, Ir and Co complex catalysis of hydrocarboxylation).

B. R. James, in *Biomimetic Chemistry,* D. Dolphin, C. McKenna, Y. Murakami, I. Tabushi, eds., Advances in Chemistry Series, No. 191, American Chemical Society, Washington, DC (1980), p. 253 (Activation of O_2 by group-VIII complexes).

G. P. Chiusoli, *Pure Appl. Chem., 52* (1980) 635 (Organic syntheses catalyzed by group-VIII complexes).

J. K. Stille, K. S. Y. Lau, *Acc. Chem. Res., 10* (1977) 434 (Oxidative-addition mechanisms of organic halides to group-VIII complexes).

i) Cobalt and Rhodium

G. W. Parshall, *Catal. Rev., 23* (1981) 107 (Commercial reactions catalyzed by soluble Co and Rh complexes).

P. Pino, *J. Organomet. Chem., 200* (1980) 223 (Hydroformylation of olefins by Co and Rh catalysts).
D.W. Meek, in *Inorganic Compounds with Unusual Properties*, R.B. King, ed., Advances in Chemistry Series, No. 150, American Chemical Society, Washington, DC (1976), p. 335 (Co(I) and Rh(I) polyphosphine complexes).

ii) Rhodium and Iridium

R.S. Dickson, *Organometallic Chemistry of Rhodium and Iridium,* Academic Press, New York (1983).

A.H. Janowicz, R.A. Periana, M.J. Buchanan, C.A. Ovak, J.M. Stryker, M.J. Wax, R.G. Bergman, *Pure Appl. Chem., 56* (1984) 13 (Oxidative addition of Rh and Ir complexes to aliphatic C-H bonds).
P.M. Maitlis, *Coord. Chem. Rev., 43* (1982) 377 (New Rh and Ir reactions).
J.A. Davies, F.R. Hartley, *Chem. Rev., 81* (1981) 79 (Pt-metal catalysts with weak donor ligands).
D. Forster, *Adv. Organomet. Chem., 17* (1979) 255 (Mechanisms of catalytic carbonylation of methanol by Rh and Ir complexes).
P.M. Maitlis, in *Inorganic Compounds with Unusual Properties – II,* R.B. King, ed., Advances in Chemistry Series, No. 173, American Chemical Society, Washington, DC (1979), p. 31 (η^5-Me_5C_5Rh and Ir catalysts for olefin and arene hydrogenation).
P.M. Maitlis, *Acc. Chem. Res., 11* (1978) 301 (Rh and Ir pentamethylcyclopentadienyls).
D.M. Roundhill, in *Fundamental Research in Homogeneous Catalysis,* Vol. 2, Y. Ishii, M. Tsutsui, eds., Plenum Press, New York (1978), p. 11; Vol. 3, p. 11 (Rh- and Ir-cluster catalyzed oxidation of ketones and CO).
S.C. Tripathi, S.C. Srivistava, R.P. Mani, A.K. Shrimal, *Inorg. Chim. Acta, 17* (1976) 257 (Rh and Ir carbonyls).
P.M. Maitlis, *Ann. N.Y. Acad. Sci., 33* (1971) 87 (Pentamethylcyclopentadienyl Rh and Ir derivatives).

iii) Cobalt

Gmelin Handbuch der Anorganischen Chemie. Organocobalt Compounds. Part 1. *Mononuclear Compounds* (New Supplement Series, Vol. 5) (1973); Part 2. *Polynuclear Compounds* (New Supplement Series, Vol. 6), Springer Verlag, Berlin (1973).

P.J. Toscano, L.G. Marzilli, *Prog. Inorg. Chem. 31* (1984) 105 (Co-C bonds in vitamin-B12).
J.E. Sheats, G. Hlatky, *J. Chem. Educ., 60* (1983) 1015 (Co sandwich compounds).
K.L. Brown, in *B-12*, Vol. 1, D. Dolphin, ed., Wiley, New York (1982), p. 245 (Organocobalt complexes).
T. Funabiki, *Rev. Inorg. Chem., 4* (1982) 329 (Olefin complexes of Co(I)).
M. Orchin, *Acc. Chem. Res., 14* (1981) 259 ($HCo(CO)_4$).
D. Forster, A. Herschman, D.E. Morris, *Catal. Rev., 23* (1981) 89 (Mechanism of olefin-hydrocarboxylation catalysis by Co).

G. Palyi, G. Varadi, I.T. Horvath, *J. Mol. Catal., 13* (1981) 61 (Co-carbonyl activation of CO and acetylenes).

R.L. Funk, K.P.C. Vollhardt, *Chem. Soc. Rev., 9* (1980) 41 (Co-mediated co-oligomerizations of hexa-1,5-dienes).

A.W. Johnson, *Chem. Soc. Rev., 9* (1980) 125 (Vitamin B-12).

E.V. Leonova, *Russ. Chem. Rev. (Engl. Transl.), 49* (1980) 147 (Co carbaboranes).

T. Toraya, S. Fukui, in *Biomimetic Chemistry,* D. Dolphin, C. McKenna, Y. Murakami, I. Tabushi, eds., Advances in Chemistry Series, No. 191, American Chemical Society, Washington, DC (1980), p. 139 (Vitamin B-12 coenzyme in the diol-dehydrase system).

J. Halpern, in *Biomimetic Chemistry*, D. Dolphin, C. McKenna, Y. Murakami, I. Tabushi, eds., Advances in Chemistry Series, No. 191, American Chemical Society, Washington, DC (1980), p. 165 (Mechanisms of B-12-dependent rearrangements).

Y. Murakami, in *Biomimetic Chemistry,* D. Dolphin, C. McKenna, Y. Murakami, I. Tabushi, eds., Advances in Chemistry Series, No. 191, American Chemical Society, Washington, DC (1980), p. 179 (Vitamin B-12 models with macrocyclic ligands).

R.G. Bergman, *Acc. Chem. Res., 13* (1980) 113 (Binuclear dialkylcobalt complexes).

H.-F. Klein, *Angew. Chem., Int. Ed. Engl., 19* (1980) 362 (Trimethylphosphine complexes of Co as homogeneous-catalysis models).

J.E. Sheats, *J. Organomet. Chem. Libr. 7* (1979) 461 (Cobaltocene, cobalticinium salts and other sandwich compounds).

G.A. Ozin, *Coord. Chem. Rev., 28* (1979) 117 (Naked Co clusters as olefin chemisorption models).

E.V. Leonov, V.K. Syundyukova, in *Methody Elementoorganicheskoi Khimii. Kobal't, Nikel', Platinovye Metaly,* A.N. Nesmeyanov, K.A. Kocheshkov, eds., Nauka, Moscow (1978), p. 7 (Organocobalt compounds, comprehensive review in Russian).

G. Schmid, *Angew. Chem., Int. Ed. Engl., 17* (1978) 392 (Tetrahedral carbonyl-Co clusters).

H. Bönnemann, *Angew. Chem., Int. Ed. Engl., 17* (1978) 505 (Co-catalyzed pyridine synthesis from alkynes and nitriles); *12* (1973) 964 (Allycobalts).

A.I. Scott, *Acc. Chem. Res., 11* (1978) 29 (Biosynthesis of vitamin B-12).

G.L. Geoffroy, R.A. Epstein, in *Inorganic and Organometallic Photochemistry,* M.S. Wrighton, ed., Advances in Chemistry Series, No. 168, American Chemical Society, Washington, DC (1978), p. 132 (Photoinduced Co-carbonyl declusterification).

G.N. Schrauzer, *Angew. Chem., Int. Ed. Engl., 16* (1977) 233 (Enzymatic reactions dependent upon corrins and coenzyme B-12); *15* (1976) 417 (Reactions of Co in corrins and vitamin B-12 model compounds).

P.L. Pauson, I.V. Khand, *Ann. N.Y. Acad. Sci., 295* (1977) 2 (Use of Co-carbonyl acetylene complexes in organic synthesis).

D. Seyferth, *Adv. Organomet. Chem., 14* (1976) 97 (Carbonfunctional alkylidynetricobalt nonacarbonyl complexes).

R.H. Abeles, D. Dolphin, *Acc. Chem. Res., 9* (1976) 114 (Vitamin B-12 coenzyme).

B.M. Babior, *Acc. Chem. Res., 8* (1975) 376 (Cobalamin-dependent rearrangement mechanisms).

A.I. Scott, *Tetrahedron, 31* (1975) 2639 (Biosynthesis of vitamin B-12).

R.S. Dickson, P.J. Frazer, *Adv. Organomet. Chem., 12* (1974) 323 (Compounds from alkynes and Co carbonyls).

T.L. Brown, *Acc. Chem. Res., 7* (1974) 408 (Co-59 NQR).

B.R. Penfold, B.H. Robinson, *Acc. Chem. Res., 6* (1973) 73 (Tricobalt carbon).

H. Bönnemann, *Angew. Chem., Int. Ed. Engl., 12* (1973) 964 (Allylcobalts).

D.G. Brown, *Prog. Inorg. Chem., 18* (1973) 178 (Vitamin B-12).

D. Dodd, M.D. Johnson, *J. Organomet. Chem., 52* (1973) 1 (Organic compounds of Co).

E. V. Leonova, N. S. Kochetkova, *Russ. Chem. Rev. (Engl. Transl.)*, *42* (1973) 278 (Cobalticene reactions).
J. M. Pratt, P. J. Craig, *Adv. Organomet. Chem.*, *11* (1973) 331 (Preparation and reactions of organocobalt complexes).
H. Bönnermann, G. Wilke, *Angew. Chem., Int. Ed. Engl.*, *85* (1973) 1024 (The allylcobalt system).
R. Morassi, I. Bertini, L. Sacconi, *Coord. Chem. Rev.*, *11* (1973) 343 (Co(II) five-coordinated complexes).
J. M. Wood, D. G. Brown, *Struct. Bonding (Berlin)*, *11* (1972) 47 (Vitamin B-12 enzymes).
L. F. Elsom, J. D. Hunt, A. McKillop, *Organomet. Chem. Rev., A*, *8* (1972) 135 (Co(II) catalysis in Grignard reactions).
G. Costa, *Pure Appl. Chem.*, *30* (1972) 335 (Reactivity of metal-carbon σ-bond in cobalt chelates).
G. Costa, *Coord. Chem. Rev.*, *8* (1972) 63 (Organometallic derivatives of cobalt chelates).
G. Palyi, . Piacenti, L. Marko, *Inorg. Chim. Acta Rev.*, *4* (1970) 109 (Methinyltricobalt enneacarbonyl compounds).
A. Bigotto, G. Costa, G. Mestroni, G. Pellizer, E. Reisenhover, L. Nardin-Stefani, A. Puxeddu, G. Tauzher, *Inorg. Chem. Acta Rev.*, *4* (1970) 41 (Vitamin B-12 models).
A. J. Chalk, J. F. Harrod, *Adv. Organomet. Chem.*, *6* (1968) 119 (Catalysis by cobalt carbonyls).
R. F. Heck, *Adv. Organomet. Chem.*, *4* (1966) 243 (Alkyl- and acylcobalt tetracarbonyls).

iv) Rhodium

V. A. Pavlov, E. I. Klabunovskii, *Izv. Akad. Nauk SSSR, Ser. Khim.*, (1983) 2015 (Asymmetric hydrogenation by Rh phosphines).
H. Brunner, *Angew. Chem. Int. Ed. Engl.*, *22* (1983) 897 (Rh-catalyzed enantioselective hydrosilylation).
L. Marko, J. Bukos, in *Aspects of Homogeneous Catalysis*, Vol. 4, R. Ugo, ed., Reidel, Dordrecht (1981), p. 145 (Rh-catalyzed hydrogenations).
F. H. Jardine, *Prog. Inorg. Chem.*, *38* (1981) 64 (Chlorotris(triphenylphosphine)rhodium(I) reactions).
H. Siegel, W. Himmele, *Angew. Chem., Int. Ed. Engl.*, *19* (1980) 178 (Rh-catalyzed hydroformylation).
J. Halpern, A. S. C. Chan, D. P. Riley, J. J. Pluth, in *Inorganic Compounds with Unusual Properties-II*, R. B. King, ed., Advances in Chemistry Series, No. 173, American Chemical Society, Washington, DC (1979), p. 16 (Cationic Rh phosphines).
P. Kvintovics, B. Heil, L. Marko, in *Inorganic Compounds with Unusual Properties-II*, R. B. King, ed., Advances in Chemistry Series, No. 173, American Chemical Society, Washington, DC (1979), p. 26 (Aromatic hydrogenation by Rh phosphines).
L. Marko, *Pure Appl. Chem.*, *51* (1979) 2211 (Rh-phosphine homogeneous catalysts).
D. A. Slack, in *Fundamental Research in Homogeneous Catalysis*, Vol. 3, M. Tsutsui, ed., Plenum Press, New York (1979), p. 983 (P-31 NMR on Rh catalysts).
V. S. Khandkarova, in *Methody Elementoorganicheskoi Khimii. Kobal't, Nikel', Platinovye Metaly*, A. N. Nesmeyanov, K. A. Kocheshkov, eds., Nauka, Moscow (1978), p. 360 (Organorhodium compounds, comprehensive review in Russian).
M. A. Bennett, in *Fundamental Research in Homogeneous Catalysis*, Vol. 2, Y. Ishii, M. Tsutsui, eds., Plenum Press, New York (1978), p. 93 (Oxidative addition to Rh(I)).
I. Ojima, in *Fundamental Research in Homogeneous Catalysis*, Vol. 2, Y. Ishii, M. Tsutsui, eds.,

Plenum Press, New York (1978), p. 181 (Asymmetric hydrogenation and hydrosilylation with chiral Rh complexes).
B. R. James, in *Transition-Metal Hydrides,* R. Bau, ed., Advances in Chemistry Series, No. 167, American Chemical Society, Washington, DC (1978), p. 122 (Asymmetric hydrogenation by Rh hydrides with chiral ligands).

v) Iridium

R. Crabtree, *Acc. Chem. Res., 12* (1979) 331 (Organoiridium compounds in catalysis).
V. S. Khandkarova, in *Methody Elementoorganicheskoi Khimii. Kobal't, Nikel', Platinovye Metally,* A. N. Nesmeyanov, K. A. Kocheshkov, eds., Nauka, Moscow (1978), p. 465 (Organoiridium compounds, comprehensive review in Russian).

i) Group VIII: Nickel, Palladium, Platinum

E. Müller, O. Bayer, H. Meerwein, K. Ziegler, eds., *Methoden der Organischen Chemie (Houben-Weyl),* Vol. XIII/9, *Metallorganische Verbindungen der VII. und VIII. Nebengruppe des Periodensystems: Ni, Pd, Pt,* by H. F. Klein, E. Langer, H. Segnitz, K. von Werner, G. Thieme Verlag, Stuttgart (1985).

E. Sappo, A. Tiripicchio, P. Braunstein, *Chem. Rev., 83* (1983) 203 (Ni triad alkyne-carbonyl clusters).
L. S. Hegedus, *Mod. Synth. Methods, 3* (1983) 61 (Group-VIII metals in organic synthesis).
E. Negishi, *Acc. Chem. Res., 15* (1982) 240 (Pd and Ni-catalyzed cross-couplings).
E. Negishi, *Pure Appl. Chem., 53* (1981) 2333 (Bimetallic catalysts containing Ni and Pd in organic synthesis).
E. Uhlig, D. Walthe, *Coord. Chem. Rev., 33* (1980) 3 (Syntheses with electron-rich Ni triad complexes).
G. P. Chiusoli, *Pure Appl. Chem., 52* (1980) 635 (Organic syntheses catalyzed by group-VIII complexes).
M. Kumada, *Pure Appl. Chem., 52* (1980) 669 (Ni and Pd-catalyzed cross couplings of organometals with organic halides).
R. J. Mureinik, *Coord. Chem. Rev., 25* (1978) 1 (Solvent paths and dissociated intermediates in square-planar complex substitutions).
J. K. Stille, K. S. Y. Lau, *Acc. Chem. Res., 10* (1977) 434 (Oxidative addition mechanisms of organic halides to group-VIII complexes).
J. F. Harrod, A. J. Chalk, in *Organic Synthesis via Metal Carbonyls*, Vol. II, I. Wender, P. Pino, eds., Wiley, New York (1977), p. 673 (Group VIII-catalyzed hydrosilylation).
D. M. Roundhill, *Adv. Organomet. Chem., 13* (1975) 273 (Hydrido- Ni, Pd and Pt complexes).
A. Peloso, *Coord. Chem. Rev., 10* (1973) 123 (Kinetics of Ni, Pd and Pt complexes).
J. H. Nelson, H. B. Jonassen, *Coord. Chem. Rev., 6* (1971) 27 (Mono-olefin and -acetylene complexes of Ni, Pd, Pt).
J. R. Miller, *Adv. Inorg. Chem. Radiochem., 4* (1962) 133 (Stereochemistry of Ni, Pd and Pt, including some organometallic compounds).

i) Nickel

Gmelin Handbuch der Anorganischen Chemie. *Organonickel Compounds*, Part 1: *Mononuclear Compounds* (New Supplement Series, Vol. 16) (1975) (Index for Parts 1 and 2 (New Supplement Series, Vol 18 (1975).

P.W. Jolly, G. Wilke, *The Organic Chemistry of Nickel,* Academic Press, New York, Vol. 2, *Organic Synthesis* (1975).

Gmelin Handbuch der Anorganischen Chemie. Organonickel Compounds, Part 2: *Mononuclear Compounds* (Concluded). Polynuclear Compounds (New Supplement Series, Vol. 17), Springer Verlag, Berlin (1974).

P.W. Jolly, G. Wilke, *The Organic Chemistry of Nickel,* Academic Press, New York, Vol. 1, *Organonickel Complexes* (1974).

H.-F. Klein, *Angew. Chem., Int. Ed. Engl., 19* (1980) 362 (Trimethylphosphine complexes of Ni as models for homogeneous catalysis).

K. Nag, A. Chakravorty, *Coord. Chem. Rev., 33* (1980) 87 (Mono-, tri- and tetravalent Ni).

B. Bogdanovic, *Adv. Organomet. Chem., 17* (1979) 105 (Ni catalyzed olefin oligomerization).

A.C.L. Su, *Adv. Organomet. Chem., 17* (1979) 269 (Ni catalyzed codimerization of ethylene and butadiene).

G.A. Ozin, *Coord. Chem. Rev., 28* (1979) 117 (Naked Ni clusters as olefin chemisorption models).

C. St.-Joly, in *Inorganic Compounds with Unusual Properties-II* R.B. King, ed., Advances in Chemistry Series, No. 173, American Chemical Society, Washington, DC (1979), p. 152 (Ni(II)-d^8-complexes).

J.J. Eisch, K.R. Im, in *Inorganic Compounds with Unusual Properties-II,* Advances in Chemistry Series, No. 173, American Chemical Society, Washington, DC (1979), p. 195 (Ni(0) cleavages and oligomerizations).

G.P. Chiusoli, G. Salerno, *Adv. Organomet. Chem., 17* (1979) 195 (Applications of organonickel compounds in organic synthesis).

M. Itoh, A.B. Kunz, in *Fundamental Research in Homogeneous Catalysis,* Vol. 3, M. Tsutsui, ed., Plenum Press, New York (1979), p. 73 (MO calculations on Ni carbonyls).

R.H. Grubbs, A. Miyashita, in *Fundamental Research in Homogeneous Catalysis*, Vol. 3, M. Tsutsui, ed., Plenum Press, New York (1979), p. 151 (Nickelacycloalkane reactions).

R. Noyori, in *Inorganic Compounds with Unusual Properties-II,* R.B. King, ed., Advances in Chemistry Series, No. 173, American Chemical Society, Washington, DC (1979), p. 307 (Ni(0)-catalyzed reaction of strained rings).

F.S. Denisov, in *Methody Elementoorganicheskoi Khimii. Kobal't, Nickel', Platinovye Metaly,* A.N. Nesmeyanov, K.A. Kocheshokov, eds., Nauka, Moscow (1978), p. 128 (Organonickel compounds, comprehensive review in Russian).

H. Takaya, in *Fundamental Research in Homogeneous Catalysis,* Vol. 2, Y. Ishii, M. Tsutsui, eds., Plenum Press, New York (1978), p. 221 (Ni-catalyzed reaction of bicyclobutanes with olefins).

K.W. Barnett, *J. Organomet. Chem., 78* (1974) 139 (The chemistry of nickelocene).

K. Fischer, K. Jones, G. Wilke, *Angew. Chem., Int. Ed. Engl., 12* (1973) 943 (Nickel-olefin complexes).

K. Fischer, K. Jonas, P. Misbach, R. Stubba, G. Wilke, *Angew. Chem., Int. Ed. Engl., 12* (1973) 943 (The Ni effect).

E.V. Leonova, N.S. Kochetkova, *Russ. Chem. Rev. (Engl. Transl.), 42* (1973) 278 (Nickelocene reactions).

R. Morassi, I. Bertini, L. Sacconi, *Coord. Chem. Rev., 11* (1973) 343 (Ni(II) five-coordinated complexes).
D.R. Fahev, *Organomet. Chem. Rev., A, 7* (1972) 245 (σ-Organonickel compounds).
H. Buchholz, P. Heimbach, H.J. Hey, H. Selbeck, W. Wiese, *Coord. Chem. Rev., 8* (1972) 129 (Nickel-catalyzed cycloadditions of 1,3-dienes).
M.F. Semmelhack, *Org. React., 19* (1972) 115 (Formation of C–C bonds via π-allylnickel compounds).
G.P. Chiusoli, in *Aspects of Homogeneous Catalysis,* Vol. 1, R. Ugo, ed., Manfredi, Milan (1970), p. 77 (Stereoselectivity in Ni-induced organic syntheses).
P. Heimbach, in *Aspects of Homogeneous Catalysis*, Vol. 2, R. Ugo, ed., Reidel, Dordrecht (1970), p. 79 (Ni-catalyzed cyclooligomerization of alkenes and dienes).
G. Wilke, *Pure Appl. Chem., 17* (1968) 179 (Organonickel compounds).
G.N. Schrauzer, *Adv. Organomet. Chem., 2* (1964) 1 (Organometallic chemistry of nickel).

ii) Palladium and Platinum

V.A. Maksakov, S.P. Gubin, *Koord. Khim., 10* (1984) 689 (Ligand activation on clusters).
M.N. Bochkarev, *Sov. J. Coord. Chem. (Engl. Transl.), 10* (1984) 673 (Covalent polynuclear transition and non-transition metal compounds).
A.L. Balch, *Comments Inorg. Chem., 3* (1984) 51 (Odd oxidation states of Pd and Pt).
U. Belluco, R.A. Michelin, P. Uguagliati, B. Crociani, *J. Oranomet. Chem., 250* (1983) 565 (Mechanisms of nucleo- and electrophilic attack on organo-Pd(II) and Pt(II) complexes).
I.S. Akhrem, N.M. Chistovalova, M.E. Volpin, *Russ. Chem. Rev. (Engl. Transl.), 52* (1983) 953 (Si–C bond breaking by Pd and Pt complexes).
V.V. Bashilov, V.I. Sokolov, O.A. Reutov, *Izv. Akad. Nauk SSSR, Ser. Khim.,* (1982) 2069 (Zero-valent Pd and Pt with Hg(II); syntheses of organo-Pd and Pt).
P.M. Maitlis, *Chem. Soc. Rev., 10* (1981) 1 (η^5-Cyclopentadienyl and η^6-arene protecting ligands for Pt metal complexes).
S. Otsuka, *J. Organomet. Chem., 200* (1980) 191 (Pd and Pt complexes of bulky phosphines).
H.C. Clark, *Pure Appl. Chem., 50* (1978) 43 (Isocyanide, carbene and related chemistry of Pd and Pt).
S.C. Tripathi, S.C. Srivastava, R.P. Mani, A.K. Shrimal, *Inorg. Chim. Acta, 17* (1976) 257 (Pd and Pt carbonyls).
R. Jira, W. Frieslebeu, *Organomet. React., 3* (1972) 1 (Olefin oxidation with group-VIII noble-metal compounds).
F.R. Hartley, *Organomet. Chem. Rev., A, 6* (1970) 119 (Starting materials for the preparation of organometallic complexes of Pd and Pt).
F.R. Hartley, *Chem. Rev., 69* (1969) 799 (Olefin and acetylene complexes of Pd and Pt).

iii) Palladium

J. Tsuji, *Organic Synthesis with Palladium Compounds,* Springer Verlag, Berlin (1980).
P.M. Henry, *Palladium-Catalyzed Oxidation of Hydrocarbons,* Reidel, Dordrecht (1980).
B.M. Trost, *Organopalladium Intermediates in Organic Synthesis* (Tetrahedron Reports on Organic Chemistry, D.H.R. Barton, J.E. Baldwin, W.D. Ollis, T. Stephen, eds)., Vol. 4, Ch. 32, Pergamon Press, Oxford (1978).

P. M. Maitlis, *Organic Chemistry of Palladium*, Academic Press, New York (1971): Vol. 1, *Metal Complexes;* Vol. 2, *Catalytic Reactions.*

V. I. Sokolov, *Pure Appl. Chem., 55* (1983) 1837 (Asymmetric cyclopalladation in enantioselective synthesis).
O. M. Tempkin, L.G. Bruk, *Russ. Chem. Rev.(Engl. Transl.), 52* (1983) 206 (Pd(I) complexes in coordination and catalysis).
I. V. Kozhernikov, *Russ. Chem. Rev. (Engl. Transl.), 52* (1983) 244 (Palladium (II) catalysis of oxidative coupling of olefins).
B. Aakermark, J. E. Backvall, K. Zetterberg, *Acta Chem. Scand., Ser. B, 36* (1982) 577 (Nucleophilic addition to Pd olefin, π-allyl and σ-alkyls).
J. Tsuji, *Pure Appl. Chem., 54* (1982) 197 (π-Allylpalladium catalysis).
R. F. Heck, *Pure Appl. Chem., 53* (1981) 2323 (Pd-catalyzed synthesis of conjugated polyenes).
J. Tsuji, *Pure Appl. Chem., 53* (1981) 2371 (Pd-catalysis in natural-products synthesis).
B. M. Trost, *Acc. Chem. Res., 13* (1980) 385 (Pd-catalyzed allylic alkylations).
P. M. Maitlis, *J. Organomet. Chem., 200* (1980) 161 (Acetylene and cyclobutadiene complexes).
J. Tsuji, *Adv. Organomet. Chem., 17* (1979) 141 (Pd-catalyzed reactions of butadiene and isoprene).
R. F. Heck, *Acc. Chem. Res., 12* (1979) 146 (Pd-catalyzed reactions of organic halides with olefins).
A. Z. Rubezhov, in *Methody Elementoorganicheskoi Khimii, Kobal't, Nikel', Platinovye Metaly,* A. N. Nesmeyanov, K. A. Kocheshkov, eds., Nauka, Moscow (1978), p. 566 (Organopalladium compounds, comprehensive review).
B. M. Trost, *Tetrahedron, 33* (1977) 2615 (Organopalladium intermediates in organic synthesis).
P. M. Maitlis, *Acc. Chem. Res., 9* (1976) 93 (Pd(II)-induced oligomerization of acetylenes).
P. M. Henry, *Adv. Organomet. Chem., 13* (1975) 363 (Pd-catalyzed organic reactions).
I. Moritani, Y. Fujiwara, *Synthesis,* (1974) 524 (Aromatic substitution of alkenes by Pd salts).
P. M. Henry, *Acc. Chem. Res., 6* (1973) 16 (Pd(II)-catalyzed exchange and isomerism).
J. Tsuji, *Acc. Chem. Res., 6* (1973) 8 (Pd-catalyzed butadiene additions).
R. F. Heck, *Top. Curr. Chem., 16* (1971) 221 (Addition-elimination reactions of Pd compounds with olefins).
P. M. Henry, *Ann. N. Y. Acad. Sci., 172* (1971) 483 (Mechanisms of Pd(II) catalyzed reactions of olefins).
R. Hüttel, *Synthesis,* (1970) 225 (Palladium salts and complexes in preparative organic chemistry).

iv Platinum

U. Belluco, *Organometallic and Coordination Chemistry of Platinum,* Academic Press, New York (1974).
Gmelin Handbuch der Anorganischen Chemie. Platinum. Part D: *Complex Compounds of Platinum with Neutral Ligands.* Springer Verlag, Berlin (1957) (Including organoplatinum compounds).

G. M. Whitesides, R. H. Reamey, R. L. Brainard, A. N. Izumi, T. J. McCarthy, *Ann. N. Y. Acad. Sci., 415* (1983) 56 (Organoplatinum-metal hydride reactions).
R. J. Puddephatt, in *Inorganic Chemistry Toward the 21st Century,* M. H. Chisholm, ed., ACS Symposium Series, No. 211, American Chemical Society, Washington, DC (1983), p. 353 (Platinacyclobutane chemistry).

F. G. A. Stone, *Inorg. Chim. Acta, 50* (1981) 33 ($Pt(0)(cod)_2$ in heterometallic cluster synthesis).
F. G. A. Stone, *Acc. Chem. Res., 14* (1981) 318 (Ligand-free Pt compounds).
H. C. Clark, *J. Organomet. Chem., 200* (1980) 63 (The author's organoplatinum work).
R. J. Puddephatt, *Coord. Chem. Rev., 33* (1980) 149 (Platinocyclobutanes).
P. I. Stroselskii, Yu. I. Solov'ev, *Sov. J. Coord. Chem. (Engl. Transl.), 4* (1978) 1123 (Platinum-olefin complexes).
A. A. Rubezhov, in *Methody Elementoorganicheskoi Khimii, Kobal't, Nikel', Platinovye Metaly*, A. N. Nesmeyanov, K. A. Kocheshkov, eds., Nauka, Moscow (1978), p. 627 (In Russian) (Organoplatinum compounds, comprehensive review).
D. M. Roundhill, in *Fundamental Research in Homogeneous Catalysis,* Vol. 2, Y. Ishii, M. Tsutsui eds., Plenum Press, New York (1978), Vol. 3, p. 11 (Pt-cluster catalyzed oxidation of ketones and CO).
M. Green, F. G. A. Stone, in *Transition-Metal Hydrides,* R. Bau, ed., Advances in Chemistry Series, No. 167, American Chemical Society, Washington, DC (1978), p. 111 (μ-H-Pt hydrosilylation catalysts).
R. Romeo, P. Uguagliatti, U. Belluco, *J. Mol. Catal., 1* (1976) 325 (Platinum-olefin complexes).
M. H. Chisholm, *Trans. N. Y. Acad. Sci., 36* (1974) 657 (Organic chemistry of Pt).
M. M. Chisholm, H. C. Clark, *Acc. Chem. Res., 6* (1973) 202 (Pt-induced carbonium ions).
J. S. Thayer, *Organomet. Chem. Rev., A, 5* (1970) 53 (Organoplatinum(IV) compounds).
U. Belluco, G. Deganello, R. Pietropaulo, P. Uguagliati, *Inorg. Chim. Acta Rev., 4* (1970) 7 (Pt(II) complexes with group-IV donors).
U. Belluco, B. Crociani, R. Pietropaolo, P. Uguagliati, *Inorg. Chim. Acta Rev., 3* (1969) 19 (Complexes of Pt(II) with unsaturated hydrocarbons).
R. J. Cross, *Organomet. Chem. Rev., 2* (1967) 97 (σ-Complexes of Pt(II) with hydrogen, carbon and other elements of group IV).

j) Group IB: Copper, Silver, Gold

E. Müller, O. Bayer, H. Meerwein, K. Ziegler, eds., *Methoden der Organischen Chemie (Houben-Weyl),* Vol. XIII/1, *CH-Acidität. Metallorganische Verbindungen der I. Gruppe des Periodensystems: Li, Na, K, Rb, Cs, Cu, Ag, Au,* by G. Bähr, P. Burba, H. F. Ebel, A. Luttringhaus, U. Schollkopf, G. Thieme Verlag, Stuttgart (1970).

A. M. Sladkov, I. R. Gol'ling, *Russ. Chem. Rev. (Engl. Transl.), 48* (1979) 868 (Organocopper and silver acetylide reactions).
G. van Koten, J. G. Noltes, in *Fundamental Research in Homogeneous Catalysis*, Vol. 3, M. Tsutsui, ed., Plenum Press, New York (1979), p. 963 (Dynamics of aryl group-IB clusters).
J. G. Noltes, *J. Organomet. Chem., 200* (1975) 177 (Group-IB aryl clusters).
M. I. Bruce, *J. Organomet. Chem., 44* (1972) 209 (Carbonyl chemistry of the group-IB metals).
A. M. Sladkov, L. Yu, Ikhin, *Russ. Chem. Rev., 37* (1968) 748 (Copper and silver acetylides in organic synthesis).

i) Copper

Gmelin Handbuch der Anorganischen Chemie. New Supplement Series. Organocopper Compounds, Part 2, Springer Verlag, Berlin (1983).
G. H. Posner, *An Introduction to Synthesis Using Organocopper,* Wiley, New York (1980).

J.G. Noltes, *Philos. Trans. R. Soc. London, Ser. A,* (1982) 308 (Organocopper clusters).
M. Ito, A.B. Kunz, in *Fundamental Research in Homogeneous Catalysis,* Vol. 3, M. Tsutsui, ed., Plenum Press, New York (1979), p. 73 (MO calculations on Cu carbonyls).
C. Kutal, P.A. Grutsch, in *Inorganic Compounds with Unusual Properties-II,* R.B. King, ed., Advances in Chemistry Series, No. 173, American Chemical Society, Washington, DC (1979), p. 325 (Cu(I) sensitization of olefin photoreactions).
G.A. Ozin, *Coord. Chem. Rev., 28* (1979) 117 (Naked Cu clusters as olefin chemisorption models).
J.F. Normant, *Pure Appl. Chem., 50* (1978) 709 (Stoichiometric vs. catalytic use of Cu(I) salts in main-group organometallic synthesis).
A. Camus, N. Marsich, G. Nardin, L. Randaccio, *Inorg. Chim. Acta Rev., 23* (1977) 131.
H.O. Howe, *Acc. Chem. Res., 9* (1976) 59 (Li-organocuprate additions).
J.F. Normant, *J. Organomet. Chem. Libr., 1* (1976) 219 (Organocopper reagents in organic synthesis).
G.H. Posner, *Org. React., 22* (1975) 253 (Substitutions by organocopper reagents).
T. Saegusa, Y. Ito, *Synthesis,* (1975) 291 (Synthesis of cyclic compounds from Cu isonitrile complexes).
P.E. Fanta, *Synthesis,* (1974) 9 (Ullmann synthesis of biaryls).
A.E. Jukes, *Adv. Organomet. Chem., 12* (1974) 215 (The organic chemistry of copper).
A.N. Nesmeyanhov, in *Methody Elementoorganicheskoi Khimii. Podgrupy Medi, Skandiya, Titana, Vanadiya, Khroma i Margantsa, Lantanoidy i Aktanoidy,* A.N. Nesmeyanov, K.A. Kocheshkov, eds., Nauka, Moscow (1974), p. 11 (Organocopper compounds).
T. Kaufmann, *Angew. Chem., Int. Ed. Engl., 13* (1974) 291 (Oxidative coupling via organocopper compounds).
M. Goshaev, O.S. Otroshchenko, A.S. Sadykov, *Russ. Chem. Rev. (Engl. Transl.), 41* (1972) 1046 (the Ullman reaction).
J.F. Normant, *Synthesis,* (1972) 63 (Organocopper(I) compounds and organocuprates in synthesis).
G.H. Posner, *Org. React., 19* (1972) 1 (Conjugate-addition reactions of organocopper reagents).

ii) Silver

Gmelin Handbuch der Anorganischen Chemie. Silver. Part B. Section 5. *Organosilver Compounds. Organosilver Salts.* Springer Verlag, Berlin (1975).
A.N. Nesmeyanov, in *Methody Elementoorganicheskoi Khimii. Podgrupy Medi, Skandiya, Titana, Vanadiya, Khroma i Margantsa, Lantanoid i Aktanoidy,* A.N. Nesmeyanov, K.A. Kocheshkov, eds., Nauka, Moscow (1978), p. 61 (In Russian) (Organosilver compounds, comprehensive review).
L.A. Paquette, *Acc. Chem. Res., 4* (1971) 280 (Catalysis of strained-bond rearrangements by silver ions).
C.D.M. Beverwijk, G.J.M. Van der Kerk, A.J. Leusink, J.G. Noltes, *Organomet. Chem. Rev., A, 5* (1970) 215 (Organosilver chemistry).
C. Blomberg, *Chem. Tech. (Amsterdam), 25* (1970) 837 (Organosilver chemistry).

iii) Gold

Gmelin Handbuch der Anorganischen Chemie, Au. Organogold Compounds (by H. Schmidbaur), Springer Verlag, Berlin (1980).

R. J. Puddephatt, *The Chemistry of Gold,* Elsevier, Amsterdam (1978) (Chapter 7: Organogold chemistry).

G. K. Anderson, *Adv. Organomet. Chem., 20* (1982) 39 (Organogolds).
K. I. Grandberg, *Russ. Chem. Rev. (Engl. Transl.), 51* (1982) 394 (Organogold(I)).
R. Uson, A. Laguna, J. Vicente, *Synth. React. Inorg. Metal-Org. Chem., 7* (1977) 463 (Arylgold chemistry).
H. Schmidbaur, *Angew. Chem., Int. Ed. Engl., 15* (1976) 728 (Organogold chemistry).
A. N. Nesmeyanov, E. G. Perevalova, *Izv. Akad. Nauk SSSR, Ser. Khim.,*(1974) 1124 (Organogold complexes).
B. Armer, H. Schmidbaur, *Angew. Chem., Int. Ed. Engl., 9* (1970) 101 (Organogold chemistry).

Chapter 11

a) General Reviews

J. Falbe, *New Synthesis with Carbon Monoxide*, Springer Verlag, Berlin (1983).
R. P. Houghton, *Metal Complexes in Organic Chemistry*, Cambridge University Press, Cambridge (1979).
I. Wender, P. Pino, eds., *Organic Synthesis via Metal Carbonyls,* Vol. II, Wiley, New York (1977).

F. G. A. Stone, *Angew. Chem., Int. Ed. Engl., 23* (1984) 89 (Isolobal principles).
D. C. Bradley, S. H. Langer, *Chem. Rev., 81* (1981) 109 (Immobilized carbonyl catalysts).
H. Alper, *Pure Appl. Chem., 52* (1980) 607 (Carbonyls in catalysis and synthesis).
H. Behrens, *Adv. Organomet. Chem., 18* (1980) 2 (Metal carbonyls in liquid ammonia).
F. Calderazzo, R. Ercoli, G. Natta, in *Organic Syntheses via Metal Carbonyls,* I. Wender, P. Pino, eds., Wiley-Interscience, New York (1978), p. 1 (Comprehensive review).
J. K. Burdett, *Coord. Chem. Rev., 27* (1978) 1 (Matrix-isolation studies on transition-metal carbonyls and related species).
F. S. Wagner, in *Kirk-Othmer, Encyclopedia of Chemical Technology,* 3rd. Ed. J. Wiley, New York (1978), Vol. 4, p. 794 (Metal carbonyls, technical aspects).
W. Nagata, M. Yoshioka, *Org. React., 25* (1977) 255 (Hydrocyanation of conjugated carbonyls).
R. B. King, in *Methodicum Chimicum*. Vol. 8. *Preparation of Transition-Metal Derivatives*, K. Niedenzu, H. Zimmer, eds., Academic Press, New York (1976), p. 421 (Metal carbonyls).
F. A. Cotton, *Prog. Inorg. Chem., 21* (1976) 1 (Metal carbonyls).
S. C. Tripathi, S. C. Srivastava, R. P. Mani, A. K. Shrimal, *Inorg. Chim. Acta, 17* (1976) 257 (Rh, Ir, Pd, Pt carbonyls and derivatives).
G. R. Dobson, *Acc. Chem. Res., 9* (1976) 300 (Ligand exchange in octahedral carbonyls).
H. Behrens, *J. Organomet. Chem., 94* (1975) 139 (The chemistry of metal carbonyls, the life and work of W. Hieber).
R. K. Sheline, J. L. Slater, *New Synth. Methods, 3* (1975) 203 (Spectral evidence for lanthanoid- and actinoid-carbonyl compounds).
S. C. Tripathi, S. C. Srivastava, R. P. Mani, A. K. Shrimal, *Inorg. Chim. Acta, 15* (1975) 249 (Ru and Os carbonyls and their substituted derivatives).

R. A. Sokolik, in *Methody Elementoorganicheskoi Khimii. Tipy Metalorganicheskikh Soedinenii Perekhodnykh Metallov*, A. N. Nesmeyanov, ed., Nauka, Moscow (1975), p. 19 (Transition-metal carbonyls, comprehensive review in Russian).
B. H. Robinson, *MTP Int. Rev. Sci. Inorg. Chem., Ser. Two, 6* (1975) 1 (Metal-carbonyl complexes).
S. C. Tripathi, S. C. Srivastava, R. P. Mani, *J. Sci. Ind. Res., 33* (1974) 570 (Group VI-metal carbonyl complexes).
M. I. Bruce, *J. Organomet. Chem., 44* (1972) 209 (Carbonyl chemistry of the group-IB metals: Cu, Ag, Au).
M. Ryang, S. Tsutsumi, *Synthesis* (1971) 55 (Organic synthesis by metal carbonyls).
E. W. Abel, F. G. A. Stone, *Quart. Rev., 24* (1970) 498 (Chemistry of metal carbonyls: synthesis and reactivity).
W. Hieber, *Adv. Organomet. Chem., 8* (1970) 1 (Forty years of metal-carbonyl research).
E. W. Abel, F. G. A. Stone, *Quart. Rev., 23* (1969) 325 (the chemistry of metal carbonyls: structural considerations).
M. E. Strem, *Ann. N. Y. Acad. Sci., 145* (1967) 123 (Metal carbonyls: their synthesis and utility).
M. Bigorgne, J. Benard, *Rev. Chim. Miner., 3* (1966) 831 (Some aspects of the chemistry of metal carbonyls).
J. C. Hilleman, *Pre. Inorg. React., 1* (1964) 77 (Preparation of metal carbonyls).
E. W. Abel, *Quart Rev., 17* (1963) 133 (General review on metal carbonyls).
J. Chatt, P. L. Pauson, L. M. Venanzi, in *Organometallic Chemistry*, H.A. Zeiss, ed., Rheinhold, New York (1960), p. 468 (Metal carbonyls and related compounds).

b) Polynuclear Metal Carbonyls (Clusters)

B. F. G. Johnson, ed., *Transition-Metal Clusters,* Wiley, New York (1980).

N. K. Eremenko, E. G. Mednikov, S. P. Gubin, *Sov. J. Coord. Chem. (Engl. Transl.), 10* (1984) 617 Carbonylphosphine clusters of Pd and Pt).
J. S. Bradley, *Adv. Organomet. Chem., 22* (1983) 1 (Carbidocarbonyl clusters).
H. Vahrenkamp, *Adv. Organomet. Chem., 22* (1983) 169 (Metal-cluster reactions).
E. Sappo, A. Tiripicchio, P. Braunstein, *Chem. Rev., 83* (1983) 203 (Alkyne-carbonyl clusters of the Fe, Co and Ni triads).
M. D. Curtis, in *Inorganic Chemistry Toward the 21st Century,* M. H. Chisholm, ed., ACS Symposium Series, No. 211, American Chemical Society, Washington, DC (1983), p. 241 (Converting high- and low-valent clusters).
A. J. Deeming, *J. Mol. Catal., 21* (1983) 25 (Small oxy ligands for metal clusters).
R. Ugo, R. Psari, *J. Mol. Catal., 20* (1983) 53 (Carbonyl clusters in CO catalysis).
M. I. Bruce, *J. Organomet. Chem., 257* (1983) 417 (Heterometallic clusters).
E. L. Muetterties, M. J. Krause, *Angew. Chem., Int. Ed. Engl., 22* (1983) 135 (Catalysis by clusters).
J. Lewis, B. F. G. Johnson, *Pure Appl. Chem., 54* (1982) 97 (Os- and Ru-carbonyl clusters).
A. J. Carty, *Pure Appl. Chem., 54* (1982) 1131 (Cluster-bound acetylides).
K. D. Kaesz, C. B. Knobler, M. A. Andrews, G. Van Buskirk, R. Szostak, C. E. Strouse, Y. C. Lin, A. Mayr, *Pure Appl. Chem., 54* (1982) 131 (Organic transformations on metal clusters).
B. F. G. Johnson, J. Lewis, *Adv. Inorg. Chem. Radiochem., 24* (1981) 225 (Clusters).
M. Tachikawa, E. L. Muetterties, *Prog. Inorg. Chem., 28* (1981) 203 (Metal-carbide clusters).
A. Simon, *Angew. Chem., Int. Ed. Engl., 20* (1981) 1 (Condensed metal clusters).

R. C. Schoening, J. L. Vidal, R. A. Fiato, *J. Organomet. Chem., 206* (1981) C43 (Formyl cluster complexes).

B. F. G. Johnson, R. E. Benfield, *Top. Stereochem., 12* (1981) 253 (Stereochemistry of carbonyl clusters).

E. L. Muetterties, *Catal. Rev., 23* (1981) 69 (Molecular metal clusters).

J. Evans, *Chem. Soc. Rev., 10* (1981) 159 (Metal-carbonyl clusters and supported catalysts).

M. C. Manning, W. C. Trogler, *Coord. Chem. Rev., 39* (1981) 89 (Electronic structures of transition-metal clusters).

M. Tachikawa, R. L. Geerts, E. L. Muetterties, *J. Organomet. Chem., 213* (1981) 11 (Heteronuclear metal-carbide clusters).

F. G. A. Stone, *Acc. Chem. Res., 41* (1981) 318; *Inorg. Chim. Acta, 50* (1981) 33 (Pt(0) compounds in the synthesis of clusters).

D. C. Bailey, S. H. Langer, *Chem. Rev., 81* (1981) 109 (Relation of clusters to supported metal catalysts).

S. D. Jackson, P. B. Wells, R. Whyman, P. Worthington, *Catal., 4* (1981) 75 (Metal-cluster catalysts).

H. D. Kaesz, *J. Organomet. Chem., 200* (1980) 145 (Clusters).

G. L. Geoffroy, *Acc. Chem. Res., 13* (1980) 469 (Mixed-metal clusters).

A. Müller, R. Joster, F. A. Cotton, *Angew. Chem., Int. Ed. Engl., 19* (1980) 875 (Trinuclear clusters of the early transition metals).

P. Chini, *J. Organomet. Chem., 200* (1980) 37 (Large metal-carbonyl clusters).

E. L. Muetterties, *J. Organomet. Chem., 200* (1980) 177 (Cluster chemistry).

D. M. P. Mingos, *Pure Appl. Chem., 52* (1980) 705 (Theory and structures of clusters).

E. L. Muetterties, T. N. Rhodin, E. Band, C. F. Brucker, W. R. Pretzer, *Chem. Rev., 79* (1979) 91 (Clusters and surfaces).

R. Ugo, in *Fundamental Research in Homogeneous Catalysis,* Vol. 3, M. Tsutsui, ed., Plenum Press, New York (1979), p. 579 (Supported metal-carbonyl clusters).

M. Moskovits, *Acc. Chem. Res., 12* (1979) 229 (Clusters and heterogeneous catalysis).

A. P. Humphries, H. D. Kaesz, *Prog. Inorg. Chem., 25* (1979) 146 (Cluster hydrides).

J. W. Lauher, *J. Am. Chem. Soc., 100* (1978) 5305 (Bonding in metal-cluster carbonyls; predicted geometry and stoichiometry).

E. Band, E. L. Muetterties, *Chem. Rev., 78* (1978) 639 (Mechanisms of metal-cluster rearrangements).

G. A. Ozin, *Coord. Chem. Rev., 28* (1978) 117 (Co, Ni and Cu naked clusters).

E. L. Muetterties, in *Fundamental Research in Homogeneous Catalysis,* Vol. 2, Y. Ishii, M. Tsutsui, eds., Plenum Press, New York (1978); p. 1 (Metal-cluster catalyzed hydrocarbon activation).

M. S. Wrighton, in *Inorganic and Organometallic Photochemistry,* M. S. Wrighton, ed., Advances in Chemistry Series, No. 168, American Chemical Society, Washington, DC (1978), p. 189 (Photocatalyzed reactions of alkenes with silanes using trinuclear metal-carbonyl catalyst precursors).

G. Schmid, *Angew. Chem., Int. Ed. Engl., 17* (1978) 392 (Tetrahedral Co clusters).

B. F. G. Johnson, J. Lewis, D. Pippard, *J. Organomet. Chem., 160* (1978) 263 (Activation of transition-metal carbonyl clusters).

D. M. Roundhill, in *Fundamental Research in Homogeneous Catalysis,* Vol. 2, Y. Ishii, M. Tsutsui, eds., Plenum Press, New York (1978), p. 11; Vol. 3, p. 11 (Cluster-catalyzed oxidation of ketones and CO).

D. Seyferth, in *Transition-Metal Organometallics in Organic Synthesis*, Vol. II, H. Alper, ed., Academic press, New York (1978), p. 1 (Clusters as synthetic reagents).

G. L. Geoffroy, R. A. Epstein, in *Inorganic and Organometallic Photochemistry,* M. S.

Wrighton, ed., Advances in Chemistry Series, No. 168, American Chemical Society, Washington, DC (1978), p. 132 (Photoinduced declusterification of Co carbonyls).
J. M. Basset and R. Ugo, in *Aspects of Homogeneous Catalysis,* Vol. 3, R. Ugo, ed., D. Reidel, Dordrecht (1977) (Molecular clusters).
P. Chini, B. T. Heaton, *Top. Curr. Chem., 71* (1977) 1 (Tetranuclear carbonyl clusters).
H. Vahrenkamp, *Struct. Bonding (Berlin), 32* (1977) 1 (Transition-metal clusters with organic ligands).
D. Seyferth, *Adv. Organomet. Chem., 14* (1976) 97 (Alkylidynetricobalt nonacarbonyl clusters).
P. Chini, G. Longoni, V. G. Albano, *Adv. Organomet. Chem., 14* (1976) 285 (High-nuclearity metal-carbonyl clusters).
K. Wade, *Adv. Inorg. Chem. Radiochem., 18* (1976) 1 (Cluster chemistry).
P. J. Vergamini, G. J. Kubas, *Prog. Inorg. Chem., 21* (1976) 261 (Organometallic sulfur clusters).
B. E. Penfold, B. H. Robinson, *Acc. Chem. Res., 6* (1973) 73 (Tricobaltcarbon clusters).
H. D. Kaesz, *Chem. Br., 9* (1973) 344 (Metal-carbonyl clusters).
R. B. King, *Prog. Inorg. Chem., 15* (1972) 288 (Transition metal cluster compounds, including carbonyls).
G. Palyi, F. Piacenti, L. Marko, *Inorg. Chim. Acta Rev., 4* (1970) 109 (Methynyl tricobalt enneacarbonyls).
P. Chini, *Pure Appl. Chem., 24* (1970) 489 (Polynuclear metal carbonyl compounds).
R. D. Johnson, *Adv. Inorg. Chem. Radiochem., 13* (1970) 471 (Transition-metal clusters with π-acid ligands).
B. P. Biryukov, Yu. T. Struchkov, *Russ. Chem. Rev. (Engl. Transl.), 39* (1970) 789 (Polynuclear metal carbonyls).
P. Chini, *Inorg. Chim. Acta Rev., 2* (1968) 31 (Closed metal-carbonyl clusters).
B. Penfold, *Perspect. Struct. Chem., 2* (1968) 71 (Stereochemistry of metal-cluster compounds).
M. C. Baird, *Prog. Inorg. Chem., 9* (1968) 1 (Polynuclear metal carbonyls).
F. A. Cotton, *Quart. Rev., 20* (1966) 389 (Transition-metal compounds containing clusters of metal atoms).

c) Other Metal Carbonyl Derivatives Containing Metal-Metal Bonds

F. A. Cotton, R. A. Walton, *Multiple Bonds Between Metal Atoms,* Wiley, New York (1982).
M. H. Chisholm, ed., *Reactivity of Metal-Metal Bonds,* ACS Symposium Series, No. 155, American Chemical Society, Washington, DC (1981).

J. Holton, M. F. Lappert, R. Pierce, P. I. W. Yarrow, *Chem. Rev., 83* (1983) 135 (Hydrocarbonyl- or hydrocarbon-bridged binuclear complexes).
A. Poe, *Chem. Br., 19* (1983) 997 (Kinetic behavior of metal-metal bonded carbonyls).
M. I. Bruce, *J. Organomet. Chem., 257* (1983) 417; *242* (1983) 147 (Heterometal-metal bonds).
R. Poilblanc, *Inorg. Chim. Acta, 62* (1982) 75 (Bridged homobimetal complexes).
H. Werner, *Adv. Organomet. Chem., 19* (1981) 155 (Metal-metal bonded complexes).
R. B. King, *Acc. Chem. Res., 13* (1980) 243 (Alkylaminodifluorophosphines: novel bidentate ligands for stabilizing low metal oxidation states and metal-metal bonded systems).
M. H. Chisholm, in *Inorganic Compounds with Unusual Properties-II,* R. B. King, ed., Advances in Chemistry Series, No. 173, American Chemical Society, Washington, DC (1979), p. 396 (W and Mo triple bonds).

A. N. Nesmeyanov, M. I. Rybinskaya, L. V. Rybin, *Russ. Chem. Rev. (Engl. Transl.)*, *48* (1979) 213 (Binuclear Fe carbonyls with nitrogen bridges).

J. L. Templeton, *Prog. Inorg. Chem.*, *26* (1979) 212 (Metal-metal bonds of order four).

F. A. Cotton, *Acc. Chem. Res.*, *11* (1978) 225 (Multiple metal-metal bonds).

M. H. Chisholm, F. A. Cotton, *Acc. Chem. Res.*, *11* (1978) 356 (Metal-metal Mo and W triple bonds).

H. Vahrenkamp, *Angew. Chem., Int. Ed. Engl.*, *17* (1978) 370 (The metal-metal bond).

W. C. Trogler, H. B. Gray, *Acc. Chem. Res.*, *11* (1978) 232 (Electronic spectra and photochemistry of quadruple M-M bonds).

G. A. Ozin, *Catal. Rev.*, *16* (1977) 191 (Very small metallic and bimetallic clusters).

R. B. King, *Coord. Chem. Rev.*, *20* (1976) 155 (η^5-Me_5C_5-carbonyls with metal-metal bonds).

E. D. Kundig, M. Moskovits, G. A. Ozin, *Angew. Chem., Int. Ed. Engl.*, *14* (1975) 292 (Transition-metal atoms forming binuclear complexes).

F. A. Cotton, *Chem. Soc. Rev.*, *4* (1975) 27 (Multiple metal-metal bonds).

A. N. Nesmeyanov, M. I. Rybinskaya, L. V. Rybin, V. S. Kaganovich, *J. Organomet. Chem.*, *47* (1973) 1 (Binuclear complexes).

B. P. Biryukov, Yu. T. Struchkov, *Russ. Chem. Rev. (Engl. Transl.)*, *39* (1970) 789 (Metal-metal bonds in complexes and polynuclear carbonyls).

W. Jehn, *Z. Chem.*, *9* (1969) 170 (Metal-carbonyl derivatives with M-M bonds).

N. S. Vyazankin, G. A. Razuvaev, O. A. Kruglaya, *Organomet. Chem. Rev.*, *3* (1968) 323 (Organometallic derivatives with M-M′ bonds, including carbonyls).

d) Structure, Bonding, and Spectra of Metal Carbonyls

P. S. Braterman, *Metal-Carbonyl Spectra*, Academic Press, New York (1975).

T. A. Albright, *Tetrahedron*, *38* (1982) 1339 (Applied MO approach to structure and reactivity).

D. M. Hoffmann, R. Hoffmann, *Inorg. Chem.*, *20* (1981) 3543 (Theory of molecular A-frames).

P. S. Pregosin, *Annu. Rev. NMR Spectrosc.*, *11 A* (1981) 227 (C-13 NMR of group-VIII complexes).

J. C. Green, *Struct. Bonding (Berlin)*, *43* (1981) 37 (Gas-phase photoelectron spectra of d- and f-block organometallics).

F. P. Netzer, *Appl. Surf. Sci.*, *7* (1981) 289 (Auger spectra).

R. Colton, M. J. McCormick, *Coord. Chem. Rev.*, *31* (1980) 1 (μ_2-Bridging carbonyls).

A. H. Cowley, *Prog. Inorg. Chem.*, *26* (1979) 46 (UV photoelectron spectroscopy).

J. K. Burdett, *Coord. Chem. Rev.*, *27* (1978) 1 (Matrix-isolation studies).

S. F. A. Kettle, *Top. Curr. Chem.*, *71* (1977) 111 (Vibrational spectra of metal carbonyls).

J. J. Turner, J. K. Burdett, R. N. Perutz, M. Poliakoff, *Pure Appl. Chem.*, *49* (1977) 271 (Matrix photochemistry of metal carbonyls).

S. Aime, L. Milone, *Prog. Nucl. Mag. Res.*, *11* (1977) 3 (Dynamic ^{13}C-NMR spectroscopy of metal carbonyls).

W. L. Jolly, *Top. Curr. Chem.*, *71* (1977) 149 (X-ray photoelectron spectra of metal carbonyls).

D. J. A. Connor, *Top. Curr. Chem.*, *71* (1977) 71 (Thermochemical studies of organotransition metal carbonyls and related compounds).

P. W. Hickmott, M. Cais, a. Modiano, *Annu. Rev. NMR Spectrosc.*, *6C* (1977) 1 (NMR data on metal carbonyls).

M. H. Chisholm, S. Godleski, *Prog. Inorg. Chem.*, *20* (1976) 299 (C-13 NMR).

P. S. Braterman, *Struct. Bonding (Berlin)*, *26* (1976) 1; *10* (1972) 57 (Spectra and bonding).

R.J. Haines, *Chemsa (South Africa), 2* (1976) 22 (Carbon-monoxide bonding to metals).
L.H. Jones, B.I. Swanson, *Acc. Chem. Res., 9* (1976) 128 (Bonding in metal carbonyls).
M. Bigorgne, *J. Organomet. Chem., 94* (1975) 161 (Vibrational spectra of metal carbonyls).
L.J. Todd, G. Wilkinson, *J. Organomet. Chem., 77* (1974) 1 (C-13 NMR).
B.E. Mann, *Adv. Organomet. Chem., 12* (1974) 135 (C-13 NMR).
D.F. Shriver, *Chem. Br., 8* (1972) 419 (Bonding patterns for the carbonyl ligand).
S.F.A. Kettle, I. Paul, *Adv. Organomet. Chem., 10* (1972) 199 (IR intensities of metal-carbonyl stretches).
B.P. Biryukov, Yu.T. Struchkov, *Itogi Nauki, Kristallokhim., 7* (1971) 142 (Structural chemistry of metal carbonyls).
B.P. Biryukov, Yu.T. Struchkov, *Itogi Nauki, Kristallokhim., 5* (1969) 148 (Structural chemistry of metal carbonyls).
L.M. Haines, M.H.B. Stiddard, *Adv. Inorg. Chem. Radiochem., 12* (1969) 53 (Vibrational spectra of metal carbonyls).

e) Reactions

G.L. Geoffroy, M.S. Wrighton, *Organometallic Photochemistry,* Academic Press, New York (1980).
V. Balzani, V. Carassiti, *Photochemistry of Coordination Compounds,* Academic Press, New York (1970).

K.A.M. Creber, K.S. Chan, J.K.S. Wan, *Rev. Chem. Intermed., 5* (1984) 37 (Free radical and radical ions in carbonyl-organometal photochemistry).
G.K. Anderson, R.J. Cross, *Acc. Chem. Res., 17* (1984) 67 (Carbonyl insertions into square-planar complexes).
D.J. Darensbourg, R.A. Kudaroski, *Adv. Organomet. Chem., 22* (1983) 129 (Activation of CO by metal complexes).
D. Walther, E. Dinjus, J. Sieler, *Z. Chem., 23* (1983) 237 (Activation of CO by metal complexes).
M.D. Johnson, *Acc. Chem. Res., 16* (1983) 343 (Bimolecular homolytic displacement of transition-metal complexes from carbon).
J.V.N.V. Prasad, C.N. Pillai, *J. Organomet. Chem., 259* (1983) 1 (Carbometalation).
J. Halpern, *Acc. Chem. Res., 15* (1982) 332 (Carbon-hydrogen bonds from reductive elimination).
M.T. Reetz, *Angew. Chem., Int. Ed. Engl., 21* (1982) 96 (Lewis-acid induced α-alkylation of carbonyls).
M.S. Wrighton, J.L. Graff, R.J. Kazlaukas, J.C. Michener, C.L. Reichel, *Pure Appl. Chem., 54* (1982) 161 (Photogeneration of reactive species).
J.M. Kelly, *Photochemistry, 11* (1982) 259 (Photochemistry of metal carbonyls).
V.I. Manov-Yuvenskii, B.K. Nefedov, *Russ. Chem. Rev. (Engl. Transl.), 50* (1981) 470 (Carbonylation of nitro compounds).
E- Negishi, *Pure Appl. Chem., 53* (1981) 2333 (Catalyzed cross-coupling reactions).
F. Basolo, *Inorg. Chim. Acta, 50* (1981) 65 (Mechanisms of carbonyl-ligand replacement).
D.J. Darensbourg, M.Y. Darensbourg, R.R. Burch, Jr., J.A. Froelich, M.J. Incorvia, in *Inorganic Compounds with Unusual Properties--II,* R.B. King, ed., Advances in Chemistry Series, No. 173, American Chemical Society, Washington, DC (1979), p. 106 (Oxygen-exchange and ligand substitution in Cr carbonyls).

R. Eisenberg, D. E. Hendricksen, *Adv. Catal., 28* (1979) 79 (Binding and activation of CO and its homogeneously catalyzed reactions).

G.L. Geoffroy, R.A. Epstein, in *Inorganic and Organometallic Photochemistry,* M.S. Wrighton, ed., Advances in Chemistry Series, No. 168, American Chemical Society, Washington, DC (1978), p. 132 (Photo-induced declusterification of Co carbonyls).

S.G. Davies, M.L.H. Green, D.M.P. Mingos, *Tetrahedron, 34* (1978) 3047 (Nucleophilic addition to organotransition-metal cations containing unsaturated hydrocarbon ligands).

R. Noyori, *Ann. N. Y. Acad. Sci., 295* (1977) 225 (Iron carbonyls in organic synthesis).

M.S. Wrighton, *Top. Curr. Chem., 65* (1976) 37 (Mechanisms of photochemical reactions).

H. Alper, *J. Organomet. Chem. Libr., 1* (1976) 305 (Carbonyls as reagents for organic synthesis).

G.R. Dobson, *Acc. Chem. Res., 9* (1976) 300 (Trends in reactivity for ligand exchange of octahedral metal carbonyls).

D.F. Shriver, *J. Organomet. Chem., 94* (1975) 259 (Basicity and reactivity of metal carbonyls).

J. Lewis, B.F. Johnson, *Pure Appl. Chem., 44* (1975) 43 (Reactions of polynuclear carbonyls with simple unsaturated molecules).

M.S. Wrighton, *Chem. Rev., 74* (1974) 401 (Photochemistry of metal carbonyls).

D.F. Shriver, A. Alich, *Coord, Chem. Rev., 8* (1972) 15 (The reaction of metal carbonyls with Lewis acids-carbon and oxygen-bonded CO).

S.W.J. Price, in *Comprehensive Chemical Kinetics,* Vol. 4, *Decomposition of Inorganic and Organometallic Compounds,* C.H. Bamford, C.F.H. Tipper, eds., Elsevier, Amsterdam (1972), p. 197 (Decomposition of metal alkyls, aryls, carbonyls and nitrosyls).

D. Shriver, *Acc. Chem. Res., 3* (1970) 231 (Transition-metal basicity).

M. Ryang, *Organomet. Chem. Rev., A, 5* (1970) 67 (Metal carbonyls as stoichiometric reagents in organic synthesis).

J. Halpern, *Acc. Chem. Res., 3* (1970) 386 (Oxidative-addition reactions).

E. Koerner von Gustorf, F.W. Grevels, *Top. Curr. Chem., 13* (1969) 366 (Photochemistry of metal carbonyls).

R.J. Angelici, *Organomet. Chem. Rev., 3* (1968) 173 (Kinetics and mechanisms of substitution reactions in metal-carbonyl complexes).

W. Strohmeier, *Top. Curr. Chem., 10* (1968) 306 (Kinetics and mechanisms of exchange and substitution reactions in metal carbonyls).

H. Werner, *Angew. Chem., Int. Ed. Engl., 7* (1968) 930 (Kinetics of substitution reactions in metal carbonyls).

D.A. Brown, *Inorg. Chim. Acta Rev., 1* (1967) 35 (Substitution reactions of metal-carbonyl compounds).

W. Strohmeier, *Angew. Chem., Int. Ed. Engl., 3* (1964) 730 (Photochemical substitutions on metal carbonyls and their derivatives).

f) Metal-Carbonyl Anions

J.P. Collman, *Acc. Chem. Res., 8* (1975) 342 (Disodium tetracarbonylferrate, a transition-metal analogue of Grignard reagents).

R.B. King, *J. Organomet. Chem., 100* (1975) 111 (Carbonyl-anion syntheses).

J.E. Ellis, *J. Organomet. Chem., 86* (1975) 1 (Metal-carbonyl anions).

E.A. Sullivan, T.F. Jula, *Int. Lab.,* (1974) 51 (Disodium tetracarbonylferrate as reagent in organic synthesis).

P. Chini, A. Cavalieri, S. Martinengo, *Coord. Chem. Rev., 8* (1972) 3 (Mixed cluster-carbonyl anions containing Ni and Co).

R. B. King, *Acc. Chem. Res., 3* (1970) 417 (Applications of metal-carbonyl anions in the synthesis of organometallic compounds).

M.I. Bruce, F.G.A. Stone, *Angew. Chem., Int. Ed. Engl., 7* (1968) 835; 747 (Nucleophilic reactions of metal-carbonyl anions with fluorocarbons).

R.B. King, *Trans. N.Y. Acad. Sci., 28* (1966) 889 (Metal-carbonyl anions).

R.B. King, *Adv. Organomet. Chem., 2* (1964) 157 (Reactions of alkali-metal derivatives of metal carbonyls).

W. Hieber, W. Beck, G. Braun, *Angew. Chem., 72* (1960) 795 (Anionic metal-carbonyl complexes).

g) Metal-Carbonyl Hydrides

R. Bau, ed., *Transition-Metal Hydrides,* Advances in Chemistry Series, No. 167, American Chemical Society, Washington, DC (1978).

E.L. Muetterties, *Transition-Metal Hydrides,* M. Dekker, New York (1971).

D.S. Moore, S.D. Robinson, *Chem. Soc. Rev., 12* (1983) 415 (Transition-metal hydride complexes).

G.L. Soloveichik, B.M. Bulychev, *Russ. Chem. Rev. (Engl. Transl.), 52* (1983) 72 (Bi-transition-metal hydrides).

L.F. Dahl, *Ann. N.Y. Acad. Sci.,* (1983) 415 (Structure and bonding in transition-metal hydrides).

G.E. Toogood, M.G.H. Wallbridge, *Adv. Inorg. Chem. Radiochem., 25* (1982) 267 (Ti- and V-group metal hydrides).

G.L. Soloveichik, B.M. Bulychev, *Russ. Chem. Rev. (Engl. Transl.), 51* (1982) 507 (Mono-transition metal hydrides).

R.G. Teller, R. Bau, *Struct. Bonding (Berlin), 44* (1981) 1 (Structural data for transition-metal hydrides).

G.L. Geoffroy, *Prog. Inorg. Chem., 27* (1980) 123 (Photochemistry of transition-metal hydrides).

H.D. Kaesz, *J. Organomet. Chem., 200* (1980) 145 (Hydridometal complexes and clusters).

R. Bau, R.G. Teller, S.W. Kirtley, T.F. Koetzle, *Acc. Chem. Res., 12* (1979) 176 (Structures of hydride complexes).

A.P. Humphries, H.D. Kaesz, *Prog. Inorg. Chem., 25* (1979) 145 (Hydrido-transition metal cluster complexes).

G.K. Reddy, N.H.M. Gowda, *J. Ind. Chem. Soc., 54* (1977) 289 (Hydride, carbonyl and hydridocarbonyl complexes of platinum-group metals).

J. Schwartz, *J. Organomet. Chem. Libr., 1* (1976) 461 (Transition-metal hydrides for organic synthesis).

J.P. McCue, *Coord. Chem. Rev., 10* (1973) 265 (Transition-metal hydrides).

H.D. Kaesz, R.B. Saillant, *Chem. Rev., 72* (1972) 321 (Hydride complexes of transition metals, including carbonyl hydrides).

B.D. James, *Rec. Chem. Prog., 31* (1970) 199 (Metal-carbonyl hydrides).

J. Chatt, *Science, 160* (1968) 723 (Hydride complexes of transition metals).

M.L.H. Green, D.J. Jones, *Adv. Inorg. Chem. Radiochem., 7* (1965) 115 (Hydride complexes of transition metals).

A.P. Ginsberg, in *Transition Metal Chemistry,* R.L. Carlin, ed., New York (1965), Vol. 1, p. 111 (Hydride complexes of transition metals, including carbonyl hydrides).

h) Metal-Carbonyl Cations

E.W. Abel, S.P. Tyfield, *Adv. Organomet. Chem., 8* (1970) 117 (Metal-carbonyl cations).
M.H.B. Stidard, *Rev. Chim. Miner., 3* (1966) 801 (Carbonyl complexes of transition metals in positive oxidation states).

i) Metal-Carbonyl Halides

R. Colton, *Coord. Chem. Rev., 6* (1971) 269 (Steric effects in Mo and W halocarbonyls).
M.W. Anker, R. Colton, I.B. Tomkins, *Rev. Pure Appl. Chem., 18* (1968) 23 (Substituted halocarbonyls of group-VI transition metals).
F. Calderazzo, in *Halogen Chemistry*, V. Gutmann, ed., Academic Press, New York (1967), Vol. 3, p. 383 (Halogenometal carbonyls).

j) Thiocarbonyls

S. Rajan, *J. Sci. Ind. Res., 38* (1979) 648 (Thiocarbonyl complexes).
P.V. Yaneff, *Coord. Chem. Rev., 23* (1977) 183 (Thiocarbonyl complexes of transition metals).
G. Gattow, W. Behrendt, in *Topics in Sulfur Chemistry,* Vol. 2, A. Senning, ed., G. Thieme Verlag, Stuttgart, (1977), p. 1 (Carbon sulfide and its complex chemistry).
I.S. Butler, *Acc. Chem. Res., 10* (1977) 359 (Transition-metal thiocarbonyls and selenocarbonyls).
I.S. Butler, *J. Organomet. Chem., 66* (1974) 161 (Activation of CS_2 by transition-metal complexes).

k) Cyanides and Isocyanides

L. Malatesta, F. Bonati, *Isocyanide Complexes of Metals,* Wiley-Interscience, New York (1969).

E. Singleton, H.E. Oosthuizen, *Adv. Organomet. Chem., 22* (1983) 209 (Metal-isocyanide complexes).
S.J. Bryan, P.G. Huggett, K. Wade, J.A. Daniels, J.R. Jennings, *Coord. Chem. Rev., 44* (1982) 149 (Acetonitrile complexes).
Y. Yamamoto, *Coord. Chem. Rev., 32* (1980) 193 (Zero-valent transition-metal complexes of organic isocyanides).
K.R. Mann, H.B. Gray, in *Inorganic Compounds with Unusual Properties-II,* R.B. King, ed., Advances in Chemistry Series, No. 173, American Chemical Society, Washington, DC (1979), p. 225 (Photochemistry of binuclear Rh(I) isocyanides).
U. Schöllkopf, *Pure Appl. Chem., 51* (1979) 1347; *Angew. Chem., Int. Ed. Engl., 16* (1977) 339 (α-Metallated isocyanides in organic synthesis).
H.C. Clark, *Pure Appl. Chem., 50* (1978) 43 (Isocyanides of Pd(II) and Pt(II)).
T. Saegusa, Y. Ito, *Synthesis*, (1978) 291 (Cyclic compounds via Cu isonitriles).
B.N. Storhoff, H.C. Lewis, Jr., *Coord. Chem. Rev., 23* (1977) 1 (Organonitrile complexes).
S.J. Lippard, *Prog. Inorg. Chem., 21* (1976) 91 (Mo complexes with cyanide and isocyanide ligands).

S. Cenini, G. LaMonica, *Inorg. Chim. Acta, 18* (1976) 279 (Organic azides and isocyanates as sources of nitrene (imino) species).
L. P. Yur'eva, in *Metody Elementoorganicheskoi Khimii. Tipy Metallorganicheskikh Soedinenii Perekhodnykh Metallov*, A.N. Nesmeyanov, K.A. Kocheshkov, eds., Nauka, Moscow (1975), p. 162 (In Russian) (Isonitrile complexes of transition metals).
F. Bonati, G. Minghetti, *Inorg. Chim. Acta Rev., 9* (1974) 95 (Isocyanide complexes).
D. Hoppe, *Angew. Chem., Int. Ed. Engl., 13* (1974) 789 (α-Metallated isocyanides in organic synthesis).
P. Rigo, A. Turco, *Coord. Chem. Rev., 13* (1974) 133 (Cyanide-phosphine complexes).
P.M. Treichel, *Adv. Organomet. Chem., 11* (1973) 21 (Transition metal-isocyanide complexes).
Y. Yamamoto, H. Yamazaki, *Coord. Chem. Rev., 8* (1972) 225 (Isocyanide insertion and related reactions).
A. Volger, in *Isonitrile Chemistry*, I. Ugi, ed., Academic Press, New York (1971), p. 217 (Coordinated isonitriles).
L. Malatesta, *Prog. Inorg. Chem., 1* (1959) 283 (Isocyanide complexes of metals).

I) Carbene Derivatives

H. Fisher, F.R. Kreissi, U. Schubert, R. Hof, K.H. Dotz, K. Weiss, eds., *Transition Metal Carbene Complexes*, Verlag Chemie, Weinheim (1984).
F. G. A. Stone, in *Chemistry for the Future*, H. Grünewald, ed., Pergamon Press, Oxford (1984), p. 149 (Bridging carbenes).
U. Schubert, *Coord. Chem. Rev., 55* (1984) 261 (Transition-metal carbene structures).

M. I. Bruce, A. G. Swincer, *Adv. Organomet. Chem., 22* (1983) 59 (Vinylidene and propadienylidene (allenylidene) metal complexes).
J.E. Hahn, *Prog. Inorg. Chem., 31* (1983) 205 (Bridging alkylidenes).
W.A. Herrmann, *Pure Appl. Chem., 54* (1982) 65 (Methylene bridges).
C.P. Casey, Jr., in *Reactive Intermediates*, Vol. 2, M. Jones, Jr., R.A. Moss, eds., Wiley, New York (1981), p. 135 (Carbene complexes as intermediates).
F.J. Brown, *Prog. Inorg. Chem., 27* (1980) 1 (Stoichiometric reactions of metal-carbene complexes).
R.R. Schrock, *Acc. Chem. Res., 12* (1979) 98 (Alkylidene complexes of Nb and Ta).
H.C. Clark, *Pure Appl. Chem., 50* (1978) 43 (Carbenes of Pd(II) and Pt(II)).
E.O. Fischer, U. Schubert, H. Fischer, *Pure Appl. Chem., 50* (1978) 857 (Selectivity and specificity in chemical reactions of carbene and carbyne complexes).
A. Nakamura, *Pure Appl. Chem., 50* (1978) 37 (Enantioselective chiral carbene complexes).
O.M. Nefedov, A.I. Dyachenko, A.K. Prokofev, *Russ. Chem. Rev. (Engl. Transl.), 46* (1977) 941 (Carbene complexes).
C.P. Casey, *J. Org. Chem., 33* (1976) 189 (Metal-carbene complexes in organic synthesis).
C.P. Casey, *J. Organomet. Chem. Libr., 1* (1976) 397 (α-Anions of metal-carbene complexes in organic synthesis).
C.P. Casey, in *Transition-Metal Organometallics in Organic Synthesis*, H. Alper, ed., Academic Press, New York (1976), p. 189 (Metal-carbene complexes).
D. Seyferth, *Adv. Organomet. Chem., 14* (1976) 97 (Alkylidene-tricobalt nonacarbonyl clusters).
K. Dötz, *Naturwissenschaften, 62* (1975) 365 (Transition-metal carbene complexes in synthetic organic chemistry).

M.F. Lappert, *J. Organomet. Chem., 100* (1975) 139 (Nucleophilic carbene complexes).
E.O. Fischer, *Angew. Chem., Int. Ed. Engl., 86* (1974) 651 (Carbene and carbyne-metal complexes).
R.B. King, *Ann. N.Y. Acad. Sci., 239* (1974) 171 (Dicyanovinylidene metal complexes).
D.J. Cardin, B. Cetinkaya, M.J. Doyle, M.F. Lappert, *Chem. Soc. Rev., 2* (1973) 99 (Carbene complexes).
B.R. Penfold, B.H. Robinson, *Acc. Chem. Res., 6* (1973) 73 (Tricobalt carbon).
D.J. Cardin, B. Cetinkaya, M.F. Lappert, *Chem. Rev., 72* (1972) 545 (Carbene complexes).
F.A. Cotton, C.M. Lukehart, *Prog. Inorg. Chem., 16* (1972) 487 (Carbenoid complexes).

m) Arene, Olefin and Polyolefin Derivatives

G. Deganello, *Transition-Metal Complexes of Cyclic Polyolefins,* Academic Press, New York (1979).
M. Herberhold, *Metal π-Complexes.* Vol. 2, *Complexes with Mono-Olefinic Ligands.* Part 1. *General Survey*, Elsevier, Amsterdam (1972); Part 2, *Specific Aspects* (1974).
E.O. Fischer, H. Werner, *Metall π-Complexes*, Vol. 1, *Complexes with Di- and Oligo-olefinic Ligands*, Elsevier, Amsterdam (1966) (Englisch translation of *Metal π-Komplexe mit di- und oligo-olefinischen Liganden*, Verlag Chemie, Weinheim (1963)).

I. Omae, *Coord. Chem. Rev., 51* (1983) 1 *Angew. Chem., Int. Ed. Engl., 21* (1982) 889 (Intramolecular coordination by diolefins).
M.I. Rybinskaya, *Pure Appl. Chem., 54* (1982) 145 (α-Carbenium stabilization in olefin-Fe complexes).
P.W. Jolly, R. Mynott, *Adv. Organomet. Chem., 19* (1981) 257 (NMR characterization of olefin and allyl complexes).
P.L. Pauson, *Pure Appl. Chem., 49* (1977) 839 (Aromatic transition-metal complexes).
W. Schurig, *Chem. Ztg., 101* (1977) 173 (Selectivity and stereochemistry of olefin π-complex formation).
A.J. Birch, I.D. Jenkins, in *Transition-Metal Organometallics in Organic Synthesis,* Vol. I, H. Alper, ed., Academic Press, New York (1976), p. 1 (Metal-olefin complexes).
G. Henrici-Olivé, S. Olivé, *Top. Curr. Chem., 67* (1976) 107 (Olefin insertion in transition-metal catalysis).
A.J. Birch, J.D. Jenkins, *J. Org. Chem., 33* (1976) 1 (Transition-metal complexes of olefinic compounds).
S.D. Ittel, J.A. Ibers, *Adv. Organomet. Chem., 14* (1976) 33 (Coordination of unsaturated molecules to transition metals).
W.E. Silverthorn, *Adv. Organomet. Chem., 13* (1975) 48 (Arene-transition metal chemistry).
M.I. Rybinskaya, in *Methody Elementoorganicheskoi Khimii. Tipy Elementoorganicheskikh Soedinenii,* A.N. Nesmeyanov, ed., Nauka, Moscow (1975), p. 217 (π-Complexes of monoolefins, comprehensive review).
R.D.W. Kemmitt, *MTP Int. Rec. Sci., Inorg. Chem., Ser. Two, 6* (1975) 295 (Olefin-metal π-complexes).
G. Deganello, P. Uguagliati, L. Calligaro, L. Sandrini, F. Zingales, *Inorg. Chim. Acta, 13* (1975) 247 (Polyolefin-carbonyl derivatives of Fe, Ru, Os).
T. Inglis, *Inorg. Chim. Acta Rev., 7* (1973) 35 (Pseudo-allyl carbonyls).
F.R. Hartley, *Chem. Rev., 73* (1973) 163 (Thermodynamic data for olefin complexes).
D.I. Hall, J.H. Ling, R.S. Nyholm, *Struct. Bonding (Berlin), 15* (1973) 3 (Chelating olefin-group V ligands).

I. I. Kritskaya, *Russ. Chem. Rev. (Engl. Transl.)*, *41* (1972) 1027 (Organic ligands for transition metals).
R. D. Kemmitt, *MTP Int. Rev. Sci., Inorg. Chem., Ser. One, 6* (1972) (Olefin complexes).
J. H. Nelson, H. B. Jonassen, *Coord. Chem. Rev., 6* (1971) 27 (Olefin complexes of Ni, Pd and Pt).
M. Cais, in *The Chemistry of Alkenes,* S. Patai, J. Zabicky, eds., Vol. 2, Interscience, New York (1970), p. 335 (Alkene complexes of transition metals).
J. F. Bielleman, H. Hemmer, J. Levisalles, in *The Chemistry of Alkenes*, S. Patai, J. Zabicky, eds., Vol. 2, Interscience, New York (1970), p. 215 (Alkene complexes of transition metals as reactive intermediates).
G. Paiaro, *Organomet. Chem. Rev., A, 6* (1970) 319 (Optical activity in olefin complexes).
F. R. Hartley, *Chem. Rev., 69* (1969) 799 (Olefin complexes of Pd and Pt).
H. W. Quinn, J. H. Tsai, *Adv. Inorg. Chem. Radiochem., 12* (1969) 217 (Olefin complexes of transition metals).
R. Jones, *Chem. Rev., 68* (1968) 785 (Metal π-complexes of substituted olefins).
E. O. Fischer, H. Werner, *Angew. Chem., Int. Ed. Engl., 2* (1963) 80 (Metal-complexes with di- and oligo-olefinic ligands; later expanded into a book).
M. A. Bennett, *Chem. Rev., 62* (1962) 611 (Olefin and acetylene metal complexes).
R. G. Guy, B. L. Shaw, *Adv. Inorg. Chem. Radiochem., 4* (1962) 78 (Olefin, acetylene and π-allylic complexes of transition metals).

n) Nitrosyl Derivatives

R. D. Feltham, J. H. Enemark, *Top. Stereochem., 12* (1981) 156 (Structures of metal nitrosyls).
R. Eisenberg, C. D. Meyer, *Acc. Chem. Res., 8* (1975) 26 (Coordination chemistry of NO).
D. Sutton, *Chem. Soc. Rev., 4* (1975) 443 (Azodiazenato nitrosyl analogues).
K. G. Caulton, *Coord. Chem. Rev., 14* (1975) 317 (Nitrosyl syntheses).
J. H. Enemark, R. D. Feltham, *Coord. Chem. Rev., 13* (1974) 339 (Metal-nitrosyl chemistry).
N. G. Connelly, *Inorg. Chim. Acta Rev., 6* (1972) 47 (Metal nitrosyls).

o) Derivatives Containing Transition-Nonmetal Bonds

E. Colomer, R. J. P. Corriu, *Top. Curr. Chem., 96* (1981) 79 (Si- and Ge-transition metal bonds).
K. Jonas, *Adv. Organomet. Chem., 19* (1981) 97 (Alkali-metal transition-metal complexes).
G. A. Razuvaev, *J. Organomet. Chem., 200* (1980) 243 (Polynuclear σ-bonded metal organometallics).
M. N. Bochkarev, *Russ. Chem. Rev. (Engl. Transl.), 49* (1980) 800 (Heterometallic-chain organometallics).
A. Bonny, *Coord. Chem. Rev., 25* (1978) 229 (Group-IV derivatives of iron-triad carbonyls).
J. A. Zubieta, J. J. Zuckerman, *Prog. Inorg. Chem., 24* (1978) 251 (Tin-transition metal structures).
N. N. Greenwood, I. M. Ward, *Chem. Soc. Rev., 3* (1974) 231 (Metal-boron bonding).
F. Höfler, *Top. Curr. Chem., 50* (1974) 129 (Si-transition metal bonds).

C.S. Cundy, B.M. Kingston, M.F. Lappert, *Adv. Organomet. Chem., 11* (1973) 253 (Si-transition metal or Si-C-transition metal bonds).
B.Y.K.Ho, J.J. Zuckerman, *J. Organomet. Chem., 49* (1973) 1 (Tin-transition metal structures).
H.G. Any, P.T. Lau, *Organomet. Chem. Rev., A, 8* (1972) 235 (Si-transition-metal bonds).
G. Schmid, *Angew. Chem., Int. Ed. Engl., 9* (1970) 819 (Boron-metal carbonyl compounds).
U. Belluco, G. Deganello, R. Pietropaulo, P. Uguagliati, *Inorg. Chim. Acta Rev., 4* (1970) 7 (Group IV-Pt(II) complexes).
E.H. Brooks, R.J. Cross, *Organomet. Chem. Rev., A, 6* (1970) 277 (Group IV-transition-metal derivatives).
B.P. Biryukov, Yu.T. Struchkov, *Russ. Chem. Rev. (Engl. Transl.), 39* (1970) 789 (Metal-metal bonds).
N.E. Kolobova, A.B. Antonova, K.N. Anisimov, *Russ. Chem. Rev. (Engl. Transl.), 38* (1969) 822 (Group-IV element derivatives of metal carbonyls).
A.G. MacDiarmid, Y.L. Baay, J.F. Bald, A.D. Berry, A.K. Gondal, A.P. Hagen, M.A. Nasta, F.E. Saalfeld, M.V. McDowell, *Pure Appl. Chem., 19* (1969) 431 (Silicon derivatives of Mo, Fe, Co carbonyls).
G.A. Razuvaev, N.S. Vyzankin, *Pure Appl. Chem., 19* (1969) 353 (Metal carbonyl derivatives with M–Si and M–Ge bonds).
J.F. Young, *Adv. Inorg. Chem. Radiochem., 11* (1968) 92 (Metal-carbonyl derivatives of group-IV elements).

p) Miscellaneous Derivatives

C.A. McAuliffe, W. Levason, *Phosphine, Arsine and Stibine Complexes of the Transition-Elements,* Elsevier, Amsterdam (1979).
C.A. McAuliffe, ed., *Transition-Metal Complexes of Phosphorus, Arsenic and Antimony Ligands,* MacMillan, New York (1973).

H. Schmidbaur, *Angew. Chem., Int. Ed. Engl., 22* (1983) 980 (Phosphorus-ylide complexes).
W.R. Cullen, J.D. Woollins, *Coord. Chem. Rev., 39* (1981) (Ferrocenylphosphine and arsine complexes).
P.D. Smith, B.R. James, D.H. Dolphin, *Coord. Chem. Rev., 39* (1981) 31 (Metal porphyrin complexes with axial-carbon ligands).
G. van Koten, K. Vrieze, *Recl. Trav. Chim. Pays-Bas, 100* (1981) 129 (α-Diimine complexes).
S.G. Murray, F.R. Hartley, *Chem. Rev., 81* (1981) 365 (Thio-, seleno- and telluroether complexes).
I. Omae, *Coord. Chem. Rev., 32* (1980) 235 (Complexes with a phosphorus-donor ligand).
W.A. Nugent, L. Haymore, *Coord. Chem. Rev., 31* (1980) 123 (Complexes with organoimido (NR) ligands).
I. Omae, *Chem. Rev., 79* (1979) 287 (Intramolecular nitrogen donors).
I. Omae, *Coord. Chem. Rev., 28* (1979) 97 (Intramolecular sulfur ligands).
C.A. Tolman, *Chem. Rev., 77* (1977) 313 (Phosphorus ligands).
J. Chatt, *J. Organomet. Chem., 100* (1975) 17 (Mononuclear dinitrogen complexes).
D.G. Holah, A.N. Hughes, K. Wright, *Coord. Chem. Rev., 15* (1975) 239 (Cyclic phosphine complexes).
D. Sellmann, *Angew. Chem., Int. Ed. Engl., 13* (1974) 639 (Dinitrogen complexes).
P. Rigo, A. Turco, *Coord. Chem. Rev., 13* (1974) 133 (Cyanide-phosphine complexes).

A. D. Allen, R. O. Harris, B. R. Loescher, J. R. Stevens, R. N. Whiteley, *Chem. Rev., 73* (1973) 11 (Dinitrogen complexes).
R. J. Clark, M. A. Busch, *Acc. Chem. Res., 6* (1973) 246 (Carbonyl-phosphorus trifluoride complexes).
R. J. Angelici, *Acc. Chem. Res., 5* (1972) 335 (Carbamoyl and alkoxycarbonyl complexes).
A. J. Carty, *Organomet. Chem. Rev., A, 7* (1972) 191 (Organometallic derivatives containing N–N bonded ligands).
R. B. King, *Acc. Chem. Res., 5* (1972) 177 (Polytertiary phosphine complexes of metal carbonyls).
W. Levanson, C. A. McAuliffe, *Adv. Inorg. Chem. Radiochem., 14* (1972) 173 (Transition-metal complexes containing bidentate phosphine ligands).
M. Kilner, *Adv. Organomet. Chem., 10* (1972) 115 (Nitrogen groups in metal carbonyls).
W. Beck, *Organomet. Chem. Rev., A, 7* (1971) 159 (Metal-fulminate complexes).
J. F. Nixon, *Adv. Inorg. Chem. Radiochem., 13* (1970) 363 (Metal-carbonyl trifluorophosphine complexes).
E. W. Abel, B. C. Crosse, *Organomet. Chem. Rev., 2* (1967) 443 (Sulfur-containing metal carbonyls).
R. G. Hayter, *Prep. Inorg. React., 2* (1965) 211 (Sulfur- and phosphorus-bridged metal carbonyls).

Chapter 12

i) η^3-Allyl Complexes

D. L. Dolgoplosk, K. L. Makovetskii, E. I. Sinayakova, C. K. Sharaev, *Polimerizatsiya Dienov Pod Vliyaniem π-Allilnykh Kompleksov*, (Polymerization of Dienes with π-Allyl Complexes) Nauka, Moscow (1968) (In Russian).

I. Omae, *Coord. Chem. Rev., 53* (1984) 261 (Intramolecular coordination by an allyl donor).
B. Bosnich, P. B. Mackenzie, *Pure Appl. Chem., 54* (1982) 189 (Asymmetric catalytic allylic alkylation).
J. Tsuji, *Pure Appl. Chem., 54* (1982) 197 (π-Allylpalladium catalysis).
M. Julemont, Ph. Teyssie, in *Aspects of Homogeneous Catalysis*, Vol. 4, R. Ugo, ed., D. Reidel, Dordrecht (1981), p. 99 (η^3-Allyl catalysis of diolefin reactions).
H. Werner, *Adv. Organomet. Chem., 19* (1981) 155 (Allyl bridges in metal-metal bonded complexes).
P. W. Jolly, R. Mynott, *Adv. Organomet. Chem.,19* (1981) 257 (NMR characterization of allyl complexes).
B. M. Trost, *Acc. Chem. Res., 13* (1980) 385 (Allylic alkylations catalyzed by Pd).
R. B. King, in *The Organic Chemistry of Iron*, E. A. Koerner von Gustorf, F. W. Grevels, I. Fischler, eds., Academic Press, New York (1978), p. 463 (Fe allyls).
I. A. Dolgoplosk, E. J. Tinyakov, S. J. Beilin, G. N. Bondarenko, O. P. Parenago, N. N. Stefanovskaya, V. M. Frolov, O. K. Sharaev, V. A. Yakovleva, *Izv. Akad. Nauk SSSR, Ser. Khim.,* (1977) 2598 (Review of Russian research on π-allyl complexes of Ti, V, Cr, Fe, Co, Ni).
G. P. Chiusoli, L. Cassar, in *Organic Synthesis via Metal Carbonyls,* Vol. II, I. Wender, P. Pino, eds., Wiley, New York (1977), p. 297 (π-Allyl metal carbonyls in organic synthesis).

L. S. Hegedus, *J. Organomet. Chem. Libr., 1* (1976) 329 (π-Allyl complexes in organic synthesis).

F. L. Bowden, R. Giles, *Coord. Chem. Rev., 20* (1976) 81 (Formation of π-allyl complexes from allenes).

I. I. Kritskaya, in *Methody Elementoorganicheskoi Khimii. Tipy Metallorganicheskikh Soedinenii Perekhodnykh Metallov* (Methods of Organoelement Chemistry. Types of Transition Metal Organometallic Compounds) A. N. Nesmeyanov, K. A. Kocheshkov, eds., Nauka, Moscow (1975), p. 734 (π-Allyl metal complexes: a comprehensive review in Russian).

J. Powell, *MTP Int. Rev. Sci., Inorg. Chem., Ser. Two, 6* (1975) 189 (Allyl complexes of the 2nd and 3rd row transition elements).

N. Rösch, R. Hoffmann, *Inorg. Chem., 13* (1974) 2656 (Geometry and bonding of transition-metal complexes with allyl groups as the only ligands).

G. Coutois, L. Miginae, *J. Organomet. Chem., 69* (1974) 1 (Reactivity of allylic organometallics).

H. L. Clarke, *J. Organomet. Chem., 80* (1974) 155 (Binary π-allyl-metal compounds $M(allyl)_n$, $n = 2, 3, 4$).

I. Inglis, *Inorg. Chim. Acta Rev., 7* (1973) 35 (Pseudo-allyl metal-carbonyl complexes).

R. Baker, *Chem. Rev., 73* (1973) 487 (π-Allyl metals in organic synthesis).

B. L. Shaw, H. A. Stringer, *Inorg. Chim. Acta Rev., 7* (1973) 1 (Transition-metal allene complexes).

M. F. Semmelhack, *Org. React., 19* (1972) 115 (Formation of C–C bonds via π-allylnickel compounds).

W. Keim, in *Transition Metals in Homogeneous Catalysis,* G. N. Schrauzer, ed., M. Dekker, New York (1971), p. 59 (π-Allyl catalysts).

L. A. Fedorov, *Russ. Chem. Rev., (Engl. Transl.), 39* (1970) 655 (NMR spectra of metal-allyl complexes).

P. Weimbach, P. W. Jolly, G. Wilke, *Adv. Organomet. Chem., 8* (1970) 29 (π-Allylnickel intermediates in organic synthesis).

U. Birkenstock, H. Bonnemann, B. Bogdanovic, D. Walter, G. Wilke, in *Homogeneous Catalysis*, B. J. Luberoff, ed., Advances in Chemistry Series, No. 70, American Chemical Society, Washington, DC (1968), p. 250 (π-Allylnickel compounds as homogeneous catalysts).

M. I. Lobach, M. D. Babitskii, V. A. Kormer, *Russ. Chem. Rev. (Engl. Transl.), 36* (1967) 476 (π-Allyl complexes of transition metals).

G. Wilke, E. Bogdanovic, P. Hardt, P. Heimbach, W. Keim, M. Kroner, W. Oberkirsch, K. Tanaka, E. Steinrucke, D. Walter, H. Zimmermann, *Angew. Chem., Int. Ed. Engl., 5* (1966) 151 (π-Allyl-metal complexes).

M. L. H. Green, P. L. I. Nagy, *Adv. Organomet. Chem., 2* (1964) 325 (Allyl-metal complexes).

E. O. Fischer, H. Werner, *Z. Chem., 2* (1962) 174 (Transition-metal complexes with π-allyl and π-enyl ligands).

ii) Carbyne-Metal Complexes

F. G. A. Stone, in *Chemistry for the Future*, H. Grünewald, ed., Pergamon Press, Oxford (1984), p. 149 (Bridging carbyne groups).

D. Seyferth, in *Transition-Metal Organometallics in Organic Syntheses*, Vol. II. H. Alper, ed., Academic Press, New York (1978), p. 1 (Alkyne complexes and clusters as synthetic reagents).

C. Huttner, A. Frank, E. O. Fischer, *Isr. J. Chem., 15* (1977) 133 (X-ray structures of carbyne complexes).

E.O. Fischer, *Adv. Organomet. Chem., 14* (1976) 1 (Carbyne complexes).
D. Seyferth, *Adv. Organomet. Chem., 14* (1976) 97 (Alkylidynetricobalt nonacarbonyl clusters).
E.O. Fischer, U. Schubert, *J. Organomet. Chem., 100* (1975) 59 (Carbyne complexes).
E.O. Fischer, *Angew. Chem., Int. Ed. Engl., 13* (1974) 651 (Carbene and carbyne complexes: Nobel lecture).
G. Palyi, F. Piacenti, L. Marko, *Inorg. Chim. Acta Rev., 4* (1970) 109 (Methynyltricobalt enneacarbonyl compounds).

Chapter 13

i) General: η^4-Diene Complexes

S.P. Gubin, A.V. Golounin, *Dieny i Ikh "Pi"-Kompleksy* (Dienes and Their π-Complexes), Nauka, Sibirskoe Odelenie, Novosibirsk (1983).

D. Hoppe, *Nachr. Chem. Tech. Lab., 30* (1982) 706, 711 (1,3-Dienetricarbonyliron).
M.F. Semmelhack, *Pure Appl. Chem., 53* (1981) 2379 (Nucleophilic addition to diene-metal complexes).
G. van Koten, K. Vrieze, *Recl. Trav. Chim. Pays-Bas, 100* (1981) 129 (Diazabutadiene complexes).
A.J. Birch, *Ann. N. Y. Acad. Sci., 33* (1980) 107 (Tricarbonyl-(cyclohexadiene)iron complexes).
A.J. Pearson, *Acc. Chem. Res., 13* (1980) 463 (Fe diene tricarbonyl syntheses).
R.B. King, in *The Organic Chemistry of Iron*, Vol. 1, E.A. Koerner von Gustorf, F.W. Grevels, I. Fischler, eds., Academic Press, New York (1978), p. 525 (Diene complexes).
S.P. Gubin, in *Methody Elementoorganicheskoi Khimii. π-Kompleksy Perekhodnykh Metallov s Dienami, Arenami, Soedineniya s σ-Svyazyu M–C* (Methods of Organoelement Chemistry. π-Complexes of Transition Metals with Dienes, Arenes, Compounds with σ-M–C Bonds), A.N. Nesmeyanov, K.A. Kocheshkov, eds., Nauka, Moscow (1976), p. 7 (A comprehensive review on butadiene, cyclohexadiene, η^4-cyclooctatetraene, η^4-cyclopentadiene complexes in Russian).
P. Dowd, *Acc. Chem. Res., 5* (1972) 242 (Trimethylenemethane).
F. Weiss, *Quart. Rev., 24* (1970) 278 (Trimethylenemethane and related α,α′-disubstituted isobutenes).
R.C. Kerbar, D.J. Ehntholt, *Synthesis,* (1970) 449 (Fulvene complexes).
R. Pettit, G.F. Emerson, *Adv. Organomet. Chem., 1* (1964) 1 (Diene-iron carbonyl complexes).
R. Pettit, G.F. Emerson, J. Mahler, *J. Chem. Educ., 40* (1963) 175 (Chemistry of diene-iron tricarbonyl complexes).

ii) Cyclobutadiene Complexes

G. Deganello, *Transition-Metal Complexes of Cyclic Polyolefins,* Academic Press, New York (1979).
M.P. Cava, M.J. Mitchell, *Cyclobutadiene and Related Compounds,* Academic Press, New York (1967).

T. Bally, S. Masamune, *Tetrahedron, 36* (1980) 343 (Cyclobutadiene).

P. M. Maitlis, *J. Organomet. Chem., 200* (1980) 161 (Palladium cyclobutadienes).

A. Efraty, *Chem. Rev., 77* (1977) 691 (Cyclobutadiene metal complexes; comprehensive review).

L. V. Rybin, in *Methody Elementoorganicheskoi Khimii. Tipy Metallorganicheskikh Soedinenii Perekhodnykh Metallov* (Methods of Organoelement Chemistry. Types of Metal-Organic Compounds of Transition Metals), A. N. Nesmeyanov, K. A. Kocheshkov, eds., Nauka, Moscow (1975), p. 542 (Comprehensive review on cyclobutadiene-metal complexes in Russian).

R. Pettit, *J. Organomet. Chem., 100* (1975) 205 (Reactions of cyclobutadiene complexes in organic synthesis).

P. M. Maitlis, K. W. Eberius, in *Nonbenzenoid Aromatics,* J. P. Snyder, ed., Academic Press, New York (1971) (Cyclobutadiene-metal complexes).

R. Pettit, *Pure Appl. Chem., 17* (1968) 253 (Review of the early results).

L. Watts, R. Pettit, in *Werner Centennial*, G.. Kauffman, ed., Advances in Chemistry Series, No. 62, American Chemical Society, Washington, DC (1967), p. 549 (Chemistry of cyclobutadiene-iron tricarbonyl).

P. M. Maitlis, *Adv. Organomet. Chem., 4* (1966) 95 (Review of the early cyclobutadiene-complex chemistry).

R. Criegee, *Angew. Chem., Int. Ed. Engl., 1* (1962) 519 (Attempts to synthesize tetramethylcyclobutadiene).

Chapter 14

J. C. Johnson, *Metallocene Technology,* Noyes Data Corp., Park Ridge, NJ (1973).

E. Neuse, H. Rosenberg, *Metallocene Polymers,* M. Dekker, New York (1970).

M. Rosenblum, *Chemistry of the Iron-Group Metallocenes: Ferrocene, Ruthenocene, Osmocene,* Wiley, New York (1965).

L. Sacconi, in *Chemistry for the Future*, H. Grünewald, ed., Pergamon Press, Oxford (1984), p. 143 (Multidecker sandwich compounds).

K. Jonas, *Pure Appl. Chem., 56* (1984) 63 (Reactive organometallics from metallocene).

P. N. Gaponik, A. L. Lesnikovich, Yu. G. Orlik, *Russ. Chem. Rev. (Engl. Transl.), 52* (1983) 294 (Ferrocene-based organo-metallics).

H. Werner, *Angew. Chem., Int. Ed. Engl., 22*, (1983) 932 (Electron-rich semisandwich bases).

M. F. Lappert, A. Singh, *J. Organomet. Chem., 239* (1982) 133 (Bis-η^5-cyclopentadienyllanthanoid(III) chlorides).

R. N. Grimes, *Pure Appl. Chem., 54* (1982) 43 (Metallocarboranes in organic synthesis).

A. G. Davies, *Pure Appl. Chem., 54* (1982) 23 (Photolytic reactions of cyclopentadienyl-metal compounds).

I. Omae, *Coord. Chem. Rev., 42* (1982) 31 (Intramolecular coordination in cyclopentadienyls).

C. Simionescu, T. Lixandru, L. Tătaru, I. Mazilu, M. Vata, *J. Organomet. Chem., 238* (1982) 363 (Ferrocene monomers and polymers).

D. W. Macomber, W. P. Hart, M. D. Rausch, *Adv. Organomet. Chem., 21* (1982) 1 (Functionally substituted metal cyclopentadienes).

H. Werner, *Adv. Organomet. Chem., 19* (1981) 155 (Metal-metal bonded complexes with bridging cyclopentadienyls).

G. P. Pez, J. N. Armor, *Adv. Organomet. Chem., 19* (1981) 1 (Titano- and zirconocene).
W. R. Cullen, J. D. Woollins, *Coord. Chem. Rev., 39* (1981) 1 (Ferrocene-containing metal complexes).
P. M. Maitlis, *Chem. Soc. Rev., 10* (1981) 1 (η^5-Cyclopentadienyl protecting groups for Pt-metal complexes).
H. Werner, *J. Organomet. Chem., 200* (1980) 335 (Triple-decker sandwiches).
E. V. Leonova, *Russ. Chem. Rev. (Engl. Transl.), 49* (1980) 147 (Co complexes of carbaboranes).
G. B. Shul'pin, *Russ. Chem. Rev. (Engl. Transl.), 49* (1980) 645 (Diastereotropy of substituted π-cyclopentadienyls).
B. E. R. Schilling, R. Hoffmann, D. L. Lichtenberger, *J. Am. Chem. Soc., 101* (1979) 585 (Bonding and structure of $C_5H_5M(CO)_2I$ complexes).
R. N. Grimes, *Coord. Chem. Rev., 28* (1979) 47 (Metal-sandwich complexes of cyclic planar and pyramidal boron ligands).
P. M. Maitlis, in *Inorganic Compounds with Unusual Properties-II*, R. B. King, ed., Advances in Chemistry Series, No. 173, American Chemical Society, Washington, DC (1979), p. 31 (η^5-Me_5C_5-Rh and Ir catalysts for olefin and arene hydrogenation).
W. E. Watts, *J. Organomet. Chem. Libr., 7* (1979) 401 (Ferrocenyl-carbocations).
J. E. Sheats, *J. Organomet. Chem. Libr., 7* (1979) 461 (Cobaltocenes).
N. S. Kochetkova, Yu. K. Krynkina, *Russ. Chem. Rev. (Engl. Transl.), 47* (1978) 486 (Applications of cyclopentadienyl transition-metal complexes).
J. E. Bercaw, in *Transition-Metal Hydrides*, R. Bau, ed., Advances in Chemistry Series, No. 167, American Chemical Society, Washington, DC (1978), p. 136 ((η^5-$C_5Me_5)_2ZrH_2$).
P. M. Maitlis, *Acc. Chem. Res., 11* (1978) 301) (Pentamethylcyclopentadienyl Rh and Ir complexes).
E. Maslowsky, Jr., *J. Chem. Educ., 55* (1978) 276 (Structures of metal-cyclopentadienyl derivatives).
P. C. Bharara, V. D. Gupta, R. C. Mehrotra, *J. Organomet. Chem. Libr., 5* (1977) 259 (Cyclopentadienyl-metal complexes).
H. Werner, *Angew. Chem., Int. Ed. Engl., 16* (1977) 1 (Varieties of sandwich complexes).
C. B. Hunt, *Educ. Chem., 14* (1977) 110 (Metallocenes, the first 25 years).
P. L. Pauson, *Pure Appl. Chem., 49* (1977) 839 (Aromatic transition-metal complexes – the first 25 years).
R. N. Grimes, *Organomet. React. Synth., 6* (1977) 63 (Metallocarborane reactions).
R. G. Sutherland, W. J. Pannekoek, C. C. Lee, *Ann. N. Y. Acad. Sci., 295* (1977) 192 (Hydrogenation of aromatic ligands).
C. U. Pittman, Jr., B. Surynarayanan, Y. Sasaki, in *Inorganic Compounds with Unusual Properties*, R. B. King, ed., Advances in Chemistry Series, No. 150, American Chemical Society, Washington, DC (1976), p. 46 (Semiconducting ferrocene polymers).
P. J. Vergamini, G. J. Kubas, *Prog. Inorg. Chem., 21* (1976) 261 (Cyclopentadienyl-metal sulfide clusters).
J. W. Lauher, M. Elian, R. H. Summerville, R. Hoffmann, *J. Am. Chem. Soc., 98* (1976) 3219 (Triple-decker sandwiches).
R. B. King, *Coord. Chem. Rev., 20* (1976) 155 (Pentamethylcyclopentadienyl-metal complexes).
K. D. Warren, *Struct. Bonding (Berlin), 27* (1976) 45 (Ligand-field theory of metal-sandwich complexes).
M. Elian, M. M. L. Chen. D. M. P. Mingos, R. Hoffmann, *Inorg. Chem., 15* (1976) 1148 (Bonding in M-C_5H_5 complexes).
J. W. Lauher, R. Hoffmann, *J. Am. Chem. Soc., 98* (1976) 1729 (Structure and chemistry of bis $(C_5H_5)ML_n$ complexes).

K. Schlögl, H. Falk, in *Methodicum Chimicum*, Vol. 8, *Preparation of Transition-Metal Derivatives*, K. Niedenzu, H. Zimmer, eds., G. Thieme Verlag, Stuttgart (1976), p. 469 (Ferrocene); H. Rosenberg, p. 500 (Sandwich compounds, metallocenes).

T. V. Nikitina, in *Methody Elementoorganicheskoi Khimii. Tipy Metalloorganicheskikh Soedinenii Perekhodnykh Metallov*, A. N. Nesmeyanov, K. A. Kocheshkov, eds., (Methods of Organoelement Chemistry. Types of Organometallic Compounds of Transition Metals), Nauka, Moscow (1975), p. 585 (Monocyclopentadienyl-metal derivatives, extensive review in Russian); E. G. Perevalova, T. V. Nikitina, p. 687 (Bis-cyclopentadienyl-metal derivatives, extensive review in Russian).

G. R. Knox, W. E. Watts, *MTP Int. Rev. Sci., Inorg. Chem., Ser. Two, 6* (1975) 219 (π-Cyclopentadienyl-metal complexes).

G. Wilkinson, *J. Organomet. Chem., 100* (1975) 273 (Fe sandwich-compounds – the first four months).

P. E. Gaivoronskii, N. V. Lamin, *Russ. Chem. Rev. (Engl. Transl.). 43* (1974) 466 (Mass spectra of π-organotransition-metal complexes).

S. A. R. Knox, F. G. A. Stone, *Acc. Chem. Res., 7* (1974) 321 (Synthesis of pentalene-metal complexes).

K. W. Barnett, *J. Organomet. Chem., 78* (1974) 139 (Chemistry of nickelocene).

A. N. Nesmeyanov, N. S. Kochetkova, *Russ. Chem. Rev. (Engl. Transl.), 43* (1974) 710 (Practical uses of ferrocene and its derivatives).

G. B. Shulpin, M. I. Rybinskaya, *Russ. Chem. Rev. (Engl. Transl.), 43* (1974) 710 (Ferrocenophanes).

D. O. Cowan, C. LeVanda, J. Park, F. Kaufmann, *Acc. Chem. Res., 6* (1973) 1 (Mixed-valence ferrocene chemistry).

G. Davidson, *Organomet. Chem. Rev., A, 8* (1972) 303 (Vibrational spectra of π-organotransition-metal complexes).

E. G. Perevalova, T. V. Nikitina, *Organomet. React., 4* (1972) 163 (Reactions of bis(cyclopentadienyl)metal compounds).

K. W. Barnett, D. W. Slocum, *J. Organomet. Chem., 44* (1972) 1 (Cyclopentadienyl metal complexes of Cr, Mo, W).

D. W. Slocum, C. R. Ernst, *Adv. Organomet. Chem., 10* (1972) 79 (Electronic effects in metallocenes and related systems).

B. Floris, C. Illuminati, G. Ortaggi, *Coord. Chem. Rev., 8* (1972) 39 (Electron-donor properties of ferrocene).

M. D. Rausch, *Pure Appl. Chem., 30* (1972) 523 (Metal-cyclopentadienyl complexes).

B. P. Biryukov, Yu. T. Struchkov, *Russ. Chem. Rev. (Engl. Transl.), 39* (1970) 789 (Metal-metal bonds in π-complexes).

R. C. Kerber, D. J. Entholt, *Synthesis*, (1970) 449 (Transition-metal complexes of fulvenes).

D. W. Slocum, C. R. Ernst, *Organomet. Chem. Rev., A, 6* (1970) 337 (Metallocene homoannular electronic effects).

D. E. Bublitz, K. E. Rinehart, *Org. React., 17* (1969) 1 (Synthesis of substituted ferrocenes and related metallocenes).

A. N. Nesmeyanov, *Pure Appl. Chem., 17* (1968) 211 (Organic chemistry of ferrocenes).

E. W. Neuse, *Adv. Macromol. Chem., 1* (1968) 1 (Ferrocene polymers).

H. J. Lorkowski, *Top. Curr. Chem., 9* (1967) 207 (Ferrocene as a building unit in polymers).

K. Schlögl, *Top. Stereochem., 1* (1967) 39 (Stereochemistry of metallocenes).

W. E. Watts, *Organomet. Chem. Rev., 2* (1967) 231 (Ferrocenophane systems).

R. Schlögl, *Top. Curr. Chem., 6* (1966) 479 (Stereochemistry of metallocenes).

A. N. Nesmeyanov, E. G. Perevalova, *Ann. N. Y. Acad. Sci., 125* (1965) 67 (Organic chemistry of ferrocene).

R. L. Pruett, *Prep. Inorg. React., 2* (1965) 187 (Synthesis of cyclopentadienyl-metal carbonyls).

M. D. Rausch, J. M. Birmingham, *Ann. N. Y. Acad. Sci.*, *125* (1965) 57 (Synthesis of metallocenes).
J. M. Birmingham, *Adv. Organomet. Chem.*, *2* (1964) 365 (Synthesis of cyclopentadienyl-metal compounds).
W. F. Little, *Survey Prog. Chem.*, *1* (1963) 133 (Metallocenes).
M. D. Rausch, *Can. J. Chem.*, *41* (1963) 1289 (Metallocene chemistry – a decade of progress).
L. Plesske, *Angew. Chem., Int. Ed. Engl.*, *1* (1962) 301 (Ring substitutions in ferrocene).
M. D. Rausch, *J. Chem. Educ.*, *37* (1960) 568 (Cyclopentadienyl compounds of metals).
G. Wilkinson, F. A. Cotton, *Prog. Inorg. Chem.*, *1* (1959) 1 (Cyclopentadienyl-metal compounds, early results).
E. O. Fischer, H. P. Fritz, *Adv. Inorg. Chem. Radiochem.*, *1* (1959) 56 (Review of early results in metallocene chemistry).
E. O. Fischer, *Angew. Chem.*, *67* (1955) 475 (Classical review on early results in metallocene chemistry).
P. L. Pauson, *Quart. Rev.*, *9* (1955) 391 (Classical review on metallocenes by one of the discoverers of ferrocene).

Chapter 15

G. Deganello, *Transition-Metal Complexes of Cyclic Polyolefins*, Academic Press, New York (1979).
H. Zeiss, P. J. Wheatley, H. J. S. Winkler, *Benzenoid-Metal Complexes, Structural Determinations and Chemistry*, Ronald Press, New York (1965).

E. L. Muetterties, J. R. Bleeke, E. J. Wacherer, T. Albright, *Chem. Rev.*, *82* (1982) 499 (Arene-metal complexes).
P. M. Maitlis, *Chem. Soc. Rev.*, *10* (1981) 1 (η^6-Arenes as protecting ligands for Pt-metal complexes).
M. F. Semmelhack, *Pure Appl. Chem.*, *53* (1981) 2379 (Nucleophilic addition to arene complexes).
M. L. H. Green, *J. Organomet. Chem.*, *200* (1980) 119 (Transition-metal atom synthesis of zerovalent arenes).
G. Jaouen, in *Transition-Metal Organometallics in Organic Synthesis*, Vol. II, H. Alper, ed., Academic Press, New York (1978), p. 65 (Arene complexes in organic synthesis).
K. H. Pannell, B. L. Kalsotra, C. Parkanyi, *J. Heterocycl. Chem.*, *15* (1978) 1057 (Heterocyclic π-complexes of transition metals).
A. N. Artemov, *Khim. Elementoorg. Soedin.*, (1977) 17 (Organo-metallic derivatives of benzenechromium tricarbonyl).
R. G. Sutherland, *J. Organomet. Chem. Libr.*, *3* (1977) 311 (π-Arene-π-cyclopentadienyliron cations).
M. F. Semmelhack, *Ann. N. Y. Acad. Sci.*, *295* (1977) 36 (Arene-metal complexes in organic synthesis).
G. Jaouen, *Ann. N. Y. Acad. Sci.*, *295* (1977) 58 (Use of arene chromium tricarbonyl complexes in organic synthesis).
M. F. Farona, *Organomet. React. Synth.*, *6* (1977) (Catalysis by arene-metal tricarbonyls).
M. F. Semmelhack, *J. Organomet. Chem. Libr.*, *1* (1976) 361 (Arene complexes in organic synthesis).

W.E. Silverthorn, *Adv. Organomet. Chem., 13* (1975) 47 (Arene-transition-metal chemistry).
K.J. Klabunde, *Angew. Chem., Int. Ed. Engl., 14* (1975) 287 (Preparation of arene-metal complexes by reactions with metal atoms).
T.A. Stephenson, *MTP Int. Rev. Sci., Inorg. Chem., Ser. Two, 6* (1975) 287 (Metal compounds of six-electron ligands).
P.L. Pauson, *Pure Appl. Chem., 49* (1975) 309 (Aromatic transition-metal complexes, the first 25 years).
M.A. Bennett, in *Rodd's Chemistry of Carbon Compounds,* 2nd ed., Vol. 13, Part B, S. Coffey, ed., Elsevier, Amsterdam (1974), p. 357 (Aromatic compounds of transition metals).
M.E. Volpin, Yu. N. Novikov, in *Top. Nonbenzenoid Aromatic Chem., 1* (1973) 269 (Graphite as an aromatic ligand).
M.D. Rausch, *Pure Appl. Chem., 30* (1972) 523 (Metal-arene chemistry).
H. Werner, *Top. Curr. Chem., 28* (1972) 141 (Reactions of aromatic-metal complexes).
M.R. Churchill, *Prog. Inorg. Chem., 11* (1970) 53 (Azulene complexes).
M. Tsutsui, *Trans. N.Y. Acad. Sci., 30* (1968) 658 (Arene π-complex chemistry).
R.I. Pruett, *Prog. Inorg. Chem., 2* (1965) 187 (Arene metal carbonyls).
K. Plesske, *Angew. Chem., Int. Ed. Engl., 1* (1962) 301, 347 (Ring-substitution in arene-metal complexes, early results).
E.O. Fischer, H.P. Fritz, *Angew. Chem., 73* (1961) 353 (π-Complexes of benzene with transition metals).
E.O. Fischer, H.P. Fritz, *Adv. Inorg. Chem. Radiochem., 1* (1959) 56 (Compounds of aromatic ring systems with metals, early results).
G. Wilkinson, F.A. Cotton, *Prog. Inorg. Chem., 1* (1959) 1 (Arene metal compounds, early results).

Chapter 16

G. Deganello, *Transition-Metal Complexes of Cyclic Polyolefins,* Academic Press, New York (1979) (Dealing with polyolefins with more than six-carbon atoms in a ring and including cycloheptatriene, cyclooctatetraene, etc.).

L.A. Paquette, *Tetrahedron, 31* (1975) 2855 (Cyclooctatetraene derivatives).
T.A. Stephenson, *MTP Int. Rev. Sci., Inorg. Chem., Ser. Two, 6* (1975) 287 (Metal complexes containing seven-electron organic ligands).
E. Ashworth, M.L.H. Green, J. Knight, *J. Less-Common Met., 36* (1974) 213 (Cycloheptatrienyl-molybdenum chemistry).
A. Streitwieser, Jr., *Proc. Rare-Earth Res. Conf.,* 11th, J.M. Haschke, H.A. Eick, eds., NTIS, Springfield, VA, *1* (1974) 278 (Uranocene and related chemistry).
A. Streitwieser, Jr., in *Topics in Nonbenzenoid Aromatic Chemistry*, Vol. 1 (1973) 221 (Uranocene: f-orbital aromatic system derived from cyclooctatetraene).
E. Pietra, *Chem. Rev., 73* (1973) 293 (Seven-membered conjugated carbo- and heterocyclic compounds and their homoconjugated analogues in metal complexes).

Chapter 17

E. Sappo, A. Tiripicchio, P. Braunstein, *Chem. Rev., 83* (1983) 203 (Alkyne-carbonyl clusters of the Fe, Co and Ni triads).

R. Nast, *Coord. Chem. Rev., 47* (1982) 89 (Metal-alkynyl coordination).

D. Seyferth, in *Transition-Metal Organometallics in Organic Synthesis,* Vol. II, H. Alper, ed., Academic Press, New York (1978), p. 1 (Alkyne complexes and clusters as synthetic reagents).

A.C. Hopkinson, in *The Chemistry of the Carbon-Carbon Triple Bond,* S. Patai, ed., J. Wiley, New York (1978), Part 1, p. 75 (Metal complexes of acetylenes).

M. Simonetta, A. Gavezzotti, in *The Chemistry of the Carbon-Carbon Triple bond,* S. Patai, ed., J. Wiley, New York (1978), Part 1, p. 36 (Acetylene-metal bonding).

R.B. King, *J. Ind. Chem. Soc., 54* (1977) 169 (Transition-metal derivatives of macrocyclic alkadiynes).

K.P.C. Vollhardt, *Acc. Chem. Res., 10* (1977) 1 (Transition-metal catalyzed acetylene cyclizations).

R.B. King, *Ann. N.Y. Acad. Sci., 295* (1977) 135 (Carbon-carbon triple bond dichotomy in alkynes using metal carbonyls).

P.L. Pauson, I.U. Khand, *Ann. N.Y. Acad. Sci., 295* (1977) 2 (Applications of cobalt-carbonyl acetylene complexes to organic synthesis).

S. Otsuka, A. Nakamura, *Adv. Organomet. Chem., 14* (1976) 245 (Acetylene complexes in homogeneous catalysis).

P.M. Maitlis, *Acc. Chem. Res., 9* (1976) 93 (Palladium oligomerization of acetylenes).

L.P. Yur'eva, in *Methody Elementoorganicheskoi Khimii. Tipy Metalloorganicheskikh Soedinenii Perekodnykh Metallov,* A.N. Nesmeyanov, K.A. Kocheskov, eds., (Methods of Organoelement Chemistry. Types of Organic Compounds of Transition Metals), Nauka, Moscow (1975), p. 384 (Acetylenic π-complexes of transition metals, extensive review in Russian).

R.S. Dickson, P.J. Fraser, *Adv. Organomet. Chem., 12* (1974) 323 (Compounds derived from alkynes and carbonyl complexes of Co).

L.P. Yur'eva, *Russ. Chem. Rev. (Engl. Transl.), 43* (1974) 95 (Oligomerization of acetylenes on transition-metal complexes).

E. Müller, *Synthesis* (1974) 761 (Transition-metal 1,4-, 1,5-, 1,6- and 1,7-diyne reactions).

F.R. Hartley, *Chem. Rev., 73* (1973) 163 (Thermodynamic data for acetylene complexes).

P.M. Maitlis, *Pure Appl. Chem., 30* (1972) 427 (Oligomerization of acetylenes induced by metals of the Ni triad).

R.D. Kemmit, *MTP Int. Rev. Sci., Inorg. Chem., Ser. One, 6* (1972) 227 (Olefin and acetylene complexes of transition metals).

F.R. Hartley, *Angew. Chem., Int. Ed. Engl., 11* (1972) 596 (Metal-acetylene bonding).

G. Marr, B.W. Rockett, *Educ. Chem., 8* (1971) 13 (Alkyne complexes of transition metals).

J.H. Nelson, H.B. Jonassen, *Coord. Chem. Rev., 6* (1971) 27 (Acetylene complexes of Ni, Pd, Pt).

F.R. Hartley, *Chem. Rev., 69* (1969) 799 (Acetylene complexes of Pd and Pt).

H. Hübel, in *Organic Syntheses via Metal Carbonyls,* I. Wender, P. Pino, eds., Interscience, New York (1968), p. 273 (Organometallic derivatives from metal carbonyls and acetylenes).

F.I. Bowden, A.B.P. Lever, *Organomet. Chem. Rev., 3* (1968) 227 (Transition-metal chemistry of acetylenes).

M.A. Bennett, *Chem. Rev., 72* (1962) 611 (Acetylene complexes of transition metals).

R.G. Guy, B.I. Shaw, *Adv. Inorg. Chem. Radiochem., 4* (1962) 78 (Acetylene complexes of transition metals).

Chapter 18

R. Taube, *Comments Inorg. Chem., 3* (1984) 69 (Stabilization of low oxidation state σ-organo-transition metals by lithium coordination).

M. Green, *Mech. Inorg. Organomet. React., 2* (1984) 271 (Metal-alkyl bond formation and fission; oxidative addition and reductive elimination).

J.D. Atwood, T.S. Janik, M.F. Pyszezek, P.S. Sullivan, *Ann. N.Y. Acad. Sci., 415* (1983) 259 (Catalytic carbonylation-decarbonylation of alkyl complexes).

J.A. Morrison, *Adv. Inorg. Chem. Radiochem., 27* (1983) 293 (Trifluoromethyl) derivatives of the transition metals).

W.A. Herrmann, *J. Organomet. Chem., 250* (1983) 319 (Methylene bridges).

M. Green, *Mech. Inorg. Organomet. React.* (1983) 211, 340 (Metal-alkyl bond fission and formation).

J.D. Chappell, D.J. Cole-Hamilton, *Polyhedron, 1* (1982) 739 (Transition-metal metallocycles).

R. Nast, *Coord. Chem. Rev., 47* (1982) 89 (Coordination by metal alkynyls).

V.E. L'vovskii, E.A. Fushman, F.S. D'yachkovskii, *Russ. J. Phys. Chem. (Engl. Transl.), 56* (1982) 1864 (Reactivity of first-row transition metal-carbon σ-bonds).

C. Floriani, *Pure Appl. Chem., 54* (1982) 59 (Metal-carbon bond formation from small molecules and one-carbon functional groups).

T.C. Flood, *Top. Stereochem., 12* (1981) 37 (Reaction and stereochemistry of transition-metal carbon σ-bonds).

A. Krief, *Tetrahedron, 36* (1980) 2531 (σ-Heterosubstituted organometallics in synthesis).

E.J. Kuhlman, J.J. Alexander, *Coord. Chem. Rev., 33* (1980) 195 (CO insertion into transition-metal-carbon σ-bonds).

R. Bergman, *Acc. Chem. Res., 13* (1980) 113 (Binuclear dialkyl-cobalt complexes).

R.J. Lagow, J.A. Morrison, *Adv. Inorg. Chem. Radiochem., 23* (1980) 178 (Trifluoromethyl-organometallic synthesis).

H.C. Clark, *J. Organomet. Chem., 200* (1980) 62 (Fluorocarbons of metals).

E. Maslowsky, Jr., *Chem. Soc. Rev., 9* (1980) 25 (Methyl metals).

D.St. Black, W.R. Jackson, J.M. Swan, in *Comprehensive Organic Chemistry*, D. Barton, W.D. Ollis, eds., Vol. 3, D.N. Jones, ed., Pergamon Press, Oxford (1979), p. 1127 (Transition metal-carbon σ-bonds).

M.I. Lobach, V.A. Kormer, *Russ. Chem. Rev. (Engl. Transl.), 48* (1979) 758 (Diene insertion into transition metal-ligand bonds).

F.L. Bowden, L.H. Wood, in *The Organic Chemistry of Iron*, Vol. 1, E.A. Koerner von Gustorf, F.W. Grevels, I. Fischler, Academic Press, New York (1978), p. 345 (Fe-C σ-bonds).

B. Gorewit, M. Tsutsui, *Adv. Catal., 27* (1978) 227 (σ-π Rearrangements and their role in catalysis).

M.F. Lappert, *Pure Appl. Chem., 50* (1978) 703 (Unusual metal alkyls).

H. Schmidbaur, *Pure Appl. Chem., 50* (1978) 19 (σ-Ylid derivatives of transition metals).

R. Taube, H. Drevs, D. Steinborn, *Z. Chem., 19* (1978) 425 (Stable σ-organo-compounds of transition metals).

F. Calderazzo, *Pure Appl. Chem., 50* (1978) 49 (Synthesis and reactivity of C-bonded transition elements).

M.D. Johnson, *Acc. Chem. Res., 11* (1978) 57 (Reactions of electrophiles on σ-bonded organo-transition metals).

M.I. Bruce, *Angew. Chem., Int. Ed. Engl., 16* (1977) 73 (Cyclometallation reactions).

R. Uson, A. Laguna, J. Vicente, *Synth. React. Inorg. Metalorg. Chem., 7* (1977) 463 (σ-Aryl-gold derivatives).

R. Fields, *Ann. Rev. NMR Spectrosc.*, *7* (1977) 1 (F-19 NMR of fluoroalkyl- and fluoraryl-transition-metal derivatives).

M. Tsutsui, N. Ely, R. Dubois, *Acc. Chem. Res.*, *9* (1976) 217 (σ-Bonded organic derivatives of f-elements).

R. Fields, *Fluorocarbon Related Chem.*, *3* (1976) 308 (Polyfluorinated aliphatic derivatives of the transition elements).

J. Dehand, M. Pfeffer, *Coord. Chem. Rev.*, *18* (1976) 327 (Cyclometallated compounds).

P.J. Davidson, M.F. Lappert, R. Pearce, *Chem. Rev.*, *76* (1976) 219 (Metal σ-hydrocarbonyl, mostly homoleptic).

R.R. Schrock, G.W. Parshall, *Chem. Rev.*, *76* (1976) 243 (σ-Alkyl and aryl complexes of transition metals).

M.F. Lappert, in *Inorganic Compounds with Unusual Properties*, R.B. Kind, ed., Advances in Chemistry Series, No. 150, American Chemical Society, Washington, DC (1976), p. 256 (The bulky alkyl ligand $(Me_3Si)_2CH$-).

B.L. Booth, *MTP Int. Rev. Sci., Inorg. Chem., Ser. Two*, *6* (1975) 137 (Complexes containing metal-carbon σ-bonds).

H. Gilman, *J. Organomet. Chem.*, *100* (1975) 83 (Perfluoro-organometallics).

M.C. Baird, *J. Organomet. Chem.*, *64* (1974) 289 (Transition metal-carbon σ-bond scission).

G. Wilkinson, *Angew. Chem., Int. Ed. Engl.*, *13* (1974) 664 (Stable alkyl-transition-metal compounds: Nobel lecture).

P.J. Davidson, M.F. Lappert, *Acc. Chem. Res.*, *7* (1974) 209 (Stable, homoleptic metal alkyls).

A. Wojcicki, *Adv. Organomet. Chem.*, *12* (1974) 32 (SO_2-Insertion into transition-metal-carbon σ-bonds).

D.V. Ioffe, M.I. Mostova, *Russ. Chem. Rev. (Engl. Transl.)*, *42* (1973) 56 (Dual reactivity of arylmethyl organometallics).

P.S. Braterman, P.J. Cross, *Chem. Soc. Rev.*, *2* (1973) 271 (Stability of σ-metallorganic compounds).

A. Wojcicki, *Adv. Organomet. Chem.*, *11* (1973) 87 (CO Insertion into transition-metal-carbon σ-bonded compounds).

D.G.H. Ballard, *Adv. Catal.*, *23* (1973) 263 (σ-Transition-metal organometallics as catalysts for polymerization of vinyl monomers).

D.R. Fahey, *Organomet. Chem. Rev., A*, *7* (1972) 245 (Organo-nickel σ-derivatives).

W.R. Cullen, *Adv. Inorg. Chem. Radiochem.*, *15* (1972) 323 (Fluoroalicyclic derivatives).

W.T. Reichle, in *Characterization of Organometallic Compounds*, M. Tsutsui, ed., Wiley-Interscience, New York (1971), Vol. 2, p. 653 (σ-Bonded organometallics).

R.S. Nyholm, *Quart. Rev.*, *24* (1970) 1 (Perfluoro-ligands for transition metals).

S.C. Cohen, A.G. Massey, *Adv. Fluorine Chem.*, *6* (1970) 185 (Polyfluoro-aromatic derivatives of metals).

T. Chivers, *Organomet. Chem. Rev., A*, *6* (1970) 1 (Polychloroorganic derivatives of metals).

G.W. Parshall, *Acc. Chem. Res.*, *3* (1970) 139 (Intramolecular or *ortho*-metallation).

M.F. Lappert, J.S. Poland, *Adv. Organomet. Chem.*, *9* (1970) 397 (σ-Heterodiazoalkane reactions with metals).

W. Kitching, C.W. Fong, *Organomet. Chem. Rev., A*, *5* (1970) 281 (SO_2 and SO_3 insertions into M-C bonds).

M. Tsutsui, M. Hancock, J. Ariyoshi, M.N. Levy, *Angew. Chem., Int. Ed. Engl.*, *8* (1969) 410 (σ-π Rearrangements of transition-metal compounds).

I.I. Kritskaya, *Russ. Chem. Rev. (Engl. Transl.)*, *35* (1969) 167 (Formation and stabilization of σ-M-C bonds of transition metals).

A.M. Sladkov, L.Yu. Ukhin, *Russ. Chem. Rev. (Engl. Transl.)*, *37* (1968) 748 (Cu and Ag acetylides).

M.J. Bruce, F.G.A. Stone, *Prep. Inorg. React., 4* (1968) 177 (Fluorocarbon complexes of the transition metals).

G.W. Parshall, J.J. Mrovca, *Adv. Organomet. Chem., 7* (1968) 157 (σ-Alkyl and σ-aryl derivatives of transition metals).

M.I. Bruce, F.G.A. Stone, *Prep. Inorg. React., 4* (1968) 177 (Fluoroalkyl and -aryl derivatives).

R.F. Heck, *Adv. Organomet. Chem., 4* (1966) 243 (Alkyl- and acyl-cobalt derivatives).

R.D. Chambers, T. Chivers, *Organomet. Chem. Rev., 1* (1966) 279 (Pentafluorophenyl derivatives of metals).

P.M. Treichle, F.G.A. Stone, *Adv. Organomet. Chem., 1* (1964) 43 (Fluorocarbon derivatives of the metals).

H.C. Clark, *Adv. Fluorine Chem., 3* (1963) 19 (Perfluoroalkyl derivatives of metals).

R. Nast, *Angew. Chem., 72* (1960) 26 (Complex acetylides of transition metals).

F.A. Cotton, *Chem. Rev., 55* (1955) 551 (Alkyls and aryls of transition metals: early review).

Subject Index

Acetylenes
- acidity 52
- coordination to metals 350
- hydrosilylation of 120
- metal derivatives
  aluminum 99, 102
  cadmium 73
  lead 174
  lithium 48
  magnesium 64
  mercury 75
  sodium 51
- metallation of 48, 51
- reactions with metal carbonyls 292, 293, 350ff, 356
- transition-metal complexes 349

Acrolein, nickel complex 261
Acrylonitrile, nickel complex 261
Actinides
- σ-alkyls, aryls 360, 364
- allylic complexes 272
- cyclooctatetraene complexes 348
- cyclopentadienyls 308, 314

Alkoxyboranes 89
Alkoxysilanes 121
Allene, reaction with iron carbonyls 276
π-Allylic ligand
- bridging 270
- molecular orbitals of 22

π-Allylic metal compexes 269
Allylidene-metal complexes 255
Aluminum, organo-, derivatives, 97
–, halides 102
–, hydrides 103
–, reactions 99
–, reagents in the synthesis of organo derivatives of antimony 206, arsenic 187, boron 87, gallium 104, lead 175, tin 154

Aminoboranes 89
Aminogermanes 147
Aminosilanes 162
Aminostannanes 162
Anion radicals, naphtalene 52
Anions
- carborane 16
- metal alkyl 360, 361
- metal carbonyl 239
- organoaluminum 98, 101
- organoantimony 205, 208, 210, 214
- organoarsenic 189
- organoberyllium 57
- organocadmium 71
- organochromium 367
- organocobalt 371
- organocopper 374
- organogallium 107
- organogermanium 148
- organogold 374, 375
- organolead 177
- organomagnesium 63
- organomanganese 369
- organoplatinum 373
- organorhenium 369
- organothallium 112
- organotin 158
- organouranium 365
- organozinc 68

Antimony, organo-, derivatives 205
–, acids 211
–, alkoxy 212, 213
–, halides 209
–, hydrides 208
–, hydroxides 214
–, nitrogen containing, 215
–, oxides 214
–, thio derivatives, 214

Arsabenzene 191
Arsenic, organo-, derivatives 183
–, acids 193, 197
–, halides 192
–, hydrides 191
–, nitrogen containing 202

Arsenic, organo-, derivatives/continued/
–, oxygen containing, 195
–, sulfides 199
–, thioacids 201
Arsines
– cyclic 204
– linear 204
– primary 191
– secondary 191
– tertiary 186
Arsinic acids 193, 197
Arsonic acids 193, 197
Arsonium salts, tetraorgano-, 188
Arsoranes, pentaorgano- 190
Association, see Polymerization
Azaferrocene 329
Azulene, manganese complex 331

Benzene
– bonding to metals 24
– mercuration of 78
– metal complexes 335, 336
– molecular orbitals of 22
Beryllium, organo-, derivatives 56
–, classification 56
–, preparation 58
–, structure 59
Biological activity 5
Bio-organometallic chemistry 5
Bismabenzene 218
Bismuth, organo-, derivatives 216
–, halides 218
–, oxygen containing 219
Bismuthonium, tetraorgano- 217
Bond energies 12
Bonding in
– acetylene-metal complexes 350
– beryllium dimethyl 16
– π-complexes 23
– lithium alkyls 17
– metal-carbene complexes 19
– metal carbonyls 18
– metal-olefin complexes 20, 21
– organoaluminum compounds 15
– uranocene 348
Bonding to metals, of
– acetylene 350
– benzene 24
– cyclopentadiene 24
– cylooctatetraene 348
Bonds (in organometallic compounds)
– covalent (sigma) 8, 12
– dative 8
– ionic 7
– delocalized 8, 16
– electron deficient 8, 14
Borazarenes 90
– metal complexes 329
Borazines 88, 90
– metal complexes 343
Borinic acids 88
Boron, organo-, derivatives 82
–, acids 88
–, classification 82
–, halides 87
–, preparation 84
–, reactions 71
Boronic acids 88
Borosiloxanes 134
Boroxines 88
Butadiene
– coordination to metals 256
– molecular orbitals of
– metal complexes with chromium 288, copper 267, iron 266, 276, 288, manganese 259, 265, 274, 288; molybdenum 288, rhodium 291, silver 267, vanadium 288

Cadmium, organo-, derivatives 71
–, halides 73
–, preparation 72
–, reactions 51, 72
Calcium, organo-, derivatives 66
Carbanions 7, 11
Carbenes
– bonding to metals 19
– metal complexes 252
Carbofunctional organosilanes 124
Carbon monosulfide, metal complexes 249
Carbon monoxide
– bonding to metals 18, 229
– metal complexes (carbonyls) 227
– molecular orbitals 18
Carboranes
– bonding 17
– metal complexes 306, 332
– preparation 92
– structure
Carbosilanes 124, 140

Carbynes
– metal complexes 282
Chromium
– acetylene derivatives 351
– σ-alkyls, aryls 360, 366, 367
– allyl complexes 270
– benzene complexes 337, 351
– borabenzene complexes 343
– butadiene complexes 288
– carbene complexes 253
– carbonyl anions 239, 241, 242
– carbonyl halides 247, 248
– carbonyl hydride 244
– carbonyls 228, 233
– carbyne complexes 282, 283
– cycloheptatriene complexes 340, 341, 345, 346
– cycloheptatrienyl derivatives 346
– cyclooctatetraene complexes 304, 342, 348
– cylopentadienyls 308, 310, 317, 318, 346, 348
– hexacarbonyl 231
– isocyanide complexes 251
– olefin complexes 258, 263
– phosphocine complexes 343
– pyridine complexes 342
– selenocarbonyl complexes 250
– thiocarbonyl complexes 249
– trimethylenemethyl complexes 305
vinylidene complexes 255
Chromocene 317
– preparation 317
– reduction of 273
Classification of
– π-ligands 224
– organo-Al compounds 97
– organo-As compounds 183
– organo-B compounds 82
– organo-Be compounds 56
– organometallic compounds 5
– organo-Si compounds 116
Cobalt
– acetylene derivatives 352, 353, 354, 355
– alkyls, aryls 361, 370, 371
– allylic complexes 277
– benzene complexes 340
– butadiene complexes 291
– carbonyl anions 239, 241, 242, 243
– carbonyl cations 247
– carbonyl halides 248
– carbonyl hydride 244, 245
– carbonyls 228, 231, 232, 233, 236
– carborane complexes 333
– cyclobutadiene complexes 293, 295, 297
– cylclohexadiene complexes 266, 302
– cyclooctadiene complexes 260
– cyclooctatetraene complexes 266
– cyclopentadiene complexes 298
– cyclopentadienone complexes 357
– cyclopentadienyls 308, 324, 325
– cyclopropenyl complexes 281
– ethylene complexes 257
– fulvene complexes 299
– isocyanide complexes 251
– olefin complexes 260, 264
– quinone complexes 357
– silacyclopentadiene complexes 299
– tris(allyl) complexes 270
Cobaltocene
– preparation 324
– reactions 260, 291
Complexes of
– germylenes 152
– organo-Al compounds 101
– organoarsines 205
– organo-B compounds 84
– organo-Be compounds 57
– organo-Hg compounds 76
– organo-Li compounds 49
– organo-Mg compounds 63
– organo-Sn compounds 157, 171
– organo-Zn compounds 172
– silylenes 143
– stannylenes 172
Copper
– σ-alkyls, aryls 361, 373, 374
– butadiene complexes 267
– carbonyl cations 247
– carbonyl halides 248
– carbonyls 228, 234, 236
– cyclohexadiene complexes 267
– cyclooctatetraene complexes 262
– cyclopentadienyls 308, 327
– isocyanide complexes 251
– norbornadiene complexes 262, 265
– olefin complexes 262
Covalent bonds in organometallic compounds 8, 12
Cubane-type tetramers 57, 60, 69, 70, 73, 79, 103

Cyclic dimers 57, 60, 63, 64, 69, 73, 90, 102, 106, 107, 108, 111, 113, 139, 151, 159, 169
Cyclic tetramers 102, 106, 107, 131, 137, 139, 150, 195, 200, 203, 233, 248
Cyclic trimers 15, 57, 60, 70, 71, 79, 83, 84, 90, 92, 103, 106, 108, 111, 131, 137, 139, 150, 151, 169, 195, 200, 203, 233
Cycloarsines 204
Cyclobutadiene
– complexes, general 291
– complexes with cobalt 293, 295, 296, 357, iron 292, 293, 294, 295, 356, molybdenum 293, 296, nickel 179, 291, 296, niobium 292, palladium 296, rhodium 293, 295, 357, titanium 292, vanadium 292, 293, 296
– molecular orbitals of 22
Cyclobutadieneiron tricarbonyl 294, 356
– aromatic reactions 294
– reactions with carboranes 306
Cyclododecadiene-1,6
– rhodium complex
Cyclogermanes 148, 149
Cycloheptadiene-1,3, metal complexes 302
Cycloheptadienyl, metal complexes 331, 332
Cycloheptatriene-1,3,5
– bonding to metals 25
– metal complexes 303, 304, 345
Cycloheptatrienyl
– bonding to metals 25
– metal complexes 345
Cyclohexadiene-1,3
– bonding to metals 256
– complexes with cobalt 302, copper 267, iron 301, manganese 259, 265, 301, molybdenum 300, osmium 302, platinum 266, rhodium 266, ruthenium 302, tungsten 300, 301, vanadium 300
Cyclohexadienyl
– metal complexes 330
Cyclononatriene-1,4,7, molybdenum complex 265
Cyclooctadiene-1,5
– complexes with chromium 263, gold 267, iron 263, manganese 265, molybdenum 263, nickel 264, palladium 264, platinum 264, rhodium 264, ruthenium 264
Cyclooctatetraene
– bonding to metals 25, 256, 348
– complexes with actinides 348, cobalt 266,
chromium 304, 342, copper 262, iron 303, lanthanides 348,
– molybdenum 304, 342, nickel 266, 279, platinum 264, rhodium 266, ruthenium 304, titanium 292, tungsten 304, 342, uranium 348
– metal complexes (general) 341, 347
Cyclooctatriene-1,3,5
– metal complexes 341
– molybdenum complex 263
Cyclopentadiene
– bonding to metals 24
– magnesium derivatives 64
– reaction with cobalt carbonyls 324, iron carbonyls 297, 320, 322
– rhenium complexes 259
– sodium derivatives 51, 52
– tetrahapto derivatives of cobalt 298, iridium 298, iron 297
Cyclopentadiene
– metal complexes 298, 357
Cyclopentadienyl-
– arsenic 187
– beryllium 60
– bonding to metals 24, 27
– bridging 280
– iridium 110
– lead 181
– metal complexes 307
– molecular orbitals of 22
– tin derivatives 170
– zinc derivatives 68
Cyclopentadienylmetal carbonyls
– bonding 23
– mononuclear 311
– neutral and ionic 309
– polynuclear 312
Cyclopentadienylmetal halides 310
Cyclopentadienylmetal isocyanides 251
Cyclopentene, rhenium complex 259
Cyclopropenyl, triphenyl-, cation
– metal complexes 280
Cyclosilanes 128
Cyclosilazanes 137
Cyclosilazoxanes 134, 137
Cyclosiloxanes 131, 142
Cyclosilthianes 139
Cyclostannazanes 168

Dative bonds in organometallic compounds 8, 18
Delocalization of charge 11
Delocalized bonds 8, 16
Dewar benzene, chromium complex 262, 263
Dicyanovinylidene complexes 252
Dimerization of
– organo-Al compounds 14, 102, 103
– organo-Be compounds 57, 61
– organo-Ga compounds 106, 108
– organo-In compounds 111
– organo-Mg compounds 63
– organo-Sn compounds 159, 166
– organo-Tl compounds 113
– trimethylaluminum 14
– triphenylaluminum 15
Direct synthesis of
– carbosilanes 140
– organo-Al compounds 99, 102
– organo-As compounds 193
– organo-Be compounds 61
– organo-Cd compounds 72
– organochlorosilanes 119
– organo-Hg compounds 77
– organo-Li compounds 46
– organo-Mg compounds 63
– organo-Na compounds 51
– organo-Pb compounds 175
– organo-Sb compounds 206, 209
– organo-Sn compounds 154
Dissociation energies 13
Disilene 127
p-Divinylbenzene, iron complex 290
Divinylboranes, iron complexes 306

Effective atomic number 19
Electron deficient bonds 8, 14
18-Electron rule 19, 23
Electronegativity 9
Electronic structure of transition metals 223
β-Elimination 12, 359
Environmental organometallic chemistry 5
Ethylene
– bonding to metals 20
– complexes with iridium 261, iron 259, manganese 258, nickel 261, osmium 260, palladium 261, platinum 225, rhenium 259, rhodium 261, ruthenium 260, silver 262
– molecular orbitals 20
– reacitions with ferrocene and lithium 260
Exchange
– hydrogen-metal 47
– lithium-halogen 47
– metal-metal 47

Ferrocene
– acylation 320
– alkylation 320
– aromatic properties 321
– discovery 5
– mercuration 78, 320
– preparation 320
– reaction with ethylene and lithium 260
– structure 321
Ferrocenyllithium, reactions 321
Fluxional
– π-allylic complexes 271
– iron-cyclopentadiene complexes 369
Free radicals
– organogermanium 153
– organosilicon 143
– organotin 169, 173
Fulvenes
– bonding to metals 256
– metal complexes 299
Fumaronitrile, nickel complex 261
Furane
– lithiation 47
– mercuration 78

Gallium, organo-, derivatives 104
–, preparation 104
–, reactions 105
Germanium, organo-, derivatives, 143
Germylenes 152
Gold
– σ-alkyls, aryls 361, 362, 374, 375
– carbonyls 228, 234, 236
– carbonyl halides 248
– cyclopentadienyl complexes 308, 328
– ethylene complexes 257
– isocyanide complexes 251
– metallocycles 375
Grignard reagents
– association 64
– intermediates in the synthesis of organo derivatives of aluminum 98, antimony 206, arsenic 186, 198, beryllium 58, boron 84, cadmium 73, gallium 104, germa-

nium 143, lead 174, mercury 74, 77, silicon 118, 119, thallium 112, 113, tin 153
– preparation 63
– reactions 66
– structure 64

Hafnium
– σ-alkyls, aryls 360
– π-allylic complexes 270
– cyclopentadienyl derivatives 308, 310, 315
Hapto, prefix 25
Heterocarboranes 94
Heterocyclic derivatives of aluminum 100, antimony 206, 208, 213, arsenic 186, 191, 192, 193, 203, bismuth 218, boron 83, 85, 90, germanium 145, 150, 151, lead 176, mercury 76, silicon 124, tin 155
Heterocyclosilazanes 138, 162
Heteroleptic compounds 12, 360
Heterosiloxanes 134
Hexafluorocyclopentadiene, cobalt complex 260
Homoleptic compounds 12, 360
Hybridization of atomic orbitals
– in Group-II elements 55
– in Group-III elements 81
– in Group-IV elements 115
Hydroalumination 98
Hydroboration 85, 86
Hydrosilylation 120
Hydrostannation 158
Hydrocarbons, acidity of 51

Indium, organo-, derivatives 109
Industrial organometallic chemistry 5
Infrared spectroscopy
– of metal carbonyls 18
– use of 33
Ionic bonds in organometallic compounds 7, 9
Iridium
– acetylene complexes 352
σ-alkyls, aryls 371
– butadiene complexes 291
– carbonyl anions 239, 241
– carbonyl cations 247
– carbonyl halides 248
– carbonyl hydrides 244
– carbonyls 228, 232, 236,
– cyclopentadiene complexes 298
– cyclopentadienyls 308, 326
– ethylene complexes 261
– isocyanide complexes 251
– norbornadiene complexes 264
olefin complexes 261, 264
– tris(allyl) complexes 270
Iron
– acetylene complexes 352, 354, 356
– σ-alkyls, aryls 361, 369, 370
– allylic complexes 275, 276
– benzene derivatives 330, 339
– butadiene complexes 266, 276, 288, 289
– carbonyl anions 239, 240, 241, 242, 243
– carbonyl cations 246, 247
– carbonyl halides 247
– carbonyl hydrides 244, 245
– carbonyls 228, 230, 232, 233, 236
– carborane complexes 306, 332
– cyclobutadiene complexes 292, 293, 294, 295, 306, 356
– cycloheptadienyl complexes 331
– cycloheptatrienyl complexes 346
– cyclohexadiene complexes 301, 302
– cyclohexadienyl complexes 330, 331
– cyclooctatetraene complexes 303
– cyclopentadiene complexes 297
– cyclopentadienone complexes 299, 350, 357
– cyclopentadienyls 297, 308, 310, 320–323, 350, 356, 369, 375
– ethylene complexes 259, 260
– fulvene complexes 299
– isocyanide complexes 251
– olefin complexes 259, 260, 264
– pentacarbonyl 231
– pentadienyl complexes 328
– phosphole complexes 300
– silacyclopentadiene complexes 294
– silacyclohexadiene complexes 302, 328
– thiocarbonyl complexes 249
– thrimethylenemethyl complexes 305
– tris(allyl) complexes 270
Isocyanide, metal complexes 250
Isoleptic compounds 360

Kharasch reagent 261, 264, 296

Laboratory techniques 30
Lanthanides
– σ-alkyls, aryls 360, 362, 364
– allylic complexes 272

– cyclooctatetraene complexes 348
– cyclopentadienyls 308, 310, 313
Lead, organo-, derivatives 173
–, alkoxides 179
–, amides 179
– carboxylates 178
– halides 177
– hydrides 178
– hydroxides 178
– preparation 174
– reactions 51, 176
π-Ligands 224
Ligand-transfer reactions 296
Literature of organometallic chemistry 35
Lithium, organo-, derivatives
–, preparation 46
–, reagents in the synthesis of organo derivatives of antimony 206, arsenic 186, beryllium 58, boron 84, cadmium 72, 73, carboranes 95, germanium 143, 145, lead 174, mercury 74, silicon 118, thallium 112, tin 153, zinc 68, 70
–, structure 16, 48
Lithium alkyls, structure 16
Lithium-halogen exchange 47

Macrocyclic diacetylenes, reactions with metal carbonyl derivatives 295, 353, 356, 357
Magnesium, organo-, derivatives 62
–, preparation 62
–, halides 63 (see also Grignard reagents)
–, reactions 63, 66
Manganocene 319
Manganese
– acetylene complexes 352
– σ-alkyls, aryls 361, 368, 369
– azulene complexes 331
– benzene complexes 338
– butadiene complexes 265, 274, 289, 290
– carbonyl anions 239, 241, 242
– carbonyl cations 246, 247
– carbonyl hydride 244
– carbonyls 228, 231, 232, 235, 247, 248
– carborane complexes 333
– cyclohexadiene complexes 265, 301
– cyclopentadienyl complexes 308, 319, 376
– ethylene complexes 259
– isocyanide complexes 251
– norbornadiene complexes 265
– olefin complexes 259
– thiocarbonyl complexes 249
– vinylbenzene complexes 290
– vinylidene complexes 255
Manganocene 319
Mass spectroscopy 33
Mercapto derivatives of aluminum 104, boron 84, 91, lead 180, silicon 123, tin 164
Mercuration 78
Mercury, organo-, derivatives 74
–, halides 77
–, intermediates in the synthesis of organo derivatives of aluminum 98, arsenic 192, boron 87, cadmium 72, gallium 104, lithium 46, sodium 51, zinc 67
–, preparation 74
–, reactions 79
–, structure 75
Metal-atom synthesis 32
σ-Metal-carbon bond, compounds of transition metals 359
Metal-carbonyl anions 239
cations 246
halides 247
hydrides 243
Metal carbonyls
– binary 228
– binuclear 228, 232
– bonding in 18
– heteronuclear 229, 238
– isocyanide derivatives 251
– mononuclear 228, 231
– polynuclear 232
– preparation 234
– reactions 238
– structure 229
– trinuclear 232
Metal-isocyanide complexes 251
Metal-isocyanide halides 251
Metal-Metal exchange 47
Metal-olefin bond 20
Metal selenocarbonyls 250
Metal thiocarbonyls 249
Metal-vapor synthesis and reactions
– chromium and pyridines 342
– iron and benzene 302
– iron and butadiene 288
– nickel and olefins 257
– titanium, vanadium, iron, cobalt and cycloheptatriene 346
Metalation 47, 48, 95

Metalating reagents 47, 52
Metallocarboranes 94
Metallocycles, containing transition metals 365, 366, 367
Molecular orbitals of
– π-allyl ligands 22
– benzene 22
– butadiene 22
– carbon monoxide 18
– cyclobutadiene 22
– cyclopentadienyl 22
– ethylene 20
Molybdenum
– acetylene derivatives 351, 353
– σ-alkyls, aryls 360, 368
– allyl complexes 270
– benzene complexes 337
– borazine complexes 343
– butadiene complexes 288, 289
– carbene complexes 254
– carbonyl anions 239, 241, 242
– carbonyl halides 248
– carbonyls 228, 235
– cyclobutadiene complexes 293
– cycloheptatriene complexes 341, 345
– cyclohexadiene complexes 300
– cyclooctatetraene complexes 304, 342, 346
– cyclopentadienyl complexes 308, 310, 317, 318
– cyclopropenyl complexes 281
– hexacarbonyl 231
– isocyanide complexes 251
– olefin complexes 258, 263
– phosphorine complexes 343
– pyridine complexes 342
– thiocarbonyl complexes 249
– trimethylenemethyl complexes 305
– vinylidene complexes 255
Molybdocene 318

Nickel
– acetylene complexes 352, 353, 355
– σ-alkyls, aryls 361, 371, 372
– allylic complexes 278
– bis(allyl) complexes 270
– carbonyl anion 239, 242
– carbonyl halide 248
– carbonyl hydride 244
– carbonyls 228, 234, 236
– cyclopentadiene complexes 291, 295, 296
– cyclooctatetraene 266, 279
– cyclopentadienyl complexes 308, 326, 327
– cyclopropenyl complexes 280, 281
– ethylene complexes 257, 261
– isocyanide complexes 251
– metallocycles 372, 373
– olefin complexes 261, 264
– tetracarbonyl 231
– thiocarbonyl 249
Nickelocene
– preparation 326, 327
– reactions 261, 278
Niobium
– acetylene complexes 292, 351
– σ-alkyls, aryls 360, 366
– benzene complexes 337
– carbene complexes 254
– carbonyl anions 239
– cyclobutadiene complexes 292
– cyclooctatetraene complexes 348
– cyclopentadienyl complexes 308, 310, 316, 348, 351
– tris(allyl) 270
Niobocene 316
Norbornadiene, complexes of copper 262, 265, 267, iron 263, manganese 259, 265, ruthenium 264

Olefins
– bidentate 262
– bonding to metals 20, 256
– bridging 265
– metal π-complexes 255
– monodentate 257
Organometallic chemistry
– definition 3
– history 4
Organometallic compounds
– classification 5
Organosilicon compounds 116
Osmium
– acetylene complexes 352
– carbene complexes 253
– carbonyl anions 239, 242
– carbonyl cations 246, 247
– carbonyl halides 247, 248
– carbonyl hydrides 244, 245
– carbonyls 228, 232, 233, 236
– cyclopentadienyl complexes 308, 324
– ethylene complexes 260

– pentacarbonyl 231
Osmocene 324

Palladium
– acetylene complexes 352, 354
– σ-alkyls, aryls 361, 372
– allylic complexes 270
– bis(allyl) complexes 270
– carbonyl cations 247
– carbonyl halide 248
– carbonyls 228, 236
– carborane complexes 332
– cyclobutadiene complexes 296
– cyclopentadienyl complexes 310, 327
– ethylene complexes 257, 261
– isocyanide complexes 251
– metallocycles 373
– olefin complexes 261, 269
Pentadienyl metal complexes 328
Phosphaferrocene 329, 330
Phosphinoborines 84, 92
Phosphosiloxanes 134
Phospholes, metal π-complexes 300, 329
Platinum
– acetylene complexes 352
– σ-alkyls, aryls 316, 372, 373
– bis(allyl) complexes 270
– butadiene complexes 266
– carbene complexes 253, 254
– carbonyl anions 239
– carbonyl cations 247
– carbonyl halides 248
– carbonyls 228, 236
– cyclohexadiene complexes 266
– cyclopentadienyl complexes 308, 327
– ethylene complexes 255, 256
– isocyanide complexes 251
– metallocycles 373
– olefin complexes 261, 264, 266
Plumbylenes 181
Polyarsines 204
Polycarbosilanes 140
Polygermanes 148
Polymerization of
– cyclosiloxanes 132, 133
– dimethylberyllium 16
– lithium alkyls 16, 48
– organoaluminum compounds 103
– organoberyllium halides 61
– organocadmium derivatives 73
– organogallium derivatives 106
– organoindium derivatives 111
– organolead derivatives 177
– organothallium derivatives 112, 114
– organotin alkoxides 161
  halides 157
  hydroxides 159
Polymetallosiloxanes 135
Polysilanes
– cyclic 128
– linear 129
Polysilazanes 136
Polysiloxanes 129
– cyclic 131
– linear 133
– synthesis 130
Polysiloxymetalloxanes 135
Polystannanes 164
Polystibines 216
Purification of solvents 29
Pyridine, metal carbonyl complexes 312

Raman spectroscopy 33
Redistribution reactions of
– arsenic compounds 192
– cadmium compounds 73
– germanium derivatives 146, 152
– mercury compounds 75, 77
– organotin derivatives 154, 156
– polysiloxanes 132, 133
– polysilthianes 139
– zinc compounds 139
Rhenium
– acetylene complexes 352
– σ-alkyls, aryls 361, 368, 369
– allyl complexes 272, 274
– carbonyl anions 239, 241, 242
– carbonyl cations 246, 247, 248
– carbonyl hydride 244, 245
– carbonyls 228, 232, 236
– carborane complexes 332
– cyclopentadienyl derivatives 319
– ethylene complexes 259
– isocyanide complexes 251
– olefin complexes 259
Rhodium
– acetylene complexes 352
– σ-alkyls, aryls 371
– allylic complexes 277
– butadiene complexes 264, 291
– carbonyl anions 239, 242, 243
– carbonyl halide 248

- carbonyl hydride 244
- carbonyls 228, 232, 233, 234, 236
- cyclobutadiene complexes 293, 295, 357
- cyclopentadienyl derivatives 308, 310, 325, 326
- ethylene complexes 261
- isocyanide complexes 251
- metallocycles 371
- olefin complexes 264
- tris(allyl) complexes 270, 277

Ruthenium
- acetylene complexes 352, 355
- allylic complexes 277
- benzene complexes 339, 340
- carbonyl anions 239
- carbonyl cations 247, 248
- carbonyl hydrides 244, 245
- carbonyls 228, 232, 236
- cyclooctatetraene complexes 304
- cyclopentadienyl complexes 308, 323
- ethylene complexes 260
- isocyanide complexes 251
- pentacarbonyl 231
- selenocarbonyl complexes 250
- thiocarbonyl complexes 249

Ruthenocene 323

Scandium
- σ-alkyls, aryls 360, 364
- allylic complexes 272
- cyclooctatetraene complexes 347
- cyclopentadienyl derivatives 308, 310, 313

Schlenk tubes 30
Semibridging bonds of carbon monoxide 230, 311
Separation techniques 32
Silabenzene 126
Silacyclobutane 125
Silacyclopropanes 125
Silacyclohexadiene, iron complex 328
Silacyclopentadiene, metal complexes 299
Silaethylenes 141, 142
Silanes 121
Silanols 122
Silazanes 122, 136
Silicon, organo-, derivatives 116
Silirenes 125
Siloxanes 130
Silver
- σ-alkyls, aryls 361, 374
- butadiene complexes 267
- carbonyl cations 247
- carbonyls 228, 234, 236
- cyclopentadienyl derivatives 308, 327
- ethylene complexes 262
- isocyanide complexes 251

Silylenes 142
Sodium, organo-, derivatives 51
Solvents, purification of 29
Spectroscopic methods 33
Stability
- kinetic 12, 13
- hydrolytic 13
- thermal 13
- thermodynamic 12

Stannanes 164
Stannazanes 167
Stannylenes 169
Stibabenzene 208
Stibines 206
Stibonium, tetraorgano- 207
Stretching frequency of coordinated and free CO molecule 18
Structure of
- π-complexes 22
- metal-carbonyl anions 241
- metal carbonyls 18, 229
- metal-olefin complexes 20
- organoaluminum compounds 16
- organoberyllium compounds 16
- organotin halides 157

Synthesis of organometallic compounds, laboratory techniques 29

Tantalum
- acetylene complexes 351
- σ-alkyls, aryls 360, 366
- benzene complexes 337
- carbene complexes 254
- carbonyl anions 239
- carbonyls 228, 235
- cyclopentadienyl complexes 308, 351
- tris(allyl) complexes 270

Technetium
- benzene complex 339
- carbonyl anion 239,
- carbonyl cations 247, 218
- carbonyl hydride 244, 246
- carbonyls 228, 232, 236
- cyclopentadienyl complexes 308, 319

– ethylene complexes 259
– olefin complexes 259
Tetraalkyllead 174
Tetracene, iron-carbonyl complex 276
Tetracyanoethylene complexes with chromium 258, molybdenum 258, tungsten 258
Tetramethylallene, iron complex 290
Tetraorganosilanes 118
Tetraphenylcyclobutadiene, metal complexes 292, 293, 294, 295, 296
Thallium, organo-, derivatives 111
Thiadiborolene, metal complexes 306
Thiolates, see Mercapto derivatives
Thiophene, mercuration of 78
Thiophene dioxide, iron complexes 300, 343
Thorium
– σ-alkyls, aryls 360, 365
cyclooctatetraene complexes 348
– cyclopentadienyl complexes 308, 310, 314
Three-center bonds 15
– in dimethylberyllium 16
– in methyllanthanide derivatives 15
– in trimethylaluminum 15
Tin, organo-, derivatives 153
–, alkoxides 160
–, amides 162
–, halides 156
–, hydrides 158
–, hydroxides 159
–, oxides 166
–, preparation 153
–, reactions 155
–, reagents in the synthesis of organoarsenic derivatives 187, organoboron derivatives 85, 87
–, selenides 168
–, sulfides 168
Titanium
– acetylene complexes 350
– σ-alkyls, aryls 360, 365
– allylic complexes 270, 272
– benzene complexes 336
– carbonyls 228, 234
– cyclobutadiene complexes 292
– cycloheptatrienyl complexes 346
– cyclooctatetraene complexes 292, 347, 348
– cyclopentadienyl complexes 308, 310, 314, 315, 347, 350
– metallocycles 366
Titanocene derivatives 315
Toxicity, organobismuth compounds 216
–, organomercury compounds 76
Transition metals
– atomic orbitals 21
– electronic structure 223
– σ-metal-carbon bond derivatives 359
Transmetallation 47
Trimethylaluminum, dimeric structure 14
Trimethylenemethyl, metal complexes 304, 305
Triphenylaluminum, dimeric structure 15
Triple-decker sandwich complexes 326, 333, 347
Trisilylamine 137
Tristannylamine 167
Tungsten
– acetylene complexes 351
– σ-alkyls, aryls 360, 368
allyl complexes 270
– benzene complexes 337
– butadiene complexes 289
– carbene complexes 253
– carbonyl anions 239, 241, 242
– carbonyl halides 248
– carbonyls 228, 235
– cycloheptatriene complexes 345
– cyclohexadiene complexes 300, 301
– cyclooctatetraene complexes 304, 342
– cyclopentadienyl complexes 308, 310, 318
– hexacarbonyl 231
– isocyanide complexes 251
– olefin complexes 258, 263
– phosphorine complexes 343
– thiocarbonyl complexes 249
Tungstocene 318

Unsymmetrical bonding of carbon monoxide 229
Uranium
– σ-alkyls, aryls 360, 364, 365
– cyclopentadienyl complexes 308, 314
– cyclooctatetraene complex 348
Uranocene 348

Vanadium
– acetylene complexes 292, 351
– σ-alkyls, aryls 360, 366
– allylic complexes 272
– benzene complexes 337, 357

– butadiene complexes 288
– carbonyl anions 239
– carbonyl hydrides 244
– carbonyls 228, 234, 235, 242
– cyclobutadiene complexes 292
– cycloheptatriene complexes 341, 345
– cycloheptatrienyl complexes 346
– cyclohexadiene complexes 300
– cyclohexadienyl complexes 330
– cyclopentadienyl complexes 308, 310, 315, 316, 350
– cyclopropenyl complexes 281
– tris(allyl) complexes 270
Vanadocene 315
Vinylbenzene, iron complex 290
Vinylidene metal complexes 255

X-ray diffraction 34

Yttrium, cyclopentadienyl complex 308, 310
Zinc, organo-, derivatives 67
–, cleavage by alkali metals 51
–, halides 70
–, hydrides 71
–, preparation 67
–, reactions 68
Zirconium
– σ-alkyls, aryls 360, 365
– allylic complexes 270, 272
– cyclopentadienyl derivatives 308, 310, 315
Zirconocene derivatives 315.